CW01543467

Medical Biostatistics

Third Edition

Chapman & Hall/CRC Biostatistics Series

Chapman & Hall/CRC Biostatistics Series

Chapman & Hall/CRC Biostatistics Series

Medical Biostatistics

Third Edition

Abhaya Indrayan

CRC Press
Taylor & Francis Group
Boca Raton   London   New York

CRC Press is an imprint of the
Taylor & Francis Group, an **informa** business

A CHAPMAN & HALL BOOK

CRC Press
Taylor & Francis Group
6000 Broken Sound Parkway NW, Suite 300
Boca Raton, FL 33487-2742

© 2013 by Taylor & Francis Group, LLC
CRC Press is an imprint of Taylor & Francis Group, an Informa business

No claim to original U.S. Government works

Printed in the United States of America on acid-free paper
Version Date: 20120601

International Standard Book Number: 978-1-4398-8414-0 (Hardback)

Library of Congress Cataloging-in-Publication Data

Indrayan, Abhaya, 1945-
   Medical biostatistics / Abhaya Indrayan. -- 3rd ed.
     p. ; cm. -- (Chapman & Hall/CRC biostatistics series)
   Includes bibliographical references and index.
   ISBN 978-1-4398-8414-0 (hardback : alk. paper)
   I. Title. II. Series: Chapman & Hall/CRC biostatistics series (Unnumbered)
   [DNLM: 1. Biostatistics--methods. 2. Biometry--methods. WA 950]

610.72'7--dc23

2012019378

Visit the Taylor & Francis Web site at
http://www.taylorandfrancis.com

and the CRC Press Web site at
http://www.crcpress.com

Contents

List of Figures

List of Tables

Preface

Biostatistical aspects are receiving increased emphasis in medical books, medical journals, and pharmaceutical literature, yet there is a lack of appreciation of biostatistical methods as a medical tool. This book arises from the desire to help biostatistics earn its rightful place as a medical, rather than a mathematical, subject. Medical and health professionals may then perceive biostatistics as their own discipline instead of an alien discipline. A book that effectively focuses on the statistical aspects of medicine with a medical perspective is clearly needed. To enhance focus, this book is titled *Medical Biostatistics*. Prefix "medical" precludes fishes and plants that a purist might include under the genre of "bio"statistics.

Variation is an essential, and perhaps the most enjoyable, aspect of life. But the consequent uncertainties are profound. Thus, methods are needed to measure the magnitude of uncertainties and to minimize their impact on decisions. *Biostatistics is the science of management of uncertainties in health and medicine.* Beginning with this premise, this book provides a new orientation to the subject. This theme is kept alive throughout the text. I have tried to demonstrate that biostatistics is not just statistics applied to medicine and health sciences but is *two* steps further, providing tools to manage some aspects of medical uncertainties.

The primary target audiences are students, researchers, and professionals of medicine and health. These include clinicians who deal with medical uncertainties in managing patients and want to practice evidence-based medicine; research workers who design and conduct empirical investigations to advance knowledge, including research workers in pharmaceutical industry who search new regimens that are safer and more effective yet less expensive and more convenient; and health administrators who are concerned with epidemiological aspects of health and disease.

Although the text is tilted to the viewpoint of medical and health professionals, the contents are of sufficient interest to a practicing biostatistician and a student of biostatistics as well. They may find some sections very revealing, particularly the heuristic explanations provided for various statistical methods.

The boundary between epidemiology methods and biostatistics is thin, if at all. This book does not limit itself to the conventional topics of confidence intervals and tests of significance. It discusses at length study designs, measurement of health and diseases, clinimetrics, and quality control in medical setup. The text fosters the thought that medicine has to be individualized yet participatory. It tries to develop pathways that can achieve this through biostatistical thinking. Emphasis is laid on the concepts and interpretation of the methods rather than on theory or intricacies. Theoretical development is intentionally de-emphasized and applications increasingly emphasized. A large number of real-life examples are included that illustrate the methods and explain the medical meaning of the results. Many statistical concepts are repeatedly explained in different contexts to bring home the point, keeping the requirement of the target audience in mind.

In the process of projecting biostatistics as a medical discipline, it is imperative to place less emphasis on mathematical aspects. But the essential algebra that is needed to communicate and understand some statistical concepts is not ignored. In fact, the second half of the book makes liberal use of notations. An attempt is made to strike an even balance. Medical and health professionals, who are generally not well trained in mathematics, may find the language and presentation very conducive. Equations and formulas are separately identified and manual calculations are described for the fundamentals, but the emphasis is on the use of computers for advanced calculations. Software illustrations for intricate methods are provided in Appendix C of this book.

The book is fairly comprehensive and incorporates a large number of statistical concepts used in medicine and health. The contents are more than an introduction and less than an advanced treatise. References have been provided for further reading. A medical or a health professional should be able to plan and carry out an investigation by oneself on the basis of this text and intelligently seek the help of an expert biostatistician when needed. Medical laboratory professionals, scientists in basic medical sciences, epidemiologists, public health specialists, nutritionists, and others in health-related disciplines may also find this volume useful. The text is expected to provide a good understanding of the statistical concepts required to critically examine the medical literature. The material is suitable for use in preparation for professional examinations such as that for membership in the College of Physicians. The content is also broad enough to cover an undergraduate biostatistics course for medical and health science students.

I am thankful to the reviewers worldwide who have examined the book microscopically and provided extremely useful suggestions for its improvement while also finding the first edition as "probably the most complete book on biostatistics" and the second edition as "almost encyclopedic in breadth." This edition incorporates most of these suggestions. The second edition increased the coverage and now the third edition increases the depth. Some details left out earlier have been included to provide more intelligible reading. Yet, many important techniques continue to be sidetracked in this text. This illustrates my escape from discussing complexities as the book is designed primarily for medical professionals.

The sequence of chapters may not look natural to statisticians because their thoughts follow mathematical continuum but may look natural to medical and health professionals whose biostatistics needs are for problem solving.

I am confident that the book would be found as the most comprehensive treatise on biostatistical methods. In the process, I realize I am undertaking the risk involved in including elementary- and middle-level discussions in the same book. I would be happy to receive feedback from readers.

Abhaya Indrayan

Data sets in the Examples in this text are available in Excel for ready download at http://MedicalBiostatistics.synthasite.com. Use these data sets to rework some of the examples of your interest and to do further analysis where needed.

Summary Tables

TABLE S.1

Methods to Compute Some Confidence Intervals

Parameter of Interest	Conditions	95% CI
Proportion (π)	Large n, $p \neq 0$ and $p \neq 1$	Equation 12.11
	Small n, any p	Figure 12.4
	Any n, $p = 0$ or 1 (bound)	Table 12.4
Mean (μ)	Large n, σ known, almost any underlying distribution	Equation 12.14
	Small n, σ known or unknown, underlying non-Gaussian	Table 12.5 (CI for median)
	Any n, σ unknown, underlying Gaussian	Equation 12.15
	Large n, σ unknown, underlying non-Gaussian	Equation 12.15
	Small n, σ known, underlying Gaussian	Equation 12.14
Median	Gaussian distribution	Equation 12.18
	Non-Gaussian conditions	Table 12.5
Difference ($\pi_1 - \pi_2$)	Large n_1, n_2—Independent samples	Equation 12.20
	Large n_1, n_2—Paired samples	Equation 12.23
Difference ($\mu_1 - \mu_2$)	Independent samples	
(σ unknown)	Large n_1, n_2—Any underlying distribution	Equation 12.21
	Small n_1, n_2—Underlying Gaussian	Equation 12.21
	Paired samples	Same as for one sample after taking the difference
Relative risk	Large n_1, n_2—Independent samples	Equation 14.4
	Large n_1, n_2—Paired samples	Same as for OR
Attributable risk	Large n_1, n_2—Independent samples	Same as for $\pi_1 - \pi_2$
	Large n_1, n_2—Paired samples	Equation 14.12
Number needed to treat	Large n_1, n_2—Independent samples	Section 14.1.3
Odds ratio	Large n_1, n_2—Independent samples	Equation 14.18
	Large n_1, n_2—Paired samples	Equation 14.21
Regression coefficient	Large n	Section 16.3.1
Regression line	Large n	Section 16.3.1
Logistic coefficient	Large n	Section 17.2.2

TABLE S.2

Statistical Procedures for Test of Hypothesis on Proportions

Parameter of Interest and Setup	Conditions	Main Criterion	Equation/Section
Small-Sized Tables			
One dichotomous variable	Independent trials		
	Any n	Binomial	Use Equation 13.1
	Large n	Gaussian Z	Equation 13.3
One polytomous variable	Independent trials		
	Large n	Goodness-of-fit chi-square	Equation 13.5
	Small n	Multinomial	Use Equation 13.6
Two dichotomous variables (2×2)	Two independent samples		
	Large n	Chi-square or Gaussian Z	Equation 13.8 or 13.9
	Small n	Fisher exact	Equation 13.11
	Detecting a medically important difference—Large n	Gaussian Z	Equation 13.10
	Equivalence test	TOSTs	Section 13.2.3
	Matched pairs		
	Large n	McNemar	Equation 13.12
	Small n	Binomial	Equation 13.13
	Crossover design		
	Large n	Chi-square	Section 13.2.2
	Small n	Fisher exact	Equation 13.11
Bigger Tables, No Matching	**The Case of Small n Not Discussed in This Book**	**Large n Required**	
Association	$2 \times C$ tables	Chi-square	Equation 13.15
Trend in proportions	$2 \times C$ tables	Chi-square for trend	Equation 13.16
Dichotomy in repeated measures	Many related 2×2 tables	Cochran Q	Equation 13.18
Association	$R \times C$ tables	Chi-square	Equation 13.15
Association	Three-way tables		
	Test of full independence	Chi-square	Equation 13.19
	Test of other types of independence (log–linear models)	G^2	Three-way extension of Equation 13.22
$I \times I$ **table**	Matched pairs	McNemar–Bowker	Section 13.3.2
Stratified	Stratified into many 2×2 tables	Mantel–Haenszel chi-square	Equation 14.26

TABLE S.3

Procedures for Test of Hypothesis on Relative Risk (RR) and Odds Ratio (OR)

Parameter of Interest and Setup	Conditions	Main Criterion	Equation/Section
Relative and Attributable Risks	**The Case of Small n Not Discussed in This Book**	**Large n Required**	
ln(RR)	Two independent samples	Gaussian Z or chi-square	Equation 14.5 or 13.8
RR	Matched pairs	As for OR	Section 14.2.2
		Gaussian Z or McNemar	Equation 14.22 or 14.23
	Stratified	Mantel–Haenszel chi-square	Equation 14.26
AR	Two independent samples	Chi-square or Gaussian Z	Equation 13.8 or 13.9
	Matched pairs	McNemar	Equation 13.12
Odds Ratio	**The Case of Small n Not Discussed in This Book**	**Large n Required**	
ln(OR)	Two independent samples	Chi-square	Equation 13.8
OR	Matched pairs	Gaussian Z or McNemar	Equation 14.22 or 14.23
	Stratified	Mantel–Haenszel chi-square	Equation 14.26

TABLE S.4

Statistical Procedures for Test of Hypothesis on Means or Locations

Setup	Conditions	Main Criterion	Equation/Section
One sample	Comparison with prespecified—Gaussian		
	σ known	Gaussian Z	Section 15.1.1
	σ not known	Student t	Equation 15.1
Comparison of two groups	Paired—Gaussian	Student t	Equation 15.3
	Paired—Non-Gaussian		
	Any n	Sign test	Equation 15.17a–c
	$5 \le n \le 19$	Wilcoxon signed-ranks W_S	Equation 15.18a
	$20 \le n \le 29$	Standardized W_S referred to Gaussian Z	Equation 15.18b
	$n \ge 30$	Student t	Equation 15.3
	Unpaired—Gaussian		
	Equal variances	Student t	Equation 15.6a
	Unequal variances	Student t	Equation 15.6b
	Unpaired—Non-Gaussian		
	n_1, n_2 between (4, 9)	Wilcoxon rank-sum W_R	Equation 15.19
	n_1, n_2 between (10, 29)	Standardized W_R referred to Gaussian Z	Equation 15.20
	$n_1, n_2 \ge 30$	Student t	Equation 15.6a or 15.6b
	Crossover design Gaussian	Student t	Section 15.1.3
	Up-and-down trial		Section 15.1.4
	Detecting medically important difference	Student t	Equation 15.23
	Equivalence tests	Student t	Section 15.4.2
Comparison of three or more groups	One-way layout Gaussian Non-Gaussian	ANOVA F	Equation 15.8
	$n \le 5$	Kruskal–Wallis H	Equation 15.21
	$n \ge 6$	H referred to chi-square	Equation 15.21
	Two-way layout Gaussian Non-Gaussian (one observation per cell)	ANOVA F	Section 15.2.2
	$J \le 13$ and $K = 3$	Friedman S	Equation 15.22a or 15.22b
	$J \le 8$ and $K = 4$	Friedman S	Equation 15.22a or 15.22b
	$J \le 5$ and $K = 5$	Friedman S	Equation 15.22a or 15.22b
	Larger J, K	S referred to chi-square	Equation 15.22a or 15.22b

TABLE S.4 (continued)

Statistical Procedures for Test of Hypothesis on Means or Locations

Setup	Conditions	Main Criterion	Equation/Section
	Multiple comparisons Gaussian		
	All pairwise	Tukey D	Equation 15.15
	With control group	Dunnett	Section 15.2.4
	Few comparisons	Bonferroni	Section 15.2.4
Repeated measures	Gaussian		Section 15.2.3

TABLE S.5

Methods for Studying the Nature of Relationship

Dependent Variable (y)	Independent Variables (xs)	Method	Equation/ Section
Quantitative[a]	Qualitative	ANOVA	Section 15.2
Quantitative	Quantitative	Quantitative regression	Chapter 16
Quantitative	Mixture of qualitative and quantitative	ANCOVA	Section 16.3.2
Qualitative (dichotomous)	Qualitative or quantitative or mixture	Logistic	Sections 17.1 and 17.2
Qualitative (polytomous)	Qualitative or quantitative or mixture	Logistic—any two categories at a time	Section 17.3.2
	Quantitative	Discriminant	Section 19.2.3
Survival	Groups	Life table	Equation 18.8
		Kaplan–Meier	Equation 18.10
		Log–rank	Section 18.3.1
Hazard ratio	Mixture of qualitative and quantitative	Cox model	Section 18.3.2

Note: Large n required, particularly for tests of significance. Exact method for small n not discussed in this book.

[a] Quantitative are variables on metric scale without any broad categories. Fine categories are admissible.

TABLE S.6

Main Methods of Measurement of Strength of Relationship between Two Variables

Types of Variables	Measure	Equation/Section
Both qualitative		
Binary categories	OR and several others	Section 17.5.1
Polytomous	Phi-coefficient	Equation 17.7a
categories—nominal	Contingency coefficient	Equation 17.7b
	Cramer V	Equation 17.7c
	Proportional reduction in error	Equation 17.8
Polytomous	Kendall tau, Goodman–Kruskal	Section 17.5.1
categories—ordinal	gamma, Somer d	
Dependent qualitative and	Odds ratio	Section 17.1
independent quantitative		
Dependent quantitative and	R^2 from ANOVA	Equation 17.9
independent qualitative		
Both quantitative	η^2 from regression	Equation 16.7
For multiple linear	R^2 from regression	Use Equation 16.7
For simple linear	r	Equation 16.17
For monotonic	r_S	Equation 16.19
For intraclass	r_I	Equation 16.20 or 16.21
Agreement		
Qualitative	Cohen kappa	Equation 17.10
Quantitative	Limits of disagreement	Section 16.5.2
	Intraclass correlation	Equation 16.20 or 16.21

TABLE S.7

Multivariate Methods in Different Situations (Large *n* Required)

Nature of the Variables	Objective	Types of Variables	Statistical Method	Section
A dependent set and an independent set	Relationship	Both quantitative	Multivariate multiple regression	Section 19.2.1
	Equality of means of dependents	Dependent quantitative and independent qualitative	MANOVA	Section 19.2.2
Dependent is one of many groups	Classify subjects into known groups	Independent quantitative	Discriminant analysis	Section 19.2.3
All variables interrelated (none is dependent)	Discover natural clusters of subjects	Qualitative or quantitative or mixed	Cluster analysis	Section 19.3.1
	Identify underlying factors that explain the interrelations	Quantitative	Factor analysis	Section 19.3.2

Note: Situations not mentioned in Tables S.1 through S.7 are not discussed in this book.

Frequently Used Notations

I have tried to restrict the mathematical expressions to a minimum but notations have been used so that clarity and generalizability do not suffer. Some notations have been used for more than one quantity. The following list may help you to understand the text more easily. The list is not exhaustive. Some notations that have been sparingly used in specific contexts are not included.

α	level of significance
$1 - \alpha$	confidence level
β	probability of Type-II error
β_k	regression coefficient in the population for the kth regressor
$1 - \beta$	power
χ^2	chi-square
ε	relative precision
η^2	coefficient of determination
κ	Cohen kappa
λ	logistic of π
Λ	Wilks criterion
$\lambda(t)$	hazard at time t
μ	population mean
μ_j, μ_k	population mean in the subscripted group ($j = 1, 2, \ldots, J$; $k = 1, 2, \ldots, K$)
ν	degrees of freedom
π	population proportion or probability
π_{rc}, π_k	π in the subscripted group ($r = 1, 2, \ldots, R$; $c = 1, 2, \ldots, C$; $k = 1, 2, \ldots, K$)
ρ	product–moment correlation coefficient in the population
ρ_s	rank correlation coefficient in the population
σ	population standard deviation
Σ	sum
φ	phi-coefficient
A	antecedent characteristic
A, B, C, D	frequencies in a 2×2 table—retrospective matched pairs
a, b, c, d	frequencies in a 2×2 table, particularly in case of relative risk and odds ratio
a_k	upper end point of the kth interval ($k = 1, 2, \ldots, K$)
a_{km}	factor loading of mth factor on kth variable ($m = 1, 2, \ldots, M$; $k = 1, 2, \ldots, K$)

b	sample regression coefficient in simple linear regression
bar (‾) over a variable	mean of that variable, such as $\bar{x}$ and $\bar{y}$
b_k	sample regression coefficient for the kth regressor
b_{mk}	factor score coefficient of mth factor for kth variable
C	cumulative frequency, number of columns in a contingency table, contingency coefficient, complaints or symptoms complex, contribution to log-likelihood, number of controls per case
D	discriminant function, disease
d	difference, discriminant value, Euclidean distance
D_k	kth discriminant function
dot ($\bullet$)	sum over the corresponding group
E	residual, Naperian base
E_k, E_{rc}, E_{rcl}	expected frequency in the subscripted group
e_t	expectation of life at age t in a life table
F	ANOVA criterion
F	function
f_k	frequency in the kth group
F_m	mth factor ($m = 1, 2, ..., M$)
H	Kruskal–Wallis criterion
H_0	null hypothesis
H_1	alternative hypothesis
hat(^) over a parameter or a variable	estimated or predicted value of the parameter or of the variable
I	sampling interval, smoking index
J	number of groups, number of independent variables
K	number of groups, number of dependent variables
L	number of layers in a contingency table, likelihood, precision, number of years lived (in life table)
L_0	likelihood under H_0
L_1	likelihood under the model
M	number of factors, number of observers, number of methods, number of items
max	maximum value among the columns
N	number of subjects in a population
n	size of sample
n_k	number of subjects in the kth group in a sample
O	Outcome
O_k, O_{rc}, O_{rcl}	observed frequency in the subscripted group ($k = 1, 2, ..., K$; $r = 1, 2, ..., R$; $c = 1, 2, ..., C$; $l = 1, 2, ..., L$)
P	probability, particularly of Type-I error
p	proportion in the sample, estimated probability

$P(-)$	negative predictivity
$P(+)$	positive predictivity
Q	Cochran statistic
Q_1	first quartile
Q_3	third quartile
R	number of rows in a contingency table
r	product–moment correlation coefficient in a sample
R^2	square of the coefficient of multiple correlation
r_I	intraclass correlation coefficient in a sample
R_{ij}	rank of Y_{ij} (in case of nonparametric methods)
r_s	Spearman's rank correlation coefficient in a sample
s	sample standard deviation
$S(-)$	specificity
$S(+)$	sensitivity
S_1, S_2	Friedman criterion
s_d	standard deviation of difference in a sample
s_p	pooled estimate of the standard deviation
t	Student t
T	test, number of time points ($t = 1, 2, \ldots, T$)
t_v	t-value at v degrees of freedom
W_R	Wilcoxon rank-sum criterion
W_S	Wilcoxon signed-ranks criterion
x, y	variable values
$X_{[1]}$	ordered value of x at ith rank
x_i, x_k	ith or kth observation, midpoint of an interval
x_{ij}, y_{ij}, y_{ijk}	ith observed value of the variables x or y in the jth or (j,k)th group ($i = 1, 2, \ldots, n; j = 1, 2, \ldots, J; k = 1, 2, \ldots, K$)
Z	Z-score or a standardized Gaussian variable
z	specific value of Z
z_α	the value such that $P(Z \geq z_\alpha) = \alpha$

1

Medical Uncertainties

The human body has a mechanism to adjust itself to minor variations in the internal as well as external environment. Perspiration in hot weather, shivering in the cold, increased respiration during physical exercise, excretion of redundant nutrients, replacement of lost blood after hemorrhage, and decrease in the diameter of the pupil in bright light are examples of this mechanism. This is a continuous process that goes on all the time in our body and is referred to as **homeostasis**. *Health could be defined as the dynamic balance of body, mind, and soul when homeostasis is going on perfectly well.* The greater the capacity to maintain internal equilibrium, the better is the health. Perhaps, human efficiency is optimal in this condition. Sometimes infections, injury, nutritional imbalances, stress, etc., become too much for this process to handle, and external help is needed. Medicine can be defined as the intervention that tries to put the system back on track when aberrations occur.

Health differs greatly from person to person and periodically differs in the same person from time to time. The variations are so prominent that no two individuals are ever exactly alike. Differences in facial features and morphologic appearance help us to identify people uniquely. But more important for medicine are the profound variations in physiologic functions. We all know that measurements such as hemoglobin level, cholesterol level, and heart rate differ from person to person even in perfect health. Variations occur not only between individuals but also periodically within individuals. Diurnal variations in body temperature, blood pressure (BP), and blood glucose levels are normal. In addition, states such as shock, anger, and excitement temporarily affect most of us and also have the potential to produce long-term sequelae. In the presence of such large variation, it is not surprising that a response to a stimulus such as a drug can seldom be exactly reproduced even in the same person. Uncertainties resulting from these variations are an essential feature of the practice of medicine and deserve recognition. If they do not prevent us from taking decisions in daily life, why should they in such important aspect as health?

Role of biostatistics: Medicine is not just ingesting a drug. It involves close interactions with the patient. More often than not, a large number of steps are taken before arriving at a treatment regimen. The patient's history is reviewed; measurements such as weight, BP, and heart rate are recorded; physical examination is carried out; and investigations such as the electrocardiogram (ECG), x-ray studies, blood glucose measurements, and stool examination are done. In passing through these steps, the patient sometimes encounters many observers and many instruments. Variations among them contribute their share to the uncertainties in clinical practice. The assessment of diagnosis, treatment, and prognosis can all go wrong. To highlight the large magnitude of these uncertainties, details of various contributing factors are provided in Section 1.1. These details show how profound the uncertainties are and how important it is to delineate them and contain their effect. The role of statistics is precisely this. Statistics is the **science of management of uncertainties**—a tool to measure them and minimize their impact on decisions.

Biostatistics comprises statistical methods that are used to manage uncertainties in the field of medicine and health. Although the bio part of biostatistics should stand for all biological sciences, it has become convention to apply the term biostatistics to statistical applications in only medical and health sciences.

Since the uncertainties are glaring, one wonders how medicine has been successful, sometimes very successful, in giving succor to mankind. The silver lining is that a trend can still be detected among these variations, and following this trend yields results within clinical tolerance in most cases. The term clinical tolerance signifies that the medical intervention may not necessarily restore the system to its homeostatic level but tends to bring it closer to that level so that the patient feels better, almost cured. Also note the emphasis on "most cases." Positive results are not obtained in all cases, nor is this expected. But a large percentage of cases respond to medical intervention. Thus, the statement is doubly probabilistic. As we proceed, I hope to demonstrate how statistical medical practice is and what can be done to delineate and minimize the role of uncertainties and thus increase the efficiency of medical decisions.

The explanation of statistics would not be complete without describing two usages of this term. The meaning given in the preceding paragraphs is valid when the term is used in the singular. A more common use, however, is in the form of a plural. *Numerical information is called statistics.* It is in this sense that the media use this term when talking about football statistics, income statistics, or even health statistics.

This chapter: This chapter attempts to highlight uncertainties present in all setups of health and disease. Details of uncertainties in day-to-day clinical practice are described in Section 1.1. However, it is in the case of medical research setup that many uncertainties requiring statistical subtleties emerge. Some of these are described in Section 1.2. Biostatistics is often associated with community health and epidemiology—and this association is indeed strong. Although an epidemiological perspective will be visible throughout this book, some aspects of health planning and evaluation are specifically discussed in Section 1.3. Section 1.4 provides an outline of the methods discussed in various chapters of the book for managing these uncertainties.

1.1 Uncertainties in Health and Disease

The state of health is a result of an intricate interaction of a large number of factors. Thus, this is an extremely complex phenomenon. Many aspects of this complexity are not fully understood, and most of what is understood seems beyond human control. The most common source of uncertainty in medicine is the natural biologic variability between and within individuals. Variations between laboratories, instruments, observers, etc., are factors that further uncertainty. Details of various sources of uncertainties are as follows.

1.1.1 Uncertainties due to Intrinsic Variation

Body temperature and plasma glucose level are everyday examples of medical parameters that are evaluated against their normal values. The need to define and use such normals arises from the realization that variations do exist and it is perfectly normal for them to occur in healthy subjects. Such variations can be due to a number of factors. The following is a list of sources of commonly occurring intrinsic variability. The list is restricted to those that have profound effect. The sources listed are not necessarily exclusive of one another. In fact, in practice, the overlap can be substantial.

1.1.1.1 Biologic Variability

Age, gender, birth order, blood group, height, and weight are among the biological factors that occur naturally in a health setup. Health parameters of children are quite different from those of adults. Almost all kinds of measurements—anatomical, physiological, or biochemical—differ from age to age. For example, levels of BP seen in subjects of age, say, 20 years can be hardly applied to subjects of age 60 years. Similarly, assessment of the health of males based on, say, hemoglobin level or a lung function is not on the same scale as that of females. Monitoring body weight is much more important in growing children than in adults. Mumps has a different prognosis when it affects a postpubertal male. Thus, sometimes even the interaction among biological factors is important.

Biological variability is seen not only between subjects but also within subjects. Examples of diurnal variation in body temperature, BP, and blood glucose levels have already been cited. Menstrual cycles in women are accompanied by many other physiologic changes that are periodic in nature. A person may respond exceedingly well at one time but fail desperately at another time. Variations in physiological parameters and their role in determining health status at a particular point of time are rarely understood or even fully appreciated, although they may, in some cases, be the most dominant contributors. Fear and anxiety can cause marked alterations in physiological functions. All these variations contribute significantly to the spectrum of uncertainty in health and medicine.

1.1.1.2 Genetic Variability

African Blacks, Chinese Mongoloids, Indian Aryans, and European Caucasians differ not only in morphological features and anatomical structure but also in physiological functions. They may vary with regard to vital capacity and blood group composition. Aberrations such as thalassemia, Down's syndrome, and color blindness are of course entirely genetic. Sickle cell anemia, muscular dystrophy, and hemophilia A are genetic in origin. Many diseases with multifactorial etiology such as hypertension and diabetes have a genetic component. Because of genetic variability in the populations, the clinician has to be wary of the possibility of a genetic influence on signs and symptoms on the one hand, and the rate of recovery on the other.

1.1.1.3 Variation in Behavior and Other Host Factors

Whereas our anatomy and physiology are traceable mostly to hereditary factors, pathology is caused mostly by our own behavior and the environment. Environmental influences are discussed in the next paragraph.

The emergence of human immunodeficiency virus (HIV) positivity has brought sexual behavior into focus. Almost all sexually transmitted infections (STIs) originate from aberrant sexual relationships. Smoking is seen as an important factor in several types of carcinomas and in many conditions affecting the heart and lung. Heavy drinking for many years can affect the bones, kidneys, and liver. A sedentary lifestyle can cause spondylitis and coronary diseases.

Nutrition is probably the most dominant factor that controls the body's defense mechanism. This is determined by the consumption of the right food and awareness of this requirement. Enormous variation in awareness impacts prevention, cause, and treatment of many diseases in an unpredictable way. Scabies and dental caries thrive from lack of personal hygiene.

Some personality traits can affect the risk of hypertension and other cardiovascular diseases. A positive attitude helps ward off some ailments and assists in early recovery when struck. Stress compromises one's ability to respond to even an established treatment. All these individual factors greatly vary from person to person. Susceptibility and response are the result of a large number of interacting factors. The nature of this interaction also can vary widely between individuals, contributing to the spectrum of uncertainty.

1.1.1.4 Environmental Variability

Climatic factors sometimes determine the type and virulence of pathogens. Pollution and global warming are now acquiring center stage in the health scenario with a predilection for grim consequences. Flies, mosquitoes, and rodents are the carriers of many deadly diseases. Many gastrointestinal disorders are waterborne. The relationship between insanitation and disease is easy to discern.

An environment of tension and stress may substantially alter one's ability to cope with, say, infections. Love, affection, and prayers of the family and others sometimes do wonders in the recovery of a patient, perhaps by providing innate strength to fight.

Availability of appropriate and timely medical help has a tremendous impact on the outcome. Thus, health infrastructure plays an important role. The health culture of people with regard to utilizing the services also varies from population to population.

All these environmental factors need due consideration while dealing with health or disease, at the individual as well as at the community levels. Their effect on different people is not uniform. Some subjects tend to be affected more than others due to variability in host–environment interaction.

1.1.1.5 Chance Variability

Let us go a little deeper into the factors already listed. Aging is a natural process, but its effect is more severe in some than in others. When exposed to heavy smoking, some people develop lung cancer, others do not. Despite consuming same water with deficient iodine, some people do not develop goiter, whereas some do—that too of varying degree. The incubation period differs greatly from person to person after the same exposure. Part of such variation can be traced to factors already mentioned, such as personality traits, lifestyle, nutritional status, and genetic predisposition. However, these known factors fail to explain the entire variation. Two patients apparently similar, not just with regard to the disease condition but also for all other known factors, can respond differently to the same treatment regimen. Even susceptibility levels sometimes fail to account for all the variation. The unknown factors are called **chance**. Sometimes, the known factors that are too complex to comprehend or too many to be individually considered are also included in the chance syndrome. In some situations, chance factors could be very prominent contributors to uncertainties and in some situations they can be minor.

1.1.1.6 Sampling Fluctuations

Much of what is known in medicine today has been learned through accumulated experience. This empiricism is basic to most medical research. This aspect is discussed in Section 1.2.1. Beware that experience is always gained by observing a fraction of subjects and never all subjects. The knowledge that chills, fever, and splenomegaly are common in malaria is based on what has been observed in cases over a period. However, the cases actually observed or studied are *not all* that occurred in the world. Only a fraction of these cases, called a sample, has been studied. Similarly, saying that the normal albumin level is 56%–75% of total serum proteins is based on the levels seen in *some* healthy subjects.

One feature of samples is that they tend to provide a different picture with repeated sampling. This is called sampling fluctuation or sampling error.

This error is not a mistake but indicates only a variation. Sampling fluctuation is discussed in Chapter 3. The objective of mentioning all this here is to point out that sampling fluctuations themselves are a source of uncertainty. A clinician constantly deals with biopsies and samples of blood, urine, and sputum. Despite overt presentation, these samples may turn out to be negative. Fine-needle aspiration cytology (FNAC) takes out a very small sample of tissue, and this particular sample may provide a normal picture even when proliferation is present. The cells in the sample material may be unaffected. Sampling fluctuation is one of the many reasons that necessitate repeat investigations in some cases.

1.1.2 Natural Variation in Assessment

Minimal variations in repeat measurements occur despite all precautions. If this can happen with fixed parameters such as length of a dining table, it could definitely happen with biological parameters that are pliable anyway. A purist can call them errors rather than natural variation. These variations can be minimized by exercising more care but cannot possibly be eliminated.

1.1.2.1 Observer Variability

Barring some clear-cut cases, clinicians tend to differ in their assessment of the same subject. Interpretation of x-ray films is particularly notorious in this respect. Disagreement exists concerning such simple tools as a chart for assessing growth of children (see, e.g., [1]). There is wide variation in the estimates of the number of diabetes cases existing in the world today. Physicians tend to differ in grading a spleen enlargement. One physician may consider a fasting blood glucose level of 136 mg/dL in a male of age 60 years sufficient to warrant active intervention, but another might opt just to monitor. Variation in BP readings due to differences in hearing acuity or in interpretation of Korotkoff sounds is on record. Some clinicians are more skillful than others in extracting information from the patient and in collating pieces of information into solid diagnostic evidence. An additional factor is the patient–doctor equation. Because of the confidence of a patient in a particular clinician, the concerned clinician is able to secure much better information. Such variability on the part of the observer, researcher, or investigator is a fact of life and cannot be wished away. It is inherent in humans and represents a healthy feature rather than anything to be decried. Efforts are made from time to time to reconcile and come to a consensus. Indeed, such a consensus is reached on many occasions. Yet, many issues remain unresolved and new ones keep cropping up, and they continue to contribute to the spectrum of uncertainty.

1.1.2.2 Variability in Treatment Strategies

The physician basically is a healer. Variability in the treatment strategies of different physicians is wide, and outcomes are accordingly affected. A large number of systems of medical care are available. Examples are allopathy, homeopathy, Ayurveda, and Unani. Then, there are acupunctures, magnetic therapies, etc. There are specialists in each system. They evaluate various medical parameters in different ways and come to different conclusions. Within the modern system, physicians tend to prefer one treatment modality relative to others. For example, some emphasize lifestyle changes for treating hypertension, whereas others depend primarily on drugs. For a condition such as acute cystitis of the urinary tract in women, the management strategies differ widely from physician to physician. There is no consensus. Such variation could have special significance when conclusions are drawn based on subjects from different hospitals or from a cross section of a population.

1.1.2.3 Instrument and Laboratory Variability

Blood pressure of an individual measured by mercury sphygmomanometer is often found different from that obtained with an electronic instrument. The weight of children on a beam balance and on a spring balance may differ. Apart from such simple cases, laboratories too tend to differ in their results for splits of the same sample. Send aliquots of the same blood sample to two different laboratories for a hemogram, and be prepared to receive different reports. The difference could be genuine due to sampling fluctuation or could be due to differences in chemicals, reagents, techniques, etc., in the two laboratories. Above all, the human element in the two laboratories may be very different. Expertise may differ, and the care and attentiveness may vary. Differences occur despite standardization, although even the standard may differ between laboratories in some cases.

Inter- and intraindividual variation and variation due to observer, instrument, sample, etc., are just one of the components responsible for uncertainties in medical decisions. Another component comes from factors such as incomplete information, imperfect instruments, and poor compliance with the regimen.

1.1.2.4 Imperfect Tools

A clinician uses various tools during the course of practice. Examples are signs, symptoms, syndrome, physical measurements, laboratory and radiological investigations, and intervention in the form of medical treatment or surgery. Besides their skills in optimally using what is available, the efficiency of clinicians depends on the validity and reliability of the tools they use. **Validity** refers to the ability to measure correctly what a tool is supposed to measure, and **reliability** means consistency in repeated use. Sensitivity, specificity, and predictivities are calculated to assess validity. Reliability is evaluated in terms of measures such as Cohen kappa and Cronbach alpha. Details of some of these measures are given in this book. In practice, no medical tool is 100% perfect. Even a computed tomography (CT) scan can give a false-negative or false-positive result. A negative histologic result for a specimen is no guarantee that proliferation is absent, although in this case positive predictivity is nearly 100%. The values of measurements such as creatinine level, platelet count, and total lung capacity are indicative rather than absolute, that is, they mostly estimate the *likelihood* of a disease. Signs and symptoms seldom provide infallible evidence. Because all these tools are imperfect, decisions based on them are also necessarily probabilistic rather than definitive.

1.1.2.5 Incomplete Information on the Patient

When a patient arrives in coma at the casualty department of a hospital, first steps for management are often taken without considering the medical history of the patient or without waiting for laboratory investigations. An angiography may be highly recommended for a cardiac patient, but treatment decisions are taken in its absence if the facility is not available in that particular health center. Even while interviewing a healthy person, it cannot be ensured that the person is not forgetting or intentionally not suppressing some information. Suppression can easily happen in the case of sexually transmitted diseases (STDs). An uneducated subject may even fail to understand the questions or may misinterpret them. Some investigations such as CT and magnetic resonance imaging (MRI) are expensive, and lack of funds may sometimes lead to proceeding without these investigations even when they are highly recommended. Thus, the information remains incomplete in many cases despite best efforts. Clinicians are often required to make a decision about treatment based on such incomplete information.

1.1.2.6 Poor Compliance with the Regimen

Medical ethics requires that the patient's consent is obtained before a procedure is used. Excision of a tumor may be in the interest of a patient, but this can be done only after informed consent of the patient is obtained. Patients do not always agree. When a drug treatment is prescribed, the patient may or may not follow it in its entirety. Noncompliance can be due to circumstances beyond the control of the patient, due to carelessness, or even intentional. In a community setup, if immunization coverage to the extent of 90% is required to control a disease such as polio, best efforts may fail if the public is not cooperative. A cold chain always remains a source of worry in a domiciliary drive against polio. Healthcare providers seldom have control on compliance, and this adds to uncertainty about the outcome.

1.1.3 Inadequate Knowledge

Notwithstanding claims of far-reaching advances in medical sciences, many features of the human body and mind, and their interaction with environment, are not sufficiently well known. How the mind controls physiological and biochemical mechanisms is an area of current research. What specific psychosomatic factors cause women to live longer than men is still shrouded in mystery. Nobody knows yet how to reverse hypertension that can obviate the dependence on drugs. Cancers are treated by radiotherapy or excision because procedure to regenerate aberrant cells is not known. Treatment for urinary tract infections in patients with impaired renal function is not known. Such gaps in knowledge naturally add to the spectrum of uncertainty.

1.1.3.1 Epistemic Uncertainties

Knowledge gaps are wider than generally perceived. One paradigm says that what we do not know is more than what we know. Unfamiliarity breeds uncertainty. Such uncertainties are called epistemic uncertainties. Besides incomplete knowledge, they also include (1) ignorance, for example, how to choose one treatment strategy when two or more are equally good or equally bad, such as between amoxicillin and cotrimaxazole in nonresponsive pneumonia; (2) parameter uncertainty regarding the factors causing or contributing to a particular outcome such as etiological factors of vaginal and vulvar cancer; (3) speculation about unobserved values such as effect of high levels of NO_2 ($100\,mg/m^3$) in the atmosphere; (4) lack of knowledge about exact quantitative effect of various factors such as diet, exercise, obesity, and stress on raising blood glucose level; and (5) confusion about definition of various health conditions such as hypertension—that the BP cutoff should be 140/90 or 160/95 mmHg. Epistemic uncertainty also arises from biases of the observers and errors in measurements.

Another kind of epistemic uncertainty arises from nonavailability of proper instrument. How do you measure blood loss during a surgical operation? Swabs that are used to suck blood are not standardized. In some surgeries, blood can even spill onto floor. Even a simple parameter such as pain is difficult to measure. Visual analog scale (VAS) and other instruments are just approximations. Stress defies measurement, and behavior/opinion type of variables present stiff difficulties. If the measurement is tentative, naturally the conclusion too is tentative.

1.1.3.2 Diagnostic, Therapeutic, and Prognostic Uncertainties

Diagnostic uncertainties arise because the tests or assessments used for diagnosis do not have 100% predictivity. The perfect test is not known. An ECG can give false-positive or false-negative results. None of the procedures, for example, FNAC, ultrasonography, and mammogram, are perfect for identifying or excluding breast cancer. Cystic fibrosis is difficult to evaluate.

No therapy has ever been fully effective in all cases. Therapeutic uncertainties are particularly visible in surgical treatment of asymptomatic gland-confined prostate cancer and in medical treatment of benign prostatic hyperplasia. Many such examples can be cited. In addition are substances such as combined oral pill where long-term use may increase the risk of breast, cervical, or liver cancer but reduce the risk of ovarian, endometrial, and colorectal cancer.

Prognostic uncertainties due to lack of knowledge exist in sudden severe illness. Such illness can occur due to a variety of conditions and its cause and outcome is difficult to identify. Nobody can predict occurrence or nonoccurrence of irreversible brain damage after an ischemic stroke. Prognosis of terminally ill patients is also uncertain. Method of care for women undergoing hysterectomy is not standardized.

1.1.3.3 Predictive and Other Uncertainties

Medicine is largely a science of prediction of diagnosis, prediction of outcome of treatment, and prediction of prognosis. In addition, there are many other types of medical predictions for which knowledge barriers do not allow certainty. Look at the following examples:

- Gender of a child immediately after conception
- Survival duration after onset of an end-stage serious disease
- Number of hepatitis B cases that would come up in the following year in a country with endemic affliction
- Number of people to die of various causes in future

These are examples of universal inadequacies. In addition is the inadequate knowledge of an individual physician, a nurse, or a pharmacist. First, it is difficult to recollect everything known at the time of confronting a patient. Second, some caregivers just do not know how to handle a situation although others know it well. All this also contributes to the spectrum of uncertainties.

The objective of describing various sources of uncertainty in such detail is to sensitize the reader to their unfailing presence in practically all medical situations. Sometimes they become so profound that medicine transgresses from a science to an art. Many clinicians deal with these uncertainties in their own subjective ways, and some are very successful. But most are not as skillful. To restore a semblance of science, methods are needed to measure these uncertainties, to evaluate their impact, and, of course, to keep their impact under control. All these aspects are primarily attributed to the domain of biostatistics and are the subject matter of different chapters in this book.

1.2 Uncertainties in Medical Research

The discussion in Section 1.1 is restricted mostly to the uncertainties present in day-to-day clinical problems. Biostatistics is not merely measurement of uncertainties but is also concerned with the control of their impact. Such a need is more conspicuous in a research setup than in everyday practice.

All scientific results are susceptible to error, but uncertainty is an integral part of medical framework. The realization of the enormity of uncertainty in medicine may be recent, but the fact is age old. Also, our knowledge about biological processes still is extremely limited. These two aspects—variation and limitation of knowledge—throw an apparently indomitable challenge. Yet, medical science has not only survived but is also ticking with full vigor. The silver lining is the ability of some experts to learn quickly from their own and others' experience and to discern signals from noise, waves from turbulence, and trends from chaos. It is due to this expertise that death rates have steeply declined in the past 50 years and life expectancy is showing a relentless rise in almost all nations around the world. Burden of disease is steadily but surely declining in most countries. The backbone of such research is empiricism.

1.2.1 Empiricism in Medical Research

Empiricism can be roughly equated with experience. An essential ingredient in almost all primary medical research is observation of what goes on naturally, or before and after a deliberate intervention. Because of the various sources of uncertainties listed earlier, such observations seldom provide infallible evidence. This can be briefly explained in the context of different types of medical research as follows. The details are given in subsequent chapters.

1.2.1.1 Laboratory Experiments

Laboratory experiments in medicine are often performed on animals but are sometimes performed on biological specimens. Experiments help us understand the mechanisms of various biological functions and of the response to a stimulus. The laboratory provides an environment where the conditions can be standardized so that the influence of extraneous factors can be nearly ruled out. To minimize the role of interindividual variation, homogeneous units (animals and biological specimens) are chosen and are randomly allocated to the groups receiving a specific stimulus. Because of the controlled conditions, an experiment can provide clear answers even when performed on a small number of subjects. Nonetheless, experiments are often replicated to get more experience and thus to strengthen confidence in the results.

1.2.1.2 Clinical Trials

Clinical trials are experiments on humans. They are mostly done to investigate new modes of therapy. Research on new diagnostic procedures also falls in this category. Clinical trials are carried out meticulously involving heavy investment. Since variation between and within subjects occurs due to a large number of factors, it is quite often a challenge to take full care of all of them. Epistemic uncertainty also plays a role. It is imperative in this situation that the rules of empiricism are rigorously followed. This means that the trial should be conducted in controlled conditions so that the influence of extraneous factors is minimized, if not ruled out. Also, the trials should be conducted with a sufficient number of patients so that a trend, if any, can be successfully detected.

1.2.1.3 Surgical Procedures

A new surgical procedure is extensively studied regarding its appropriateness before initiating its trial. Abundant precautions are taken in this kind of research, and each case is intensively investigated for days. Because of such care, an operation found successful in one patient is likely to be successful in another similar patient when the same precautions are taken. Surgeons can afford to be extra cautious because the new procedure will seldom be used on a large number of cases until sufficient experience is gained. They generally have the opportunity to study each case in depth before performing a new type of surgery. Perhaps response of tissues to a surgery is less variable than of physiological functions to a medication. Despite all this, failures do occur, as in the case of organ transplantation, and uncertainties remain prominent in this kind of research also.

Nonetheless, there are surgical trials that are comparable with medical trials. Evaluation of transurethral incision against transurethral resection of prostate in benign prostatic hyperplasia can be done similar to a drug trial. The same kind of uncertainties exists in this setup, and strategies such as randomization and blindness can be used to minimize their impact. A trial for comparing methods of suturing can also be done just as a regular clinical trial.

1.2.1.4 Epidemiological Research

In epidemiological research, association or cause–effect relationships between etiologic factors and outcomes are investigated. Research on factors causing cancers, coronary artery disease, and infections occurring in some but not in others is classified as epidemiological. This can also help us understand the mechanism involved. A very substantial part of modern medical research is epidemiological. The relationship under investigation is influenced even more by various sources of uncertainty in this kind of research and a conclusion requires experience gained from a large number of cases. Statistical methods again are needed to separate clear signals from chance fluctuations.

1.2.2 Elements of Minimizing the Impact of Uncertainties on Research

The sources of intrinsic variation listed in Section 1.1.1 are mostly beyond control, but their impact can still be managed. Other sources of uncertainty also contribute, but investigations can be designed such that their influence on decisions is minimized. These designs are discussed later in detail. Elementary concepts are given in the following paragraphs as illustration. The quality of decisions in the long run can also be enhanced by devising and using improved medical methods. Both require substantial statistical inputs.

The real challenge in research is thrown by epistemic uncertainties that arise from inadequate medical knowledge. Statements are sometimes made without realizing that they are assumptions. Gastric ulcer was thought to be caused by acidity till it was established that the culprit is *Helicobacter pylori* in many cases. Thus, even fully established facts should be continuously evaluated and replaced by new ones where needed.

1.2.2.1 Proper Design

A clinical trial aims to evaluate the efficacy of one or more treatment procedures—generally different drugs or different dosages of the same drug—relative to one another. The "another" could be "no treatment" or "existing treatment" and is called the **control**. Among the precautions sometimes taken is the **baseline matching** of the subjects in various groups so that the known sources of variability have less influence on the outcome. Another very effective strategy is **randomization**, which equalizes the chance of the presence of different sources of uncertainty in various groups, including the unknown sources. The techniques of observation and measurement are standardized and uniformly implemented to minimize the diverse influence of these techniques on the outcome. If identifiable sources of uncertainty still remain uncontrolled, they are taken care of at the time of analysis by suitable adjustments. Appropriate statistical methods help in arriving at a conclusion that has only a small likelihood of being wrong.

These preliminaries are stated in the context of clinical trials, but other medical investigations, be they in a community, in a clinic, or in a laboratory, have the same basic structure and require similar statistical inputs. Descriptive research, such as to estimate the magnitude of the problem or to delineate normal levels of, say, a hematological parameter in a specific population, also requires similar care in selection of subjects and similar quality control of the instruments. The investigations into cause and effect or association also need similar inputs to minimize the influence of uncertainty and thus to increase the reliability of the conclusions.

1.2.2.2 Improved Medical Methods

Better drugs cannot be introduced without enough convincing evidence but health practices such as respiratory exercises and low-calorie diet are many times introduced, such as for reducing cardiovascular risk, without sufficient trials. Many researchers consider this anomalous since health practices tend to be adopted unnoticed on large scale and can silently endanger life and health of many people if the practice turns out to be harmful. Or, it can give false assurance if not effective. Some argue that such health practices can only benefit—thus cause no alarm. This might be true but science requires evidence.

Although the health of each individual is important, and clinical practice must use the best available methods, research endeavors generally require especially improved methods that are more accurate and more exact. This makes medical research an expensive proposition. Compromise on improved methods can substantially affect the quality of research. If such improved methods are not available, research may have to be redirected to devise such methods.

Incomplete knowledge about a patient's condition can be overcome by research into new methods that are quicker, safer, and more accurate and which can be performed more easily. To fill gaps in medical knowledge, research into more exact delineation of factors responsible for specific conditions of ill health and their mechanism is required. All this will help devise strategies to minimize the uncertain space.

Some epistemic uncertainties can be minimized by using an appropriate scoring system. Inadequacies in medical tools such as diagnostic tests can be minimized only through research on newer, more valid, and more reliable tools. Compliance with prescribed regimens can be improved by devising regimens that are simple to implement, less toxic, and more effective. Instrument and observer variability can be controlled by adhering to strict standards and thorough training. Thus, improving the methods can minimize the uncertainties arising from these deficiencies. Research into these requires scientific investigations so that the conclusions arrived at are valid as well as reliable. Proper design of the investigation helps to achieve this aim.

1.2.2.3 Analysis and Synthesis

Because of the uncertainties involved at every stage of a medical investigation, the conclusion can seldom be drawn in a straightforward manner. In almost all cases, the data obtained are carefully examined to find the answers to the questions initially proposed. For this, it is generally necessary that the data be collated in the form of tables, charts, or diagrams. Some summary measures are also chosen and computed to draw inferences. Because of the inherent variations in the data and because only a sample of the subjects is investigated rather than the entire target population, some special methods are required to draw valid conclusions. These methods collectively are called **techniques of statistical inference**. These techniques depend on the type of questions asked, design of the study, kind of measurements used, number of groups investigated, number of subjects studied in each group, etc. These techniques are the primary focus of this book and are discussed in detail in various chapters. All data processing activities, beginning with data exploration and ending with drawing inferences, are generally collectively called **statistical analysis**. The role of this analysis is to help draw valid and reliable conclusions. The term analysis probably comes from the fact that the total variability in the data is broken into its various components, thus helping to sieve clear signals or trends from noise-like fluctuations.

Although statistical analysis is acknowledged as an essential step in empirical research, the importance of synthesis is sometimes overlooked. Synthesis is the process of combining and reconciling varied and sometimes conflicting evidence. The findings of an investigation do not often match those in another investigation. Diabetes, smoking habits, and BP levels were found to be significant factors of mortality in Italy in one study but not in other studies in the same country [2]. Prevalence of hypertension in India was found to range widely from 0.36% to 30.92% in a general population of adults [3]. These differences occur for a variety of reasons such as genuine population differences; sampling fluctuation; differences in definitions, methodology, and instruments; and differences in the statistical methods used. A major scientific activity is to synthesize these varying results and arrive at a consensus. The discussion part of most articles published in medical journals tries to do such a synthesis. The objective of most review articles is basically to present a holistic view after reconciling the varying results in different studies. In addition, techniques such as meta-analysis seek to combine evidence from different studies. These synthesis methods too are primarily statistical in nature and are important for medical research.

1.3 Uncertainties in Health Planning and Evaluation

Most of the discussion so far was focused on clinical and research aspects. But medical care is just one component of the health-care spectrum. Prevention of disease and promotion of health are equally important. At the individual level, prevention is in terms of steps such as immunization, changes in lifestyle, use of fertility control methods, and improved personal hygiene. Health education is basic to all prevention and much more efficiently done at the community level than at the individual level. All these are geared to meet the specific needs of the population. These needs vary widely from population to population, area to area, and time to time, depending on the perceived need, level of infrastructure, urgency, etc. A predominantly pediatric population with a high prevalence of infectious diseases requires entirely different services than an aging population with mostly chronic ailments. These variations compel the use of statistical methods in health planning. Health situation analysis is the first step in planning the services. When a health program is implemented, the administrator always wants to know how well it is running. Sometimes, midcourse corrective steps are taken to put the program back on track. Then, the final outcome is measured in terms of the impact the program has had on the community. All these exercises have a substantial biostatistical component.

1.3.1 Health Situation Analysis

The quality and type of health services required for a community depend on the size of the community, its age–sex structure, and prevalence of various conditions of health and ill health in different sections of the community. The health services also depend on culture, traditions, perception, socioeconomic status of the population, existing infrastructure, etc. Variations and uncertainties are prominent in these aspects too. All these need to be properly assessed to prepare an adequate plan. This assessment is called health situation analysis. The basic aim of this analysis is to provide the baseline information. Since the situation can quickly change because of either natural growth of the population or interventions, a time perspective is always kept in view in this analysis.

1.3.1.1 *Identification of the Specifics of the Problem*

Generally, the broad problem requiring health action is already known before embarking on the health situation analysis and only the specifics are to be identified. It seems ideal to talk about both good and bad aspects of health, but, in practice, a health plan is drawn to meet the needs as perceived by the population. These needs are obviously related to adverse aspects of health rather than to positive aspects. These are identified from the complaints received by various social and medical functionaries. The functionaries could be mass media, such as newspapers, magazines, and television, or others such as political organizations, voluntary agencies, and medical practitioners. Sometimes a survey is required, and sometimes an expert group is set up to identify the specifics of the problem. A community-level program is planned to meet those needs that are felt by a sizable section of the population. Individual needs tend to be ignored in this scheme.

1.3.1.2 *Size of the Target Population*

Statistics attaches special meaning to the term population. A program on control of nutritional deficiencies may be restricted to the population of children younger than 5 years, a program on senile cataract restricted to the population of the aged, and a program on dental caries may include general population of age 5 years and above. An assessment of the size of such a population is necessary for two reasons. First, the magnitude of the services to be provided would depend on this size. Second, population is the denominator for many rates such as incidence and prevalence rates and birth and death rates. For the infant mortality rate, the denominator is the population of live births. For the general fertility rate, the denominator is the population of women of reproductive age. Such rates are basic to health situation analysis because they delineate the magnitude and help in measuring the impact of a program. A correct denominator is as essential for accurate assessment as the correct numerator, and obtaining them accurately can be very difficult in some situations.

The **census**, carried out periodically in most countries, is the most important source for the count of the general population. Distribution by age, gender, rural or urban area, education level, etc., is also available in most census reports. Great care is generally taken to ensure that the enumeration is complete and accurate. Thus, the counts provided by census are accepted almost without question. But there are two limitations. First, such population censuses are mostly done once in 10 years. The changes from year to year are not uniform. In areas where registration of births and deaths is not complete and migration is unchecked, the count for intermediary years is estimated by assuming that the rate of change each year is the same as the average over the decade. Second, the census is restricted to a small number of characteristics of the population. For example, a special survey may still be required to find the income distribution. This distribution may be needed as the denominator for infectious diseases, such as tuberculosis and leprosy, and for chronic diseases, such as diseases of the cardiovascular system. These diseases are strongly influenced by socioeconomic status, of which income is an important component.

1.3.1.3 Magnitude of the Problem

Assessment of the magnitude of a health problem is the core of health situation analysis. Simply stated, all it requires is the count of persons with different kinds or with different grades of the problem along with their background information. This background helps in dividing subjects into relevant groups that may be etiologically important. Grades of the problem are obtained either by direct numerical measurement, such as birth weight in grams and plasma triglyceride in milligrams per deciliter, or in terms of categories, such as none, mild, moderate, and severe as in the case of disabilities. Such gradation is fraught with uncertainties and consequent risks—thus, a caution can always be advised. The number of subjects with various grades of the problem could provide the information for various indicators that measure morbidity, mortality, and fertility. A composite index, covering multiple aspects, can also be calculated to assess magnitude of the problem.

One prerequisite for accurate assessment of the magnitude of a problem is standardized definition. Because of the differences in the population characteristics, in the perception of the investigators in grading severity, in the tools used, etc., the definitions sometimes become blunt. Chronic diarrhea [4], malnutrition [5], and hypertension in children [6] are some examples of conditions where controversies exist. To minimize uncertainty, it is necessary that the relevant terms are fully defined and uniformly used by all the investigators involved in the exercise.

The magnitude of different health problems in a community can be assessed in terms of morbidity as well as in terms of premature mortality. Assessment of both may be required to develop a plan to combat the problem.

A series of tools such as personal interview, physical examination, laboratory investigation, and imaging may be required to assign morbidity to its correct diagnostic category. Lack of perfection in these tools can be monitored by using the concepts of validity and reliability, and steps can be taken to keep it in check. Incomplete response or nonresponse can introduce substantial bias in the findings. Uncertainties remain prominent.

Health is not merely absence of morbidity and premature mortality. Situation analysis for planning a program on fertility control requires the count of births in different birth orders and the age of the women at which various orders of birth take place. It may also require investigation of age at menarche, marriage, and menopause. Breast-feeding and postpartum amenorrhea provide additional dimensions of the problem. All these aspects come under neither morbidity nor mortality, yet they are important for assessing reproductive health. Thus, morbidity and mortality will not necessarily cover all aspects of health. In any case, they rarely cover social and mental aspects.

1.3.1.4 Health Infrastructure

Assessment of the functionally available health infrastructure is an integral part of health situation analysis. This includes facilities such as hospitals, beds, and health centers; staff in terms of doctors, nurses, technicians, etc.; supplies such as equipments, drugs, chemicals, and vehicles; and most of all their timely availability at the functional level so that they can be effectively used. Other components of the health infrastructure are software such as training facilities in newer techniques; a well-defined duty schedule so that the staff knows what to do and when; motivational inputs so that the staff is not found lacking in eagerness to provide service, etc. All these vary from situation to situation. These aspects sometimes play a greater role in the success or failure of a health program than the physical facilities. Creating facilities is one thing, and their adequate utilization is another. Even good facilities can remain underutilized if the public and the patients are ignorant about them, the facilities are too expensive, cultural barriers prevent the population from using them, fear sets in for some reason, or the staff is not fully accountable. The accurate assessment of such factors is difficult.

Although the availability of health infrastructure can be accurately assessed, their utilization pattern creates an uncertain environment. A sample survey may be required and biostatistical methods would be needed to come up with an adequate assessment of the utilization pattern by different segments of population.

1.3.1.5 Feasibility of Remedial Steps

Health situation analysis is also expected to provide clues to deal with the problem. The magnitude of the problem in different segments of the population can provide an epidemiological clue based simply on excess occurrence in specific groups. Clues would also be available from the literature or from the knowledge and experience of experts in that type of situation. Not all clues necessarily lead to remedial steps. Their feasibility as an effective tool also needs to be assessed. Consider, for example, maternal complications that arise more frequently in the third and subsequent births. Avoiding those births could certainly be a solution, but many segments of the society in which such births are common may not accept the suggestion. Thus, other alternatives such as increased spacing, better nutrition, and early checkups have to be explored. Although early and frequent checkups in the antenatal period are highly desirable, poor availability of facilities may put them out of bounds for most people. Provision of a quick information system and availability of well-equipped ambulances is desirable to handle accident and trauma patients, but constraints of funds may not allow setting up an efficient system. Assessment of all such aspects is a part of the exercise of health situation analysis. Uncertainty is an integral part of all these assessments.

1.3.2 Evaluation of Health Programs

Evaluation of a health program has two distinct components. First is assessing the extent to which objectives of the program have been achieved. This requires that the objectives are measurable. For example, if the objective is to control blindness, it is always desirable to state the magnitude present at the time the program begins and the level of reduction expected at the end of the program or in each year of its implementation. If this level is mentioned for each age, gender, urban and rural area, and socioeconomic group, then the task of the evaluators becomes simpler. A second and more important component of evaluation is to identify the factors responsible for that kind of achievement—good or bad—and to measure the relative importance of these factors. Thus, the evaluation might reveal that the objectives were unrealistic considering the prevalent health situation and the program inputs. Supplies may be adequate but the population may lack the capacity to absorb all of this due to cultural and economic barriers. The target beneficiaries may not be fully aware of the contents and benefits of the program. Or, the program may fall short of the health needs as perceived by the people. Among other factors that could contribute to partial failure are errors in identification of the target beneficiaries, inadequate or ineffective supervision, and lack of expertise or motivation. Thus, *evaluation is the appraisal of the impact as well as of the process.*

Just as any other investigation, evaluation passes through various stages, each of which contains an element of uncertainty. Because of inter- and intraindividual variations in receivers as well as in providers, some segments could benefit more than the others despite equal allocation of resources. Tools to implement the program are never perfect. Supplies may be lacking because of a variety of bottlenecks, such as in storage or in transport. Some officials are more motivated than others and work better despite fewer supplies. If a program, for example, involves physical exercise by the patients and envisages screening the population for coronary heart disease to exclude these cases, then imperfect validity and imperfect reliability of the screening criteria introduce uncertainty. A program against infections may have to depend on an enzyme-linked immunosorbent assay (ELISA) test that itself could be false negative in some cases and false positive in some others. Statistical methods are required to manage these uncertainties.

1.4 Management of Uncertainties: About This Book

The discussion in the previous sections may have convinced you that uncertainties are present in practically all medical situations and they need to be properly managed. Management in any sphere is a complex process, more so if it concerns phenomena such as uncertainties. *Management of uncertainty requires a science that understands randomness, instability, and variation.* Biostatistics is the subject that deals specifically with these aspects. Instead of relating it to conventional statistical methods such as test of hypothesis and regression, biostatistical methods are presented in this book as a medical necessity to solve some medical problems. In doing so, I deliberately avoid mathematical intricacies. The attempt is to keep the text of this book light and enjoyable for the medical fraternity so that biostatistical methods are perceived as a delightful experience and not a burden. Statistical jargon is avoided and a human face is projected. For better communication, the text is sometimes in an interactive mode. Sometimes the same concept, such as different types of bias, is explained many times in different contexts in various chapters to bring home the point. For example, crossover trials are discussed for experiments, clinical trials, analysis for proportions, and analysis of means in four different chapters. Such repetition is deliberate. It is desirable too from the viewpoint of medical and health professionals who are the target audience, although statisticians might not appreciate such repetitive presentation.

A new science establishes itself by demonstrating its ability to resolve issues that are intractable within the perimeters of present knowledge. The effort in this book is to establish biostatistics as a science that helps manage medical uncertainties in a manner that no other science can do.

As already stated, uncertainties in most medical setups are generally intrinsic and can seldom be eliminated. But their impact on medical decisions can certainly be minimized, and it can be fairly ensured that the likelihood of a correct decision is high and that of a wrong decision is under control. Section 1.2.2 briefly described some elementary methods to minimize the impact of uncertainties on conclusions of a research investigation.

This minimization is indeed a challenge in all setups—clinical practice, community health care, and medical research. Measurement and consequent quantitation are a definite help. This can lead to mathematics, perhaps intricate calculations, but effort in this book is to describe the methods without using calculus. I also try to avoid complex algebra and provide heuristic explanations instead. Even the high school algebra-based material is gradually introduced after the first few chapters. This may be friendlier to medical and health students and professionals, who generally have less rigorous training in mathematics. Real-life examples, many of which are from the literature, should provide further help in appreciating the medical significance of these methods. The methods and their implications are explained fully in narrative form, which should plug much of the gaps in the existing biostatistics literature. For example, Section 15.4 demystifies many statistical processes. Readers may find this kind of friendly explanation as another distinctive feature of this book.

1.4.1 Contents of the Book

The process to keep a check on the impact of uncertainties on decisions begins from the stage of conceptualizing or encountering a problem. Identification of the characteristics that need to be assessed; definitions to be used; methods of observation, investigation, and measurements to be adopted; and methods of analysis and of interpretation are all important. This book devotes many of the subsequent chapters to these aspects.

1.4.1.1 Chapters

Because of a very different orientation of this book, the organization of chapters too is different from what you conventionally see in biostatistics books. A statistician may find the text disjointed because of a new sequence, but medical and health professionals, who are the target audience, may find a smooth flow. The organization of chapters is as follows:

Basics such as types of medical studies and tools of data collection (Chapter 2).

Design for medical investigations so that the conclusions remain focused on the questions proposed to be investigated, including sampling (Chapter 3), observational studies (Chapter 4), medical experiments (Chapter 5), and clinical trials (Chapter 6).

Numerical and graphical methods for describing variation in data (Chapters 7 and 8). These methods help in understanding the salient features of the data and in assessing the magnitude of variation. Study of variation helps generate awareness about the underlying uncertainties.

Methods of measurement of various aspects of health and disease in children, adolescents, adults, and the aged, including in a clinical setup (Chapter 9). These help to achieve quantitation and consequently some sort of exactitude. They include the nature of the reference values so commonly used in medical practice; measurement of uncertainty in terms of probability, particularly in the context of diagnosis, prognosis, and treatment; and assessment of the validity of medical tests in terms of sensitivity–specificity and predictivities.

Further on, quantitative aspects of medicine under the rubrics of clinimetrics and evidence-based medicine. These include various indexes and scoring systems (Chapter 10) that help introduce exactitude to largely qualitative characteristics.

Indicators used for measuring the level of health of a community (Chapter 11). This includes a discussion of some composite indices such as disability-adjusted life years (DALYs).

The need and rationale for confidence intervals and of tests of statistical significance in view of the sampling fluctuations (Chapter 12). These methods assign probabilities to various types of right and wrong decisions based on samples.

Methods for taking decisions despite the presence of uncertainties, particularly with regard to assessing that difference in proportions of cases in two or more groups is real or has arisen due to chance (Chapter 13), and the assessment of the magnitude of difference in terms of relative risk and odds ratio (Chapter 14).

Methods for testing statistical significance regarding difference in means of two or more groups including analysis of variance (ANOVA) (Chapter 15).

Whether or not a relationship between two or more variables really exists and, if so, the nature of the relationship and measurement of its magnitude. These include usual quantitative regression (Chapter 16) and logistic regression (Chapter 17).

Duration of survival and how it is influenced by antecedent factors (Chapter 18).

Multivariate methods such as multivariate analysis of variance (MANOVA), discriminant functions, cluster analysis, and factor analysis (Chapter 19). Only the basic features of these methods are discussed. The objective is to describe medical situations in which such methods can be advantageously used. The intricacies of these methods are not fully described. Thus, this chapter would not provide skills to use these methods but would provide knowledge about situations in which they can and should be used.

Statistical methods for assessing quality of medical care and of medical tools (Chapter 20). This includes measures of validity and reliability of instruments, as well as assessment of robustness of results through methods such as sensitivity analysis and uncertainty analysis.

Fallacies that so commonly occur in statistical applications to health and medicine. I wrap up this book with a quite extensive discussion of such fallacies and the corresponding remedies (Chapter 21).

1.4.1.2 Limitations and Strengths

Biostatistics these days is a highly developed science. It is not possible to include all that is known, not even all that is important, in one book. My attempt in this text is to cover most of what is *commonly* used in medicine and health. Computers have radically changed the scenario in that even very advanced and complex techniques can be readily used when demanded by the nature of the problem and of the data. Yet, there are some basics. Their understanding can be considered crucial for appropriate application. Within the constraints set for this book, the discussion is restricted mostly to basics and sometimes intermediaries. Among the topics left out are infectious disease modeling [7], analysis of time series [8], and bioassays [9]. Many others are not included. Nevertheless, the book covers most topics that are included in an undergraduate and graduate biostatistics course required for medical and health students. It also includes the statistical material generally required for membership examinations of many learned bodies, such as the College of Physicians. Above all, the book should be a useful reference to medical and health professionals for acquiring enough knowledge to plan and conduct different kinds of investigational studies on groups of subjects. It incorporates methods that are considered essential armament for medical research. It would also help medical researchers to critically interpret medical literature, which is becoming increasingly statistical. The first edition of this book was acclaimed as "probably the most complete book on biostatistics we have seen," second edition "encyclopedic in breadth," and the endeavor in this edition is to make it even more complete. A comprehensive index of terms is provided at the end, which will make it easier for you to locate the text of your choice.

As a medical or health professional, you may or may not do any research yourself but would always be required to interpret the research of others in this fast-growing science. Thus, the discussion in this book could be doubly relevant—first for appreciating and managing uncertainties in medical practice and second to correctly interpret the mind-boggling medical research now going on. Just how much help is provided by the knowledge of biostatistical method in understanding the articles published in reputed journals? The results of an investigation are surprising. A reader knowing mean, standard deviation, etc., has statistical access to 58% of the articles.

Adding *t*-tests and contingency tables increases comprehension to 73% [10]. The contents of the book should also be useful to those believing in evidence-based medicine. They search available evidence and interpret it for relevance in managing a patient. Biostatistical methods contribute not just to probabilistic thinking but also encourage critical and logical thinking. An avid reader would find sprinkling in the text of this book of how logic is used for interpreting evidence for improving clinical practice.

The thrust throughout this book is to try not to lose focus on medical and health aspects of the methods. The chapters often start with a real-life medical example, identify its statistical needs, and then present the methods. This is done to provide motivation for the methods discussed. Chapter titles too are mostly problem oriented rather than method oriented.

I have been amid medical professionals for long and am aware of their math phobia. I also appreciate the emphasis of some medical professionals on diagnosis and treatment as an art rather than a science because it depends very much on clinical acumen and individual equation between the patient and the doctor. This text incorporates some of these concerns also.

1.4.1.3 New in the Third Edition

The second edition of this book has been microscopically examined by the reviewers across the world and a large number of suggestions have emerged. The third edition incorporates many of those suggestions. This edition contains a completely revised and enlarged chapter on survival analysis including elements of Cox model. This edition incorporates more detailed discussion on (1) receiver operating characteristic (ROC) curves, (2) assessing equivalence, (3) repeated measures ANOVA, and (4) area under the concentration curve. A new section on meta-analysis has been added that includes forest plot and funnel plot. Among new topics introduced are clinical trial designs with interim appraisals (adaptive designs, stopping rules, and sample size reestimation), noninferiority margin, Likert scale, STROBE statement, half life, graphical and analytical methods for checking Gaussianity of data, radar graph, dietary indices, a measure of health inequality, Poisson distribution, Cochran Q-test, McNemar–Bowker test, CI for number needed to treat, measures of ordinal association, Dunnett test, adjusted R^2, choosing form of regression, ridge and splines, path analysis, an alternative approach to agreement assessment, classification trees, Six Sigma, and limitations of statistical models. Many of these are discussed at an introductory level, just to apprise the reader of the situations where these methods can be used. In the absence of these topics, the second edition looked incomplete.

This edition contains **software illustrations**. These are provided in SPSS (except one on ROC based on MedCalc) in Appendix C for relatively intricate statistical methods such as Tukey test, repeated measures ANOVA, stepwise regression, curvilinear regression, analysis of covariance (ANCOVA), logistic regression, and survival analysis. Many of these illustrations give further details of the steps involved in actually using the method. These illustrations may help to acquire skills to do intricate analysis with the help of SPSS and skills to interpret the output. In addition, SYSTAT solutions of many examples are available on the book's website (http://www.MedicalBiostatistics.synthasite.com) where all the datasets used in examples in the book are also placed in Excel format for anyone to try. Many of these solutions also go beyond the details provided in the text.

1.4.2 Salient Features of the Text

Important terms and topics mostly appear in section and subsection titles. If not, they are in **boldface** in the text at the time when they are defined or explained. This may help you to easily spot the text on the topic of your interest. The terms or phrases that I want to emphasize in the text appear in *italics*. Whenever feasible, the name of a formula is given on the line before or the same line as the formula so that it can be immediately identified.

While general statistical procedures for different situations and their illustrations are described in the text, also included are numbered paragraphs at many places. These contain comments on the applicability of the procedures, their merits and demerits, their limitations and extensions, etc. These paragraphs are important inputs for an avid reader and should not be ignored if the intention is to acquire a critical understanding of the procedure. I have tried to arrange them in a crisp, brief format separately for each point of comment.

A large number of examples illustrate the procedure discussed in each chapter. These are given with a title in bold with an indent. Many times these examples explain important aspects of the procedure, which might not be adequately discussed in the text. Thus do not ignore the examples. They are part of the learning material.

1.4.2.1 System of Notations

Although the notations used in this text are explained at their first use, an explanation of the *system* of our notations may provide considerable help in quick understanding of the contents.

A measurement or an observed value of a quantitative variable is denoted by x or y. The number of subjects studied is denoted by n. The indexing subscript for this is the letter i, Thus, x_i is the measurement obtained on the ith subject. If the heart rate of the third patient in a trial is 52/min, then $x_3 = 52$. The values $x_1, x_2, ..., x_n$ are sometimes referred to as n **observations**.

If the spectrum of values is divided into groups such as systolic BP into 100–119, 120–139, 140–149, 150–159, 160–179, 180–199, and 200+ mmHg, then the number of groups so formed is denoted by K. In this example, $K = 7$. Sometimes notation J is also used for the number of groups, particularly when grouping is on two different factors. The subscripts for them are lowercase k and j, respectively. The groups are not necessarily quantitative. They can also be nominal, such as male and female ($K = 2$), or ordinal, such as mild, moderate, and severe pain ($K = 3$). Thus, y_{ijk} is the ith observation in the jth group of factor-1 and the kth group of factor-2.

The probability of any event is denoted by P, but the primary use of this notation in the latter half of this book is for probability of Type-I error. In case of categorical data, π_k is the probability of an observation falling into the kth group ($k = 1, 2, \ldots, K$). The actual number of subjects in the kth group of a sample is denoted by n_k or by $O_{k\bullet}$. The proportion of subjects with a specified characteristic out of the total sample is denoted by p and, in case of groups, in the kth group by p_k. Note that the notation p is used in this book for sample proportion and P for probability. Many books and journals use p for probability such as p-value, and the same notation for sample proportion. I have tried to make a distinction.

When the subjects are cross-classified by two factors, the tabular results would generally have rows depicting classification on factor-1 and columns on factor-2. The number of groups formed for factor-1 is the same as the number of rows and is denoted by R. The number of columns formed for factor-2 is denoted by C. The indexing subscripts for them are r and c, respectively.

Note that this text denotes the number of groups by uppercase J, K, R, and C and indexing subscript is the corresponding lowercase j, k, r, and c. The only exception is subscript i for subject number that goes up to n (and not I). The latter is the convention followed by most books and this book retains this to avoid confusion.

Observed frequency in rth row and cth column of a contingency table is denoted by O_{rc}. Summation of values or of frequencies is indicated by Σ and the sum so obtained is sometimes denoted by a dot ($\bullet$) for the corresponding subscript. Thus, $O_{\bullet c} = \Sigma_r O_{rc}$ and $O_{r\bullet} = \Sigma_c O_{rc}$. In the first case, the sum is over rows and in the second over columns. The subscript of Σ is the summing index and is mentioned wherever needed for clarity. Multiplication in notations is denoted by * following the computer language but is denoted by the conventional $\times$ while working with numbers.

I find it convenient to retain the notations a, b, c, d used by many authors for the number of subjects in the four cells of a 2×2 table in the context of relative risk and odds ratio. Otherwise the notations are O_{rc}.

It is customary in statistical texts to denote the population parameters by Greek letters and the corresponding observed values in the sample by Roman letters. Thus, population mean, standard deviation, and probability are denoted by μ, σ, and π and their sample values by $\bar{x}$, s, and p, respectively. All notations are in italics including the notational subscripts except Greek letters.

1.4.2.2 Guide Chart of the Biostatistical Methods

For easy reference, a chart is provided at the beginning of the book that gives a summary of the methods presented in this text. The chart is divided into 7 tables from S.1 to S.7 and refers to the equation or expression number or the section where that method is described. The chart also refers to the method applicable for different types of data. Thus, you can go directly to the place in the book where the method for the dataset in your hand is described. This may be a very useful guide.

References

1. Vidal E, Carlin E, Driul D, Tomat M, Tenore A. A comparison study of the prevalence of overweight and obese Italian preschool children using different reference standards. *Eur J Pediatr* 2006; 165:696–700.
2. Menotti A, Seccareccia F. Cardiovascular risk factors predicting all causes of death in an occupational population sample. *Int J Epidemiol* 1988; 17:773–778.
3. Gupta R. Meta-analysis of prevalence of hypertension in India. *Indian Heart J* 1997; 49:43–48.
4. Stanton BF, Clemens JD, Ahmed S. Methodological considerations in defining chronic diarrhoea using a distributional approach. *Int J Epidemiol* 1990; 19:439–443.
5. Jelliffe DB. *Assessment of the Nutritional Status of the Community: With Special Reference to Field Surveys in Developing Regions of the World*. WHO Monograph Series No. 53. Geneva, Switzerland: World Health Organization, 1966.
6. Sinaiko AR. Hypertension in children. *N Engl J Med* 1996; 335:1968–1973.
7. Castillo-Chavez C, Blower S, vanden Driessche P, Kirschner D, Yakubu A (Eds.). *Mathematical Approaches for Emerging and Reemerging Infectious Diseases: Models, Methods and Theory*. New York: Springer, 2002.
8. Diggle PJ, Liang KY, Zeger SL. *Analysis of Longitudinal Data*, 2nd edn. Oxford, U.K.: Clarendon Press, 2002.
9. Govindarajulu Z. *Statistical Techniques in Bioassay*, 2nd edn. Basel, Switzerland: S. Karger, 2000.
10. Emerson JD, Coditz GA. Use of statistical analysis in the *New England Journal of Medicine*. *New Engl J Med* 1983; 309:709–713.

2

Basics of Medical Studies

You may have realized from the previous chapter that empiricism is the backbone of medical knowledge. Studies in various forms are constantly carried to better understand the interactions of various factors affecting health. Sometimes one or two interesting or unusual cases are extensively investigated and **case reports** on them are prepared. This mode of investigation certainly adds to knowledge, but it does not form the core of modern research. To identify a trend, generally a group of subjects is studied. A preplanned series of steps is followed to collect evidence and to draw conclusion. This chapter focuses on statistical aspects of such research studies rather than on clinical approaches. If you are not into research, you would still like to know the basics to understand the full implications of the studies reported in the medical literature that you consult periodically to update your knowledge.

This chapter: Empiricism requires that the study of a group of subjects should essentially consist of preparation of protocol; collection of observations; their collation, analysis, and interpretation; and the conclusions. The tool that really architects the study is the protocol. Features of study protocol are described in Section 2.1. The major component of a protocol is the design that contains the details of how various sources of uncertainties are proposed to be controlled. An outline of various designs for medical studies is presented in Section 2.2. This discusses how the broad objectives of the study, the choice of strategy, and the methods to implement that choice finally determine the design of the study.

The broad objective of the study could be to obtain descriptive data, for example, prevalence rate of a disease, or analytical data, which aims to investigate cause–effect type of relationships. Sampling is a necessary ingredient of all empirical studies, especially of descriptive studies. In view of its importance, sampling is discussed separately in Chapter 3. Methods of data collection and the intricacies involved are discussed in Section 2.3. A common strand in all strategies for medical studies is the control of various kinds of biases. These biases are identified and listed in Section 2.4. Their control is explained in subsequent chapters.

2.1 Study Protocol

The protocol is the backbone that supports a study in all steps of its execution. Thus, sufficient thought must be given to its preparation. On many occasions, it gradually evolves as more information becomes available and is progressively examined for its adequacy. The most important aspect of study protocol is to state the problem, objectives, and hypotheses. This deserves a separate discussion.

2.1.1 Problem, Objectives, and Hypotheses

It is often said that research is half done when the problem is clearly visualized. There is some truth in this assertion. Thus, do not shy away from devoting time in the beginning for identifying the problem, to thoroughly understand its various aspects, and to choose the specifics you would like to investigate.

2.1.1.1 Problem

A problem is a perceived difficulty, a feeling of discomfort about the way the things are, the presence of a discrepancy between the existing situation and what it should be, a question about why a discrepancy is present, or the existence of two or more plausible answers to the same question [1]. Among the countless problems you may notice, identifying one suitable for study is not always easy. Researchability, of course, is a prime consideration but rationale and feasibility are also important. Once these are established, the next important step is to determine the focus of the research. This can be done by reviewing existing information to establish the parameters of the problem and using empirical knowledge to refine the focus. Specify exactly what new the world is likely to know through this research.

The title of research by itself is not the statement of the problem. Statement of the problem is a comprehensive statement regarding the basis for selecting the problem, the details of lacunae in existing knowledge, a reflection on its importance, and comments on its applicability and relevance. The focus should be sharp. For example, if the problem area is the role of diet in cancers, the focus may be on how consumption of meat affects the occurrence of pancreatic cancer in males residing in a particular area. To further focus, the study may be restricted to only nonsmoking males to eliminate the effect of smoking. To further the depth of the study, meat can be specified as red or white. Additionally, one can also include the amount of consumption of red and white meat and its duration. The role of other correlates that promote or inhibit the effect of meat in causing cancer can also be studied. The actual depth of the study would depend on the availability of relevant subjects on the one hand and the availability of time, resources, and expertise on the other. Such sharp focus is very helpful in specifying the objectives and hypotheses, developing an appropriate research design, and conducting the investigation.

2.1.1.2 Broad and Specific Objectives

The focus of the study is further refined by stating the objectives. These are generally divided into broad and specific. Any primary medical research (Section 2.2) can have two types of broad objectives. One is to describe features of a condition such as the clinical profile of a disease, its prevalence in various segments of the population, and the levels of medical parameters seen in different types of cases. This covers the distribution part of epidemiology of a disease or a health condition, such as what is common and what is rare, and what is the trend. It helps to assess the type of diseases prevalent in various groups and their load in a community. However, a descriptive study does not seek explanation or causes, nor tries to find which group is superior to the other. Evaluation of the level of β_2 microglobulin in cases of HIV/AIDS is an example of a descriptive study. A study on growth parameters of children or estimating prevalence of blindness in cataract cases is also a descriptive study.

The second type of broad objective could be to investigate cause–effect type of relationship between an antecedent and outcome. Whereas cause–effect would be difficult to ascertain unless certain stringent conditions are met (Chapter 20), an association or correlation can be easily established. Studies that aim to investigate association or cause–effect are called analytical.

A broad objective would generally encompass several dimensions of the problem. These dimensions are spelt out in specific objectives. For example, the broad objective may be to assess whether a new diagnostic modality is better than an existing one. The specific objectives in this case could be separately stated as (1) positive and negative predictivity; (2) safety, in case of an invasive procedure; (3) feasibility under a variety of settings such as field, clinic, and hospital; (4) acceptability by the medical community and the patients; and (5) cost-effectiveness. Another specific objective could be to evaluate its efficacy in different age–sex or disease-severity groups so that the kinds of cases where the procedure works well are identified. Specific objectives relate to the specific activities, and they identify the key indicators of interest. They are stated in measurable format.

Keep the specific objectives as few and focused as possible. Do not try to answer too many questions by a single study especially if its size is small. Too many objectives can render the study difficult to manage. Whatever objectives are set, stick to them all through the study as much as possible. Changing them midway or at the time of report writing signals that enough thinking was not done at the time of protocol development.

2.1.1.3 *Hypotheses*

A hypothesis is a precise expression of the expected results regarding the state of a phenomenon in the target population. Research is about replacing existing hypotheses with new ones that are more plausible. In a medical study, hypotheses could purport to explain the etiology of diseases, prevention strategies, screening and diagnostic modalities, distribution of occurrence in different segments of population, the strategies to treat or to manage a disease, to prevent recurrence of disease or occurrence of adverse sequelae, etc. Consider which of these types of hypotheses can be investigated by the proposed study.

Hypotheses are not guesses but reflect the depth of knowledge of the topic of research. They must be stated in a manner that can be tested by collecting evidence. The hypothesis that dietary pattern affects the occurrence of cancer is not testable unless the specifics of diet and the type of cancer are specified. Antecedents and outcome variables, or other correlates, should be exactly specified in a hypothesis. Generate a separate hypothesis for each major expected relationship.

The hypotheses must correspond to the broad and specific objectives of the study. Whereas objectives define the key variables of interest, hypotheses are a guide to the strategies to analyze data.

2.1.2 Protocol Content

A protocol is the focal document for any medical research. It is a comprehensive yet concise statement regarding the proposal. Protocols are generally prepared on a structured format with an introduction, containing background that exposes the gaps needing research; a review of literature, with details of various views and findings of others on the issue including those that are in conflict; a clear-worded set of objectives and the hypotheses under test; a methodology for collection of valid and reliable observations and a statement about methods of data analysis; and the process of drawing conclusions. It tries to identify the uncertainty gaps and proposes methods to plug those gaps.

Administrative aspects such as sharing of responsibilities should also be mentioned. In any case, a protocol would contain the name of the investigator, academic qualification, institutional affiliation, and the hierarchy of advisers in the case of masters and doctoral work. The place of the study such as the department and institution should also be mentioned. The year the proposal is framed is also required. An appendix at the end includes the pro forma of data collection, the consent form, etc. The main body of a protocol must address the following questions with convincing justification.

Title

- What is actually intended to be studied and is the study sufficiently specific?

Introduction

- How did the problem arise? In what context?
- What is the need of the study—what new is expected that is not known so far? Is it worth investigating? Is the study exploratory in nature, or are definitive conclusions expected?
- To what segment of population or to what type of cases is the problem addressed?

Review of Literature

- What is the status of the present knowledge? What are the lacunae? Are there any conflicting reports? How the problem has been approached so far? With what results?

Objectives and Hypotheses

- What are the broad and specific objectives and what are the specific questions or hypotheses to be addressed by the study—are these clearly defined, realistic, and evaluative?

Methodology

- What are the subjects, what is the target population, what is the source of subjects, how are they going to be selected, how many in each group, and what is the justification? What are the inclusion and exclusion criteria? Is there any possibility of selection bias, and how this is proposed to be handled?
- What exactly is the intervention, if any—its duration, dosage, frequency, etc. What instructions and material are to be given to the subjects and at what time?
- Is there any comparison group? Why is it needed and how will it be chosen? How will it provide valid comparison?
- What are the possible confounders? How are these and other possible sources of bias to be handled? What is the method of allocation of subjects to different groups? If any blinding, how will it be implemented? Is there any matching? On what variables and why?
- On what characteristics would the subjects be assessed—what are the antecedents and outcomes of interest? When would these assessments be made? Who will assess them? Are these assessments necessary and sufficient to answer the proposed questions?

- What exactly is the design of the study? Is the study descriptive or analytical? If analytical, is it observational or experimental? An observational study could be prospective, retrospective, or cross sectional. An experiment could be carried out on biological material or animals, or human in which case it is called a trial. If experimental, is the design one way, two way, factorial, or what?
- What is the operational definition of various assessments? What methods of assessment are to be used—are they sufficiently valid and reliable? What information will be obtained by inspecting records, by interview, by laboratory and radiological investigations, and by physical examination? Is there any system of continuous monitoring in place? What mechanism is to be adopted for quality control of measurements? What is the set of instructions to be given to the assessor?
- What form is to be used for eliciting and recording the data? (Attach it as an appendix.) Who will record? Will it contain the necessary instructions?
- What is to be done in case of contingencies such as dropout of subjects or nonavailability of the kit or the regimen, or development of complications in some subjects? What safeguards are provided to protect the health of the participants? Also, when to stop the study if a conclusion emerges before the full course of the sample?
- What is the period of the study and the timeline?

Data Analysis

- What estimations, comparisons, and trend assessments are to be done at the time of data analysis? Whether the quality and quantity of available data would be adequate for these estimations, comparisons, and trend assessments?
- What statistical indices are to be used to summarize the data—are these indices sufficiently valid and reliable?
- How is the data analysis to be done—what statistical methods would be used and whether these methods are really appropriate for the type of data and for providing correct answer to the questions? What level of significance or the level of confidence is to be used? How are missing data, noncompliance, and nonresponse to be handled?

Validation of Results

- What is the expected reliability of the conclusions? What are the limitations of the study, if any, with regard to generalizability or applicability?
- What exercises are proposed to be undertaken to confirm the internal and external validity of the results?

Administration

- What resources are required, and how are they to be arranged?
- How are responsibilities to be shared between investigators, supporting units (e.g., pathology, radiology, and biostatistics), hospital administration, funding agency, etc.?

In short, the protocol should be able to convince the reader that the topic is important, and the data collected would be reliable and valid for that topic, and that contradictions, if any, would be satisfactorily resolved. Present it before a critical but positive audience and get their feedback. You may be creative and may be in a position to argue with conviction, but skepticism in science is regularly practiced. In fact it is welcome. The method and results would be continuously scrutinized for possible errors. The protocol is the most important document to evaluate the scientific merit of the study proposal by the funding agencies as well as by the accepting agencies. Peer validation is a rule rather than exception in scientific pursuits. A good research is the one that is robust to such reviews.

A protocol should consist of full details with no shortcuts, yet should be concise. It should be to the point and coherent. The reader, who may not be fully familiar with the topic, should be able to get clear answer about the why, what, and how of the proposed research. To the extent possible, it should embody the interest of the sponsor, the investigator, the patients, and the society. The protocol also is a reference source for the members of the research team whenever needed. It should be complete and easy to implement.

The protocol is also a big help at the time of writing of the report or a paper. The introduction and methods sections remain much the same as in the protocol, although in an elaborate format. The objectives as stated in the protocol help to retain the focus in the report. Much of the literature review done at the time of protocol writing also proves handy at the time of report writing.

Whereas all other aspects may be clear by themselves or would be clear as we go along in this book, special emphasis should be placed on the impartiality of the literature review. Do not be selective to include only those pieces of literature that support your hypotheses. Include those that are also inconsistent with or oppositional to your hypotheses. Justify the rationale of your research with reasons that effectively counter the opposite or indifferent view. *Research is a step in the relentless search for truth, and it must pass the litmus test put forward by conflicting or competing facts.* The protocol must provide evidence that the proposed research would stand up to this demand and would help minimize the present uncertainties regarding the phenomenon under study.

2.2 Types of Medical Studies

Medical studies encompass a whole gamut of endeavors that ultimately help to improve the health of people. Functionally, medical studies can be divided into basic and applied types. **Basic research**, also termed as pure research, involves advancing the knowledge base without any specific focus on its application. The results of such research are utilized somewhere in future when that new knowledge is required. **Applied research**, on the other hand, is oriented to an existing problem. In medicine, basic research is generally done at the cellular level for studying various biological processes. Applied medical research could be on the diagnostic and therapeutic modalities, agent–host–environment interactions, or health assessments.

I would like to classify applied medical studies into two major categories. The first category can be called **primary research** that includes analytical studies such as cohort studies, case–control studies, and clinical trials. It also includes descriptive studies such as surveys, case series, and census. The second category is **secondary research**, quite common these days, that includes decision analysis (risk analysis and decision theory), operations research (prioritization, optimization, simulation, etc.), evaluation of health systems (assessment of achievements and shortcomings), economic analysis (cost benefit, cost-effectiveness, etc.), qualitative research (focus group discussion), and research synthesis (Figure 2.1). This book contains biostatistical methods applicable to primary research, which still forms the bulk of modern medical research, and excludes much of the methods applicable to secondary research.

Any research on diagnostic, prophylactic, and therapeutic modalities or on risk assessment is empirical. Experience on one or two patients can help in special cases but generally investigation of a large group of subjects is needed to come to a definitive conclusion. To achieve this, study design is of paramount importance.

2.2.1 Elements of Design

Since medicine is an empirical science, the decisions are based on evidence. The conclusions must also stand up to the reasoning. Hunches or personal preferences have no role. The nature of evidence is important in itself but credence to this is acquired by the soundness of the methodology adopted for collecting such evidence. *Design is the pattern, scheme, or plan to collect evidence.* It is the tool by which credibility of research findings is assessed. The function of a design is to permit valid conclusion that is justified and unbiased. It should take care of confounding problems that could complicate the interpretation. Thus, it should be able to provide correct answers to

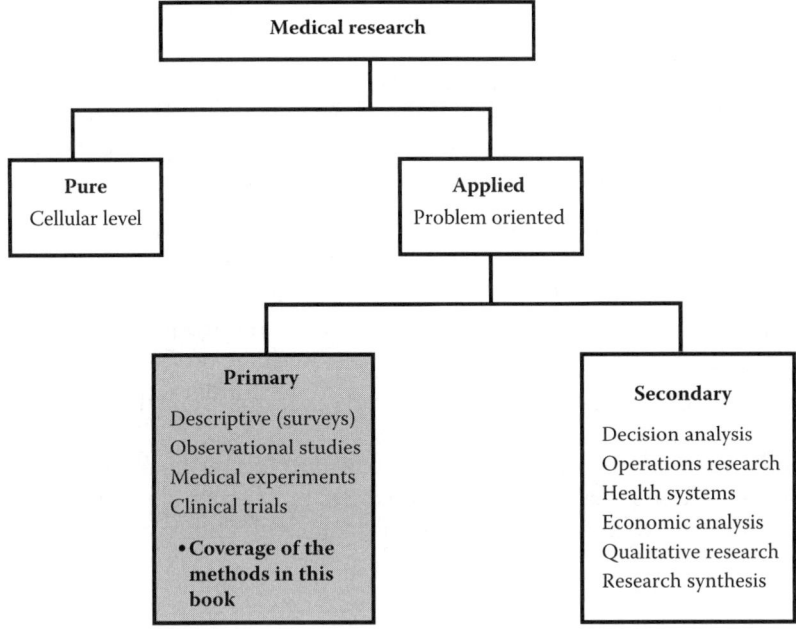

FIGURE 2.1
Types of medical studies.

the research questions. The objective of a design is to get the best out of the efforts. Various elements of design are as follows:

1. Definition of the target population: inclusion and exclusion criteria, area to which the subjects would belong, and their background information

2. Specification of various groups to be included with their relevance

3. Source and number of subjects to be included in each group with justification including statistical power or precision considerations as applicable

4. Method of selection of subjects

5. Strategy for eliciting data: experiment (animal or human) or observation (prospective, retrospective, or cross sectional)

6. Method of allocation of subjects to different groups if applicable or matching criteria with justification

7. Method of blinding if applicable and other strategies to reduce bias

8. Specification of intervention, if any

9. Definition of the antecedent, outcome, and cross-sectional characteristics to be assessed along with their validity for the study objectives

10. Identification of various confounders and the method proposed for their control

11. Method of administering various data-collecting devices, such as questionnaire, laboratory investigation, and clinical assessment, and the method of various qualitative and quantitative measurements

12. Validity and reliability of different devices and measurements

13. Time sequence of collecting observations and their frequency (once a day, once a month, etc.), duration of follow-up, and duration of study, with justification

14. Methods for assessment of compliance and strategy to tackle ethical problems

15. Strategies to handle nonresponse and other missing data

Primary feature of the design is the type of study in terms of descriptive or analytical. These have been described earlier but require some more explanation although details appear in the subsequent chapters.

2.2.2 Basic Types of Study Design

A complete chart of various relevant study formats is shown in Figure 2.2. This chart gives divisions of descriptive and analytical studies.

2.2.2.1 Descriptive Studies

Studies that seek to assess the current status of a condition in a group of people are called descriptive. These are also called **prevalence studies**. Consider the power of the following results from a 2010 National Health Interview Survey in the United States, which was a descriptive study:

1. Most (82%) U.S. children aged 17 years and under had excellent to good health.

2. Fourteen percent of children had ever been diagnosed with asthma.

3. Eight percent of children aged 3–17 years had a learning disability and 8% had attention deficit hyperactivity disorder.

The importance of descriptive studies is not fully realized but it is through such studies that the magnitude of the problems is assessed. They also help to assess what is normally seen in a population. Unfortunately, even body temperature among healthy subjects is not known with precision for many populations. Thus, there is considerable scope for carrying out descriptive studies. Such studies can provide baseline data to launch programs such as breast cancer control or rehabilitation of elderly people and can measure progress made. A descriptive study can generate hypotheses regarding the etiology of a disease when the disease is found more common in one group

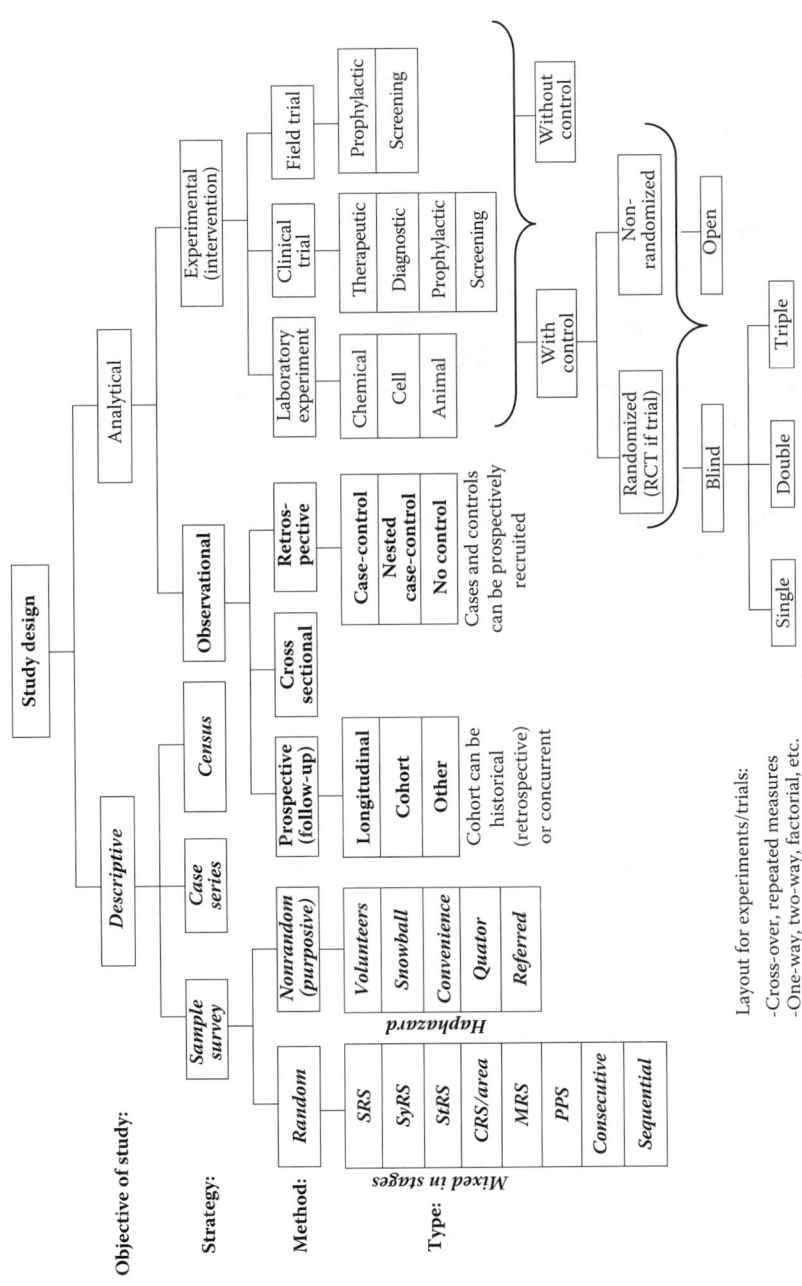

FIGURE 2.2
Various types of study designs.

than another. In some situations, a descriptive study can be designed to test a hypothesis regarding the status of a parameter such as whether at least 30% patients of abdominal tuberculosis come with the complaints abdominal pain, vomiting, and constipation of long duration or whether prevalence of non-insulin-dependent diabetes is 10% among married females of age 50 years or more whose husband is diabetic.

A case study, which generally describes features of a new disease entity, is also descriptive. It is anecdotal in nature. A story can have a definite impact, and it is easily understood. But be wary of anecdotal evidence. It can be interesting and can lead to a plausible hypothesis, but anecdotes are sometimes built around an exciting event and used by the media to sensationalize the reporting. You must verify anecdotes even for forwarding a hypothesis. A series of such cases form **case series**. They summarize common features of the cases or may highlight the variation. This too can lead to a hypothesis. Initial case series of HIV positives in San Francisco, almost exclusively among homosexual men, led to the suspicion that sexual behavior could be a cause. Surveys too are descriptive studies, although this term is generally used for community-based investigations. When repeatedly undertaken, they can reveal time trends. Complete enumeration such as a population census is also descriptive. A descriptive study generally has only one group since no comparison group is needed for this kind of study. Its design is mainly in terms of a sampling plan, which is discussed in Chapter 3. Those who are aware of statistical errors realize that no Type-I or Type-II error arises in descriptive studies unless we test a hypothesis such as for presence of correlation.

2.2.2.2 Analytical Studies

The other kind of study is an analytical study that tries to investigate etiology or cause–effect type of relationship. Determinants of a disease or of a health condition are obtained by this kind of study. Differences between two or more groups are also evaluated by analytical studies. Although the conclusions are associational, the overtones are cause–effect. A properly designed analytical study indeed can provide a conclusion regarding a cause–effect relationship. Two strategies are available for analytical studies: observation and experiment. Observational studies are based on naturally occurring events. There is no human intervention. Record-based studies are also placed in this group. Experimental studies require deliberate human intervention to change the course of events.

2.2.2.3 Basic Types of Analytical Studies

Although observational studies could be descriptive, the term is generally used for particular types of analytical studies. A study of smoking and lung cancer is a classical example of an observational study. Nobody would ever advocate or try intentionally exposing a group of persons to smoking to study its effect. Thus, experimentation in this case is not an option, at least for human beings. An observation of what is happening, has happened, or will happen in groups of people is the only approach. On the other hand, the anesthetic effect of an eutectic mixture used for local anesthesia can be safely studied in human beings because this cream has practically no toxic effect. The advantage is that it is a painless procedure, particularly suitable for children, as in electrolysis, although skin biopsy will not fall in this category. The rule of thumb is that human experiments can be carried out only for potentially beneficial modalities. Observational studies can be carried out for hazardous as well as for beneficial modalities. They can be prospective, retrospective (case–control), or cross sectional depending upon the methodology adopted to collect data. The details are provided in Chapter 4.

An initial form of observational study is **ecological study** where populations or groups are observed and not individuals. No individual data are collected and the unit of analysis too is a group. For example, the result that oral cancer is more common in countries with high prevalence of chewing tobacco is based on ecological study since the result is for countries and not for persons. You can easily see the fallacies that can occur in this setup as many unobserved factors can intervene and vitiate or concoct the result. Despite its proneness to bias, ecological study is considered a good starting point for epidemiological conclusions because of ease with which this can be carried out. This type of study mostly requires analysis of published data, and the inputs of cost and time are substantially less since no primary data are collected.

Other than observation, perhaps more convincing and scientifically more valid but ethically questionable strategy is to intentionally introduce the intervention of which the effect is proposed to be evaluated. Thus, there are human efforts to change the course of events. These are experiments, but, in medicine, the term is used generally when the units are animals or biological material. Customarily, experiments on humans are called trials.

Laboratory and animal experiments can be done in far more controlled conditions. These experiments are described in Chapter 5. Trials can be conducted in a clinic or in a community. Since humans are involved, a large number of ethical and safety issues crop up. Far more care is required in this setup, and strategies such as randomized controlled trial and blinding are advocated. The details are provided in Chapter 6.

2.2.3 Choosing a Design

In the face of so many designs shown in Figure 2.2, it could be indeed a task to choose an appropriate design for a particular setup. Issues that determine the choice would be clear when various designs are discussed in detail. A summary of my recommendations is as follows.

2.2.3.1 Recommended Design for Particular Setups

A good research strategy provides conclusions with minimal error within the constraints of funds, time, personnel, and equipment. In some situations no choice is available. For investigating the relationship between pesticides and oxidative stress, intervention in terms of deliberately exposing some people to pesticides is not an option. Observation of those who are already exposed is the only choice. For determining the efficacy and safety of a drug, intervention in terms of administering the drug in controlled conditions is a must. In this case, observational strategy is not a good option. As just indicated, the role of potentially harmful factors is generally studied by observations, but potentially beneficial factors can be studied by either strategy. The effect of garlic on cholesterol level can be evaluated by studying people who include garlic in various quantities in their diet and also by asking people who almost never took garlic to consume it for a while in specified quantity. Experiments have the edge in providing convincing results. Also, they can be carried out in controlled conditions; thus, a relatively small sample could be enough. However, experiments raise questions of ethics and feasibility. Guidelines for choosing a strategy for analytical studies can be listed as follows:

1. Should be ethically sound, causing least interference in the routine life of the subjects
2. Should be consistent, generally, with the approach of other workers in the field, and if not, the new approach should be fully justified
3. Should clearly isolate the effect of the factor under investigation from the effect of other factors in operation
4. Should be easy to implement and acceptable to the system within which the research is being planned
5. Should confirm that the subjects would sufficiently cooperate during the entire course of the study and would provide correct responses
6. Should be sustainable so that it can be replicated if required

As may be clear by now, the choice of design depends on the type of question for which an answer is sought by the study. Designs appropriate for specific research questions are given in Table 2.1. Details of these designs appear in subsequent chapters.

2.2.3.2 Choice of Design by Level of Evidence

As a health professional, you would be regularly required to update your knowledge of the fast-developing science and interpret the findings reported

TABLE 2.1

Recommended Design for Different Types of Research Questions

Question	Recommended Design
1. What is the prevalence or distribution of disease or a measurement, or what is the pathological, microbiological, and clinical profile of certain type of cases?	Sample survey
2. Whether two or more factors are related to one another (without implication of cause–effect)?	Cross-sectional study
3. Whether two or more methods agree with one another?	Cross-sectional study
4. What is the inherent goodness of a test in correctly detecting presence or absence of disease (sensitivity and specificity)?	Case–control study
5. What are the risk factors for a given outcome, or what is their relative importance?	Case–control study
6. How good is a test or a procedure in predicting a disease or any other outcome?	Prospective study
7. What is the incidence of a disease, or what is its risk in a specified group or the relative risk?	Prospective study
8. What are the sequelae of a pathological condition? And how much or how a factor affects the outcome?	Prospective study
If the factor under study is potentially harmful	Prospective study in humans; in rare situations animal experiment
If the factor under study is potentially beneficial	Randomized controlled trial for humans, and experiment for animals or biological material
9. Is an intervention really effective or more effective than the other?	Randomized controlled trial for humans, and experiment for animals or biological material

in the literature. You would want to assess the credibility of the evidence provided in different kinds of studies. If you are a researcher, the level of evidence produced in your work will be a crucial consideration.

The ultimate objective of all analytical studies is to obtain evidence for cause–effect type of relationship. This would depend on a large number of factors such as representativeness of the sample of subjects, their number in each group, control of confounders, etc., but guidelines can still be laid for the choice of design. Various types of studies according to the level of evidence are listed in Table 2.2. The comparison in this table is valid when other parameters of the study are the same. Inverted pyramid in Figure 2.3 may be more convincing. This also shows that there is no design with no chance of bias. Further details are in subsequent chapters.

TABLE 2.2

Hierarchy of Study Designs by Level of Evidence for Cause–Effect Relationship

Evidence	Type of Design	Advantages	Disadvantages
Level-1 (best)	RCT—double blind or crossover	Able to establish cause–effect and efficacy	Assumes ideal conditions
		Internally valid results	Can be done only for potentially beneficial regimen
			Difficult to implement
			Expensive
Level-2	Trial—no control/ not randomized/ not blinded	Can indicate cause–effect when biases are under control	Cause–effect is only indicated but not established
		Effectiveness under practical conditions can be evaluated	Can be done only for a potentially beneficial regimen
		Relatively easy to do	Outcome assessment can be blurred
	Prospective observational study	Establishes sequence of events	Requires a big sample
		Antecedents are adequately assessed	Limited to one antecedent
		Yields incidence	Follow-up can be expensive
Level-3	Retrospective observational study—case–control	Outcome is predefined, so no ambiguity	Antecedent assessment can be blurred
		Quick results obtained	Sequence of events not established
		Small sample can be sufficient	High likelihood of survival and recall bias
Level-4	Cross-sectional study	Appropriate when distinction between outcome and antecedent is not clear	No indication of cause–effect—only relationship
		When done on representative sample, same study can evaluate both sensitivity/ specificity and predictivity	
Level-5	Ecological study	Easy to do	Fallacies are common
Level-6	Case series and case reports	Help in formulating hypothesis	Many biases can occur
			Do not reflect cause–effect

Note: Laboratory experiments excluded.

2.3 Data Collection

After deciding on the design of the study, the next step is the collection of data. This is done by "measuring" the past or the present of the subjects by different methods. The measurement is not necessarily quantitative; it can

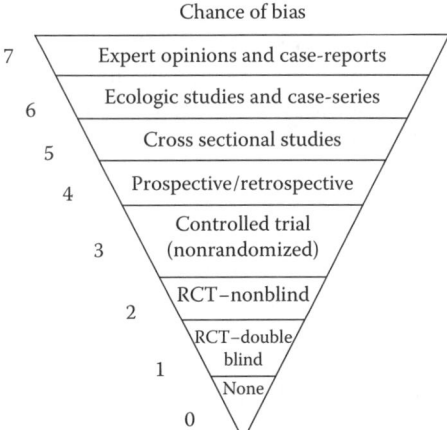

FIGURE 2.3
Chance of bias in different types of study.

be qualitative also. The details of various types of measurements are given in Chapter 7, but note now that the purpose of all measurements is to get a handle on uncertainties. Measurements help to provide exactitude concerning the state of the characteristics of the subjects under study that cannot be attained in any other way.

2.3.1 Nature of Data

The nature of data varies from investigation to investigation and within an investigation from item to item. Data could be factual, knowledge based, or a mere opinion. Data can be obtained by observation, interview, or examination. This section discusses these aspects first before moving on to various tools of data collection used in medical studies.

2.3.1.1 Factual, Knowledge-Based, and Opinion-Based Data

Data are factual when objectively measured. A subject's gender is factual and can seldom be wrong. Age and income are factual measurements but may not be known or may be wrongly reported. Extra care is needed when obtaining data on such characteristics. Height, weight, blood pressure, serum glucose level, etc., can be factually recorded. Disease signs are factual but their correct observation depends on the acumen of the attending clinician. Generally speaking, data on most but not on all of the current physical state of a subject can be factually obtained. Symptoms such as severity of pain are rarely factual because they are mostly based on the perception of the patient, which can change quickly even when the intensity of pain does not change.

A patient's history with regard to complaints in the past can sometimes be obtained from records, but otherwise would depend on the awareness and alertness of the subject. Response to interview largely depends on the knowledge and cooperation of the subject. Sometimes knowledge per se is assessed, as on transmission of hepatitis B infection. This is done to prepare a plan to combat the problem at the community level. Note that it is not easy to assess the level of knowledge on a factual basis.

Data of the third kind are based on opinion. Hopes, aspirations, and fears also fall in this category. These generally reflect attitudes. Fertility control activities in many developing countries are going on at a slow pace because a favorable opinion of the populace, sometimes even the government, is lacking. Consent of the patient for a new procedure also depends on this attitude. Some persons are very favorably disposed to trying a new modality, whereas others believe in sticking to convention. A 7-point scale from most favorable (+3) to most unfavorable (−3) can be used to measure opinion, with zero for indifferent opinion.

It is easy to advocate that data should be factual but obtaining such data is not always easy. Aspects such as knowledge and opinion are important ingredients, particularly in the prevention and control of diseases such as sexually transmitted infections. In the case of treatment, the patient's knowledge—right or wrong—about, for example, side effects or complications of surgery is an important consideration in determining the outcome. Also, opinion can sometimes be important in faithful implementation of a regimen.

2.3.1.2 Method of Obtaining the Data

Medical data are mostly obtained by observation, by interview, or by examination. Data on some behavioral variables are obtained by observation only. This may be important in case of psychosomatic disorders. The primary methods in a clinic are interview and examination. An interview may or may not reveal the full truth. Data obtained by examination are most reliable in the sense that they can be largely believed to be correct. This includes laboratory and imaging investigations, which form the core of diagnostic tools these days. Some of these methods are very expensive to adopt or just not available in a particular setup, and the clinician may have to resort to interview and physical examination or to a test of lower cost.

Although factual data based on examination are often given more weight, knowledge, opinion, and complaints as revealed by interview and observation, particularly for symptoms, have an important place in the practice of medicine. All efforts should be made to obtain valid as well as reliable data when these methods are used. Validity refers to the ability to measure correctly what is intended to be measured. Fever with chills is not a valid indicator of malaria because it is present in some other conditions as well. Similarly, childless after 10 years of marriage is not a valid basis for pronouncing infertility. Also, increased level of thyroid-stimulating hormone (TSH) alone is not enough to conclude hypothyroidism. Reliability refers to the ability to obtain the same value in repeated measurements. Body temperatures are not likely to be different when repeatedly measured in a patient at the same time, but responses concerning pain intensity can change on repeated questioning. The measurement of the former is more reliable than that of the latter. A detailed discussion on these concepts is presented in Chapter 20.

2.3.2 Tools of Data Collection

The tools of data collection include the different instruments used to collect data. These can be existing records, questionnaires and schedules, interviews and examinations, laboratory and radiological investigations, etc. For example, a patient with liver disease can be identified by asking the individual whether the disease has already been diagnosed, by looking at the health records, by clinical examination, or by carrying out some biochemical tests. Visual acuity in a subject can be assessed just by noting the power of the spectacles being worn, if any, or by a proper visual examination by an optometrist. Each method of eliciting information has merits and demerits in terms of validity and reliability on the one hand, and cost and time on the other. These factors have to be assessed in the context of the problem in hand and of the resources available. This section contains some guidelines.

2.3.2.1 Existing Records

Almost all hospitals and clinics maintain fairly good records of the patients served by them. A civil registration system may have records of births and deaths in a community along with cause of death in some countries, which, when combined with information on age, sex, occupation, etc., can be a useful resource. Some countries have a system of health centers where a comprehensive record of each family is maintained. Then, there are records of ad hoc surveys done by different agencies for specific purposes. Many individuals also keep a record of their health parameters based on periodic examinations and investigations.

Use of existing records has some demerits:

1. Records tend to be incomplete. Information on some subjects may not be available and some pieces of information on some other subjects may be missing. Conditions on which no physician was consulted may never appear in the records. Records may not contain information about items that may look vital for the study in hand but were not perceived to be so when the recording was done. Whereas the medical information may be accurate in many cases, the demographic information on the patients may not be so because attending physicians often do not pay sufficient attention to the accuracy of this information. In the case of a civil registration system on births and deaths, some events may not be registered at all—a problem that

could be very serious in some developing countries. The cause of death in such a system of registration may or may not be medically certified.

2. If the records are handwritten, as in many hospitals across the world, they may even be difficult to decipher.

3. Hospital records seldom represent the target population. Some hospitals cater to a specific ethnic, economic, or geographical group. Quite often, only severe cases go to the hospital. If the study objective demands that mild cases also be included, such records may not be adequate.

4. If only published records are accessible, publication may be late, making the information obsolete. Sometimes only selective records are published or only summaries are available.

5. Records maintained by different hospitals or other agencies or patients are seldom in a uniform format. They may lack comparability, which may make pooling difficult. They are rarely comparable across different sources or across time because of differences in diagnostic tools, strategies, expertise, etc. Records over 100 years may show false changes just because the names of medical conditions have changed (e.g., dropsy to congestive heart failure), the definitions have changed (e.g., isolated systolic hypertension from systolic blood pressure in excess of 160 mmHg then to 140 mmHg now), or the diagnostic methods have improved (e.g., introduction of digital methods of imaging).

Despite all these problems, records are and should be used for medical studies wherever feasible because they are the cheapest source of data. They preexist and so are not likely to be biased. In fact, the first attempt in all studies should be to explore the available records, published or not. These records can be compared with a road that is full of bumps and holes but is still passable by a careful driver [2]. By evaluating aspects such as extent of underreporting and the population that the records can reasonably represent, and by linkage of various records of the same person at different places, it may still be possible to put together a usable picture for specific segments of the target population. In cases such as suicides and accidental deaths, records may be the only source to provide a health profile of the victim. Properly maintained records may be vital in a study of trends in causes of early mortality.

A good example of a record-based study is clinico-pathological profile of cases such as ovarian tumors reported in a hospital over the past 20 years. Features such as age, parity, laterality, and size and stage of tumor can be studied by retrieving the records. Such a descriptive study can be the basis to subsequently plan an analytical study regarding the risk factors such as for sex cord tumors against other ovarian tumors.

2.3.2.2 Questionnaires and Schedules

A questionnaire contains a series of questions to be answered by the respondents. This could be self-administered or could be administered by an interviewer. In the case of the former, the education and the attitudes of the respondent toward the survey can substantially influence the response. In the case of the latter, the skill of the interviewer can make a material difference. The interviewer may have to be trained on how to approach a subject; how to obtain cooperation; when to prompt, probe, pause, or interject, etc. Some details of these aspects are described in simple language by Hepburn and Lutz [3].

The term schedule is used when it contains a list of items on which information is collected. The information can be obtained by observation, by interview, or by examination. Whether a questionnaire or a schedule is used, sufficient space is always provided to record the response. A list of possible or expected responses can be given to choose from against the questions or items. Then the form is called **close ended**. In the list of choices, the last item can be "others (specify)," to make the list exhaustive. If needed, carry out a pilot study and ensure that all common responses are listed so that responses under "others" do not exceed, say, 10% of the total responses. If they exceed, examine the "specify" component of "others" and make a new category of the frequently cited responses.

The list of responses should be designed in such a manner that no choice is forced on the respondent. They should also be preferably mutually exclusive so that one and only one response is applicable in each case. Depending upon the question or item, one of the responses could be "not applicable." For example, the question of results of laboratory investigation does not arise in cases where laboratory investigation was not required. This should be distinguished from cases in which the investigation was required but could not be done for some reason. The degree of accuracy of the measurements must be specified to help the surveyor to record accordingly. For example, age can be recorded as on the last birthday, weight to the nearest kilogram, and serum glucose level to the nearest $5\,mg/dL$. The responses are sometimes precoded for easy entry into the computer. When the response is to be recorded verbatim, the question or item is called **open ended**. This can present difficulty during interpretation and analysis of data but can provide additional information unforeseen earlier. Also, such open-ended responses can provide information regarding the quality of response.

Open-ended questions can provide results different from the results of close-ended questions. Suppose a question for parents is, "What is the most important thing for the health of their children?" more than 60% might say nutrition when this alternative is offered on a list. Yet, only less than 10% may provide this answer when no list is presented.

Framing questions is a difficult exercise because the structure of the sentence and the choice of words become important. Some words may have different meanings to different people. In any case, it is evident that the questions must be unambiguously worded in simple language. They must make sense to the respondent. The sequence must be logical. The length of the questionnaire should be carefully decided so that all the questions can be answered in one sitting without a feeling of boredom or burden. It is always helpful for the interviewee as well as for the interviewer to divide the questionnaire into sections. Use of features such as italics, boldface, underline, and capitals can help to clarify the theme of the question.

Respondents tend to be careless toward the tail end of the questionnaire if it is unduly long and not interesting. A large number of questions may be a source of attrition for the subjects and can increase nonresponse. Thus, nonessential questions must be avoided. Ask yourself how much you lose if a particular question is not asked. If the answer is negligible, then delete the question, otherwise do not do so. All the questions should be short, simple, nonoffending, and corroborative. They should be such that the respondent is able to answer.

One aspect of some questions is the rating scale used for providing the answer. For a question such as "how fit are you after one week of a surgery?" many may rate themselves +3 if the scale is from −5 to +5 but not 8 if the scale is from 0 to 10.

Sometimes, methods such as **Rasch analysis** are used to develop and evaluate a questionnaire or a schedule. The objective of Rasch analysis is to increase the chance of more able person to get any item correct compared to less able person and to increase the chance of getting a more difficult item correct by more able person compared to less able person. However, the procedure has limitations as it depends on who and how many respondents attempt different items. Also, it generally assumes that all items have equal correlation with the latent trait intended to be assessed. For further details, see Bond and Fox [4].

The questionnaire or schedule must always be accompanied by a statement of the objectives of the survey so that the respondent becomes aware. Easy-to-follow instructions for recording responses and explanatory notes where needed are always helpful. Special care may have to be taken concerning possible memory lapse if a question or item requires recall of an event. Whereas serious events such as accidents and myocardial infarctions are easy to recall even after several years, mild events such as episodes of fever or of diarrhea may be difficult to recall after a lapse of just 1 month.

There is also an effect of attitude of the patient toward certain ailments. An older person may perceive poor vision or infirmity as natural and may not report it at all. A smoker might similarly ignore coughing. On the other hand, mild conditions may be exaggeratedly reported depending on the disturbance they create in the vocational pursuits of the patient. A vocalist may be worried more about the vocal cords than a fractured hand. Thus, interview responses measure the person's perception of the problem rather than the problem itself. Some conditions with a social stigma such as venereal diseases and impotence may not be reported despite their perception as important.

Many medical studies require physical examination of the subject and investigations in the laboratory or in the radiology unit. The information so received will most likely be recorded on a structured form or a pro forma containing the items on which the information is required. The recording should be accurate after following the correct and full procedure as required in the protocol, but what needs to be emphasized is the legibility of writing—more so because most of the schedules for such purposes would be open ended. A computer-generated form, which allows direct entry of data without the help of intermediaries, is a big advantage in this context. However, the quality of information depends on the cooperation of the patient, the skills of the doctor, and the validity of the instruments. Ensuring all of these can be an uphill task in a given situation.

2.3.2.3 Likert Scale

For psychometric assessment of opinion, belief and attitude, a specific kind of tool is popular, called Likert scale. For a particular construct such as satisfaction from hospital services, a set of items is developed so that all aspects of that construct can be assessed. For hospital services, these items could be on competence of doctors, humanitarian nursing services, proper diagnostic facilities, etc. A declarative statement such as "I am happy with the nursing services in this hospital" is made in each item and the respondent is asked to specify his level of agreement generally on a 5-point scale such as strongly disagree, disagree, indifferent (neutral), agree, and strongly agree. This is called Likert scale.

You can have as many items in your questionnaire as is necessary but a 10-item assessment looks enough to cover different aspects of a construct without burdening the respondent. Sometimes, many more items are devised for discussion by the group of stakeholders who select "good" items. If the survey is repetitive, item analysis [5] of the responses of previous survey is done to delete the redundant items and possibly replace with new ones. Internal consistency can be assessed by a measure such as Cronbach alpha as discussed later in this book. Construct validity of the items is assessed by factor analysis, also discussed later in the book.

Response for each item is graded on the same scale—not necessarily 5-point—it could be 7-point, 9-point, or any other scale. Attempt is made that the options from strongly disagree to strongly agree are equally spaced so that they can be legitimately assigned scores of 0–4, 0–6, or 0–8 depending upon the number of options. Instead, these scores can be −2 to +2, −3 to +3, −4 to +4, etc. Various scoring patterns can give different results and the choice is yours so long as you can justify it.

Options for some items are reversed so that "strongly agree" is the first option and "strongly disagree" the last option. Or, some items are negatively framed. This helps in eliciting a well-thought response instead of consistently selecting same response like "agree" for most items. At the time of analysis, such reversed items and corresponding scores are rearranged in proper order.

Analysis is generally done for the total score. Add score of all the items and get total score for each respondent. This is called **Likert score** of that respondent for the construct under study. For a 10-item questionnaire and each item on 5-point scale (0–4), the total score for each respondent will range from 0 to 40. This can be analyzed just as any other numerical measurement. Second analysis can be focused on each item. If item 3 is on satisfaction with diagnostic facilities, and you have $n = 50$ responses on 0–4 scale for this item, you can calculate median score to get an assessment of satisfaction with diagnostic facilities, or to compare it with, say, nursing services. You can thus find which component of the construct is adequate and which need strengthening.

2.3.3 Pretesting and Pilot Study

It is generally considered essential that all questionnaires, schedules, laboratory procedures, etc., are tested for their efficacy before they are finally used for the main study. This is called pretesting. Many unforeseen problems or lacunae can be detected by such an exercise, and the tools can be accordingly adjusted and improved. This pretesting can reveal whether the items of information are adequate, are feasible and clear, the space provided for recording is adequate, the length of the interview is within limits, the instructions are adequate, etc. This also serves as a rehearsal of the actual data collection process and helps to train the investigators. Sometimes pretesting is repeated to standardize the methodology for eliciting correct and valid information.

A study on a small number of subjects before the actual study is called a pilot study. This simulates the actual study. This provides a preliminary estimate of the parameters under investigation. This preliminary estimate may be required for the calculation of sample size as per the details given later in this book. If the phenomenon under study has never been investigated earlier, the pilot study is the only way to get a preliminary estimate of the parameter. A pilot study may also provide information on the size of the clusters and sampling units at various stages that could be important in planning cluster or multistage sampling for large-scale surveys.

Remember that pilot study is not done to investigate statistical significance of result. It only provides a preliminary estimate of the relevant factors for designing a full study where prior data are not available. In a rare case, when the effect size is really large and pilot study carried out with proper selection of subjects, the results may even be conclusive, although many would question the conclusion based on a pilot study.

2.4 Nonsampling Errors and Other Biases

Different samples contain different sets of individuals and they can provide different results. This is called **sampling error** although this is not an error. The better term is sampling fluctuation. Routine statistical methods are designed to provide dependable results despite this error. More tricky are nonsampling errors that occur due to various types of biases in concepts, sample selection, data collection, analysis, and interpretation. These indeed are errors. Most common among nonsampling errors is nonresponse.

2.4.1 Nonresponse

The inability to elicit full or partial information from the subject after inclusion in a study is termed nonresponse. This can happen due to the subject turning noncooperative, relocation, injury, death, or any such reason. The opportunity for nonresponse is particularly present in follow-up studies, but even in the case of one-time evaluation, the subject may refuse to answer certain questions or may not agree to submit to a particular investigation or examination even when prior consent has been obtained.

Nonresponse has two types of adverse impacts on the results. The first is that the ultimate sample size available to draw conclusions reduces, and this affects the reliability of the results. This deficiency can be remedied by increasing the sample size corresponding to the anticipated nonresponse. The second is more serious. Suppose you select a sample of 3000 out of 1 million. But if only 250 respond out of 3000, your survey could be severely biased. These responders could be conformers or those with strong views. If not biased, a sample of 250 is not too bad to provide a valid estimate or to test a hypothesis in most situations. Mostly the nonresponding subjects are not random segment but are of specific type such as seriously ill cases who do not want to continue in the study or very mild cases who opt out after feeling better or some such segment. Their exclusion can severely bias the results. A way out is to take a subsample of the nonrespondents and undertake intensive efforts for their full participation. Assess how these subsample subjects are different from the regular respondents and adjust the results accordingly. A provision for such extra efforts to elicit responses from some nonrespondents should be made at the time of planning the study.

Experience suggests that some researchers fail to distinguish between nonresponse and zero value or absence of characteristic. Take care that this does not happen in the data you are examining as evidence for practice or in the data you are recording for research.

This book later discusses methods such as imputation for missing values and intention-to-treat analysis that can partially address the problem arising from nonresponse. But no analysis, howsoever immaculate, can replace the actual observation. Thus, all efforts should be made to ensure that non-response is minimal if not altogether absent. Strategies for this should be devised at the time of planning the study, and all necessary steps should be taken.

2.4.2 Variety of Biases to Guard Against

Medical study results often become clouded because some bias is detected after the results are available. Therefore, it is important that all sources of bias are considered at the time of planning a study, and all efforts are made to control them.

2.4.2.1 List of Biases

Various sources of bias are as follows. These are not mutually exclusive. In fact, the overlap is substantial. Some of the biases in this list are collection of many biases of similar type. If all these are stated separately, the list may become unmanageable. These are described in brief here. The details are provided later in the context where they predominantly occur.

1. **Bias in concepts:** This occurs due to lack of clarity about the concepts to be used in the proposed study. Lack of clarity gives an opportunity to the investigators to use subjective interpretation that can vary from person to person. Sometimes, the logic used can be faulty and sometimes the premise of the logic itself can be incorrect. For example, it is generally believed for body mass index and blood pressure that the lower the better. In fact, their very low values are also associated with increased morbidity and mortality. Ignoring stress as a factor in causation of a disease such as diabetes just because stress is so difficult to measure also comes under this bias.

2. **Definition bias:** The study subjects should be sharply defined so that there is no room for ambiguity. For example, if the cases are of tuberculosis, specify that these would be sputum positive, Mantoux positive, radiologically established, or some combination. A blurred definition gives room to the assessor to use his own interpretation that can affect the validity of the study.

3. **Bias in design:** This bias occurs when the case group and control group are not equivalent at baseline, and differentials in prognostic factors are not properly accounted for at the time of analysis. Concato et al. [6] report another type of bias in designs for prostate cancer detection where the study groups comprised asymptomatic men who received digital rectal examination, prostate-specific antigen was screened, and transrectal ultrasound was performed. There was no control group with "no screening." Thus, the effectiveness of screening could not be evaluated.

4. **Bias in selection of subjects:** The subjects included in the study may not truly represent the target population. This can happen either because the sampling was not random or because the sample size is too small to represent the entire spectrum of subjects in the target population. Studies on volunteers always have this kind of bias. Selection bias can also occur because the serious cases have already died and are not available with the same frequency as the mild cases (**survival bias**). Bidzan et al. [7] had an initial group of 158 older patients with mild cognitive impairment (MCI) but conclusions are drawn on 52 who were available at follow-up 5 years later. Similar bias can occur in selection of cases with highly variable incubation periods such as AIDS. (See also length bias.)

5. **Bias due to concomitant medication or concurrent disease:** Selected patients may suffer from other apparently unrelated conditions, but their response might differ either because of the condition itself or because of medication given concurrently for that condition.

6. **Instruction bias:** When there are no instructions or when unclear instructions are prepared, the investigators use discretion and this can vary from person to person and from time to time.

7. **Length bias:** A case–control study is generally based on prevalent cases rather than incident cases. Prevalence is dominated by those who survive for a longer duration. And these patients are qualitatively different from those who die early. Thus, the sample may include disproportionately more of those who are healthier and survive longer. The conclusions cannot be generalized to those who have less survival time. The disease profile can also differ since more cases would be those in which disease progression is slow. Those suffering from an aggressive form of the disease would be missed because of the rapid progression of the disease and early death.

8. **Bias in detection of cases:** Error can occur in diagnostic or screening criteria. For example, a laboratory investigation done properly in a hospital setting is less error prone compared to one carried out in a field setting where the study is actually done. In a prostate cancer detection study, if prostate biopsies are not performed in men with normal results after screening, true sensitivity and specificity of the test cannot be determined.

9. **Lead time bias:** All cases are not detected at the same stage of the disease. With regard to cancers, some may be detected at the time of screening, for example, by Pap smear, and some may be detected when the disease starts clinically manifesting. But the follow-up is generally from the time of detection. This difference in "lead time" can cause systematic error in the results.

10. **Bias due to confounder:** This bias occurs due to failure in taking care of the confounders. Then, any difference or association cannot be fully ascribed to the antecedent factors under study.

11. **Bias due to epistemic factors:** Efforts can be made to control only those factors that are known. But there may be many factors unknown that can affect the results. These epistemic factors can bias the results in very unpredictable ways.

12. **Contamination in controls:** Control subjects are generally those that receive placebo or regular therapy. If these subjects are in their homes, it is difficult to know if they have received some other therapy that can affect their status as controls. In the prostate cancer detection project reported by Concato et al. [6] and discussed in preceding paragraphs, the control subjects are those who are under routine care. But some of these may be screened outside the study and treated. Thus, their survival rate would not be sufficiently "pure" to be compared with the survival of those who were screened by the test procedures. In a field situation, contamination in a control group can occur if the control group is in close proximity with the unblinded test group and learns from the experience of the latter. The neighboring area may not be the test area of the research but some other program may be going on there that has spill-over effect on the control area.

13. **Berkson's bias:** Hospital cases when compared to hospital controls can have bias if the exposure increases the chance of admission. Thus, cases in a hospital will have disproportionately higher number of subjects with that exposure. Cases of injury in motor vehicle accidents have this kind of bias.

14. **Bias in ascertainment or assessment:** Once the subjects are identified, it is possible that more care is exercised by the investigators for cases than for controls. This can also occur when subjects belonging to a particular social group have records but others have to depend on recall. Sometimes this is also called **information bias.**

15. **Interviewer bias or observer bias:** Interviewer bias occurs when one is able to elicit better responses from one group of patients (say, those who are educated) relative to the other kind (such as illiterates). Observer bias occurs when the observer unwittingly (or even intentionally) exercises more care about one type of responses or measurements such as those supporting a particular hypothesis than those opposing the hypothesis. Observer bias can also occur if the observer is, for example, not fully alert when listening to Korotkoff sounds while measuring blood pressure or not being able to properly rotate the endoscope to get an all round view of, say, the duodenum in a suspected case of peptic ulcer.

16. **Instrument bias:** This occurs when the measuring instrument is not properly calibrated. A scale may be biased to give a higher reading than the actual or lower than the actual such as a mercury column of a sphygmomanometer not being empty in the resting position. The other possibility is inadequacy of an instrument to provide a complete picture, for example, an endoscope not reaching the site of interest, thereby giving false information. An example is mentioned earlier of +3 on −5 to +5 Likert scale may be more frequent than +8 on 0 to 10 scale.

17. **Hawthorne effect:** If subjects know that they are being observed or being investigated, their behavior and response can change. In fact, this is the basis for including a placebo group in a trial. Usual responses of subjects are not the same as when under a scanner.

18. **Recall bias:** There are two types of recall bias. One such bias arises from better recall of recent events than those that occurred a long time ago. Also, serious episodes are easy to recall than mild episodes. The second type of bias arises when cases suffering from a disease are able to recall events much more easily than the controls if they are apparently healthy subjects.

19. **Response bias:** Cases with serious illness are likely to give more correct responses regarding history and current ailments compared with controls. This is not just because of recall but because the former keep records meticulously. Some patients such as those suffering from sexually transmitted diseases (STDs) may intentionally suppress sexual history and other information because of the stigma attached to these diseases. Injury history may be distorted to avoid legal consequences. If the subjects are able to exchange notes, the response to questions might alter, in some cases might even be uniform. An unsuspecting illness, death in the family, or any such drastic event may produce an extreme response. Response bias also comes under **information bias**.

20. **Bias due to protocol violation:** It is not uncommon in a clinical trial that some subjects do not receive the full intervention or the correct intervention, or some ineligible subjects are randomly allocated in error. This can bias the results.

21. **Repeat testing bias:** In a pretest–posttest situation, the subjects tend to remember some of the previous questions and they may remove previous errors in posttest—thus doing better without the effect of the intervention. The observer may acquire expertise the second or third time to elicit the correct response. Conversely, fatigue may set in with repeat testing that could alter the response. It is widely believed that most biological measurements have a strong tendency toward the mean. Extremely high scorers tend to score lower in subsequent testing, and extremely low scorers tend to do better in a subsequent test.

22. **Midcourse bias:** Sometimes the subjects after enrollment have to be excluded if they develop an unrelated condition such as an injury or become so serious that their continuation in the trial is no longer in the interest of the patient. If a new facility such as a health center is started or closed for the population being observed for a study, the response may alter. If two independent trials are going on in the same population, one may contaminate the other. An unexpected intervention such as a disease outbreak can alter the response of those who are not affected.

23. **Self-improvement effect:** Many diseases are self-limiting. Improvement over time occurs irrespective of the intervention, and it may be partially or fully unnecessarily ascribed to the intervention. Diseases such as arthritis and asthma have natural periods of remission that may look like the effect of therapy.

24. **Digit preference:** It is well known that almost all of us have a special love for digits zero and five. Measurements are more frequently recorded ending with these digits. A person aged 69 or 71 is very likely to report one's age as 70 years. Another manifestation of digit preference is in forming intervals for quantitative data. Blood glucose level categories would be 70–79, 80–89, 90–99, etc., and not 64–71, 72–79, etc. If digit zero is preferred, 88, 89, 90, 91, and 92 can be recorded as 90. Thus, intervals such as 88–92, 93–97, and 98–102 are better to ameliorate the effect of digit preference and not the conventional 85–89, 90–94, 95–99, etc.

25. **Bias due to nonresponse:** As already discussed in detail, some subjects refuse to cooperate, suffer an injury, die, or become untraceable. In a prospective study, there might be some dropouts for various reasons. Nonrespondents make two types of effects on the responses. First, they are generally different from those who respond, and their exclusion can lead to biased results. Second, nonresponse reduces the sample size that can decrease the power of the study to detect differences or associations.

26. **Attrition bias:** The pattern of nonresponse can differ from one group to the other in the sense that in one group more severe cases drop out, whereas in another group mostly mild cases drop out. In a rheumatoid arthritis databank study, attrition during follow-up was high in patients of young age, who were less educated and were of non-Caucasian race [8].

27. **Bias in handling outliers:** No objective rule is available to label a value as outlier except a guideline that the value must be far away from the mainstream values. If the duration of hospital stay after a particular surgery is mostly between 6 and 10 days, some researchers would call 18 days as outlier and exclude it on the suspicion of

being a wrong recording, and some would consider it right and include it in their calculation. Some would not exclude any outlier, however different it might be. Thus, the results would vary.

28. **Recording bias:** Two types of errors can occur in recording. The first arises due to the inability to properly decipher the writing on case sheets. Physicians are notorious for illegible writing. This can happen particularly with similar looking digits such as 1 and 7 and 3 and 5. Thus, the entry of data may be in error. The second arises due to the carelessness of the investigator. A diastolic level of 87 can be wrongly recorded as 78, or a code 4 entered as 6 when memory is relied upon, which can fail to recall the correct code. Wrongly pressing adjacent keys on the computer keyboard is not uncommon either.

29. **Bias in analysis:** This again can be of two types. The first occurs when gearing the analysis to support a particular hypothesis. For example, while comparing pre- and postvalues, for example, hemoglobin (Hb) levels before and after weekly supplementation of iron, the increase may be small that will not be detected by comparison of means. But it may be detected when evaluated as a proportion of subjects with levels <10 mg/dL before and after iron supplementation. The second can arise due to differential interpretation of P-values. When $P = 0.055$, one researcher may refuse to say that it is significant at 0.05 level and the other may say that it is marginally significant. Some researchers may change the level of significance from 5% to 10% if the result is to their liking.

30. **Bias due to competing cofactors:** Some factors influence results synergistically or antagonistically when relevant cofactors are present. If this is not properly taken into account, the effect of an intervention can be under- or overestimated.

31. **Prevalence–incidence bias:** This occurs when effects of risk factors on prevalence that could be a function of the duration of disease can be mistaken for effects on disease occurrence.

32. **Bias due to lack of power:** You will soon notice that statistical tests are almost invariably used to check the significance of differences or associations. The power of these tests to detect difference or association depends to a large extent on the number of subjects included in the study—the sample size. If the study is conducted on a small sample, even a big difference cannot be detected, leading to a false-negative conclusion. When conducted on an appropriate number of subjects, the conclusion can change.

33. **Interpretation bias:** This arises from the tendency among some research workers to interpret the results in favor of a particular hypothesis ignoring the opposite evidence. This can be intentional or unintentional.

34. **Reporting bias:** Researchers are human beings. Some can create a report such that it gives a premonitory result even though it is based on evidence. It is easy to suppress contradictory evidence by not talking about it.

35. **Bias in presentation of results:** Scales in graphs can be chosen such that a small change looks like a big change or vice versa. The second is that the researcher may merely state the inconvenient findings that contradict the main conclusion but does not highlight them in the same way as the favorable findings are done.

36. **Publication bias:** Many journals are much too keen to publish reports that give a positive result regarding efficacy of a new regimen compared with the negative trials that did not find any difference. If a vote count is done on the basis of the published reports, positive results would hugely outscore the negative results, although the fact may be just the reverse.

2.4.2.2 Steps for Minimizing Bias

The purpose of describing various types of biases in so much detail is to create awareness to avoid or at least minimize them. Everything possible should be done to keep them under control so that you do not have to give an explanation such as "hand of God" in a soccer game. The following steps can be suggested to minimize bias in the results in a research setup. All steps do not apply to all the situations. Adopt the ones that apply to your setup. Details of some of these steps are in subsequent chapters.

1. Develop an unbiased scientific temperament by realizing that you are in the occupation of relentless search for truth.

2. Specify the problem to the minutest detail.

3. Assess the validity of the identified target population and the groups to be included in the study in the context of objectives and the methodology.

4. Assess the validity of antecedents and outcomes for providing correct answer to your questions. Beware of epistemic uncertainties arising from limitation of knowledge.

5. Evaluate the reliability and validity of the measurements required to assess the antecedents and outcomes, as also of the other tools you plan to deploy.

6. Carry out a pilot study and pretest the tools. Make changes as needed.

7. Identify all possible confounding factors and other sources of bias, and develop an appropriate design that can take care of most of these biases if not all.

8. Choose a representative sample, preferably by a random method.

9. Choose an adequate size of sample in each group.

10. Train yourself and your coworkers in making correct assessments.

11. Use matching, blinding, masking, and random allocation as needed.

12. Monitor each stage of research, including periodic checking of data.

13. Make determined efforts to minimize nonresponse and partial response.

14. Double check the data and rectify errors in recording, entries, etc.

15. Analyze the data with proper statistical methods. Use standardized or adjusted rates where needed, perform the stratified analysis, or use mathematical models such as regression to take care of those confounders that could not be ruled out by design.

16. Interpret the results in an objective manner based on evidence.

17. Report only the evidence-based results, enthusiastically but dispassionately.

18. Exercise extreme care in drafting the report and keep comments or opinions separate from evidence-based results.

Bias and other aspects of design can be very adequately taken care of if you could imagine yourself presenting the results a couple of years hence to a critical but friendly audience [9]. Consider what your colleagues would question or advise at that time, and their reaction when you conclude that the results are significant or if you conclude that the results are not significant. Can there be noncausal explanations of the results? Are there any confounding factors that have been missed? Whether chance or sampling error could be an explanation? Such considerations will help you to develop a proper design and to conduct the study in an upright manner.

References

1. Fisher AA, Foreit JR. *Designing HIV/AIDS Intervention Studies: An Operations Research Handbook.* New York: Population Council, 2002, p. 8.

2. Lutz W. *Community Health Surveys—A Practical Guide for Health Workers: 3 Using Available Information.* International Epidemiological Association, 1983, p. 10.

3. Hepburn W, Lutz W. *Community Health Surveys—A Practical Guide for Health Workers. 5 Interviewing and Recording.* International Epidemiological Association, 1986.

4. Bond TG, Fox CM. *Applying the Rasch Model: Fundamental Measurement in Human Sciences,* 2nd edn. Mahwah, NJ: Lawrence Erlbaum, 2007.

5. Shin S-H. *A Polytomous Nonlinear Mixed Model for Item Analysis: Item Calibration Using SAS NLMIXED and PARSCALE.* Saarbrücken, Germany: VDM Verlag, 2009.

6. Concato J, Peduzzi P, Kamina A, Horwitz RI. A nested case-control study of the effectiveness of screening for prostate cancer: Research design. *J Clin Epidemiol* 2001; 54:558–564.
7. Bidzan L, Pachalska M, Bidzan M. Prediction of clinical outcome in MCI. *Med Sci Monit* 2007; 13:CR398–CR405.
8. Krishnan E, Murtagh K, Bruce B, Cline D, Singh G, Fries JF. Attrition bias in rheumatoid arthritis databanks: A case study of 6346 patients in 11 databanks and 65,649 administrations of the Health Assessment Questionnaire. *J Rheumatol* 2004; 31:1320–1326.
9. Elwood M. Forward projection—Using critical appraisal in the design of studies. *Int J Epidemiol* 2002; 31:1071–1073.

3

Sampling Methods

The concept of sampling is neither new nor unfamiliar in everyday life. A cook examines a few grains of rice to find whether nearly all of them are properly cooked or not. In medicine, study of blood, urine, stool, and sputum samples and biopsy specimens is common. This chapter deals primarily with subjects, mostly human beings but sometimes laboratory animals, who have great intrinsic biologic variability.

This chapter: The chapter begins with an explanation in Section 3.1 of terms and concepts that are typical to sampling. This also includes a discussion on the advantages and limitations of sampling. The aim of the sampling method is to choose a fraction that adequately represents the entire spectrum of the target subjects. The purpose of sampling is to extrapolate the results to a substantially larger population. Various common methods of random sampling that could meet this aim are discussed in Section 3.2. Some other methods, including nonrandom methods, are presented in Section 3.3.

The literature on sampling methods is extensive. For statistical details of these methods, see Thompson [1]. Even though the description in this chapter is brief, it is sufficient to explain the basics of sampling methods. The notations used here are those that are conventionally used in texts that deal with sampling and are sometimes different from those used in later chapters of this book.

3.1 Sampling Concepts

As explained earlier, in statistics the term "population" has a special meaning. It can be understood as the target group from which sample subjects are chosen. In a descriptive study of acute respiratory infection (ARI) in a country, the target population could be all existing cases of ARI in that country. For a cervical cancer control program, the target population could be all married women of age more than 40 years. For studying risk factors of enlarged prostate, the population of interest could be all men of age 50 years and above in an area. Cost and logistic considerations seldom allow study of all the subjects in a population. Therefore, sampling becomes a natural choice. Nevertheless, all subjects in the target population can also be investigated. Then, such a study is called complete enumeration or **census**.

Even if all existing cases are surveyed, there is no guarantee that the results would apply to future cases. Thus, the concept of population in the context of medicine is more hypothetical than real. When all existing cases are indeed included, it still remains a sample considering that future cases are not included. Medical empiricism implies that the findings on the existing cases are used for future cases. Sampling is a prerequisite for this paradigm. However, if the objective is to find the prevalence of diabetes mellitus in the year 2008 among females of age 40 years and above residing in a particular city, complete enumeration is possible. Similarly, if a complete registry of cancer cases in a defined population is available, perhaps sampling is not needed for assessing the existing situation.

A futuristic perspective brings in the concept of **universe**. A universe could be larger than the target population that has implications for the result. Future cases are part of the universe but not of the population.

3.1.1 Advantages and Limitations of Sampling

Although there are many advantages of sampling, there are sometimes limitations too. Both advantages and limitations are described in this section. These are clearly understood by acknowledging the unfailing presence of sampling fluctuations in sample-based investigations.

3.1.1.1 Sampling Fluctuations

One sample from a population in all probability will be different from the second sample. Some or all individuals in the second sample may be new. Thus, the results obtained from one sample may not match those from another sample. This variability is included in the sources of uncertainties enumerated earlier and is called sampling fluctuation. If a sample of 300 healthy males of age 60–64 years from a population has an average systolic blood pressure (BP) of 142 mmHg, it is quite possible that another sample of 300 from the same population yields an average of 139 mmHg. But how likely is it that the average in a new sample will be as low as 128 mmHg or as high as 155 mmHg? The magnitude of this "error" depends primarily on three factors: (1) The method of sampling: The subjects should be selected in such a manner that a wide spectrum has adequate representation. Then, repeated samples may give nearly the same picture. (2) The size of the sample: When the sample includes a large number of subjects, the picture obtained from one sample is not likely to be very different from that of another sample of the same size because both tend to be fair representatives of the population. This cannot be said for small samples. (3) Variability among the subjects in the population: If the cholesterol level differs widely from person to person, then obviously the samples would reflect the same variability.

Only the first factor is discussed in this chapter; the other two are deferred to later chapters.

3.1.1.2 Advantages of Sampling

The advantages of sampling can be listed as follows:

1. May be the only feasible method for collection of relevant data in some cases. Samples of blood, urine, semen, and biopsies are everyday examples in medicine. Complete enumeration is not feasible for such items.
2. Lower cost and less demand on personnel because of smaller coverage in samples relative to the total population.
3. Higher speed with which the results can be obtained.
4. Because of the relatively small number of subjects in the sample, more reliable information can often be collected by deploying better trained personnel, by adopting technologically more accurate methods, and by close supervision.

3.1.1.3 Limitations of Sampling

Despite these clear advantages, the results obtained from a sample cannot always be accepted. The following limitations may be noted:

1. Sampling necessarily entails an argument from the fraction to the whole. The validity of this argument depends on the representativeness of the sample. Not all samples are representative, although methods are available that make it *likely* to happen.
2. When information is required for small segments containing few individuals, sampling may fail to provide sufficiently precise information on them.
3. Sometimes a complete count is needed anyway, as for a diagnosis and outcome profile of cases admitted to a hospital each year. In such cases, sampling is unnecessary.
4. Inclusion of some subjects in the sample and exclusion of others for a study may cause a feeling of discrimination among the subjects. Caution, some groundwork, and persuasion may be required.

3.1.2 Some Special Terms Used in Sampling

Before various sampling methods, it is necessary to understand some terms.

3.1.2.1 Unit of Inquiry and Sampling Unit

Different kinds of units are used in sampling. The **unit of inquiry** is the subject on which the information is obtained. A **sampling unit** is the one that is used for selection of the subjects. In a community survey on protein energy malnutrition, the sampling unit could be a family but the unit of inquiry could be a child younger than 5 years. One sampling unit can have multiple units or no unit to inquire on. Sometimes the sampling is done in stages, such as selection of some hospitals in the first stage, then wards or departments within the selected hospitals in the second stage, and then patients in the selected wards in the third stage. The unit of inquiry could be a patient, but sampling units are multiple in this kind of sampling.

3.1.2.2 Sampling Frame

The list of all sampling units in the target population is called the sampling frame. The units are chosen from this frame. The units must be mutually exclusive and the frame should be an exhaustive list. If there is a study on hypertensives and diabetics, a person who has both cannot be listed twice. The list may separately include those who have both diseases, those who have only one of the two diseases, and those who have neither of these two diseases. Inclusion and exclusion criteria must be fully known to the sampler. Preparation of the frame requires precise definition of the unit as well as of the population. For example, if the unit is a case of sexually transmitted disease (STD), all the STDs should be specified. State whether a case of hepatitis B infection would be included or not. Also state whether the cases would be those that have frank disease or cases with subclinical infection would also be included. The criteria of diagnosis of each should be stated, that is, the diagnosis is based on clinical evidence alone or a laboratory confirmation is also required. In some situations, a map of the study area serves as the frame. In this case, geographical entities are selected instead of individuals or families. People living or available in the selected areas constitute the sample.

3.1.2.3 Parameters and Statistics

Most medical studies draw conclusions on the basis of the sample estimate of either mean or proportion or difference. Sometimes the ratio of two means or of two proportions is of interest. Such quantities of interest are called parameters when calculated for the entire population and are called statistics when calculated for the sample. Note that the term statistics is being used here in the plural sense. The objective of sampling is to provide values of the statistics that are adequate estimates of the parameters. This adequacy is generally measured in terms of lack of bias and in terms of the standard error (SE) of the estimate. An estimator is called **unbiased** when its average value over all possible samples from a population is the same as the value of the parameter. This property increases the likelihood that the estimate in the long run approaches the right target. The SE quantifies the variation expected from sample to sample. The details appear in later chapters.

3.1.2.4 Sample Size

The number of subjects in the sample is called the sample size. It is denoted by n. The number of subjects in the entire target population is denoted by N. The ratio n/N is called **sampling fraction**.

Perhaps, the most frequent question faced by a practicing statistician concerns an adequate number of units for inclusion in the sample. This apparently simple question has many facets. Just as a physician cannot prescribe a therapeutic regimen without knowing some details of a patient's ailment, a statistician cannot answer questions on sample size unless at least some basic information is supplied to him. This involves information on issues such as the magnitude of interindividual variation, the degree of precision required from the sample, and the least confidence one would like to have in the results. Each of these issues requires details that are provided in some of the following chapters. Therefore, the question of determination of sample size is postponed until Chapter 12.

3.1.2.5 Nonrandom and Random Sampling

A sample of volunteers for phase I of a clinical trial is a nonrandom sample. This kind of sampling is adopted particularly when large number of people of the type actually required for the study is not available. Nonrandom samples inhibit generalizability but can still be useful in some situations as discussed later in this chapter.

A sample is called random when inclusion or exclusion of a particular eligible subject depends on chance and cannot be predicted in advance. However, as shortly illustrated, the chances are not necessarily equal for all subjects for inclusion in the sample. For this reason, it is sometimes prudent to call it a **probability sample**.

Random selection is just a strategy to get a representative sample. If representativeness can be achieved by any other method, the same can be adopted. The larger the sample relative to the size of the population, the more is the chance of it being representative—random or not.

3.2 Common Methods of Random Sampling

A large number of methods are available for choosing a random (or probability) sample. However, only some are commonly used.

3.2.1 Simple Random Sampling

When the scheme is such that each unit of the population has the *same chance* of being included in the sample, it is called simple random sampling (SRS). A rigorous definition is that all possible samples of the same size have an equal chance of selection. As already stated, it is customary in sampling to denote the size of the sample by n and the size of the population by N. SRS is like picking n slips from a lot containing N look-alike slips numbered 1 through N. A more scientific method is to use random numbers. These can be very easily generated on a computer.

SRS seems easy but can turn out to be difficult to implement. A prerequisite is the availability of the sampling frame. Preparing this in many cases can be a very expensive exercise. If a study involves several hospitals, each hospital will have a list of its own patients but a joint list of the patients in all participating hospitals may not be available at one place. In a domiciliary study, it may be difficult to prepare a list of families in an area. The next problem is that the randomly selected units can be physically very far apart residing in different areas, admitted to different hospitals, or being attended to in clinics in different locations. Note that such physical divergence may or may not add to the representativeness of the sample. There is no guarantee that SRS will adequately represent different segments of the target population. It is possible that the sample happens to include cases of serious or moderate severity but none of mild severity. Or all cases can be adults with no or very little representation of children. If adequate representation of such segments is required, the method of selection should be stratified sampling.

Example 3.1: Mean and SD in SRS Relative to the Values in the Population

Consider waist–hip ratio (WHR) and triglyceride (TG) data in a population of 100 male hypertensive subjects (Table 3.1). An SRS of size 16 may be the subject numbers 3, 18, 21, 31, 33, 45, 49, 53, 59, 62, 66, 71, 78, 79, 91, and 96. These are marked by an asterisk in the table. The method of calculation of mean and standard deviation (SD) is given later in Chapter 7, but note for now that the sample mean ($\bar{x}$) TG level of these 16 subjects is 153.9 mg/dL and sample SD (s) is 15.8 mg/dL. The population mean (μ) for all the 100 subjects is 155.5 mg/dL and population SD (σ) is 17.4 mg/dL. The values of statistics in this sample are slightly lower than the values of the corresponding population parameters. This can happen with any sample.

3.2.2 Stratified Random Sampling

A big drawback of SRS is that it can fail to give adequate representation to one or more subgroups of interest. For example, in a study on the relationship of maternal complications with parity, it is necessary that women of different parities are included in the sample. The definition of the sampling unit in this case could be a currently pregnant woman reporting in a particular group of antenatal clinics. An SRS of size, say, 60 women in this case can yield a sample in which by chance parity 4 is not represented at all or inadequately represented. Note that the objective of the study is not met unless all parities have adequate representation. The procedure therefore should be to first divide the frame by parity status such as 1, 2, 3, 4, 5, and 6+ and then draw an independent SRS of size, say, 10 from each division. Such a division of the frame is called **stratification** and each division a **stratum**.

TABLE 3.1

Serum TGs in mg/dL and WHR in a Population of 100 Male Hypertensive Subjects

Subject Number	WHR	TG	Subject Number	WHR	TG	Subject Number	WHR	TG
1	1.21	152	35	1.17	151	68	0.93	148
2	0.97	145	36	1.41	167	69	1.27	162
3*	1.52	183	37	0.80	131	70	1.33	178
4	0.83	138	38	0.99	128	71*	1.09	140
5	1.06	147	39	1.24	142	72	1.07	138
6	1.15	167	40	1.07	162	73	1.14	169
7	1.44	178	41	1.13	148	74	1.20	128
8	1.23	160	42	0.95	152	75	1.27	139
9	1.17	180	43	1.26	158	76	1.07	172
10	0.98	128	44	1.06	150	77	0.92	136
11	1.89	197	45*	1.17	157	78*	0.97	148
12	1.41	162	46	0.84	136	79*	0.98	169
13	1.32	158	47	0.91	147	80	1.32	173
14	1.04	150	48	0.90	162	81	1.48	185
15	1.35	175	49*	1.07	157	82	1.27	196
16	0.76	161	50	1.04	149	83	1.08	147
17	0.81	151	51	1.73	195	84	1.08	173
18*	0.90	148	52	1.39	187	85	1.38	175
19	1.44	159	53*	1.08	151	86	1.40	184
20	1.20	167	54	1.17	162	87	1.01	160
21*	1.00	142	55	1.10	148	88	1.02	151
22	1.37	138	56	1.12	138	89	1.07	145
23	1.12	149	57	1.57	167	90	1.15	162
24	1.29	162	58	1.31	175	91*	1.18	171
25	1.08	158	59*	0.94	167	92	0.90	162
26	1.34	149	60	0.85	145	93	0.87	128
27	1.61	182	61	0.75	132	94	0.95	149
28	0.73	120	62*	0.66	125	95	0.96	160
29	0.88	138	63	1.53	192	96*	0.84	134
30	1.17	129	64	1.91	169	97	0.88	158
31*	1.01	138	65	1.02	152	98	1.15	167
32	0.84	141	66*	1.07	165	99	1.20	162
33*	0.93	167	67	0.94	129	100	1.17	139
34	0.92	129						

WHR			TG	
≤0.89	14 subjects		≤159	57 subjects
0.90–1.09	38 subjects		≥160	43 subjects
≥1.10	48 subjects			

Population mean ($n = 100$) = 1.127

Population SD ($n = 100$) = 0.238

Population mean ($n = 100$) = 155.5 mg/dL

Population SD ($n = 100$) = 17.4 mg/dL

* In the sample.

In Table 3.1, the strata could be WHR categories, given at the bottom of the table. One may decide how many units to be selected from different strata—they need not be equal. This procedure of choosing the sample is called stratified random sampling (StRS). The characteristic chosen for stratification is either the one suspected to affect the variable under study or the one that makes groups of interest for which separate results are required. After this, the sample would adequately represent the stratifying characteristic but not necessarily other factors that may be of consequence. For example, in a study on diabetes mellitus, the subjects may be stratified by levels of plasma glucose, but the sample may still not be representative for obesity, age–gender, coexisting diseases, patient cooperation, etc. All these can affect the prognosis or the outcome. Thus, care is always needed in extrapolation of results even after adopting stratified sampling.

Example 3.2: TG in WHR Strata

Evidence has accumulated that higher WHR may be a health risk. In Table 3.1, the subjects are divided into thin, normal, and obese according to WHR ≤ 0.89, $0.90 \leq$ WHR ≤ 1.09, and WHR ≥ 1.10, respectively. With this categorization, the SRS of 16 subjects in Table 3.1 happens to contain 2 thin, 11 normal, and 3 obese subjects. However, the actual numbers of thin, normal, and obese in the population according to this categorization are 14, 38, and 48, respectively. The obese are clearly underrepresented in the sample. If you divide the population in the first instance and then take a sample, the result could be very different. A commonly adopted strategy in this case is to take a **proportionate sample** from each stratum. In this case, 16 out of 100 gives a sampling fraction of 0.16. Applying this to the strata gives a sample of two from the first stratum, six from the second stratum, and eight from the third stratum. These, when randomly drawn by me, are subject numbers 28 and 62 from the first stratum; 5, 21, 33, 47, 76, and 95 from the second stratum; and 8, 54, 58, 63, 69, 73, 85, and 99 from the third stratum. You may get different subjects. The means are now calculated separately for each stratum. These means for TG levels of subjects in my sample are 122.50, 155.83, and 169.62, respectively. These are multiplied by the respective stratum size and divided by the population size, that is,

$$\bar{x}_{st} = \frac{\Sigma N_k \bar{x}_k}{N},$$

where
  N_k is the size of the kth stratum
  $\bar{x}_k$ is the mean obtained for kth stratum

In this case,

$$\bar{x}_{st} = \frac{(122.50 \times 14 + 155.83 \times 38 + 169.62 \times 48)}{(14 + 38 + 48)} = 157.8.$$

The stratum size is now the "weight" for calculation of the mean. A similar formula is required to calculate the sample SD, but this is outside the scope of this text.

SIDE NOTE: This sample mean for StRS is slightly more than the population mean of TG in this example. It is natural for such a difference to arise in any sampling. But now the sample has adequate representation of each WHR category.

In case of proportionate samples as in Example 3.2, the probability of selection of each subject is nearly the same in different strata. The advantage with this is that the sample becomes self-weighting. This can be explained as follows.

Proportionate sample implies that the sample from each stratum is in the same proportion as in the population. This means,

$$\frac{N_k}{N} = \frac{n_k}{n} \quad \text{for all the } k \text{ (i.e., for each stratum).}$$

When this is substituted, we get

$$\bar{x}_{st} = \frac{\Sigma n_k \bar{x}_k}{n}.$$

This can be easily seen to be the same as the usual sample mean. Thus, in the case of proportionate samples, the mean for the stratified sample can be calculated just as a usual unweighted mean is calculated. This property is called **self-weighting** of the sample.

Nonetheless, proportionate samples are not a requirement for stratified sampling. If five subjects are selected from each stratum in Example 3.2, probability of selection is 5/14, 5/38, and 5/48 for subjects in the first, second, and third strata, respectively. The unequal probabilities do not render StRS invalid. Many medical investigations use equal samples from strata of different sizes. It is important to realize that the probabilities for different units can be unequal despite random samples and they impose a restriction of the type illustrated by the formula of $\bar{x}_{st}$. However, in this case, the estimate of mean in different strata would have different precision and it is difficult to justify mixing estimates of varying precision for obtaining the combined estimate.

3.2.3 Multistage Random Sampling

In studies that involve a population of large size, it is sometimes helpful to draw samples in stages. If the subjects spread all over a state are the target, you may select a small number of districts or counties in the first stage (first stage units are called **primary sampling units**); then some blocks, colonies, or hospitals in the second stage from the selected districts or counties; and finally the subjects from the selected colonies or hospitals. Thus, there are sampling units of various sizes. When sampling is done in stages from bigger to smaller units within the units selected at the previous stage, it is called multistage sampling. When selection in each stage is random, this becomes multistage random sampling (MRS). In a study to find the prevalence of smoking in females of age 20 years and above in a particular state with, say, a million families, you may, for example, first select 4 counties by the random method, then 12 census blocks within each selected county and 50 families within each selected block. All females of age 20+ years in the selected families could be the unit of inquiry although the sampling units are counties, blocks, and families. Some families may have two or more units of inquiry and some none at all, but most may have just one. If there are many families with two or more eligible females, this can produce a clustering effect as discussed in the next section.

In the preceding example, a total of $4 \times 12 \times 50 = 2400$ families could be in the sample. This looks like an extremely small number compared with a total of a million families in the state. Yet this could provide a fairly precise estimate of the prevalence of smoking among the females of age 20+ years in the state.

If an SRS of 2400 families is chosen out of a million instead of an MRS, the selected families may be scattered all over the state, say, in 200 census blocks. A block may have to be visited for just one family. This could mean a substantially higher cost of travel by the survey team and loss of time. In the case of MRS in this example, only 4 counties need to be visited and the survey work will concentrate in 12 blocks within each county. Thus, the major advantage of MRS is the reduced cost and saving of time because of less travel. The second advantage is that a full sampling frame of the smaller units is not required. In this example, the frame required is the list of all counties in the state, the list of blocks in the *selected* counties, and the list of families in the *selected* blocks only. In the case of SRS, on the other hand, the frame will be the list of all families in the entire state. Preparation of this frame in itself could be a major exercise in some situations. The third advantage is that, in most practical situations, the smaller sample chosen by MRS may be sufficient to achieve good precision relative to SRS.

Example 3.3: MRS for Studies on Hypercholesterolemia

Milias et al. [2] studied prevalence of self-reported hypercholesterolemia in Greek adults. In this nationwide survey, 5003 adults of age 18–74 years were enrolled by multistage sampling. Full details are not given but they state that the multistage sampling was based on age–sex distribution of Greek population according to the 2000 census. However, the stages are not identified. The accompanying table on the age–sex distribution of the sample and population indicates that the age representation in the sample was nearly the same as in the population.

SIDE NOTE: The reporting made in this article has a couple of lessons. First, the sampling is stated as multistage but stages are not identified. They may have selected few districts in the first stage, few counties in the second stage, and few families in the third stage. Second, the authors use the term "study population" instead of study sample for selected 5003 adults. Such misuse of the term population is quite common in medical literature. Third, the proportionate representation of age indicates that possibly age stratification has been used but not explicitly stated.

3.2.4 Cluster Random Sampling

When the size of the primary sampling units is not large, that is, when they generally contain a small number of subjects, then it is sometimes advisable that these units are not sampled further. All the elements in the selected primary units are then surveyed. This tends to increase the total number of subjects in the sample without a corresponding increase in the cost. Since many subjects in close proximity are included in the sample, the travel time and cost are saved. When this is done, it is convenient to understand a primary unit as a **cluster**. If a population comprises a total of N clusters, then n clusters out of N are randomly selected. If the ith cluster has M_i subjects, then a total of ΣM_i elements of these n clusters are investigated. This is called cluster random sampling (CRS). This sampling gives extremely good results when the units within the clusters are heterogeneous. The method is used also in settings with large units but then they are divided into small clusters.

Besides the fact that CRS helps to increase the size of the sample and thus the precision without a corresponding increase in the cost, the other advantage relative to SRS is again that the sampling frame of the elements is not required. The only frame required is the list of clusters. This type of sampling is also very easy to administer. Since survey of subjects within the cluster is quick due to close proximity, CRS is sometimes considered a **rapid assessment method**. The World Health Organization (WHO) recommended this kind of sampling to estimate the percentage of children immunized in a community, particularly in developing countries. Their recommended strategy is 30 clusters of size 7 each—called **30 × 7 sampling**. Bennett et al. [3] have given some very useful details of this methodology. The methodology has become popular and is used in many other setups, more for convenience than statistical propriety. The 30 × 7 methodology is appropriate when the anticipated prevalence is around 50% and the precision required is ±10% points. For other setups, a different size of cluster and a different number of clusters may be needed. Since immunization level in most countries has reached 70% or 80%, this methodology has become outdated for immunization surveys. However, such CRS is used as a rapid assessment method for assessing the status of other aspects of health as in estimating prevalence of poor vision in old age (see Example 3.4).

As will be discussed later, a disadvantage of CRS is that the elements within a cluster tend to be similar to one another and produce a clustering effect.

This is also called the **design effect**. This effect reduces the chances of getting the full spectrum of subjects in the sample. To compensate this, a larger sample may be required relative to SRS. However, it sometimes happens that even a large sample chosen by CRS is less expensive to investigate than a small sample chosen by SRS. Intercluster comparison is not valid in the case of CRS.

It is customary in the case of CRS that the population is divided into clusters of equal or approximately equal size, although this is not a prerequisite. The size of cluster—that it should have 20, 30, or 50 subjects—is generally determined by administrative convenience, such as the ability of a team to finish the survey in one day. However, the number of clusters would depend on the statistical requirement of reliability of the estimate.

> **Example 3.4: Cluster Sampling for Assessing Prevalence of Poor Vision in Old Age**
>
> For a survey on prevalence of poor vision (visual acuity < 6/36 in the better eye with corrective glasses if any) in persons of age 50 years and above (50+) in a district with half a million population, suppose 20 clusters of size 30 each are selected as per the following scheme:
>
> a. A list of census blocks is prepared along with the population of each, and this is cumulatively added.
> b. Since the sampling fraction is one cluster per 25,000 population, one number less than or equal to 25,000 is randomly selected. Then, 25,000 is sequentially added every time in a systematic fashion and thus a sample of 20 numbers is obtained. Twenty blocks containing the chosen 20 numbers are selected from the list made in (a). These blocks are now in the sample.
> c. Home visits are made from a geographically random point in each of the selected blocks, and the first 30 persons of age 50+ residing in contiguous houses are listed and examined for visual acuity. This gives one cluster of 30 subjects from each selected block.

The scheme in Example 3.4 is similar to the one recommended by WHO for surveys to assess immunization coverage in developing countries but is not exactly the same. Note the following features of the CRS in Example 3.4:

1. The frame required is only the list of census blocks, which in any case is generally available. No listing of households or of persons of age 50+ is required.
2. The selection of blocks is on the basis of the size of the population. This is inherent in step (b). Blocks with a larger population have a higher chance of being included in the sample. This is called sampling with **probability proportional to size** (PPS). Details of this method are given later in this chapter. The size in this case is the population in the blocks and the subjects are the persons of age 50+.

It is reasonable to expect that this age group would have nearly the same proportion in each census block. The PPS sampling makes the estimate self-weighting, and the usual sample proportion becomes a statistically valid estimate of the population proportion. Weighted calculations, as done for StRS (Example 3.2), are not required.

3. Starting from a geographically random point ensures random sampling but may require a map of each of the selected blocks.

4. The houses to be visited are contiguous. This should make the survey substantially faster. Because of this property, it is valid to consider it as a rapid assessment method. It is up to the investigators to define contiguity. These could be houses lined up on both sides of a street or houses whose entrance is closest to the previously selected houses. This must be defined before the survey begins.

5. The total number of clusters of size 30 would be extremely large in a district that has half a million population. A sample of 20 clusters is relatively exceedingly small. This would be generally the case in CRS.

6. All of the first 30 persons of age 50+ selected for the survey may not be available and some may not cooperate in examination of their vision. In practice, the nonresponse could be large and may have to be tackled separately.

3.2.5 Systematic Random Sampling

Example 3.4 on CRS contains a scheme in which the first unit was randomly selected and the others automatically included on the basis of the **sampling interval**, $I = N/n$. (In that example, this is 25,000.) Such a scheme is called systematic random sampling (SyRS). The units are numbered 1 to N in any order. If I is not an integer, then the integer part is taken. The first unit is selected at random from the first I units. Suppose this is the rth unit. Then, the subsequent units in the sample are $(r + I)$th, $(r + 2I)$th, etc. Thus, the first selected unit determines the entire sample.

Let there be $N = 101$ subjects in the target population out of which $n = 8$ are proposed to be selected by the systematic method. The sampling interval is $I = 101/8$ and its integer part is 12. If the randomly selected unit (subjects in this case) out of the first 12 is the 9th, then the remaining units have numbers 21, 33, 45, 57, 69, 81, and 93.

Although the systematic method works well when the sampling interval is an integer, two kinds of anomalies can occur in other cases: (1) The last few units may never have a chance to be included in the sample. In the preceding example, if the first selected number is the maximum 12, then the last number in the sample is 96. Thus, unit numbers 97–101 can never be selected. (2) In some cases, the sample size may increase by one. In this example, if the first randomly selected unit is the 4th, then the last would be the 100th. The sample then would contain nine units instead of the stipulated eight. However, this is not serious when n is large, say, more than 30. A way out of this mess is **circular sampling**, in which the start is made from a random number between 1 and N, and every Ith thereafter is selected in a cyclical manner, that is, the $(N + 1)$th unit becomes the first unit again. This is done until n units are selected.

SyRS has the following merits: (1) It is very easy to execute. When the patients are coming to a clinic, every eligible kth after the random first can be easily pulled out for inclusion in the study. It is also very quick relative to SRS, where each unit is to be selected at random. (2) SyRS does not need a full frame— knowledge about the total number of units in the population is enough. If you know that a clinic generally gets $N = 275$ patients a month and a sample of size $n = 30$ is to be chosen from among the patients coming in a particular month, then $I = 9$. If the first randomly selected patient is the 7th, then the others in the sample are the 16th, 25th, etc. These can be chosen according to the sequence of arrival and no list of patients is required for sampling. (3) In some cases, it can yield a much more representative sample than SRS. The chosen units are equally spaced until the end, so all sections of the population are likely to be represented. If a sex clinic runs 6 days a week with Mondays and Thursdays reserved for teenagers, Tuesdays and Fridays for male adults, and Wednesdays and Saturdays for female adults, then SyRS would automatically give proportional representation to each of the three groups of patients when the sampling includes all the days.

SyRS is not without demerits: (1) The SyRS could yield a biased sample if a periodicity or trend is hidden in the subjects as you go from 1 to N. (2) As in the case of SRS, this method can fail to give adequate representation to some specific groups of interest.

Example 3.5: Systematic Sampling for Evaluating Risk Factors of Folate Deficiency among Adolescents

Vitolo et al. [4] describe results of a study in Sao Leopoldo in Brazil for evaluating risk factors of folate deficiency among adolescents. In the first stage, they selected census regions, then random blocks and street corners where sampling would start, and finally houses were sampled systematically one in three. All individuals of age 10–19 years in the selected houses constituted the sample.

In this study, systematic sampling was done at the last stage to select the houses. In most large-scale studies, such multistage sampling is used, and systematic sampling can be at one or more stages.

SIDE NOTE: The authors stated they employed cluster sampling. Actually they did not use the cluster method at any stage of sampling. Such improper use of the statistical terms is quite common in medical literature.

3.2.6 Choice of Method of Random Sampling

A visual comparison of some methods of random sampling is in Figure 3.1. The method of choice is always SRS unless it is difficult to adopt.

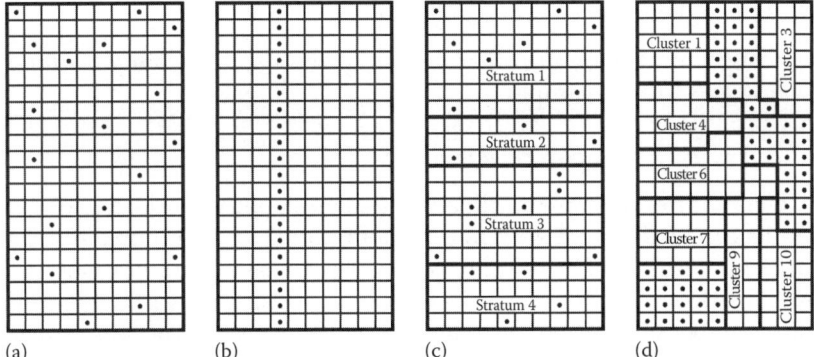

FIGURE 3.1
Visual illustration of some random sampling methods: (a) simple, (b) systematic, (c) stratified, and (d) cluster.

The estimate provided by SRS is generally the most precise. Some problems it can create are the following:

1. Nonavailability of sampling frame
2. Widely dispersed subjects in the sample making the approach difficult
3. Less representation of specific groups that are important and must be adequately represented
4. Obtaining so many distinct random numbers may be difficult

The first two problems can be handled either by cluster or by multistage sampling. The answer to the third is stratified sampling, and the remedy of the fourth is systematic sampling, because it requires only one random number. The systematic method also does not require the full frame. In any case, with wide availability of computers, generating distinct random numbers is not a problem.

Although small-scale studies may use selection by only one of the preceding methods, medium- and large-scale studies would generally be based on a mix of two or more methods.

3.3 Some Other Methods of Sampling

The common methods of random sampling described in Section 3.2 are the basic methods used in most medical setups. Their applicability would be clearer as we go along and describe various designs to control the impact of uncertainties in subsequent chapters. In addition, there are other random and nonrandom methods of sampling that are used in typical situations.

3.3.1 Other Random Methods of Sampling

The most prominent among other random methods of sampling are the PPS sampling and inverse sampling.

3.3.1.1 Probability Proportional to Size Sampling

The remarks after Example 3.2 illustrate that probability of selection of different units can be different even in a random sample. Equal probability helps in achieving self-weighting and consequently easy computations. What should one do if the sampling units themselves greatly vary in their size? If a certain number of counties are to be selected from a state for, say, estimation of prevalence of smoking among adolescents, a self-weighting scheme can be achieved by assigning the probability of selection to each county in proportion to its population. This is called PPS sampling and is primarily used for community-based surveys. The major advantage is a simpler method for calculating mean, variance, etc.

> **Example 3.6: PPS Sampling for Assessing Dental Caries Prevalence**
>
> Siegal et al. [5] report a survey among children of age 3–5 years enrolled in Ohio Head Start programs. They selected head start centers by using PPS sampling, where size was the number of children enrolled.
>
> The survey found a prevalence of 38% of dental caries in 3–5 year old head start children, and 28% had at least one untreated decayed tooth. This high prevalence is consistent with other reports for the state of Ohio.

Naturally, the PPS sampling can be implemented only when the size of the unit is known. If the objective is to estimate prevalence of smoking among adolescents by sampling some counties, the size actually is the population of adolescents in a county. However, the total population can be used as a **surrogate** assuming that the proportion of adolescents is nearly same in each county. The size of a hospital can be measured by patients attended, number of beds, or number of occupied beds. Once the size is determined, the PPS sampling can be performed as illustrated in Example 3.7.

> **Example 3.7: Illustration of Selection of a PPS Sample**
>
> Consider a state with 15 counties from whom 4 are to be selected by PPS sampling. For this, a list is made in any predetermined manner with population size of each county. This size is cumulated as shown in Table 3.2.
>
> The total size of all counties together is 75. From this, four are randomly selected as usual. If these are numbers 11, 15, 58, and 62, the corresponding counties in selection are numbers 3, 6, 10, and 12, which gave rise to the selected numbers in cumulation. Incidentally, only one is a large size county (no. 6) in this sample, but the sampling scheme has provided better chance to the bigger counties.

TABLE 3.2

Illustration of PPS Sampling in Example 3.7

County No.	1	2	3[a]	4	5	6[a]	7	8	9	10[a]	11	12[a]	13	14	15
Size (population in thousands)	3	7	1	1	2	12	6	5	18	4	2	1	3	8	2
Cumulated size	3	10	11	12	14	26	32	37	55	59	61	62	65	73	75
Selected numbers			11			15				58		62			

[a] Selected counties.

PPS sampling is often done with cluster sampling as illustrated in Example 3.4. The size could be the population in thousands, number of hospitals, population of females in reproductive age group in hundreds, or any other size relevant for the problem.

3.3.1.2 Area Sampling

If you are interested in sampling households for a community-based survey when their listing is not available, a way out is to select areas with the help of a map. If a district is to be covered by a survey, it can be divided into, say, 12 areas in a manner considered appropriate, and 3 of them can be selected randomly. In case of multistage sampling, the selected three areas can be subsampled randomly for a predefined number of households for which the household listing would be needed, or else a predetermined number of clusters of households can be selected using the map.

The advantage of area sampling is that the survey work is restricted to the selected areas, and traveling time is saved. The disadvantage is that one area may have a high density and many more households may be present than in another area. Even if geographical areas are equal, the actual size in terms of inquiry units may be widely different, causing unbalanced sample and biased estimates.

Note that the selection of districts or counties is also a form of area sampling but this term is not used for sampling such well-defined units.

3.3.1.3 Inverse Sampling

When the sampling frame of relevant subjects is not available, one can continue sampling until a predetermined number of subjects meeting the criteria are included. This is called inverse sampling. This method is used when available units are in random order. In an outpatient clinic, all kinds of patients come ostensibly in random order. If the objective is to select 35 cases of patients suffering from kidney diseases, all those coming in are filtered one by one and the first 35 who meet the criteria are selected.

3.3.1.4 Consecutive Subjects Attending a Clinic

A medical study is easy to carry out in consecutive eligible cases attending a particular clinic. So long as the date of start is predetermined, which has no link with the type of cases that might come, this kind of sample is as good as SRS, provided each eligible case is faithfully included with no bias. This is so because the sequence of arrival of cases can be genuinely considered random and can be expected to represent a fair cross section of the target population. This argument seems valid for cases coming to a clinic, but consecutive cases residing in contiguous houses in an area can have similar socioeconomic and health status. This clustering effect can render sampling of contiguous houses different from SRS.

Opposed to this, if the cases arriving on, say, Mondays, Wednesdays, and Fridays are selected, they can have an inbuilt bias. The day a patient reports in a hospital may be determined by the availability of a particular physician who might have special skills, if not for treating, but for making friends with patients and building up a rapport for extracting intricate information.

Remember that random selection is not an end but a means to obtain a representative sample. Adequate representation is the objective and randomization is a method. If any other method provides an equally unbiased sample, it is as good. However, statistical theory is built around random samples.

3.3.1.5 Sequential Sampling

In sequential sampling, the eligible subjects from the target population are selected one by one in a random manner and assessed. Further sampling is stopped as soon as a reliable result one way or the other is available. This method of sampling is not so popular in medicine.

3.3.2 Nonrandom Methods of Sampling

In medicine, a subject is sometimes precisely defined such as a menopausal woman of age less than 60 years suffering from ovarian cancer, still smoking, and a resident of California. Not many may exist that meet such specific criteria, and they would be difficult to locate from hospital records or otherwise. The question of random sampling hardly applies to such a situation. In many other situations, a nonrandom sample is taken just for convenience. This is also called a **purposive sample**. It can be used for exploratory studies. Various purposive samples can be described as follows.

3.3.2.1 Convenience Samples

Subjects who are easily available or who can easily submit to the study form a convenience sample. Sample of **volunteers** is the most prominent example. Volunteers have a definite place in medical studies, particularly in phase I of a clinical trial. Results of a phase I trial may not be amenable to generalization, but such studies do tend to indicate the tolerated dose and major side effects. Volunteer studies do provide an indication whether or not the drug or the intervention is worth pursuing further.

Another convenience sample is of a **captive population** such as medical students who can be easily persuaded by the faculty to be subjects of the study. For example, they may be persuaded to undergo lung function testing before and after an exercise or can be asked to fill in an anonymous questionnaire on sexual behavior. Medical students may be physically and mentally healthier than other youths. Thus, the results cannot be generalized to the population. Students in one particular school may not represent the students in other schools because of typicalities associated with each school.

A third type of convenience sample in medicine is **referred cases**. They are easily available and may also easily agree to participate. But referred cases almost invariably are complicated cases and do not represent the general class of patients. They may not even represent complicated cases because sometimes referral also is done of noncomplicated cases for a particular investigation or examination that another center specializes in.

Telephone sampling is an effective instrument in the developed countries as almost every household has a phone, and the listing is available separately for homes and commercial establishments (white and yellow pages). Any of the methods of probability or nonprobability sampling can be used to select telephone numbers. But keep in mind four rather severe limitations: (1) Many households have two or more phone numbers. (2) Some households without telephones can never appear in the telephone sampling. This is a serious limitation for areas where many households are without phones. (3) Some households suppress their telephone numbers and their numbers are not listed. (4) Attrition (nonresponse) rate is generally higher in telephone surveys than in personal interviews. This may increase the bias. However, in telephone sampling, a replacement is easy to make.

3.3.2.2 Other Types of Purposive Samples

Stigmatized subjects such as injecting drug users and people visiting commercial sex workers are difficult to locate. However, they generally operate in a network. If one such subject is identified, he can identify three or four others known to him. These, in turn, can help locate another three or four each, and so on. This type of sampling is called **snowball sampling**.

The **Delphi method** is a procedure by which experts are brought to a consensus in stages by gradually eliminating the isolated differential opinion. Such a gathering of experts can be termed as a **Delphi sample**. The consensus arrived at may or may not be shared by the experts who did not participate in the exercise or whose opinion was eliminated after being found not in line with those of the majority.

The term **quota sampling** is used for selecting a predecided number of subjects without recourse to random selection. Quota sampling is generally used after dividing the subjects into relevant groups such as selecting 50 obese postmenopausal and 30 nonobese women for their enzyme levels.

Whosoever is easily available is included. If one group is complete, recruitment for the other groups continues.

Another purposive method for which no specific name is given is selecting one or more schools considered to be representative of, say, undernourished children and another group of schools representative of well-nourished children. The selection is not random but purposive that meets the requirement. Within schools, the selection of children may be random or nonrandom. This kind of sampling is quite popular.

Sometimes a combination of two or more methods of purposive sampling is used. This is called **haphazard sampling**.

By describing these nonrandom methods, I am not advocating or justifying these methods. These definitely have extremely limited utility. Besides difficulty in identifying the target population these samples represent, nonrandom samples cannot be used for statistical inference such as tests of significance and confidence intervals. At the same time, they can sometimes provide useful leads for further studies that can be based on random samples. Nevertheless, large-sized samples, even when nonrandom, may still represent the cross section of the corresponding population and can be adequate for statistically valid inferences.

References

1. Thompson SK. *Sampling*, 2nd edn. New York: Jossey Bass, 2002.
2. Milias GA, Panagiotakos DB, Pitsavas C, Xenaki D, Panagopoulos G, Stefandis C. Prevalence of self-reported hypercholesterolemia and its relation to dietary habits in Greek adults: A national nutrition and health survey. *Lipids Health Dis* 2006; 5:5.
3. Bennett S, Woods T, Liyanage WM, Smith DL. A simplified general method for cluster-sample surveys of health in developing countries. *World Health Stat Q* 1991; 44:98–106.
4. Vitolo MR, Canal Q, Campagnolo PD, Gama CM. Factors associated with risk of low folate intake among adolescents. *J Pediatr (Rio J)* 2006; 82:121–126.
5. Siegal MD, Yeager MS, Davis AM. Oral health status and access to dental care for Ohio head start children. *Pediatr Dent* 2004; 26:519–525.

4

Designs for Observational Studies

Nature is a great experimenter. Many changes occur naturally, requiring no human intervention. Some people are exposed to iodine-deficient water because of environmental conditions, and some women have low hemoglobin (Hb) levels because of their nutritional status. The study of such naturally occurring events can be invaluable in studying cause–effect relationship with conviction. Such a study can be based on records, on actual observations, or a combination of both. This type of study is generally categorized as an **observational study**, also sometimes referred to as an epidemiological study. Since there is no deliberate human intervention (such as a drug) in this setup, such studies carry little risk of harm to the subjects or the society, although they are invasive with regard to time and privacy of the respondents. An observational study is generally conducted for specific groups such as those with high disease prevalence; thus, extrapolation to the general population is not immediate. Many observational studies are done in a hospital setup rather than in communities.

This chapter: Observational studies might look simple to carry out but they provide valid results only when carried out with full precaution. This chapter describes various formats in which an observational study can be carried out, the merits and demerits of these formats, and the precautions needed for the results to be valid. Some basic concepts of observational studies are explained in Section 4.1. Section 4.2 describes prospective studies that include cohort and longitudinal formats, Section 4.3 describes retrospective studies that include case–control and nested case–control formats, Section 4.4 describes cross-sectional studies that provide a snapshot of the situation, and Section 4.5 describes comparative features of these designs.

4.1 Some Basic Concepts

Following basic concepts about observational studies will help in better appreciation of this format of studies.

4.1.1 Antecedent and Outcome

The aim of an observational study is to extract information from events that have already occurred or are occurring. The investigator, however, has the liberty to decide the kind of subjects to include in the study and the kind of subjects to exclude. The more judicial the choice, the better is the study. An essential feature of all analytical studies is that there must be at least one antecedent characteristic under investigation and at least one outcome on which the subjects are monitored. Antecedents and outcomes are also chosen by the investigator, although the choice is guided by the objectives of the study.

An observational study of the effect of preexisting maternal anemia on gestation term (preterm or full term) can be done in a variety of ways. To understand these, note first that maternal anemia is an antecedent factor in this example and gestation is the outcome. An antecedent is a precursor such as an exposure (e.g., sex with a subject suffering from a sexually transmitted disease [STD]) or a risk factor (high cholesterol) suspected to affect the outcome. The other names for antecedent are cause, predisposing factor, and determinant.

An outcome could be a health state, recovery, side effect, or death. It can also be pregnancy, prognostic complication, disappearance of a symptom, or any other event of interest. This is the consequence and must necessarily occur after the occurrence of the antecedent. Other terms for this are effect and result. An outcome could be positive such as recovery to full health or could be negative such as death.

Three different ways that an observational study can be carried out are prospective, retrospective, and cross-sectional, depending upon the starting point being an antecedent, outcome, or none at all, respectively. These three methods are depicted in Figure 4.1. Instead of looking at the direction of antecedent–outcome relationship, sometimes a study is labeled prospective or retrospective depending on the time frame. This can create confusion and should be guarded against. Later in this chapter is an example of a study on past events that is prospective.

There could be setups where the antecedent and outcome are not easily identified. Gender and blood group are genetically determined simultaneously and none is antecedent to the other. Anxiety can cause disease and vice versa. Thus, the direction of relationship may not be clear in some cases. Further details are given later in this chapter.

You may like to make an interesting statistical distinction between a factor and a response. A factor is a characteristic that is prefixed and used for selection of subjects. Response is what you elicit from the subjects. The information on response is available after you have collected the data. This kind of distinction helps in explaining prospective, retrospective, and cross-sectional studies, but the term factor is used in several other contexts also.

4.1.2 Confounders

A confounder or confounding factor is an antecedent characteristic that can be described as a possible explanation of the outcome in addition to

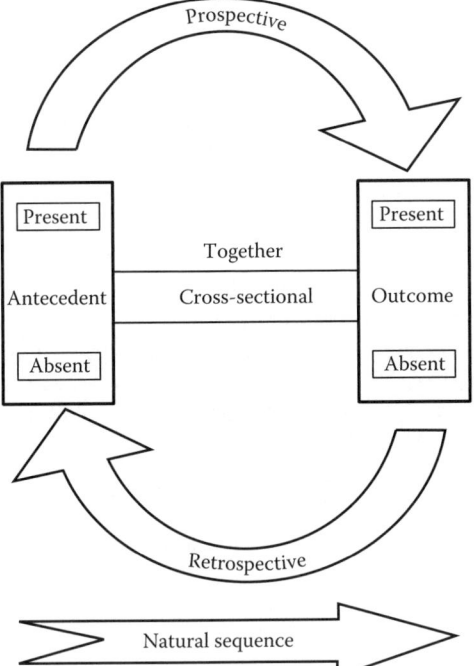

FIGURE 4.1
Schematic depiction of three methods of observational studies.

the one under investigation. Thus, it is an extraneous factor that plays spoil sport. It can also be understood as the one that is related to the antecedent as well as to the outcome but is not in the causal chain under consideration. It mixes the association of disinterest with the ones of interest and blurs the causal pathway.

In a study on smoking and hypertension, one confounder is obesity (see Figure 4.2). Smokers tend to be obese and hypertension is also related to obesity. In other words, obesity can also be at least a partial explanation for hypertension in addition to smoking and it is not in the causal chain under study. The effect of smoking and obesity on hypertension cannot be disaggregated unless obese and nonobese subjects are separately studied. The second confounding factor in this example is age. As age increases, the lifelong burden of smoking increases for smokers, and the chance of developing hypertension also increases because of age-related arterial

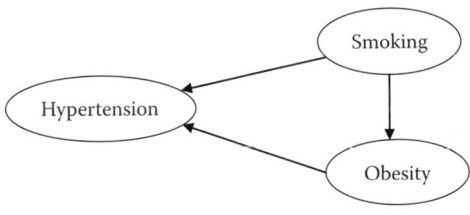

FIGURE 4.2
Obesity as a confounding factor for smoking and hypertension.

changes. Thus, again, age as a possible explanation should be ruled out. This can be done either by developing a proper design or by performing a suitable statistical analysis. This book contains plenty of information on these two methods.

Other names for confounding factors are **prognostic factors** and **concomitant variables**. The concept as encountered in different setups is explained further later in this book. Obviously, confounders should be identified before data are collected. Previous research, clinical insight, and clarity about the disease process can help in this identification. One easy method of identifying confounders is to draw a list of all possible factors that might influence the outcome of interest. The list may be based on your own knowledge, wisdom of seniors, or review of literature. Out of this list, choose the ones that would be studied as risk factors or antecedents for their role in the outcome. Whatever remains in the list are confounders for that outcome.

Note, however, that the confounders so identified would be restricted to those that are known. The knowledge could be incomplete and the list may not be comprehensive. If so, epistemic uncertainties would remain in the results. Nothing can be done to remove this lacuna except to expand the horizon and look for factors that are not in the conventional domain.

4.1.3 Effect Size

You may be interested in count or percentage such as how many cured or improved or in quantities such as average improvement in pain score or in glycated Hb after a therapy. The parameter of interest can also be degree of relationship between two or more factors, risk or relative risk (RR) of disease, etc. The generic term for such outcome-based parameters is **effect size**. This answers "how much" for the outcome of interest. This could also be probability, mean, odds ratio (OR), strength of association, effect of one factor on another, or any other parameter of interest. Sometimes effect size is measured in terms of standard deviation (SD) such as 0.7SD.

4.2 Prospective Studies

The most desirable format for studying the relationship between maternal anemia and gestation is that antenatal women with different levels of Hb are selected and followed up for timing the gestation period. Antecedents are assessed before the outcome in this format in the sequence they naturally occur. The outcome is assessed afterward. The antecedent is a factor and the outcome is a response in this format. A study following such a format is called a prospective study. Subjects are generally identified as exposed or unexposed to a risk factor, or to a health protector, or a health promoter. Since any outcome occurs after the antecedent, some follow-up is inbuilt in this format. Thus, this is also called a **follow-up study**.

Exceptional prospective studies can be cited that do not require any follow-up in actual time frame. In a study on the effect of profession on smoking habits, a cohort of people joining different professions in one particular year can be followed up for a 10-year period. This would be a standard prospective study. But the effect can also be studied by selecting people who have been in different professions for nearly 10 years and noting their present smoking habit at the end of the 10-year period in the profession. This also is a prospective study since the direction of the study is from antecedent (in this case, profession) to the outcome (in this case, smoking), but there is no follow-up of the subjects in a conventional sense.

Analysis of data from prospective studies is mostly done in terms of RR and attributable risk (AR). These are discussed in Chapter 14.

4.2.1 Variations of Prospective Studies

Prospective is an umbrella term that includes various types of studies based on follow-up of the subjects. Prospective study can be a cohort study with concurrent or historical cohort and can also be a longitudinal study.

4.2.1.1 Cohort Study

A cohort is a group of subjects that share a common base and observed forward in time. In the case of a usual prospective study, the subjects can be enrolled continuously and leave the study abruptly, whereas, in the case of cohort, enrollment, joining, or leaving in the middle is rarely admissible. In a study on use-effectiveness of oral contraceptive pills [1], the users joined the group when they started using the pills and left when they stopped using them. They could not join 2 or 3 months after starting the pill. All subjects of a cohort do not have to start from the same calendar time, but they all start from the time of occurrence of the same event.

The term cohort generally connotes substantial time gap between exposure and outcome, and the observation spans all or most of this period. There could be a cohort of children born in a particular year followed up for growth pattern or a cohort of adults residing in an area at a time followed up for diet–exercise and occurrence of coronary events. There could be a cohort of smokers and a matched cohort of nonsmokers residing in an area at a particular time followed up for 20 years for the development of chronic obstructive pulmonary disease (COPD). Many other examples can be cited. Such cohorts are called **concurrent cohorts** when exposed and unexposed groups are followed up in future.

If the group identified for the study is the one that has been recently exposed to a risk factor, it is called an **inception cohort**. For example, an inception cohort of early rheumatoid arthritis can be studied for assessing predictive factors of orthopedic surgery.

It is not necessary that the follow-up chronology is in the future; it could be in the past also. Sokal et al. [2] carried out a cancer risk study in the year 1992 on the basis of the records of women sterilized with transcervical quinacrine hydrochloride pellets in Chile between 1977 and 1991. Traceable women were also interviewed. Despite being based on past records, it is not a retrospective study since the direction of investigation is from antecedent to outcome. A **historical cohort** or **retrospective cohort** is a group of subjects with exposure in the past, investigated for development of an outcome. Terms such as **retrospective follow-up** and **historical prospective** are also used for this kind of methods. This requires that past records are fully available.

Example 4.1: A Cohort Study on Relation of Level of Cognition with Birth-Weight Parameters

Richards et al. [3] report findings of a 53-year follow-up of 1946 birth cohort, initially consisting of 5362 children of nonmanual and agricultural workers, and a random sample of one-in-four of manual workers selected from all single and legitimate births that occurred in England, Scotland, and Wales during 1 week in March 1946. (Note the rigorousness with which the specifications are stated.) The cohort was studied on 21 occasions between birth and age 53 years, when information about sociodemographic factors and medical, cognitive, and psychological function was obtained by interview and examination at each point of contact. They concluded that birth weight and postnatal growth are independently associated with level of cognition at different ages. In this case, the main outcome of interest was level of cognition and the antecedents were birth weight and postnatal growth. The outcome was repeatedly measured over the period so that the cognition achieved at different ages could be studied.

SIDE NOTE: Postnatal growth may be a function of birth weight, but, as far as cognition was concerned, this study found that the two act independently.

Example 4.2: A Retrospective Cohort Study for Identifying High-Risk Patients Undergoing Emergency Surgery

Neary et al. [4] carried out a retrospective cohort study of all nonelective general and orthopedic surgical procedures performed on a total of 1869 patients in a hospital in the United Kingdom during the calendar year 2000. Outcomes were identified from various related hospital databases, and case notes of those who died were reviewed. Note that the study was from antecedent to outcome, but the subjects belonged to a past period.

The study found that increasing age, size of operation, and American Society of Anesthesiologists (ASA) grade were significantly associated with higher risk of death within 1 year. The authors concluded that a simple scoring system could be used to identify high-risk patients among those who required nonelective surgery. Such patients could be targeted for interventions for reducing the risk of death.

SIDE NOTE: The conclusion reached is the same as anticipated by common sense. Yet the study has value, first for documenting the evidence and second by linking it to the scoring system. The study would hold greater value had the RR quantified and if its confidence interval (CI) provided.

4.2.1.2 Longitudinal Study

Another version of the prospective study is a longitudinal study in which repeated observations or measurements are serially made at several points of time on the same group of subjects, particularly over a long period. When a cohort of low-birth-weight children born in the year 2004 is followed up every year for growth, development, anthropometry, biochemical profile, pathological conditions, etc., then it is a longitudinal study. The Framingham Heart Study is among the most popular longitudinal studies, still ongoing for more than 50 years, and has spilled over to the offsprings of the subjects of the study (see, e.g., [5]). The study on cognitive level in children mentioned in Example 4.1 is also a longitudinal study.

The primary objectives of a longitudinal study are twofold: (1) to study the trend of values over a period and (2) to study the changes over the baseline or over the previously observed value. The time points in either case may be minutes such as after anesthesia in surgical patients or may be in years such as for effect of birth weight on lipid profile in adulthood. The time intervals may be regular or irregular; this does not change the basic characteristic of the study.

4.2.1.3 Repeated Measures Study

Sometimes a fine distinction is made between longitudinal study and repeated measures study. Both are prospective. In the case of repeated measures, all subjects are measured at prefixed points of time such as 1, 3, 6, and 12 months after treatment. In the case of longitudinal study, the time points for different subjects may differ such as time 1, 3, 6, and 12 months for one subject and 2, 4, 6, and 8 months for another subject depending on availability and convenience of the subject or the assessor.

Prospective, cohort, and longitudinal are not mutually exclusive terms. A cohort study can be longitudinal comprising observations at several points of time or can have only two assessments: one at the beginning and the other at the end. Some use the prospective study as a synonym to the cohort study [6], but this is not entirely correct. A study of antenatal women for birth outcome, coming to a clinic staggered over a period of time at different gestations, is a prospective study but would not be called a cohort study.

4.2.2 Selection of Subjects for a Prospective Study

An important consideration in the selection of subjects for any study is the feasibility of obtaining data on them. This means that the subjects must be approachable and cooperative. For a prospective study, especially, accurate and complete information must be available on them so that they can be correctly classified into exposed and nonexposed groups, and the effect of other characteristics on the outcome can be properly assessed. It is natural to expect that the subjects included in the study truly represent the target population. Thus, the target population must be clearly defined. It could be, for example, patients attending a particular group of diabetes clinics who are observed for development of retinopathy or those exposed to a carcinogen in a particular district. Note the geographic limitation associated with the definition of a population.

In a prospective study, one risk factor will be of primary interest but other risk factors to be concurrently studied also need to be properly identified. For example, in a study of maternal complications, these could be parity, nutrition status, Hb level, and the nature of natal care. Decide in this case whether the study is to be restricted to women who are currently pregnant or will include past pregnancies also.

A basic feature of a prospective study is that the incidence of outcome such as disease or cure is evaluated in those subjects who are exposed. It is often helpful to study a parallel group, also called a control, which is not exposed, so that a proper comparison can be made.

4.2.2.1 Comparison Group in a Prospective Study

Proper selection of the comparison group enhances the validity of conclusions from a prospective study. Quite often the control group comes from within the cohort, in whom some subjects are naturally exposed and some are not. For valid comparison, the exposed and unexposed groups must be similar at baseline, particularly with regard to the factors that can influence the outcome. If the objective is to study the effect of recently acquired central obesity on the electrocardiogram (ECG) changes over time, factors such as age, gender, personality traits, stress conditions, and smoking need to be matched between the study group (with central obesity) and the control group (without central obesity). If complete matching is not possible, as would generally happen in practice, statistical methods are used to do the required adjustment at the time of analysis. Some of these methods are presented in later chapters. Such an adjustment can become incomprehensible if done for a large number of factors.

An external group can be used for comparison in some situations. Adequate number of nondiabetics may not be available in a diabetes clinic for assessing the development of coronary incidents. External controls can be included in such a situation; however, they should come from the same milieu and should preferably be matched for all the factors except the exposure. In a rare situation when an appropriate external group is also not available, comparison can be done with the outcome rates in the general population. For example, incidence of birth defects in babies born to women of age 45 years or more can be compared with that in the births to women of child-bearing age in the general population. The actual control group in this setup should be births to women of age less than 45 years, but a separate incidence of birth defects in them may not be easily available. The incidence in births to women of age less than 45 years may not be much different from that in all women of child-bearing age since births after that age are rare. However, in many situations, the rate in the general population is not comparable with the rate in the unexposed group and a great degree of precaution is required in using such a general group as control.

In some prospective studies, it is useful to have multiple groups for comparison. For example, the effect of profession on smoking habits can be investigated by including several categories of profession in the same study. In this study, none would be the control group in the conventional sense. In another setup, subjects with different exposure levels can also be chosen for follow-up that would also provide multiple groups for comparison of the outcome.

It is sometimes impossible to find a group that is completely nonexposed. An example is exposure to dichloro-diphenyl-trichloroethane (DDT). Even people in remote locations, such as Canada's Baffin Island, harbor traces of DDT. In such cases, the comparison effectively would be between the less exposed and the more exposed.

4.2.3 Potential Biases in Prospective Studies

A large number of biases are listed in Chapter 2. Biases typically occurring in a prospective study setup are the following.

4.2.3.1 Selection Bias

A prospective study group is generally a sample from a target population of subjects. Selection bias is said to have occurred when the study group has a different composition with regard to etiologic factors such as heredity, age, gender, nutrition status, and addictions compared with the composition in the population. Studies on volunteers or on clinic subjects almost invariably suffer from such a bias. A method of selection that has a random component is considered insulation against such bias. But such selection fails to take cognizance of special bias that can result from sources such as improper definition of the population. In a study on causes of psychiatric illness in old age, if the subjects are those who are single and of age 70 years or above at the time of enrollment, the bias occurs because some with severe illness may have already expired before attaining the age of 70 years.

4.2.3.2 Bias due to Loss in Follow-Up

A major task in prospective studies is to accomplish successful follow-up of all the subjects. Loss occurs due to change in residence to an unknown or remote address, unrelated death, severe illness other than the one under study so that the required investigations cannot be done, loss of motivation of the patient to cooperate, fault developing in machines such as treadmill whose rectification takes time, absence of a trained technician, etc. In a clinic-based follow-up, when the patients are advised to report at periodic intervals, some may not come on the required day, and one or two follow-ups may be missed. Such loss constitutes a threat, first because the size of the group shrinks and second because those lost are seldom a random subgroup. They are generally typical, for example, subjects who are seriously ill and are not hopeful of living long, or those who are mildly affected and who consider continuation in the study not worth taking a risk. If the rate of disease and the rate of severity are different in the subjects who have discontinued, it would affect the validity of the results. Where possible, a random subsample of the discontinued subjects should be investigated intensively to evaluate their characteristics versus the characteristics of those who have not discontinued. If they are really different either with respect to outcome or even with respect to the baseline information collected at the time of first contact, then an adjustment may be required at the time of analysis to remove the effect of such bias. When nonrespondents could not be contacted despite best efforts, the baseline information can still be used for adjustment. Some methods of such adjustment are discussed in a later chapter.

4.2.3.3 Assessment Bias and Errors

Sometimes the exposure or the accompanying baseline information and sometimes the outcome are not correctly recorded. This can occur due to bias of the observer or of the recording clerk who may unwittingly classify a subject into a particular category of interest supporting one's individual hypothesis. Observer bias can occur if the subjects with disease are evaluated more intensively and more carefully than those without disease. Blinding is presented in a later chapter as a means of safeguarding against this bias but such blinding is rarely feasible in studies requiring long follow-up.

Incorrect information can also arise for the following reasons. These can occur in any setup but are typically more common in a prospective study setup.

1. Human error in correctly assessing the condition of the patient. This can be due to either carelessness or lack of expertise of the observer. The physician may lack competence, and the recording clerk may lack training or motivation. Assessment during the later part of a longitudinal study may be less accurate as fatigue sets in or more accurate due to learning effect.

2. Inaccurate reports from the laboratory arising from use of nonstandardized techniques or chemicals or from use of faulty instruments.
3. Doubtful validity of the diagnostic or screening test.

All such biases and errors in assessment can be reduced simply by being more careful and by using precise instruments, measurements, and classification criteria that have been pretested for their validity.

4.2.3.4 Bias due to Change in the Status

In a prospective study on central obesity and coronary artery diseases (CADs), it is possible that some subjects of the nonobese group become obese while the follow-up is still in progress. In a study on effect of smoking, a nonsmoker at the initial stage may start smoking in the middle of the study, or a smoker may quit smoking. Exclusion of such cases is one option but is feasible only when their number is small. A long-term cohort is also affected by the environmental changes such as introduction of a new drug in the market that can influence the conclusion regarding incidence of the disease under study.

4.2.3.5 Confounding Bias

As already explained, confounding is said to exist when two or more factors move together. They compete with the hypothesized risk factor as an explanation for the observed response. Egg eaters generally have raised lipid levels. If these subjects are found to be at higher risk of CAD, it is difficult to say that this is due to eating eggs per se or is contributed, at least partially, by the raised lipid level that can occur for other reasons as well. To evaluate their separate effects, special care should be taken to include egg eaters with normal lipid profile and noneaters with raised lipid levels. If unforeseen confounding occurs, a **poststratification** can be tried after the data are collected. This would indicate if a sufficient number of cases are available in each group after stratification. If there were very few cases in one or more groups, the validity of the results would suffer.

4.2.3.6 Post Hoc Bias

A study can incidentally indicate an interesting relationship that was not anticipated earlier and was not a part of the protocol. Such findings should be regarded only as hypotheses for future studies and no conclusion should be drawn on the basis of the current study.

4.2.3.7 Validity Bias

Some studies try to develop criteria to distinguish subjects with greater risk of disease from those with less risk. If sufficient distinguishing features are detected, then these criteria could indeed be developed. Such criteria may work wonderfully well on the group from which they have been derived but they need to be externally validated on another group of similar subjects. The details are provided in Chapter 20. Although statistical principles say that criteria based on representative sample should work nearly equally well on another sample from the same population, evidence of external validation is considered essential before such criteria are accepted. This validation could be done on another sample from the same population or even on a sample from a different population. The latter, of course, provides evidence that the results could be valid for the other populations too.

The identification and resolution of bias is primarily a matter of epidemiological judgment. The success of a prospective study often depends on the care taken by the investigator in recognizing and correcting these biases. Although some bias can be handled at the time of data analysis by using appropriate statistical techniques, the success of these techniques depends on validity considerations, particularly on the adequate number of cases in groups and subgroups. This cannot be ensured beforehand in this situation. Precautionary steps are preferable wherever feasible.

4.2.4 Merits and Demerits of Prospective Studies

The main merits and demerits of prospective studies are as follows.

4.2.4.1 Merits of Prospective Studies

The advantages of prospective studies will be clearer after the details of other designs—retrospective and cross-sectional—of observational studies are outlined. Nevertheless, it is evident that the temporal sequence between exposure and disease can be more easily established by prospective studies. Risk of outcome such as of disease or the chance of getting cured can be directly measured. The incidence of an outcome cannot be assessed by any other method. Prospective studies are particularly well suited for assessing the effect of rare exposure such as of a specific chemical or of a new therapeutic modality because the cohort is expected to start with an adequate number of subjects. These studies allow for examination of multiple effects of a single exposure. If the outcome is death and the records are not adequate, a prospective design is the only choice.

4.2.4.2 Demerits of Prospective Studies

There are several demerits too. Sometimes an outcome, such as carcinoma, may take years to appear after the exposure. It may occur only in a small percentage of subjects, which could mean a very large cohort to obtain an adequate number of subjects having disease. Thus, prospective studies tend to be heavy on time and resources. In some situations, the natural course of the disease or the characteristics of the subjects may change during the follow-up period. As already stated, obesity, dietary pattern, and smoking can all change in a long-term follow-up. In a prospective study, subjects know that they are being observed and this awareness may change their behavior and outcome. This is called the **Hawthorne effect**, as listed in Section 2.4.2.1. If the study spans several years, it is difficult to maintain motivation and retain the trained staff. Supervision may also lose sharpness. Notwithstanding these difficulties, prospective studies are technically the most correct designs because they move in the natural direction from exposure to outcome.

4.3 Retrospective Studies

The second format for examining gestational age in relation to maternal anemia is that births with different gestation periods are chosen and anemia status of the mothers during the antenatal period is retrieved from records. The first assessment in this format is outcome, and antecedents are subsequently assessed for each type of known outcome. This is called a **retrospective study**. Note the temporal difference between a retrospective study and a prospective study.

Investigation in the reverse direction—from outcome to the antecedents—may seem unnatural but is generally considered more efficient. Note the quickness with which a study in this format can be carried out. There is no need to wait for the outcome to develop or not. The outcome is already known and the antecedent is obtained either from records or inquiry. The cases, and in most studies, controls, are assembled, and information regarding their past exposure or risk factors is collected. The cases can arise or can be recruited in future such as of breast cancer coming to a clinic now on, yet the study is technically retrospective so long as it investigates antecedents for a known outcome.

Only subjects for whom the outcome is already known can be studied with a retrospective design. The outcome is the factor in this format and the antecedent, which is elicited in retrospective studies, is the response.

Many studies do not proceed from the outcome to the antecedent, yet are termed as retrospective in the medical literature. This usage indicates the time frame and not the etiologic sequence. For example, Hilska et al. [7] report an analysis of 150 patients with primary proximal colon cancer in Finland who were operated upon during 1981–1990. But the outcome measure was a 5-year survival rate. The study still is from the antecedent (colon cancer) to the outcome (survival). It is a prospective study going by the terminology used in this text and could be called a retrospective cohort as already explained, but not a retrospective study.

A retrospective study can also be conducted in the current time frame in special situations. For example, women with preeclampsia can be assessed for their present nutrition level and parity as risk factors in the hope that the level of such risk factors was the same before the occurrence of preeclampsia and contributed to the condition. The direction of investigation in this study is from the outcome to the antecedent.

The first step in a retrospective study is to identify and define the outcome of interest. This could be negative, such as disease or death, or positive, such as relief or a specified minimum reduction in cholesterol level. The criteria for diagnosis of a disease and its severity must be fully specified. For example, in cancer cases, the stage of disease must be specified and, of course, the affected site. Depending on the amount of information available, it is sometimes useful to keep a separate track of the definite from suspected cases.

4.3.1 Case–Control Design

The dominant format of a retrospective study is case–control in which subjects with and without disease are investigated for past exposure. Those suffering from the disease or have the health condition of interest are called **cases**, and those without that particular health condition are called **controls**. All case–control studies are retrospective, but all retrospective studies are not case–control. Investigating past history of cases of myocardial infarction (MI) is a retrospective study, but there may not be any controls. Then, it is not a case–control study. The control group provides a legitimate basis for attributing differences in the two groups to the antecedents. Thus, a case–control setup is considered a natural format for retrospective studies.

Sometimes an adequate number of cases are not available at the time of initiation of the study and the cases accruing in next few years are included. If the rate of cases with hypothyroidism in a hospital is 40 per year and a minimum of 100 are proposed to be included, then the recruitment will naturally extend to 2½ years. Such recruitment over a protracted period cannot be regarded as a prospective design though the study may sound like it has been going on for a long time.

In a prospective study, a separate comparison group is not necessary and it might automatically arise from within. However, in a case–control study, as the name suggests, a control group for comparison is essential. The term control is used generically for any reference group against which the case group is compared. In comparing patients of MI with those of stroke for risk factors, the group with primary interest is the case group and the other is the control group, although this is also a group with disease. In a study on efficacy of diagnostic tools, if the interest is in comparing ultrasound images with tomography images, the former could be the control group and the latter the case group. Note that cases are their own controls in this situation. Since the control group is not necessarily "without disease," it is sometimes prudent to call this as a **case–referent study**.

The term associated with outcome is risk and the corresponding term for antecedents is odd. In place of RR, OR is computed in case–control studies. Details of RR and OR are given in Chapter 14. The estimate of OR from case–control studies in most practical situations is a good approximation to the RR obtained from prospective studies. However, case–control studies cannot provide an estimate of the risk or incidence that prospective studies can.

Example 4.3: A Case–Control Study on Fruit and Vegetable Consumption and Risk of Some Colorectal Cancers

Protective effect of consumption of fruits and vegetables (F/V) on some cancers is pretty much known, but this protection could be differential for some specific sites of colorectal cancers (CRCs) opposed to the others. A case–control study on 834 cases of CRC and 939 controls conducted between 2005 and 2007 in Western Australia [8] found that risk of proximal colon cancer and rectal cancer was not associated with intake of total F/V. However, a significant negative association of distal colon cancer was seen with total F/V and total vegetable intake. An increased risk for CRC was associated with intake of fruit juice.

SIDE NOTE: (1) The results suggest that different F/V confer differential risks for different types of CRC. (2) The study used food frequency questionnaire as a tool to collect data. This tool is discussed in Chapter 11.

A case–control study can be nested, prevalent, sequential, or two phase (for details, see Armitage and Colton [9]). Of these, the nested design is quite common. Following are some details of a nested design.

4.3.1.1 Nested Case–Control Design

A design could combine features of cohort and case–control studies. Consider a cohort of persons of age 40–44 years who are followed up for 15 years for cataract development. Now the persons who develop cataract become cases for investigation of those risk factors that could not be studied in the cohort setup. This would require matched controls. They can come either from the same cohort among those who did not develop cataract or may come from outside. Note that cohort studies start with one or two specific antecedents, but a case–control format allows investigation of a large number of antecedents. Thus, new hypotheses can be examined. The baseline data are already available from the cohort study, and these data may be free of recall bias. This type of design is called a nested case–control design since the case–control setup is nested within a cohort. Example 4.4 describes a situation where a nested case–control design can be useful. The advantages of a case–control design were derived from within the cohort that was being followed already.

Example 4.4: A Nested Case–Control Study on Inverse Association between Body Mass Index and Ovarian Cancer

A prospective study was conducted in France to investigate any association between body mass index (BMI) and ovarian cancer [10]. Information on anthropometry, demographic characteristics, medical history, and lifestyles was obtained at the time of recruitment of subjects. (Note the advantage of availability of a lot of information in this setup.) Women diagnosed with primary, invasive epithelial ovarian cancer ($n = 122$) diagnosed 12 months or later after recruitment served as cases. (n was still large.) Two controls for each case, matched for menopausal status, age, and date of recruitment, were randomly chosen from the same cohort.

SIDE NOTE: Appropriate logistic regression showed an inverse association between BMI and ovarian cancer risk, that is, for increasing quartiles of BMI, the OR exhibited a decreasing trend. Such a dose–response type of relationship is one of the many indications that the relationship could be causal.

4.3.2 Selection of Cases and Controls

Appropriate selection of cases and controls is important for any analytical study but has special significance for case–control studies. The observations made in Section 4.2 regarding the selection of study groups for cohort studies generally apply to case–control studies as well. Other considerations are as follows.

4.3.2.1 Selection of Cases

The source of cases with the disease or any other outcome of interest can be hospital inpatients, patients seen as outpatients, cases identified in a survey, cases listed in records of a health facility, etc. Hospital-based studies may be biased for, say, subjects from upper socioeconomic class, whose nutritional status may be different, or for any other factor influencing hospital intake. Population-based studies can become very expensive. Each, however, has relevance in particular situations.

A case could be either newly diagnosed (incident) or an already existing (prevalent) subject with the disease. Inclusion of prevalent cases, particularly for chronic disease, can easily increase the sample size, but care is required at the time of interpretation of results because the factors determining the duration of disease can be important. Mild cases or those physically strong may survive longer. In addition, cases surviving for long are more likely to have recall lapse. Newly diagnosed cases do not have any such problem and are thus preferable.

4.3.2.2 Selection of Controls

Selection of an appropriate control group is critical to the success of the case–control study. Control subjects, for example, could be patients from the same hospital but suffering from another disease, or peer workers, neighbors, relatives, and friends if that does not mask the relationship. But they should come from the same setting. Sometimes multiple controls are studied when they are available in abundance and are inexpensive and quickly elicited. This helps to increase the reliability without corresponding increase in cost. In any case, the choice should be such that all confounders are ruled out as much as possible. A comprehensive discussion on selections of controls in case–control studies is available in a series of articles by Wacholder et al. [11–13].

4.3.2.3 Sampling Methods in Retrospective Studies

First, the sample size must be adequate that can represent the entire spectrum of subjects and can provide reliable results. Only then the results are generalizable.

Cases included should ideally be representative of all persons with the specified outcome, but many case–control studies are carried out on a nonrandom sample. Random selection is especially important for descriptive studies but probably not so important for analytical studies. Experience suggests that the relationship between antecedent and outcome can be adequately assessed despite a nonrandom sample in many situations as long as the bias is under check. The basic requirement is the baseline equivalence of the cases and controls.

At the same time, these studies too involve estimation of a parameter such as OR, finding CI, and testing of hypothesis. These statistical procedures do require a random sample of the subjects. Whenever feasible, indeed a random sample should be taken. See Example 4.5 for a nested case–control study that uses stratified sampling to assess the effect of serum selenium levels on cancer mortality.

Example 4.5: Stratified Sample for a Nested Case–Control Study

Serum selenium level is widely suspected to affect cancer mortality. But the results across studies are not coherent. Belgium has a system to follow up each patient till death. A stratified (for sex—male and female) random sample of 201 cancer deaths of age 25–74 years out of a total of 343 during a 10-year period was studied for their selenium level as well as some other factors [14]. Three controls were selected for each case and these were matched for age and gender. Thus, a total of 603 controls were also studied. Serum selenium level was found a significant predictor of cancer mortality in males but not in females.

In Example 4.5, sex stratification of the subjects helped to come up with a conclusion that is different for males than for females. Thus, the stratification strategy paid off in this case. Also note that the investigations proceed from outcome (cancer death) to an antecedent (serum selenium level), and thus the study is retrospective in nature. Since controls were also investigated, it is a case–control study. It is nested because follow-up of each person is routinely done in Belgium and cases are chosen from this follow-up. Controls were easily available and choosing three controls per case helped to increase the reliability of results without the corresponding increase in cost.

On the flip side is the sample of 201 cancer deaths out of 343 and the claim that it is a random sample. Such 60% sample is not a norm: one can legitimately wonder why all 343 could not be included in the study. Had all these been investigated, it would still be a sample in the sense that they occurred in a 10-year period. Previous deaths and future deaths would still not be incorporated.

4.3.2.4 Confounders and Matching

Ideally, the control group should include exactly similar subjects except for the disease and the antecedent factors not under investigation so that it is parallel in a true sense. Ensuring exact similarity may not be feasible, but it is easy to understand that the cases and controls should be matched with respect to all those factors that do not fall into the set of hypothesized risk factors.

For finding whether a conventional STD per se has a role in hepatitis B infection, cases would be hepatitis B positives and controls would be hepatitis B negatives but they should be of same age, same gender, having similar sexual behavior (such as multiple partner sex), similar glycemia levels, etc. These are the confounding factors in this study.

The results of case–control studies are much more valid if there is one-to-one matching between cases and controls, that is, each case should have a corresponding one **matched control**. These are also called matched pairs and the process is called matching. The attempt should be to simulate an identical twins situation. The purpose is to be able to conclude that any difference between the cases and controls is attributable to the antecedent under study and to no other factor. Thus, the role of confounders is minimized. In the case of perfect one-to-one matching, the statistical analysis should consider two groups paired and not independent samples.

It is an uphill task to match more than two or three confounders. Generally, matching stops at age and sex, which are confounders in almost every medical setup. If matching of many confounders is required, an acceptable but less valid procedure is **group matching** (also called frequency matching). Under this scheme, controls are matched with the cases on average or with regard to pattern of presence of the confounding factors. If obesity is a confounding factor, and if 35% of cases are obese, then nearly the same percentage of controls should also be obese for group matching. Since finding controls with many matching characteristics is difficult, some confounders may have to be adjusted at the time of analysis.

In addition to selection, matching should also be in ascertainment. Consider whether the cases and controls are likely to respond to the questionnaire in a similar manner, and no bias is likely to creep in due to the differential pattern of responses not related to the factors under study. The controls must be assessed with the same keenness and with the same methodology as that for the cases. Similar pro forma and procedures should be used as far as possible. If cases are being interviewed in a clinic, the controls should also be interviewed in a clinic. Controls may be healthy subjects, but interviewing them in home can alter the response. Cases may be more motivated but try to extract the same cooperation from the controls as well. Controls should be able to provide an equally correct estimate of the rate of occurrence of antecedents in subjects without the disease. Wherever possible, consult records because they are likely to be far less biased than verbal responses. This is feasible only when the records are complete.

I have made out a case for matching for all the factors that can affect the outcome except the ones under study. The last qualifier is the actual operational clause. This is sometimes ignored. The real question is what to do with factors that affect the outcome as well as the risk factors under study—earlier called confounders. Matching for these can sometimes result in **overmatching** and biased results toward no effect are obtained. For example, in a study on effect of postmenopausal estrogens on uterine cancer, matching on uterine bleeding can present invalid findings since uterine bleeding is associated with uterine cancer also. Any matching on such outcome-related variable can distort the results. See also Example 4.6.

Example 4.6: Overmatching

Marsh et al. [15] describe a study of relation between radiation exposure and mortality from leukemia in workers at a nuclear reprocessing plant. The matching factors in this study were site, sex, work status, age (within 2 years), and date of entry (within 2 years). Note that risk of leukemia varies with these factors and they looked like valid factors for matching. Date of entry was considered necessary as the risk of leukemia changes with calendar time. Examination of data since 1950 shows that the radiation dose steeply declined after 1980. Matching for date of entry unwittingly also matched for radiation exposure, which was the antecedent under study. Such overmatching obscured the relationship between radiation dose and risk of leukemia. Overmatching reduced the statistical significance.

SIDE NOTE: Likewise, exposure-related variables that can make risk factor distribution similar in cases and controls should not be matched.

4.3.3 Merits and Demerits of Case–Control Studies

Before using the case–control strategy, consider the following merits and demerits.

4.3.3.1 Merits of Case–Control Studies

The basic advantage of a case–control design is that the long period of follow-up is avoided. This can drastically reduce the cost. This design is particularly well suited for outcomes with a long gestation period. Since the subjects are selected on the basis of their outcome or disease status, this design allows beginning with an adequate number of diseased subjects. This is particularly advantageous for a rare disease such as leukemia. A case–control design also allows investigation of a wide spectrum of potential etiologic factors, and thus a range of hypotheses can be investigated.

4.3.3.2 Demerits of Case–Control Studies

A major problem in a case–control study can arise since both exposure and disease have already occurred at the time of initiation of the study. This makes it particularly susceptible to bias. Since the presence or absence of disease is already known, bias could very easily creep into the reporting or recording of the exposure information. In addition, cases may easily recall past events since they are sick or affected, whereas controls may fail to do so. In the case of a prospective study, events as they unfold can be accurately observed. Such an opportunity is not available in a case–control design. Dependence is entirely on the reporting done by the subject or on the records. Recall lapse can occur and the records may be incomplete. The biases mentioned in the previous section for prospective designs are generally applicable to case–control designs as well, except those arising from the time gap between exposure and outcome.

If the outcome is death, cases are no longer available. Such a study necessarily has to be based on records, which may be incomplete and not in the required format. Thus, think twice before deaths are considered as cases in a case–control study.

4.4 Cross-Sectional Studies

The third format for a gestation–anemia study is that a group of deliveries are chosen irrespective of maternal anemia and gestational age, and both are elicited. This is a cross-sectional study since both antecedent and outcome are observed at the same time. Both are responses in this setup and none is a factor. The presence or absence of either antecedent or outcome is not a consideration at the time of selection in this kind of design, but the objective still is to investigate the association. Although this kind of format is more appropriate for descriptive studies, such as for estimating the prevalence of a health outcome, it is also appropriate to generate hypothesis regarding etiology.

In some setups, the distinction between antecedent and outcome is blurred. Considering cleft lip and thalassemia in children, neither is a known cause of the other, yet dependence of one on the other can be investigated for generating a hypothesis. Such studies are analytical in this sense and do not remain purely descriptive. For studying the association between blood group and gender, if male and female subjects are selected and tested for their blood group, it could be difficult to categorize it either as a prospective or as a retrospective study though it might merit to be called a case–control study if one gender is regarded as case and the other as control. When the antecedent and the outcome are not identifiable, the study may be called descriptive rather than analytical. It would only indicate the blood group profile in the two genders without aspersions on the exposure–outcome type of relationship. Such a study would continue to be of a descriptive type until one is identified as an antecedent or at least as a suspected antecedent for the purpose of the investigation.

A study on evaluation of concordance between two or more methods is also cross-sectional. Cross-sectional studies are more appropriate for assessing the relationship between fairly stable entities, such as gender and hypertension, which do not change during the course of the study. Note that a cross-sectional study provides a one-time snapshot of the status of the characteristics and not a long-term perspective.

In a cross-sectional study, confounding factors can also be investigated at the same time. For example, in a study on determinants of the increase in serum cholesterol with age in adults, a group of subjects could be elicited for sex, diet, BMI, etc., in addition to age and serum cholesterol level. All measurements would be considered valid for the date of investigation although assessment of diet in this case could be based on the food intake during previous 3 days. The investigation might have the objective of determining the role of sex and obesity in increasing the cholesterol level with age, but the design is such that the role of sex, diet, and cholesterol level in obesity can also be investigated with the same validity. Statistically, any characteristics in a cross-sectional study can be considered dependent on the rest; the only restriction is the plausibility or justifiability of the relationship obtained.

4.4.1 Selection of Subjects for a Cross-Sectional Study

A cross-sectional study is categorized as analytical because it also intends to investigate associations and differences. However, the validity of conclusions would depend on the proportional representation of various levels of responses. Thus, it is important that a cross-sectional study is done on a randomly selected sample of adequate size, obtained from the target population without recourse to any consideration of presence or absence of antecedent or outcome. If this is compromised, the conclusions based on a cross-sectional study could be biased.

The first step for sampling for a cross-sectional study is, as usual, to identify the target population for which the results would generalize. Then decide which sampling method would be appropriate. For age- and sex-related disease such as hypertension and diabetes, stratification by age and sex might be useful. For a community-based study in a large population, a multistage sampling involving selection of counties and households can be adopted, or a cluster sampling might be more convenient. In a clinical setup where subjects come in queue, systematic sampling could be adopted.

4.4.2 Merits and Demerits of Cross-Sectional Studies

The following paragraphs describe the demerits of cross-sectional designs before describing the merits because for such studies demerits should be considered first when investigating antecedent–outcome relationship.

4.4.2.1 Demerits of Cross-Sectional Studies

Cross-sectional studies give a snapshot view and cannot measure risk. They may turn out to be a poor choice in situations in which the antecedent, the outcome, or both are rare. If the outcome of interest is testicular cancer or the antecedent under investigation is exposure to synthetic estrogen, then a cross-sectional study is not appropriate. An extremely inadequate number of subjects with the characteristics of interest may render the entire exercise futile.

There are other demerits too. The analysis can certainly be extrapolated to evaluate the net relationship between any two characteristics, keeping the others constant, but it should be clear that a cross-sectional design investigates the presence and not the appearance of the condition. Transient cases or rapidly fatal cases may inevitably remain underrepresented in this kind of design. The causes that determine the appearance are confounded with those influencing the duration of the disease, and it may be difficult to draw a clear inference about either set of causes. Dropouts or migrants tend to be excluded in a cross-sectional study.

The other serious difficulty in a cross-sectional study is that it cannot be ensured that the antecedent has actually preceded the outcome. This might have important implications for a causal inference. A firm conclusion on cause–effect can rarely be drawn from cross-sectional studies, and this is a major limitation for such studies to be truly analytical. The concept of a control group is not relevant to cross-sectional studies. Nevertheless, the biases earlier mentioned for cohort studies are largely applicable to such studies as well. However, information bias or memory lapse could be practically absent in this setup.

Caution is required in interpreting the results of a cross-sectional study. Such a study might reveal, for example, that the prevalence of hypercholesterolemia increases with age, but the fact could be that it is not age induced but is due to the changes in diet pattern of younger subjects resulting from increased awareness of the harmful effects of a high-cholesterol diet. Such awareness was less common 30 years ago and practically absent 50 years ago. A high-cholesterol diet consumed by older subjects for a long time when they were young because of lack of awareness might still have a carryover effect that persists despite a change to a low-cholesterol diet. Cross-sectional studies fail to take care of such confounders.

4.4.2.2 Merits of Cross-Sectional Studies

It follows from the preceding discussion that the cross-sectional design is particularly well suited for acute conditions with a short latent period or for chronic diseases that are stable and nonfatal. This design can be recommended for situations in which the distinction between antecedent and outcome is blurred. In an earlier example on association of sex with blood group, there is no antecedent. In disease–anxiety syndrome, disease can cause anxiety and anxiety can cause disease—thus either could be an antecedent. Also, a cross-sectional study is generally considered a rapid and inexpensive way to provide clues for further and more valid investigations.

For analysis of data from a cross-sectional study, as mentioned earlier, any one of the characteristics that can plausibly be considered as the outcome can be considered dependent on the others. Logistic regression or usual multiple regression (Chapters 16 and 17) can be used to find the joint or net effect of each of the independent factors in the model. In the case of cross-sectional studies, the assessment is generally made in terms of OR. Sometimes the prevalence rate ratio (PRR) is used.

> **Example 4.7: A Cross-Sectional Study on Diabetes and Sleep Apnea in Obese Subjects**
>
> A study was conducted on 137 extremely obese subjects (mean BMI = 46.9 kg/m^2) and their diabetes status and obstructive sleep apnea (OSA) were elicited [16]. Thus, this is a cross-sectional study. Among subjects with normal glucose tolerance, 33% had OSA. This was 67% in prediabetic subjects and 78% in type 2 diabetes patients. Thus, the association between OSA and diabetes status in extremely obese subjects was clear. This continued to be so after age, sex, BMI, etc., were adjusted.
>
> **SIDE NOTE:** This cross-sectional study excluded the role of various confounders, yet the conclusion rightly is of association and not cause–effect.

4.5 Comparative Performance of Prospective, Retrospective, and Cross-Sectional Studies

Caution is the bottom line for results obtained from any observational study. Because of a large number of confounding factors in this setup, some of which may be obscure and beyond redemption, a firm conclusion could be difficult. Results from such studies are many times considered suggestive and not conclusive. The confidence level increases when the same result is obtained in a variety of settings in different studies.

TABLE 4.1

Three Designs for Studying the Relationship between Maternal Anemia and Term of Gestation

Gestation (Outcome)	A Maternal Anemia (Antecedent)			B Maternal Anemia (Antecedent)			C Maternal Anemia (Antecedent)		
	Yes	No	Total	Yes	No	Total	Yes	No	Total
Preterm			?	$O_{1\bullet}$					?
Full-term			?	$O_{2\bullet}$					?
Total	$O_{\bullet 1}$	$O_{\bullet 2}$	n	?	?	n	?	?	n
	Prospective			Retrospective			Cross-sectional		

Now that the details of various observational studies have been described, a comprehensive view of the relative features of three types of designs can be presented. The structure of data obtained from these three types of observational studies is in Table 4.1. Note that column totals are fixed a priori in prospective studies and row totals in case–control studies. Only the grand total is fixed a priori in cross-sectional studies. Statistically, the hypothesis under test is for homogeneity of column frequencies in prospective studies, homogeneity of row frequencies in prospective studies, and association in cross-sectional studies. Statistical methods for these setups are discussed in later chapters.

Comparative features of the three types of study designs are summarized in Table 4.2. Their performance with regard to certain criteria is compared in Table 4.3. From an analytic point of view, assessment of the antecedent–outcome relationship is important. Thus, the last two criteria in Table 4.3 are crucial. The following brief contains some repetition of what has been already stated.

4.5.1 Performance of Prospective Studies

More than one outcome such as absence of side effects, recovery, and duration of survival can be simultaneously investigated in a prospective study. The greatest strength of a prospective study is that it gives more accurate results since the events are observed as they occur. The time sequence of events can be studied. The prospective design is the only methodology by which true incidence can be estimated.

A prospective study is time consuming and expensive but it is suitable for rarely observed antecedents. Generally, a large number of subjects with a particular antecedent are required so that a reasonable number of cases with the desired outcome are available at the end of the study for reliable conclusions. However, the study can have dropouts during follow-up, which can introduce bias as explained earlier. During the course of the follow-up

TABLE 4.2

Comparative Features of Case–Control, Cohort, and Cross-Sectional Designs

Item	Cohort (or Prospective)	Case–Control (or Retrospective)	Cross-Sectional
Main antecedent	Mostly known at the time of recruitment but in cohort of general population may be assessed as a baseline after recruitment	Elicited	Elicited (the distinction between antecedent and outcome may be blurred)
Outcome	Elicited after the assessment of antecedents	Already present and known	Elicited
Recruitment of subjects	On the basis of the antecedent	On the basis of the outcome	Neither outcome nor antecedents are considered
Definition of a case	Subject with the specified antecedent	Subject with the specified outcome	Any subject in the defined population
Definition of a control	Subject without the specified antecedent	Subject with outcome other than specified	No control is required
Measure of disease frequency	Incidence	None	Prevalence
Samples required	One cohort of exposed and sometimes one cohort of unexposed	One group of cases and one group of controls	One sample from population
Direction of investigation	Forward—into the outcome	Backward—into the antecedent	Cross-sectional situation as it exists

period, characteristics of the subjects can change such as cessation of smoking or an obese subject becoming thin.

4.5.2 Performance of Retrospective Studies

Retrospective study, generally a case–control study, has inverse properties. It can be accomplished with relatively a small number of subjects in less time and resources. A retrospective study is efficient for rare outcomes because it can begin with a sufficient number of cases. It can simultaneously evaluate many causal hypotheses. It is efficient also in the evaluation of interaction between different risk factors. A case–control study allows easy control of confounders. All these advantages accrue because a large number of affected cases are available in this format.

On the downside, a recall lapse is common in a case–control study. Differentials such as the ability of cases to recall events easily than controls can cause additional bias. It can also be biased because only those who already have had the required outcome can be included. Many severe cases

TABLE 4.3

General Performance Comparison of Case–Control, Cohort, and Cross-Sectional Designs

Criteria	Cohort (or Prospective)	Case–Control (or Retrospective)	Cross-Sectional
Cost and time	High	Low	Low
Number of subjects required	Large	Small	Large
Suitability for rare exposures	Good	Poor	Poor
Suitability for rare outcomes	Poor	Good	Poor
Spectrum of etiologic factors that can be investigated	Small	Large	Large
Spectrum of outcome factors that can be investigated	Large	Small	Large
Recall lapse and other biases	Not likely	Very likely	Not likely
Completeness of information	High	Low	Full, but only cross-sectional
Dropouts	More	Less	None
Changes in the characteristics of the subjects over time	More likely	Less likely	None
Assessment of temporal relationship	Good	Difficult	Not possible
Suitability for assessment of sensitivity and specificity	No	Yes	Yes, if the sample is representative
Suitability for assessment of predictivities	Yes	No	Yes, if the sample is representative
Evaluation and control of confounders	Poor	Good	Fair
Assessment of risk	Direct by RR	Indirect by OR	Approximate by PRR
Assessment of cause–effect relationship	Good	Fair	Poor

may have already died and cannot be a part of this type of study. A case–control format will not be able to establish the sequence of events.

4.5.3 Performance of Cross-Sectional Studies

A cross-sectional format is rarely used for an analytical study because it fails to provide a good assessment of cause–effect type of relationship, particularly when it is not clear what the antecedent is and what the outcome is. Between peptic ulcer and milk consumption, either could be a cause of the other. A major disadvantage is that this format fails to include severe cases that tend to be rapidly fatal.

A cross-sectional study is a good tool to generate hypotheses, which can be subsequently tested by a case–control or a prospective study. A cross-sectional study is certainly as good as a descriptive study. It is quick and easy to complete. It starts with a reference population, so generalization is immediate when based on a genuine random sample of sufficiently large size. Repeated cross-sectional studies are good for detecting changes in known risk factors besides, of course, the time trend of the disease.

As explained in a later chapter, it is many times helpful to obtain sensitivity/specificity or predictivity of a test procedure as indicators of its validity. The former requires a case–control study and the latter a prospective study. Sometimes a cross-sectional study is used to calculate both types of indices. When the subjects are real representatives of the target population, the percentage of subjects with disease and without disease would be nearly the same as in the target population. The test positives and test negatives would also be in representative proportion. But they will be prevalent cases and not incident cases, nor will they be "incidence" of antecedents. As explained later, prevalence may be in proportion to incidence under stable conditions. In such a situation, cross-sectional studies can be used to obtain both types of indices.

4.5.4 Reporting Results of Observational Studies—STROBE

Experience suggests that some authors do not give full details of the study they carried out. A group of researchers has developed a check list of items for STrengthening the Reporting of OBservational studies in Epidemiology, called STROBE. This check list generally applies to all the three formats of observational studies—a few items are specific to each of prospective, retrospective, and cross-sectional studies. Most of these items are included in the preceding text on these studies. Some others not included or need reinforcement are as follows.

The study design should be included in the title itself of the report. In the methods section, describe setting, locations, and dates, including period of enrolment, period of exposure, duration of follow-up, and the time when data were collected. Provide details of the selection of participants with matching criteria if applicable. Justify the study size and describe the flow so that the number of subjects available at each stage of the study is clear. Explain how missing data were handled, if any. In the results section, give unadjusted rates as well as confounder-adjusted rates with 95% CI. Mention all the analysis you did and state those, if any, that you carried out but do not form part of the present report. Discuss the generalizability and limitations.

References

1. Indrayan A, Bagchi SC, Verma V. Medico-social factors contributory to dropouts in a rural cohort of oral contraceptors. *J Fam Welf* 1972; 18:65–75.
2. Sokal DC, Zipper J, Guzman-Serani R, Aldrich TE. Cancer risk among women sterilized with transcervical quinacrine hydrochloride pellets, 1977 to 1991. *Fertil Steril* 1995; 64:325–334.

3. Richards M, Hardy R, Kuh D, Wadsworth MEJ. Birthweight, postnatal growth and cognitive function in a national UK birth cohort. *Int J Epidemiol* 2002; 31:342–348.

4. Neary WD, Foy C, Heather BP, Earnshaw JJ. Identifying high-risk patients undergoing urgent and emergency surgery. *Ann R Coll Surg Engl* 2006; 88:151–156.

5. Coviello AD, Zhuang WV, Lunetta KL et al. Circulating testosterone and SHBG concentrations are heritable in women: The Framingham Heart Study. *J Clin Endocrinol Metab* 2011; 96:E1491–E1495.

6. Last JM (Ed.). *A Dictionary of Epidemiology*, 4th edn. Oxford, U.K.: Oxford University Press, 2001, pp. 33, 145.

7. Hilska M, Gronroos J, Collan Y, Laato M. Surgically treated adenocarcinomas of the right side of the colon during a ten-year period: A retrospective study. *Ann Chir Gynaecol* 2001; 90 (Suppl 215):45–49.

8. Annema N, Heyworth JS, McNaughton SA, Iacopetta B, Fritschi L. Fruit and vegetable consumption and the risk of proximal colon, distal colon, and rectal cancers in a case-control study in Western Australia. *J Am Diet Assoc* 2011; 111:1479–1490.

9. Armitage P, Colton T (editors-in-chief). *Encyclopedia of Biostatistics*. Chichester, U.K.: John Wiley & Sons, 1998, pp. 503–540.

10. Lukanova A, Toniolo P, Lundin E et al. Body mass index in relation to ovarian cancer: A multi-centre nested case-control study. *Int J Cancer* 2002; 99:603–608.

11. Wacholder S, McLaughlin JK, Silverman DT, Mandel JS. Selection of controls in case-control studies: I. Principles. *Am J Epidemiol* 1992; 135:1019–1028.

12. Wacholder S, Silverman DT, McLaughlin JK, Mandel JS. Selection of controls in case-control studies: II. Types of controls. *Am J Epidemiol* 1992; 135:1029–1041.

13. Wacholder S, Silverman DT, McLaughlin JK, Mandel JS. Selection of controls in case-control studies: III. Design options. *Am J Epidemiol* 1992; 135:1042–1050.

14. Kornitzer M, Valente F, de Bacquer D, Neve J, de Backer G. Serum selenium and cancer mortality: A nested case-control study with age- and sex-stratified sample of Belgian adult population. *Eur J Clin Nutr* 2004; 58:98–104.

15. Marsh JL, Hutton JL, Binks K. Removal of radiation dose response effects: An example of over-matching. *Br Med J* 2002; 325:327–330.

16. Fredheim JM, Rollheim J, Omland T et al. Type 2 diabetes and pre-diabetes are associated with obstructive sleep apnea in extremely obese subjects: A cross-sectional study. *Cardiovasc Diabetol* 2011; 10:84.

5

Medical Experiments

Essential ingredients of an experiment are a purported manipulable cause and an anticipated effect. The relationship is speculative—if already confirmed there is no need for an experiment. An experiment is a procedure to verify or falsify a causal relationship by introducing the purported cause and observing the effect. This definition requires that a cause must be hypothesized and its possible outcomes visualized in advance. Observations are geared to measure this anticipated outcome.

Broadly speaking, the occurrence or nonoccurrence of different maternal complications in anemic and nonanemic women is an experiment albeit performed by nature, so is the occurrence of goiter of various grades in areas with iodine deficiency in water. When a rare or unique opportunity is available to observe or to study the effects of specific events as a result of naturally occurring changes, it is called a **natural experiment**. For example, the tsunami of 2005 provided a rare opportunity to study the health consequences of such a disaster. John Snow's classical discovery that cholera is a waterborne disease was the outcome of a natural experiment. Snow identified two mixed populations, alike in many important respects but different in the sources of water supply to their households. The large difference in the occurrence of cholera among these two populations gave a clear indication that cholera is a waterborne disease. This was demonstrated in the year 1854, long before the advent of the bacteriological era. Such natural experiments generally come under the domain of observational studies as described in the preceding chapter.

This chapter concerns experiments with *human intervention* for changing the course of events. They are therefore called **intervention studies**. By exercising control on the extraneous factors that can affect the outcome, experiments are the most direct method to study the cause–effect relationship. Experimental evidence is generally more compelling than that available from observational studies.

A medical experiment can be carried out in a laboratory, clinic, or community. The subjects for experiment in the clinic or community are human beings, and such an experiment is generally termed a **trial**. Laboratory experiment, on the other hand, may involve inanimate entities such as physical forces or chemicals: in the context of medicine, laboratory experiments are generally conducted on biological material or animals. Laboratory experiments often provide important clues to the potential of the intervention for formulating into a therapeutic agent. When successful, they pave the way for human studies. Thus, such experiments have a special place in medical studies. Even if you do not plan to conduct an experiment yourself, the details given in this chapter will help you to better understand and interpret the results of experiments conducted by others.

This chapter: Trials conducted on human beings are discussed in Chapter 6. This chapter is restricted to experiments on biological material and animals. In such experiments, relatively more maneuverability is available and stricter control of extraneous factors is possible. Section 5.1 discusses basic features that include principles of experimentation, advantages, and limitations. Section 5.2 is about various designs of experiments such as one-way, two-way, factorial, and repeated measures designs. These designs are easy to adopt for experiments on biological material and animals, but they have wider applicability to human studies also. Since clinical trials have a large number of other issues for discussion in Chapter 6, the current chapter describes various designs that could be used for animal experiments as well as for human trials. You should read both Chapters 5 and 6 to get a comprehensive picture of various designs of experimental studies done on animals and humans. Section 5.3 is about choice and sampling of experimental units.

5.1 Basic Features of Medical Experiments

Sometimes passive observation is not enough. You may have tried twisting the tail of your dog to see how he reacts. There is always a curiosity to explore the consequences of our actions. The first basic feature of any experiment is manipulation—the stimulus. The natural course of events is sought to be changed by human intervention. Although an experiment can be carried out for an intervention whose mechanism of action is still unknown, cause–effect inference is complete only when a biological explanation is available.

In the context of drug development, an experiment in the laboratory is performed first to develop a molecule that has desirable biochemical properties—thus has propensity to be beneficial in promotion of health, prevention of sickness, or treatment of disease—and second to establish that these properties are indeed present. This exercise requires enormous inputs of theoretical knowledge about various compositions. Once such a compound is obtained, its properties are investigated. The compound is modified as needed and the drug development starts. A formulation is prepared that could ultimately take the shape of a drug. This formulation is first tried on animals that could somewhat simulate human conditions. Trying a new drug on human subjects without establishing its efficacy and safety in laboratory and animal setups is considered unethical and is almost never allowed.

The second basic feature of scientific experimentation is the meticulous control over the experimental conditions that helps to draw inference on cause–effect type relationships. This is done by following established principles of experimentation over the units of experiment and the process, as described next. The **unit of medical experiments** is the biological specimen or animal that receives or does not receive the intervention and is observed for the anticipated changes.

The third basic feature of an experiment is that it is replicable. Since the experimental conditions are standard and the intervention well defined, it is not difficult for others to repeat the experiment and verify the results.

Although these features and other aspects are discussed in this chapter in the context of experiments on biological samples and animals, they are, by and large, applicable to clinical trials as well.

5.1.1 Statistical Principles of Experimentation

The purpose of statistical principles is to provide adequate control over the experimental conditions so that the effect of uncertainties is minimized and a valid conclusion can be drawn. These include control group, randomization, and replication. Blinding and matching are more relevant for human trials and these are discussed in the next chapter. Nonstatistical principles include ethics. For ethics in animal experiments, see Monamy [1].

5.1.1.1 Control Group

An important source of bias in all medical studies is the Hawthorne effect. In the context of experiments, this says that the experimental units tend to behave differently when they know that they are being observed. The same occurs when an inert procedure is used that otherwise should not produce any change. To account for this psychological effect, an experiment is almost invariably carried out by including a control group that does not receive the active ingredient of intervention but receives null stimulus or an inert substance such as saline injection or a placebo. In some experiments, the control group could receive the existing or prevalent intervention, while the test group receives the new intervention. For example, in an experiment on a new drug that could relieve pain, the control could receive the existing drug after pain is induced in both groups of mice by some mechanism.

When a separate group of units receives control regimen, this is called **parallel** control group. Some experiments are done without a parallel control group. Measuring tail-flick latency in mice to heat exposure before and after an analgesic does not have a parallel control. Such an experiment is called **before–after study**, where each unit serves its own control. Thus, there is a control in this setup also, although it is not a separate group.

Importance of parallel controls is described in Chapter 4 in the context of case–control studies. Parallel control is desirable in an experiment so that the *net effect* of the intervention can be evaluated. Although the need of a control is not debated, what actually is required for control is a subject of debate.

Units in the control group should be equivalent to the experimental units at baseline. Since medical experiments are on biological material or on animals, this is not as much a restrictive requirement. For an experiment on the shape of red blood cells at various ionic strengths, various groups of cells including the control would be equivalent so long as they belong to the same genetic stock. It is because of such baseline equivalence and standardized conditions in the laboratory that the experiments are able to provide compelling evidence of the presence or absence of a cause–effect type of relationship. This chapter contains many such examples.

5.1.1.2 Randomization

Randomization is a standard scientific procedure now almost universally adopted in medical experiments. This envisages random allocation of units to different stimuli, one of which could be null—the control. Even though the units in all groups should be homogenous to begin with, randomization is advocated for providing chance to the residual or unforeseen heterogeneity to equally divide among various groups. *The test group should be no different than the control group except for the intervention itself.* Then, any difference from control in the outcome can be attributed fairly to the intervention. The effect of chance, arising from including a limited number of units in the experiment, would be small when such equality is assured. Randomization also tends to remove the possible bias of the investigator in allocating units to different stimuli. In addition, randomization is a necessary ingredient for validity of the statistical methods needed to analyze the data.

Randomization apparently looks easy but actually can be difficult to implement. For example, assigning the even numbered animals in sequence to the test group and the odd to the control group has potential to cause bias because it can be easily tempered. Assigning one group of mice to the test and the other to the control can also be biased because the first group may be of specific type and the second group of another type. If an experiment is to be conducted on 30 mice with 15 each in the test and the control group, genuine randomization is best achieved by computer-generated random numbers. The website *www.randomization.com* does this very efficiently and tells you which numbers will be in group-1 and which in group-2. If there are more than two groups, random allocation can be made accordingly by using the simple procedure provided at this site.

Randomization tends to work well in the long run but may fail for a particular experiment. The standard practice is to compare the baseline of the two groups after allocation to confirm that they indeed are equivalent. In case of experiments on mice, they can be compared for age, sex, weight, or any other characteristics that might affect the outcome. Details are given later but statistically this equivalence can be confidently inferred only when each group has a reasonably large number of units. Also, beware that such equivalence is for averages or proportions in the two groups and not for one-to-one matching.

Average equivalence is enough for most experiments that aim to compare the average outcome in one versus the other group or groups.

5.1.1.3 Replication

Replicability is a basic feature of any experiment, and it is through replication that considerable confidence is added when the same results are obtained. Replication by another worker in a different setting helps to confirm that the results were not dependent on local conditions that were unsuspectingly assisting or hindering the outcome. Such replication also provides evidence for the robustness of the results.

Sometimes the investigator himself wants to replicate the experiment to be on firm footing. This is particularly desirable when the response varies widely from subject to subject. The variability across subjects remains unaltered, whereas replication reduces variability in experimental results. If the results between replicates are similar, their reliability naturally increases.

Replication fulfills an important statistical requirement. As would be explained later, they help quantify the random error in the experimental results. Replication is also an effective instrument in reducing the effect of this error. **Random errors** occur due to imprecision of the process—in methods/measurements, such as staining, magnification, rotation, counting, and timing, and in the environment, such as room temperature, humidity, or minor variation in chemicals, and reagents—or due to the varying care adopted by different observers. All these are those already listed as sources of uncertainties in Chapter 1.

5.1.2 Advantages and Limitations of Experiments

Despite being the most direct method of investigating cause–effect type of relationships, experiments do have some limitations. But first, the advantages.

5.1.2.1 Advantages

As already stated, the basic advantage of an experiment is its ability to provide clear evidence of the effect of an intervention when properly designed and performed in standard laboratory conditions. An antecedent–outcome relationship can be obtained if present even when the underlying mechanism is not clear—although in this case it will not be explained as cause–effect. Thus, a mere opportunity of an experiment generates awareness about complications involved in a medical relationship. An experiment is a good portal to test existing knowledge of the possible confounders, and it also opens up the possibility of considering unknown confounders that are in epistemic domain. Thus, an experiment can broaden the possibilities of new explanations and new ways to solve a problem, howsoever limited.

Experiments on biological samples and sometimes on animals can also be done for harmful procedures and substances. Such an opportunity does not exist for human trials. Nobody would deliberately expose people to a pesticide to find how it affects health. This can be done with animals within the perimeter of ethics. Experiments that require sacrifice can also be done on small-sized animals (see Example 5.1).

Example 5.1: An Animal Experiment on Hormone Replacement Therapy

To determine whether continuous or cyclic hormone replacement therapy (HRT) is better, Sun et al. [2] conducted an **experiment** on 142 Sprague–Dawley rats that were randomly divided into seven groups. Besides normal estrous and ovariectomized controls, the other five groups received treatments imitating clinical regimen with different combinations. The rats were sacrificed, and mitotic index and proliferating cell nuclear antigen (PCNA) were the outcome measures. Note the relevance of animal experimentation in this setup.

SIDE NOTE: The results suggest that continuous regimen was better than the cyclic regimen in postmenopausal HRT in rats.

5.1.2.2 Limitations

The most serious limitation of an experiment is that it is carried out in near-ideal conditions that do not exist in the practical world. Thus, the results sometimes do not reproduce when applied in actual conditions. An experiment is necessarily context specific and generalization is difficult. A fully internally valid experimental result may have low external validity. Experiments on biological material and animals rarely provide results that can be used on humans, although they do provide evidence one way or the other to proceed to human experimentation. Thus, they indeed provide a valid base for clinical trials.

As stated earlier, experiments are sometimes done without understanding the biological processes involved in the anticipated relationship between antecedents and the outcome. In that case, the results may remain uninterpretable.

5.2 Design of Experiments

An experiment must be organized such that it can answer the question of interest clearly and efficiently. This process is called the design of experiment. It includes features such as what factors to investigate, what levels of various factors are important, how many groups of units are needed, how to allocate groups to various interventions including control, etc. Validity of experimental results is directly affected by the structure of the design and its correct implementation. For this reason, attention to experimental design is extremely important. It is an important component of study protocol also.

A design is easy to construct and a constructed design is easy to understand when a few terms used in this context are clear. First, a **factor** is an *independent* characteristic whose levels are set by the experiments. In an experiment on mice, mice can be divided by strain (Sprague–Dawley, Wistar, Donru, etc.) to see the effect in different strains or by weight into light, moderate, and heavy to study the effect in different weight categories. The factor is strain in the first case and weight category in the second case. This usage of the term factor is essentially the same as explained earlier.

Levels of the factor connote the strata into which a factor is divided. When there are three weight categories of mice as in the preceding paragraph, weight has three levels. The weight levels light, moderate, and heavy have gradient, but that is not necessary. Strains of mice do not have any such gradient, but they are still called levels. If four strains are considered for the experiment, it has four levels. The design will greatly depend on how many levels of various factors need to be studied to adequately answer the research question.

An essential factor in an experiment is the intervention. It has a minimum of two levels—intervention present and intervention absent (control). If there are three dose levels such as 0, 5, and 10 mg/kg of body weight of mice, the intervention has three levels.

The last term in the context of experiment is the **response**. This also was discussed earlier and has the same meaning here. Response is the outcome over which the experimenter has no control. It can be qualitative such as positive change, no change, and negative change or can be quantitative such as time taken in recovery or the magnitude of decrease in the level of pain.

Nuances of qualitative and quantitative characteristics are discussed in a later chapter but note here that the statistical analysis of data heavily depends on such nature of the response variable.

Haphazard collection of data without a design can be expensive and time consuming, requiring many runs, and yet not provide the focused conclusions you are looking for. Confounding may make it difficult to distinguish the effect of various factors. Thus, design is important. A well-designed experiment can (1) easily spot where the differences exist, (2) provide more reliable and focused answers to your questions, (3) reduce cost by avoiding wastage, and (4) provide insight regarding the patterns that would be otherwise difficult to discern. Following are some of the popular designs used in medical investigations.

5.2.1 Classical Designs: One-Way, Two-Way, and Factorial

Classical is my term that you may not find in the literature for experimental designs. One-way, two-way, etc., are the basic structures on which more complex designs can be constructed. I am calling them classical designs.

5.2.1.1 One-Way Design

A *K*-way design considers *K* factors together. Thus, one-way design is the simplest with only one factor. In the experimental setup now under consideration, this one factor is the intervention itself—conventionally called treatment. There are a minimum of two levels of this factor, viz., intervention present and intervention absent. Intervention can have many levels but the design remains one way so long as the experimental units are not subdivided by inherent factors such as age and sex. If there were three levels of intervention such as continuous HRT, cyclic HRT, and no HRT, the experimental units, say mice, would be randomly divided into three groups to receive these three treatments. The age of the mice or their strain will not be a consideration in a one-way design; however, mice in all the groups can be of uniform age and of the same strain.

Statistically, a one-way design is called **completely randomized design**. It is also called one-way classification. The data arising from any *K*-way design including one-way are analyzed by using the statistical technique called analysis of variance, popularly known as ANOVA, when the outcome is quantitative. This technique helps to find whether or not any difference in the average outcome among groups is likely to be beyond the realm of sampling fluctuation, and thus qualifies to be called real. Multiple comparisons procedure is subsequently used to find which group or groups have a really different outcome. Such statistical methods are a very potent tool to rule out a major part of empirical uncertainty in the results.

5.2.1.2 Two-Way Design

In the HRT example cited for one-way, now introduce one additional factor, that is, the mice are normal estrous with ovary or ovariectomized. This factor has these two levels. With three levels of treatment, continuous HRT, cyclic HRT, and no HRT, and two levels of estrous status, this is a 3×2 experiment. With a total of 6 subgroups and 10 mice in each, this may look like as shown in Table 5.1.

TABLE 5.1

Two-Way Design (3×2) with 10 Mice in Each Group (Mice Number in Each Group Shown after Random Allocation)

Factor-1	Factor-2	
	Normal Estrous	**Ovariectomized**
Continuous HRT	3,7,8,14,18,30,31,37,48,52	1,5,10,11,15,20,27,42,43,49
Cyclic HRT	4,9,13,21,32,39,44,47,50,57	6,17,26,33,34,40,51,56,58,59
No HRT (control)	12,22,23,25,36,41,45,53,55,60	2,16,19,24,28,29,35,38,46,54

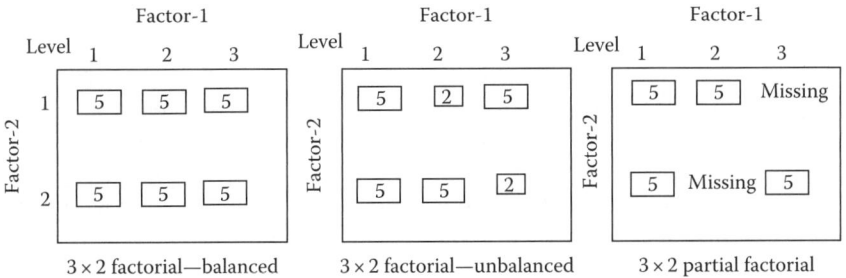

FIGURE 5.1
Balanced and unbalanced factorial and partially factorial designs (in box is the number of subjects).

Note that mice are allocated to each level of each factor so that there are six subgroups. In this example, each subgroup has the same number of mice, but that is not a requirement for any *K*-way design. Only that equal numbers make the analysis and interpretation a lot easier. A design with equal numbers is called **balanced** and otherwise an **unbalanced design** (Figure 5.1).

Grouping of mice into normal estrous and ovariectomized in this example is done to obtain homogeneous mice within each group. Conventionally, in statistics, this is called blocking and a two-way design as **randomized block design**. Sometimes, a one-way experiment is repeated two or three times to increase the level of confidence in the findings. This also is blocking, where the blocking factor is time. However, take care that the conditions such as reagents, calibration, and environment do not change in such repetitions. However, such terms are getting out of vogue, which were initially coined for agricultural experiments.

The response in this experiment is the mitotic index. The two-way design allows study of response variation between levels of factor-1 as much as between levels of factor-2. It is possible in this case that the difference in mitotic index between continuous and cyclic HRT is not the same in estrous mice as in the ovariectomized mice. This is called interaction. In view of the importance of interaction in medical experiments, this is explained in detail in the next paragraph.

5.2.1.3 Interaction

Biological interaction is generally understood to occur when two or more factors are simultaneously needed to produce or enhance (or retard) the effect. Statistical interaction in the context of designs is departure from the additive model. This means that the simultaneous effect of two factors is different from the sum total of their individual effects. Interaction is a term that belongs to *K*-way ($K \geq 2$) experiment but does not restrict to experiments on biological samples and animal experiments. In fact, it is better explained by examples of human experimentation.

The interaction explained in the next paragraph is not a pharmacological interaction between drugs but is a statistical interaction between the *effects* of a regimen when administered in conjunction with another regimen.

Some factors work more effectively when other conducive factors are also present. Iron supplementation is more effective in increasing Hb level when folic acid is also given. Their combined presence is much more effective than the sum total of their individual effects. This is a positive interaction and is called **synergism**. Since aspirin can reduce the beneficial effect of angiotensin-converting enzyme inhibitors in patients with heart failure, they possibly have negative interaction. This is called **antagonism**. Most interactions cannot be classified into any of these two categories. Osteoporosis is more severe in older women than older men (Figure 5.2a). Thus, age and gender interact for severity of osteoporosis, and perhaps they have no interaction for total lung capacity (TLC). When they are plotted, the decline in TLC with age runs almost parallel in men and women (Figure 5.2b). A similar pattern of response for various levels of factors, except for a nearly constant difference, indicates absence of interaction. Then, they are called factors with **additive effect**. If the responses are not parallel, interaction is said to be present. Epidemiologically, interaction is called effect modification.

Note that the term interaction is used for the product of two or more antecedent factors and not for the relationship between an antecedent and an outcome. Differential effect of various levels of an antecedent on outcome is not called an interaction.

Even when interaction is absent and the factors' effects are **additive**, a two-way design gives different results than two one-way designs—one for each factor. In any case, a two-way design is needed to explore whether interaction is present.

Although methods exist that can evaluate interaction in factors even when only one subject receives each combination of factor levels, generally it is evaluated when more than one experimental unit is assigned to each factor

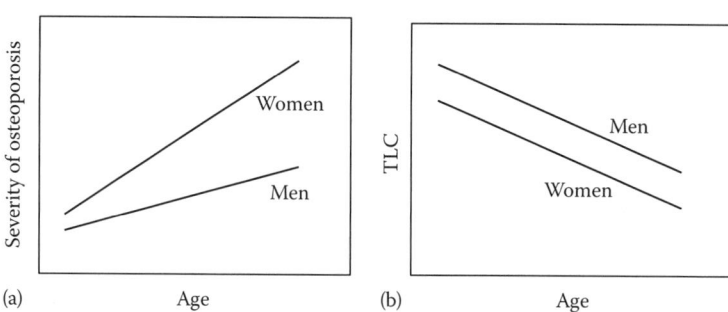

(a)   Age   (b)   Age

FIGURE 5.2
(a) Interaction present and (b) interaction absent.

level combination. In Table 5.1, 10 mice are assigned to each combination of factor levels, and thus interaction can be assessed. More than one subject for each combination of factor levels is essentially the same as replication described earlier.

5.2.1.4 K-Way and Factorial Experiments

Extension of two-way design to three or more factors is easy to imagine. In the HRT on mice example, add the factor of age (young and old mice) and the experiment becomes three-way. However, three or more factors are seldom considered together because the experiment becomes complex and the interpretation, particularly of interaction, becomes difficult. For this reason, three-way or four-way experiments are rarely seen in medical literature.

The name factorial experiment sounds like a misnomer. When all possible combinations of the levels of various factors are included in the experiment, it is called (full) factorial. This design answers the question what *combination* of levels of various factors is most effective. It is good to study interactions. A factorial design is easy to implement when prior knowledge is available regarding the levels of various factors that should be simultaneously considered. The 3×2 experiment described in Table 5.1 is a factorial experiment. All possible six combinations are included. An experiment with K-factors, each with C levels, is called a C^K factorial experiment. A simpler format is K-factor at two levels each—a 2^K factorial experiment. See Example 5.2 for a 3^2 experiment.

Factorial experiment can be implemented only when each level of one factor can be administered with each level of the other factors. In case of drugs, for example, it is possible that high dose of one drug cannot be given with the high dose of the other drug. In that case, factorial experiment cannot be done.

Example 5.2: A 3^2 Factorial Experiment

In the case of potentially therapeutic or anesthetic agents, it is common to try to evaluate combinations of various dosages of two or more formulations. In an experiment on anesthetic drug A (e.g., xylocaine), dosages under experiment could be 0 (none), 1, and 2 mg, and for drug B (e.g., bupivacaine), dosages could be 0, 0.5, and 1 mg for adult rabbits. The possible combinations in this experiment are given in Table 5.2.

TABLE 5.2

Combinations in a 3^2 Factorial Experiment

	Group								
	1	2	3	4	5	6	7	8	9
Dosage of drug A (mg)	0	0	0	1	1	1	2	2	2
Dosage of drug B (mg)	0	0.5	1	0	0.5	1	0	0.5	1

All combinations are included in a factorial experiment, relevant or not. In the example of HRT in mice, if control (no HRT) is studied only for normal estrous and not for ovariectomized mice as one might be tempted to do, the experiment is no longer fully factorial. If there is prior information that 2 mg of drug A in combination of 1 mg of drug B in Example 5.2 is harmful, this combination will be unethical and omitted. Such designs are called **partially factorial** or **quasi-factorial** (Figure 5.1). Example 5.3 illustrates an application of partially factorial design.

Example 5.3: A Partially Factorial Design for Testing Anticonvulsant Drugs

Carbamazepine and lamotrigine are anticonvulsant drugs, which may have a role in treatment of epilepsy and bipolar disorders. For experiments on mice, they require a vehicle such as methylcellulose.

The effect of these drugs is often studied on various parameters in mice. One among these is oxidative stress measured by malondialdehyde (MDA), glutamine synthase (GS), etc. For an example of such an experiment, see Pavone and Cardile [3].

An experiment on carbamazepine (factor-A), lamotrigine (factor-B), plus the vehicle (factor-C) has these three factors. With given and not given as two levels of each of these factors, a full-factorial experiment requires $2 \times 2 \times 2 (=2^3) = 8$ groups. Although the vehicle alone can be given, A and B cannot be administered without the vehicle. Thus, the combinations available are O (control), C, AC, and BC. The combination of the two drugs with vehicle (ABC) can be tried but the objective was not to study this combination.

In this experiment, only four groups were feasible or relevant against the required eight for a full-factorial experiment. Thus, this is a partially factorial experiment.

The statistical analysis of a full-factorial experiment is relatively easy and interpretation is even easier than of a partially factorial experiment. Thus, statisticians tend to advise full-factorial design and avoid partial setup. I would also advise so. However, in conditions where one or more particular combinations of levels of factors are undesirable or irrelevant, a partially factorial experiment can be designed and executed. Statistical analysis of this design will be tough but can still be done with the help of modern software. The main casualty in such designs is the interaction of two or more factors because that can be rarely evaluated. Nonetheless, partially factorial designs can still be devised if a particular interaction is important to study.

5.2.2 Some Unconventional Designs

Popular among unconventional designs are repeated measures design and crossover design. Again, these are described later in the context of laboratory experiments, but these too have extensive application in human experimentation.

5.2.2.1 Repeated Measures Design

In place of incorporating a separate control group, it is sometimes prudent to assess the outcome variable before and after the intervention in the same unit. In a bacterial colony, the bacteria could be counted initially and then provided a specific favorable environment such as sucrose in the medium. As already stated, this is a before–after experiment, and the simplest form of repeated measures design. A parallel control group is not necessary in this setup, although some experiments may have such a group as well for observing a trend in them also after placebo.

A carcinogenic marker such as AgNOR can be measured just before administration of a carcinogen in mice and after 1, 2, 5, and 10 days. This can be done separately for the control group also that receives the placebo. The changes are noted in the experimental and the control group. Experiments on most anesthetic agents follow such a repeated measures design. This helps to study the short-term outcome separately from that emerging in the long term, and the trend can be studied. Exercise caution in this setup so that fatigue does not set in and the readings at later instances are taken with the same earnestness as in the earlier phase.

The follow-up in repeated measures design is done at predefined intervals. The trend of the response can be assessed by repeated measures. The focal outcome is the difference in the response trend between the groups under experiment, although statistically one or two particular features of the trend are examined, such as the rate of rise and time to reach to the peak response.

5.2.2.2 Crossover Design

Crossover design too has pronounced applications to human experiments. Thus, the following discussion is mixed for laboratory and clinic setups. A crossover experiment on mice for drug abuse liability measured by lower back brain stimulation reward (BSR) thresholds is in Example 5.4.

Medicine is an incomplete science as it is today; many drugs work for a limited period. They have to be administered again, else the disease returns with the same intensity. Some drugs do have some carryover effect but in others there is practically none. When all traces of their effect vanish, say, 2 or 3 days after the last ingestion, the same subject can be used again to try the other drug, provided the condition of the subject reverses to what it was. This is called the steady state. In the case of humans, epilepsy, migraine, and end-stage renal failure are examples of such diseases. In some patients, hypertension and diabetes require daily medication, otherwise the disease returns to the baseline. In the case of animals, the experiment could be for the effect of cocaine versus saline on behavior in terms of distance traveled in 30 min following the injection separately in long-sleep and short-sleep mice. In this experiment, some mice may receive cocaine on day-1 and saline on day-2 and others saline on day-1 and cocaine on day-2. The process of trying two regimens on the same subject after providing a time

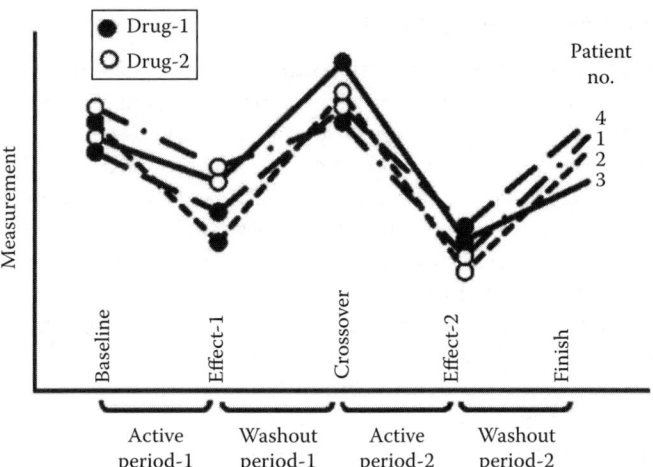

FIGURE 5.3
Representation of crossover design.

gap for complete disappearance of the carryover effect of the first regimen is called crossover (Figure 5.3).

You can see that crossover is more suitable for quantitative outcomes rather than therapeutic (yes/no) response. Nonetheless, it can be used for quantal response as well. Suitability of crossover also depends on the regimen. Quickly reversible regimens are obviously more suitable than long-acting regimen such as steroids. Crossovers are difficult to implement for multidose trials.

An important prerequisite of a successful crossover experiment is the absence of carryover effect. The carryover effect need not be just physical but can also be psychological if that affects the pharmacological outcome. The time gap required for carryover effect to vanish is called **washout period**. This generally should be at least four times the half-life of the regimen. **Half-life** is the period at which half of the peak concentration of drug is available in the body and the other half is eliminated or utilized by the body. Nothing should occur during the washout period that can affect the outcome. Absence of carryover effect can be statistically tested in a crossover design.

In a crossover experiment, the subjects are divided into two equal groups by randomization—one group receives drug A followed by drug B and the other receives drug B followed by drug A. Allocation can be done by tossing a coin or by any other random method. It is desirable to continue observation till the end of the washout period of the second drug. This provides confidence regarding the drug exiting the system.

The advantage of crossover is the availability of natural controls because the same experimental unit is used twice. This tends to reduce the impact of experimental error. The benefit of randomization is also available. Fewer subjects are required in this design relative to a parallel design with a separate control group. It is not necessary to recruit all subjects at a particular point of time. Period has a relative meaning. Some patients may have completed both periods before others are even recruited.

There are many cautions for using crossover design. Besides the basic requirement of no carryover effect, crossover is recommended when the effect of the intervention can be assessed quickly so that the patient can be crossed over without a large time gap to deny opportunity for confounders to appear. If baseline condition can affect the outcome, all subjects in the trial should be homogenous. Crossover is not a suitable design for cyclic diseases such as asthma and arthritis because they have natural remission periods. As for any design, there should not be any concomitant medication or comorbidity that can affect the outcome. Note that in this design, the subjects must stay in the experiment twice as long as otherwise needed in an experiment with parallel control group. This can cause higher dropouts. However, periods in the case of crossover are not necessarily calendar periods. It is possible that one subject receives first-period treatment in the month of June and second period in the month of July, and the second subject is recruited in August for the first-period treatment. Also, the entire trial should not be modified midway in any manner such as for dosage, period, and sequence because that could complicate the analysis. Crossover should not be used when substantial dropout is anticipated because then the statistical analysis becomes complex and the reliability steeply declines.

In some situations, a subject is used for multiple crossovers. A comprehensive text on crossover design is by Senn [4].

Example 5.4: Crossover Design for an Animal Experiment on Drug Abuse Liability

In rat, the ability of drugs to lower BSR thresholds often correlates well with high abuse reliability. Gill et al. [5] implanted bipolar electrodes into the medial forebrain bundle of male mice for assessing BSR thresholds after intraperitoneal alternative saline and cocaine in a crossover design.

For a human example, consider a representative group of 90 patients with essential hypertension. To study the effect of different drugs, they can be given atenolol, propranolol, and acebutolol for 6 days each after a gap of 3 days in between as a washout period for the drug to exit the system and for the disease to reappear at the same level. In this example, one patient is used three times. The order in which a patient receives the drugs is randomized. If the conditions are favorable, you can have four or five regimens in a trial.

Particularly in a crossover and in repeated measures design without parallel control in general, the results are obtained by comparing the average within subject responses at different time points. In the usual experiments with a separate control group, the results are obtained by comparison of averages across groups. Crossover and before–after experiments have smaller variability because the subjects are the same, and such experiments require fewer subjects to detect the same effect than the design with parallel control group.

5.2.2.3 Other Complex Designs

The designs described in this chapter are the basic formats that are generally used for medical experiments. However, in some situations, it might be appropriate to mix features of two or more designs. For example, a crossover design can be planned for the young and the old subjects or for males and females, introducing a factor that is not considered in the basic crossover experiment. As explained earlier, a design is partially factorial or incomplete factorial when one or more combinations are missing. It is called unbalanced when the number of units in each group is not the same. In Table 5.1, if 10 normal estrous mice receive continuous HRT but only 8 ovariectomized receive this therapy, the design is unbalanced. The experimenter may have started with 10 but 2 may have expired or may not be available to record the outcome.

Further details of various designs of experiments are given by Montgomery [6].

5.3 Choice and Sampling of Units for Laboratory Experiments

An experiment necessarily requires exposing units to an intervention whose effect is unknown. If the effect is known, there is no need to do that experiment. Any investigation originates from uncertainty—the environment in an experiment is doubly uncertain because of the intervention. A suitable design certainly helps in controlling some of these uncertainties, but choice of experimental units, method of selection, and the size of experiment are equally important. The first two aspects are discussed in the following paragraphs, and the size is discussed later for different setups including experiments. For size, note at this stage that laboratory experiments are conducted in fairly standard conditions, where many sources of uncertainty are under control—thus, they can provide equally reliable results on the basis of a relatively small sample.

5.3.1 Choice of Experimental Unit

As mentioned earlier, an experiment can be carried out on biological material, on animals, and later on human beings. Common sense ethics dictates that the preference is lowest biological entities and then progressively higher ones when the lower entities fail to meet the requirement. Thus, your efforts must be directed first to identify a suitable material such as tissue, cell, blood sample, serum, and body fluid for experiment. For example, for effect of temperature on erythrocyte sedimentation rate (ESR), the experimental unit for exposure to various temperatures would obviously be blood sample.

Material may be already available in a laboratory, which is collected for the investigation required for the patient but is preserved as record. In that case, no fresh consent is required. The investigation would be unlinked anonymous and the person concerned does not come into the picture. See Example 5.5 for such an experiment.

Example 5.5: Biofilm-Forming Ability of an Organism

The twin organisms *Acinetobacter calcoaceticus* and *Acinetobacter baumannii* can be isolated and identified from clinical samples such as tracheal tips, wound swabs, urine, and blood. Biofilm-forming ability of the complexes can be studied under different conditions such as different media strength, room temperature, and pH value of the medium. Biofilm-forming ability is measured by optical density value. This property may be related with antimicrobial susceptibility pattern and the site of isolation of strains. Thus, the results of such an experiment can help devise coatings or other mechanisms that inhibit biofilm-forming ability of these bacteria in humans.

In some situations, however, caution may be needed in conducting experiments on biological material. A great debate is going on regarding stem cell research, particularly when they are obtained from human embryos, even though such embryos pile up in some hospitals.

Moving beyond to live animals, preference is given to small animals such as mice, guinea pigs, and rabbits. Ethics for them is slightly relaxed compared to bigger animals such as monkeys and dogs. The choice is primarily influenced by suitability of the animal for the disease under consideration and the possibility of the implication of results for humans. Mice are favorite because they are small, prolific, vertebrate, mammals, and get diseases such as diabetes and cancer. They mimic human conditions well. Whatever species and strains you use, they must be sensitive so that the effect if any can show up. No group should be in a different room or in different shelf because environment may be different.

Before embarking upon animal experimentation, consult the local ethics that is generally quite well documented. The issues regarding administering harmful substances to animals, their torture, and sometimes sacrifice are more concerned with ethics than statistics. See books such as by Wolfensohn and Lloyd [7] for animal experimentation ethics and care.

5.3.2 Sampling Methods in Laboratory Experiments

For sampling of biological material, when identified and linked to a human being, the situation is back to sampling of individuals. For these, the same method of sampling as described in Chapter 6 for clinical trials should be followed. For experiments on unlinked and anonymous material, like those stored for record, sampling may be needed when more than adequate numbers meeting the criteria are available. If the experiment is on wound swabs for behavior of an organism under manipulated conditions, a random sample of swabs with that organism can be taken when a large number is available.

Laboratory experiments on animals also do not raise much concern about sampling methods. Generally, experimental animals of one particular species of isogenic strains are chosen, and they can be easily considered as representative of their population. Sensitive, specific, and uniform material helps in reducing the size of the sample. However, whenever factors such as age, gender, and weight can affect findings, the animals should be either stratified for independent experiment on each stratum or only one particular stratum of animals should be investigated. For example, to study the effect of middle turbination resection on facial growth of rabbits, the animals must be of the same age. For telomerase activity, which is implicated in all immortalization and carcinogenesis, age and sex of rats are important because they affect the level of this activity. Also in this case, the strain of rats can affect the findings. Thus, separate experiments should be done on samples of rats of different strains when relevant.

5.3.3 Choosing a Design of Experiment

The question of interest should be precisely defined so that a design can be devised that can adequately answer that question. An elaborately conceived design will fail if it answers a wrong question. Thus, always devote some time in brooding over the problem and converting it to an investigable question. This can be a big help in devising a right design. A precise question can help to decide what specific factors to investigate, and what and which levels of these factors should be included. Narrow them down to as small a list as you can without sacrificing the utility of the experiment. If you are in doubt on any particular factor even after consulting the literature and experts, it is better to include it in the experiment and check if it is helpful in explaining the response. Alternatively, compare the cost of ignoring a factor with the cost of inclusion, and do what looks more efficient.

Although the appropriate conditions for each design are mentioned at the time of discussing the design, the summary in Table 5.3 may be helpful to get an overall view.

5.3.4 Pharmacokinetic Studies

Experiments have a special place in pharmacokinetic studies. When a drug is ingested, it takes time to reach to the system. Conventional measures in

TABLE 5.3

Design Appropriate for Various Experimental Conditions

Outcome of Interest	Conditions	Design
Effect of a stimuli or doses of stimuli in homogenous group of units	No interest in stratified analysis	One-way
Effect of two or more treatments in the same group of subjects	No carryover effect and no natural periods of remission	Crossover
Average outcome for combination of levels of two or more factors, including interaction	All combinations of factor levels are relevant and feasible; interest is in stratified analysis if the units are heterogeneous	Factorial—it can be two-way or multiway depending on the number of factors
Average outcome for each level of each factor but interactions not of much interest	Some combinations of levels of one or more factors not relevant or not feasible	Partially factorial
Trend over time	—	Repeated measures

pharmacokinetic studies are time to reach peak concentration (T_{max}), time taken to return to the initial level, the peak level attained (C_{max}), and the response level after a specific time gap.

Another parameter of interest in pharmacokinetic studies is half-life about which a mention was made earlier in the context of washout period. An ingested drug generally reaches its peak concentration in the system fairly quickly. Then, it starts to decay. **Half-life** is the time taken from the peak concentration to decrease by half. The decrease could be in its pharmacological activity instead of concentration.

Half-life is obtained by fitting a regression (discussed in a later chapter) of concentration on time and locating the time point where concentration is one-half of the peak. In most pharmacological applications, concentration decline is slow in the beginning and increasingly rapid later on, which suggests log scale. Thus, linear regression of log(concentration) versus time is obtained. If the slope of this line is b,

$$\text{Half-life: } T_{1/2} = \frac{\ln(0.5)}{b}.$$

Since the concentration is declining over time, b in this case would be negative, and so would be $\ln(0.5)$. Thus, $T_{1/2}$ would be positive. Note that the concentration after two half-lives will not be zero but will be nearly one-fourth of the peak concentration.

References

1. Monamy V. *Animal Experimentation: A Guide to the Issues,* 2nd edn. Cambridge, U.K.: Cambridge University Press, 2009.
2. Sun A, Wang J, Zhu P. How to use progestin in hormone replacement therapy: An animal experiment. *Chin Med J* (Eng) 2001; 114:173–177.
3. Pavone A, Cardile V. An in vitro study of new antiepileptic drugs and astrocytes. *Epilepsia* 2003; 44 (Suppl 10):34–39.
4. Senn S. *Crossover Trials in Clinical Research,* 2nd edn. New York: John Wiley & Sons, 2002.
5. Gill BM, Knapp CM, Kornetsky C. The effects of cocaine on the rate independent brain stimulation reward threshold in the mouse. *Pharmacol Biochem Behav* 2004; 79:165–170.
6. Montgomery DC. *Design and Analysis of Experiments,* 7th edn. New York: John Wiley & Sons, 2008.
7. Wolfensohn S, Lloyd M. *Handbook of Laboratory Animal Management and Welfare,* 3rd edn. Oxford, U.K.: Blackwell Publishing Inc., 2003.

6

Clinical Trials

More out of convention than for semantics, medical experiments on human beings are called trials. The objective is to study the cause–effect relationship between a medical intervention and a health outcome in human subjects. Since the subjects are human, a large number of issues crop up ranging from stricter ethics to profound variations. Thus, trials do need extra care.

A human experimentation cannot be done unless sufficient reasons are present. The medical community must be sufficiently uncertain about the effect of the intervention under the trial. The regimen must have passed through rigors of preclinical phases before reaching humans. These phases involve studying biochemical properties and experiments on suitable animal models as discussed in Chapter 5. It must be clearly established that the intervention is more likely to be beneficial than harmful. The motto "Do no harm" is scrupulously followed in all medical research including trials. If any unsuspecting harm is later on detected, the intervention is immediately discontinued.

By their very nature, all trials are prospective studies where antecedent is the intervention and the efficacy and safety are the outcomes. A follow-up is built into all the trials. Thus, many ideas developed earlier for prospective observational studies apply to the trial setup as well.

This chapter: The intervention in a clinical trial is generally a therapeutic agent or a modality, in which case it is called a therapeutic trial. The setting for such trials is generally a clinic visited by patients of the type required for the trial. This classical setup of clinical trials is discussed in Section 6.1. This includes crucial strategies such as randomization, matching, and blinding. Issues such as equipoise, equivalences, efficacy versus effectiveness, and design are discussed in Section 6.2. Many concepts explained in the context of therapeutic trials are applicable to setups where intervention is a diagnostic or screening procedure or a preventive strategy. These give rise to diagnostic and prophylactic trials including vaccine trials. Some of these are conducted in a clinic but many in the community in a field setup. The details of such trials are presented in Section 6.3. The discussion is mostly restricted to the statistics-related aspects.

6.1 Therapeutic Trials

The primary objective of a therapeutic clinical trial usually is to evaluate the safety and efficacy of a treatment regimen in individuals with different levels of severity of disease and of various backgrounds such as different age, sex, and nutrition status. Extreme care is required because therapeutic trials generally involve exogenous material that may have side effects and may not be beneficial at all relative to the existing modes of therapy. For this reason, before a therapeutic trial is undertaken, it is necessary to be sufficiently convinced regarding intoxicity and potentiality as a beneficial regimen. Thus, the previous phases in the laboratory must have provided unequivocal results. Since the clinical trials are precarious, they are pursued in phases, particularly for a new formulation or substance as for drug development.

6.1.1 Phases of a Clinical Trial

Basic features of various phases of a clinical trial can be described as follows. Although these phases are used in several other setups, all these phases must be conducted with convincing success for developing a new therapeutic regimen. Such phasing is sometimes waived for a therapy or its variation that is already in use for some other condition and the trial is to examine its use in a new set of conditions.

6.1.1.1 Phase I Trial

Phase I of a clinical trial is done on human volunteers to study the pharmacological properties of the regimen, such as concentration–time profile, food interaction, toxicity, and major side effects, and most of all to delineate the maximum tolerated dose. Thus, dose escalation may be done in this phase. It may not be easy to find volunteers for this phase of trial, except possibly hopeless cases who find a ray of hope in the new regimen or courageous, healthy people who agree to participate for some compensation. The compensation should be proportional to the expected discomfort and not excessive that could be frowned upon as coercive or as unnecessary inducement. Except for regimens for diseases such as cancer that compromise tolerance, healthy subjects are preferred in this phase because therapeutic efficacy is not an issue at this stage. This phase generally needs less than 20 participants for each dose. Large number is not desirable since serious ethical issues arise due to perils that volunteers may face. There is no control group in this phase.

If diseased cases are included, take care that their varying severity does not affect the outcome. Comorbidities can spoil this phase of the trial. Thus, rule out not only symptomatic diseases but also asymptomatic conditions such as low hemoglobin level and high triglyceride level so that the side effects are not unfairly attributed to the drug.

6.1.1.2 Phase II Trial

Phase II of a trial is done on patients for whom the test regimen may be eventually indicated. The objectives of this phase are to (1) get initial idea of potential clinical efficacy, (2) assess short-term incidence of side effects, (3) identify a dose schedule for various kinds of cases (such as for mild, moderate and severe cases, or for children and adults), (4) study the interaction with other drugs or effect of comorbidities, and (5) collect further pharmacological data. Phase II also establishes or refutes that the new regimen is likely to meet at least the minimum level of efficacy. If this level is not met, there is no use pursuing the regimen any further. This is a crucial phase that really establishes that the regimen is going to be useful or not. Thus, it also provides **proof of concept**. The number of participants in this phase is generally 100–200. Sometimes it is a randomized trial with a control group on the pattern of a phase III trial.

Phase II can help in learning more about the treatment regimen and about the type of patients and kind of symptoms for which the treatment is beneficial. An appropriate dose and the appropriate subjects are identified for the phase III trial. Phase II may have to be stopped early if the regimen is found beyond tolerance in patients. Failure of phase II may help to identify the problems with the regimen, which may indicate the need to go back to the basics for improving the formulation.

When interpreting results of a phase II trial, note that the efficacy and toxicity might be interdependent. Thus, the error rate may be higher than that apparently obtained by considering them independently. In this phase, comorbidities are generally not excluded because applicability would suffer. Comparison of efficacy and side effects in patients with and without comorbidities will help define the exclusion criteria for phase III trial.

6.1.1.3 Phase III Trial

The stage is set for phase III trial once the early trials establish overall safety of the regimen, its basic clinical pharmacology, its therapeutic properties, and its most important side effects. There must be a valid control group and allocation of subjects fully randomized in this phase to the **test arm** and the **control arm**. For this reason, this is called **randomized controlled trial** (RCT). Selection of cases and controls and randomization for this phase are discussed later in this section. Through such strategies, a phase III trial is expected to provide compelling evidence of the efficacy of the regimen or its lack. Safety should not be sidetracked. When benefits are explored, proper assessment of harm is equally crucial. In fact, these days, safety is an overriding consideration in many trials. Phase III is also a prerequisite to meet the regulatory standards of license.

Phase III is a large-scale trial with generally 300–1000 subjects recruited for each arm. The follow-up must be sufficiently long for efficacy and side effects to emerge and to rule out that any relief is transient. Sometimes safety is a major issue, particularly when efficacy is already exceeding 80%. For side effects, a larger sample may be needed than to evaluate efficacy. This phase is expected to provide a full picture of the clinical performance of the treatment under test. Specifically, this can include (1) exact identification of the diagnostic group that responds reasonably well and comparison of the beneficial and adverse effects with those of the existing treatment, if any; (2) an increase in patient exposure in terms of both the number of patients and the length of follow-up so that less common and late side effects can also be identified; (3) more evidence on possibility of adverse interaction with other treatment regimens with which the new treatment is likely to be prescribed; (4) the ideal dose regimen for different types of patients with regard to age, body weight, severity of disease, etc.; (5) further pharmacological studies; and (6) receptivity to the treatment regimen of communities with different medical cultures. The last objective can be achieved by conducting a multicentric trial.

Example 6.1: A Randomized Controlled Trial on Nutritional Supplementation for GIT Cancer

In Italy [1], a total of 305 patients with preoperative weight loss <10% and cancer of the gastrointestinal tract (GIT) were randomized to receive preoperative artificial nutrition supplementation, postoperative jejunal infusion (perioperative group), or no artificial nutrition (conventional group). There are three groups in this RCT including one **control** (the conventional group).

SIDE NOTE: The outcome variables were postoperative infections and length of hospital stay. Intention-to-treat analysis (Chapter 20) and differences between the groups showed that preoperative supplementation was as effective as perioperative administration, and both strategies are superior to the conventional approach.

6.1.1.4 *Phase IV: Postmarketing Surveillance*

The regimen or the drug can be released for marketing after getting the license from the regulatory authorities on the basis of successful completion of the phase III trial. However, the monitoring continues for the effects and side effects. This is called postmarketing surveillance and many consider it phase IV of clinical trial. A whole new science of **pharmacoepidemiology** has emerged to study such issues when the treatment is a drug regimen. A good reference on this subject is Strom [2].

Phase IV is generally carried out in observational study mode than a trial mode. The drug is used in a routine setup and not intentionally ingested by the experimenter. The efficacy and adverse reactions are observed as they occur and the conditions are not controlled.

Under postmarketing surveillance, all adverse reactions and other events attributable to regimen are monitored. The effectiveness is also evaluated.

Patient preferences are studied. Recent findings about tamoxifen carrying a risk of endometrial cancer and arthroscopic surgery not beneficial for osteo-arthritis of knee are results partially attributable to such surveillance.

Any evidence of compromise on safety is a sensitive issue that can be easily hijacked by the media without gathering sufficient evidence. Pharmaceutical companies may have to react swiftly to investigate any such report because its damages can be irreversible.

6.1.2 Selection of Subjects

Phase I is on volunteers about which some remarks are already made. Phase II also is mostly an uncontrolled trial and the subjects can be purposively included. Phase III is necessarily an RCT for which selection of subjects is crucial. Note, however, that all RCTs are not necessarily phase III since phases are used mostly for developing a new regimen and not in other setups. An RCT can be conducted on an existing modality on a new kind of patient or a slightly modified modality can be tried that may be more convenient or more potent. This will not be a phase III trial. The following discussion is focused on RCT—phase III or otherwise—since the issues are relevant to some other setups as well.

Validity of results of a clinical trial heavily depends on proper selection of participants in different phases. They should mirror the target group. When a control group is present as in RCT, the selected participants are randomly allocated to the test and the control arm.

6.1.2.1 Selection of Participants for RCT

When an inordinately large number of subjects are available that pass the inclusion and exclusion criteria, they must be randomly selected for inclusion in the trial. This allows generalizability. One method could be systematic (e.g., every fifth), and the second method is to include consecutive eligible patients arriving in a clinic or hospital within a specified period. Otherwise, random numbers can be used for selection. Note that this random selection is different from random allocation about which a discussion follows.

Ethical considerations such as informed consent can preselect a biased group. Some patients or some clinicians may have strong preference for a particular therapy under trial and they can refuse randomization. Some eligible patients may refuse to participate when they are told that they could be randomized for placebo or the existing therapy. Some may refuse because it is a trial and not treatment per se. Considerable efforts may be needed to keep such refusals at a minimum.

In addition, the groups should be such that there is a priori uncertainty about the efficacy of the test therapy in them. This is called **patient equipoise** and helps to ensure that the patients are homogenous. Details are given later in this chapter on such equipoise. Patient equipoise implies guarding against unwitting tendency to include subjects who are likely to benefit from the drug under trial without declaring that the trial is restricted to such specific groups. Sometimes health conscious people agree to enter a trial while, in fact, participants should be fair representatives of the class of patients that are finally targeted to benefit from the regimen in case the trial is successful. The participants should be uncertain about the outcome of the trial so that the results are unbiased.

Be extra cautious in trials on severe cases. Besides medical surveillance that these cases require, note that many such cases sometimes hardly have scope to deteriorate further in case the drug is ineffective—they can only improve. If the patients have very high blood pressure (BP), they may remain so, whereas patients with relatively low BP can easily show natural rise. When comparing cases of high BP with low BP for efficacy of a drug, the drug may be unnecessarily considered to have caused rise in low BP group.

The number of subjects should be reasonably large in each group so that the full clinical spectrum is represented and a trend, if present, can clearly emerge. This also ensures reliability of the results. It should have adequate statistical power to detect a minimum medically relevant difference.

Bias can still occur in subtle or unknown ways in an RCT despite random allocation and blinding. A major source of bias is negligence in follow-up. If the follow-up requires recalling or revisiting the patients, some may not turn up or may refuse to cooperate, some may be untraceable, and some could die from unrelated causes. Even if the outcome assessment is within the hospital stay, some can leave against medical advice. Another factor that could affect a clinical trial is the need to change the treatment modality midway if any patient develops a serious illness. In addition, there could be patients who did not follow the full regimen. This is called the **partial compliance**. Take preemptive steps to minimize such losses and plan to adjust the results if needed.

6.1.2.2 Control Group in a Clinical Trial

As in other experiments, clinical trials can have one or more treatment regimens, but a **parallel or concurrent control group** is almost invariably required except in crossover and before–after setup. These are also called **reference-controlled** trials. Real controls are those that are similar but follow the natural course of disease without any intervention. In practice, however, the reference control group is either treated with an existing regimen or administered a placebo.

The placebo is important because some subjects tend to behave or respond differently when no treatment is given compared with when a sham treatment is given. This may happen due to activation of mu-opioid reception in the brain by the *expectation* of relief [3]. The placebo should look exactly like the therapeutic agent under trial, perhaps with same taste, and should be given in a parallel dose.

There is always a question about using a placebo on patients who are known to have the disease because they need an active ingredient to cure their ailment. However, placebos can be easily used in the following situations. Some of these situations may affect randomization.

1. No standard treatment is available, that is, the existing treatment modality has very doubtful results—perhaps no better than a placebo.
2. New evidence has emerged regarding the doubtful efficacy of the standard therapy.
3. Existing regimen is too costly or is rarely available to the population at large.
4. On patients who have already been given standard treatment and have not benefited, and no second line of treatment is available for them.
5. Test regimen is an add-on to the existing regimen. This means that all patients in the trial, including those on placebo, would receive the normally prescribed therapy any way.
6. Patients refuse to accept existing therapy and are willing to be part of a trial where they know that they can receive a placebo.

In situations where these conditions are not met, a group on existing therapy can serve as control. But control is now widely considered as a scientific necessity.

The control group should undergo the same medicinal rituals, such as dietary regulations, as the treatment group. This is more easily said than done. There are procedures for which a placebo group is nearly impossible. Examples are renal dialysis and fitting of an artificial limb. If a nonparallel group is a control, then appearance, schedule of administration, discomfort, etc., may cause differential compliance. Such a trial cannot be double blind, about which a discussion is presented later in this section. Finding a strategy that minimizes bias in such cases can be a challenging task. Whatever bias creeps in will have to be tackled at the time of analysis of data, and this too may not be an easy task.

When the comparison is with an existing therapy, this group is called **active control**. Such a group too should have similar baseline and must be subjected to the same maneuvers as the test group, except for the regimen itself.

There is a debate whether surgical trials need a group with sham surgery as placebo. Perhaps, evidence is not enough that sham surgery has the same psychological benefits as a placebo in a drug trial. Nevertheless, a sham surgery group can be adopted for a setup where it is not too expensive and is harmless to the participants.

An apparently simple approach is the comparison of the current test group with a group previously treated with the required control regimen including placebo. This is called **historical control**. This may be derived from previous clinical trials or records such as registries and databases. The advantage is that no concurrent control is required—thus the requirement of cases reduces by half. Cost also reduces accordingly. Ethical issues regarding recruiting and exposing subjects to the control regimen are also avoided. Historical controls may be appropriate for a disease that has a relatively stable natural history, and understanding of prognostic aspects has not changed. Multiple controls in this setup may help increase the confidence if the results replicate in each group of controls.

Despite demonstrable efficacy, the results are rarely accepted as definitive when based on historical controls. The flaw is that some factors may have changed over time. Known changes can be accounted for in the interpretation of results, but there might be some obscure changes that could affect the results. Diagnostic techniques and evaluation procedures may have improved over time. Lack of randomization also compromises the credibility of results in this setup. Historical controls may not have been monitored with the same keenness as concurrent controls would.

Notwithstanding the strong argument made earlier for some kind of controls, there might be situations where they are not needed. If a treatment is being tried for a rapidly fatal disease such as tuberculous meningitis, where is the need of a control group? Saving of some cases is enough evidence of efficacy. Drugs with dramatic effects such as penicillin do not need a control. Utility of Pap smear was established without recourse to a controlled trial. However, such instances are rare.

Example 6.2: Random Sample of Control Subjects but not of the Cases in a Trial

Abnormal mammogram can cause anxiety in some women who possibly need help. Barton et al. [4] performed a trial to evaluate the effect of an educational intervention that taught the skills to cope with anxiety. The subjects were women of age 39 years or older in seven mammography sites who came for screening. Of 8543 such women, 1439 had abnormal mammogram and were included in the trial. A random sample of 1405 women was taken from the remaining 7104 women with normal mammogram. Thus, these control subjects may not be matched for baseline characteristics. The investigators possibly expected that age may not affect the response or expected that age structure of the controls will not be much different from that of cases.

SIDE NOTE: Subsequent investigations in subjects with abnormal mammogram showed that many of these were false positive. The authors concluded that immediate reading of mammograms was associated with less anxiety than educational intervention targeting coping skills, because many were, in fact, false positives.

6.1.3 Randomization and Matching

The difference in outcome between the test group and the control group can be legitimately ascribed to the intervention when the participants with and without intervention are equivalent to begin with. Randomization is a very potent tool to achieve such equivalence. This works well for trials on a large number of participants but occasionally fails for small samples. If the conditions inhibit a large trial, a suitable alternate strategy is to identify pairs of participants matched for baseline characteristics and randomly allocate one of each pair to the test arm and the other to the control arm. The details are as follows.

6.1.3.1 Randomization

Randomization is random allocation of subjects to various arms of a trial. This is considered a scientifically valid strategy and is also very popular. Some methods of random allocation are given in the next section. Randomization is insurance that works in the long run but is not a guarantee for a particular trial. The possibility of selection bias is eliminated and the groups are more likely to become more comparable. It is fair to expect that the unaccounted prognostic factors such as age, gender, and grade of disease will be distributed nearly equally, thereby achieving baseline homogeneity across groups. No group may have dominance of participants of a particular type that can favor or go against the test regimen. Note that these covariates are not affected by treatment since they are measured at baseline at the time of enrolment. This, however, should be confirmed by post hoc analysis of the baseline characteristics of the subjects in the two groups. In addition, randomization provides a basis for statistical inference so that probabilities can be legitimately assigned. Random allocation among groups, one of which is a control, is a prerequisite for an RCT.

A distinction must be made between randomization and random selection of subjects out of those eligible. The former is a strategy to achieve baseline equivalence of the groups so that the difference emerging after the intervention can be legitimately ascribed to the intervention. This helps in achieving **internal validity**. On the other hand, random selection is for achieving **external validity**—being able to extrapolate results to the subjects not included in the trial. Whenever feasible, randomly select $2n$ subjects and then randomly allocate n of these to test arm and the other n to the control arm. When a large number of eligible subjects are not available, random selection becomes inapplicable.

When comparing a new regimen with the existing, the patients may like to switch their group after randomization. This can happen when blinding is not done. Such changes after randomization may affect the comparability as discussed later for intention-to-treat analysis.

In some situations, as in a trial to compare mastectomy with chemotherapy as a management for breast cancer, it might be difficult to get patients to agree to be randomized. Also, it might be difficult to identify a group of patients that are equally good candidates for mastectomy and chemotherapy.

6.1.3.2 Matching

As already stated, randomization works well in the long run and is advocated for a setup in which an adequate number of eligible subjects who are willing to be randomized are available. If the number of cases is not as large, examine if

matched pairs are available and if the two persons forming a pair can be randomized to receive the test and the control regimen. Matched pair design may be suitable for acute rather than chronic conditions. If the number of eligible subject is even less, controls may have to come from elsewhere. In this situation, matching becomes even more important. Experiments using matching instead of randomization are called **quasi-experiments**. This term is used for all those experiments and trials where the element of randomization is missing, either because it was not feasible or for any other reason. Evidence from such experiments is not considered as strong as from randomized trials.

Ideally, all relevant characteristics that might influence the outcome, except those under study, should be matched on a one-to-one basis. This does not stop at age and sex as is sometimes done. Nonetheless, comprehensive matching for all prognostic factors may not be feasible in all situations and some constrains on conclusions may become necessary. For example, in a trial of a new oral antidiabetic drug, the subjects in the test and control group could be matched for age, sex, and, perhaps, obesity, but it may be difficult to match for genetic factors and stress conditions. These two factors can also influence the outcome. The other important prognostic factors in this case are severity of the disease and any coexisting disease. They may have to be adjusted at the time of analysis.

In addition, matching can be tried only for the known factors. There might be other factors in the epistemic domain about which nobody knows yet—an uncertainty that still remains. Note that *randomization has the advantage of giving chance to the known as well as the unknown factors to be equally distributed*.

In some situations, it is possible to simultaneously give a different treatment to known pairs such as two eyes or two limbs of the same persons. Twin studies also come under this category. Randomization can be done within each pair to determine which one will receive test regimen and which control regimen. Thus, randomization and matching can go on simultaneously—they are not mutually exclusive. If the trial is on comparison of methods such as pulse oximeter and sphygmomanometer BP readings at the same time in the two arms, many pairs would be easily available. If the trial is for treatment regime, it would be extremely difficult to find matched pairs, such as the same severity of glaucoma in the two eyes or both limbs with same degree of paralysis.

It may not be possible to find a control of age 62 years for matching with a case of the same age. In most situations, matching within ±2 years for adults is considered adequate. Such relaxation can be possibly allowed for other factors as well. In tough situations, **group matching** is done instead of one-to-one matching. This means that if 30% of cases are females, 30% of controls are also females; if 60% cases have body mass index $\leq 25\,\mathrm{kg/m^2}$, a similar percentage is in the controls. This is also called frequency matching.

In most cases, one matched control is included in the study for each case, but when controls are easy to find and are less expensive you can include two or more controls for each case. This increases the sample size and helps to increase the reliability of the conclusion.

You can see that matching may mean incurring extra cost due to baseline investigation on a large number of subjects, many of which may be discarded as unmatchable. Generalizability suffers as the control group is somewhat distorted and interactions cannot be properly assessed. Special statistical methods are required to analyze such paired data.

6.1.4 Methods of Random Allocation

Human nature is basically biased. Subjective allocation is not considered random even if the investigator justifies it by stating that he allocated without any bias. It is necessary that a scientifically accepted method of random allocation is adopted. Some of these methods are as follows.

6.1.4.1 Allocation Out of a Large Number of Available Subjects

Suppose a large number N of eligible subjects is available for the trial. Out of these, suppose n_1 are to be allocated to group-1, n_2 to group-2, n_3 to group-3, etc. This can be done by drawing lots or with the help of random numbers. Such numbers can also be readily generated on a computer. These can be one digit, two digits, three digits, etc., depending on N. Assign serial numbers 1 to N to the eligible subjects. One method of allocation that reasonably works is as follows. Pick n_1 distinct random numbers, each less than or equal to N. Subjects with these serial numbers are allocated to group-1. Pick another set of n_2 distinct random numbers less than or equal to N after excluding those that are already in group-1. Subjects with these serial numbers form group-2. The third set of n_3 random numbers, excluding those in group-1 or group-2, gives subjects for group-3. This is illustrated in the following example. Websites are available that can do this easily such as *www.randomization.com*.

Example 6.3: Random Allocation from a Large Population

Suppose $N = 125$ eligible subjects are available for a trial of a drug in two doses, 50 and 100 mg. The third group receives a placebo. Suppose also that the control group will have $n_1 = 20$ subjects and the dose groups will have 10 subjects each. That is, $n_2 = 10$ and $n_3 = 10$. One possible set of three-digit random numbers less than or equal to 125 is as follows:

Group-1 (control)	074 096 022 008 107 043 054 059 124 010 027
	089 054 119 018 006 097 032 068 029
Group-2 (dose-1)	078 004 080 095 033 045 069 022 091 114 082
Group-3 (dose-2)	001 019 120 010 027 031 087 062 082 052 071
	042 029 032

Underlined numbers are those that are excluded because they have already appeared in a previous group. Subjects assigned these numbers will form the respective groups.

6.1.4.2 Random Allocation of Consecutive Patients Coming to a Clinic

The general practice is to do a trial on consecutive patients reporting in a clinic after excluding those who are not eligible or do not provide consent. If the patients indeed come in random order and the consent does not introduce any bias, then the even-numbered patients can be assigned to one group and the odd-numbered patients to the other group. For three or more groups, the first patient can be randomly allocated to any group by, say, draw of lots and the remainder allocated in a systematic way. If there are four groups and the first patient is randomly allocated to group-3, then the allocation for the incoming patients will be group-4, 1, 2, 3, 4, 1, etc.

This sequential scheme will fail in achieving unbiased allocation if the subjects follow an unknown design in the sequence of their arrival. Also, it is difficult to enforce blinding (see next section) with this kind of allocation. A more acceptable method is to draw or generate a random number between 1 and K (both inclusive), where K is the number of groups. When an eligible subject arrives, assign him to the group bearing this number. If the random number drawn is 3, then the subject is allocated to group-3. If the random number is 1, then the subject goes to group-1, and so on. This is called **simple randomization**. This requires that the number of subjects in each group is the same.

In place of generating a random number between 1 and K, it may be convenient to generate a two-digit random number. Divide it by K and add 1 to the remainder to get the group. (This assumes that the number of groups is $K \leq 9$.) If $K = 3$ and the generated random number is 52, then assign the subject to group-2 (remainder of 52/3 is 1, and addition of 1 gives 2). In fact, a small computer program can be developed that randomly allocates subjects to the specified number of groups. This program can be used over and over to provide fresh allocation whenever a new trial begins.

A difficulty in simple randomization is that one particular group may have its full quota of subjects much before the other groups. This is called imbalance since the subjects appearing late will not have chance to be allocated to the already completed group. In that case, the process of allocation continues as follows. The subjects for whom the allocated group turns out to be the already completed group are excluded from the trial. Suppose there are three groups, and each is planned to have 30 subjects. It is possible in this scheme that the full 30 are allocated to group-2 when group-1 has only 22 and group-3 26 after allocation of 78 subjects. In this case, ignore the 79th subject if the random number for this subject is 2. This subject is not included in any group. If the random number for the 80th subject is 3, then the subject is assigned to group-3. This process continues until each group has its full quota of 30 subjects. This procedure may mean wastage of some eligible subjects, but it ensures fair allocation. It is not considered fair to assign the last few subjects to only one or two groups and deny them a chance to be theoretically in the other groups.

If there are only two groups, the allocation can also be made by toss-ing a coin, but this should be done before the patient physically appears. Otherwise, the patient's confidence may be shaken. It is sometimes consid-ered convenient to include patients who report to a clinic on alternate days or during a specified time of the day. This may introduce bias because some patients may choose a time and day according to the availability of a particu-lar physician, and the management of cases by this particular physician may have prognostic implications. Similarly, allocation of subjects on the basis of even and odd date of birth may apparently look random but can be misused. All such methods are called **quasi-random allocation**. As already noted, the biggest problem is that such allocations cannot be kept blind and the chances of bias in assessment remain. Random allocation can be kept open so that the concerned subjects know that they are randomized to which group. But the strategy of concealment of allocation is the gold standard.

6.1.4.3 Block, Cluster, and Stratified Randomization

A popular method is block randomization. This requires that subjects are divided into M blocks of size $2n/M$ each, where n is the stipulated size of each of the two groups. The block size must be a multiple of number of groups. For two groups, the block size can be 4 or 6 or 8 but not 5 or 7. If you have enrolled a total of 80 subjects, you can make 20 blocks of 4 subjects each. Within each block, allocate two subjects at random to group-1 and the other two to group-2. Thus, you can have one of the following allocations:

$$(1,1,2,2), (1,2,1,2), (1,2,2,1), (2,2,1,1), (2,1,1,2), (2,1,2,1)$$

When you allocate randomly, one of these six is randomly chosen. This method is called block randomization.

An advantage of block randomization is that the possibility of one group becoming full before the other is ruled out. But the difficulty is that you know that the fourth subject after first three going to groups 1,1,2 must go to group-2. Thus masking is difficult. For this, several random block sizes are advocated that are concealed from the investigators. Just as simple ran-domization, this method too does not eliminate the possibility of one group getting subjects with different baseline covariates than the other.

For a large trial, particularly in a community, such as for evaluating the impact of special education on sexual behavior in school adolescents, you can randomly allocate 5 schools out of participating 10 to receive the educa-tion and the other 5 to serve as control. This is called cluster randomization. Schools serve as cluster.

Another method, though rarely used, is stratified randomization. This ensures that the subjects with important covariates are equally distributed to the groups. If your study is on wonder dose that controls blood sugar level for 1 month, and if you know that the effect could be different in males of age <50 years than females of age ≥50 years, you may want to divide the enrolled subjects as <50 M, ≥50 M, <50 F, and ≥50 F so that each of these strata is adequately represented, and then divide them equally to group-1 and group-2.

6.1.5 Blinding and Masking

As mentioned earlier, there is a tendency for the subjects to respond differently depending on whether they are in the treatment group or in the control group. To control this bias, two precautions are taken. First is called blinding of the subjects and the assessors, and the second is called masking of the regimen. These two are different procedures as explained next although many workers mix up these two terms.

6.1.5.1 Blinding

Trials with no blinding are called **open trials**. Everybody knows who is receiving what. When the patients do not know that they are receiving control or test therapy, then this is called **single blinding**. This eliminates the possibility of patients psychologically changing their response when they know that they are in the control group. They may feel discriminated against if the allocation is open. Also, patients who know that they are receiving a new regimen may either exhibit increased anxiety or have favorable expectations. Bias resulting from all these is due to the Hawthorne effect mentioned earlier. In addition, patients are more likely to seek adjunct intervention in an open trial and more likely to dropout.

If the treating clinician is aware of the upcoming allocations, he could exclude patients he considers unsuitable for that treatment. This exclusion after allocation can introduce bias. During the course of the trial, a clinician who knows the group assignment is more likely to administer co-intervention and more likely to adjust dose. If the assessor knows that a particular subject is a case (in the test group) or is a control, this may well affect the way questions are asked, investigations done, or interpretations made. Thus, it is desirable that the assessor also is kept blind. If he also does not know that the patient belongs to the test group or control group, then this is called **double blinding**. This removes possible bias of the physicians or nurses involved in patient assessment—it at least mitigates any subconscious influence of the assessor on the outcome assessment. Such precaution is an important criterion for validity of the results of a trial. Double-blind RCT is considered a **gold standard** to assess the efficacy of a new regimen.

Blinding is easy to implement by assigning codes to the treatment and the control. The record that a subject has received treatment or control is kept with a third person. Sometimes even the data analyst becomes interested in particular findings and can gear the analysis and interpretation accordingly. To avoid this, the analyst is also kept blind about the codes. The codes are broken only after the data analysis is complete. This makes the trial **triple blind**.

Rigid coding systems, such as code X for the treatment and code Y for the placebo, should be avoided because breaking the code for one patient breaks it for the rest of the trial. It is often satisfactory to use sealed envelopes bearing the serial number on the outside and containing the treatment allocation inside. Details of how blinding was actually implemented should be stated when reporting a trial. Merely stating that blinding was done is not enough. In fact, such details should be given in the protocol itself.

Morality issues are attached to blinding because the information is withheld from the participants who would be keen to know what they are getting in case an unusual side effect appears or unusual recovery occurs. If the doctor does not know, he may not be able to take remedial measures if anything happens to the detriment of the participant. Breaking the code must be axiomatic if the care providers consider it necessary.

Blinding is easier said than done. There are situations where blinding is not feasible. For assessing the outcomes such as quality of life, readmissions, and falls after hip surgery, blinding is just not possible if one maneuver is keeping the patients in hospital for a specified number of days, and the other is early discharge and home rehabilitation. In most surgical interventions, control has to be another kind of surgery, and not a placebo. As mentioned earlier, a sham surgery may be unethical many times because it exposes a patient to surgical risks. In either case, it is extremely difficult to enforce blinding in a surgical trial. The patient can be kept blind after proper consent but the surgeon definitely knows. However, a mechanism can possibly be developed wherein all assessments subsequent to the operations are done by another surgeon who does not know and cannot decipher whether the patient belongs to the test surgery or the control surgery.

May I remind that blindness refers to the patients and the assessors? To implement blinding faithfully, it is necessary to have a referee who keeps the code and assigns subjects to test or control group according to a predevised plan such as random allocation. Many times the term blinding is used to include masking, although masking is different.

6.1.5.2 Masking

Masking is the collection of steps that makes the test and the control regimen difficult to distinguish by the subjects and the assessors alike. There is a natural curiosity in participants and assessors to decipher the concealment. Thus, continuous vigilance is required. Masking ensures that the allocation remains concealed throughout the trial.

The top ingredient of masking is that the placebo or the control regimen has exactly the same physical properties—packaging, labeling, handling, color, size, shape, smell, and possibly taste. They must be administered in an undifferentiated fashion. If one regimen is once-a-day (OD) and the other twice-a-day (BD), the OD group should be given a placebo second dose to give an identical look.

The next ingredient of masking is that the control subjects must pass through the same medical rigmarole in terms of physical and laboratory assessments, diet, change of wards or beds, duration and frequency of examination, and the attention paid to the complaints, so that there is no scope for deciphering the group to which the patient belongs and of altered response due to differential procedures. Once blinding is done, this kind of masking will naturally follow. Yet, masking is the arrangements made to ensure that the identity of groups is not revealed till the trial is over. In fact, this can improve compliance and retention of the subjects by clearly demonstrating that groups are being treated equally. While trying to implement a perfect masking, beware of regimen-specific complaints such as bradycardia in those receiving beta-blockers. Such complaints can still unmask the code. A very careful strategy may have to be devised in some situations so that the bias is minimized if not eliminated.

A protocol should provide complete details of how masking would be done. Otherwise the audience remains skeptical. They must be convinced that masking would remain in effect until all opportunities of bias have passed.

6.2 Issues in Clinical Trials

The complexity of clinical trials is increasing as new factors affecting the outcome are being discovered. A large number of issues crop up that need to be considered for conducting a proper clinical trial. For a detailed account of these issues, see Piantadosi [5]. Clinical trials are difficult to implement and are becoming very expensive. This section contains a brief account of only those issues that are concerned with biostatistics.

6.2.1 Outcome Assessment

A trial may be immaculately planned and executed, but the conclusion will heavily depend on how outcomes are defined and assessed.

6.2.1.1 Specification of End Points or Outcome

Efficacy is always related to a particular outcome. Terms such as recovery and discharge are vague outcomes. They must be specified in terms of measurements such as glomerular filtration rate for kidney diseases, in terms of images such as x-ray for dislocated joint, or in terms of any such objective criterion. Also, the duration after which the outcome is to be assessed should be specified—within a day, within a week, or what. This applies to death also. Everybody dies, but if a death occurs 3 months after a surgery, should this be ascribed to the surgery? Follow-up period for different outcomes of interest must also be fully specified.

There are other issues as well relating to the outcomes. The actual interest may be in cardiovascular outcomes, but for expediency, change in BP level can be considered as a surrogate end point. Large cohorts and long follow-up are expensive—surrogates tend to make them expedient. For example, microalbuminuria is a promising surrogate of renal protection in many cases. However, do not use surrogates indiscriminately. Examine first whether they are indeed valid markers for the hard end point you are looking for. The surrogate should accurately assess not only the benefit but also the harm or the lack of it. A correlate may not be a suitable surrogate.

Outcomes can also be assessed at prespecified interim stages. This can help in discontinuing a trial if confirmed results are available one way or the other. Desired efficacy may be proved or unacceptable severe side effects may appear. Sample size can also be reassessed. However, mid-appraisal can unblind the study.

6.2.1.2 Causal Inference

The current and previous chapters maintain throughout that properly conducted experiments—whether in the laboratory or in the clinic—provide compelling evidence for or against cause–effect relationship between the antecedent and the outcome. The antecedent could be the regimen under trial, and outcome may be any that indicates efficacy. However, there are limitations: compelling evidence—yes; indisputable evidence—no.

An antecedent is not considered a cause unless it has plausible biological explanation for the effect. As mentioned earlier, the trials can still be conducted without full knowledge of the biological relationship, and this relationship can be explored later after the trial is over. In that case, the relationship is interpreted as a mere association and not cause–effect.

A perfectly valid trial in terms of randomization, control, and blindness may still provide results that are difficult to apply elsewhere. First, the selected subjects may not be representative of the target population; second, the conditions under which a trial is done may be too restrictive; and third, different trials may give different results. These differences propel synthesis techniques, such as meta-analysis, that allow a more reliable conclusion regarding causality [6].

RCTs are one of the many ways, perhaps the most dependable, to establish the properties of a regimen. But there might be alternate methods also of arriving at credible answers. They all should match. In addition, all such trials are plagued with epistemic uncertainties—unrecognized factors may affect the outcome. A positive aspect of varying results of different trials on the same regimen is that they raise awareness about the uncertain domain and bring humility to our endeavors.

6.2.1.3 Side Effects

A regimen should not be assessed only in terms of its benefits or efficacy. Except possibly those that alter lifestyle, no intervention is without risk of side effects and toxicity. Thus, the benefit must be seen in relation to the possible risk. This has special relevance to potentially hazardous drugs. In some situations, safety is more important than efficacy. For an account of benefit–risk assessment of various drugs, see Korting and Schafer-Korting [7].

There is practically no intervention without side effects. I can subjectively say a health-threatening side effect appearing in less that 1 is 10,000 users can be considered extremely rare and a side effect appearing in more than 1 in 10 users too common to tolerate. In between frequencies can be called rare, uncommon, common, etc. Of course, this will depend on the nature of side effect. Some of the side effects may be preexisting or occur in any case in a person or even a group of persons, but some could be attributed to the regimen. How can this attribution be achieved? Methods are available that help to categorize side effects into those that are definitely due to the regimen, possibly due to the regimen, and unlikely due to the regimen. For this, one procedure is to withdraw the treatment and administer it again, and see if the side effect disappears and recurs. Withdrawl of treatment and readministration may be occuring naturally in some cases.

6.2.1.4 Effectiveness versus Efficacy

Trials are generally done in ideal conditions that do not exist in practice. For this reason, these are more fully called **explanatory trials**. The actual performance of the regimen in practice may differ. Efficacy of a treatment is what is achieved in a trial that simulates optimal conditions, and effectiveness is what is achieved in practical conditions when the treatment is actually prescribed. For clarity, the latter is sometimes called use-effectiveness. Effectiveness could be lower than efficacy because of lack of compliance of the regimen, inadequate care, nonavailability of drugs, etc. These do not occur in a trial. Experience suggests that nearly three-fourths of the patients do not adhere to or persist with prescriptions. Thus, patients and maneuvers adopted during a trial do not translate their results for patients at large. Generally, such external validity of the trials is not high. But explanatory trials do establish the potency of a regimen to effect a change.

6.2.1.5 Pragmatic Trials

A regimen may have 90% efficacy but what it is worth if it is extremely difficult to implement! Effectiveness under practical conditions has brought pragmatic trials into focus. The patients recruited for this kind of trial are not homogeneous as in a regular clinical trial but reflect variations that occur in real clinical practice. Strategies such as randomization and control are also not structured. Instead patients receiving existing regimen or no treatment serve as control. Because of a large number of intervening factors in this setup, the interpretation could be difficult. Statistically, the standard deviation could be relatively large. For details of pragmatic trials, see Roland and Torgerson [8].

In the absence of blinding and placebos in a pragmatic trial, the results could be biased because of Hawthorne effect. The expectation of participants could be favorable or unfavorable and that will determine the actual bias. If patients are allowed to choose a treatment as could occur in practice, a further bias may creep in. Causal inference, which says that the effect is due to certain regimen, suffers. Thus, a better approach would be to do pragmatic trial for assessing usefulness in real-life situation *after* efficacy in ideal conditions is established.

6.2.2 Various Equivalences in Clinical Trials

The objective in a clinical trial may be to establish superiority, equivalence, or noninferiority of the test regimen compared with an existing regimen. Equivalence can be therapeutic equivalence or bioequivalence.

6.2.2.1 Superiority, Equivalence, and Noninferiority Trials

Superiority in any case would be an objective when comparing with placebo, many trials are also conducted to show that a new regimen has better efficacy than the existing regimen. These are called superiority trials. Superiority is established when the efficacy improves by at least a predefined margin. Choosing this margin can be nerve wrecking in some cases as it should be fully justified. This margin obviously would be more when comparing with placebo than when comparing with an active regimen. If a new treatment for systemic lupus erythematosus has 36% efficacy and the existing treatment has 30% efficacy, is this gain of 6% good enough to switch to the new treatment? This margin is determined by clinical considerations and not by statistical considerations. For example, when comparing Venae Sectio with modified Seldinger Technique for totally implantable access ports, this margin could be a 15% higher success rate—80% versus 95% [9]. The incremental benefits of newer regimens are shrinking as the technology is advancing. Thus, the superiority margin is progressively decreasing. Statistical procedure is different for testing such superiority than for equivalence or noninferiority.

> **Example 6.4: Superiority Trial for a Surgical Technique of Duodenum Preserving Pancreatic Head Resection**
>
> Koninger et al. [10] have described the protocol of a superiority RCT that compared end points such as duration of surgical procedure and quality of life at 12 months for partial resection of the pancreatic head without transection of the organ and visualization of the portal vein (Berne procedure) against resection with dissection of the pancreas from the portal vein. A reduction of at least 1 h in operating time in the test group was considered clinically important for superiority.

SIDE NOTE: Sample size was determined for adequate power to detect the clinically important difference just stated. The size actually increased because many subjects could not receive the intended procedure in this trial due to its complexity. Blinding was not feasible because of different surgeries, but the patients were randomly allocated. Track was kept of the adverse events.

Equivalence trials aim to show that the effects differ by no more than a medically unimportant specified margin. These trials consider the possibility of lesser efficacy as well as of higher efficacy. If the specified margin is 3% and if the existing therapy has 70% efficacy, for equivalence the efficacy of test therapy should be between 67% and 73%. If the efficacy is either more than 73% or less than 67%, the test regimen is not equivalent. Equivalence trials need two-tailed statistical tests and confidence intervals, which are discussed in a later chapter.

Equivalence trials need far more care than the usual comparative trials. The kind of incentive available in showing that a difference exists probably is not available in equivalence trials. Thus, equivalence trials lack natural internal checks. Inclusion and exclusion criteria should also be strict otherwise something like concomitant medication can tilt results toward equivalence.

In a therapeutic setting, the interest generally is in noninferiority rather than equivalence. If a new regimen has the potential to be at least as effective as the existing regimen but is cheaper or more convenient, interest would be in noninferiority. Noninferiority trials aim to show that the effect of the new regimen is not worse than the existing regimen by more than a specified margin, and therefore the new regimen can be advocated. Statistically, they need one-tailed procedures just as the superiority trials although the critical tail in noninferiority is on the left side. For superiority, it is on the right side of the distribution. For a short and crisp discussion about noninferiority trials, see Snapinn [11]. He argues that noninferiority and superiority can be assessed in the same clinical trial without statistical penalty.

Example 6.5: A Noninferiority Trial for Treating Major Depression

A total of 251 adult outpatients with acute major depression were randomized to receive hypericum ($n = 125$) and paroxetine ($n = 126$) in a noninferiority trial of hypericum extract WS5570 [12]. The primary outcome measure was a change in the Hamilton depression scale from baseline to day 42. Since hypericum extract is better tolerated and considered useful in cases with high risk of chronicity, it was helpful to find after the trial that herbal hypericum extract is at least as effective as synthetic paroxetine and is better tolerated. Paroxetine has proved efficacy in patients with depression of any severity.

The concepts of superiority, equivalence, and noninferiority mostly are applicable mostly in comparison to another regimen, which could be the existing standard therapy or some other regimen. Thus, the control in this setup is generally an active control and not a placebo. The regimen under test should be sufficiently efficacious; otherwise equivalence can occur when both are equally ineffective. Also, it is commonly known that superiority, equivalence, and noninferiority trials generally require higher number of subjects.

6.2.2.2 Therapeutic Equivalence and Bioequivalence

Equivalence trials have two distinct dimensions of equivalence. The kind of equivalence mentioned in the preceding paragraphs is for efficacy. This is called therapeutic equivalence and considers only the success rate as the end point. Many therapeutic equivalence trials are done for alterative dose schedule or method of administration. For example, alendronate 35 mg once weekly may be therapeutically equivalent to 5 mg daily for prevention of osteoporosis. Innovative and generic formulations of beclomethasone dipropionate may be therapeutically equivalent in adult patients of moderate to severe asthma. Both give nearly the same efficacy, that is, the efficacy does not differ by more than a clinically unimportant difference. Baker [13] argues that different mesalamine products may not be therapeutically equivalent because they differ with regard to where the drug is released in the intestinal tract, and this may affect the outcome.

Only the end point is considered for therapeutic equivalence, whereas bioequivalence considers the entire course of the recovery of patients. When the course of the disease or the improvement pattern over a period of time is the same for two regimens, they are considered bioequivalent. This requires pharmacological studies and the comparison may be in terms of peak concentration, time to reach its peak, half-life, area under the curve, etc. Sometimes pharmacokinetic studies on bioavailability are done to study bioequivalence. Popovic et al. [14] studied spline function for bioequivalence of verapamil single 240 mg orally standard retard tablet with single 5 mg intravenous dose.

Many consider bioequivalence to imply therapeutic equivalence. However, the reverse is obviously not true. Distinction should also be made between bioequivalence at individual patient level and average bioequivalence in groups of patients. If the two regimens under comparison produce different responses in individuals, this is ignored in average bioequivalence, whereas in individual bioequivalence, this interaction is an important consideration. Average bioequivalence would imply that either of these regimes can be prescribed. Individual bioequivalence would mean that the patient can be switched from one regimen to the other in the midst of the ongoing treatment. Bioequivalence studies can be done on healthy subjects as well for some regimens using crossover strategy. For statistical details of bioequivalence, consult Patterson and James [15].

6.2.3 Designs for Clinical Trials

Generally, selection of subjects, randomization, and blinding are considered adequate to specify the design of a clinical trial. However, in some situations, the design described in the context of laboratory experiments such as one-way, two-way, factorial, and, particularly, crossover are used for clinical trials. In addition, there are designs such as N-of-1, up-and-down, and sequential that are specific to a clinical trial setup.

6.2.3.1 One-Way, Two-Way, and Factorial Designs

All of these designs have been discussed in the previous chapter in the context of laboratory experiments. As mentioned there, they have applications to human studies as well, although probably they are not as frequently used in a clinical trial setup.

An elementary design is **before–after**. One can measure oxidative stress enzyme levels (e.g., superoxide dismutase) in Parkinson's disease, introduce an intervention to reduce this stress, and measure again. Subjects are their own control. There is no separate (parallel) control group, and thus in that sense, this is an **uncontrolled trial**. Problem with before–after design is that the effect, at least partially, could be due to psychological (placebo) reasons and not due to the drug. The two are inextricably mixed and it is impossible in this kind of design to separate the drug effect from the psychological effect. Thus, this design is used only when parallel controls are not feasible.

If there is a parallel control group, which also is subjected to before–after measurements, a commonsense method to arrive at causative relationship is **difference-in-differences**. The before–after difference in the test group is compared with the before–after difference in the control group. If this difference in differences is substantial, the conclusion of the effect of the test regimen can be safely reached.

All clinical trials that compare the test regimen with controls without any other consideration have effectively a one-way design with treatment as factor at two levels. In case, there are dosages under experiment or more than one test regimen under trial, the levels of this factor are more than two. Comparison of surgery, radiotherapy, hormone therapy, and watchful waiting in prostate cancer has four levels. If the patients are divided by duration of disease also, such as <1, 1–2, and 2+ years, this becomes a two-way design with 4 × 3 levels. But this stratification should be done before the commencement of trial and not afterward. If there are subjects in each combination, for example, with surgery but <1 year duration of disease, with radiotherapy for <1 year duration, etc., this is a (fully) factorial experiment. If the number of subjects for each combination of levels is the same, it is a **balanced design**—if not, an **unbalanced design**. If there is no case, for example, in the group with watchful waiting and duration of disease 2 years or more, the design is **partially factorial**. Such distinctions are important to choose the right method of statistical analysis. These explanations are the same as provided in the preceding chapter (Figure 5.1). Note that factorial design is good to study interactions.

6.2.3.2 Crossover and Repeated Measures Designs

These designs too have been discussed in the previous chapter on experiments but consider now the clinical trial applications.

The effect of antihypertensive drugs generally lasts for a limited period, and the disease bounces back to the original level soon after discontinuation. In the end-stage renal failure, urea clearance diminishes and dialysis has to be done repeatedly. In such situations, the same subject can be used again for the other regimen. The subjects become their own control. This helps to remove much of the variability between groups occurring in a parallel control setting. Crossover is relatively easy when the treatment can be rapidly started and stopped. There should not be any carryover effect after the washout period. However, crossover strategy with random allocation of the order of therapy and adequate washout period as discussed earlier is essential for valid results from such self-control designs. Crossover design cannot be adopted if cure or death is the outcome. In addition, regulatory agencies seldom approve crossover because of the possibility of carryover effect even if demonstrated to be absent. Also, side effects and tolerability are difficult to study in a crossover setup since the administration is for a short period. Thus, crossover is not recommended for phase III trials. But this design can be easily used in phase II, particularly for identifying the right dosage. The conventional crossover is two regimens–two periods–two sequences strategy. For further details, see Senn [16].

Repeated measure is an extension of the before–after strategy, where after-assessments are serially done at several instances to study the trend. Bioequivalence studies necessarily use repeated measures design, because only then the pharmacokinetics and the course of disease can be studied. The basic feature of repeated measures is longitudinal follow-up, although this can be short as in the case of trials on anesthetic agents or can last for years as for quality of life in surgical interventions.

Example 6.6: A Repeated Measures Design for Behavioral Stress Cues in Family-Centered Care of Premature Infants

A convenience sample of 114 premature infants (and their parents), divided into two groups, received routine intensive care and additional individualized, developmentally supportive family-centered care, respectively [17]. The infants were repeatedly assessed, among others, for stress cues. The developmentally supportive group had lower cues and needed less sedatives/narcotics and vasopressors. Complication rates and parental satisfaction were not different between the groups.

SIDE NOTE: The aforementioned example is based on the abstract of the publication that could not give the details. The instances of follow-up (repeated measures) are not clear, neither is it clear if all infants were assessed at the same age or not. A big statistical problem in this study is the convenience sample and nonrandom allocation of the infants to the intervention and control. Thus, both internal validity and external validity are questionable.

6.2.3.3 N-of-1, Up-and-Down, and Sequential Designs

Classical format of RCTs described earlier is considered the gold standard, but there can be problems in their implementation. Some of these problems can be addressed by alternate designs as listed next.

RCTs assess the average situation in the test group with that in the control group. Averages can be deceptive. There might be a treatment that works in a very specific type of case, which may not be many in the RCT groups. The disease itself may be rare or unusual such as orthostatic hypotension and narcolepsy, for which enough cases are not available for conducting an RCT. In such situations, a trial on a single patient can be considered for individual efficacy. However, the methodology is not firmly established yet.

N-of-1 trial is done on one patient when the patient is eager to cooperate and the disease causes low grade morbidity. It is a variation of crossover strategy with several pairs of treatment periods. The test regimen is given in one period and the control treatment in the other period. The order of these two treatments is randomized within each pair of periods separately. Such trials can be conducted only for chronic stable conditions with reversible symptoms and quantifiable outcome.

Obviously, a trial is done when efficacy is in doubt. This is always so for a new regimen. If a patient is reluctant to comply with the existing regimen because it is not effective in his case or because of side effects, he might agree to participate in an N-of-1 trial. This trial provides a unique opportunity to test a new regimen for a particular patient.

As in the case of crossover trials, a suitable regimen for N-of-1 trial is the one that can be rapidly started and stopped. There should not be any carry-over effect after the washout period. A minimum of three crossover pairs are advised. Blinding helps minimize patient and observer bias.

Example 6.7: N-of-1 Trial on a Series of Patients

Sometimes N-of-1 trials are conducted on a series of patients. Woodfield et al. [18] conducted such trials for quinine efficacy in 13 patients of repeated skeletal muscle cramps of the leg. Those patients were already prescribed quinine. Following a 2-week washout period, each patient received three blocks of 4-week treatment by quinine sulfate and matched placebo with a randomized crossover design. Only 10 patients completed the trial—three showed significant benefit in cramp occurrence, six showed nonsignificant benefit, and one showed no benefit. Thus, the results were equivocal and uncertainty remained even after this trial.

The second strategy could be the up-and-down method. This is used for calibrating the dose. Up-and-down trial involves a series of patients with the rule of an increase in dose following negative response and decrease in dose following positive response. The trial can be carried out on one patient also, provided the validity conditions mentioned for N-of-1 trials are met. The first test should

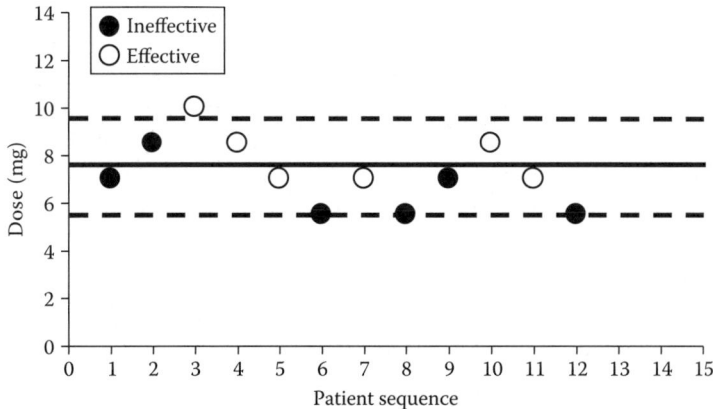

FIGURE 6.1
A trial with up-and-down method.

be performed at a dose close to your guess of the median dose. The method is also known as **sensitivity experiment**, also as Dixon and Massey method.

Consider a trial starting with 7 mg of intrathecal hyperbaric bupivacaine for an anesthetic effect in lower limb surgeries that failed to produce a satisfactory response in a particular patient. If the increment fixed in advance is 1.5 mg, the next patient will get 8.5 mg. If that also failed, the next will get 10 mg. If that succeeds, the next will now get a lower dose, 8.5 mg. If that also succeeds then next will get 7 mg. Success will decrease the dose and failure will increase the dose each time by the fixed margin (Figure 6.1).

The trials can be conducted on as many patients as decided in advance. Once the data are available for successive doses and success–failure information, they can be analyzed to estimate least mean effective dose. Confidence interval can also be obtained. The statistical method is given in Chapter 15. This type of trial cannot be used to estimate other locations such as ED_{95}— the dose effective in 95% cases. But one drug can be compared with the other for the least mean effective dose. In the just cited example, hyperbaric bupivacaine can be compared with isobaric bupivacaine and the one with lower dose can be recommended.

Some clinicians may find the up-and-down method very convincing for identifying a minimum critical dose that is effective. Although the dose in this method is increased or decreased by a fixed margin, the up-and-down method helps estimate the exact mean dose that could be in-between those tried.

The other design sometimes used in clinical trials is the sequential design. In this design, one subject (or a pair of subjects in case of matched controls) is added to each group at a time and the cumulative results are analyzed to come to one of the three decisions: (1) the treatment is effective, (2) the treatment is not effective, and (3) the evidence is not yet enough to support either (1) or (2). In the case of (3), the trial is done on one more subject in each group. This sequence goes on until a decision (1) or (2) is reached. The trial can be terminated midway if limitation of resources does not allow further continuation. The objective is to avoid wastage in case the evidence one way or the other can be obtained with a smaller number of subjects. But this can unblind the trial. Sequential trials are discussed by Whitehead [19].

The designs discussed in this section are still evolving and they still have to be properly understood. As experience accumulates, their merits and demerits will emerge. The method of statistical analysis too is not fully standardized for N-of-1 trial, up-and-down method, and largely for sequential designs.

6.2.3.4 Choosing a Design for a Clinical Trial

After describing so many types of designs, it could be expedient to provide a guideline for choosing an appropriate design. The following is the priority list depending upon the level of evidence, but the actual choice depends on the constraints imposed by the conditions of the trial and the resources available:

1. Whenever feasible, choose a random sample of eligible subjects from the target population. Divide these subjects randomly into the test and control groups. Blind the subjects and the observers about allocation, and make arrangements that this remains concealed until the results are available.
2. If random selection is not possible, choose the available subjects who meet the inclusion and exclusion criteria and allocate them randomly to the test and control groups. Blinding is desirable wherever feasible.
3. If random allocation is not feasible, match the cases and controls for their baseline characteristics. Use this strategy for small samples even if randomization is feasible.
4. If matching too is not feasible, use before–after strategy, that is, assess the subjects before intervention and after intervention. This strategy can be used in other setups as well.
5. If baseline information is difficult to assess, use existing information on baseline of similar subjects.

Level of evidence obviously is also adversely affected by (1) an unrepresentative sample, (2) improper randomization and lack of blinding that reduce the validity, (3) the small size of the trial that diminishes reliability and reduces power to detect difference even when present, (4) nonresponse that affects validity, (5) historical control or lack of control group that affects effect attribution, and (6) enrolling patients at different points of time that can affect homogeneity. The design must be so chosen that it can take care of these problems and provide reliable and valid evidence.

6.2.4 Designs with Interim Appraisals

Industry estimates suggest that nearly one-half of clinical trials fail. Such huge failure places heavy burden on those who invest in the development of new regimens. Costs are rapidly escalating anyway as the investigations are becoming intricate. Such considerations have obligated researchers to devise innovative designs that allow trials to mend during the course itself or stop if the analysis of interim data provides adequate evidence of desired efficacy or of futility. Futility is concluded if the regimen has little chance of meeting the minimum efficacy target. In cases where the decision one way or the other is not reached, the trial runs its full course, although in this case also modifications can be allowed to increase the likelihood of reaching to a definitive conclusion. Under this paradigm, you can decide to stop or modify the trial at interim stages for nonstatistical considerations also such as poor quality of investigations, slow enrolment, bigger dropout, unacceptable compliance, and increased cost. Design with interim appraisal can be grouped into two broad types: one, that allows early stopping but no other modification, and two, that allows other adaptations.

6.2.4.1 Designs with Provision to Stop Early

Consider huge savings in terms of patient exposure and time and management cost if you are able to stop a trial early. If sufficient evidence of desired efficacy is available, besides savings, you may be able to introduce the regimen early and earn revenue, initiate follow-up study for further development of the regimen, and meet the competition, and such other indirect benefits may accrue due to expediency. On the flip side is that early stopping could be unnecessarily guided by transitory effect and may fail to consider secondary end-points or some subgroups. If you design to stop early, ensure that no such criticism is mounted.

Simplest is to decide beforehand at the time of planning that the data accrued at one prespecified stage, say, when proportion p of the trial is complete, would be evaluated to decide whether to continue or stop. This is a **two-stage design** and allows only one interim analysis. Proportion p could be in terms of the number of subjects, in terms of length of follow-up, or any such criterion. For example, if the total trial is planned to follow up all n subjects for a period of $T = 12$ months and the appraisal is made at 8 months, then $p = 8/12$. Whether appraisal is at completion of p^*n subjects or at p^*T months and a decision to continue or stop is made, the Type-I error and power are affected. These important statistical concepts are discussed later in this book. In a clinical trial setup, Type-I error corresponds to probability of accepting a poor regimen, and power corresponds to accepting a good regimen. Adjustments are required to preserve them. Procedures for this are given in a later chapter while discussing sample size for clinical trials.

In place of two stages, you may like to plan to examine the data accrued at several stages. In the extreme case, one can think of analyzing the data sequentially after each pair of observations (one subject for test group and one for control group) and decide to continue or stop. This is difficult and not considered logistically practical for clinical trials. Generally, the data are analyzed at equispaced stages, such as four stages each after completing one-fourth of the trial. This is called **group sequential design**. This design is accepted by most regulatory agencies and is efficient in comparison to adaptive designs. Adaptive designs are described in a short while.

Group sequential designs are initially planned to have a large number of subjects so that relatively small gain can be detected although the actual target may not be to detect such a small gain. Commitment of large sample up front may be scary to some researchers but that is a feature of group sequential designs. If interim analysis at any stage provides clear evidence of that efficacy, the trial is stopped prematurely and a decision is made with reduced sample size. If the evidence emerges that there is hardly any chance of achieving the desired minimum efficacy, it is futile to continue and the trial is stopped. Else the trial continues.

Total number of stages and stopping criteria are specified in advance in the design itself. Minimum clinically relevant gain must also be specified a priori. Stopping rules are devised in such a manner that the level of significance and power are least affected. In this approach, the sample size or design is not modified. This is done in adaptive design.

6.2.4.2 Adaptive Designs

Suppose after due consideration of available knowledge, you plan a trial on 1000 subjects in each arm with 1 year follow-up. After 2 months into the trial, you find from formal or informal interim analysis that your anticipations at planning stage were incorrect and the trial will give you confirmatory results about efficacy (or lack of it) in just 6 months (or that you need to extend it to 16 months to get adequate number with the desired end point); only 700 would suffice (or 1500 would be needed); doses you are trying are too high (or too low) and you need to have a middling dose; a concomitant treatment is needed; or a particular subgroup such as males of age 70+ years needs to be excluded. Sometimes even the baseline information on the enrolled subjects may indicate that modifications in the design are needed. For example, this may tell you that expected kind of subjects are not being enrolled and eligibility criteria need to be changed. In such situations, you would not like to waste resources by sticking to the original plan in the hope of wonders, instead would like to adapt the trial to the realities revealed by actual experience.

Adaptive designs allow flexibility to redesign the trial midstream, guided by the interim data. This can make the trial more efficient by saving time, money, and patients. The difficulty is that the adaptive methodology is still evolving as this could involve many features of the design and possibly can be done at several stages of an ongoing trial. Adaptation is planned in advance by anticipating different scenarios as they unfold in an ongoing trial. Thus, it can handle only limited issues and not those that were unforeseen or ignored.

Despite clear advantages of adaptive trials, they have been rarely used so far. There is a confusion of what all should be considered for adaptation, how to reestimate the sample size under different adaptations, and how to handle logistic problems. The analysis becomes computationally complex. Specially tailored softwares are coming up. Nonetheless, the adaptive strategy is being explored with enthusiasm by drug companies, regulators, and researchers. As the experience accumulates, adaptive trials can be better designed and will have wider acceptability. An adaptive trial can quickly move from phase II to phase III since lessons learnt are already incorporated. This can expedite the process of product development. For sure, this strategy can address frustration arising in conventional structured designs when the trial gives negative results and you wish that if something can be done differently so that the results would not be so disappointing. Adaptive trials can soon become industry standard as the problems are sorted out.

Among various adaptations, statistically most relevant is to reestimate the sample size on the basis of the actual effect size found at interim stages. Reestimation requires intricate statistical inputs as this is done to preserve the level of significance and the power. Such adaptation does not cause much of ethical problems, rather seems to enhance ethics by keeping provision to stop the trial early in case convincing evidence of efficacy or of futility appears. Care is taken that such appraisal does not undermine the integrity and validity of the trial.

Adaptive designs incorporate practical considerations of possibly not getting things right first time when design is prepared. They generally start with relatively small sample but with built-in opportunity to escalate the trial after seeing the data at interim stages.

Such interim analysis has potential to unblind an otherwise blind trial. Unblinding is necessary to assess whether the treatment arm is giving evidence of sufficient efficacy relative to the control. Steps are specified at the time of the design as to who will be unblinded (such as Data Safety and Monitoring Board), and how the investigators and the subjects will continue to remain blinded for trial to go unhindered. The person/team unblinded for interim analysis must be independent of the team of investigators and should make only indirect statement such as "the trial will continue for 8 more months" or "will enroll 130 more subjects."

For more details of designs with interim appraisals, see Chow and Chang [20].

6.2.5 Biostatistical Ethics for Clinical Trials

Trials have the potential to harm the participants and the society since an unproven modality or intervention is used although the perception may be that it is beneficial. In an eagerness to complete the trial, the findings may become statistically biased. Then, the entire trial is questionable. Following precautions can be advised.

6.2.5.1 Equipoise

Uncertainty is considered a moral prerequisite for trials. Equipoise is espoused as the essence of the **uncertainty principle**. Among various equipoises discussed in the context of clinical trials, medically most important is **clinical equipoise**. This is the collective uncertainty among clinicians about the efficacy of the regimen under trial. For example, this exists for the use of vasopressin for management of septic shock. Lilford [21] cites the example of amniocentesis and chronic villous sampling for such clinical equipoise, although in his opinion the latter is twice as risky for miscarriage. These two are suitable candidates for comparison in a trial. Similar equipoise exists between stenting and endarterectomy for carotid restenosis. Optimal route (intramuscular vs. subcutaneous) of administration of influenza and pneumococcal vaccines in elderly patients is also in debate.

When the concept is stretched further, it not only means uncertainty but also that the two arms of a trial are likely to result in *equal* efficacy on an a priori basis. It connotes equal uncertainty for the positive and negative outcomes of the trial. Practically, though, genuine uncertainty is enough without pressing for equal uncertainty. Clinical equipoise is the condition under which clinicians as a group would not object for their patients to participate, and patients may rationally accept randomization. This equipoise also insulates against prejudiced assessment of the patients by the investigators. Previous evidence of benefit of a treatment may be flawed but can disturb the equipoise [22]. Sometimes a trial is terminated early when overwhelming evidence emerges and the equipoise is disturbed.

The second is **patient equipoise**, referred earlier. This is the uncertainty among the participants regarding the outcome of the trial. Primarily this is a statistical requirement for psychological homogeneity of the subjects at baseline.

The third is the **personal equipoise** of the clinician so that he does not feel uncomfortable about his own views and about his patients. A particular clinician may have a strongly positive or a very bitter feeling about a regimen even though clinical equipoise in terms of collective uncertainty may exist. A clinician who is convinced that one treatment is better than another for a particular patient cannot ethically agree to randomization. Personal equipoise may be difficult to achieve, but efforts can be made by discussing evidence regarding the underlying uncertainties and try to convince him that equipoise indeed exists.

Cheng et al. [22] have discussed these three kinds of equipoise in the context of a trial for melioidosis, but the concepts are explained very well.

6.2.5.2 Ethical Cautions

The other important consideration is the ethics of conducting *a trial* on subjects some of whom could be sick. The treatment being under test itself is an indication that its utility is doubtful. Important ethical considerations are as follows:

1. Is the treatment regimen under test reasonably safe?
2. Is there sufficient information that the treatment is likely to be beneficial? This, however, runs counter to the equipoise just discussed.
3. Have the subjects been informed about the potential benefits and possible side effects, and their consent obtained?
4. Is it ethical to use a placebo on some subjects who are sick? This point has been discussed earlier.
5. Is it ethical to allocate the subjects randomly among various groups to receive different treatments? Are sufficient precautions built-in to take immediate action in case an adverse reaction develops?
6. Is it proper for a trial to be blind in any way?

The basic theme of all these considerations is that the science cannot compromise the interest of the individual subjects without taking them into full confidence.

I would like to add that a trial is also unethical if proper statistical methods are not used. The entire effort of conducting a well-designed trial can go waste, including inconvenience to the patients, if the analytical methods are substandard. This, in fact, amounts to misuse of resources.

6.2.5.3 Statistical Considerations in a Multicentric Trial

A multicentric trial is necessarily a large trial that requires not only resources but also coordination. The objectives of such a trial generally are to evaluate the efficacy of the test regimen in varying conditions and to check if consistent results are obtained when uniform protocol is adopted. If yes, the data can be pooled and a result with greater reliability obtained, being based on a much larger sample after pooling. Bigger accrual of cases is among the primary objectives of multicentric trials. The generalizability also increases. If the results across centers differ despite uniform protocol, perhaps a new hypothesis can be developed to explain the differences.

Although varying conditions is an essential ingredient of multicentric trial, they tend to exacerbate due to varying interpretation of the protocol requirements. The difficulty arises when such variations are not acknowledged and go on inadvertently, finally damaging the comparability. Thus, these trials need periodic review—much more than single-center trials. Reviews such as by Data Safety and Monitoring Board help in quality assurance and uniformity across sites. Complex regimens that require intensive training may not be appropriate for multicentric trials.

6.2.5.4 Multiple Treatments with Different Outcomes in the Same Trial

In some trials, several doses of a drug, or two or three different drugs, are investigated simultaneously. This provides an opportunity to do comparative studies of different dosages or drugs, or other regimens, in uniform conditions. But this raises additional issues as well. The comparison is easy if the outcome of interest is the same. If the outcome is recovery for one treatment and arrest of the disease for the other, then the statistical analysis becomes complicated. Even when the interest is uniform in recovery, the ancillaries such as speed of relief, cost of the treatment, and adverse side effects would necessarily become part of the comparison. The investigator must anticipate these variations and decide beforehand on how these would be reconciled to come to a decision. This is a clinical exercise rather than a statistical exercise.

6.2.5.5 Size of the Trial

There is increasing awareness of the need to evaluate each intervention thoroughly as it moves from the laboratory to the clinic. With advances in science, it is becoming difficult to formulate a regimen that can produce a very substantial improvement over an existing regimen. Thus, even a small difference is now considered clinically important. As explained later, detection of a small difference requires a bigger trial. More and more trials these days are larger in size. For example, the Collaborative Low-dose Aspirin Study in Pregnancy (CLASP) was a trial that involved more than 9000 women recruited in 222 centers in 17 countries [23]. This was a prophylactic trial. Prophylactic and screening trials can be very large. There is an example later in this chapter of an oral cancer screening trial involving more than 150,000 participants. In therapeutics, a multicentric trial on nicardipine retard in dementia involved more than 6000 patients [24].

Small trials have not lost relevance. They are important at least in making a beginning. They can also help in reaching a definitive conclusion where the performance of the new regimen is substantially better than that of the existing regimen. A medium-sized trial would generally include 300 cases and an equal number of controls. A trial on as low as 30 subjects in each group can be conducted for training purposes, such as for Master's dissertation. The actual number would depend on a variety of considerations such as the magnitude of the difference to be detected, the expected variability in the response by different subjects, confidence required in the result, etc. Details are given in Chapter 12.

Also beware of what is now popularly known as Lasagna Law. This says that the eligible subjects rapidly disappear from the clinic as you begin the trial and the number actually available may be much less than anticipated. They appear again after the trial is over.

6.2.5.6 Compliance

An intervention study requires active participation and cooperation of the study subjects. After agreeing to participate, some subjects may deviate from the protocol for a variety of reasons. These include developing side effects, forgetting to take their medication, or simply withdrawing their consent after randomization. Analogously, in a trial of a surgical therapy, those who were randomized to one group may choose to obtain alternative treatment on their own initiative. In addition, there may be instances in which participants cannot comply, such as when the condition of a randomized patient rapidly worsens to a point where continuation in the trial becomes contraindicated. Consequently, the problem of achieving and maintaining high compliance is an issue that needs consideration in the design and conduct of all clinical trials. A more acute statistical problem is differential compliance in the comparison groups. This difference may be related to prognostic factors such as occurrence of nausea in the treatment group but not in the control group. If so, the comparability may be lost despite full randomization and random selection of subjects from the target population.

Some participants in a trial may become noncompliant despite all reasonable efforts. Try to obtain complete information on these subjects. Adjust the results in accordance with the outcome observed in those who are similar to the ones who dropped out.

The other problem is lack of compliance with the protocol by the observers or the investigators. This problem can be severe if many observers are involved. Multiple observers are natural in a multicentric trial, and therefore such trials are especially vulnerable to such bias. More intensive efforts in terms of training and periodic assessment of the methods and the procedures followed by different observers may be helpful in minimizing this bias.

The analysis of data from clinical trials is done similar to the analysis for laboratory experiments. This is presented in Chapters 13 and 14 for proportions and in Chapter 15 for means.

6.2.6 Reporting Results of a Clinical Trial

After spending so much resource on planning and conducting a clinical trial, it is important to properly report it to the fraternity. Two popular guidelines are available. One is called CONsolidated Standards Of Reporting of Trials (CONSORT) that concentrates on content and format of a conventional report and the other emphasizes open access. They are not mutually exclusive but complement each other.

6.2.6.1 CONSORT

A good trial such as an RCT must also be reported in a format that could be appreciated by the readers. In view of the significance of such reporting, CONSORT have been formulated, which are revised periodically. The basic

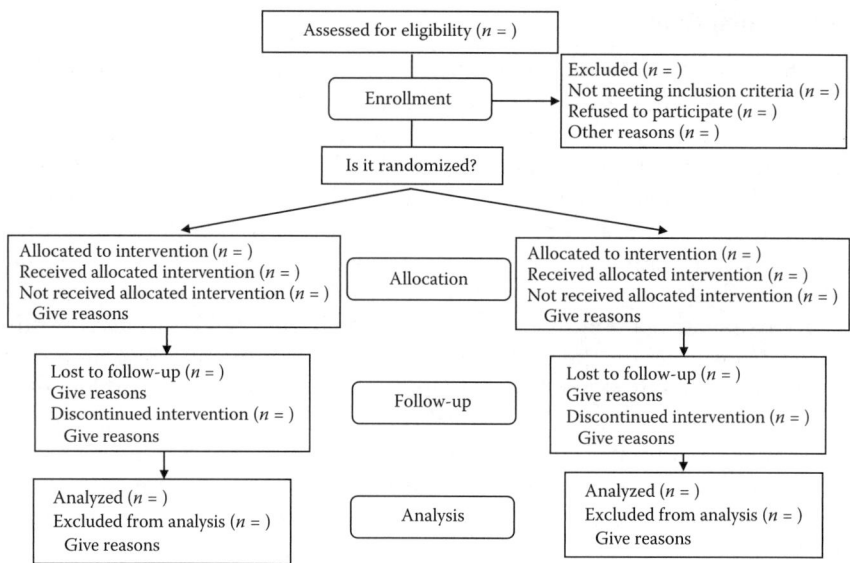

FIGURE 6.2
The CONSORT 2010—flowchart.

features of these standards are the same as already stated, namely, the report of the trial should indicate why the study was undertaken, should include scientific background and explanation of rationale, structured review of all pertinent literature not leaving out the opposite view, selection of subjects and sample size, allocation of subjects and blinding, baseline data, transparency regarding analytic methods including for missing data, noncompliance, etc. [25]. The revised CONSORT statement is designed to help minimize confusion and promote clarity in reporting the methods and results. This comprises a 25-item checklist and a flow diagram (Figure 6.2) to help ensure clear reporting of key elements of clinical trials. As of the year 2011, the flow diagram does not include information on blinding, although it contains information on randomization. Moreover, there is no exclusive mention of the number of cases actually followed up, although this can be deduced. CONSORT guidelines do not require much information on statistical errors. The statement outlines the minimum requirement and you can always add more.

6.2.6.2 Registration of Trials and Open Access

The conventional medium for publication of clinical trial results is a peer-reviewed journal. Although there are many merits in such publications, they suffer from restricted access, fixed format, and limited space. Thus, many trials, particularly those with negative findings, remain obscure. Also, trials that are partially reported because of limited space can accentuate bias.

It is now widely realized that the researchers have an ethical obligation to make the full results of human research public. Several innovations are quickly occurring to meet this need. The World Health Organization is leading an international effort to promote registration of clinical trials at the time of initiation. International Committee of Medical Journal Editors (ICMJE) has issued its own registration requirements [26]. These requirements include submission of the complete protocol that would make it obligatory not just to publish but also not to miss out any inconvenient findings. This may also enhance the quality of trial, because protocol has to follow a standard format, and reduce the deviation from the protocol, which now would have to be explained. Selective reporting and suppression of unfavorable findings can be easily detected. This also requires public sharing of the data (except confidential and proprietary information) soon after publication of the results. The data can be posted on the individual websites for public viewing.

ICMJE is for publication of trial results in regular journals, but Wager [27] argues that peer review has a poor record in detecting incorrect or fabricated data, the interpretation could be subjective, and claims can be unfounded. The author makes a passionate plea for open access as advocated by Public Library of Science (PLoS). If the results are freely accessible on public funded websites, the cost and delays can be reduced. Journals can publish reviews and critiques of those results and can also provide interpretation for different audiences such as researchers, clinicians, and patients. The journal may also relax its norms and not preclude articles based on web-posted results.

BioMed Central publishes a large number of open-access journals. Among these is the *Journal of Negative Results in BioMedicine*. This is in realization of the need of professionals to know about negative trials as much as about positive trials.

6.3 Trials Other than for Therapeutics

Trials other than for therapeutics include diagnostic, prophylactic, field, and vaccine trials. These terms are not mutually exclusive. The general principles for these trials are the same as for therapeutic trial but there are subtle differences in details.

6.3.1 Clinical Trials for Diagnostic and Prophylactic Modalities

Diagnostic trials are almost invariably conducted in clinics, whereas prophylactic trials can be conducted in clinic as well as in the field in a community. Field-based trials are discussed in the next section.

TABLE 6.1

Results of a Diagnostic Trial

Results of the Diagnostic Procedure under Trial	Actual Disease		Total
	Present	Absent	
Positive	a	b	$a + b$
Negative	c	d	$c + d$
Total	$a + c$	$b + d$	n

6.3.1.1 Diagnostic Trials

The intervention in a diagnostic trial is not a therapeutic agent but a procedure that can change the diagnosis and thus the course of the treatment. Thus, they have the potential to improve decision making and patient management.

From the ethics point of view, noninvasive procedures such as measuring BP and weight do not cause much anxiety except for time, cost, and inconvenience to the patient, but an invasive procedure such as endoscopy has the potential to cause harm to the health of the patient. More care is required for trial on an invasive procedure.

Most RCTs have a parallel control group, but many diagnostic trials are self-controlled. For comparison of prostate-specific antigen levels and ultrasound images for prostate cancer, both procedures would be done on the same set of patients. Agreement between the two can be evaluated, but to find which is better, a gold standard is needed. MRI can be compared with arthroscopy as the reference for detection of meniscal ruptures. When a reference is available, the diagnostic efficacy in terms of sensitivity, specificity, and predictability can be obtained. For a diagnostic trial on electrodiagnostic consultations for stenosis, Haig et al. [28] considered consensus among clinicians as the gold standard for clinical syndrome of lumbar spinal stenosis (Example 6.8).

When a group of suspected cases is available meeting specific inclusion and exclusion criteria, the diagnosis would be established by a gold standard in some of them. The diagnostic procedure under trial would also be used on all the suspected cases and some would be found positive. This would yield a 2 × 2 table of the type shown in Table 6.1. In this table, $(a + c)$ cases were found to have the disease by gold standard and $(a + b)$ by the procedure under trial. Various validity parameters can be estimated using the computational methods described in Chapter 9.

Example 6.8: Electrodiagnostic Testing for Lumbar Spinal Stenosis

Haig et al. [28] report a prospective, masked, controlled diagnostic trial to determine the sensitivity and specificity of electrodiagnostic consultation for the clinical syndrome of lumbar spinal stenosis. A total of 150 suspected cases were included. Out of these, clinical consensus on diagnosis was reached for 55 cases. This was considered gold standard. Electrodiagnostic measurements such as paraspinal EMG score and composite limb and paraspinal filtration score were evaluated for their sensitivity and specificity.

6.3.1.2 Prophylactic Trials in Clinics

Prophylaxis is a procedure that promotes health or controls primordial factors that adversely affect health. A prophylactic trial is generally conducted in the field where a community is involved. But it can be conducted in a clinical setup also. An example is prophylactic nasal continuous positive airway pressure after major vascular surgery. Amnioinfusion for meconium-stained amniotic fluid in labor at the time of childbirth is a prophylactic procedure that can be tried for specific types of births. As all other trials, prophylactic trials are also restricted to a particular segment of the subjects. For example, for a prophylactic drug such as aspirin to reduce cardiovascular events, the participants may be vulnerable people of age 50 years or more.

Principles of clinic-based prophylactic trials are the same as for therapeutic trials. Perhaps, a prophylactic procedure is somewhat insulated against harmful effects, thus slight relaxation in ethics may allow stricter control over confounding factors. Community-based trials in the field have a different setup, whether for therapeutic modality, for prophylactic agent, or for a screening procedure. These are discussed next.

6.3.2 Field Trials for Screening, Prophylaxis, and Vaccines

Field conditions, where a segment of population is involved, are different from clinical conditions. Many confounders such as severity condition of the patient can be controlled in the clinic but would be difficult to control in the field. In any case, field trials are done for modalities that can be used on mass scale on the general population or its segment. Thus, they can have wider health policy implications that clinical trials seldom have.

In public health, field trials are sometimes done with the health facility as the unit of experiment. The intervention could be training to the peripheral workers such as for Pap smear to find if that improves the case detection rate against control area where no such training was given. Issues in such trials may be slightly different from what were discussed in this section for individual-based trials.

6.3.2.1 Screening Trials

Screening is quite in vogue for cancers. The Prostate, Lung, Colorectal, and Ovarian Cancer Screening Trial initiated in 1992 in the United States has enrolled more than 150,000 participants [29]. Nearly half are randomly assigned to the intervention (screening) and the other half remain as control. Whether screening helps in reducing cancer mortality is yet to be seen.

Many collateral benefits emerged though. For example, it was found that chest radiograph abnormalities not suspicious for lung cancer are common, and prostate volume and age are independently associated with increased prostate-specific antigen in men undergoing screening. Second example is a randomized mass screening trial for abdominal aortic aneurysm. Ten-year results show 73% reduction in mortality by this aneurysm in those screened as compared to those not screened [30].

Note how screening trials look like based on mass screening but the procedures used are hospital based. Sankaranarayanan et al. [31] report a community-based trial wherein all persons of age 35 years or older were screened for oral cancer in intervention villages and not screened in the control villages. The screening was done three times at 3-year intervals for signs of oral cancer. There were nearly 60,000 eligible subjects in the intervention group and nearly 55,000 in the control group. The villages were cluster randomized rather than individual randomized to receive or not receive the intervention. The difficulty was, as in most field trials, low compliance when referred for confirmatory examination that has to be done in a hospital. In this trial, compliance was lower than 70%.

6.3.2.2 Prophylactic Trials in the Field

A prophylactic trial in the field could be for a strategy such as lifestyle changes for coronary disease or could be for vitamin intake—even drugs that are stipulated to prevent occurrence or recurrence of adverse events. Giving vitamin A supplements to infants and young children to improve their retinol level is an example of such intervention. Although there is a fine distinction between preventive and prophylactic measures, I am including both into the prophylactic category. Such trials have tremendous value in policy formulation, in saving lives, and in improving health, but they do not receive that kind of attention.

You can see that a prophylactic trial is not necessarily conducted in the general population. Vitamin A trial cited in the preceding paragraph is for children. A trial on educational campaign for responsible sexual behavior may target adolescents. Another trial on hematinic supplementation may target antenatal women of low socioeconomic stratum.

> **Example 6.9: Community Trial on Insecticide-Treated Clothes for Protection against Malaria**
>
> A total of 198 refugees were selected in Kenya by multistage cluster sampling and were divided almost equally to receive the intervention and placebo by cluster randomization [32]. The intervention was insecticide-treated personal clothes and linen, and the placebo in this case was plain water treatment. Double blinding was done for malaria parasite smear. The odds of malaria infection in the intervention group were reduced by about 70%.

6.3.2.3 Vaccine Trials

Vaccine trials are conducted in phases as therapeutic trials but need even more precaution. The need for extra care arises from the applicability of vaccines to a large segment of populations who are not sick but are at risk, as opposed to therapeutics that is applied only to patients and administered under close supervision. A feature of vaccines is immunogenicity, which might be an important consideration in some diseases, in addition to protective efficacy. In others, duration of protection may be important. Quality and quantity of immune responses required for protection against infection and against development of disease are scientific challenges. In the case of HIV, for example, there would be a vaccine that inhibits HIV infection, and there could be a vaccine that inhibits or retards development of disease—AIDS—in those already infected.

In view of the complexities involved in vaccine trials, an additional phase called **phase IIB** is sometimes advocated. This is also called "test of concept" phase. Aim of phase IIA could be to establish the schedule of administration for different age-groups as it would be most likely a factorial experiment with dose level as one factor and age-group as the second factor. Thus, four phases are required instead of the usual three. The objective of phase IIB is to evaluate whether the vaccine has any (>0%) efficacy at all. In a phase III trial for vaccines, this objective shifts generally to at least 30% efficacy. The participants in phase IIB are not necessarily representative of the target population. For phase III, a representative sample is strongly indicated. Phase IIB also assesses the operational efficiency, whereas the objective of phase III is to produce compelling evidence of efficacy for licensure from regulatory agencies.

Phase III trial for a vaccine has to be a large-scale trial so that adequate numbers developing the disease, particularly in the control group, are available. The total number of subjects may run into thousands and the follow-up too may go up to several years. Since phase III is an expensive trial for vaccines, phase IIB becomes a highly desirable proposition to indicate whether or not to proceed to phase III. Phase IIB, however, increases the time frame because this too can take at least a couple of years.

6.3.3 Issues in Field Trials

Since field trials are generally done on apparently healthy people for screening, prophylaxis, or prevention, they raise societal questions beyond individual choices. The trial must be consistent with the local social ethics and cultural norms. For example, a trial on fish-based diet may not be accepted by the vegetarian segment, and they might even object to propagation of such a diet. Pacific islanders may not like to be screened for obesity because many of them feel very comfortable with it. In a population where sex is taboo, a trial on education for responsible sexual behavior among teenagers in the context of HIV may be frowned upon.

Field conditions are affected by a large number of factors and they can be seldom controlled. Thus, the results point to effectiveness rather than efficacy.

At the same time, field trials offer some advantages as well. Results of a trial in field conditions can be easily adopted for action because it is already constrained by the actual field conditions.

6.3.3.1 Randomization and Blinding in Field Trials

Field trials are generally large scale and individual randomization becomes difficult. Thus, groups of participants such as schools or clinics are randomized instead of the subjects. This is called cluster randomization as already explained. In an oral cancer screening trial referred to earlier, villages were randomized to receive or not receive the intervention. However, randomization remains a strong ingredient of field trials as much as for clinical trials.

The same cannot be stated for blinding. Field trials are very difficult to be blinded except in some typical situations. Lwegaba [33] reports a single-blinded field trial on educational material for tobacco prevention. The trial was conducted on school students and the schools were distantly located so that blinding could hold on.

6.3.3.2 Designs for Field Trials

Experimental design for field trial can be one-way, two-way, factorial or partial factorial, or repeated measures but is rarely a crossover. Trial on different salt iodine concentrations for assessing urinary iodine excretion is a one-way design. The unit of experiment in this case would be a family and they can be randomized to receive different concentrations such as 5, 20, and 30 mg/kg. With one control, this factor has four levels in this trial.

Iron and folic acid are known to help raise hemoglobin level in anemic women. But the utility of their supplementation through salt in a population is not fully known. Since anemia is mostly a condition in the underprivileged section of the population, the trial can be conducted in this section. It may be interesting in this trial to see the effect of salt fortified by iron alone, salt fortified by folic acid alone, and salt fortified by both. With control, this is a 2×2 factorial experiment. If folic acid alone is not considered worth a try, and three groups are included, the experiment becomes partially factorial.

References

1. Gianotti L, Braga M, Nespoli L, Radaelli G, Beneduce A, Di Carlo V. A randomized controlled trial of preoperative oral supplementation with a specialized diet in patients with gastrointestinal cancer. *Gastroenterology* 2002; 122:1763–1770.
2. Strom BL (Ed.). *Pharmacoepidemiology*, 4th edn. New York: John Wiley & Sons, Inc., 2005.
3. Zubieta JK, Bueller JA, Jackson LR et al. Placebo effects mediated by endogenous opioid activity on mu-opioid receptors. *J Neurosci* 2005; 25:7754–7762.

4. Barton MB, Morley DS, Moore S et al. Decreasing women's anxieties after abnormal mammograms: A controlled trial. *J Natl Cancer Inst* 2004; 96:529–538.
5. Piantadosi S. *Clinical Trials: A Methodologic Perspective*, 2nd edn. New York: Wiley Interscience, 2005.
6. Cooper HM. *Research Synthesis and Meta Analysis: A Step-by-Step Approach*, 4th edn. Thousand Oaks, CA: Sage Publications, Inc., 2009.
7. Korting HC, Schafer-Korting M (Eds.). *The Benefit/Risk Ratio: A Handbook for the Rational Use of Potentially Hazardous Drugs*. Boca Raton, FL: CRC Press, 1998.
8. Roland M, Torgerson DJ. Understanding controlled trials: What are pragmatic trials? *BMJ* 1998; 316:285.
9. Knebel P, Frohlich B, Knaebel HP et al. Comparison of Venae Sectio vs. modified Seldinger Technique for totally implantable access ports: Portas-trial [ISRCTN: 52368201]. *Trials* 2006; 7:20.
10. Koninger J, Seiler CM, Wente MN et al. Duodenum preserving pancreatectomy in chronic pancreatitis: Design of a randomized controlled trial comparing two surgical techniques [ISRCTN: 50638764]. *Trials* 2006; 7:12.
11. Snapinn SM. Noninferiority trials. *Curr Control Trials Cardiovasc Med* 2000; 1:19–21.
12. Szegedi A, Kohnen R, Dienel A, Kieser M. Acute treatment of moderate to severe depression with hypericum extract WS 5570 (St John's wort): Randomised controlled double blind non-inferiority trial versus paroxetine. *BMJ* 2005; 330:503.
13. Baker DE. Therapeutic equivalence of mesalamine products. *Rev Gastroenterol Disord* 2004; 4:25–28.
14. Popovic J, Mitic R, Sabo A, Mikov M, Jakovljevic V, Dakovic-Svajcer K. Spline functions in convolutional modeling of verapamil bioavailability and bioequivalence II: Study in healthy volunteers. *Eur J Drug Metab Pharmacokinet* 2006; 31:87–96.
15. Patterson S, James B. *Bioequivalence and Statistics in Clinical Pharmacology*. Boca Raton, FL: CRC Press, 2006.
16. Senn S. *Crossover Trials in Clinical Research*, 2nd edn. Chichester, U.K.: John Wiley & Sons, 2002.
17. Byers JF, Lowman LB, Francis J et al. A quasi-experimental trial on individualized, developmentally supportive family-centered care. *J Obstet Gynecol Neonatal Nurs* 2006; 35:105–115.
18. Woodfield R, Goodyear-Smith F, Arroll B. N-of-1 trials of quinine efficacy in skeletal muscle cramps of the leg. *Br J Gen Pract* 2005; 55:181–185.
19. Whitehead J. *The Design and Analysis of Sequential Clinical Trials*, 2nd edn. New York: John Wiley & Sons, 1997.
20. Chow S-C, Chang M. *Adaptive Design Methods in Clinical Trials*. New York: Chapman & Hall, 2006.
21. Lilford RJ. Equipoise is not synonymous with uncertainty (Letter). *Br Med J* 2001; 323:574.
22. Cheng AC, Lowe M, Stephens DP, Currie BJ. Ethical problems of evaluating a new treatment for melioidosis. *Br Med J* 2003; 327:1280–1282.
23. Anonymous. CLASP: A randomized trial of low-dose aspirin for the prevention and treatment of preeclampsia among 9364 pregnant women. CLASP (Collaborative Low-dose Aspirin in Pregnancy) Collaborative Group. *Lancet* 1994; 343:619–629.

24. Gonzalez-Gonzalez JA, Lozano R. A study of tolerability and effectiveness of nicardipine retard in cognitive deterioration of vascular origin (Spanish). *Rev Neurol* 2000; 30:719–728.

25. Schulz KF, Altman DG, Moher D. CONSORT 2010 statement: Updated guidelines for reporting parallel group randomized trial. *J Pharmacol Pharmacother* 2010; 1:100–107.

26. De Angelis C, Drazen JM, Frizelle FA et al. Clinical trial registration: A statement from the International Committee of Medical Journal Editors. *N Engl J Med* 2004; 351:1250–1251.

27. Wager E. Publishing clinical trial results: The future beckons. *PLoS Clin Trials* 2006; 1:e31.

28. Haig AJ, Tong HC, Yamakawa KS et al. The sensitivity and specificity of electrodiagnostic testing for the clinical syndrome of lumbar spinal stenosis. *Spine* 2005; 30:2667–2676.

29. Oken MM, Marcus PM, Hu P et al. Baseline chest radiograph for lung cancer detection in the randomized Prostate, Lung, Colorectal and Ovarian Cancer Screening Trial. *J Natl Cancer Inst* 2005; 97:1832–1839.

30. Lindholt JS, Juul S, Fasting H, Henneberg EW. Preliminary ten-year results from a randomised single centre mass screening trial for abdominal aortic aneurysm. *Eur J Vasc Endovasc Surg* 2006; 32:608–614.

31. Sankaranarayanan R, Mathew B, Jacob BJ et al. Early findings from a community based, cluster-randomized, controlled oral cancer screening trial in Kerala, India: The Trivandrum Oral Cancer Screening Study Group. *Cancer* 2000; 88:664–673.

32. Kimani EW, Vulule JM, Kuria IW, Mugisha F. Use of insecticide-treated clothes for personal protection against malaria: A community trial. *Malar J* 2006; 5:63.

33. Lwegaba A. Field trial to test and evaluate primary tobacco prevention methods in clusters of elementary schools in Barbados. *West Indian Med J* 2005; 54:283–291.

7

Numerical Methods for Representing Variation

The preeminent role of empiricism in medicine is to make observations. Observations on a group of subjects give rise to data. When data on, say, 200 patients with duodenal ulcer are available, how does one make sense out of them? One possibility is a graphical representation of the data. This is discussed in the next chapter. More commonly, a numerical summarization is done. Two initial steps for this are as follows: (1) Make a summary of the data in such a manner that none of their important features is lost. (2) Calculate a few summary values that can adequately represent location and scatteredness in the data. Does the representation in the literature really give you the right picture of their data? The methods discussed in this chapter may provide you the answer.

This chapter: The methods for summarizing data depend on the types of measurement— quantitative, qualitative, categorical, etc. These types are described in Section 7.1. One way to numerically summarize data is to prepare data tables. The methods for this are presented in Section 7.2. A simple method to express the gist of a set of data is by stating the proportions and rates of occurrence of different events. This is discussed in Section 7.3. Other summary statistics such as mean and standard deviation (SD) are discussed in Sections 7.4 and 7.5, respectively.

7.1 Types of Measurement

Blood glucose level and urea clearance are measured in terms of quantities, whereas the blood group of a person is a quality recorded as O, A, B, or AB. Age can be measured in days and hours but is often categorized in years as (0–4), (5–14), (15–49), etc. Disease severity and extent of malnutrition are quantities but are generally measured as none, mild, moderate, and serious. Site of malignancy is also a measurement in a statistical sense but is recorded as oral, lung, abdomen, breast, etc. Thus, there are a large variety of measurements. Statistical methods to study variation depend on the type of measurements. These types can be grouped in a variety of ways. A majority of them are described here.

7.1.1 Nominal, Metric, and Ordinal Scales

A scale is an instrument on which the characteristics are measured. It can be quantitatively calibrated in the usual sense or can be qualitative. The following types can be identified.

7.1.1.1 Nominal Scale

Not all measurements are necessarily in terms of quantities. Complaints and site of cancer are qualities but are still considered *measurement* in a statistical sense. The categories in such cases are only names and the scale of such a measurement is called nominal. All space-related measurements such as organ affected, site of lesion, and place of occurrence of disease are nominal and so are attributes such as race and blood group. These names do not have any specific order. Thus, there is no notion of less than or more than in this kind of scale, and the only valid comparison is of equality or inequality. Gender is either male or female and none is higher or more than the other. Diagnosis of liver disease as hepatitis, cirrhosis, or malignancy is nominal and so is the nature of a handicap such as visual, speech, orthopedic, mental, etc. These variables are genuinely categorical, and the only way to associate numbers with them is by way of assigning a **code** to each category. Note, however, that codes for nominal categories are not metric scores—a category receiving code 4 is not twice the category receiving code 2. These codes cannot be treated as quantities and cannot be added, subtracted, multiplied, or divided. Codes are not the same as scores.

When the assessment of a characteristic is in terms of only two categories such as yes/no, present/absent, or favorable/unfavorable, these are called **dichotomous categories.** The corresponding variable is called a **binary**. Recording gender as male or female is the most glaring example. If the number of categories is more than two, such as cirrhosis, hepatitis, and malignancy for liver disorders, these are called **polytomous categories.** Statistical methods are simpler for dichotomous than for polytomous categories. If a statistician advises you to collapse three or more categories into two for the sake of convenience during calculations, you should agree only if this does not reduce relevance and the operational value of the conclusion is not compromised. Clinical relevance should not be sacrificed for statistical expediency. Computers have largely obviated the need to compromise on this aspect.

7.1.1.2 Metric Scale

At the other end of the spectrum are characteristics that can be exactly measured in terms of a quantity. Duration of a disease, hemoglobin (Hb) level, heart rate, and parity are examples of such characteristics. These are said to be measured on a metric scale. Often these are recorded in categories such as years of age in (0–4), (5–14), (15–49), (50–69), and (70+) groups.

Such categorization tends to ordinalize the metric scale and results in loss of information. Yet, it is preferred in some situations, as discussed later in this section.

There is a tendency in health and medicine to develop a **scoring system** for the soft data such as the degree of severity of disease. This introduces a metric scale for soft data and definitely helps to achieve better exactitude. Such scores are statistically satisfying but sometimes lose clinical relevance. If you are using a scoring system, ensure that it adequately represents varying grades of the observations. Assigning a number should not be at the cost of unjustified air of accuracy.

Sometimes the metric scale is divided further into interval and ratio scales. In an **interval scale**, there is no absolute zero, for example, body temperature. A temperature of 105°F cannot be interpreted as only 5% higher than 100°F. Similarly, it is incorrect to say that a person with an intelligence quotient (IQ) of 160 is twice as intelligent as a person with an IQ of 80. Differences matter but ratios are irrelevant. This is so for many measurements in medicine that do not start from zero. Plasma glucose level, blood pressure (BP), heart rate, all are examples of this kind of measurement. On the other hand, in a **ratio scale**, a zero point can be meaningfully designated. It is correct to say that the duration of survival of 6 years is twice as much as 3 years, and parity 3 is thrice as much as parity 1. However, in this case also, their medical interpretation may not be based entirely on a proportional factor. The fine distinction between interval and ratio scales in most cases is not required for managing uncertainties. Therefore, I refrain from giving much attention to this aspect in this book.

The disadvantages of measurements on the metric scale are that, in many cases, such measurements are relatively more difficult, more time consuming, and more expensive. A large number of parameters may have to be considered together to say that the extent of burns is 78%, while it is easy to say on visual inspection alone that the burns are extensive. However, such shortcuts invariably lack objectivity. Whenever feasible, prefer the metric scale to the ordinal scale. It is always wise to ensure that the metric measurements you are using are indeed valid and reliable.

7.1.1.3 Ordinal Scale

There are certain characteristics that should be measured in terms of quantity, but the nonavailability of a good instrument compels measurement in terms of what is called an ordinal scale. Disease severity when measured as none, mild, moderate, serious, or critical; likelihood of the presence of a particular disease when measured as ruled out, unlikely, doubtful, likely, and confirmed; and self-perception of health from very bad to very good on, say, a 7-point scale are examples of such characteristics. These are inherently metric but an ordinal scale is more convenient in such measurements. The main reason for nonavailability of a metric scale is that many of these characteristics are multifactorial. For example, disease severity depends on signs and symptoms and measurements such as BP and plasma glucose levels, radiological assessment, etc.

Sometimes a device to measure a characteristic is easily available but is not adopted because such a level of accuracy is not needed. Smoking can be measured as the number of cigarettes smoked, but categories such as none, light, moderate, and heavy seem to serve the purpose sufficiently well in many clinical situations. Age can be measured in terms of years, but categorization into child, adult, and old may be adequate in some situations.

The basic advantage of using an ordinal scale rather than a metric scale is convenience in eliciting, recording, and reporting. The use of any sophisticated device is avoided. Ordinal categories are often easier to comprehend than metric categories. In the process, however, valuable and accurate information is lost and the analysis of data is rendered less efficient. Metric measurements are amenable to a host of mathematical manipulations that are not possible with ordinal measurements. Thus, prefer hard measurements such as BP level instead of grade of hypertension and prostate volume instead of grade of enlargement. If the efforts required for hard measurements are enormous, such as in measuring the size of the brain, or when no metric scale is available, use an ordinal scale. However, beware of anomalies in some ordinal categories. What is mild for me may be moderate for you. Very rarely are these terms strictly defined.

Between nominal and ordinal, there might be measurements that are **semiordered**. Classification of malignancy as definitely absent, probably absent, uncertain, probably present, and definitely present is an example of a semiordered scale. These categories are partly nominal and partly ordinal.

Quite often, numerals are associated with ordinal categories such as 0 for none, 1 for mild, 2 for moderate, and 3 for serious. These numbers are then subjected to all sorts of algebraic calculations. Such calculations are valid only when the moderate degree is considered two times the mild degree and the serious degree is considered three times the mild. These numerals also assume that the difference between mild and no disease is the same as that between serious and moderate disease. In practice, this may not be so. Thus, caution is required in assigning numerals to ordinal categories and in drawing conclusions when based on calculations involving such numbers. Note that these are scores and not codes.

7.1.1.4 Grouping of a Metric Scale (Categorizing Continuous Measurements)

Exact measurements on the metric scale are indeed statistically preferable to ordinal measurements. The irony is that sometimes circumstances force grouping of metric data into categories even after exact data are obtained. The weight of a woman may be recorded to the nearest kilogram but may have to be categorized into 5 kg intervals such as (40–44), (45–49), etc. Data reported in this manner are called **grouped data**, and the process is commonly referred to as categorizing continuous variables and the groups are called **class intervals**. The reasons for doing this may be one or more of the following:

1. Consider a dataset containing systolic BP levels of 1200 persons. The only effective way to present these in a report is by using groups such as (100–109), (110–119), (120–129), etc., and stating the number of subjects in each such group. This saves space and at the same time makes the data more intelligible. Storage of grouped data may take only about half a page or 1 kB of space, whereas storage of 1200 individual values may take four pages or 8 kB of space. Such grouping also makes the data more sensible while 1200 ungrouped values may be difficult to comprehend.

   Groups such as (0–4) and (5–9) for age assume that age is noted in terms of *completed* years or age last birthday. The interval (5–9) actually means 5 to less than 10 years and can also be written as (5–10) years. It is customary in such statistical grouping that the upper end of the interval is considered to belong to the next interval. Mathematically, these are written as [0–5), [5–10), etc., but this kind of exact notation is seldom used in practice. Wherever the intervals are continuous in this text, the convention of (0–4), (5–9), etc., is followed.

2. It is well known that the end digit is predominantly 0 or 5 in many data values. This happens either because of approximation done by subjects themselves at the time of inquiry, such as stating one's age as 45 years instead of the more exact 44, or because of the observer's bias such as in recording a systolic 130 mmHg instead of the exact 132. Intervals (105–114), (115–124), (125–134), etc., or (108–112), (113–117), (118–122), etc., would dilute the effect of such digit preference. In another setting, suppose waist and hip sizes are measured without sufficient care and could be in error of up to 5 mm. Grouping of waist–hip ratio in intervals (0.7–0.8), (0.8–0.9), (0.9–1.0), etc., would minimize the effect of such errors and the purpose of assessing central obesity could still be adequately achieved despite errors, provided they are minor.

   The preceding two reasons are valid for grouping at the stage of reporting or analysis, but sometimes even the recording is done in a grouped form. This is done for the following reasons:

3. Eliciting a woman's age and anybody's income is sometimes considered improper. Some people prefer to keep such information confidential. Stating them in a grouped form may be more acceptable. The exact value remains confidential, yet data available are in a usable form.

4. Many clinicians are accustomed to think in terms of anemia present or absent and its degree as mild, moderate, or severe in place of exact Hb or hematocrit values. Thus, they sometimes prefer grouped values. Two or more measurements can also be simultaneously

considered in this kind of grouping. Categorization of growth of a child into excessive, normal, retarded, and dismal depends not only on height and weight but also on the age of reaching different milestones of development. Such multifactorial grouping is sometimes more relevant for the practice of medicine.

5. In an experiment on lethal dose of a drug in mice, it is much easier to observe each morning and record the number of dead mice than to keep a continuous watch and note the exact time of death. In this case, the survival time would be available in 24 h categories. Serum glucose level is measured in units of 5 mg/dL because the analyzer in some cases is so calibrated. Thus, 5 mg/dL categories are inadvertently formed. Greater accuracy may be redundant in this case. If better accuracy is needed, cost and efforts may substantially increase.

Whenever data are available in an exact form, statistical analysis should be done using the exact data. Grouping in this case renders analysis less efficient in the sense that some important features of the data may fail to emerge. Statistical inference becomes less efficient as the methods assume that the values in grouped data are flat within each interval. This is against the factual position since discontinuity is imposed by categorization across interval boundaries. Interval 160–169 of systolic BP forgets that 168 mmHg is more than 162 mmHg and the difference is more than between 168 and 170, which now belong to separate categories. Cut-points of intervals are mostly arbitrary and different cut-points can give different results. In some cases, though, this loss can be compensated by increasing the sample size. A larger sample size helps to capture a better spectrum of values even when data are grouped. In fact, in cases in which grouped data can be rapidly obtained at a substantial saving, more reliable results can be obtained by investigating a larger sample within the same cost. In such a situation, grouped data on metric measurements can be rightly advocated. However, it is important that the number of groups and width of intervals are appropriately chosen so that the essential features of data are not compromised and the relevance is not lost. One example of misuse of grouping can be found in an old Indian Council of Medical Research [1] report that gives the mean weight of children of age 0–3 months as 4.5 kg. Since the weight at 3 months is nearly double the weight at birth, the mean in this interval has little practical utility.

7.1.2 Other Classifications of the Types of Measurement

Generally speaking, a characteristic that tends to vary from subject to subject or from unit to unit is called a **variable**. All biological characteristics—age, gender, birth order, body temperature, body mass index (BMI), duration of survival, disease severity, etc.—are variables. Characteristics on the metric scale already have a numeric outcome, such as birth order 3. Characteristics on the nominal and ordinal scale do not have this feature.

Even though gender is male or female, we can assign a code such as 0 for male and 1 for female. Similarly, disease severity in none, mild, moderate, and serious categories can be assigned scores 0, 1, 2, and 3, respectively, with limitations as stated earlier. When such a numeric assignment is done, the characteristic becomes variable in a statistical sense. Mathematical formulations then become easy, although caution is advised because codes cannot be added or subtracted and scores may not exactly measure the characteristic.

7.1.2.1 Discrete and Continuous Variables

Some variables can take only a small number of values. Gender has only two possible values and severity of disease has four values as seen in the previous paragraph. Some others can take any of a large number of values such as systolic BP ranging from 100 to 200 mmHg. BP can be 132.7 mmHg if an instrument giving such accuracy is available. On the other hand, parity of a woman can be 1 or 2 but never 1.6. A variable that can take only a finite, generally small, number of values in a range is called discrete. Number of deaths in a hospital in a day, blood group, and diagnosis are other examples of discrete variables.

It is common in medicine that discrete variables take only nonnegative *integer* values, but the definition does not require it to be so. Shoe size can be $7, 7\frac{1}{2}, 8, 8\frac{1}{2}$, etc., but is still discrete. It can take only these four values between 7 and $8\frac{1}{2}$. Compare this with a variable such as age. If needed, age can be measured accurately as 7.2613 years. Although recording of age in terms of completed years is often considered adequate, particularly for adults, theoretically age can take infinite number of values between 7 and $8\frac{1}{2}$. Thus, this is not considered a discrete variable.

A variable that can take an infinite number of values in a range is called continuous. Age, cholesterol level, body temperature, enzyme level, and weight are examples of continuous variables. In practice, it is redundant to measure them to several decimal places. For example, it does not help to say that the Hb level is 14.038 g/dL. Weight is generally measured to the nearest kilogram, BP to the nearest millimeter of mercury and Hb level to one-tenth of a gram per deciliter. It can be argued that such approximation makes the variable discrete but, in practice, the continuous character is not lost as long as the measurement is sufficiently accurate. Inversely, variables such as heart rate, platelet count, and respiration rate are, in fact, discrete yet are considered continuous because of the large number of possible values. *Only variables that can take a small number of values, say, less than 10, are generally considered discrete.* Others can be treated as continuous for practical purposes even when they are theoretically discrete.

Classification of variables into discrete or continuous is important because statistical methods for the two types of variables are different. The importance of such classification will become clear as you proceed further.

7.1.2.2 Qualitative and Quantitative Data

As stated earlier, a characteristic becomes a variable in a statistical sense when numerals are assigned to its categories or to its measurements. This does not mean that all the variables are quantitative. It is important to keep the distinction between values and codes and between a quality and a quantity. This text often uses the terms "qualitative data" and "quantitative data" to maintain the distinction. The former includes nominal, ordinal, and categorical data (when quantity is disregarded) and the latter includes uncategorized metric data.

Uncertainties are better managed when the measurement is exact. Whereas exactitude is inbuilt in quantitative measurements, the qualities also need to be fully specified for correct measurement. If the variable is a symptom such as cough, then its intensity, frequency, and duration may have to be considered. If the variable is a disease such as coronary artery disease (CAD), the electrocardiogram (ECG) findings, frequency and duration of angina, and homocysteine level may be relevant. Even for an apparently simple qualitative variable such as socioeconomic status, the measurement could be in terms of income, education, and occupation. The classification of people into black, white, and brown may not be always straightforward. Blindness needs to be qualified by a cutoff point for visual acuity (VA). Nonetheless, there are certain qualitative variables such as gender and mortality that can be directly measured without further specification in almost all cases.

Medical practice has a preponderance of qualities over quantities. Use of quantitative variables such as BP, body temperature, glucose level, and Hb level is, of course, common, but at some stage in patient management they tend to be interpreted as qualities such as high, borderline, normal, and low. Management of a subject with a total leukocyte count (TLC) of $2.1 \times 10^9/L$ is generally the same as that of a subject with a TLC of $2.2 \times 10^9/L$. Thus, the exact count may not be all that necessary in some cases. This is not to suggest that exact measurements are not important. They definitely are and are always preferable because it is only through such quantities that borderline cases and the trend in terms of improvement or deterioration can be identified. Yet, the fact is that medicine still is based mostly on qualitative assessment. The general practice is to use exact metric for research and use grades for day-to-day patient management.

7.1.2.3 Stochastic and Deterministic Variables

Variables can be categorized on the basis of yet another feature. Some measurements are considered known for subjects. The others are subsequently obtained. The former are factors and the latter responses, as explained in Chapter 4. Responses are considered stochastic because they are subject to chance fluctuation. They cannot be exactly predicted. Factors are considered deterministic because they are known beforehand and are not subject to chance fluctuation. If you are measuring fasting plasma glucose level of obese and nonobese people, obesity status is already known and is not subject to any fluctuation. It is deterministic. Plasma glucose level is stochastic in this setup. If you have people of known plasma glucose level and you measure their cholesterol level, the glucose level is deterministic and cholesterol level is stochastic. In a prospective study, antecedents are known and thus deterministic, whereas in a retrospective study the outcome is already known and is deterministic. In a prospective study, the outcome is stochastic and in a retrospective study, antecedents are stochastic. However, in a cross-sectional study, both are stochastic, as none is known beforehand.

Most of the inferential statistical methods discussed in this book apply to stochastic rather than to deterministic variables. Noninferential methods can be used for deterministic variables. *The role of the deterministic variables is to provide help in coming to a more valid conclusion concerning the stochastic variables.*

7.2 Tabular Presentation

Tables and graphs are just about the only way that data on a large number of subjects can be presented. Different kinds of tables are discussed in this section. Graphs require a more elaborate discussion and deserve a full chapter. The next chapter deals only with graphs.

Not all tables contain data. Tables 4.2 and 4.3 on the comparison of various types of study designs have no data. The discussion in the present section is restricted to data-based tables. These may contain the number of subjects or frequency in different groups or the frequency of different characteristics in a group of people (such as Tables 7.1 and 7.2). Some tables contain other types of information such as mean, survival rate, or dosages in different groups (Table 5.2). Tables may also contain random numbers or statistical calculations.

TABLE 7.1

Distribution of 1000 Subjects Coming to a Cataract Clinic by Age, Gender, and VA in the Worse Eye

Age Group (Years)	VA ≥ 6/60			6/60 < VA ≤ 1/60			VA < 1/60			Total		
	M	F	P	M	F	P	M	F	P	M	F	P
−49	11	8	19	37	32	69	12	10	22	60	50	110
50–59	18	21	39	69	73	142	13	16	29	100	110	210
60–69	25	21	46	183	142	325	42	47	89	250	210	460
70–79	10	11	21	54	44	98	26	25	51	90	80	170
80+	3	4	7	9	14	23	8	12	20	20	30	50
Total	67	65	132	352	305	657	101	110	211	520	480	1000

M, male; F, female; P, person.

TABLE 7.2

Distribution of Multiple Births (Twins, Triplets, and Quadruplets) by Gender

Number of	Number of Male Children					
Female Children	0	1	2	3	4	Total
0	×	×	17	5	0	22
1	×	25	6	2	×	33
2	12	3	0	×	×	15
3	4	0	×	×	×	4
4	1	×	×	×	×	1
Total	17	28	23	7	0	75

Thus, there is a large variety of tables. Of these, the frequency tables have special relevance in a statistical context and need detailed discussion. These are of two types: contingency tables and multiple response tables.

7.2.1 Contingency Tables and Frequency Distribution

Categories are called **mutually exclusive** when only one of them is applicable to one subject. While measuring BMI, categories such as −14, 15–24, 25–34, and 35+ kg/m² are mutually exclusive because a person's BMI can be in only one of these categories. These are **exhaustive** too because no BMI can be beyond these categories. When a group of subjects are classified into such mutually exclusive and exhaustive categories, you obtain a contingency table. For example, a table giving the number of children with severe, moderate, and mild forms of bronchiolitis, attended to in a clinic, is a contingency table. Note that categories are necessary for representation of frequencies in a contingency table form. For a variable like heart rate (per minute), the categories could be (60–64), (65–69), (70–74), etc. For family size, each number by itself could be a category.

When the number of variables is two or more, a cross-classification can be done. A contingency table is called a one-way, two-way, or K-way table depending upon the number of variables on which the subjects are cross-classified. Table 7.1 is a three-way table in which 1000 subjects coming to a cataract clinic are divided by age, gender, and the VA in the worse eye. Totals collapse one or more characteristics and yield two-way tables, and further collapsing yields one-way tables. The last column of Table 7.1 is a one-way classification of subjects by age and the bottom row totals provide the two-way classification by VA and gender. When totals and column P, which is the sum of M and F, are excluded, this table has 30 cells. The number of subjects in a cell is called the **cell frequency**. Since VA has three categories, age five categories, and gender two categories, the **order of the table** is 3 × 5 × 2, which makes a total of 30 possible categories. Cells obtained by totals are not counted.

In Table 7.1, gender is on nominal scale while age and VA are metric but categorized into groups. Irrespective of the scale, all contingency tables represent what is called a frequency distribution.

A table continues to be a contingency table when percentages are mentioned in place of the cell frequencies as long as the total n is stated somewhere. When n is known, the percentages can be readily converted to the respective frequencies.

You will need these basic concepts to understand medical literature, when trying to analyze data from your own clinic, and when trying to do research.

7.2.1.1 Empty Cells

It is not uncommon in medical data that some cells in a contingency table have no frequency or zero frequency. It is important for the purpose of analysis to distinguish between observed zeros and structural zeros. An **observed zero** is one where some frequency could have occurred but happens to be zero in the sample. This is not much of a problem and is treated just like any other small frequency. A **structural zero** occurs when it is just not possible to have any subject in the cell.

Consider a study on multiple births—twins, triplets, and quadruplets. The births are classified by gender. The distribution may be as displayed in Table 7.2. There are some observed zeros in this table. In addition, note the × sign in some cells where no frequency is possible. Such tables are called **incomplete tables**. Special methods are required to analyze such tables. For a summary of such methods, see Agresti [2]. This book excludes discussion on such incomplete tables, but observed zeros are admissible in the methods discussed in this book.

7.2.1.2 Problems in Preparing a Contingency Table on Metric Data

A problem frequently encountered in preparing a contingency table on all continuous and most metric variables is in deciding the number and width of intervals. When age in years is divided into (0–4), (5–14), (15–49), (50–69), and (70+) categories, the number of intervals is five and they are all unequal. Reporting of systolic BP (in millimeters of mercury) is mostly done in equal groups (120–129), (130–139), (140–149), etc., and of diastolic BP in (70–74), (75–79), (80–84), etc. The choice mostly depends on commonsense evaluation of the utility of such groups in conveying the basic features of the data. Generally, the number of such groups should be between four and eight. More groups are better for describing the variability, but this should also not take away the advantage of compactness of a contingency table. In the case of Table 7.1, VA can be divided into (6/6–6/9), (6/9–6/18), (6/18–6/60), (6/60–3/60), (3/60–1/60), and (1/60–PL+) and (PL–), where PL is for perception of light. That certainly would give a better view of the subjects but would also make the table more clumsy and less intelligible. Also, for cataract management, such detailed categorization may not be necessary.

All contingency tables give the distribution of subjects over various values of a measurement. Thus, these describe a **frequency distribution**. In this distribution, the values of a continuous variable would be grouped such as for age and VA in Table 7.1. For a discrete variable, the values may appear as such, for example, 0, 1, 2, 3, 4, and 5+ for parity. The last category in this case is also a group. Ordinal and nominal groups may appear as such without any numeric assigned. Note the following:

1. A one-way contingency table describes a univariate distribution, a two-way table a **bivariate distribution**, and a three-way table a trivariate distribution. Table 7.1 contains a trivariate distribution of cases coming to a cataract clinic by age, gender, and VA.

2. Calculating **cumulative frequencies** is sometimes useful. These are obtained by sequentially adding the frequencies in the successive groups. In Table 7.1 cumulative frequencies up to the age 60 and 70 years are 320 (32%) and 780 (78%), respectively.

3. A table may sometimes need explanatory footnotes. Be liberal in providing such explanations where needed. Clarity should not be sacrificed for brevity.

7.2.2 Multiple Response Tables and Other Features

All frequency tables are not necessarily contingency tables. Consider the data in Table 7.3 on the number of cases of abdominal tuberculosis with major symptoms [3]. This is a classification of the cases but the categories are neither exclusive nor exhaustive. They are not exclusive because one patient can have two or more complaints: a patient may have vomiting as well as constipation. This is called a multiple response. The categories are not exhaustive either because only major symptoms are listed. Patients with other symptoms may be present but are not included in this table. When the categories are exhaustive and multiple responses are present, the sum total of frequencies would necessarily exceed the total number of subjects. In Table 7.3, the sum is more even when the categories are not exhaustive. To avoid confusion, the total number of subjects in the case of multiple response is called a **base** in place of total. Percentages are generally calculated using this base.

TABLE 7.3

Cases of Abdominal Tuberculosis with Major Symptoms

Symptom	Number of Cases	Percent
Pain in abdomen	126	89.4
Vomiting for a long time	85	60.3
Constipation for a long time	57	40.4
Total cases (base)	141	100.0

Source: Adapted from Das, P. et al., *Am. J. Proctol.*, 26, 75, 1975.

TABLE 7.4

Conversion of Cases of Abdominal Tuberculosis
with Major Symptoms into a Contingency Table

Group of Major Complaints	Number of Cases	Percent
Pain, vomiting, and constipation	44	31.2
Pain and vomiting, no constipation	37	26.2
Pain and constipation, no vomiting	10	7.1
Vomiting and constipation, no pain	2	1.4
Pain, no vomiting, no constipation	35	24.8
Vomiting, no pain, no constipation	2	1.4
Constipation, no pain, no vomiting	1	0.7
Other symptoms	10	7.1
Total	141	100.0

Table 7.3 is not a contingency table but can be converted to a contingency table when additional information is available. One way to do it is suggested in Table 7.4. Note that the categories are now mutually exclusive and exhaustive. Perhaps, this table provides more useful clinical information.

7.2.2.1 Features of a Table

Each frequency table or contingency table must contain the total number of subjects, group wise if groups are present. Confusion first arises while calculating percentages. The base for percentages should be a predetermined number. In a case–control study, the relevant percentage is of those possessing a particular antecedent out of the total cases and controls separately. If 28 out of 80 breast cancer cases, and 21 out of 80 matched controls report age at menarche <12 years, do not calculate percentage of 28 and 21 out of the total 49. Instead calculate percentage of each out of 80. In a prospective study, the percentage should be based on the exposed and unexposed totals that have been followed up. In a cross-sectional study, the percentage should be calculated out of the grand total (and not row totals or column totals) because the grand total is the prefixed number.

For calculating the number of decimals in a percentage, use the rules described in Chapter 21. All percentages must add up to 100 except in cases of multiple response. Table 7.5 illustrates both kinds of percentages. The table also illustrates the placing of explanatory matter in footnotes. Use symbols *, †, ‡, §, and ¶ in that order to link footnote with the table contents. Else use superscripts a, b, c, d, etc. Codes, abbreviations, symbols, inconsistencies, obscure information, etc., should be explained in footnotes. If data are not original, give the source in footnote without using a connecting symbol.

TABLE 7.5

Postnatal Complications in Women

Postnatal Complications[a]	Number of Women	Percentage
Excessive bleeding	30	7.25
Urinary tract infection	52	12.56
Convulsions	68	16.43
Foul discharge	56	13.53
Others	39	9.42
Total with complications (base)	209	50.48
No complication	205	49.52
Total	414[b]	100.00

[a] Multiple response.
[b] Total women are 427, but complication information is not available for 13 women.

Example 7.1: Illustration of Features of a Table

Table 7.5 contains hypothetical data from a hospital. I am not providing any further information. See if you find the table self-explanatory or not. Note that this is not a contingency table since multiple response is allowed. The complications are not mutually exclusive: two or more complications can be simultaneously reported by a single patient.

7.2.3 Other Types of Statistical Tables

Among other statistical tables are those that give rates, ratios, *P*-values, correlations, etc. Many such tables appear in the medical literature that you may have to properly interpret for the benefit of your medical practice. Sometimes you may like to prepare a table yourself, particularly for research. Several such statistical tables appear throughout this book. Their features and interpretation are also described.

7.2.3.1 What Is a Good Statistical Table?

If you are frequently reading medical literature, or writing papers or reports on the basis of a research, you may find that some tables are not sufficiently clear. An important feature of a good statistical table is that it should tell a worthwhile and intelligible story. Although a table is generally accompanied by a text, it should be able to stand alone and explain the data by itself. The table title should be short but complete, and column and row headings self-explanatory. If needed, footnotes should be provided. A very large table tends to be confusing and a very short table may look incomplete. Thus, a table should be just adequate to provide complete information.

7.3 Rates and Ratios

A numerical method for representing uncertainties is by way of computing a rate or a ratio. These are calculated for qualitative data only. If the
measurement is metric, it can be converted to a qualitative form by suitable grouping. For example, subjects having Hb level less than 10 g/dL can
be said to suffer from anemia and those having diastolic BP greater than
90 mmHg as having hypertension. When such qualitative conversion is
done, the rates and the ratios can be computed for quantitative data also.
For this, an understanding of the concepts of proportion, rate, and ratio is
a prerequisite.

7.3.1 Proportion, Rate, and Ratio

Proportion, rate, and ratio are distinct but interrelated concepts. The following details may help to understand similarities and dissimilarities
among them.

7.3.1.1 Proportion

Because of variation, some subjects in a group would possess the quality
or the attribute under investigation and others would not. The simplest
way to express this variation is by stating the percentage of subjects with
the target attribute, for example, 30% of myocardial infarction (MI) cases
are obese and 40% are smokers, treatment A is effective in 85% of the cases
with a particular disease, or 2% of people of age 50 years and above in
an area have kidney disease. All these are proportions when calculated
per person. In these examples, the proportions are 0.30, 0.40, 0.85, and
0.02, respectively. These proportions are valid indicators only when they
are based on a sufficiently large n. If one out of a total of three accidental
deaths is a female in an area in a year, it is incorrect to say that the female
proportion is 0.33.

It is convenient to express percentages instead of proportions in most
cases. However, if the disease or the health condition is rare, this could be
expressed per thousand or even per million. For example, one may say that
12 per 1000 exposed to long-term air pollution suffer from asthma or that
5 per million are deaf and dumb in a particular population. The proportions
in these examples are 0.012 and 0.000005, respectively.

A quantity can *increase* by more than 100% but cannot *decrease* by more
than 100%. If a quantity is measured in percent—saying that it increased by
7 percentage points is not the same as it increased by 7%. Increase from 53%
to 60% is 7 percentage points but it is 13% of the baseline 53%.

7.3.1.2 Rate

You may have come across urea clearance rate as something like 70 mL/ min in healthy adults. Rate is a measure of the frequency of occurrence of a phenomenon. It counts the number of events occurring in a specific period as per unit of time. Time is an essential element of this concept—per day, per month, per year, etc. Since this frequency can change over time, rate is time specific. If 8 deaths occurred per 1000 population in a particular year, this becomes the death rate for that year. If on average two new cases of cirrhosis appeared per 1000 population in the year 2012 in an area, we call it the incidence rate per 1000 for that year. When the variation from time to time or from year to year is not large, the reference to the point of time can be omitted. If death rate from accidents remains fairly constant at about 14 per million population year after year in an area, then a general statement could be made without mentioning any particular year. But new discoveries and advances in technology can alter any rate even when the ground situation remains the same. Thus, exercise caution in comparing rates over time.

Essential components of a rate are (1) a numerator, (2) a denominator, (3) a specification of duration, and (4) a multiplier such as percent or per thousand. The multiplier helps to convert an awkward-looking fraction to a convenient and nice number. The denominator is generally the number of subjects in the group *at risk* for the events counted in the numerator. Measures such as birth rate per thousand are exceptions because the children born are not part of the entire population, yet it is called a rate. Since the essential feature of a rate is the frequency of occurrence over time, it is legitimate to call it a birth *rate*.

The main purpose of computing a rate is to be able to compare two or more groups, times, places, etc. The denominators are population, patients, subjects, etc., which vary periodically and geographically. Rate makes the denominator uniform, and thus allows valid comparison. However, a rate is a valid tool only for stable conditions over the period for which it is calculated. If the population varies within that period, such as the beginning of the year and the end of the year, it is conventional to use the average population, say, at the middle of the year. Here in lies the catch. If a city has a population of 10,000 at the beginning of the year and 6000 perish due to a severe earthquake in the month of February, the midyear population would be nearly 4000 (some would die due to routine causes and some births may take place). In that case, the death rate, when based on the midyear population, for that year would be nearly 6000 × 1000/4000 or 1500 per 1000 population, or 1.5 per person. This rate would be entirely different if earthquake deaths occur in the latter half of the year. This dramatic example illustrates that the midyear or average population can be fallacious if deaths, for that matter any event for which rate is required, do not occur regularly throughout the period at nearly a uniform rate. The concept of rate is not applicable when such unusual spikes or dips occur.

7.3.1.3 Ratio

Broadly speaking, a ratio is one quantity relative to another. It can be expressed as *a:b* or as *a/b*. In a broad sense, all rates too are ratios because there is a numerator and there is a denominator. In practice, the usage of the term ratio is restricted to a situation in which the numerator is not a part of the denominator, nor the denominator part of the numerator. Both are separate and distinct entities. The ratio of white blood cells (WBCs) to red blood cells (RBCs) is 1:600. The sex ratio in a general population is computed as number of females per 1000 males and the dependency ratio as the number of dependents (say, of age –15 and 65+ years) in a population to the working (15–64 years) population. Doctor–population ratio, bed–population ratio, waist–hip ratio, acid output basal to maximal ratio, and albumin–globulin ratio are the other examples.

7.4 Central and Other Locations

This section deals with quantitative data only. Suppose there are data on duration of hospitalization of a large number of cases of a particular disease. One way to represent variation in this duration, of course, is to prepare a table showing the numbers of cases with different durations. A step further is to be able to say that the most common duration is 8 days and that duration ranged between 5 and 10 days in nearly 90% of the cases. The former represents a central value and the latter scatter or dispersion. These two in many situations give a fairly good idea of the entire dataset. Sometimes one is interested in a noncentral quantity such as the duration of hospitalization seen in two-thirds of the cases. This is a quantile, one form of which is the percentile used in growth charts. Quantiles are used to measure locations other than the central value.

7.4.1 Central Values: Mean, Median, and Mode

A value nearly in the center, or commonly observed, can be considered as a representative of the central value. The arithmetic mean, commonly called an average, is a very popular measure of central value. If $x_1, x_2, ..., x_n$ are n observations in our sample, the sample mean is $\bar{x} = \Sigma x_i/n$. The median is the middle value, which is obtained as the $((n + 1)/2)$th value if n is odd, after arranging in ascending order, and average of $n/2$th and $((n/2) + 1)$th if n is even. The mode is the most common value.

Example 7.2: Calculating Mean, Median, and Mode

The following are the data on cases of acute polymyositis of the back in 38 women.

Duration of immobility (days):

7	5	9	7	36	4	6	7	5	8	3	6	5
7	8	10	7	14	10	9	4	6	11	9	6	5
8	8	6	7	5	5	12	3	5	9	10	7	

Mean = 7.9 days

Median = Average of the 19th and 20th values after rearranging
in ascending order

$$= \frac{(7+7)}{2} = 7 \, days$$

Mode = 5 and 7 days, occurring in seven patients each

A distribution containing two modes such as in Example 7.2 is called a **bimodal distribution**. Age distribution in Hodgkin's disease and leukemia is bimodal with one (smaller) peak around 20 years and the other (bigger) peak at around 60 years.

Mean is denoted by $\bar{x}$ for sample and by μ for population. No such universally accepted notations are available for median and mode.

7.4.1.1 Understanding Mean, Median, and Mode

Visual display can help to further understand the meaning of mean, median, and mode. Various graphs are discussed in the next chapter, but here a simple plot of points explains the meaning of these measures. Figure 7.1 displays the duration of hospital stay (days) after a surgery for 30 patients. The mean is the center that balances the beam on which the points are plotted. This balancing is a very useful feature of this central value. The median is such that half of the dots are less than or equal to the median. Although it is not so clear in Figure 7.1 because of the discrete data, half of the values will be more than the median and the other half less. This may or may not happen with the mean. The mode clearly refers to the peak.

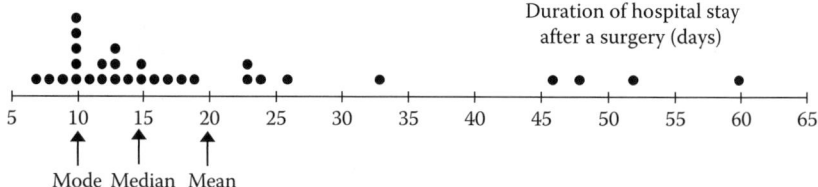

FIGURE 7.1
Schematic of mean, median, and mode.

7.4.1.2 Calculation in Case of Grouped Data

When the exact values are available, the calculations must invariably be based on such values. However, for reasons given earlier, the data are sometimes available in grouped form. In that case, the calculations become approximate. In many applications, midpoint of the class intervals containing the required central value is considered adequate. However, the best approximations are obtained as follows.

Let the intervals in ascending order be (a_0, a_1), (a_1, a_2), ..., (a_{K-1}, a_K) and their midpoints $x_k = (a_{k-1} + a_k)/2$, $k = 1, 2, ..., K$. Note that these are continuous intervals. Let the number of subjects or the frequency in the first interval be f_1, the second interval f_2, and so on. Then,

$$\text{Grouped data: Mean} = \frac{\Sigma f_k x_k}{n}, \quad \text{where } n = \Sigma f_k. \tag{7.1}$$

$$\text{Grouped data: Median} = a_{m-1} + \frac{n/2 - C}{f_m} * h_m, \tag{7.2}$$

where
- f_m is the frequency in the interval containing the $(n/2)$th observation (called the median interval)
- a_{m-1} is the lower limit of the median interval
- C is the cumulative frequency preceding the median interval
- h_m is the width of the median interval

$$\text{Grouped data: Mode} = a_{M-1} + \frac{f_M - f_{M-1}}{2 f_M - f_{M-1} - f_{M+1}} * h_M, \tag{7.3}$$

where
- f_M is the frequency in the interval with the highest frequency (modal class)
- f_{M-1} and f_{M+1} are the frequencies in the preceding and succeeding intervals
- a_{M-1} is the lower limit of the modal class
- h_M is the width of the modal class

This considers adjacent frequencies to determine the peak.

Example 7.3: Mean, Median, and Mode in Grouped Data

Group the data in Example 7.2 and obtain the continuous intervals by subtracting and adding 0.5 to the end points as in Table 7.6. Now,

$$\text{Mean} = \frac{(11 \times 4 + 16 \times 7 + 8 \times 10 + 2 \times 13 + 1 \times 36)}{38} = 7.8 \, \text{days}.$$

TABLE 7.6

Grouping of Data in Example 7.2 on Duration of Immobility in Cases of Acute Polymyositis

Group (a_{k-1}, a_k)	2.5–5.5	5.5–8.5	8.5–11.5	11.5–14.5	36	Total
Midpoint (x_k)	4	7	10	13	36	
Frequency (f_k)	11	16	8	2	1	38
Cumulative frequency	11	27	35	37	38	

For the median, the interval containing the 19th and 20th observations is (5.5–8.5) days. Thus, $f_m = 16$, $a_{m-1} = 5.5$, and $C = 11$. Also, $h_m = 3$.

$$\text{Median} = 5.5 + \frac{19 - 11}{16} \times 3 = 7.0 \, \text{days.}$$

Note that 19 is away from cumulative 11, and it is reasonable to divide this difference proportionately.

For the mode, the interval containing the highest frequency is again (5.5–8.5) days. Thus, $f_M = 16$, $a_{m-1} = 5.5$, $f_{M-1} = 11$, and $f_{M+1} = 8$.

$$\text{Mode} = 5.5 + \frac{16 - 11}{32 - 11 - 8} \times 3 = 6.7 \, \text{days.}$$

SIDE NOTE: The continuous intervals formed in Example 7.3 are a natural consequence of the observed durations. The duration of immobility would rarely be exactly 6 days or exactly 8 days. When it is noted as 6 days, it is likely to be anywhere between 5.5 and 6.5 days. If the duration is noted in terms of completed days, then 6 days is really between 6 and 7 days. In that case, the continuous intervals would be (3–6), (6–9), etc., and the mean, median, and mode would change accordingly.

The values of the mean, median, and mode sometimes change because of grouping. The values obtained on the basis of the ungrouped data in Example 7.2 are exact. The approximation in Example 7.3 occurs because all observations in an interval are assumed to be at its midpoint. The magnitude of error depends mostly on the width of the intervals. You may wish to try another grouping and see how the values of the mean, median, and mode are affected.

7.4.1.3 Which Central Value to Use?

If you want to tell a patient one duration of immobility generally seen in cases of acute polymyositis of the back, would you use the mean, median, or mode? The mean is the popular choice because it is simple to calculate and easy to understand. It certainly should be preferred except in cases in which it can mislead. Add to the data in Example 7.2 two more cases with duration 32 and 47 days. There is already a case of immobility of 36 days. The mean now becomes 9.5 days. Because 31 out of now 40 patients have immobility for 9 days or less, 9.5 days cannot be a representative central value. The mean is vitiated due to **outliers**. Outliers are unusually high or unusually low values that do not go along with the other values. Whenever such values are present, the mean is not a good choice. In such instances, use the median because it is not affected by unusually high or unusually low values. For these 40 durations, the median is the average of the 20th and 21st values in ascending order. This again is $(7 + 7)/2 = 7$ days.

Use the mode when the interest is specifically in the most common value or the "typical" value. Consider the incubation period of cholera from exposure to the appearance of symptoms. This generally ranges from 1 to 4 days, but the most common (mode) is 2 days. This mode has special significance in this case because it can help to estimate a peak occurrence if the exposure is from a common source. Thus, hospitals can be geared-up to handle the cases. In Example 7.2, there are two modes—5 days and 7 days—each seen in nearly one-fifth of the patients. *The mode can be multiple but the mean and median are always unique.* The practical utility of the mode is sometimes lost when it is more than one except when it is genuinely bimodal. In case of nominal data, mode is an appropriate central value.

My guideline for choosing an appropriate central value is as follows: Always prefer mean to represent the central value except when (1) outliers are present (use the median then) or (2) the interest is specifically in the most common value (use the mode then). If multiple modes are present, use the mean. You will soon see that in a large number of situations in medicine and health, the opportunity to choose may not arise because the three central values tend to be the same. Nonetheless, use the following precautions:

1. The mean and other measures of centrality are sometimes misused. One misuse occurs when they are cited without the accompanying variability of the underlying values. The mean of the fasting plasma glucose level could be 105 mg/dL when the range is 90–110 mg/dL and also when the range is 60–200 mg/dL. The same means in these two cases have entirely different implications. Mean must always be accompanied by a measure of variation. Various measures of variability are presented in Section 7.5.

2. Mean can be ridiculous when outliers are present. If 8 out of 10 persons in a sample do not take alcohol and 2 persons take 120 mL each per day, it is unwise to say that the average intake per person is

24 mL. As as extreme example, consider the absurdity of average number of arms in human beings = 1.99999, since some are without arms and some have only one arm.

3. Mean, median, and mode are much more meaningful when they are based on relatively homogenous groups. For example, the mean survival period of assorted patients admitted in a hospital could be an inappropriate measure of efficiency since there would be cases of hernia with minimal risk of death and cases of peritonitis with grave prognosis. The survival period could be widely different.

4. The number of values on which a mean is based must always be stated. Mean of values in two or three subjects has little meaning. The mean of 5 subjects may be the same as of 200 subjects, but the two means have very different reliability.

5. The other common misuse occurs when a mean or median is applied to one particular case in place of a group. If the median survival time after detection of a malignancy is 3 years, it does not mean that a person with this malignancy is likely to survive for 3 years. Even when the term "likely" is added, the statement does not become valid. The only correct statement is that nearly half survive for 3 years or less and the other half for more than 3 years.

6. Arriving at a conclusion in medicine based on mean alone without considering other correlates could be misleading. If an antihypertensive drug is able to reduce diastolic BP by 10 mmHg on average after 1 week of use by patients with hypertension, other considerations such as side effects, cost, and convenience of intake cannot be ignored altogether while evaluating the usefulness of the drug.

7. Most scientific applications would be careful about the misuses just mentioned, but popular media such as television channels and newspapers sometimes pick up average values and report them without sufficient care about the variability attached to them. Thus, the public is not properly informed. This can cause all sorts of problems regarding the perception of a disease or in understanding the gravity of a health problem.

8. Mean, median, and mode are proportionately affected by change of origin and scale. For example, if you add 6 to each value of x, mean also will be 6 more. If you multiply each value of x by 3, mean also will be three times. That is,

$$\text{Mean}(a + bx) = a + b * \text{mean}(x).$$

If you have measured body temperature in Fahrenheit and subsequently decide to convert this to Celcius, you do not need to calculate mean again. Since $C = 5/9 * F - 160/9$, (mean in °C) = $5/9 \times$ (mean in °F) − 160/9.

7.4.1.4 Geometric Mean

There are at least two other types of mean, although they are rarely used. The first is as follows:

$$\text{Geometric mean: GM} = (x_1 * x_2 * \cdots * x_n)^{1/n}. \qquad (7.4)$$

For the geometric mean (GM), the n values are multiplied and then their nth root is taken. Note that $\log \text{GM} = (1/n) * \Sigma \log x_i$. Thus, GM is the antilogarithm of the arithmetic mean of logarithms. In the case of grouped data, this becomes

$$\text{Grouped data: } \log \text{GM} = \left(\frac{1}{n}\right) * \Sigma f_k \log x_k \qquad (7.5)$$

where x_k is the midpoint of the kth ($k = 1, 2, ..., K$) interval as in Equation 7.1 and $n = \Sigma f_k$. GM can be calculated only when all xs are positive. GM can be very different from arithmetic mean. For example, GM of 4, 16, and 64 is 16 whereas arithmetic mean of these values is $84/3 = 28$.

Situations in which logarithmic transformation of values are helpful in achieving certain desirable statistical properties are discussed later in this book. In those cases, GM can be used. One common use of this measure is in tissue attenuation correction for gastric emptying time. The GM of gastric counts is considered the **gold standard** for this correction. Gastric counts generally possess a multiplicative feature so that GM is more appropriate than the arithmetic mean. Other common use of GM is for measuring antibody titer and gamma radiation count. These are also multiplicative phenomena rather than additive phenomena.

7.4.1.5 Harmonic Mean

Sometimes rates are stated in a reciprocal manner. An example is the average population served per doctor. If this is 1000 for the rural area in a district and 500 for the urban area, what is the average for the district as a whole? Even if the populations in rural and urban areas are equal, the average is not 750. Consider the data in Table 7.7.

TABLE 7.7

Population Served per Doctor in Rural and Urban Areas:
Equal Rural–Urban Population

Area	Population Served per Doctor	Population	Number of Doctors
Rural	1,000	50,000	50
Urban	500	50,000	100
Total		100,000	150

When rural and urban areas are combined, the average population served per doctor is 100,000/150 = 667. It is not 750. This can be computed as follows:

$$\text{Average population served per doctor} = \frac{50,000 + 50,000}{(50,000/1,000) + (50,000/500)} = 667.$$

This is called the harmonic mean (HM). If the populations in rural and urban areas are unequal (Table 7.8), then the average population served per doctor is 100,000/125 = 800. In other words,

$$\text{Average population served per doctor} = \frac{75,000 + 25,000}{(75,000/1000) + (25,000/500)} = 800.$$

The general procedure for grouped data is as follows:

$$\text{Grouped data: Harmonic mean} = \frac{\Sigma f_k}{(f_1/x_1) + (f_2/x_2) + \cdots + (f_K/x_K)}, \qquad (7.6)$$

where x_k is the value in f_k subjects. In case of categorized metric data, x_k is the midpoint of the kth category ($k = 1, 2, \ldots, K$). For **ungrouped data,**

TABLE 7.8

Population Served per Doctor in Rural and Urban Areas:
Unequal Rural–Urban Population

Area	Population Served per Doctor	Population	Number of Doctors
Rural	1,000	75,000	75
Urban	500	25,000	50
Total		100,000	125

each $f_k = 1$ for all k and $K = n$. HM is the reciprocal of the mean of the reciprocals. Note the following:

1. GM gives relatively small weight to the large values, and HM gives relatively large weight to the small values. This may become necessary in some cases as illustrated earlier. However, the use of these two averages is rare and rightly so. They are difficult to calculate and understand and are not desirable unless really needed.

2. GM and HM cannot be computed if one or more value is zero or negative.

7.4.2 Other Locations: Quantiles

Quantiles are the values of the variable that divide the total number of subjects into ordered groups of equal size. Names such as percentiles, deciles, quintiles, quartiles, and tertiles are used when the numbers of divisions are 100, 10, 5, 4, and 3, respectively. In general, the values dividing subjects into s equal groups may be called s-tiles. The total number of s-tiles is $(s - 1)$. The procedure for their computation is described next.

7.4.2.1 Quantiles in Ungrouped Data

Quantiles in ungrouped data are obtained simply in the following manner:

$$p\text{th } s\text{-tile} = \left(p * \frac{n}{s} \right)\text{th value in ascending order of magnitude}, \qquad (7.7)$$

where n is the total number of subjects. For $n = 200$ subjects,
35th percentile = $(35 \times 200/100)$ = 70th value,
7th decile = $(7 \times 200/10)$ = 140th value,
3rd quintile = $(3 \times 200/2)$ = 120th value,
1st quartile = $(1 \times 200/4)$ = 50th value, and
2nd tertile = $(2 \times 200/3)$ = 133rd value,

in ascending order of magnitude.

Note that percentiles are very different from percentages. For example, 90th percentile of esters in cholesterol could be 58%. This means that 90% people have esters less than or equal to 58%. Also note that highest is 99th percentile (better than 99%). No one is ever 100th percentile. Another feature worth noting for all quantiles is that quantile of $(x + y) \neq$ quantile of x + quantile of y. This needs to be understood in the context of mean since mean$(x + y)$ = mean(x) + mean(y).

7.4.2.2 Quantiles in Grouped Data

In case of grouped data, the calculation is based mainly on the quantile interval—the interval containing the required quantile.

$$\text{Grouped data: } p\text{th } s\text{-tile} = a_{p-1} + \frac{p*n/s - C}{f_p} * h_p, \qquad (7.8)$$

where

a_{p-1} is the lower limit of the quantile interval; the quantile interval is the one containing the $(p*n/s)$th observation in order of magnitude

C is the cumulative frequency until the quantile interval

f_p is the frequency in the quantile interval

h_p is the width of the quantile interval

Percentiles can be denoted by P_1, P_2, P_3, etc., deciles by D_1, D_2, D_3, etc., and quartiles by Q_1, Q_2, and Q_3.

The other method for obtaining approximate quantiles in case of grouped data is graphical. Various graphs are discussed in the next chapter, but here a simple line diagram does the job. For graphical calculation of quantiles, the cumulative percentages of subjects are plotted against the upper end of the data intervals. For the age distribution of subjects in Table 7.1, this plot is shown in Figure 7.2. Note that the cumulative percentages for age –49, –59, –69, –79, and beyond 79 are 11, 32, 78, 95, and 100, respectively. This plot of cumulative frequencies (in percentage or absolute number) is called an **ogive**. To obtain a pth s-tile, draw a horizontal line at $100p/s\%$ and read the value on the x-axis where this horizontal line intersects the percent-based ogive. Figure 7.2 shows the 40th percentile and the 3rd quartile.

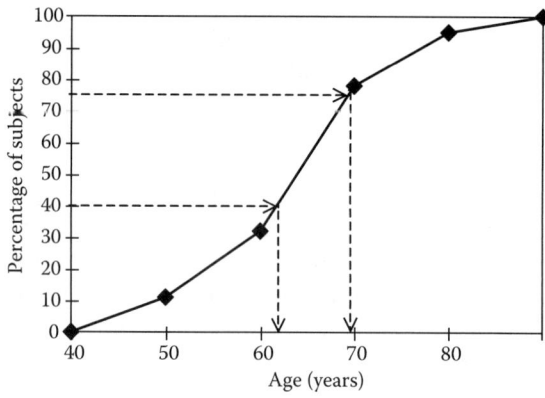

FIGURE 7.2

Approximate calculation of some quantiles from ogive.

Example 7.4: Calculation of Various Quantiles for Grouped Data

Consider the duration of immobility data in Table 7.6. The 2nd tertile in this dataset is $2 \times 38/3 = 25$th value in ascending order. This is 8 days. Also,

$$85\text{th percentile} = \left(85 \times \frac{38}{100}\right)\text{th or 32nd value in ascending order} = 10\,\text{days}$$

The same data are grouped in Example 7.3. The 25th value is in the interval (5.5–8.5) days. Thus, for the 2nd tertile, $a_{p-1} = 5.5$, $C = 11$, $f_p = 16$, and $h_p = 3$. Therefore,

$$2\text{nd tertile (grouped data)} = 5.5 + \frac{25 - 11}{16} \times 3 = 8.1\,\text{days}.$$

Similarly,

$$85\text{th percentile (grouped data)} = 8.5 + \frac{32 - 27}{8} \times 3 = 10.4\,\text{days}.$$

You may wish to obtain quantiles from an ogive also of the type shown in Figure 7.2 and compare the results with the present values. The calculations in the case of grouped data are, in any case, approximate for quantiles just as for the mean, median, and mode.

7.4.2.3 Interpretation of Quantiles

All calculations can be done with the help of computers, but the method of computation helps in understanding quantiles and their proper interpretation. (Later, I avoid giving methods of computation for many statistical methods when they become complex and concentrate on usage and interpretation instead.) A pth s-tile is the value that is more than $100p/s\%$ of observations but less than (or equal to) the other $100(1 - p/s)\%$ observations. In Example 7.4, 2nd tertile = 8 days implies that two-thirds of the patients have immobility for 8 days or less. If the 97th percentile of weight of 2-year old boys in a county is 15.2 kg, it implies that 97% of such children in that county weigh 15.2 kg or less. The other 3% have higher weight.

Figure 7.3 shows the duration of hospital stay after surgery for 30 patients. The data are the same as in Figure 7.1. See how quartiles are determined. They divide the total number of subjects into four equal groups in terms of frequency. These frequencies may not be exactly equal in the case of the discrete data depicted in Figure 7.3 but would be equal in the case of really continuous data. Other quantiles have similar interpretations.

Quantiles are sometimes used for an objective categorization of the subjects. It may be easier to say in some situations that the bottom one-third values are low, the middle one-third are medium, and the top one-third are

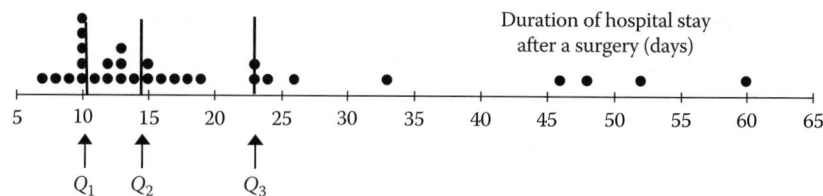

FIGURE 7.3
Schematic of quartiles.

high. In the case of data on the duration of immobility in Example 7.2, one may say that the duration of immobility up to 6 days experienced by one-third of the patients is low, that between 6 and 8 days is medium, and that above 8 days is high.

A popular use of quantiles is in grading scores. In educational testing, it is common to consider the scores in the top decile as excellent because only 10% of the subjects have these high scores. In medicine too, scoring is sometimes used to grade the severity of disease (see Chapter 10). A condition with a score less than the first quartile (scored by the least affected 25% of cases) can be considered mild, one with a score between the first and second quartiles as moderate, that between the second and third quartiles as serious, and that beyond the third quartile as very serious. Such nomenclature may or may not agree with the prognosis because the categorization is based on statistical rather than on medical considerations. The argument is that a condition with a score, for example, worse than that of 75% of cases can be safely called very serious. When the categorization is based on quantiles and not on prognosis, the interpretation is relative and not absolute. It is in this sense that percentiles are used in growth charts.

Quantiles are the *points* that divide the total number of subjects into equal groups, but it is colloquially popular to say, for example, that the scores are in the top decile. This assumes that the decile is a range. This is not technically accurate but is accepted as long as the meaning is clear.

Example 7.5: A Good Use of Percentiles for Assessing the Risk of Heart Disease

There is growing evidence that children with low birth weight but with rapid gain in weight and increase in BMI after the age of 1 year are at greater risk of coronary heart disease later in life, particularly male children [4]. Rapid gain means that a child who was at lower percentile for birth weight reaches to a higher percentile at, say, age 8 years. A boy with birth weight in the bottom 10% (less than 10th percentile) who rapidly gains weight to reach the top 25% (beyond 75th percentile) at age 5 or 10 years is at much greater risk of coronary heart disease. This is called **crossing the centiles** and illustrates a very apt application of the concept of percentiles.

7.5 Measuring Variability

Some measurements such as body temperature do not vary much across healthy individuals whereas others such as cholesterol level are highly variable. Thus, dispersion (scatter) is important. See Figure 7.4 for distributions that differ in location and scatter. These curves would be more clear after reading Chapter 8.

Consider systolic BP measured for two groups of five persons each:

Group-1: 134, 132, 124, 132, and 128 mmHg

Group-2: 110, 140, 118, 150, and 132 mmHg

Both groups have the same mean (130 mmHg) and the same median (132 mmHg). Despite this equality in groups with regard to central values, the groups are very different. The BP values in group-1 vary from 124 to 134 and in group-2 from 110 to 150. Their scatter or dispersion is different. Thus, representation of data merely by a central value is incomplete. It needs to be accompanied by a measure of dispersion as well.

Can we devise a method to measure the dispersion by a single value? One such measure of variability is **range**. This is defined as the difference between the maximum and minimum values in a group. The range in group-1 is 134 − 124 = 10 mmHg and in group-2 is 150 − 110 = 40 mmHg. Although useful in many cases, range gives a completely distorted picture when outliers are present. If the systolic BP of the last person in group-1 is 158 mmHg in place of 128, the range jumps to 158 − 124 = 34 mmHg. This range disregards the fact that the other four readings are fairly close to one another and that they do not have much dispersion. A modification of range is

$$\text{Semi-interquartile range:} \frac{(Q_3 - Q_1)}{2}, \tag{7.9}$$

where
Q_1 is the first quartile
Q_3 is the third quartile

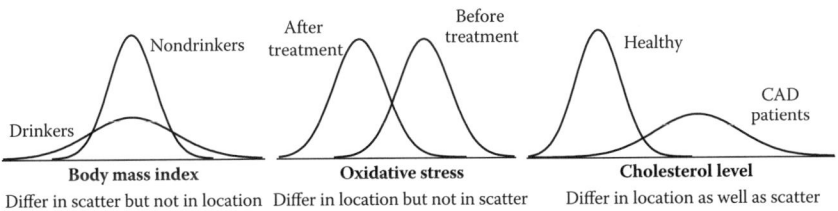

FIGURE 7.4
Distributions differing in location and scatter.

Quartiles are not much affected by outliers but they also do not use all values in their entirety. All quantiles, including quartiles, use ranks rather than the values themselves. We need a measure that uses all the values in the data.

7.5.1 Variance and Standard Deviation

Dispersion, scatter, and variability all connote the same phenomenon. Its magnitude depends on the extent of differences each value has from each of the others. Instead of calculating so many differences, it is convenient to compute the difference of each from a central value, namely, the mean. Let the sample observations be $x_1, x_2, ..., x_n$ and the mean $\bar{x}$. The difference $(x_i - \bar{x})$, where $i = 1, 2, ..., n$, is called the deviation from the mean. An average of these deviations could be a measure of dispersion, but this would always be zero. Some of these deviations would be positive and some negative. One way out is to ignore the sign, get absolute values, and calculate the average $\Sigma |x_i - \bar{x}|/n$. This is called the **mean deviation**. Absolute values are mathematically difficult to handle. It is easy to get rid of the minus sign by squaring the deviations. The average of squared deviations $(x_i - \bar{x})^2$ is $\Sigma(x_i - \bar{x})^2/n$. This is called the **variance** and is a very useful and popular measure of dispersion. For population variance, use μ in place of $\bar{x}$. Population variance is denoted by σ^2.

Variance has been extensively studied and has been found to be a very adequate measure. The only difficulty is that the unit of the variable values is also squared (e.g., square meters for area). The original unit is retrieved by taking the square root. The quantity obtained as the positive square root of the variance is called the **standard deviation**. It is seen in case of samples that the denominator $(n - 1)$, in place of n, gives better estimate of the population SD in the long run. Population SD is calculated with n in the denominator.

7.5.1.1 Variance and Standard Deviation in Ungrouped Data

As already explained, the formulas for ungrouped data can be stated as follows:

$$\text{Ungrouped data: Sample variance: } s^2 = \frac{\Sigma(x_i - \bar{x})^2}{n-1}, \tag{7.10}$$

$$\text{Sample SD: } s = \sqrt{\frac{\Sigma(x_i - \bar{x})^2}{n-1}}. \tag{7.11}$$

SD is invariant under change of origin. It is not affected by addition or subtraction of each value by a constant. That is, SD of 5, 10, 15, and 20 is the same as of 6, 11, 16, and 21. But it is affected by scale. If you divide each value of x by 4, the SD also will be one-fourth. In general,

$$SD(a + bx) = b * SD(x).$$

Since SD is heavily dependent on mean $\bar{x}$, it can give weird result when mean is not an adequate central value. Thus, avoid using SD for data where mean is not appropriate such as when outliers are present.

The calculations for SD are illustrated in Example 7.6.

Example 7.6: Standard Deviation in Two Groups with Diverse Dispersion

Calculation of the variance and SD of the systolic BP given earlier for the two groups of subjects are shown in Table 7.9. SD in group-2 is more than four times the SD in group-1. This can be legitimately used to conclude that the variation in group-2 is nearly four times than in group-1. Many other uses of SD are described in later sections.

See Figure 7.5 to get an idea of what variance measures. Three medical measurements are shown. The serum iron varies in this figure from 50 to $170\,\mu g/L$ in healthy subjects. It has a large variance relative to systolic BP, which varies between 115 and $140\,mmHg$. Serum sodium varies within a narrow range of $135-144\,mEq/L$. The variance is even smaller in this measurement.

TABLE 7.9

Calculation of Variance and SD in Two Disparate Groups

Group-1			Group-2		
SysBP (x)	Deviation $(x - \bar{x})$	Square $(x - \bar{x})^2$	SysBP (y)	Deviation $(y - \bar{y})$	Square $(y - \bar{y})^2$
134	4	16	110	−20	400
132	2	4	140	10	100
124	−6	36	118	−12	144
132	2	4	150	20	400
128	−2	4	132	2	4

Total: $\Sigma(x_i - \bar{x}) = 0$; $\Sigma(x_i - \bar{x})^2 = 64$ $\qquad$ $\Sigma(y_i - \bar{y}) = 0$; $\Sigma(y_i - \bar{y})^2 = 1048$

Sample variance: $\dfrac{\Sigma(y_i - \bar{y})(x_i - \bar{x})^2}{n-1} = \dfrac{64}{4} = 16$ $\qquad$ $\dfrac{\Sigma(y_i - \bar{y})^2}{n-1} = \dfrac{1048}{4} = 262$

Sample SD: $\sqrt{\dfrac{\Sigma(x_i - \bar{x})^2}{n-1}} = \sqrt{16} = 4$ $\qquad$ $\sqrt{\dfrac{\Sigma(y_i - \bar{y})^2}{n-1}} = \sqrt{262} = 16.2$

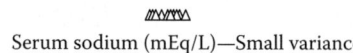

Serum sodium (mEq/L)—Small variance

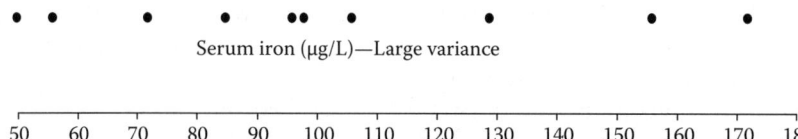

Systolic BP (mmHg)—Medium variance

Serum iron (μg/L)—Large variance

| 50 | 60 | 70 | 80 | 90 | 100 | 110 | 120 | 130 | 140 | 150 | 160 | 170 | 180 |

FIGURE 7.5
Schematic of differential variance in three measurements.

The value of the variance (or of SD), just as of the mean, depends on the unit of measurement. It would be high if the unit was microgram per liter instead of milligram per liter. The SD of serum iron values in healthy subjects may be 25 when measured in microgram per liter but would be only 0.025 when measured in milligram per liter.

7.5.1.2 *Variance and Standard Deviation in Grouped Data*

Let there be K intervals $(a_0, a_1), (a_1, a_2), \ldots, (a_{K-1}, a_K)$ and their midpoints $x_k = (a_{k-1} + a_k)/2, k = 1, 2, \ldots, K$. Let the frequency in the kth interval be f_k. Then, for

$$\text{Grouped data—Sample variance: } s^2 = \frac{\Sigma f_k (x_k - \bar{x})^2}{n-1}, \tag{7.12}$$

and

$$\text{Grouped data—Sample SD: } s = \sqrt{\frac{\Sigma f_k (x_k - \bar{x})^2}{n-1}}, \tag{7.13}$$

where $\bar{x} = \Sigma f_k / n$ and $n = \Sigma f_k$.

As in the case of the mean, this also involves approximation in that all observations in an interval are assumed centered on the midpoint. This works fairly well when the width of the intervals is not large.

7.5.1.3 *Variance of Sum or Difference of Two Measurements*

Mean has the property that mean($x + y$) = mean(x) + mean(y). But variance or SD does not have this feature. SD of sum or difference of the values is not the sum or difference of SDs. In a specific situation when x and y are linearly independent, variance($x + y$) = variance(x) + variance(y). In general,

$$\text{variance}(x + y) = \text{variance}(x) + \text{variance}(y) + 2 * \text{covariance}(x, y),$$

where covariance$(x, y) = \Sigma(x_i - \bar{x})(y_i - \bar{y})/(n-1)$. This is the sum of product of deviations and is applicable when each value of x has correspondingly one value of y. Note that $(x + y)$ can be calculated only in such paired case. For variance of $(x - y)$, the sign before covariance becomes negative. For linearly independent x and y, covariance becomes zero. Covariance is explained further in Chapter 16 in the context of correlation.

Example 7.7: Pulse Pressure as Difference and its Variance

Pulse pressure is the difference between systolic pressure (SysBP) and diastolic pressure (DiasBP). A value around 45 mmHg is considered normal. A genuine low value such as less than 25 mmHg may occur in shock or aortic stenosis. A high value such as more than 60 mmHg may indicate stiffness of major arteries or a leak in the aortic valve. Thus, the pulse pressure itself is important medical parameter irrespective of actual values of systolic and diastolic level.

Suppose 12 persons have systolic and diastolic levels as shown in Table 7.10. Pulse pressure is also shown. Their mean and SDs are also

TABLE 7.10

Illustration of Relation of Variance($x - y$)
with Variance(x) and Variance(y)

	SysBP x	DiasBP y	Pulse Pressure $x - y$
	116	72	44
	124	82	42
	145	89	56
	139	87	52
	140	93	47
	118	77	41
	136	85	51
	127	79	48
	129	84	45
	110	67	43
	115	79	36
	128	83	45
Mean	127.25	81.42	45.83
Variance	124.20	52.08	29.24
Covariance(x, y)	73.52		

shown. Note that mean$(x - y) = $ mean$(x) - $ mean(y), but variance$(x - y) \neq$ variance$(x) - $ variance(y). In this case, covariance$(x, y) = 73.52$—thus, variance of pulse pressure $= 124.20 + 52.08 - 2 \times 73.52 = 29.24$. This is the same as obtained in Table 7.10 by directly computing the variance of differences.

7.5.1.4 Measuring Variation in Skewed and Nominal Data

Mean and, consequently, SD are inappropriate measures in case of highly skewed data. In that case, one reasonable measure of variability is

$$\text{Inter-quartile range: IQR} = (Q_3 - Q_1),$$

where Q_1 and Q_3 are the first and third quartiles. This range covers middle 50% of values.

In case of nominal scale, index of qualitative variation is used in place of SD. This index is given by

$$\text{Variation ratio: VR} = 1 - \frac{f_m}{n}, \tag{7.14}$$

where f_m is the frequency in the modal class.

If

No disease = 10%,
Mild disease = 40%,
Moderate disease = 30%,
Serious disease = 15%, and
Fatal disease = 5%,

then

$$\text{VR} = 1 - \frac{40}{100} = 0.60.$$

If most cases are concentrated, the VR is small.

7.5.2 Coefficient of Variation

An SD of 5 mmHg for systolic BP readings is small, but an SD of even 3 g/dL for Hb level is large. This is because the SD has to be assessed in relation to the values themselves or their mean. If the mean systolic BP level of the subjects under study is 132 mmHg, then SD = 5 mmHg is only 3.8% of the mean. If the mean Hb level is 15 g/dL, SD = 3 is 20% of the mean. The latter SD is surely higher, relatively. Thus, SD by itself cannot be used to *compare*

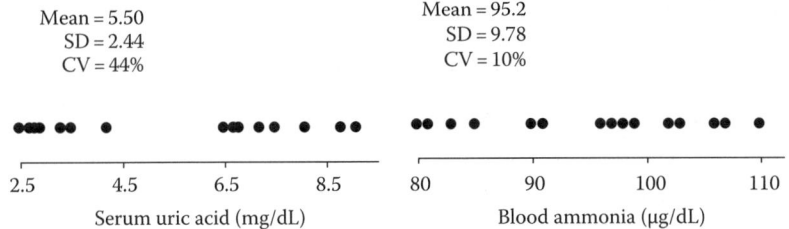

FIGURE 7.6
Less SD but higher CV of serum uric acid levels relative to blood ammonia levels.

variability in two different kinds of variables. Its ratio with the mean can be used. This gives rise to the following measure:

$$\text{Coefficient of variation: CV} = \frac{\text{SD}}{\text{Mean}}. \tag{7.15}$$

Note that the units of measurement (g/dL or mmHg or any other unit) cancel out in such a ratio, which makes the CV independent of units.

Coefficient of variation (CV) can be used to compare variability in different variables that otherwise is not comparable. For example, it can be concluded on the basis of CV that the variation between healthy individuals in thermoregulation (oral, rectal, and skin temperatures) is small and that in renal functions (urinary flow, creatinine clearance, urea clearance, etc.) is high. In the former case, the SD does not generally exceed 1% of the mean value, but in the latter case, the SD could be as much as 50% of the respective averages. Cardiovascular functions (systolic/diastolic BPs, heart rate, etc.) are somewhere in between with SD around 5%–15% of the mean.

Figure 7.6 shows plots of value of serum uric acid and blood ammonia in a sample of subjects. They are plotted in a manner that the variability looks equal. It all depends on the scale chosen on the x-axis. In these subjects, the level of serum uric acid is between 2.5 and 9.1 mg/dL. The SD is 2.44 mg/dL, and the CV is 44%. Against this, the level of blood ammonia ranges from 80 to 110 µg/dL but the SD is only 9.78 µg/dL. The CV is only 10%. Thus, variability is much lower in blood ammonia than in serum uric acid. The graphs could be deceptive.

Example 7.8: CV and Brainstem Function

The R-R interval in the ECG can be used to evaluate brainstem function. This interval is highly variable in healthy subjects. The R-R interval is measured several times in a subject in one ECG, and the mean and SD are obtained. Thus, CV can be obtained for each subject separately. Nezu et al. [5] studied 18 children with severe brain damage and 22 controls. They obtained the results as shown in Table 7.11.

TABLE 7.11

Mean CV of R-R Intervals in Cases with Different Conditions

Condition	n	Mean CV of R-R Intervals (%)
Neurologically normal controls	22	5.56
Patients complicated with respiratory insufficiency (RI)	10	2.19
Patients with severe athetotic cerebral palsy (SA)	8	11.30
Patients with brain death (included in RI group also)	4	1.00–1.29

The authors did not investigate the cause of the higher CV in the severe athetotic cerebral palsy (SA) group but concluded on the basis of the extremely low CV in patients with brain death that the CV of the R-R interval may be useful for quantitative evaluation of severe neurological deficit. In general, the CV of R-R intervals is also reduced in the aged and possibly also in subjects suffering from essential hypertension.

Chang et al. [6] have described a useful application of CV. They measured the nuclear diameter of cancer cells by ocular micrometry and calculated the CV. Recurrent cases and cases in which death occurred had higher CV than their respective complements. They concluded that the extent of variation as measured by CV of nuclear diameter of cancer cells offers a prognostic adjunct to standard clinical and histological analysis.

Although CV is unit free, it is affected by change of origin and scale since mean and SD are affected. Thus, it can give misleading result for data on interval scale where zero has no substantive meaning and a constant can be added or subtracted. Otherwise also, if some values are negative and others positive, mean could be small or close to zero. In this case, CV will undesirably inflate. Thus, use CV for ratio scale data only where zero is defined and values generally positive. Since mean and SD are not appropriate for highly skewed data, use CV = IQR/Median for this setup. In case of minor skewness, the same CV as defined earlier can be used.

References

1. Indian Council of Medical Research. *Growth and Physical Development of Indian Infants and Children.* Tech Rep Ser No. 18. New Delhi, India: Indian Council of Medical Research, 1972.
2. Agresti A. *Categorical Data Analysis.* New York: John Wiley & Sons, 1990, pp. 244–250.
3. Das P, Kumar P, Gupta CK, Indrayan A. Clinical patterns of abdominal tuberculosis. *Am J Proctol* 1975; 26:75–86.

4. Eriksson JG, Forsen T, Tuomilehto J, Osmond J, Barker DJ. Early growth and coronary heart disease in later life: Longitudinal study. *Br Med J* 2001; 322:949–953.
5. Nezu A, Kimura S, Kobayashi T, Osaka H, Uehara S. Coefficient of variation of R-R intervals in severe brain damage. *Brain Dev* 1996; 18:453–455.
6. Chang IC, Kuo SH, How SW. Coefficient of variation of nuclear diameters as a prognostic factor in papillary thyroid carcinoma. *Anal Quant Cytol Histol* 1991; 13:403–406.

8

Presentation of Variation by Figures

Tools such as graphs, diagrams, charts, and maps are commonly used for visual display of data. They are generally referred to as figure in the literature, although this term also includes nondata-based pictures such as of a bone or of a lesion.

Visual display is considered a powerful medium for communication and for understanding the basic features of a set of data. Salient features are often easily brought forth by an appropriately drawn figure. Also, the impression received from such a figure seems to be more vivid and lasts longer than the impression from numeric data. The main function of a figure is to provide perception and cognition of the basic features of the data to the viewer. However, for this an appropriate diagram must be chosen and correctly drawn. A large number of methods are available, and the choice is not always easy.

This chapter: Generally speaking, a graph is a figure drawn to a scale. The scale could be on a horizontal axis or on a vertical axis or both. Graphs are used to display the relationship of variables or to display the pattern of their distribution. A very useful graph is a frequency curve, which is discussed in Section 8.1. The term graph is disappearing from medical literature and the term diagram is preferred instead. Bars or lines (Section 8.2) are essentially graphs but are colloquially called diagrams. Factually, a diagram is a figure depicting data not necessarily to a scale. The schematic depiction of three methods of observational studies in Figure 4.1 is an example of this kind of diagram. There are many others in this book.

Electrocardiograms (ECGs), electroencephalograms (EEGs), and chymographs are examples of data-based figures commonly used in the practice of medicine. They are far too specialized to be discussed in this text. But there are others not so specialized diagrams such as growth charts, dendrograms, and partograms that are typically used in health and medicine. They are discussed in Section 8.3.

A chart in the context of medicine is a figure that shows the interrelationship of various factors. But the factors in the case of a chart are generally stated in the form of text. Refer to Figure 2.1 on types of medical studies for an example. However, use of this term to denote a graph is not uncommon. A growth chart is an example that, in fact, is a graph.

In medicine, the term map is generally used for a trace of a body part, such as a gene map of the human chromosome or a map of a lesion. Statistically, a map is a conventional geographic map showing the boundaries and location of areas. Epidemiological maps are used to effectively display the interregional variation in incidence and prevalence rates of various diseases. Details of charts and maps are given in Section 8.4.

8.1 Graphs for Frequency Distribution

Table 8.1 contains a frequency distribution of subjects, who are attending a hypertension clinic, by their level of total serum cholesterol. Note that the categories are not equal. Levels up to 200 mg/dL are considered safe and 200–239 mg/dL are not considered in the danger zone. They form one category each. Higher values cause alarm. For this reason, subsequent intervals are smaller, thereby allowing adequate monitoring. The last category puts all readings from 340 to 399 mg/dL in one group. (No subject had a level of 400 mg/dL or more in this group.) Such a distribution can be graphically depicted in the following ways.

8.1.1 Histogram and Its Variants

A histogram is a set of contiguously drawn bars showing a frequency distribution. The bars are drawn for each group (or interval) of values such that the area is proportional to the frequency in that group. The variable values are plotted on the horizontal (x) axis and the frequencies are plotted on the vertical (y) axis. This diagram has certain variants.

TABLE 8.1

Distribution of Subjects Attending a Hypertension Clinic by the Total Serum Cholesterol Level

Cholesterol Level (mg/dL)	Number of Subjects (f)	Percentage
−199	3	3.7
200–239	13	15.9
240–259	16	19.5
260–279	17	20.7
280–299	24	29.3
300–319	6	7.3
320–339	0	0
340–399	3	3.7
Total	82	100.0

8.1.1.1 Histogram

A histogram for the data in Table 8.1 is shown in Figure 8.1a. Note the follow-ing for these data:

1. The first is an open interval because the lower end point is not men-tioned. For the purpose of drawing the histogram, it is assumed to be 160–199, that is, of the same width as the next interval.

2. Thus, the first and the second intervals have width 40 each, whereas all others except the last have width 20. For the *area* of the bar to rep-resent frequency, the height of the bar for the first two intervals is half of the frequency in these intervals. This suitably adjusts for the double width of these intervals.

3. The interval 320–339 has no subject, and this is represented by a blank space.

4. The last interval has width 60. This is three times the width of most of the other intervals. To adjust this width, the height of the bar for this interval is one-third of the frequency in that interval.

The vertical axis can represent percentage instead of frequency. This will affect the scale but not the shape of a histogram. The following two conditions are prerequisites for a histogram to be a valid representation: (1) Characteristics to be represented on the horizontal axis must be on a metric scale, preferably a continuous variable such as cholesterol level in Table 8.1. The categories must be numeric intervals and must be mutu-ally exclusive and exhaustive. Ordinal and nominal scales are not rep-resented by a histogram, nor are multiple responses. (2) The data to be represented on the vertical axis must be either percentages that add to 100 or frequencies that add to total n. They cannot be other values such as means, rates, or ratios.

These two conditions are fairly severe and restrict the use of histogram to a very specific kind of data.

8.1.1.2 Stem-and-Leaf Plot

A variant of the histogram is a stem-and-leaf plot. This shows the actual values as in Figure 8.1b for another set of data on systolic blood pressure (BP). In this figure, the first two digits of systolic BP are considered the stem and the third digit the leaf. The shape is similar to that of a histogram except that the representation in this figure is horizontal rather than vertical. Recurrence of the same values becomes more evident in this representation, but intervals that are not multiples of 10 or are unequal are difficult to dis-play by a stem-and-leaf plot.

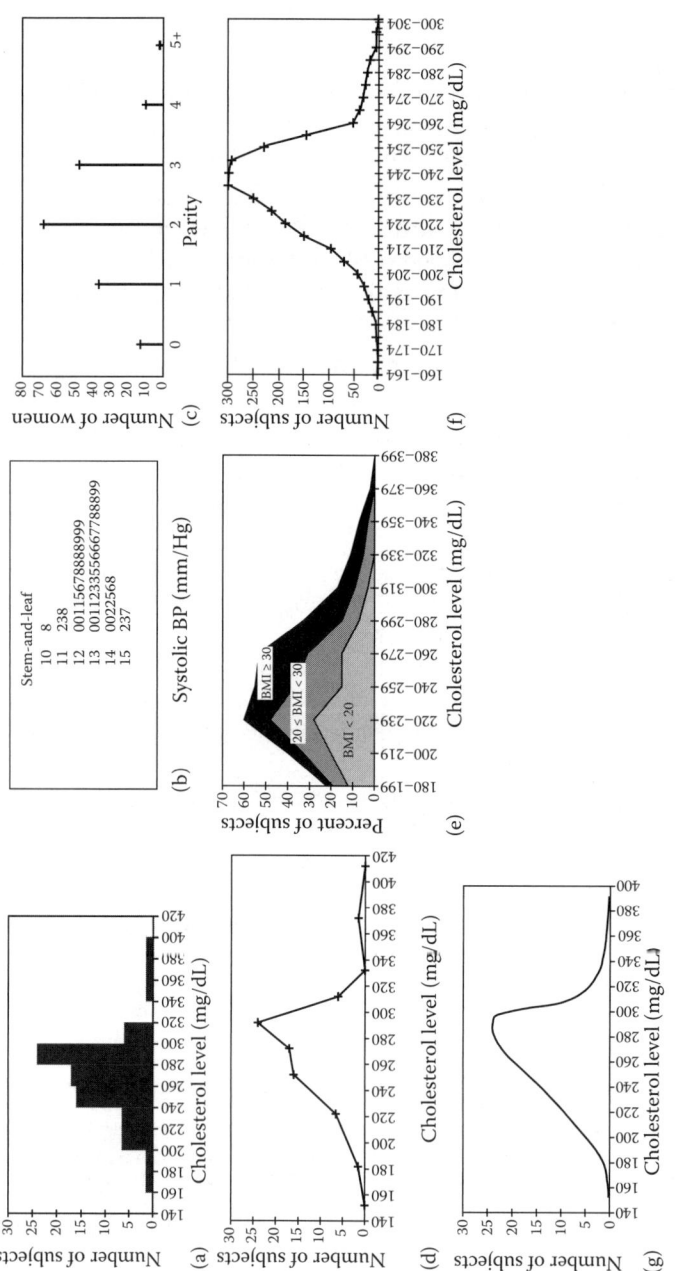

FIGURE 8.1

Different forms or representation of frequencies. (a) Histogram for cholesterol data in Table 8.1. (b) Stem-and-leaf plot for a systolic BP data. (c) Number of women with various parities attending a family welfare clinic. (d) Frequency polygon for cholesterol data in Table 8.1. (e) An area diagram for cholesterol levels in different BMI categories. (f) Polygon when $n = 3000$ and cholesterol interval $= 5\,\text{mg/dL}$. (g) Frequency curve for the cholesterol data in Table 8.1. Note the following: (1) The frequency histogram in (a) transforms to polygon in (d) and then to curve in (g). (2) The stem-and-leaf plot in (b) is horizontal version of a histogram. (3) The area diagram in (e) is showing distribution for individual BMI categories as well as the cumulative picture. (4) The polygon in (f) is taking a shape of a curve because of large n and small intervals.

8.1.1.3 Line Histogram

When the variable is really discrete with a small number of values, such as parity, the frequency can be represented by vertical lines in place of bars, as shown in Figure 8.1c. Another representation is by stacked dots as in Figure 7.1.

8.1.2 Polygon and Its Variants

This category of diagrams comprises the frequency polygon and the area diagram. Their details are given in the following.

8.1.2.1 Frequency Polygon

A polygon is a shape enclosed by straight lines. A frequency polygon is drawn only in cases in which a histogram can be drawn. Thus, the two conditions mentioned earlier apply to the polygon also. For a polygon, plot the points corresponding to the frequency (or percentage) on the midpoint of the class intervals, and join them by straight lines. This procedure is the same as joining the midpoints of the tops of the bars in a histogram. For the data in Table 8.1, the polygon is shown in Figure 8.1d. The frequency before the first interval and that after the last interval is zero each, and so lines are also drawn to join zero points. This completes the polygon as an enclosed shape.

8.1.2.2 Area Diagram

The area diagram is a variant of a polygon. In many cases, the area under the polygon has a useful interpretation, such as the total number of subjects. Figure 8.1e shows the distribution by cholesterol level of subjects of different body mass index (BMI) values in a hypertension clinic. But the BMI levels are stacked over one another. The sequential addition of the frequencies gives the cumulative distribution. In this case, more obese subjects commonly have higher cholesterol levels and the distribution changes shape from one BMI category to another.

8.1.3 Frequency Curve

Imagine the shape of a frequency polygon when the number of subjects is extremely large and the width of intervals extremely small. If the data in Table 8.1 are for 3000 cases and the cholesterol intervals are only 5 mg/dL, the polygon will tend to take the shape of a curve. This is shown in Figure 8.1f. Theoretically, a curve would be obtained when the subjects are infinite and the intervals infinitely small. This is called a frequency curve. Quite often, such a curve takes a regular shape and is best obtained by means of a mathematical equation. Such an equation seems to work well as a good approximation to the data arising in practice. Some regular shapes of frequency curves are discussed in Chapter 9.

Instead of depending entirely on mathematics, a fairly good approximation can be obtained by drawing a smooth curve closest to the polygon. A few points may be far off from the curve so drawn. The curve for the data in Table 8.1 is shown in Figure 8.1g. Note that a frequency zero in the interval 320–339 mg/dL is ignored to achieve a smooth and regular shape of the frequency curve. When an extremely large number of measurements are available, the frequency in the interval 320–339 mg/dL would not be zero.

Basic information provided by the histogram, polygon, or curve is the nature of the distribution of the subjects over various values of the variable, that is, whether they are evenly distributed or are concentrated around some value, and whether the values are widely scattered or are compact. Thus, these figures are more of an exercise in data exploration than in analysis of data.

8.2 Pie, Bar, and Line Diagrams

The histogram, polygon, and frequency curve are meant exclusively to display the frequency distribution of a variable on a metric scale, preferably continuous. Various other figures can be drawn for other kinds of data.

8.2.1 Pie Diagram

A pie is a circular diagram divided into segments, each segment representing frequency in a category. If the frequency in a category is f_k out of n, the angle of the corresponding segment is $360^* f_k/n^\circ$. It is necessary that $\Sigma f_k = n$, that is, the categories must be mutually exclusive and exhaustive. A pie cannot correctly display a multiple response.

As in the case of a histogram, percentages can replace frequencies in the case of a pie. In addition, the categories can be nominal or ordinal. Such categories are not admissible for a histogram. Grouping of continuous data, as in the case of a histogram, is also required for a pie diagram if not already present. Discrete data may or may not be grouped. This means that a pie can be drawn for all datasets valid for a histogram as well as in some other situations. When the metric data are in categories, the functional difference between a pie and a histogram is that the pie is not adequate to depict the frequency distribution over various values. In addition, pie does not provide a good representation when there are a large number of categories. An adequate histogram can still be drawn. However, pie is better in depicting the concentration of values in one category *relative* to

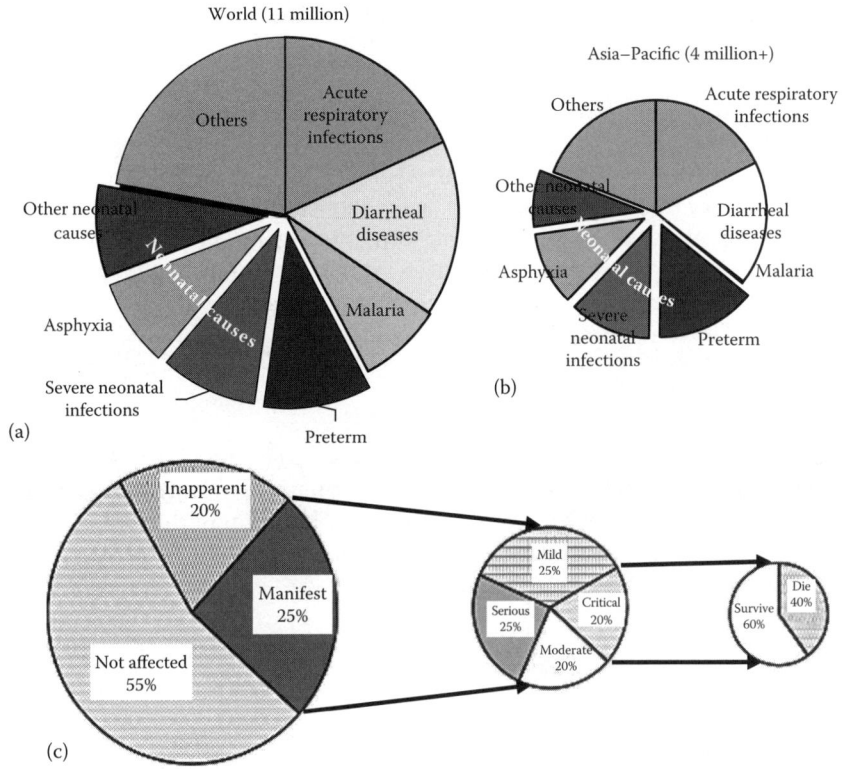

FIGURE 8.2
Pie diagrams: (a) Causes of under-five mortality—World. (b) Causes of under-five mortality—Asia–Pacific 2000–2003. (From WHO, *World Health Report 2005*, World Health Organization, Geneva, Switzerland, 2005.) (c) Spectrum of disease by connected pies. (From Indrayan, A., *Basic Methods of Medical Research*, 2nd edn., AITBS Publishers, Delhi, India, 2008.)

the other categories. See Figure 8.2a, which is a pie diagram for causes of death of children of age under 5 years in the world [1]. Acute respiratory diseases and diarrheal disease are the predominant causes. Figure 8.2b is for Asia–Pacific region. Figure 8.2c depicts the disease spectrum with the help of a series of pie diagrams that are linked to a segment of the previous pie [2]. In this diagram, manifestation of disease in mild, moderate, serious, and critical proportions is shown. Also shown is the proportion of survival and death in the case of critical manifestation of the disease.

As far as possible, each segment of pie must be labeled and must show the number or percentage of subjects it represents.

8.2.1.1 Useful Features of Pie Diagram

One useful application of pie diagram is in comparing the relative distribution of two unequal groups of subjects divided into the same categories.

Figure 8.2a is based on 11 million deaths of children of age less than 5 years in the world per year during 2000–2003. Figure 8.2b is based on 4 million such deaths in Asia–Pacific.

Higher *n* in the world is represented by a proportionate increase in the size of the pie. The segments are still comparable, and the greater predominance of malaria in the world relative to the Asia–Pacific is clear. Such comparability in groups of unequal size is rarely achieved by any other type of diagram.

Another useful feature of a pie diagram is **wedging**. When attention is to be specifically drawn to one particular category, the segment representing that category is wedged out as shown for neonatal deaths in Figure 8.2a and b. A pie can also be **exploded** so that all segments are wedged out.

8.2.1.2 Donut Diagram

Make a hole in a pie and get donut. Both have same functionality and same requirement. The total must be a meaningful quantity. Donut comparison of age distribution of males and females in cases of mild hypertension is shown in Figure 8.3. Females are nearly one-half of males as shown by size of donuts in this figure. Note how large percentage of 35–44 years in females is shown relative to the males.

8.2.2 Bar Diagram

The bar diagram is the most common form of representation of data and is indeed very versatile. If the data are mean, rate, or ratio from a cross-sectional study, then the bar may be the only appropriate diagram. It is especially suitable for nominal or ordinal categories although it can be drawn for metric categories as well.

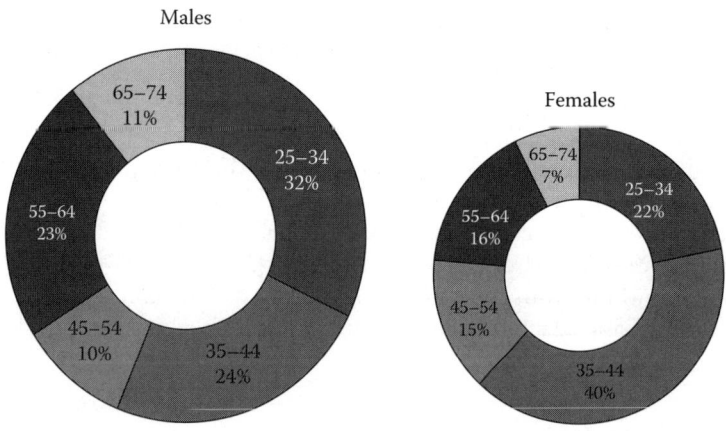

FIGURE 8.3

Donut diagrams: Age (in years) distribution (in %) of males and females in cases of mild hypertension.

Figure 8.4a represents the percentage distribution of palpable breast lumps by histopathological diagnosis in different age groups. This is called **divided bar** diagram. As age advances, the relative incidence of carcinoma increases according to these data. If the total number of women with palpable breast lumps were highest in the age group 45–49 years, then these bars would not show this information. Each bar represents the total cases in the respective age groups as 100%.

Figure 8.4b is a **multiple bar** diagram showing the mean and standard deviation (SD) of some lung functions in male workers in factories with different pollutants. Both inter- and intrafactory comparisons can be made with the help of this diagram.

The bars do not have to be vertical. Sometimes horizontal bars give a better representation. Figure 8.4c is one such diagram, simultaneously showing

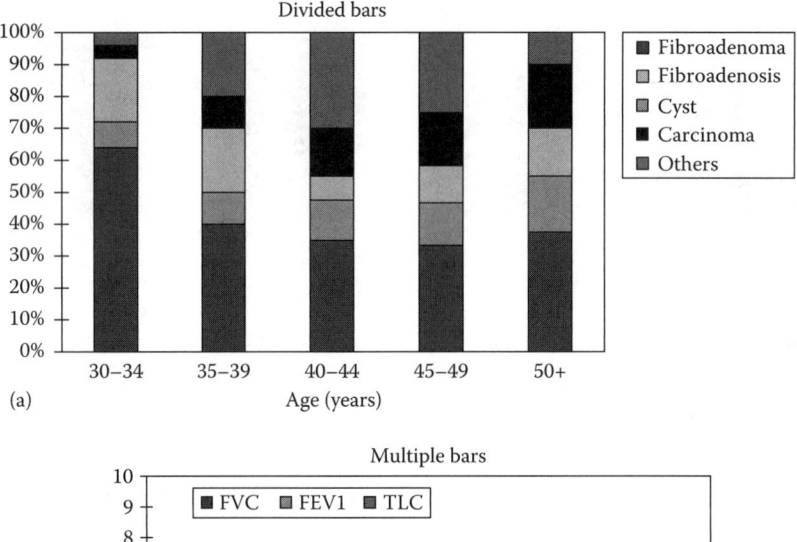

(a)

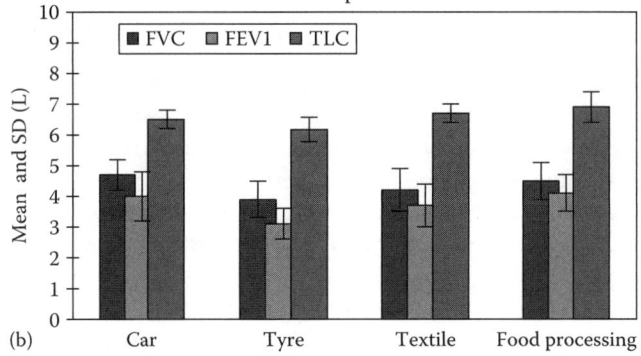

(b)

FIGURE 8.4
Different types of bar diagram: (a) Palpable breast lump by histopathological diagnosis in different age groups. (b) Mean and SD of lung functions in male workers in different types of factories.

(*continued*)

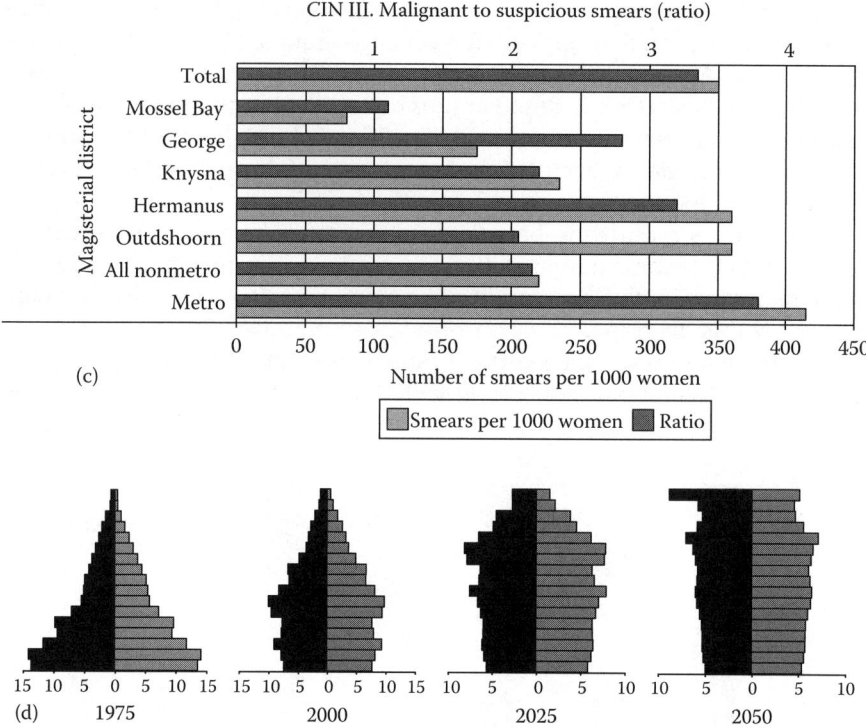

(c)

(d)   1975   2000   2025   2050

FIGURE 8.4 (continued)
(c) Ratio of malignant to suspicious smears and their screening incidence by magisterial district, Western Cape, South Africa. (From Bailie, R. and Bourne, D., Surveillance for equity in cervical cytology screening, *Int. J. Epidemiol.*, 25, 46–52, 1996, by permission of Oxford University Press.) (d) Population pyramids (UN projections for a developing country—1975, 2000, 2025, and 2050).

two pieces of information for some magisterial districts in South Africa [3]. One horizontal axis measures the number of cervical smears per 1000 women and the other the ratio of malignant to suspicious smears. Note how a bar diagram can quickly become complex. Space between bars is kept same unless one particular category is to be highlighted.

A special bar diagram used in demography is the **population pyramid**. This shows the number or percentage of males and females in different age groups in a population. Figure 8.4d shows the changing age–sex distribution of the population in a developing country over a long period. Vertical axis at the center is age, and horizontal bars on the left side are for females and on the right side for males. *The age groups must be equal (except the last) for this kind of depiction.* Note the striking difference between the shapes of the population pyramids for the different periods. The shape initially in the year 1975 is triangular with a broad base, showing predominance of the child population. Decline in population in every age group shows that deaths occur at almost every age at that stage of development. Fifty years later, the distribution of population up to the age of 60 years is nearly uniform. Practically no deaths occur in these age groups because they tend to live longer. After that the deaths are fast and frequent. As the population evolves, the women of old age groups increase. This is clearly visible in the pyramid for the year 2050.

Sometimes you would find it difficult to properly interpret a bar diagram you see in the literature because the information is incomplete or the diagram is not properly drawn. Wherever feasible, each bar should show the value it represents. If a bar shows that 42% cases of age 50 years and above reporting pain in knees had arthritis, mention 42% at the top of the bar or inside the bar so that the reader can immediately understand. Bar diagrams in Figure 8.4 have this deficiency.

8.2.3 Scatter and Line Diagrams

A scatter diagram is considered a good representation because it depicts all the observations. A line diagram is used to show trend. The details are as follows.

8.2.3.1 Scatter Diagram

A scatter diagram aims to show the variation in the values of one variable in relation to another. Simple examples are BP values in subjects of different ages and cholesterol levels in subjects with different BMIs. Different symbols can be used to distinguish between male and female subjects. If one variable is dependent on the other, the dependent variable is shown on the vertical (y) axis. This must be a metric variable in all scatter diagrams. In Figure 8.5a, the percent cholesterol esters is shown as dependent on total bilirubin in cases of cirrhosis, hepatitis, and common bile duct (CBD) obstruction. In the presence of regurgitation jaundice, the flow of bile into the duodenum may decrease, causing a reduction in the esterified form of cholesterol. Note an outlier in this figure. Scatter diagrams provide the opportunity to spot such outliers.

You can see that **outliers** are those obstinate values that are located in isolation in the graph—not fitting with the others. Many of these can be attributed to errors such as misplaced decimal, wrong method of measurement, wrong reporting, misuse of instrument and wrong record. Sometimes, rarely though, the outlier may be right and other values wrong such as when first value is actually observed and the rest are model-based estimates that are way-off. At other times, an outlier may be a genuine out-of-the-box value for a very unusual person or patient.

When the scale on the x-axis is ordinal or nominal, the scatter takes the shape of stacking of symbols over one another as shown in Figure 8.5b. Different symbols are used for different severity of jaundice in this diagram.

Already it is difficult to untangle the points and identify the pattern for cirrhosis, hepatitis, and CBD obstruction as shown in Figure 8.5a. The problem

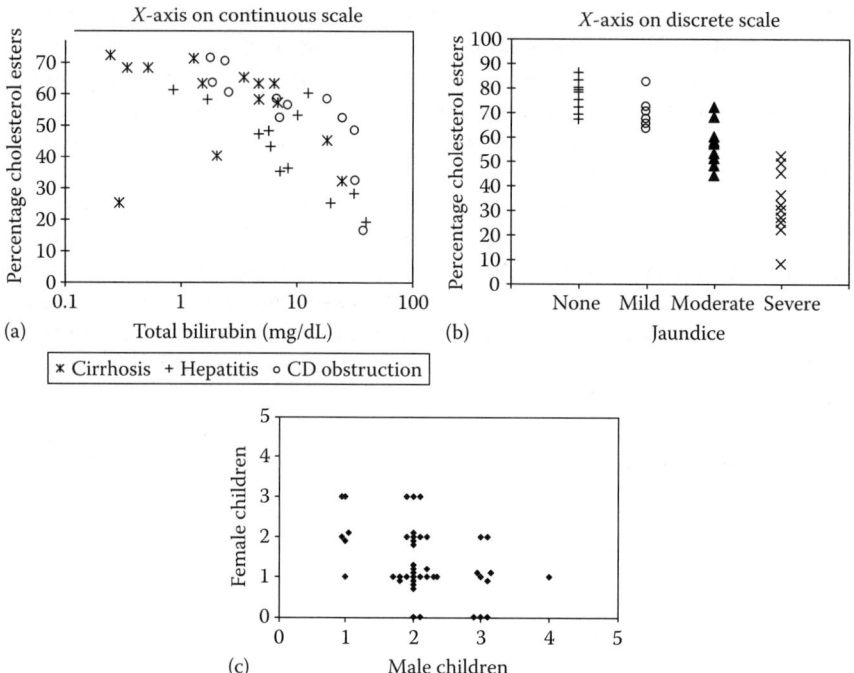

FIGURE 8.5
Three types of scatter diagram: (a) Dependence of percentage cholesterol esters on degree of jaundice measured by total bilirubin. (b) Percentage cholesterol esters in cirrhosis patients with different degrees of jaundice. (c) Use of jitter for overlapping points—male and female children at the time of sterilization in families preferring male children.

accentuates when measurements with different units are shown on right and left axes. The next problem occurs due to too many points at one place in a scatter diagram that may look like a vague blob. For such overlapping points, try jitter, which is a minor perturbation for plotting purposes only, and use small size symbols (see, e.g., Figure 8.5c).

8.2.3.2 Line Diagram

A line diagram is used to show the trend of one variable over another. Figure 8.6a shows the trend of infant mortality rate against the socioeconomic status of families in a developing country. The plotted points are joined by a line. It can be called **time plot** when x-axis is time. It continues to be called a line diagram even when the representation is by a curve instead of a straight line. The variable on the y-axis could be an average as in Figure 8.6b. Note how the points on the x-axis are adjusted for unequally spaced values. This facility is not available in a bar diagram. Such adjustment can be done when the x-axis is on a metric scale but not when it is on an ordinal scale. The question of trend does not arise in the case of nominal

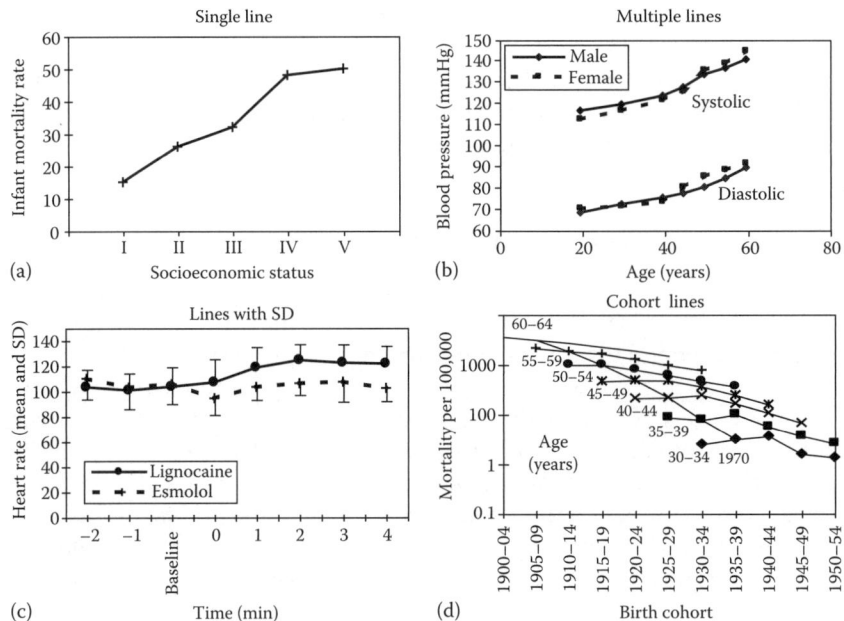

FIGURE 8.6

Different types of line diagram: (a) Infant mortality rate per 1000 live births by socioeconomic status in an area. (b) Trend of average systolic and diastolic BP over age in males and females in a general population of adults. (c) Heart rate in women with pregnancy-induced hypertension undergoing cesarean section—esmolol and lignocaine groups. (d) Trend in age-specific CHD mortality per 100,000 in Australia from 1960 to 1992 within 5-year birth cohorts for females aged 30–64 years (y-axis is on logarithmic scale). (From Wilson, A. and Siskind, V., Coronary heart disease mortality in Australia: Is mortality starting to increase among young men?, *Int. J. Epidemiol.*, 24, 678–684, 1995, by permission of Oxford University Press.)

scale because then the categories do not have any particular order. A quality control chart presented later in Chapter 20 is an entirely different type and illustrates a typical use of a line diagram. One line diagram you have already seen is the ogive given in Figure 7.2.

8.2.3.3 Complex Line Diagrams

Two or more lines can be drawn in one diagram (Figure 8.6b) but more than five lines make it very complex and difficult to understand. Another complexity arises when the SD of the variable for each point on the x-axis is shown on the y-axis by vertical lines on either side as in Figure 8.6c. This gives an indication of the variation present in the data. As explained later in Chapter 21, representation of 1 × SD on either side can be fallacious because the actual variation is much more.

An interesting representation by line diagram could be of mortality in several cohorts. Figure 8.6d shows the age-specific coronary heart disease

(CHD) mortality rate in different 5-year birth cohorts for females in Australia from 1910–1914 to 1950–1954 [4]. Note that the cross-sectional lines can be used to show the age trend in mortality in a particular group of years. One such line is drawn for the years 1970 ± 5. This is because various cohorts attain different ages in the same calendar year.

8.2.4 Choice and Cautions in Visual Display of Data

Before going on to more complex representations, it may be helpful to review and get a clear picture of situations in which one diagram is more appropriate than the others. This is shown in Table 8.2.

When the range of values on the y-axis is extremely large, examine whether or not a logarithmic scale would be more appropriate. This scale converts 10 into 1, 100 into 2, 1000 into 3, 10,000 into 4, etc. One such representation appears in Figure 8.5a where the total bilirubin on the x-axis is plotted on a log scale. If the ordinary scale, called linear, is used, the points for low levels of bilirubin would become indistinguishably close. Another log scale is on the y-axis of Figure 8.6d. One advantage of a log scale is that it can represent change relative to the previous value. Cautious interpretation is needed in this case.

Note the following points regarding various diagrams:

1. A diagram should not be too complex. A large number of relationships can be shown in one diagram, but it is advisable to restrict one diagram to not more than two relationships or not more than three variables.

TABLE 8.2

Appropriate Type of Diagram for Various Situations

Vertical Axis (Must Represent a Quantity)	Scale on Horizontal Axis			
	Metric and Continuous No Categories	Metric Categories or Discrete	Ordinal	Nominal
Frequency or percentage	×[a]	Histogram[b]	Bar	Bar
		Pie[c] in these three cases when contribution of each category relative to the others is to be shown		
Mean, rate, or ratio	×[a]	Bar	Bar	Bar
		Line in these two cases for trend		
Values of a variable	Scatter	Scatter	Scatter	Scatter

[a] Frequency, percentage, mean, rate, or ratio is not possible on y-axis when x variable is continuous and has no categories.
[b] Polygon or frequency curve.
[c] Pie diagram has no horizontal or vertical axis.

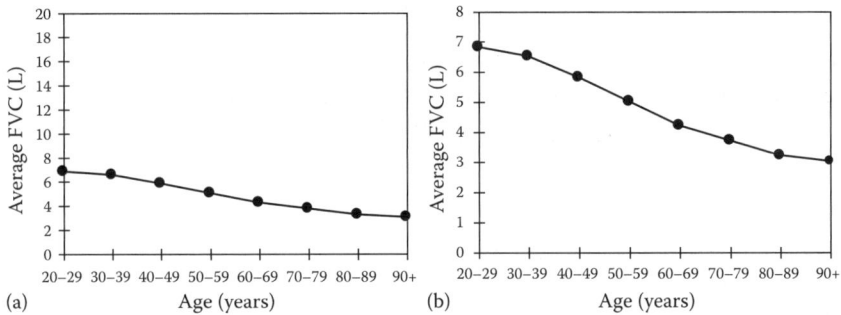

FIGURE 8.7
Effect of choice of scale on the slope or a line: (a) Relation of average FVC with age. (b) The same trend with a different scale on the *y*-axis.

2. **Choice of scale**: A diagram can be made to show a steeper or flatter relationship by choosing the scale accordingly. Figure 8.7a and b show the same relationship between age and average forced vital capacity (FVC), but FVC is plotted on different scales in the two figures, giving a different impression.

3. Scale of calibration should be clearly indicated. Choose a scale that is suitable and does not exaggerate or understate the values. For comparing two or more groups, use the same scale for both the groups.

4. In case secondary data are used to draw a figure, provide the source.

5. Figures displaying all the data points, such as a scatter diagram, are preferable because they allow readers to draw their own conclusions. However, sometimes it is not feasible to show the entire set of data. Only summary statistics such as percentages, rates, or averages are shown.

6. All graphs and diagrams must be self-explanatory, containing informative (what, where, and when) yet concise titles, legends to identify various components of a diagram, labels for axes, unit of measurement, etc.

7. Use diagrams mostly for exploring the data rather than to draw conclusions. Because the sample size *n* is important for conclusions and small or large *n* does not affect many diagrams, conclusions based on a diagram alone can be very fallacious. Details of such fallacies are given in Chapter 21.

8. As far as possible, directly label the categories, lines, etc., so that reader does not have to refer to legend or text for this purpose.

A graph or a diagram is necessarily an approximate depiction. An average 103.2 and 103.4 can be shown distinct in a table but would look the same in a graph. Thus, *graphs are good for visual display of pattern or trend but not to depict exact values.*

8.2.5 Mixed and Three-Dimensional Diagrams

After describing simple diagrams, the stage is now set to discuss more complex diagrams. Among many such diagrams that can be discussed, the following are quite commonly used in health and medicine.

8.2.5.1 Mixed Diagram

Sometimes it is advisable to simultaneously show the variation in two characteristics, each measured on the same x-axis. This definitely saves space and helps to provide a more comprehensive view. Figure 8.8a shows the age-wise prevalence of cataract blindness in a community and the percentage operated (surgical coverage). The former is shown by bars and the latter by a line. Thus, this is a mixed diagram.

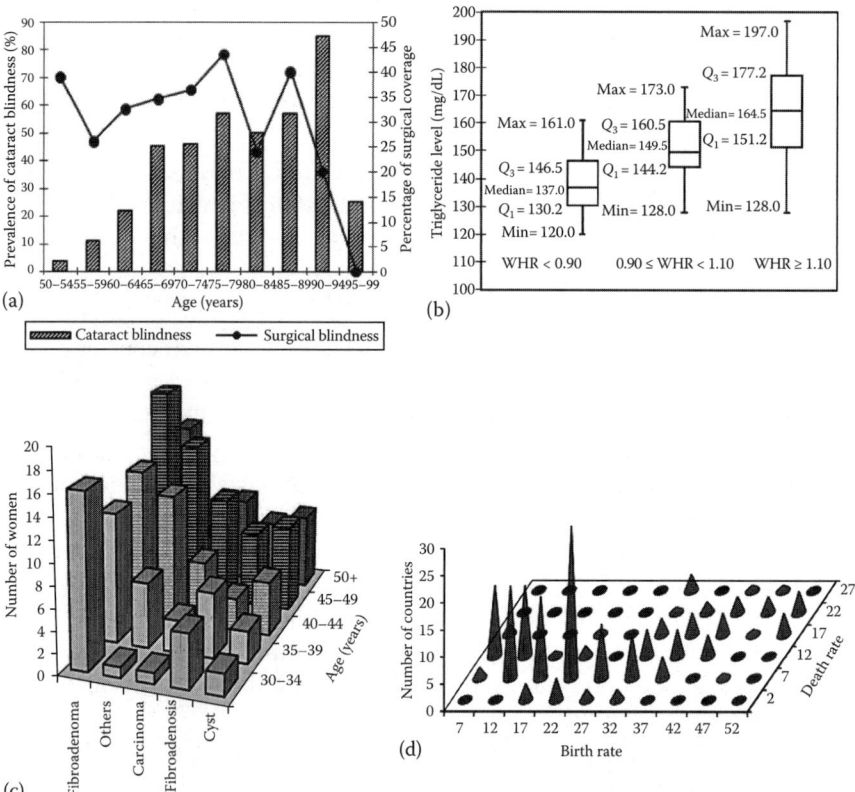

FIGURE 8.8
Some types of mixed and 3D diagrams: (a) Age-wise prevalence of cataract blindness (one or both <6/60) and the percentage of surgical coverage. (b) Box-and-whiskers plots for TGLs in different WHR categories. (c) Distribution of palpable breast cases in different age groups by histopathological diagnosis. (d) Distribution of countries in the world by birth rate and death rate, 2003.

Although both measurements on y-axis in Figure 8.8a are percentages, they are on a different scale. This itself makes reading difficult. Moreover, if the units are different, then different measurements on the y-axis can create confusion.

8.2.5.2 Box-and-Whiskers Plot

A diagram considered useful in data exploration is a box-and-whiskers plot. A box is made at the value of the median with two divisions. The lower end of the box represents the first quartile Q_1 and the upper end the third quartile Q_3. The height of the box, thus, is the interquartile range (IQR) and covers middle 50% of the values. The larger the box, the greater is the spread of the values. The width of the box is arbitrary. Vertical lines are drawn from the "minimum" to the lower end of the box (Q_1) and from the upper end of the box (Q_3) to the "maximum." These look like whiskers. Values more than three interquartile distances away from the median are considered clear outliers and marked with an asterisk, and values more than 1.5*IQR (called inner fence) but less than 3.0*IQR (called outer fence) are considered mild outliers and marked by 0. The "minimum" and "maximum" values used to draw whiskers exclude these mild and clear outliers. The diagram can be made horizontally also.

A box-and-whiskers plot (also called a **box-plot**) for the triglyceride levels (TGLs) in different waist–hip ratio (WHR) categories for the data in Table 3.1 is shown in Figure 8.8b. The values of median, Q_1 and Q_3, as well as the minimum and maximum values are also shown. In this case, there are no outliers. As just stated, a tall box indicates that the data values are widely dispersed. A short box would show that they are compact. The figure on the whole displays the location of the median and the dispersion of the data as well as the skewness. The plot for three WHR groups in Figure 8.8b shows that the dispersion of TGL is fairly spread out and is symmetric when WHR ≥ 1.1 but relatively compact and very skewed (the distance between the median and Q_1 is very different from the distance between the median and Q_3) when $0.9 \leq$ WHR < 1.1. Rising box-and-whiskers plots in this figure also shows that TGL rises as WHR increases.

8.2.5.3 Three-Dimensional Diagram

Many computer packages readily draw three-dimensional (3D) diagrams. The additional variable is shown on a third axis. A 3D representation of a relationship is a response surface of the type shown later in Chapter 16 (Figure 16.8). Another is in Figure 8.8c, which shows the distribution of palpable breast cases in different age groups by their histopathological diagnosis. This is a 3D representation of Figure 8.4a. Sometimes the bars can be replaced by spikes as in Figure 8.8d. This figure shows the distribution of countries around the world by birth rate and death rate in the year 2003. The height of the spike indicates the number of countries. Many countries had a birth rate around 22 per 1000 and a death rate around 7 per 1000 population in that year. Some countries had a birth rate exceeding 50 and some had a death rate less than 5 per 1000 population.

8.2.5.4 Biplot

A biplot is a graphical display of multivariate data. Although two or three variables are easy to depict, depiction of a larger number is difficult. How can mean plasma glucose level (both fasting and postprandial), weight, and BP (systolic and diastolic) be shown for two groups of subjects in the same figure? Several methods have been proposed that can show four or even five variables simultaneously. However, the methods are complex and the figure so obtained is difficult to interpret. Such a figure, as of now, does not seem to serve the purpose of better perception and adequate cognition that a figure is supposed to serve. Thus, further details are not given. Interested readers may see Gabriel and Odoroff [5].

8.2.5.5 Nomogram

A set of scaled lines, each representing a metric variable, arranged in a diagram such that the value of one variable can be read in relation to the values of two or more other variables, is called a nomogram. The scales on the lines are differentially calibrated so that corresponding values of the variables lie on a straight line. This obviates the need to do calculations because the values can be read by using a simple ruler. The WHO Expert Committee [6] has given a nomogram that computes BMI for a given height and weight.

Figure 8.9 shows a nomogram designed to find the number of clusters required for a survey to estimate the prevalence of a disease with specified precision when a guesstimate of prevalence is used and the cluster size or the design effect is known [7]. This was used for rapid surveys to assess the prevalence of senile cataract blindness in a community. As an illustration, a cross-sectional line is also drawn, which shows that when the guesstimate of the prevalence of blindness due to senile cataract is $p = 0.04$, the ratio of design effect to cluster size $(D/B) = 0.05$, then the number of clusters required is nearly 460 for $\alpha = 0.05$ for estimating prevalence within one-tenth of its value. The notation α is for probability of Type-I error, which is explained later in Chapter 12.

Such nomograms are used in several different applications. Lawoyin and Onadeko [8] developed a nomogram to predict the birth weight category of babies born at term on the basis of the maternal weight changes at different periods of gestation. Studd [9] constructed a nomogram that is claimed to separate normal labor from abnormal labor in primigravidas admitted at different stages of cervical dilation. Partin et al. [10] developed a nomogram based on clinical stage, Gleason score of the prostate needle biopsy, and serum prostate-specific antigen. This was designed to improve the ability to predict the pathological stage of a prostate tumor.

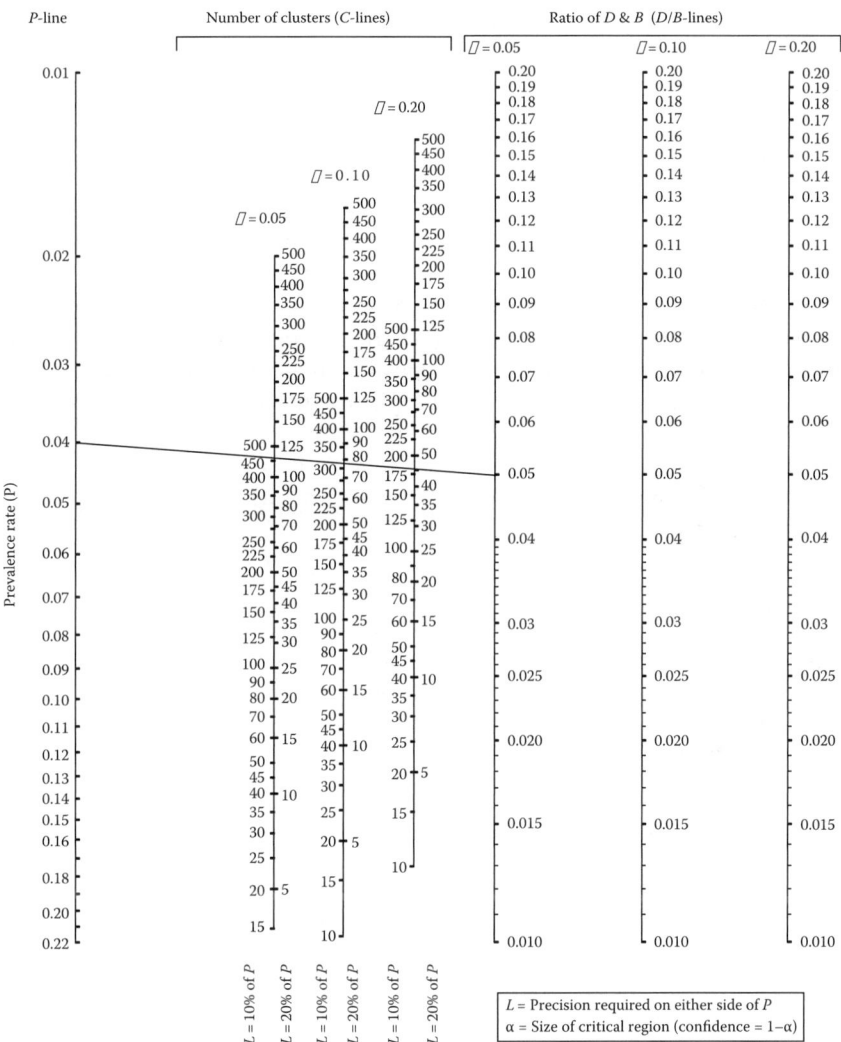

FIGURE 8.9
Nomogram: number of clusters required for rapid assessment of prevalence of a disease for different cluster sizes. (From Kumar, A. and Indrayan, A., *Int. J. Epidemiol.*, 31, 463, 2002.)

8.3 Special Diagrams in Health and Medicine

Among the most common diagrams used by clinicians is the ECG. This is a line diagram showing the cardiac electrical activity over time. Time is measured in fractions of seconds. P, Q, R, S, T, and U waves are recorded. The PR interval, duration of the QRS complex, shape of the ST segment, QT intervals, etc., are used to assess various cardiac functions. All these show variation even in healthy subjects and thus come under the domain of statistics. However, following the convention, these are kept out of the scope of this book. The same is true for many other diagrams such as the EEG, electromyogram, and electrooculogram.

At least five other types of diagrams are statistical in nature and used typically in health and medicine. The first is a growth chart, used to assess the growth and nutritional status of a child. The second is a dendrogram, used to show affinity or similarity of one entity with others. The third is an epidemic curve, which depicts the progress of an epidemic. The fourth is partogram, which is used to monitor the progress of labor in childbirth. The fifth is area under the curve diagram, which is used in pharmacokinetic study of drugs. Many of these and some others are discussed in this section.

Discussion is easy when the diagrams specially used in health and medicine are divided into those used in public health and those used in personal care and research.

8.3.1 Diagrams Used in Public Health

Three special diagrams used in public health are epidemic curve, lexis diagram, and isobars grid. If you have not come across these diagrams in medical literature, the chance is that you would see them at some point in your professional career.

8.3.1.1 Epidemic Curve

An epidemic of a disease is said to exist when the occurrence is clearly in excess of the usual occurrence. The term was originally used for infectious diseases such as influenza and cholera but is now used for conditions such as deaths in vehicular accidents.

In the case of infectious diseases, the number of cases gradually rises, reaches a peak, and then starts to decline. A graph showing the number of cases from the beginning to the end of the epidemic is called an epidemic curve. Time is plotted on the horizontal axis and the number of cases on the vertical axis. Even though this could be a bar diagram, it is interpreted as a curve because it shows trend over time. An infectious disease epidemic in a population that has a common source (such as contaminated water for cholera) would look like Figure 8.10a. A sharp peak occurs a couple of days later than the incubation period, but tapering off may not be as sharp because of varying time taken in recovery. The Bhopal gas tragedy in India and the Minamata disease outbreak in Japan caused by eating fish containing high concentrations of methyl mercury are well-known examples of common source epidemics.

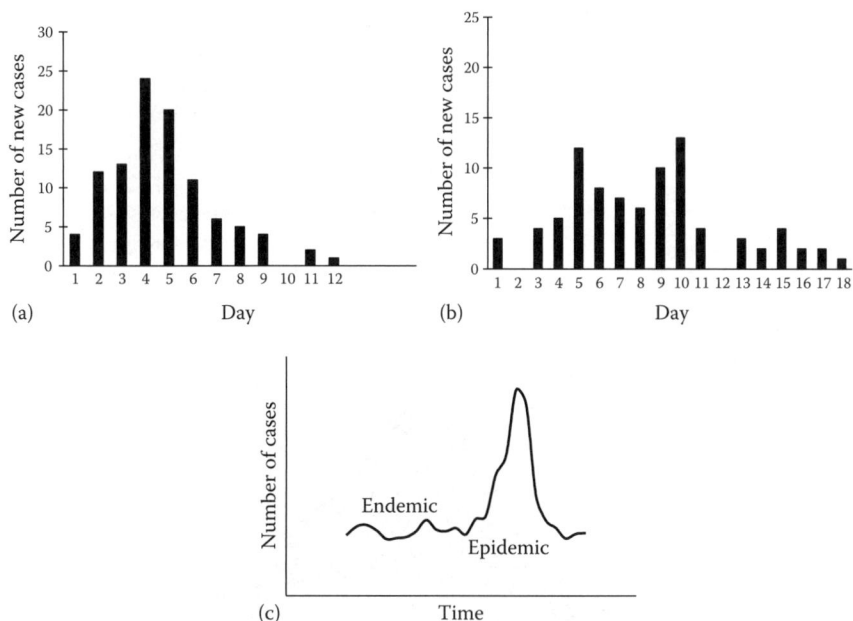

FIGURE 8.10
Epidemic curves: (a) common source, (b) propagated, and (c) endemic versus epidemic levels.

A propagated epidemic results from person-to-person transmission (e.g., of hepatitis A and polio). It shows a gradual rise, sometimes intermittent peaks, and a very gradual fall (Figure 8.10b). The speed of spread depends upon herd immunity, opportunities for contact, and the secondary attack rate.

A steep rise in cases over the endemic level is shown in Figure 8.10c. It helps to identify the time of start and finish of epidemic. This is the real epidemic *curve* and not a bar diagram.

8.3.1.2 Lexis Diagram

A very typical diagram that shows age-specific mortality in different birth cohorts is a contour-like diagram as shown in Figure 8.11c. This figure shows the lung cancer mortality in Australian males in 1950–1985 by age at death and year of death [11]. This is called a lexis diagram, although the authors of the article call it a **synoptic chart**. The data displayed in Figure 8.11c can be conventionally shown by a line diagram (Figure 8.11a) or in three dimensions by a surface chart (Figure 8.11b). The birth cohorts are indicated by broken lines in Figure 8.11a. Solid lines connect points with equal mortality rate (per 100,000 males per year). Perhaps easiest to understand is the surface in Figure 8.11b. It can be easily seen, for example, that the mortality rate at age around 70 years steeply increased during the period 1970–1980 and showed a slight decline thereafter.

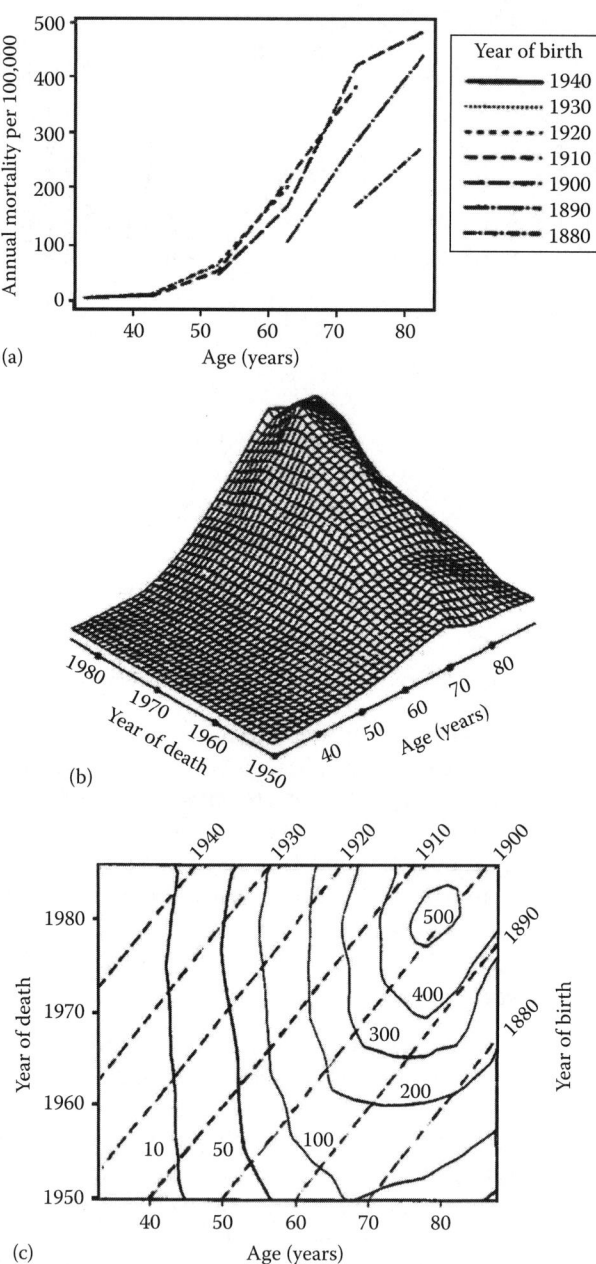

FIGURE 8.11
Mortality from lung cancer in Australian males during 1950–1985: (a) line diagram, (b) surface chart, and (c) lexis diagram. (Adapted from Jolley, D. and Giles, G.G., Visualizing age-period-cohort trend surfaces: A synoptic approach, *Int. J. Epidemiol.*, 21, 178–182, 1992, by permission of Oxford University Press.)

8.3.2 Diagrams Used in Individual Care and Research

Growth chart and partogram are typical diagrams that can help in assessing the health of individuals, and dendrogram and radar graph are examples of diagrams used in medical research. The details of these diagrams are given next.

8.3.2.1 Growth Charts

Among the dimensions of physical growth of children are weight, height, and head circumference. Each can be assessed against age or against each other. For the purpose of illustration of the type of diagram they are, see Figure 8.12a and b, which is weight-for-age charts for girls, developed by the United States National Center for Health Statistics (NCHS) and the World Health Organization (WHO).

The charts show various percentile curves based on a large number of healthy children. Thus, this technically is a line diagram but is called a chart. Fiftieth percentile is like a median that can be used as a reference, and 3rd and 97th percentile curves define the lower and upper limits for healthy growth. Space between these two percentile curves is considered road to health. Sometimes Z-scores (Chapter 9) are used instead of percentiles.

A growth chart is used for longitudinal monitoring rather than for one-time cross-sectional assessment. The *trend* should follow the same pattern as the reference curve. A flattening or declining trend relative to the reference is indication of decline in nutritional status and steeper upward trend not crossing the upper limit indicates improvement.

Further details of how these charts are used for assessing various aspects of growth are presented in the next chapter.

8.3.2.2 Partogram

A partogram is used by midwives in field conditions, especially in developing countries, for prevention of prolonged labor in childbirth. Progress of labor is recorded in a graph in terms of cervical dilation against time. Observations are recorded at regular time intervals, say, every hour. The partogram contains an alert line and an action line (Figure 8.13). Crossing the alert line is associated with fetal distress. Neonatal resuscitation is more likely if the alert line is crossed. If the action line is also crossed, the chances of stillbirths are higher. The function of a partogram is to provide early warning for detection of abnormal progress of labor. Dujardin et al. [12] found the partogram useful and efficacious in a study in Senegal. Less postpartum sepsis was reported [13] after implementation of the partogram in a multicenter trial done in Indonesia, Malaysia, and Thailand.

8.3.2.3 Dendrogram

Consider the problem of dividing a fixed number of entities, n, into a small number of homogeneous groups, K, with respect to J measurements.

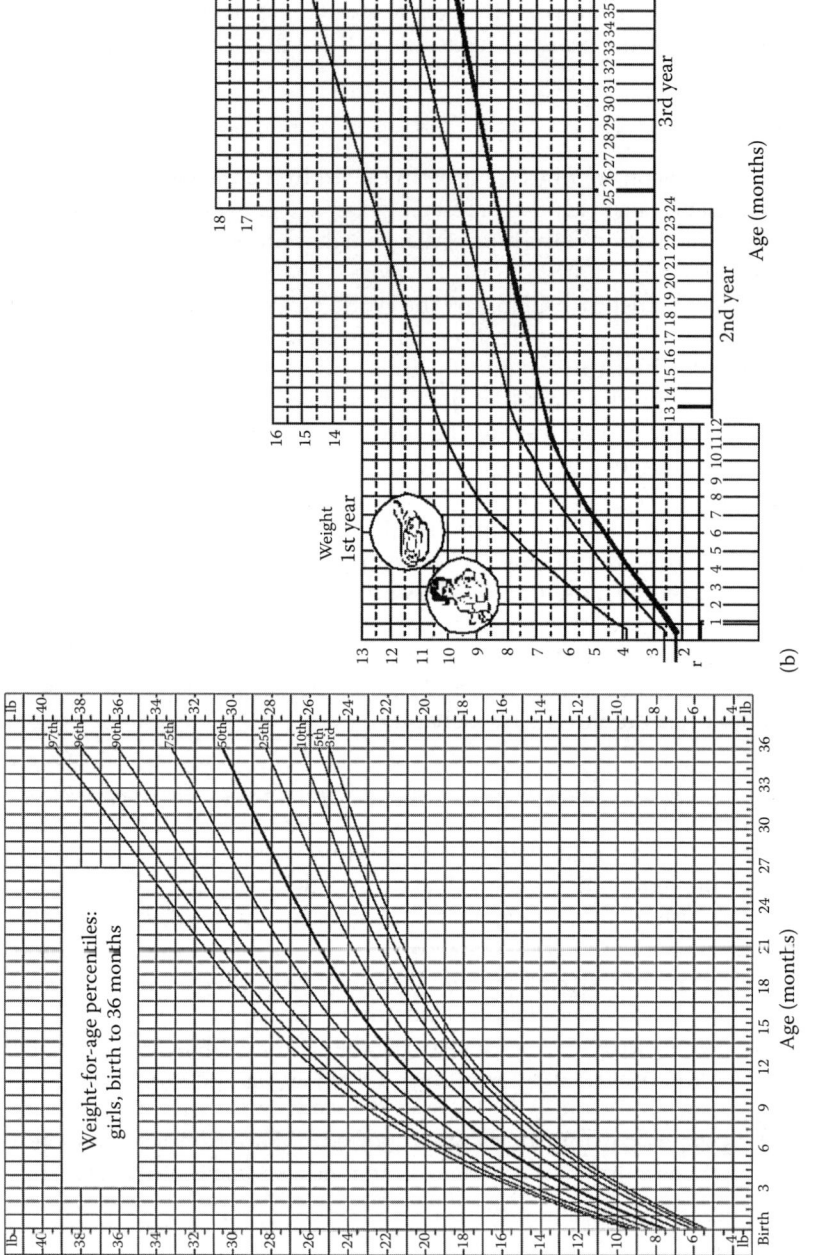

FIGURE 8.12
Weight-for-age charts (girls): (a) NCHS and (b) WHO.

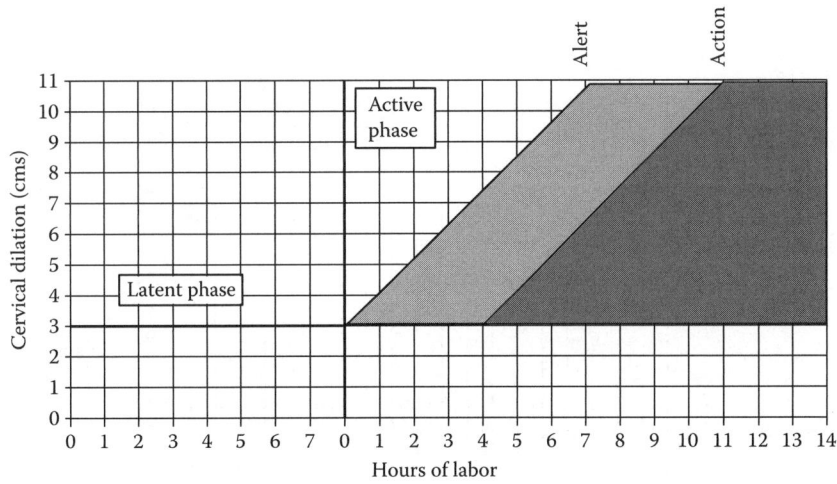

FIGURE 8.13
Partogram.

This could be like dividing various hemoglobinopathies into groups that are internally similar but distinct from others on the basis of, say, anion high-performance liquid chromatography (HPLC) values of hemoglobin (Hb) A2, Hb variant, and HbF, even Hb, red blood cell (RBC), mean corpuscular volume (MCV), mean corpuscular hemoglobin (MCH), etc. A measure of affinity or similarity is obtained, which helps to classify the entities. When the measurements are metric, an opposite, that is, distance, is generally measured. This is mostly done in terms of what is called a Euclidean distance. For present/absent types or variables, other indices of similarity are used. See Romesburg [14] for details of how this and other distance measures are obtained and how the entities are divided into homogeneous groups.

The statistical technique used for such division is known as **cluster analysis**. A large number of methods are available. One broad category of cluster methods is hierarchical in nature. Hierarchical clustering goes on merging similar entities sequentially until all *n* entities are finally merged into one big group. A dendrogram is a graphical display of the merging taking place at each stage. For example, Figure 8.14 shows the sequential merging of 20 SDS–PAGE (sodium dodecyl sulfate–polyacrylamide gel electrophoresis) groups on the basis of whole cell protein profile in strains of *Pseudomonas aeruginosa* [15]. This is one of the many forms that a dendrogram can take. The groups are rendered progressively less homogeneous as the merging takes place. If no extraneous indication of the number of homogeneous groups is available, the number can be determined statistically by observing a sudden jump in the value of a coefficient at each stage of the hierarchical process. A brief discussion of this method is in Chapter 19, which is on such multivariate methods.

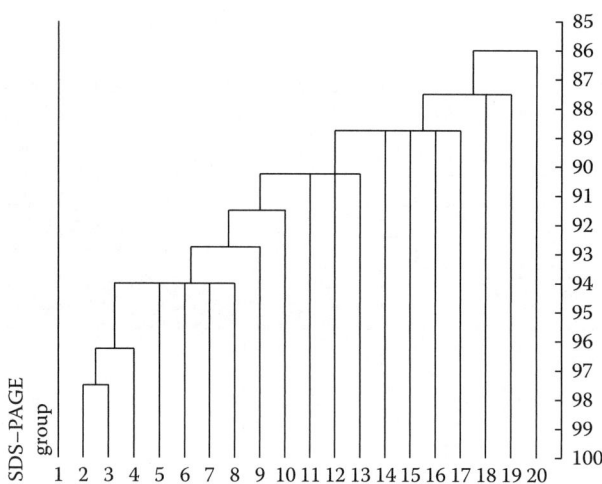

FIGURE 8.14

Dendrogram: Sequential merging of SDS–PAGE groups on the basis of whole cell protein profile in strains of *Pseudomonas aeruginosa*. (From Khan, F.G. et al., *Indian J. Med. Res.*, 104, 347, 1996. With permission of *Indian Journal of Medical Research*.)

8.3.2.4 Area under the Concentration Curve

In pharmacokinetic studies, the bioavailability of drug at different points of time is studied. In a clinical trial setup, you may be interested in response or outcome at different points of time. When the outcome, either in terms of percentage efficacy or in terms of average measurement, is plotted against time, you get a line diagram. This is generally called concentration curve. As a comprehensive measure of performance, area under this line is computed. This area can be compared between two or more groups to decide which one is better or worse than the others. See also Figure 21.4. The problem in comparing area under the concentration curve is discussed in Chapter 21.

8.3.2.5 Radar Graph

Radar graph provides a suitable presentation when performance of one group on 4–6 variables is to be compared with another group. As many axes are drawn as the number of variables and a polygon is drawn according to the values as shown in Figure 8.15a and b. This looks like a radar. The first figure compares average age, percent females, percent hypertensives, percent obese, and average hemoglobin level in cases and controls at baseline in a randomized controlled trial (RCT). The graph reveals that percent hypertensives are very different in the two groups but all other values are not much different. Second figure compares male and female diabetic patients for six measurements. Notice the marked difference in percent on diet control and with positive history. If one radar portrays general population averages for five or six variables

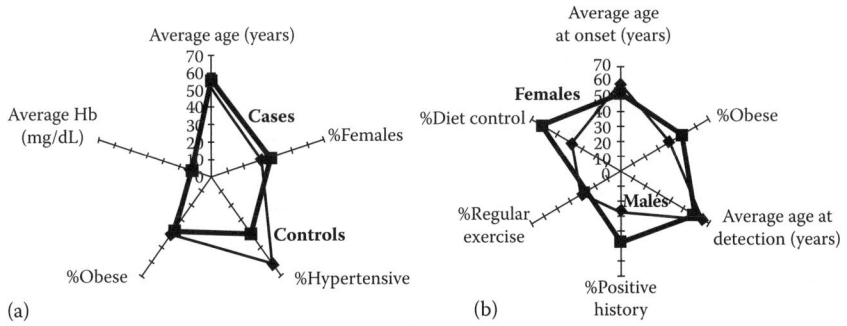

FIGURE 8.15
Radar graphs: (a) Comparison of some characteristics of cases with controls in a RCT; (b) comparison of some characteristics of male cases of diabetes with female cases.

and the other same variables for a specific group such as those suffering from cancer, such a graph can depict on which variable, if any, cancer patients exceed the general population and on which variable they have less value.

Radar graph seems like a good tool for multivariate comparisons. But only two groups can be cleanly compared. For three or more groups, this becomes too cluttered. As almost all other graphs, radar graph too can mislead depending upon the scale chosen for depiction. In Figure 8.15b, the axis for average age at detection can be stretched and then the difference between the two groups will magnify. Any axis can be stretched or compressed. The figure is most effective when all variables have same scale such as percentage.

8.4 Charts and Maps

Charts and maps are yet other types of figures used in health and medicine to illustrate relationships. These can be described as follows.

8.4.1 Charts

As mentioned earlier, charts use text, and the relationships are described with the help of boxes, arrows, and other shapes. Many intricate relationships can be easily explained with the help of such charts.

8.4.1.1 Schematic Chart

An example of such a chart is presented in Figure 8.16. It shows one possible scheme of multiphasic screening for diabetes mellitus. Another example is in Figure 2.2, which divides various medical studies into prospective, retrospective, etc. A third example is given in Chapter 10 (Figure 10.1)

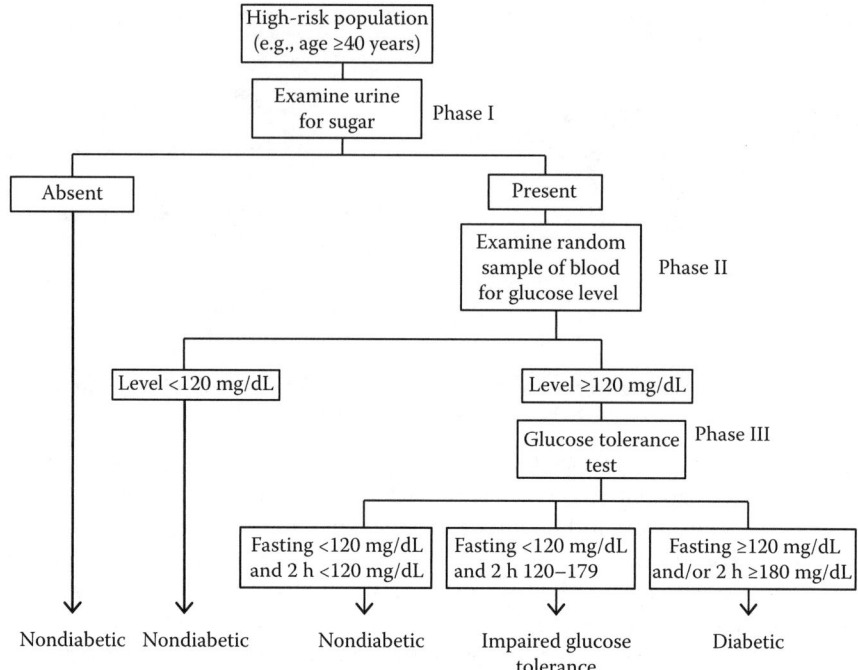

FIGURE 8.16
Chart showing a scheme of multiphasic screening for diabetes mellitus.

showing the interaction of various primordial and secondary factors causing myocardial infarction.

8.4.1.2 Pedigree Chart

Pedigree charts are commonly used for studying genetic diseases. One example is shown in Figure 8.17 [16]. The pattern in such a chart helps to identify a trait as autosomal dominant or recessive. The dominant traits (e.g., familial hypercholesterolemia) show a vertical pattern of inheritance (parents and children affected), whereas the recessive traits (e.g., beta-thalassemia) show a horizontal pattern of inheritance (siblings affected). Males are represented by squares and females by circles. Pedigree charts can help to investigate X-linked disorders such as hemophilia A and color blindness. Affected persons are represented by filled squares or circles and unaffected ones by hollow squares or circles. Carriers are represented by half-filled and half-hollow squares or circles.

8.4.2 Maps

Several types of maps can be drawn for medical data. The following are the commonly used ones.

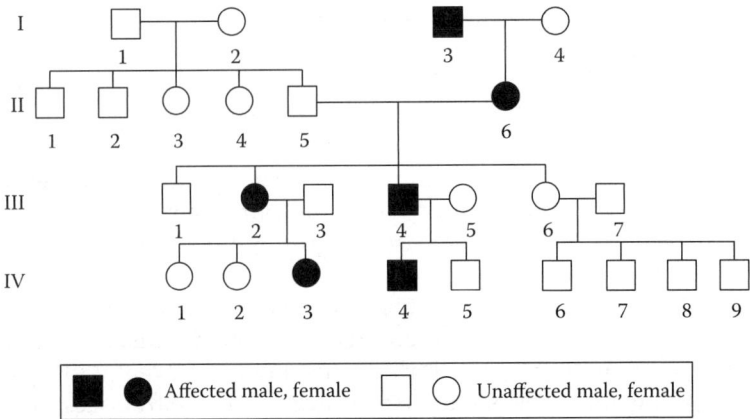

| ■ ● Affected male, female | □ ○ Unaffected male, female |

FIGURE 8.17

Pedigree chart. (From Wilson, J.D. et al. (eds.), *Harrison's Principles of Internal Medicine*, 12th edn., Vol. 1, International edition, McGraw Hill, New York, 1991. With permission of McGraw Hill Companies.)

8.4.2.1 Spot Map

Maps are a powerful medium for showing the spatial distribution of a disease or of a health condition. A dot can represent one or many cases (such as 1 dot = 10 cases). A concentration of dots in any area indicates that the incidence of disease is high in that area. Figure 8.18 is a map of India for HIV prevalence rate in persons with sexually transmitted diseases (STDs) in the year 2005 [17]. Areas with higher incidences have

FIGURE 8.18

Spot map: HIV prevalence in STD cases in India, 2005.

many dots. Such a map is called a spot map and can be used to investigate a localized outbreak of a disease.

8.4.2.2 Thematic Choroplethic Map

The second type of statistical mapping is called thematic choroplethic cartography. Areas with similar rates have the same shade and the shading becomes darker as the rate increases. Figure 8.19 shows the countries of Asia–Pacific by under-five mortality rate.

The four categories shown in Figure 8.19 are based on consideration of the statistical equivalence in rates based on cluster analysis. Quite often the categories in thematic maps are arbitrarily chosen of equal width and the number of categories also remains arbitrary. This introduces considerable subjectivity in the cognition and perception obtained from the maps. A discussion on this aspect is given by Indrayan and Kumar [18] who advocate that the categories should be natural, as dictated by the data in place of arbitrary choices. These natural categories can be identified by

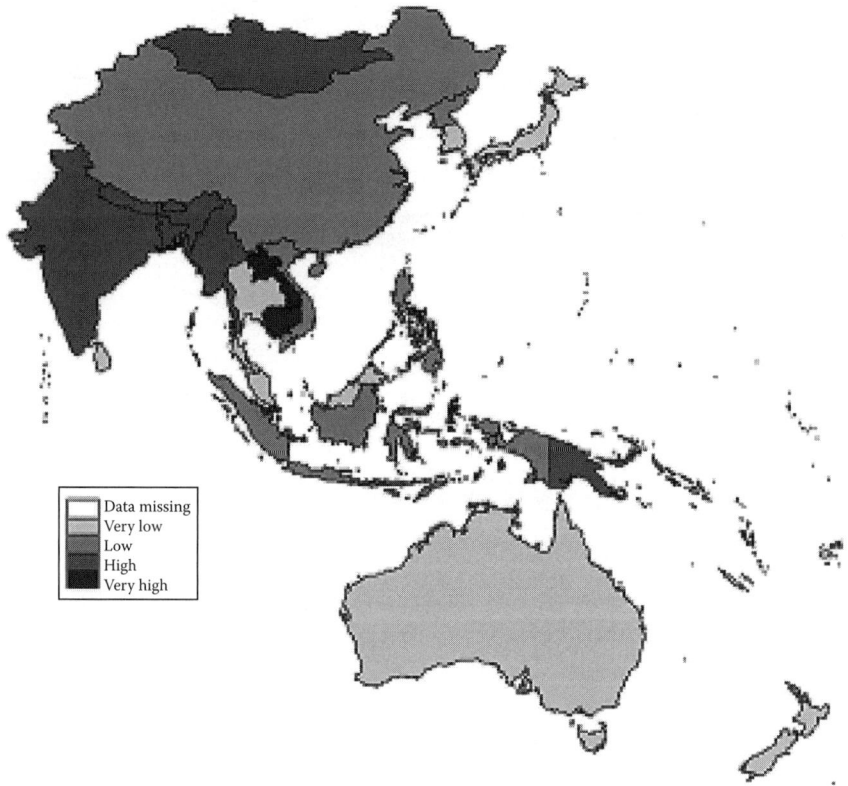

FIGURE 8.19
Thematic choroplethic map of under-five mortality rate in countries of the Asia–Pacific region.

consensus in the groups obtained by various methods of clustering and the picture thus obtained can be substantially different from the one based on equal-width categories.

8.4.2.3 Cartogram

Generally speaking, a cartogram is a map that depicts the size of various areas in proportion to the magnitude of the problem. A cartogram of 35 ECG leads is used in precordial mapping for diagnosis of combined myocardial ventricular hypertrophy. The concern in this text is mostly with epidemiological mapping. Sometimes it is convenient to show the area of a city or country in proportion to the population or the cases occurring in that city or country. The actual boundaries are approximated although efforts are made to retain the shape. An example of such a cartogram is Figure 8.20, which shows estimated mortality per million attributable to climate change by the year 2000 in 14 regions of the WHO [19]. Note how Africa and the Indian peninsula are oversized, and North and South America and Europe are drastically undersized.

I have discussed a large variety of figures used for depiction of data or relationships in health and medicine; however, the list is far from exhaustive. There are several other types of figures that have specialized applications.

Some new methods of diagrammatic representation will come up later in this text. A reference is already made to response surface, control chart, and the etiology diagram that appear in other chapters. Many other such diagrams can be made. For further details of visual display of data, see Tufte [20].

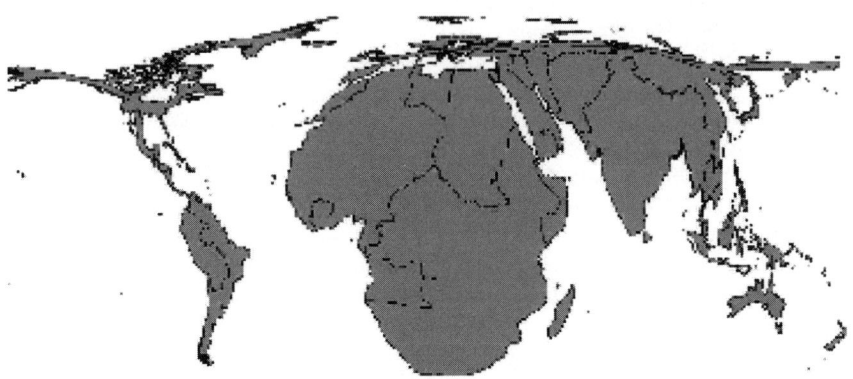

FIGURE 8.20
Cartogram: estimated mortality attributable to climate change in 14 regions of WHO. (From Patz, J., *Emerging Disease Threats from Global Ecological Change: Biological Threats Symposium,* Center for World Affairs and the Global Economy, Madison, WI, 2006.)

References

1. WHO. *World Health Report 2005*. Geneva, Switzerland: World Health Organization, 2005.
2. Indrayan A. *Basic Methods of Medical Research*, 2nd edn. Delhi, India: AITBS Publishers, 2008, p. 202.
3. Bailie R, Bourne D. Surveillance for equity in cervical cytology screening. *Int J Epidemiol* 1996; 25:46–52.
4. Wilson A, Siskind V. Coronary heart disease mortality in Australia: Is mortality starting to increase among young men? *Int J Epidemiol* 1995; 24:678–684.
5. Gabriel KR, Odoroff CL. Biplots in biomedical research. *Stat Med* 1990; 9:469–485.
6. WHO. Physical status: The use and interpretation of anthropometry. Technical Report Series No. 854, Geneva, Switzerland: World Health Organization, 1995, p. 434.
7. Kumar A, Indrayan A. A nomogram for single-stage cluster sample surveys in a community for estimation of a prevalence rate. *Int J Epidemiol* 2002; 31:463–467.
8. Lawoyin TO, Onadeko MO. A nomogram for screening low birth weight and large for gestational age babies for use in primary health care centres. *East Afr Med J* 1993; 70:746–748.
9. Studd J. Partograms and nomograms of cervical dilatation in management of primigravid labour. *Br Med J* 1973; 4 (5890):451–455.
10. Partin AW, Yoo J, Carter HB et al. The use of prostate specific antigen, clinical stage and Gleason score to predict pathological stage in men with localized prostate cancer. *J Urol* 1993; 150:110–114.
11. Jolley D, Giles GG. Visualizing age-period-cohort trend surfaces: A synoptic approach. *Int J Epidemiol* 1992; 21:178–182.
12. Dujardin B, De Schampheleire I, Sene H, Ndiaye F. Value of the alert and action lines of the partogram. *Lancet* 1992; 339:1336–1338.
13. World Health Organization Maternal Health and Safe Motherhood Programme. World Health Organization partograph in management of labour. *Lancet* 1994; 343:1399–1404.
14. Romesburg C. Cluster analysis for researchers. Lulu.com, 2004.
15. Khan FG, Rattan A, Khan IA, Kalia A. A preliminary study of fingerprinting of *Pseudomonas aeruginosa* by whole cell protein analysis by SDS-PAGE. *Indian J Med Res* 1996; 104:347–348.
16. Wilson JD, Braunwald E, Isselbacher KJ et al. (Eds.). *Harrison's Principles of Internal Medicine*, 12th edn., Vol. I, International edition. New York: McGraw Hill, 1991, p. 26.
17. National AIDS Control Organisation, India. Directory of HIV Data. http://www.nacoonline.org (last accessed on March 8, 2012).
18. Indrayan A, Kumar R. Statistical choropleth cartography in epidemiology. *Int J Epidemiol* 1996; 25:181–189.
19. Patz J. *Emerging Disease Threats from Global Ecological Change: Biological Threats Symposium*. Madison, WI: Center for World Affairs and the Global Economy, 2006.
20. Tufte ER. *The Visual Display of Quantitative Information*. Cheshire, CT: Graphics Press, 1983.

9

Some Quantitative Aspects of Medicine

This chapter is primarily concerned with individuals rather than groups. Earlier, I defined biostatistics as the science of management of uncertainties in health and disease. Basic to all this is measurement, because it is through measurement that quantities are obtained and uncertainties aptly studied. Medicine is more qualitative than quantitative, but qualitative measurements too are inherently assessed in terms of quantities. This can be explained as follows.

Why was malignancy considered an adverse health condition when it was first diagnosed in a person long ago? Not because it resulted in rapid deterioration of health in an isolated case but because several such cases were found to follow nearly the same adverse course. This can also be said about any sign–symptom syndrome that forms the sheet anchor of medical practice. Experience gained on a *group* of subjects is applied to a new individual in the hope that he too would follow the same pattern. Similarly, a treatment regimen is first tried on a *group* of subjects before it is accepted for use on individual patients. The likelihood of a response is evaluated, and then the regimen is used on new subjects. Health complaints and their relief are both qualitative characteristics at an individual level. However, the assessment depends on quantitative evaluation of their likelihood in terms of rate, proportion, or probability. Thus, qualitative measurements too are inherently assessed in terms of quantities even at an individual level.

Qualitative aspects such as complaints and relief are, and will remain, dominant features while dealing with individual patients. But there is also a realization that these are overwhelmingly subjective and depend mostly on the perceptions of the patients and the doctor. Although this perception is important in medical practice, the trend is to depend increasingly on quantitative measurements that are more reliable, objective, and exact. These measurements serve as aids and not the ends. The objective of medical and health practice is to provide relief to patients and to promote well-being, and these are more of a quality than a quantity. All measurements in medicine are carried out with this intention.

The science of medicine—generally identified with diagnosis, treatment, and prognosis—has increasingly become measurement oriented than subjective-assessments oriented, more quantitative than qualitative, and more evidence based than perception based. Evidence-based medicine is making a firm foothold, and the patients too sometimes demand evidence of the effect of the procedure that they are advised to undergo. Measurements are the cornerstone for achieving this objective. Because of the important and wide-ranging implications of medical measurements, I am dividing the discussion on it into two major parts. This chapter focuses on some specific measurements and their reliability and validity assessment, and the next chapter is on scores and risks that form the core of evidence-based medicine.

This chapter: Quantitative measurement in health and medicine can sometimes be difficult. Some epidemiological measures commonly used for assessing different aspects of health of individuals are discussed in Section 9.1. The usual practice is to evaluate various quantitative parameters of a subject against their reference values. The methods generally used to delineate such reference values as well as their implications are discussed in Section 9.2. This assessment is done in terms of probabilities because of variation and uncertainties, which are discussed in Section 9.3. Section 9.4 is on assessment of the validity of medical tests. These last two sections also apply to the qualitative measurements. Receiving operating characteristics (ROC) curves are used for assessing the overall performance of the tests. Details of these curves are presented in Section 9.5.

9.1 Some Epidemiological Measures of Health and Disease

Assessment of the level of health and disease in individuals can be very subjective. Perfect health can seldom be defined, and the meaning of well-being changes from person to person. The subject's own perception also changes from time to time. Health becomes an unattainable ideal in a true sense. This, however, does not distract scientists. Efforts are always made to measure the level of health. The focus, though, remains on lack of health rather than on its presence.

Positive health can be understood as the ability to cope with physical, biological, psychological, and social stress. This can become far too abstract. Yet, measurements such as hemoglobin (Hb) level, high-density lipoprotein (HDL), immunity level, vital capacity, and pain-bearing capacity can be possibly used to assess positive health. Same is true for low cholesterol, low sedimentation rate, and low bleeding time. Measurement of positive health has remained unexplored and requires the attention of the researchers. The profile of persons who rarely fall sick and are able to do more work than others while leading an enjoyable life can be studied to identify factors that contribute to positive health. It is possible, though, that psychological factors such as personality profile, absence of stress, and carefree attitude contribute more to positive health than physiological parameters. This is a flagship concept mooted in the first edition of this book in 2001 and has been picked up by others for exploration.

Multiplicity of measures mentioned hereafter in this section for some aspects of health raises the question of the choice. Choose the one that exactly measures the targeted aspect and possesses established reliability and validity. The chosen indicator must be unambiguous and should have biological relevance. If it is simple, inexpensive, easy to implement, and less inconvenient to the subjects, consider that a bonus in this age of technological complexity.

9.1.1 Epidemiological Indicators of Neonatal Health

Health of a child is determined from the time of conception. Some diseases such as thalassemia and sickle cell anemia have their origin in genes. Several others, such as coronary artery disease (CAD) and diabetes mellitus, possibly have a substantial genetic component. But an attempt to study the genetic profile is usually made only in cases in which a manifestation has occurred or is feared. Those apart, commonly used statistical measures of neonatal health are as follows.

9.1.1.1 Birth Weight

Although ultrasonographic measurements can be used to assess the growth of the fetus, the first measurement of physical health of a child after birth is the weight. This generally declines for a few days after birth and is then regained. If the weight immediately after birth cannot be recorded, the weight on the seventh day in many cases would be a good approximation. The normal range is 3.2–3.7 kg. A birth weight less than 2.5 kg is conventionally considered low in many countries. A low birth weight not only has been found to be associated with increased risk of early mortality but also is surmised to affect growth and development during adolescence and trigger diseases in adulthood such as CAD and diabetes. A useful index for neonatal weight is

$$\text{Ponderal index (for neonates)} = \frac{\text{Weight in g}}{(\text{Length in cm})^3} * 100.$$

A child with ponderal index 3.0 or more can be considered overweight but, in some conditions such as in maternal smoking, reduced length may also be implicated. An index between 2.5 and 3.0 is considered normal, between 2.0 and 2.5 marginal, and a child with an index less than 2.0 is classified as small for gestational age (SGA). When weight and length are both low, this index may not reveal the deficiency, but the prognosis is poorer. Such a symmetric SGA child is generally classified as intrauterine growth retarded (IUGR). This is identified by a very low weight but a nearly normal ponderal index. Some organizations do not distinguish between SGA and IUGR children, and both are identified only by low weight, generally below 10th percentile point for the gestational age, irrespective of the length of the child.

Another index of SGA is

$$\text{Birth weight ratio: BWR} = \frac{\text{Actual weight}}{\text{Expected weight for gestational age and gender}}.$$

A statistical relationship (third-degree polynomial) is available to find the expected weight for gestational age and gender [1]. The categorization is as follows:

Normal	BWR $\geq$ 0.90
Mild SGA	$0.75 \leq$ BWR < 0.90
Severe SGA	BWR < 0.75

9.1.1.2 Apgar Score

This index quantifies the neonatal prognosis and is generally measured 1 and 5 min after birth. Apgar is the name of the scientist who first proposed this index but is now also an acronym for what it measures—**a**ppearance, **p**ulse rate, **g**rimace, **a**ctivity, and **r**espiration. Skin color, heart rate, response to stimulation, muscle tone, and respiration are graded on a scale of 0–2, generally only as 0, 1, or 2. The sum of these five scores is called the Apgar score. A low score is associated with risk of disability, even death, and thus calls for immediate attention. A score of 8 or more is considered normal and 7 or less is an indication of asphyxia. An Apgar of 0–2 can be considered severe asphyxia, 3–4 moderate, and 5–7 mild asphyxia.

9.1.2 Epidemiological Indicators of Growth in Children

Physical growth of a child is assessed by anthropometric measurements such as weight, height, chest circumference, and head circumference. Mathematical models for growth in stature of children are discussed by Ledford and Cole [2].

Each of the anthropometric measurements can be independently assessed by the percentile point achieved by a child relative to the healthy children of that age and gender in the same population. Median is regarded as a reference value, and 3rd and 97th percentiles as thresholds to indicate abnormally low or abnormally high values. The norms are generally obtained for age intervals of 6 months after 2 years, that is, age 2 years, $2\frac{1}{2}$ years, 3 years, $3\frac{1}{2}$ years, etc. Smaller age intervals are desirable for age-specific norms for younger children and infants.

Interpretation of health of the children with measurements outside the 3rd and 97th percentiles can be difficult. No matter how healthy are the children who are measured to construct the chart, 3% of them have weight less than the 3rd percentile curve. Thus, even some fully healthy children may show a weight in the low category. This is an acknowledged limitation of a growth chart, but the chart is still useful. Note that low weight in such children is in a relative sense only—relative to the other 97% healthy children. Thus, a low weight does not necessarily indicate poor health in an absolute sense.

9.1.2.1 Weight-for-Age

This is the most commonly used indicator but is more effective when the trend over age for the same child is studied. This trend is compared with the trend seen in healthy subjects in that population. The assessment is population specific. There is a mention of weight-for-age chart in Chapter 8, which contains the trend seen in healthy U.S. girls. The difficulty with weight-for-age, however, is that it fails to distinguish a thin but tall child from a well-proportioned child.

9.1.2.2 Weight-for-Height and Height-for-Age

The weight-for-height index obviates the need to know age particularly if it is between 1 and 10 years and can be safely used for children when age is in doubt. This index measures the balance between weight and height (length in case of children less than 2 years). A weight less than the 3rd percentile point for a particular height indicates wasting (i.e., thinness) associated with failure to gain weight or loss of weight [3]. This is considered an indicator of acute undernourishment. Weight-for-height fails to detect abnormalities when both height and weight are affected.

You have seen the ponderal index for neonates in one of the preceding paragraphs. A general form of the ponderal index is weight/(height)^b, where b is estimated from the regression (Chapter 16) of log(weight) on log(height) separately for each age. Freeman et al. [4], for example, found that $b = 2.08$ for boys of 7 years, $b = 2.20$ for girls of 7 years, $b = 2.44$ for boys of 16 years, and $b = 1.75$ for girls of this age in the United Kingdom. This varies from population to population.

Another index for assessing growth is height-for-age. Low height-for-age is an indicator of stunting (i.e., shortness) when weight for height is normal. This is frequently associated with chronic undernourishment resulting mostly from a poor overall economic condition or repeated illness. A height-for-age chart for healthy girls is illustrated in Figure 9.1 [5]. It includes weight-for-age percentiles also in the same chart so that a combined assessment can be made.

9.1.2.3 Z-Scores and Percent of Median

The interpretation of anthropometric measurements sometimes becomes easier when the Z-score is computed. This is given by

$$Z\text{-score} = \frac{\text{Weight} - \text{Mean}}{\text{SD}}, \tag{9.1}$$

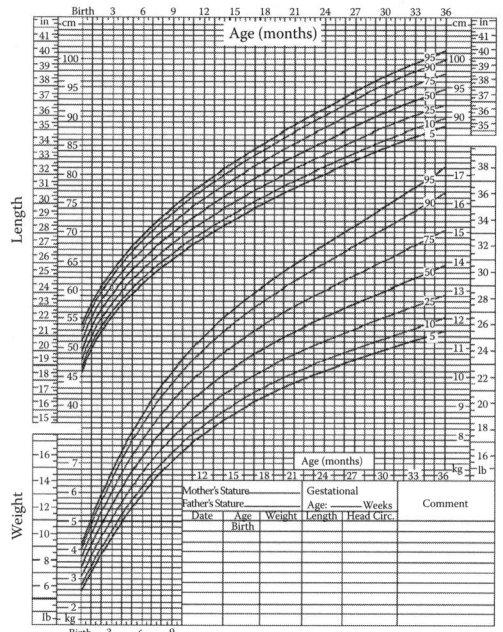

FIGURE 9.1
Height-for-age and weight-for-age chart for healthy U.S. girls.

where mean and SD are calculated for reference healthy children of that age or that height. A Z-score below −2 is considered low and below −3 exceedingly low. These Z-scores can be obtained for almost all anthropometric measurements. Justification for these thresholds will be clear after the Gaussian distribution is explained in Section 9.2.

The other index used to assess growth is percent of median. If the median weight of healthy children of height 1.10 m is 21.0 kg and the weight of a child of this height coming to a clinic is 18.5 kg, this is 18.5/21 × 100 = 88% of median. A measurement above 80% of the median is regarded as normal, between 71% and 80% as indicating undernutrition of grade I, between 61% and 70% as grade II, and 60% or less as grade III. Each population can evolve its own classification. The parents are advised of suitable corrective steps depending on the grade of undernutrition found in a child.

Example 9.1: Z-Scores and Percent of Median

Healthy children from well-to-do families were surveyed for their height, weight, age, and gender. The distribution of 450 girls by weight, taken within a week of their seventh birthday, is as follows:

Weight (kg)	14–16	16–18	18–20	20–22	22–24	24–26	26–28
No. of girls	4	36	85	182	96	40	7

These give mean = 21.1 kg and SD = 2.3 kg. A girl who is nearly 7 years old comes to a clinic from the same area. Her weight is 18.3 kg. Can she be considered underweight?

$$\text{Z-score for the girl} = \frac{18.3 - 21.1}{2.3} = -1.22.$$

$$\text{Weight as percent of median} = \frac{18.3}{21.1} \times 100 = 87.$$

A negative Z-score and weight less than 100% of median both indicate that the girl's weight is less than the average. However, because the Z-score is not less than −2 and the percent of the median is not less than 80, the weight can be regarded as within the normal variation and the girl is not classified as underweight. Thus, there is no cause for alarm. What is important in this case is the longitudinal follow-up to monitor that the pattern remains on normal trajectory.

9.1.2.4 Growth Velocity

Velocity is the rate of growth per unit of time. This is higher at the beginning of life and tapers off as age increases. There is a slight upswing, called midgrowth, around 6 or 7 years of age in some cases. A definite spurt is seen in adolescence. Velocity is indicated by the steepness of the curve and represents the incremental growth per unit of time. This indicator is also used to monitor growth and requires longitudinal measurements. A velocity less than normal for a particular age indicates failure to thrive. Conventional velocity charts involve two charts and hence are difficult to adopt in practice. A 3-in-1 weight-monitoring chart has been devised [6] for infants. It consists of conventional weight centiles complemented with extra lines called thrive lines, where the slope defines a cutoff for failure to thrive. The weight must be measured at 4-week intervals. This chart needs to be field-tested in different populations.

Weight velocity and height velocity can be used as indicators of growth in the immediate past and thus help detect acute malnutrition. A sudden decline in weight velocity (or in weight gain) in a particular child may provide better insight into the existence of a health problem than a weight-for-age measurement. Weight-for-age is relatively slow to react and slow to show that a problem exists. Height-for-age and weight-for-age are also affected by hereditary factors, particularly the size of the parents. Velocity, on the other hand, when compared with the previous velocity of the same child, is relatively independent of heredity.

9.1.2.5 Skinfold Thickness

A large part of body fat is deposited under the skin. Tools such as Harpenden calipers are used to measure skinfold thickness. Soft tissue radiography can also be used to delineate subcutaneous fat. Skinfold is generally measured at the mid-triceps, biceps, and subscapular and superiliac regions. The sum of these four measurements should not generally exceed 40 mm in boys and 50 mm in girls between 5 and 10 years of age. Widely acceptable reference data for subcutaneous fat are not yet available. But for this, the skinfold is an appropriate measure of adiposity and thus is a good indicator of obesity.

The most common site for measuring skinfold thickness is triceps. In females, this continuously increases up to the age of nearly 30 years, remains stable at around 25–30 mm between 30 and 50 years, and then declines. In contrast, in males it stabilizes quickly at age 20 years at around 15 mm [7].

Skinfold thickness in the newborns could be a marker for mother's nutritional status because this is not generally determined by genetic factors. Other anthropometric measurements such as birth weight, crown-heel length, and head circumference have a substantial genetic component.

9.1.3 Epidemiological Indicators of Adolescent Health

Vital changes occur during adolescence that set the basics of adulthood. A spurt in gain in height and weight occurs and genitals take shape. Pubic hair grow. Menarche occurs, and breasts develop in girls. Variation, as always, remains an integral part of all these developments. Relatively few physical sicknesses occur in this phase of life. Health is generally measured in terms of adequacy of physical growth and sexual maturation.

9.1.3.1 Growth in Height and Weight in Adolescence

Most countries have not developed standards for growth in height and weight during adolescence. The National Center for Health Statistics (NCHS) data, obtained for U.S. children and adolescents [5], are often used for comparison. Maximum growth in median height in 1 year occurs during the 13th year in U.S. boys (7.5 cm) and during the 11th year in girls (7.5 cm). This maximum in British children occurs around the 14th year and the 12th year, respectively [8]. Many children show slower or delayed growth mostly due to genetic and nutritional factors. A child's measurement can be compared with the NCHS chart or the chart of the child's native country (where available) to assess the progress of growth. The assessment again is in terms of the percentile achieved.

Preece and Baines [9] have developed models that can be used to evaluate height parameters such as age at takeoff, height at takeoff, velocity at takeoff, age at peak height velocity, and peak height velocity. These parameters have special relevance to the adolescent group.

Short stature in adolescent girls that persists into adulthood is associated with increased risk of adverse reproductive outcomes. Risk of low birth weight babies, cephalopelvic disproportion, dystocia, and cesarean section increases in short mothers. No specific health risk is known for short-statured boys.

In U.S. boys, body mass index (BMI) is least (median 15.3 kg/m^2) at age 6 years and in girls at age 5 years (median 15.2 kg/m^2). This rises to 23.0 kg/m^2 at age 20 years in boys and to 21.7 kg/m^2 in girls at this age [7]. Thus, BMI is not age independent in children. For internationally applicable cutoffs for well-to-do children, see Cole et al. [10].

9.1.3.2 Sexual Maturity Rating

Breast development in females, appearance of pubic hair in both boys and girls, and the development of male genitalia are graded into stages from 1 to 5, where the first is the preadolescent stage and the last is the fully matured adult stage. One can think of these stages as scores and use them as measures of extent of pubertal development. This can be related to age to find whether or not the development is on course. Clear guidelines on this are not yet available and need to be developed separately for each population depending on the rate of sexual maturation generally seen in healthy adolescents in that population.

9.1.4 Epidemiological Indicators of Adult Health

Health of an adult can be measured in several dimensions. Again, however, the discussion in this biostatistics book is restricted to some epidemiological indicators of general health and excludes particular diseases.

9.1.4.1 Obesity

Obesity has been found to be associated with risk of diseases such as hypertension, atherosclerosis, gallbladder disease, and diabetes. It is now standard practice in clinics to assess obesity and give advice accordingly. It really should be assessed by the amount of fat present in the body, but that is difficult to assess. Weight relative to height is considered as a good surrogate. The following indicators are used:

$$\text{(i) Body mass index: BMI} = \frac{\text{Weight in kg}}{\left(\text{Height in m}\right)^2}. \tag{9.2}$$

A BMI between 20 and 25 kg/m^2 is considered normal in adults. A low value (particularly if BMI < 18) indicates that the person is thin and a high value is indicative of excess fat. BMI between 25 and 30 indicates overweight. A value more than 30 is a definite indication of obesity. BMI is considered age–sex independent in adults, and nearly the same thresholds can be used for females as for males without much error although it is slightly less in females. It is lower in children. Another name for this index is Quetlet index after the Belgian scientist who first suggested it.

Recent evidence suggests that BMI < 16 is more of a risk for early mortality than BMI > 25. A disaggregated analysis has indicated that BMI = 26 or 27 has nearly the same risk as BMI = 25 kg/m²; thus, these values are no longer considered to indicate overweight.

BMI has the same basic form as the general ponderal index mentioned earlier. The exponent of the denominator is 3.0 in case of infants but progressively declines as age advances. It finally settles down to 2.0 for adults as in denominator of (i):

$$\text{(ii) Broca index for normal weight (kg)} = \text{Height in cm} - 100. \qquad (9.3)$$

This is extremely simple to understand and to use. It says that if a person's height is 165 cm then the weight should be around 65 kg.

Evenly distributed fat is probably not as harmful as accumulation around the waist. This is measured by

$$\text{(iii) Waist–hip ratio} = \frac{\text{Waist circumference}}{\text{Hip circumference}}. \qquad (9.4a)$$

This measures central obesity or abdominal obesity (or truncal obesity). The normal ranges are 0.8–1.0 for men and 0.70–0.85 for women. A large waist–hip ratio is found to be more closely associated than BMI with risk of some serious conditions such as stroke. Waist measurement obviously depends on height and cannot be considered as standalone indicator:

$$\text{(iv) Waist–height ratio} = \frac{\text{Waist circumference}}{\text{Height}}. \qquad (9.4b)$$

A value of more than 0.6 of this ratio is considered a health risk.

9.1.4.2 Smoking

Cigarette smoking is known to be a predominant killer in many countries and is gradually taking center stage in the remaining countries. It is generally measured in terms of number of cigarettes smoked per day. However, the duration of smoking is also important. A person's intensity of smoking may also vary from time to time. A measure could be the total number of cigarettes smoked so far in life. This number is given by

$$S_1 = n_1 x_1 + n_2 x_2 + \cdots + n_K x_K, \qquad (9.5)$$

where n_k ($k = 1, 2, \ldots, K$) cigarettes per day (intensity) are smoked for x_k years (duration). This is more exact than pack-years generally used for smoking. S_1 suffers from the same demerit as the pack-years: smoking 10 cigarettes a day for 25 years is the same as smoking 25 cigarettes per day for 10 years. This obviously is not always true. For lung cancer, duration may be more important, whereas for coronary disease, intensity of smoking may be more important. Despite this deficiency, pack-years continues to be the most accepted measure, so should S_1.

Now consider a multiplier p_k to indicate further variation in the intensity of smoking. Evidence suggests that p_k could be 0.15 for passive smoking (i.e., effect of passive smoking—also called second hand smoking—is 15% of the effect of active smoking), 0.67 for filter cigarettes, 5.0 for cigar, and 2.5 for pipe. Similar equivalence can be possibly postulated for smokeless tobacco such as snuff, chewing tobacco, and betel quid. With this adjustment, an elementary index of smoking is

$$S_2 = \Sigma p_k n_k x_k \quad (p_k = 1 \text{ for regular cigarette}). \tag{9.6}$$

S_2 measures modified cigarette-years of smoking and can range from 0 to 2000 or more. For many diseases, the cumulative burden of smoking is not linear but progressively slows down as the amount of smoking increases after the damage is already done. That is, the first 1000 cigarette-years are more harmful than the next 1000 cigarette-years. Logarithm is generally considered a good moderation in such situations, but this would be too severe for smoking as it would reduce 100 cigarette-years to 2 and 1000 cigarette-years to 3. I looked for a simple function of S_2 that may take a value of nearly 5 for S_2 = 100 and nearly 15 for S_2 = 1000. The values 5 and 15 are my subjective assessment of the years that must elapse after quitting for the burden of such smoking to completely disappear in many cases. The function with this feature is $\frac{1}{2}\sqrt{S_2}$. When the benefit of the time elapsed since quitting is also allowed,

$$S_3 = \frac{1}{2}\sqrt{\Sigma p_k n_k x_k} - y; \quad y \le S_2; \tag{9.7}$$

where y is the number of years elapsed since quitting in case of past smokers (for current smokers $y = 0$).

S_3 can be adjusted for the age at start of smoking. Not much evidence is available, but assume that the burden is twice as much when the age at start is 15 years as when the age at start is 30 years or more. Epistemic uncertainties are prominent for this aspect but the postulation seems plausible. Under these assumptions, the final index of smoking is

$$S = \left(3 - \frac{a}{15}\right)\left(\frac{1}{2}\sqrt{\Sigma p_k n_k x_k} - y\right). \tag{9.8}$$

where
  a is the age (years) at start of smoking (take $a = 30$ for $a > 30$)
  p_k is the intensity of smoking for n_k years ($p_k = 1$ for cigarettes)
  x_k is the number of cigarettes/cigars/etc., smoked
  y is the number of years elapsed since quitting in case of past smokers (for current smokers $y = 0$)

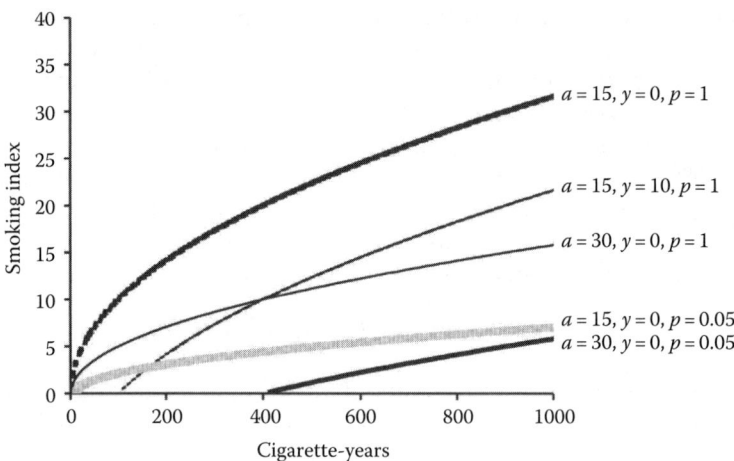

FIGURE 9.2
Smoking index for some typical values (see text).

A value less than zero is interpreted as zero. This is a comprehensive index of the present burden of smoking as it incorporates (1) the duration of smoking, (2) the quantity of smoking, (3) smoking of filter cigarettes and other forms of tobacco consumption that can be factored to cigarette smoking, (4) more damage by smoking in the beginning and progressively less burden from smoking additional pack-years in life, (5) benefit of the time elapsed since quitting, and (6) deleterious effect of starting smoking early in life. The index has an inbuilt feature to consider current smokers and ex-smokers and obviates the need to divide ever smokers into such dichotomy. The index models the entire history of smoking into a single metric. The modification of the cigarette-years in some typical conditions is shown in Figure 9.2. This index does incorporate a large number of aspects of smoking but fails to capture occasional smoking or the beneficial effect of interruption.

Example 9.2: Calculation of Smoking Index

The following is the smoking history of three persons:

Person A: Started smoking at age 12 years. Initially smoked 10 regular cigarettes for $3\frac{1}{2}$ years. Since then has been smoking 20 filter cigarettes a day for the last $17\frac{1}{2}$ years.

Person B: Started smoking at age 21 years. Smoked 12 cigarettes a day for 1 year, 15 cigarettes a day for $2\frac{1}{4}$ years, 20 cigarettes a day for $1\frac{1}{2}$ years, no smoking for 6 months, and 2 cigars a day for 1 year. He has not smoked for the past 4 years.

Person C: Never smoked but spouse smoked. When both were together, an average of five cigarettes a day were smoked. This started at age 27 years and went on for 6 years. There has been no exposure to cigarette smoke for the past 3 years.

Smoking index

Person A: $S = \left(3 - \dfrac{12}{15}\right)\left(\dfrac{1}{2}\sqrt{10 \times 3.5 + 0.67 \times 20 \times 17.50}\right) = 2.2 \times 8.21$

$= 18.1.$

An index of 8.21 increases to 18.1 because smoking started at an early age of 12 years.

Person B: $S = \left(3 - \dfrac{21}{15}\right)\left(\dfrac{1}{2}\sqrt{12 \times 1 + 15 \times 2.5 + 20 \times 1.5 + 5 \times 1} - 4\right)$

$= 1.64 \times (4.73 - 4) = 1.2.$

The present burden of smoking is small because of quitting 4 years ago.

Person C: $S = \left(3 - \dfrac{27}{15}\right)\left(\dfrac{1}{2}\sqrt{0.15 \times 5 \times 6} - 3\right)$

$= 1.2 \times (-1.9) = 0$ (negative value is to be taken as zero).

The burden was small because of passive smoking and that too vanished because of no exposure for the past 3 years.

9.1.4.3 Physiological Functions

Besides routine measurements of body temperature, heart rate, blood pressure, etc., it may be useful to measure parameters such as lung functions and cardiac output. Lung functions are measured by a variety of indicators such as expiratory volume and vital capacity. Details are available in Cole [11].

Assessment of physiological measurements as normal or abnormal is discussed later in this chapter. Borderline values present trickier problems if the terms such as high normal, marginally high, and probably abnormal are used. A level of 50 mg/dL of serum urea may be considered normal by one physician and high by another. Blood pressure 130/92 mmHg and homocysteine level 15 μmol/L are also borderline values. Thus, there is always a risk of misdiagnosis and missed diagnosis. Normality or otherwise of a measurement should be assessed after considering the values seen in healthy and sick subjects with sufficient precautions for overlapping values.

9.1.4.4 Quality of Life

Myocardial infarction (MI), breast cancer, multiple fractures, and peritoneal surgery are examples of conditions that have many survivors, but quite a few of them are not able to lead a normal life of a healthy person. The disability may be apparent such as in walking and talking or more subtle as in doing heavy work for long hours. Quality-of-life assessment is gaining importance as more and more people are able to live longer due to medical intervention but retain residual disability of one kind or the other. It is being increasingly assessed for the general population as well or for patients of various types even when there is no disability. Quality of life is also commonly used as an outcome measure in research on the relative benefits of different treatment methods.

Quality of life is generally equated with hopes and ambitions matched by experience. It involves a person's own perception and values. Note that this is quite abstract and thus is difficult to measure. Physical, psychological, and social well-being, including functionality in daily living, are generally included in a quality-of-life assessment. In the case of chronic patients, this may contain items on sleep, appetite, sexual functions, social participation, work performance, etc. It is often considered convenient to divide the quality-of-life questionnaire into domains, such as physical health and psychological well-being, and to divide a domain into facets, such as psychological well-being into negative and positive feelings, self-esteem, and memory.

Several instruments are available that claim to measure the quality of life in different kinds of subjects, particularly in patients with chronic ailments. For cardiovascular disease, a quality-of-life measure is the multidimensional index based on 35 questions on different domains of quality of life developed by Avis et al. [12]. For cancer, there is a quality-of-life questionnaire, called EORTC QLQ-C30, containing 30 items of inquiry [13]. As the quality of life is mostly the perception of the subject, the rating sometimes may be inconsistent with the actual physical condition such as tumor stage. A patient in an advanced stage of malignancy may still report a good quality of life.

For the general population, the World Health Organization has devised a quality-of-life (WHO-QOL) questionnaire with 100 items. This questionnaire is considered too detailed and difficult to answer. A brief questionnaire (WHO-QOL-BREF) with 26 items is also available. More popular is the short form with 36 items called SF-36. This measures functional health and well-being. Such questionnaires can be easily downloaded from various websites such as *www.sf-36.org*. These questionnaires may have to be adapted to the local conditions.

9.1.5 Epidemiological Indicators of Geriatric Health

Geriatric health is assuming increasing importance as the older population is rapidly rising in most countries. Some epidemiological measures of geriatric health are given here—they can be used for individual assessment.

9.1.5.1 Activities of Daily Living

In the case of old age or handicaps, the degree of disability can be measured in terms of an activities of daily living (ADL) index. Scores are assigned to the level of independence assessed on several ADL such as walking, bathing, use of toilet, and dressing. The score could range from 0 for complete dependence to, say, 4 for complete independence on each item. The sum of these scores is called ADL index (see, e.g., Katz and Akpom [14]). A disadvantage of such an index is that it is insensitive to change when the level improves on some items and deteriorates on the others.

No index, including the one just mentioned, is widely acceptable. Fourteen questions ranging from difficulty in self-care (e.g., eating or dressing) to higher level activities (e.g., carrying weights or doing housework), used by the WHO Eleven Countries Study [15], can be adapted to suit local conditions.

9.1.5.2 Mental Health of the Elderly

Population is quickly aging around the world. Physical limitations emerging from degeneration are recognized, but the mental agility also deserves attention.

Among many instruments available for measuring mental health of the elderly, one in common use is the mental health component of the quality-of-life questionnaire such as SF-36. This is no different from the tool used for the general population of adults. SF-36, however, is restricted to the functional status, including for mental health.

Insomnia and anxiety are common in old age. For these, Pittsburgh Sleep Quality Index [16] and *Diagnostic and Statistical Manual of Mental Disorders, Fourth Edition (DSM-IV)* [17], respectively, can be used. Both are for the general population and not specific to the elderly. Beck Depression Inventory is used for assessing the severity of depression symptoms. This is an old instrument but continues to be commonly used, particularly for old age people.

9.2 Reference Values

Reference values are extensively used for decisions on managing patients. It is known that the normal body temperature in humans is 98.6°F and the normal level of glucose in serum on fasting is between 75 and 115 mg/dL. When a person comes in for the first time with complaints, his measurements are evaluated against such reference values. The evaluation can be less risky and more meaningful if the basic principles of establishing such normal values are known. For this, an understanding of the distributional aspects of measurements is essential.

9.2.1 Gaussian and Other Distributions

You may wish to revisit Section 8.1 and refresh your memory about the frequency curves that depict distribution of continuous variables. Figure 8.1g shows the distribution by cholesterol level, given in Table 8.1, of subjects attending a hypertension clinic. Note that this distribution has a peak at around 285 mg/dL. This indicates that, among hypertensives, this level is more common than any other level. The mean is 267 mg/dL, the median is 270 mg/dL, and the mode is 285 mg/dL. These three are fairly far apart. The subjects are hypertensives and the shape of the distribution is not symmetrical. When subjects are healthy, many medical measurements such as cholesterol level, serum iron, and glucose level tend to follow a specific symmetric shape called Gaussian. The frequencies are high in the center and rapidly decline on either side in a fashion similar to that shown in Figure 9.3 for the serum iron data of Table 9.1. The figure is a smoothed curve. Almost all errors in measurement also follow this shape. Is it not amazing that a chaotic measurement such as error tends to follow a regular pattern? Such a distribution is also called a **normal distribution**, but I avoid using this term in this text because normal has a different meaning in medicine. It is not normal for sick people to have "normally" distributed values. If you are mostly dealing with patients rather than healthy people, you may wonder why your values do not follow a so-called normal (Gaussian) distribution.

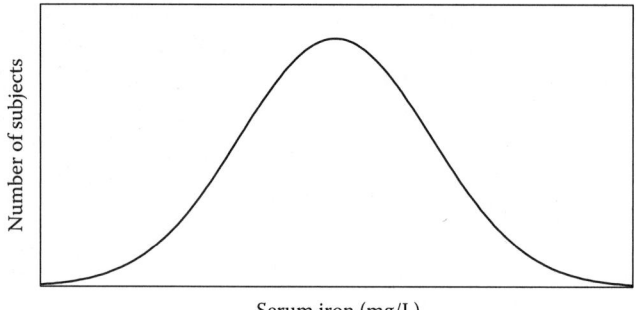

Serum iron (mg/L)

FIGURE 9.3
Distribution of serum iron in healthy subjects (smoothed curve).

TABLE 9.1

Serum Iron Level in 165 Healthy Subjects

Serum Iron (mg/L)	Number	Percent
0.50–0.69	2	1.2
0.70–0.89	12	7.3
0.90–1.09	33	20.0
1.10–1.29	68	41.2
1.30–1.49	34	20.6
1.50–1.69	13	7.9
1.70–1.89	3	1.8
Total	165	100.0

9.2.1.1 Properties of a Gaussian Distribution

A Gaussian distribution has the following properties:

Property 1: The shape is symmetric like a bell.

Property 2: The mean, median, and mode coincide.

Property 3: The limits from (mean − 2SD) to (mean + 2SD) cover the measurements of nearly 95% of subjects. These are referred to as **±2SD limits** or sometimes as **2-sigma** limits.

Another often-cited property of a Gaussian distribution is that the limits from (mean − 3SD) to (mean + 3SD) cover almost all subjects (99.7% to be exact). These 3-sigma limits are rarely used in health and medicine. An exceptional use of these limits is in Z-scores defined by Equation 9.1.

Properties 1 through 3 hold true for the population as a whole but can also be used for samples as an approximation. The approximation works well when the sample size is large. Sometimes these properties are used inversely. When these properties hold for a set of data, the distribution is considered Gaussian. This also works well in most practical situations. Gaussian distribution also has a specified form of peak, called kurtosis, though this aspect is generally ignored.

Example 9.3: Gaussian Distribution of Serum Iron Level in Healthy Subjects

In the case of the serum iron data in Table 9.1, mean = 1.207 mg/L, median = 1.199 mg/L, and mode = 1.196 mg/L. These three are nearly equal. The smoothed shape of the distribution is shown in Figure 9.3.

For these data, SD = 0.227 mg/L: Thus, the ±2SD limits are

$$(1.207 - 2 \times 0.227) \text{ to } (1.207 + 2 \times 0.227),$$

or

$$0.75 - 1.66.$$

Nearly 94% of healthy subjects mentioned in Table 9.1 have serum iron levels within these limits, whereas ±2SD should have 95%. This difference of 1% is due to sampling fluctuation. Utility of such limits in the practice of medicine is discussed later in this chapter.

9.2.1.2 Other Distributions

Gaussian distribution pervades statistical thoughts so much that it is seen afflicted by Ghost of Gauss. Although many medical measurements in healthy subjects do indeed follow a Gaussian pattern, not all do. Figure 9.4a

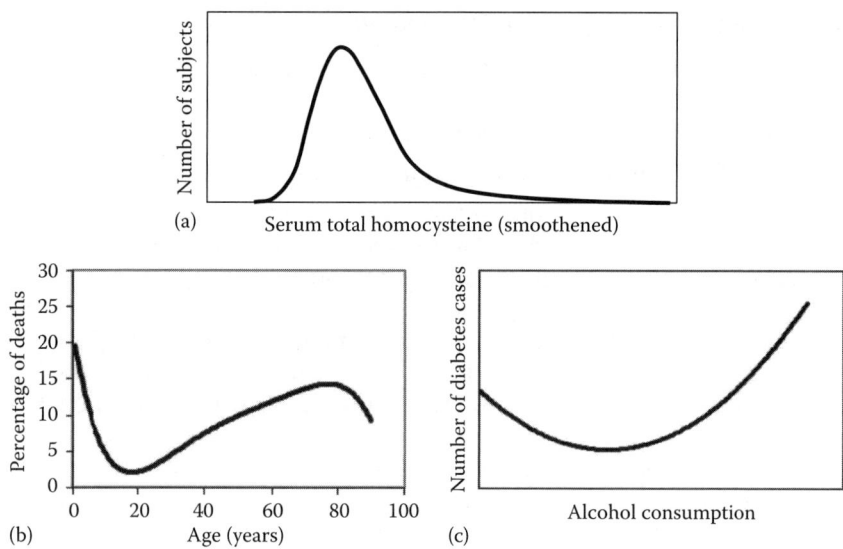

FIGURE 9.4
(a) Serum total homocysteine in 100 healthy subjects (smoothed curve). (From Arnesen, E., Refsum, H., Bonaa, K.H., Ueland, P.M., Forde, O.H., and Nordrehaug, J.E., Serum total homocysteine and coronary heart disease, *Int. J. Epidemiol.*, 24, 704–709, 1995, by permission of Oxford University Press.) (b) Distribution of deaths by age at death in a developing country—smoke pipe distribution. (c) Distribution of cases of type-2 diabetes by sustained level of alcohol consumption—J-shaped distribution.

shows the distribution of serum total homocysteine in healthy subjects in Norway [18]. This distribution has a longer tail on to the right because higher-than-mode levels are more common in healthy subjects. This is called a **right-skewed** distribution. The homocysteine level in any population is likely to follow the same pattern. Triglyceride levels in Belgian children are also found to be highly skewed to the right [19]. On the other hand, the distribution of Hb level is generally **left skewed** because lower values are commonly seen in healthy subjects. Figure 9.4b shows the age distribution of deaths in a population of a developing country. This is entirely different from a Gaussian pattern and has a shape of a smoke pipe. Thus, shapes other than Gaussian also occur in practice. Figure 9.4c illustrates a J-shaped distribution that occurs for cases of type-2 diabetes at different levels of sustained consumption of alcohol.

Most right-skewed distributions can be converted to a Gaussian form by a suitable logarithmic or square-root transformation. For a left-skewed distribution, a reciprocal transformation can be examined; sometimes it works, sometimes not.

As mentioned earlier, the distribution of most medical measurements is Gaussian in healthy subjects and skewed in sick subjects. Examples of measurements with remarkable skewness among patients are tumor markers such as carcinoembryonic antigen (CEA) (median in lung cancer 3.5ng/mL but range 1–7580). In cancer particularly, the aberration multiplies rather than adds such as becoming twice as much per unit of time. Logarithms transform it to additive and tend to give it a symmetric shape.

Another feature of a distribution is **kurtosis**. This measures the peakedness of the frequency curve. Gaussian distribution has a specific peak, which is defined as zero, and peak of all other distribution is measured against this value. A distribution with flat peak has kurtosis less than zero and called platykurtic, and a distribution with sharp peak has kurtosis more than zero and called leptokurtic. On a Likert scale from 0 to 8, if most values (say nearly 80%) are around 4 and 5, and very few lower or higher, the distribution will be platykurtic.

9.2.1.3 Checking Gaussianity: Simple but Approximate Methods

Since most statistical methods assume Gaussian distribution, you should check this when in doubt. Sometimes the biological process underlying a medical measurement provides sufficient clue that the distribution of a particular measurement is Gaussian or not. Some examples are given in preceding paragraphs. In all other cases, you will be required to make a judgment on the basis of the data you have on a sample of subjects. In this situation, you can use any of the following methods. These methods work well for large n but may fail for small n.

Easiest may be to check if properties 1 through 3 stated earlier hold. There are several other methods. Also, first few methods given next may fail to detect the deviation in peakedness (kurtosis) of the distribution though they are adequate for skewness. When n is small, you may have to use your subjective judgment. If you are calculation oriented, just calculate the coefficient of skewness. Actual procedure for calculating the coefficient of skewness is complex as it requires sum of the cubes of deviations from mean. A simple procedure is to calculate

$$\text{Coefficient of skewness} \approx \frac{\text{Mean} - \text{Mode}}{\text{SD}}. \tag{9.9}$$

This works reasonably well for unimodal (single-peak) distributions. Negative value indicates left skewness and positive value right skewness. For a symmetric distribution, this coefficient is zero. A value <-1 or $>+1$ indicates highly skewed distribution.

In a Gaussian distribution—in fact in all symmetric unimodal distributions—mean, median, and mode are equal (Figure 9.5a). In sample values, this could be approximately so. For others, note the following:

Right-skewed distribution: Mode < Median < Mean
Left-skewed distribution: Mean < Median < Mode

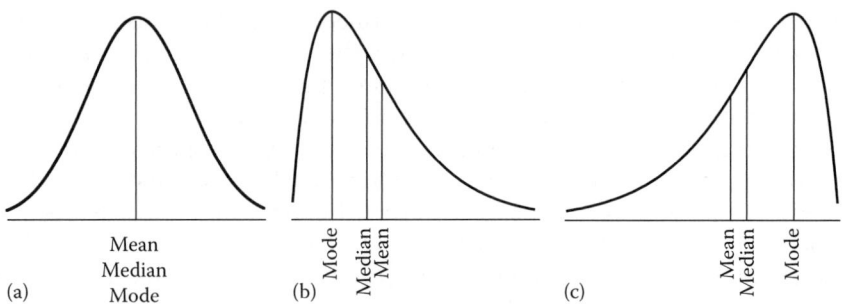

FIGURE 9.5
Location of mean, median, and mode in (a) symmetric, (b) right-skewed, and (c) left-skewed
distributions.

These are also shown in Figure 9.5b and c. Incidentally, these words appear
in a dictionary in the order seen for left-skewed distribution and reverse
in the right-skewed distribution. Also the distance between mean and
median in a dictionary is small relative to distance between median and
mode. The coefficient of skewness (Equation 9.9) is also based on such considerations. Thus, the first method to find that a distribution is symmetric
or not is to calculate the mean, median, and mode and see if they follow
any of the aforementioned patterns. In samples, the difference between
mean, median, and mode must be substantial for the distribution to be
considered skewed.

If you are graph oriented, most basic to check Gaussianity is histogram.
Draw it for frequencies in different class intervals and see if it largely follows
a bell shape or not. An alternative to histogram is stem-and-leaf plot. Second
approximate method is **quartile plot** of the type shown in Figure 9.6a through
c. For this, compute Q_1, Q_2, and Q_3 and plot them on minimum to maximum
axis. If the distance between Q_1 and Q_2 is nearly the same as between Q_2 and

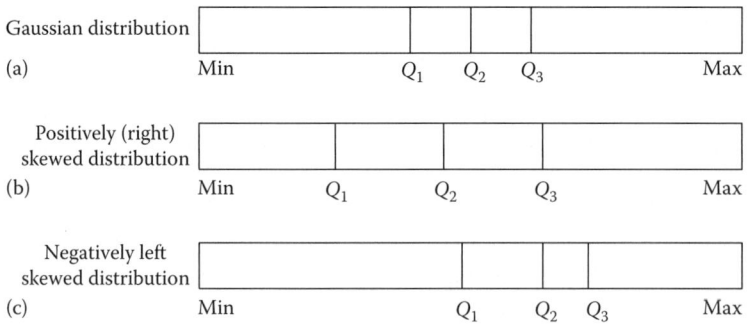

FIGURE 9.6
Pattern of quartiles in (a) symmetric (Gaussian), (b) right-skewed, and (c) left-skewed
distributions.

Q_3, you can safely assume symmetry and possibly Gaussian. If the pattern is different as in Figure 9.6b and c, the distribution is either left skewed or right skewed. If the sample size n is really large, you may like to try this type of plot with deciles instead of quartiles.

An alternative to quartile plot is the box plot of the type shown in Figure 8.8b. For symmetry, the boxes above and below median as well as the whiskers on both the sides should be nearly equal. Third graphic method is ogive. You know that ogive is the plot of cumulative frequencies against the x-values. In a Gaussian distribution, this takes the shape of a sigmoid. If the shape is substantially different, take it as an indication of non-Gaussian shape.

Whereas the methods just described are for symmetry, the following considers all aspects of Gaussianity including kurtosis. Try to plot the cumulative relative frequencies for each distinct value in your sample versus cumulative probabilities of Gaussian given in Table B.1. Values in this table are explained shortly. This is called **P-P (proportion-by-probability) plot**. If the distribution is Gaussian, this will be nearly a straight line. If substantially different, suspect that the distribution is not Gaussian. In place of P-P, you can also try **quantile-by-quantile (Q-Q) plot**. These plots observed quantile for each distinct value against expected for that value under Gaussian pattern. This should also provide nearly a straight line.

More exact methods require calculations and checking statistical significance of the departure from Gaussian. Such significance based on Anderson–Darling, Shapiro–Wilk, and Kolmogorov–Smirnov tests is discussed in Chapter 12.

9.2.2 Reference or Normal Values

The term normal is used in medicine with several different meanings. It is normal for a 70-year old person to have myopia, and it is normal for women undergoing chemotherapy for breast cancer to lose hair. Murphy [20] has given a very interesting discussion of the different uses of this term in medical literature. A moot question is whether normal is the same as ideal or optimal. If yes, how to define the optimal? It is possible that a person with 37.5°C body temperature and 149/92 mmHg BP does exceedingly well when accompanied by other corresponding physiological changes. Who knows!

Normal level of a quantitative measurement can be defined in many ways. Most will agree that normal values are those that are generally seen in healthy subjects. However, each individual has his own normal (call it self-normal) in a healthy condition and, whenever possible, the evaluation of the current condition should be made against the value normally present in that person in healthy state. If a person is known to have a BP of 110/70 mmHg when healthy, then a level of 130/80 mmHg would be considered a definite rise, sufficient to put the attending clinician on alert for contemplating some action. For example, this can happen in pregnancy-induced hypertension.

The range of plasma levels of most hormones in healthy subjects is wide. As a consequence, the level of a hormone in an individual may be halved or doubled (and thus be grossly abnormal for that person) but still be within the so-called normal range. However, the healthy level of a patient coming to a clinic for the first time is seldom known. In such cases, the patient's level is evaluated against the levels generally seen in healthy subjects in the population. Call these population-normals. These serve the purpose of reference values. Thus, reference values are the levels generally seen in healthy subjects in a population. Such references are also used to delineate the extent of allowable variation in a subject even when his own healthy levels are known. In the example just cited, because a raised BP of 130/80 mmHg is still well within the normal limits for the other healthy subjects, therapy may not start unless the complaints are severe. Most likely, there would be no such complaints. Yet a big rise is surely enough to be on alert.

This text uses the two terms—reference values and normal values—interchangeably for values generally seen in healthy subjects.

9.2.2.1 Implications of Normal Values

Reference values could be different for different segments of the population. A vital capacity normal for females would not be normal for males. A level of BP seen normally in adults would not be normal for children. The normal weight of 2-year olds in Sudan may not be the same as the normal weight of 2-year olds in Sweden. Normals may also change from time to time. Height seems to have increased all over the world during the past 50 years. Lung functions also seem to be improving. The following additional points should be noted:

1. Normal values or reference values are based on measurements of healthy subjects, preferably the healthiest segment of the population. The following are the criteria for normals to be reliable:

   a. They should be based on a sample that includes at least 200 individuals in each group (say, 200 males and 200 females) if the groupwise reference values are required. A smaller sample would be adequate only when the interindividual variability is really small.

   b. There should not be any outlier in the data.

   c. The sampling procedure must be scientific so that the sample indeed represents the entire spectrum of values present in the group.

   d. Measurements should be carefully made by trained workers using adequately tested standardized instruments and methods. The data should be available to anyone for review.

2. When the interindividual variation among healthy subjects is small, as in the case of body temperature, the normal level could be a single value (e.g., 98.6°F) instead of a range (Section 9.2.3). This single value is a representative central value and would be a mean, median, or mode. When the distribution is Gaussian, all three are the same and any one is nearly as good as another. However, the mean is preferred because of its good statistical properties, particularly its relative consistency from sample to sample. For distributions other than Gaussian, the choice again would mostly be the mean because of its easy understandability and better reliability. The median is used when the distribution is highly skewed, or when extreme values or outliers are present in the data that cannot be excluded. When interest is specifically in the most common value for some reason, the choice naturally is the mode. These guidelines are the same as those prescribed earlier in Chapter 7 for choosing an appropriate central value.

3. When the interindividual variation in healthy subjects is large, a single normal value is not sufficient and we need a range of normal values.

9.2.3 Normal Range

Even though normal is the level generally seen in healthy subjects, there always are persons with very high or very low values who are still absolutely healthy. Thus, there could be considerable overlap between normal and abnormal values. Determining a threshold that works in all situations has remained an elusive objective. The following approaches are available.

9.2.3.1 Disease Threshold

The best method to delineate normal levels is to observe people with different levels for a sustained period, and identify a threshold beyond which people start feeling the burden in some sense—not able to do work to one's full capacity, or entailing a risk for an adverse condition later in life. This is an extremely complex procedure and requires consultation from experts, who, in turn, should have full evidence for the threshold they propose. The cutoff 140/90 mmHg for BP is such a threshold. Experts have observed that a higher BP considerably increases the risk of CAD. Not many examples of this type of cut-point are available, but there would be people with level 145/92 who would be healthy and there would be people with level 136/88 and not healthy (with BP-related complaints such as headache and irritability). Thus, even this threshold does not rule out errors. This type of threshold is known as disease threshold of normal level.

9.2.3.2 Clinical Threshold

The second alternative is to compare levels of those who are in perfect health with those who are not. Since each of these groups will have a distribution

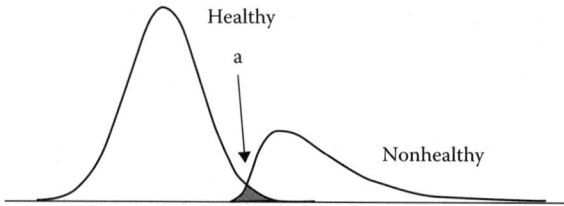

FIGURE 9.7
Pattern and overlap of measurement in healthy and nonhealthy subjects.

of its own, the situation typically will be as in Figure 9.7. This figure has the following features:

1. The number of healthy persons far exceeds the nonhealthy persons.
2. The distribution among healthy subjects is Gaussian whereas in nonhealthy persons is skewed.
3. The variation in the levels is smaller for healthy than for nonhealthy persons. Note how scatteredness in the levels of nonhealthy is relatively large.
4. In this figure, nonhealthy subjects have higher levels. This is true for many measurements such as T_3, T_4, BP, blood glucose, etc., but not for all. Higher levels of Hb, peak expiratory flow rate, HDL cholesterol, etc., indicate good health. For these measurements, the curve for nonhealthy will be on the left side.
5. There is some overlap between levels seen in healthy subjects and the levels seen in nonhealthy subjects. This is shown as the shaded area in Figure 9.7. If there is no overlap, the healthy levels and nonhealthy levels can be immediately defined. In practice, this overlap is substantial and causes problems in defining healthy levels.

Statisticians have shown that the point where the two curves intersect provides the cutoff with least number of misclassifications. This level is indicated as "a" in the figure. This is the clinical threshold that could be used to define normal levels. Indeed this is a very convincing approach but can be adopted only when the distribution in the healthy and nonhealthy groups is known and the overlap is minimal. The biggest problem in this approach is the choice of criterion to categorize a person as healthy or nonhealthy for drawing these curves. Threshold will not be known without the curves and the curves cannot be drawn without categorizing subjects as healthy and nonhealthy. Obviously external criteria are needed, and those may or may not work. Nonetheless, such clinical threshold has inbuilt provision for tolerating error of misclassification as indicated by the shaded area. Errors are not ruled out by this method also. Bigger the overlap, larger is the shaded area and higher the chances of error.

9.2.3.3 Statistical Threshold

When the distribution is Gaussian, Property 3 (Section 9.2.1) is invoked to say that (mean − 2SD, mean + 2SD) are the normal limits. They exclude nearly 2.5% of healthy subjects with extreme measurements on either side. This is arbitrary but now accepted around the world. The mean and SD are computed from measurements obtained on a large number of healthy subjects. These are statistical thresholds and popularly known as ±2SD limits. Most of the normal ranges used in medical practice are obtained in this manner. In the case of the serum iron data in Table 9.1, the normal range is 0.75–1.66 mg/L according to the calculations shown in Example 9.3. When the distributional shape is far from Gaussian, the range from 2.5th to 97.5th percentile points is considered normal instead of ±2SD limits. Not exactly these but the 3rd and 97th percentiles are illustrated in Figure 8.12a for weight measurements of girls. Note that the ±2SD limits for the Gaussian distribution are also from 2.5th to 97.5th percentile. You can therefore forget about ±2SD limits and use the percentile-based range for all measurements irrespective of the shape of the distribution. But ±2SD limits are ingrained in the minds of many clinicians and statisticians alike. One reason for this is that ±2SD limits fit well into the confidence interval and testing hypothesis strategy that are discussed in later chapters. Note the following for such statistical thresholds:

1. No matter how healthy the subjects are, there are always 2.5% healthy subjects at the lower end and another 2.5% at the upper end who will have levels outside such a normal range. This is an error but is tolerated because an error of this magnitude may always occur irrespective of the method used to establish normal limits. This error is at least quantitatively known for statistical thresholds but would not be easily known in other approaches.

2. The ±2SD limits are purely statistical. A level beyond these limits is abnormal only in the sense that such an extreme level is rare in healthy subjects. Whether this translates to medical problems is not known. However, these limits seem to be working well as an aid in most practical situations.

3. A measurement such as 106 mg/dL for fasting blood glucose level is not abnormal when the normal range is from 75 to 105; just that the chance of this value occurring in a healthy person is small—less than 2.5%. Gaussian theory stipulates that this chance reduces steeply as the measurement becomes farther and farther away from mean. A value of 104 mg/dL has nearly the same prognosis as the value 106 mg/dL. No miracle happens at the cutoff, such as 105, that would suddenly make a measurement abnormal. Nonetheless such cutoff is needed somewhere as a guideline to start suspicion. (Mean ±2SD) provides such a cutoff. But it is applicable to one type of measurement at a time. If there are five different types of measurements such as different components

of lipid profile, the chance of a healthy person labeled as healthy by such statistical criteria for all measurements together is not large.

4. Some disease entities are based almost exclusively on a single parameter. Diagnosis of anemia caused by iron deficiency is based on Hb level, hypertension on BP levels, diabetes mellitus on serum glucose level, and glaucoma on intraocular pressure. Other indications such as signs–symptoms play a minor role for classifying such diseases. Evidence exists that persons with statistically abnormal levels do have an increased risk of the concerned morbidity and mortality. An intervention, such as therapy, to bring the level back to the normal range helps to reduce this risk.

5. The normal levels, whether statistical ±2SD limits or based on the healthy–sick dichotomy, should be determined by measuring a large number of subjects according to the guidelines given in the previous section. Only then they command confidence. Normative data based on small samples can at best be indicative that need confirmation in subsequent testing. Small sample based normals can seldom be used for diagnostic or prognostic purposes.

6. The risk of misdiagnosis and missed diagnosis seems to be universally present irrespective of the procedure used to delineate reference values. As already explained, if the reference values are not statistical but based on values actually present in healthy and diseased subjects, then too some overlap is inevitable. If the diagnosis is based on clinical signs and symptoms instead of the value of a single parameter, there will also be cases with a nontypical picture. Even a composite picture jointly obtained by several measurements, investigations, and signs and symptoms can turn out false in some cases. As the information on a patient increases, the risk of error decreases, but it would rarely vanish at the diagnosis stage. This is where the acumen of a clinician comes in handy. The human brain is always superior to any technological input, particularly in the case of medicine. The answer lies in putting together the pieces of a jigsaw puzzle in as efficient a manner as possible.

9.3 Measurement of Uncertainty: Probability

You may have noticed that uncertainties in medical decisions are profound. These can be minimized but not eliminated. We must learn to live with them. Measurement of their magnitude is the first step in this direction.

An accepted measure of uncertainty is probability. The term has everyday meaning, but its computation can be nerve wrecking in some intricate cases. Mathematically speaking, an event that cannot occur, such as human male giving birth to a child, has probability zero. This is as low as it can get. An event that is certain to occur, such as death, has probability one. No probability can be negative, nor can it exceed one. Statistical definition of probability is based primarily on empiricism and thus is milder. If a woman of age 58 years has never conceived in the history of a community, the statistical probability of occurrence of such an event in that community is zero. It does not necessarily imply that the event is impossible.

This text is concerned with statistical probabilities. In simple terms, if oral cancer is seen to occur in 8% of a *large number* of habitual tobacco chewers for more than 10 years, the probability P that a randomly picked tobacco chewer of this type will get oral cancer is 0.08. An interpretation of probability is the relative frequency in a large number of cases. Thus, it measures the likelihood of an event and is complementary to uncertainty. In an etiologic investigation, this probability is referred to as risk—a term extensively used to delineate the hazards of a disease on exposure to an unfavorable factor. If the risk of oral cancer among nonchewers is 0.005, then the risk in chewers in our example is 16 times of that in nonchewers.

I do not wish to go much into the details and restrict this book to the elementary laws that are handy in computing probability in some medical situations. Later in this section, a more detailed discussion is given on the use of probability in clinical assessments.

In many medical setups, precise probabilities are difficult to obtain. Long term chance of occurrence of cancer in a person with specific trait would be rarely known. You will find some methods in this book of estimating such probability. **Imprecise probability** is a term doing round these days. Such probability could be based on your personal belief and may be fuzzy with an element of gamble. Probability is imprecise when stated in interval with lower and upper expectations such as between 10% and 15%. Such probability arises when the information is scarce, vague, or conflicting and where your preferences are of "may be" type. There is no bar in using imprecise probability where precise is not available but note that a decision based on imprecise probability can only be tentative. Remember also that output of any model cannot be more precise than the inputs. If at all, the output will be less precise.

9.3.1 Elementary Laws of Probability

Among the most elementary laws of probability are the law of multiplication, which helps to calculate probability of joint occurrence of two or more events, and the law of addition, which helps to calculate probability of one or the other event.

9.3.1.1 Law of Multiplication

CAD is more common in diabetic patients than in nondiabetic patients. We say that CAD and diabetes are associated. These are termed statistically dependent diseases although such dependence does not imply any cause–effect type of relationship. On the other hand, blindness and deafness in a person are **independent events** in the sense that occurrence of one does not increase or decrease the chance of occurrence of the other. For such independent events, the joint probability of the two occurring together in a person can be easily computed as the product of the individual probabilities. Thus,

$$P(\text{blindness and deafness}) = P(\text{blindness}) * P(\text{deafness}), \qquad (9.10a)$$

where P is the notation for probability.

Symbolically "and" is denoted by $\cap$ and is called intersection. Thus, for independent events,

$$P(A \cap B) = P(A) * P(B). \qquad (9.10b)$$

This is called the law of multiplication or **product rule of probabilities**. A useful feature of this relationship is its inversibility. If Equation 9.10b holds, then the events A and B are independent, otherwise not.

Computation of the joint probability $P(\text{blindness and deafness})$ directly, without using Equation 9.10a, requires knowledge of the percentage of subjects in which these two occur together. In many situations, this might be cumbersome to obtain compared with the two individual probabilities on the right side of Equation 9.10a. Note that the individual probability in our example is simply the prevalence rate, which would be easily available. If these are 1 per 1000 and 2 per 100,000, respectively, then the prevalence of blindness and deafness occurring together is $0.001 \times 0.00002 = 0.00000002$ or 2 per 100 million. The law of multiplication is useful for obtaining joint probability in such cases.

9.3.1.2 Law of Addition

After corrective surgery for residual deformity in multiple injuries, the recovery may be full, partial, or none. Because a patient at any particular instance can belong to only one of these categories, these are called **mutually exclusive** categories. In the case of such categories, the probability of belonging to one or the other is computed by the law of addition. That is,

$$P(\text{full or partial recovery}) = P(\text{full recovery}) + P(\text{partial recovery}). \qquad (9.11)$$

If the probability of full recovery is 0.30 and that of partial recovery 0.40, then the probability of at least some recovery is 0.70. In notation, the symbol $\cup$, called union, is used for "or." Thus, for mutually exclusive events,

$$P(A \cup B) = P(A) + P(B). \qquad (9.12)$$

Note that mutually exclusive events cannot be independent and vice versa. There is no bar on independent events occurring together, whereas mutually exclusive events cannot occur together. Because full, partial, and no recovery are the only possibilities, out of which one has to happen, these are called **exhaustive categories**.

Example 9.4: An Interesting Relationship between BMI and Skinfold Thickness

Table 9.2 is constructed from the table on the correspondence between BMI and subscapular to triceps skinfold ratio in middle-aged Cretan men given by Aravanis et al. [21].

Cutoff points chosen for each category are tertiles. These divide the subjects into three groups of equal size. Thus, nearly equal numbers in the margin are not surprising. In this case,

$$P(\text{low BMI}) = \frac{109}{331} = 0.329,$$

and

$$P(\text{low skinfold ratio}) = \frac{110}{331} = 0.332.$$

Thus, $P(\text{low BMI}) \times P(\text{low skinfold ratio}) = 0.329 \times 0.332 = 0.109$. On the other hand, the joint probability on the basis of the cell frequency, $P(\text{low BMI and low skinfold ratio}) = 56/331 = 0.169$.

These two probabilities are very different, indicating that low BMI and low skinfold ratio are not independent. The aforementioned calculations show that if one is low, there is a greater chance that the other is also low; similarly for high. Because 62/331 is much more than $(111/331) \times (107/331)$, there is a greater chance of one being high when the other is high. But there is no evidence that middle BMI occurs more frequently with middle skinfold ratio. When BMI is in the middle category, the joint probabilities of low, middle, and high skinfold ratios are 36/331 = 0.109, 40/331 = 0.121, and 35/331 = 0.106, respectively. These

TABLE 9.2

Correspondence between Skinfold and BMI in Cretan Men

BMI	Subscapular to Triceps Skinfold Ratio			Total
	Low (<1.77)	Medium (1.77–2.33)	High (>2.33)	
Low (<25.3 kg/m²)	56	43	10	109
Medium (25.3–28.7 kg/m²)	36	40	35	111
High (>28.7 kg/m²)	18	31	62	111
Total	110	114	107	331

are not very different from the product of individual probabilities $(111/331) \times (110/331) = 0.111$, $(111/331) \times (114/331) = 0.115$, and $(111/331) \times (107/331) = 0.108$. When BMI is in the middle category, skinfold ratio is thus independent and falls in any of the three categories with almost equal frequency.

Theoretically, even a mild difference in the joint probability from the product of individual probabilities suggests association. Practically, the difference should be *substantial* to conclude association. There are two other considerations. First, $n = 331$ is a sample in Example 9.4 and there is a need to be cautious about sampling fluctuation that can easily produce minor deviations. Second, a very mild association may not be medically relevant in this case.

9.3.2 Probability in Clinical Assessments

Probabilities play an important role in wide-ranging clinical activities. When a diagnosis is reached on the basis of the complaints and physical examination, this is generally only the most *likely* diagnosis. As the investigation reports become available or the response to a therapy is known, the probability changes and sometimes even the most likely diagnosis also changes. *If they do not change the probability, then there is no use for these investigations and they are a waste of effort.*

Prospects of recovery after a therapeutic intervention are almost always stated in terms of probability. When the first heart transplantation was done, the chances of success were rated as 80%. The probability of recovery of a patient from tetanus after clinical manifestation is often considered less than 60%. The probability of survival for more than 5 years after detection of acute lymphoblastic leukemia is generally assessed as less than 30% despite therapy. Thus, probabilities have extensive usage in clinical assessments.

9.3.2.1 Probabilities in Diagnosis

Consider the following for the process of diagnosis. A set of signs, symptoms, and other evidence, occurring together more often than expected by chance and generally found useful in prognosis and treatment, is given a name and the process is called diagnosis. It requires discovering clusters such that similar patients fall within the same cluster, but the clusters themselves remain relatively distinct from one another.

There are two kinds of diagnostic entities. First are those that are based mainly on the value of a set of particular measurements. As already mentioned, diagnosis of essential hypertension depends almost exclusively on levels of systolic and diastolic blood pressures, diagnosis of glaucoma on intraocular pressure, and diagnosis of diabetes mellitus on serum glucose level. Once a defined cutoff point is available, the diagnosis is immediately obtained in these cases. Nevertheless, because of the wide-ranging variability even among healthy subjects and the distinct possibility of overlap between measurements found in healthy and ill subjects, a diagnosis based on such clear-cut definitions does not always remain above board. Probabilities play a preeminent role.

The second type of diagnostic entities is essentially multifactorial. They depend on signs and symptoms and on findings of laboratory and radiological investigations. Most cardiopulmonary disorders and hepatic dysfunctions fall in this group. Even though gold standards are available in some cases, such as positive cerebral angiography in embolism, the facility to carry out such an investigation may not be immediately available. In some cases, the gold standard itself can be in error. Positive histological evidence in carcinomas is considered confirmatory, but negative findings sometimes fail to correctly exclude the disease. When diagnosis depends on a multitude of factors, the clinician's personal judgment becomes important. Childhood leukemia and abdominal tuberculosis (abdTB) are examples of such diagnoses. When a patient presents with complaints of pain in the right hypochondrium, anorexia, and dyspepsia, liver disease is suspected. Liver function tests are then done and various enzymes in the blood are estimated. They too, in many cases, fail to provide a differential diagnosis on cirrhosis, malignancy, or hepatitis. Looking at the totality of the presentation, only the most probable diagnosis is made.

Sometimes diseases are described in the literature in the form of complaints commonly associated with those diseases. The track then is from disease to complaints. The actual diagnostic process is just the reverse, from complaints to the disease. To illustrate this problem further, take an example of the use of medical records. They are sometimes important sources for updating our knowledge.

Suppose the analysis of records shows that 70% of patients with abdTB present with long-term complaints of abdominal pain, vomiting, and constipation. Then,

$$P(\text{pain, vomiting, constipation/abdTB}) = 0.70. \qquad (9.13)$$

Because of the restriction to a specific group, which is mentioned after the slash (/) sign, namely, cases of abdTB in Equation 9.13, such a probability is called **conditional probability**. Note that the probability in Equation 9.13 is of very little value to a clinician. It tells what is seen in the patients but does not tell about the presence or absence of the disease when these complaints are reported. The inverse probability

$$P(\text{abdTB/pain, vomiting, constipation}) \qquad (9.14)$$

is useful to a clinician because it gives the diagnostic value of the complaints. The difficulty, however, is that hospitals maintain records diseasewise and not complaintwise. Thus, the records to compute the probability in Equation 9.13 are easily located. This requires only screening the records of cases of abdTB with regard to those complaints. But to compute probability in Equation 9.14, records of all the cases first need to be screened, irrespective of the disease, and the cases must be separated according to reporting of these three complaints. Among these, the cases with abdTB need to be counted. This is a relatively big exercise. One way to get around the problem is to use a formula that converts $P(A/B)$ into $P(B/A)$ and discussed a little later in this chapter.

9.3.2.2 Forwarding Diagnosis

Establishing the diagnosis for a particular patient is an exercise in classification. When a patient comes with a certain presentation, several possible diagnoses may flash into mind. The clinician mentally works out the chances of each specific diagnostic category and assigns the patient to the most likely category. The diagnosis, thus, is essentially a probabilistic entity.

Until sufficient information is available to evaluate probability using the procedure given in the previous section, personal probabilities can be used. **Personal probability** is a measure of one's belief in a statement. If you believe on the basis of experience or otherwise that among the patients who present with long-standing complaints of abdominal pain, vomiting, and constipation simultaneously, only 1 in 6 is of abdTB, your personal probability of abdTB given these three complaints (and nothing else) is 1/6.

High fever, rigors, splenomegaly, and presence of parasite in the blood are the stages that progressively confirm malaria. As the information increases, the diagnosis becomes firm, and the probability of absence or presence of the disease becomes concrete. The probability depends on what information is already available. The chance part is the uncovered information. The probability of any event without availability of any particular information is called **prior probability,** and the probability after that information is available is called **posterior probability**. The latter obviously depends on the kind of information available to alter the prior probability.

9.3.2.3 Assessment of Prognosis

In prognosis, the effort is to correlate the outcome (e.g., survival with or without residual effect or death) with the antecedent state (e.g., disease severity with reference to agent, host, and environment factors). The relationship depends, among other things, on correct diagnosis, promptness of treatment, type of treatment, facilities available, cooperation of the patient, attention of medical personnel, their ability, etc. Because of the variability in all these factors, not all patients with a given similar antecedent state have the same outcome. Again, the probabilities are helpful in making a decision. As in the case of diagnosis, the probability here is also conditional on the antecedent state.

If outcome is denoted by O and antecedent state by A, the probability to be calculated is $P(O/A)$. In fact, long-term follow-up studies of patients of different types and with different severities are needed before $P(O/A)$ can be objectively evaluated. Books and other literature may carry information on such probability of a specific outcome for a particular antecedent state. In the absence of such studies, the clinician's own experience of the percentages of patients who in the past have recovered, been disabled, or died can be used. So far, the prognosis lacks much of such quantification.

9.3.2.4 Choice of Treatment

Choice of treatment is a complex process and includes single or multiple therapy, dosage and duration, surgery, etc. It is known that the progress of disease and its severity differ from patient to patient and that a patient's response to a treatment always remains uncertain. When a choice of treatment modalities is available, that modality is prescribed which is *most likely* to result in best relief for the patient according to the assessment of the treating clinician. Thus, the choice of the treatment too is a matter of probability. So is its effectiveness. This probability again is conditional on a priori information regarding patient characteristics, disease prognosis, facilities available, cost, etc. Although the efficiency, efficacy, and effectiveness of various treatment regimens are established with the help of clinical trials, in many cases, a physician is again guided by his or others' experience and uses personal probabilities to choose a particular treatment.

9.3.3 Further on Diagnosis: Bayes Rule

As mentioned earlier, the scheme of medical knowledge is sometimes such as to provide probabilities of the form P(complaints/disease), whereas the probabilities actually required in practice are P(disease/complaints). For simplicity and for generalizability, denote the set of complaints and investigation results by C and the particular disease by D. When some additional information is available, $P(D/C)$ can be obtained from $P(C/D)$ by using Bayes rule.

9.3.3.1 Bayes Rule

The cases with C and with D can be shown in a diagram as in Figure 9.8. This is called a **Venn diagram**. The shaded area is common to both and is $C \cap D$ in our notation. As explained for Equation 9.13, C/D is the portion of D that is in C. Thus,

$$P(C/D) = \frac{P(C \cap D)}{P(D)}. \tag{9.15}$$

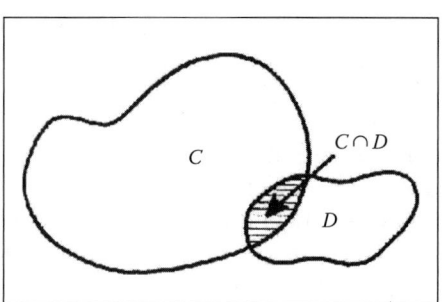

FIGURE 9.8
Venn diagram.

This implies that

$$P(C \cap D) = P(C/D)P(D). \tag{9.16}$$

Similar to Equation 9.15,

$$P(D/C) = \frac{P(D \cap C)}{P(C)}. \tag{9.17}$$

Because $D \cap C = C \cap D$, from Equations 9.16 and 9.17,

$$\text{Bayes rule: } P(D/C) = \frac{P(C/D)P(D)}{P(C)}. \tag{9.18}$$

Equation 9.18 is called Bayes rule and helps to convert $P(C/D)$ into more useful $P(D/C)$. For this conversion, two additional probabilities are required. The first is $P(D)$. This is the prior probability of the disease and is the same as the prevalence of the disease in the subjects under investigation. This can be derived from the records. The second is $P(C)$, which is the relative frequency of the complaints in the same kind of patients. Special efforts may have to be made to compute $P(C)$. In many cases, it may still be found easier to obtain $P(C)$ than $P(D/C)$ directly. Nevertheless, $P(C)$ can be easily computed by an alternative method given later in this section. Once these probabilities are available, the required probability $P(D/C)$ can be computed from Equation 9.18. This is the posterior probability after the complaints are known. Example 9.5 illustrates the calculation and shows that $P(D/C)$ can be very different from $P(C/D)$.

Example 9.5: Wide Difference between *P(C/D)* and *P(D/C)*

Suppose in a hospital 1 in 5000 patients on average is finally diagnosed as a case of abdTB. Thus, for these patients, $P(D) = 1/5000 = 0.0002$. Suppose further that the complaints of abdominal pain, vomiting, and long-term constipation are seen in 70% of cases of abdTB. Thus $P(C/D) = 0.70$. If a survey of records of that hospital also shows that these complaints are reported by nearly 1 in 1000 patients with all diseases, then $P(C) = 0.001$. Therefore, from Equation 9.18,

$$P(D/C) = \frac{0.70 \times 0.0002}{0.001} = 0.14.$$

Thus, there is only a 14% chance that a random patient reporting to that hospital with those complaints is a case of abdTB. Compare it with $P(C/D)$, which is 70%. $P(C/D)$ is high but has a very small diagnostic value when the complaints are nonspecific and can occur in some other conditions also, such as amebiasis and hepatitis in this case.

The probabilities $P(C/D)$, $P(D)$, and $P(C)$ may be available in a research setup, but a guesstimate may have to be made for everyday practice on the basis of knowledge and experience. A practitioner of long standing can have a fair idea of $P(C/D)$, $P(D)$, and $P(C)$ from his experience in his area of expertise. Example 9.5 demonstrates how medical records can be used to estimate these probabilities.

9.3.3.2 Extension of Bayes Rule

Bayes rule is more useful when it is known that the set of complaints C can occur only in K mutually exclusive categories. If the group of complaints (abdominal pain, vomiting, and constipation) are considered to occur only in abdTB (D_1), amebiasis (D_2), and hepatitis (D_3) and no other disease, and if the diseases do not really overlap, then we have $K = 3$ mutually exclusive and exhaustive categories of C. If that is so, Figure 9.9 shows that $C = (C \cap D_1) \cup (C \cap D_2) \cup (C \cap D_3)$.

From an extension of Equation 9.12, we get

$$P(C) = P(C \cap D_1) + P(C \cap D_2) + P(C \cap D_3)$$
$$= P(C/D_1) P(D_1) + P(C/D_2) P(D_2) + P(C/D_3) P(D_3) \qquad (9.19)$$

from Equation 9.16. Just as $P(C/D_1)$ and $P(D_1)$ are easily available, so would $P(C/D_2)$, $P(D_2)$, $P(C/D_3)$, and $P(D_3)$. In Example 9.5, these are the probabilities of occurrence of the complaints and the prevalence rates of amebiasis

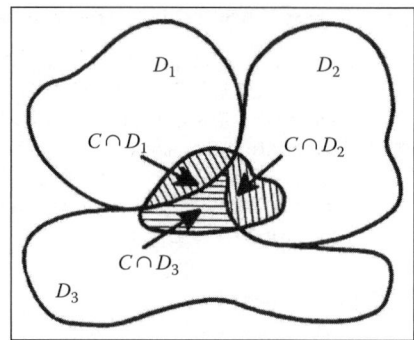

FIGURE 9.9
C as union of three intersections.

and hepatitis, respectively. Substituting Equation 9.19 in the denominator of Equation 9.18 gives for D_1,

$$P(D_1/C) = \frac{P(C/D_1)P(D_1)}{P(C/D_1)P(D_1) + P(C/D_2)P(D_2) + P(C/D_3)P(D_3)} . \qquad (9.20)$$

This is the expanded version of Bayes rule. This version obviates the need to know $P(C)$, the relative frequency of complaints in the example, which could be difficult to obtain. Example 9.6 illustrates the use of Bayes rule in another context.

Example 9.6: Caution in Using Bayes Rule

Suppose only 1 in 2000 male adults suffers from a coronary disease for which a screening tool is ECG. When a person actually has the disease, ECG is positive 98% of the time, but it is also positive 1% of the time in those who do not have the disease. What is the probability that an ECG-positive person actually has the disease?

Let D_1 = {person has the disease}, D_2 = {person does not have the disease}, and C = {positive ECG}. Then, $P(D_1) = 1/2000 = 0.0005$, $P(D_2) = 1 - 0.0005 = 0.9995$, $P(C/D_1) = 0.98$, and $P(C/D_2) = 0.01$. To find $P(D_1/C)$, from Equation 9.20, we get

$$P(D_1/C) = \frac{0.98 \times 0.0005}{0.98 \times 0.0005 + 0.01 \times 0.9995} = 0.047.$$

If the values are really as in this example, a very small $P(D_1/C)$ indicates that a positive ECG does not provide any clue to the disease. This may seem paradoxical in view of 98% positivity of the ECG in coronary cases. The difficulty is in it being positive in 1% of the subjects without the disease. And this group is very large: 1999 out of 2000 in this screening exercise. Thus, most positive results arise from errors rather than from diseased cases. A tool with much smaller error rate is required to be effective in such a situation.

SIDE NOTE: In clinics, however, an ECG will be done only for those who are otherwise suspected to have a coronary problem. And in them $P(D_1)$ would be high. $P(C/D_1)$ would also be high. Thus, $P(D_1/C)$ would be much more in a clinic setup than the one obtained in a screening setup. Then, an ECG will not be a waste of resources.

9.4 Validity of Medical Tests

Tools such as laboratory tests are an integral part of modern medicine. Fine-needle aspiration cytology (FNAC) is done to detect breast cancer, Pap smear for cervical cancer, Western blot for human immunodeficiency virus (HIV) infection, and chest x-ray for tuberculosis. Often, the sign–symptom syndrome functions as a *test* to form the basis for establishing a diagnosis. For example, generalized maculopapular rashes and fever with cough are considered indicative of measles, and prolonged acute chest pain indicates MI.

The tools used for evaluation and management of health and disease are seldom perfect. All tests are flawed to a degree. These produce correct results in many cases but fail, fully or partially, to perform well in some cases. Healthy individuals are occasionally classified wrongly as ill, sending false alarm, and some individuals who are really ill may not be detected, raising complacency. All such errors need to be controlled because there is a cost involved—cost of unnecessary treatment, cost of side effects, inconvenience, progression of disease, and even death of some patients. *The ability of a tool or of a procedure to perform its assigned function correctly is called its validity.*

A valid diagnostic test would correctly detect the presence as well as the absence of the disease. Some tests are more valid than others although they may be more expensive. Rectal sonography is considered more efficacious than prostate-specific antigen (PSA) values for detecting prostate cancer. Scintigraphy gives better results than ECG for MI. However, **gold standards** that give perfect results all the time are rare, and most of them are expensive in time and effort. No medical test is valid in absolute sense. Errors in classification such as misdiagnosis and missed diagnosis occur no matter what test is used.

Malaria is characterized by high fever with chills and rigors, splenomegaly, and a positive blood smear. How valid is this set of criteria? Can it correctly identify all the cases of malaria and can it correctly exclude all the nonmalarial cases? These two aspects are discussed next.

9.4.1 Sensitivity and Specificity

The ability of a test to give positive result in true cases of a disease is called sensitivity. Specificity is the ability to give negative result in cases where the disease is absent. Both are probabilities though many times expressed in percentage terms by multiplying by 100. These are two components of validity of a test that measure its inherent goodness. Denote presence of disease by $D+$ and absence by $D-$ and test positivity by $T+$ and test negativity by $T-$. Also let true positives be abbreviated as TP, false positives as FP, true negatives as TN, and false negatives as FN. The ability of a test to be positive when the disease is present is

$$\text{Sensitivity } S(+) \quad \text{or} \quad P(T+/D+) = \frac{\text{TP}}{\text{TP} + \text{FN}}. \tag{9.21}$$

The ability of a test to give a negative result when the disease is absent is

$$\text{Specificity } S(-) \quad \text{or} \quad P(T-/D-) = \frac{\text{TN}}{\text{TN} + \text{FP}}. \tag{9.22}$$

These are best illustrated with the help of an example.

Example 9.7: Calculation of Sensitivity and Specificity

ECG was performed on a total of 700 subjects with complaints of prolonged acute chest pain. Of these, 520 cases were earlier confirmed for the presence of MI and 180 for its absence. The results obtained are given in Table 9.3.

Note the following for ECG as a test for the diagnosis of MI in this example:

True positives (TP) = 416
False positives (FP) = 9
True negatives (TN) = 171
False negatives (FN) = 104

TABLE 9.3

MI and ECG in Cases of Acute Chest Pain

ECG	MI		Total
	Present	**Absent**	**Total**
Positive	416 (TP)	9 (FP)	425
Negative	104 (FN)	171 (TN)	275
Total	520	180	700

It is known that 520 subjects had MI and the other 180 did not. Thus,

$$\text{Sensitivity of ECG, } S(+) = \frac{416}{520} \times 100 = 80\%,$$

and

$$\text{Sensitivity of ECG, } S(-) = \frac{171}{180} \times 100 = 95\%.$$

Thus, in this example, a negative ECG seems sufficiently specific in non-MI cases but not so sensitive in the cases with MI. Many cases of MI, in this example 20%, are missed by ECG, possibly because the required elevation in ST segment does not appear.

The following measures the combination of sensitivity and specificity, when both are equally important:

$$\text{Inherent validity of a test} = \frac{(TP + TN)}{n}, \tag{9.23}$$

where n is the total number of subjects. In Example 9.7, inherent validity of ECG is $(416 + 171)/700 \times 100 = 84\%$.

9.4.1.1 Features of Sensitivity and Specificity

A difficulty with the concepts of sensitivity and specificity is that they can be evaluated only when the presence or absence of the disease is known. If the disease status is already known, where is the need for a test? Nevertheless, the concepts are useful, as shortly explained. The following points deserve attention:

1. The true diagnosis is evaluated on the basis of more refined methods, a gold standard, that may be far more difficult to adopt. Often, the real diagnosis emerges after the passage of time, for instance, on response to therapy or by autopsy. Often, a surrogate is used as gold standard such as histological evidence for cancer. If the gold itself is a bit shoddy, a good sensitivity or specificity may give a false sense of security.

2. The value of sensitivity and specificity can be changed by altering the criterion for positivity. In Example 9.7, if the creatine phosphokinase (CPK) level is also considered, then some other values of sensitivity and specificity are obtained. The exact values will depend on what threshold of CPK level is chosen to indicate MI. Different thresholds will give different values of sensitivity and specificity.

3. Generally speaking, an alteration in the criteria will either increase or decrease sensitivity but will affect the specificity in the opposite

manner. That is, an increase in sensitivity is generally accompanied by a decrease in specificity, and vice versa.

4. It is assumed in these calculations that the disease and the test can be classified as present/absent or negative/positive. In practice, you can have a category "may be" or "indeterminate" or "±." The calculations then become extremely complex and raise many additional issues that are outside the scope of the present book.

5. Genuine cases with equivocal or uninterpretable test results should be excluded. In practice, though, such results are seen to have some association with disease state. If so, the bias can creep in even when they are excluded.

6. As for almost any other criterion, values of sensitivity and specificity are valid for the type of cases actually used for calculation. If only severe cases are included, the values would not be valid for mild cases. For example, large advanced tumors are easily picked up. You may like to include appropriate proportion of cases with different spectrum of disease. But the number in each category must be sufficient.

7. Co-morbidities may also affect the values. For example, NESTROFT (naked eye single tube red cell osmotic fragility test), used for screening thalassemia in children, shows good sensitivity in patients without any other hemoglobin disorder but also produces positive result when other types of hemoglobinopathies are present.

8. The subjects without disease should be similar as encountered in practice. They should be subjects with initial suspicion because otherwise the question of testing does not arise.

9. Another condition is that prior knowledge of presence or absence of disease should not affect the performance of the test. That is, the person interpreting the test should not consider the disease status. Better still if the person is kept blind to the disease status.

10. Guard against possible interobserver variation if more than one observer is used. This is particularly so when observer abilities are important in diagnosis such as for bone density assessment through magnetic resonance imaging (MRI) by an experienced radiologist and a junior radiologist.

11. If you are studying a really rare (e.g., one in a million) disease, a positive result even with a good test is likely to be false on somebody in the remaining 999,999.

9.4.1.2 Likelihood Ratio

Sensitivity and specificity can be combined to provide another useful measure of the inherent goodness of a test. Consider (1) the chance that the test is positive in those who have the concerned disease and (2) the chance that the test is positive in those who do not have the disease. The ratio of these two chances is called the likelihood ratio of a positive test result and denoted by LR+. The first chance is measured by sensitivity and the second by (1 − specificity). Thus,

$$LR+ = \frac{Sensitivity}{1 - Specificity}. \qquad (9.24a)$$

A high likelihood indicates that the test is good in being positive where it should be. If sensitivity is 90% and specificity is 80%, then LR+ = 0.90/(1 − 0.80) = 4.5. Analogously, for negative result,

$$LR- = \frac{1 - Sensitivity}{Specificity}. \qquad (9.24b)$$

In this example, this is (1 − 0.90)/0.80 = 0.125.

Although the concepts of LR+ and LR− are for inherent qualities of the test, their interpretation is mostly in terms of likelihood of presence or absence of disease. LR+ measures the increase factor of odds of disease when the test is positive, and LR− measures the decrease factor of odds of no disease when the test is negative. In other words, posttest odds = pretest odds ∗ LR. If the pretest probability of test positivity is 10%, the odds are 1:9 or 1/9. If LR+ = 6, posttest odds = 6/9, or the probability is 6/(6 + 9) = 0.4. The chance after testing positive increases from 10% to 40%. LRs provide further insight into the inherent quality of the test. A value of 10 or more of LR+ and 0.1 or less of LR− indicates that the test is extremely good in those whose disease status is known.

9.4.2 Predictivities

Sensitivity and specificity are indicators of the validity of a test but they do not measure the diagnostic value of the test. Diagnostic value is obtained in terms of predictivities.

9.4.2.1 Positive and Negative Predictivity

The actual problem in practice is to detect the presence or absence of a suspected disease by using a test. The diagnostic value of a test is measured by the probability of actual presence of disease among those who are test positives and the probability of actual absence of disease among those who are test negatives. These indicators are kind of an inverse of sensitivity and specificity and are called positive predictivity and negative predictivity, respectively. In the literature, these are referred to as predictive value of a positive test and predictive value of a negative test, respectively. But I prefer the shorter names and hope these short names do not cause any confusion. These are also called **posttest probabilities** and measure the utility of a test in correctly identifying or correctly excluding the disease. Positive predictivity can also be understood to measure how good the test is as a marker of the disease.

In terms of notations,

$$\text{Positive predictivity } P(+) \quad \text{or} \quad P(D+/T+) = \frac{\text{TP}}{\text{TP} + \text{FP}}, \tag{9.25}$$

and

$$\text{Negative predictivity } P(-) \quad \text{or} \quad P(D-/T-) = \frac{\text{TN}}{\text{TN} + \text{FN}}. \tag{9.26}$$

Try to interpret the data in Table 9.3 inversely. Suppose now that 425 ECG-positive subjects were further investigated and 416 were found to have MI. A group of 275 ECG-negative subjects were also investigated and 104 were found to have MI. Then, the predictivity of a positive ECG for the presence of MI is

$$P(+) = \frac{416}{425} \times 100 = 98\%,$$

and the predictivity of a negative ECG for the absence of MI is

$$P(-) = \frac{171}{275} \times 100 = 62\%.$$

The diagnostic value of a test is not reflected by sensitivity and specificity but is reflected by predictivities. These also are indicators of the validity of a test. If 700 subjects in this example are considered good representatives of such cases, the specificity is high at 95%, yet the test is poor in excluding MI—only 62% are correctly ruled out. This happens in this case because many subjects with negative ECG have MI. What it tells is that the ECG can be safely used to detect the presence but not to detect the absence of MI, as far as this example is concerned.

A screening test should have high negative predictivity whereas a confirmatory test should have high positive predictivity. When both are equally important, use

$$\text{Predictive validity} = \frac{\text{TP} + \text{TN}}{\text{Total subjects tested}}. \tag{9.27}$$

This combines the two predictivities assuming that both are equally important. If they are not, an index can be devised that gives differential weight as needed. Equation 9.27 is the same as Equation 9.23 but the interpretation is different since the denominator in Equation 9.23 is the total subjects with and without disease and in Equation 9.27 is the total subjects who are test negative and test positive.

9.4.2.2 Predictivity and Prevalence

The concepts of sensitivity and specificity work in an inverse direction. They move from disease to the test. Predictivities do provide assessment in the right direction, but they are severely affected by the prevalence of disease among those tested. This is illustrated in Example 9.8. The advantage of sensitivity and specificity is that they are absolute and do not depend on prevalence.

Example 9.8: Predictivities Depend on Prevalence

Out of 700 tested, let the number with MI in Example 9.7 be changed to 300 (the prevalence is now 43%). If sensitivity and specificity remain as before at 80% and 95%, respectively, then the different numbers would be as shown in Table 9.4. Now,

$$\text{Positive predictivity} = \frac{240}{260} \times 100 = 92\%,$$

and

$$\text{Negative predictivity} = \frac{380}{440} \times 100 = 86\%.$$

These values are very different from the ones obtained earlier because of the entirely different prevalence rate. The previous prevalence rate was 520/700 = 0.74 or 74%.

From Bayes rule, positive predictivity,

$$P(+) \text{ or } P(D+/T+) = \frac{P(T+/D+)P(D+)}{P(T+/D+)P(D+)+P(T+/D-)P(D-)}$$

$$= \frac{S(+)*p}{S(+)*p+[1-S(-)]*(1-p)}, \tag{9.28}$$

TABLE 9.4

Increased Prevalence of Disease in Table 9.3

	D+	D−	Total
T+	240	20	260
T−	60	380	440
Total	300	400	700

where p is the prevalence rate per unit. When sensitivity and specificity are constants, it is easy to see from Equation 9.28 that positive predictivity increases as prevalence increases.

As with Equation 9.28, it can also be shown for negative predictivity that

$$P(-) \text{ or } P(D-/T-) = \frac{S(-)*(1-p)}{S(-)*(1-p)+[1-S(+)]*p}. \tag{9.29}$$

As prevalence increases, the negative predictivity decreases. Dependence of predictivity on prevalence arises from putting the information in its proper context. A patient with high fever and shivering might be diagnosed as influenza in Europe but malaria in West Africa [22].

The predictivities for some specific values of sensitivity and specificity and for different prevalences are shown in Table 9.5. As the prevalence increases, the positive predictivity also increases, and this increase is more pronounced when the specificity is low. Higher prevalence leads to less negative predictivity, more so when sensitivity is low.

Equations 9.28 and 9.29 also express the relationship between sensitivity–specificity and predictivities. If prevalence is known, predictivities can be obtained by using sensitivity and specificity. Herein lies the importance of these two somewhat reversed indices. Based on confirmed cases, sensitivity and specificity are easy to obtain. Use them to calculate diagnostically important positive and negative predictivities with the help of Equations 9.28 and 9.29. Direct calculation of predictivities requires follow-up studies to find the confirmed cases with and without disease among those that have

TABLE 9.5

Predictivities for Some Specific Values of Sensitivity, Specificity, and Prevalence

Sensitivity S(+)	Specificity S(−)	Prevalence	Positive Predictivity P(+) (%)	Negative Predictivity P(−) (%)
0.20	0.20	0.10	3	69
		0.50	20	20
		0.90	69	3
0.20	0.90	0.10	18	91
		0.50	67	53
		0.90	85	11
0.90	0.20	0.10	11	95
		0.50	53	67
		0.90	91	18
0.90	0.90	0.10	50	99
		0.50	90	90
		0.90	99	50

shown positive and negative tests. Such follow-up studies could be expensive, and can be avoided for predictivities when prevalence is known and sensitivity–specificity easily obtained.

9.4.2.3 Meaning of Prevalence for Predictivity

The dependence of predictivities on prevalence is to be cautiously interpreted. This prevalence is among those who are administered the test. A diagnostic test is generally administered to those who are suspected to have the disease, and in them the proportion with disease is likely to be high. This proportion is the same as prevalence in the sense used here. When this is high, it becomes difficult to correctly identify the negatives.

Another very useful interpretation of prevalence is the extent of belief or confidence that a clinician has in the presence of disease in a particular subject, earlier called personal probability. On the basis of the information available on the subject before the test, if a clinician evaluates that the chance of disease in that subject is 60%, then this has exactly the same connotation as prevalence. Thus, prevalence can also be understood as the pretest probability.

If the test is for screening purposes in an endemic area or in a high-risk population, for example, enlarged prostate in males of age 60+ years, the prevalence may still be low because the entire (high-risk) population is being screened. In this case, even a test with high positive predictivity would leave out many positive cases.

In summary, the calculation of predictivities should be done using subjects who correctly represent the proportion of diseased and nondiseased cases among those who are to be tested. For this reason, predictivities are not comparable across groups or across populations whereas sensitivity and specificity are.

Although it may be clear that sensitivity–specificity are based on retrospective (case–control) studies and direct calculation of predictivities requires prospective studies, what should be done if the study is cross-sectional? Luckily, cross-sectional is a setup that can be used to calculate both sensitivity–specificity and predictivity. There is a big rider though. *For these calculations to be valid, the cross-sectional study must be based on a sample that proportionately represents the diseased and nondiseased subjects in the target population on which the test is proposed to be used. Without this, the values arrived at could be wrong and misleading.*

9.4.2.4 Features of Positive and Negative Predictivities

The following are important features of diagnostically useful predictivities:

1. The computation of sensitivity–specificity and of predictivities assumes that a gold standard is available and has been used to confirm the presence or the absence of disease. This is not always the case. For example, pancreatic carcinoma can be confirmed only by

laparotomy if the patient is alive or by autopsy if the patient is dead. Many suspected cases or their relatives might not agree to these procedures. A gray zone is always present, and efforts are made to keep the size of this gray zone minimal. As the technology advances, new procedures are discovered, the gray zone decreases, and the calculation of validity indicators becomes more accurate. Remember that predictivity calculations, just as for sensitivity–specificity, depend on "gold" being above suspicion. As mentioned earlier, a doubtful gold will cloud the performance of the new test.

2. Another way to look at the relationship between sensitivity–specificity and predictivities is to compare the formula given in Equation 9.21 with Equation 9.26. In both these expressions, FN is extra in the denominator. The more are false negatives the lower is the sensitivity and the lower the negative predictivity. A similar comparison of formulas given in Equations 9.22 and 9.25 shows that the more are the false positives, the lower is the specificity and the lower the positive predictivity. Ironic as it may sound, positive predictivity corresponds more to specificity than to sensitivity, and negative predictivity corresponds more to sensitivity.

3. Just as in the case of sensitivity and specificity, there is also a trade-off between positive predictivity and negative predictivity. In Example 9.7, if the criterion of elevated CPK is added to the ECG, then the ability to predict cases of MI correctly will increase but, at the same time, some patients with MI may be incorrectly classified in the non-MI group.

4. For a condition such as hypertension, neither very high positive predictivity nor very high negative predictivity is required. Rosner and Polk [23] sent out a questionnaire to 30 hypertension experts to estimate the positive and negative predictivity required to diagnose hypertension. The median of the desired positive predictivity was 80% according to these experts and the median of the desired negative predictivity 77.5%. Neither is high.

5. Depending on the requirement, either a test with high negative predictivity or a test with high positive predictivity can be chosen. Screening tests, which are used at initial stages or in a community setup, should be able to correctly exclude all negative cases. Cervical cytology (Pap smear) is used for screening of cervical cancer and mammography for breast cancer. Random blood glucose level is used for screening of diabetes, and Mantoux test for tuberculosis infection. They are tests with high negative predictivity. When the test is negative, it is safe to assume in these cases that the disease is absent. In addition, a screening test must be durable to withstand constant use and considerable abuse because it tends to be used on a large number of subjects.

On the other hand, the goal of diagnosis is to ferret out cases while keeping false positives to a negligible level. False positivity can create an unmanageable backlog of cases who actually do not require medical attention. This can lead to organizational fatigue and the staff may wear out, causing loss of alertness. Thus, a diagnostic test is good if it has high positive predictivity. Histology is good to confirm malignancy, so is x-ray for bone fracture. Try to use such tests in your practice.

6. If the emphasis is not to miss any positive case, then a test with higher negative predictivity is desirable. Diseases with a grave prognosis have this feature. Malignancy is an example in which missing a case could become very expensive later on, particularly if it is in a treatable stage. When a test with high negative predictivity is used, false negatives would be less but false positives may be high. These false positives could be excluded later on by further investigations. This surely increases the burden on the medical care system but that is the price for not missing many positive cases. But there are risks too. If the initial treatment is costly or toxic to the nondiseased, then it should not be given merely on the basis of suspicion. Another risk of high false positivity, particularly for nearly untreatable diseases such as leukemia and AIDS, is that it can cause unnecessary psychological trauma in the subjects who actually are nondiseased.

7. In a research setup, for a new test, you should not be contented with the validity computations on one dataset. You should apply the test to some other groups to check the repeatability. It is only after such external validation that the values of sensitivity–specificity and predictivities can be accepted and adopted for wider application.

8. Finally, note that the calculations of sensitivity–specificity and predictivities help to measure the uncertainties. This is a step forward compared with using qualitative terms such as definite, doubtful, and suspected. The qualitative terms can mean different things to different people. Validity indices remove such subjectivity.

9.4.3 Combination of Tests

Very few tests are individually good in correctly detecting both the presence and the absence of disease. A combination of two or more tests can help to enhance the efficacy. There are two principal forms of combination: series and parallel.

9.4.3.1 Tests in Series

If a subject is considered to have a disease when he is positive by *each* of a battery of tests, then the tests are said to be in series. A subject negative by any one or more of the tests is classified as nondiseased. Thus, it is easy to classify a subject into the nondiseased group and more difficult to classify him into the diseased group. As a consequence, many with the disease are more likely to be classified as negatives. Thus, false negatives may be high and the following is likely to happen:

Tests in series	Sensitivity	Low
	Specificity	High
	Positive predictivity	High
	Negative predictivity	Low

Note the word "likely". There may be exceptions.

9.4.3.2 Tests in Parallel

Two or more tests are said to be in parallel when a positive result on any one or more tests is considered sufficient to classify a subject into the diseased group. A subject negative by both or all the tests is classified as nondiseased. In this case, some without disease are more likely to be wrongly classified into the diseased group but many without disease are correctly excluded. This gives the following:

Tests in parallel	Sensitivity	High
	Specificity	Low
	Positive predictivity	Low
	Negative predictivity	High

As in the case of tests in series, these may isolate situations where this may not happen.

Example 9.9: Parallel and Serial Tests

Ultrasound and CT scan of the pancreas were used as tests in cases of carcinoma of pancreas. The results are in Table 9.6.

Note that the tests in series help to increase the specificity (at the cost of sensitivity) and that the tests in parallel sacrifice specificity to gain sensitivity.

As in the case of a single test, the choice to arrange two or more tests in series or in parallel depends on the purpose and the nature of the disease. In the case of pancreatic carcinoma in Example 9.9, if it is considered more important not to

TABLE 9.6

Sensitivity and Specificity of CT Scan and Ultrasound in Series and in Parallel

Test	Sensitivity	Specificity
Ultrasound	80	60
CT scan	90	90
Two tests in series	72	96
Two tests in parallel	98	54

miss any positive case, as for screening, then high negative predictivity is desirable. In that case, the tests should be used in parallel. Any person positive by either or both the tests would be classified as a suspected case of pancreatic carcinoma. Many of these could be false positives and would undergo the inconvenience of further tests. Many would be excluded later by subsequent procedures.

A strategy sometimes used to detect right cases is to use highly sensitive test in the first stage that may identify nearly all suspected cases. Then, in the second stage, use a highly specific test that will filter out nearly all negative cases.

9.4.4 Gains from a Test

As mentioned earlier, predictivities can be understood as the posttest probabilities. They measure the likelihood of the disease or of no disease after the test result becomes known. The pretest probabilities are the prevalence in the target group. Assuming that the patient is a random arrival from the group on which the test is done, the chance that he has the disease is the same as the prevalence rate of the disease in that group. A test is useful only if it substantially alters the posttest probability of disease compared with the pretest probability. This difference is the gain attained by the test. This is easily appreciated as gain when the terms pretest and posttest probabilities are used instead of prevalence and predictivity.

> **Example 9.10: Gains from α-Fetoprotein Determination**
>
> Consider α-fetoprotein determination as a test for hepatoma in patients with cirrhosis. If this test has a sensitivity of 96% and specificity 88%, what is the gain by the test at different levels of prevalence?
>
> When the sensitivity is 96% and specificity 88%, the gain by the test is as given in Table 9.7 for different pretest probabilities. Posttest probabilities have been obtained with the formulas given in Equations 9.28 and 9.29.
>
> In this case, the gain in $P(+)$ is maximum when the prevalence is between 20% and 40%. When a person before the test is estimated to have nearly a 1:3 chance of the disease, the test result would be most useful in enhancing confidence in arriving at a decision about the presence of the disease. This substantially enhanced confidence may not be enough, say, less than 80%, but that is the best achievable with the help of this test for that pretest likelihood.
>
> SIDE NOTE: If you already believe before the test that the person has an extremely high chance of being positive or an extremely high chance of being negative, then the application of the test would make only a marginal difference in your belief. But, in this example, if the pretest likelihood of the absence of the disease is as low as 10%, a negative test firms up your belief to 71% that the disease is absent. That is a very substantial gain.

These gains would vary from one set of sensitivity–specificity values to another set. You may like to confirm that for $S(+) = 80\%$ and $S(-) = 74\%$, the gains are maximum when the pretest chances are nearly 50:50.

TABLE 9.7

Gain by the Test at Different Levels of Prevalence for Sensitivity 96% and Specificity 88%

Probability of Presence of Disease (%)			Probability of Absence of Disease (%)		
Pretest (Prevalence)	Posttest $P(+)$	Gain by the Test	Pretest (100 – Prevalence)	Posttest $P(-)$	Gain by the Test
10	47	37	90	99	9
20	67	47	80	99	19
30	77	47	70	98	28
40	84	44	60	97	37
50	89	39	50	96	46
60	92	32	40	94	54
70	95	25	30	90	60
80	97	17	20	85	65
90	99	9	10	71	61

9.4.4.1 When Can a Test Be Avoided?

In some situations such as in prostate biopsy, the test is invasive. It carries a risk of side effects besides cost and inconvenience. In such situations, it is possible to find whether the burden of a diagnostic test is worth pursuing or whether it is better to make therapeutic decisions without the test (if legal considerations allow). For example, one can decide that if the pretest likelihood of disease is more than 80% or less than 10% then there is hardly any need for the test. The treatment can proceed in the former case, and the patient can be advised to keep watch in the latter case and to come back if and when new problems occur or if the complaints persist.

It is possible to obtain two thresholds P_1 and P_2 such that:

If pretest probability is less than P_1, then do not test and do not start treatment.

If pretest probability is more than P_2, then treat without the test.

If the pretest probability is between P_1 and P_2, then carry out the test and be guided by the test result.

For this, the following quantities are estimated, mostly through accumulation of experience:

Benefit of appropriate therapy,

B = average net gain in utility (benefit minus risk) among the patients who are properly treated.

Loss from unnecessary therapy,

L = average net loss in utility in patients who do not have the disease but are still given treatment.

Risk of diagnostic test,

R = average loss in utility from any complication of administration of the test.

The details are avoided but it can be shown that

$$P_1 = \frac{(\text{FP rate}) * L + R}{(\text{FP rate}) * L + (\text{TP rate}) * B},$$

and

$$P_2 = \frac{(\text{TN rate}) * L - R}{(\text{TN rate}) * L + (\text{FN rate}) * B}.$$

Example 9.11: Postponing a Test in Acute Renal Failure

Acute renal failure has a grave prognosis. Thus, early diagnosis and prompt therapy are important. But early symptoms are wide and varied, and the diagnosis is seldom clear. Unnecessary therapy (dialysis or medication) can sometimes lead to complications such as embolism arid dyselectrolytemia besides, of course, the cost and inconvenience to the patient and to the hospital. Diagnostic approaches include renal biopsy, which also carries some risk. Suppose the likelihoods are as follows:

Average net gain from proper treatment, $B = 0.80$.
Average net loss due to unnecessary therapy, $L = 0.10$.
Risk of diagnostic test, $R = 0.01$.
FP rate of the diagnostic tests used = 0.25.
TP rate of the diagnostic tests used = 0.90.
TN rate of the diagnostic tests used = 0.75.
FN rate of the diagnostic tests used = 0.10.

Then,

$$P_1 = \frac{0.25 \times 0.10 + 0.01}{0.25 \times 0.10 + 0.90 \times 0.80} = 0.05,$$

and

$$P_2 = \frac{0.75 \times 0.10 - 0.01}{0.75 \times 0.10 + 0.10 \times 0.80} = 0.42.$$

If the preceding data are believed to be true, the following conclusions can be drawn. If the clinician has less than 5% suspicion in a case, then possibly the test can be avoided or at least postponed until such time as the situation becomes clearer. If the clinical condition of the patient is sufficient to assess more than 42% chance of acute renal failure, then therapy can be started without worrying about the test. If the pretest probability is between 5% and 42%, then the test results are the best guide.

Note the following:

1. An example can be constructed such that $P_1 > P_2$, that is, the procedure does not work. But this is unlikely to happen in practice.
2. This kind of approach is recommended only for tests that are risky or not widely available.

9.5 Search for the Best Threshold of Continuous Test: ROC Curve

Tests such as white blood cells (WBCs) count for appendicitis are continuous as a range of values are obtained. Tests such as presence or absence of abdominal pain are discrete—mostly dichotomous. This section deals with tests that give continuous measurements.

Thresholds that minimize the sum of false positives and false negatives can be obtained by a procedure called discriminant analysis (Chapter 19). This procedure is particularly useful when the number of possible categories is three or more, for example, cirrhosis, hepatitis, and malignancy of liver. For discriminating between two competing diagnoses, two approaches are available as discussed next. The first approach based on the sensitivity–specificity ROC curve is more popular but less valid, and the second based on predictivity ROC is a new and more valid approach.

9.5.1 Sensitivity–Specificity-Based ROC Curve

Is T_4 better or TSH better in discriminating between hyperthyroid and euthyroid cases? Comparison of the performance of two quantitative tests can be obtained by the area under the ROC curve. Conventional ROC curve is obtained by plotting sensitivity (i.e., true positive rate) versus (1 − specificity) (i.e., false positive rate) for different values of measurements as illustrated in Example 9.12. Historically, the name comes from signal detection developed during World War II. For details of ROC curve, see Sackett et al. [24].

Example 9.12: ROC for Duration of Rupture of Membrane for Cesarean Delivery

Many full-term births in a hospital require induction of labor. Induction succeeds in most cases but fails in a few. In case the induction fails, a Cesarean is done for delivery. This involves pain, time, and money but also requires mental preparedness.

It would be nice for both the woman and family as well as the attending obstetrician to anticipate a Cesarean delivery on the basis of patient characteristics. The traditional method is to compute Bishop Score based on dilatation, effacement, consistency, and position. Other parameters that influence the success of induction of labor are maternal age, parity, BMI, and amniotic fluid index. A study was carried out in $n = 166$ cases with prelabor rupture of membrane to find if the duration since rupture can help in predicting the Cesarean delivery. Although the study was prospective, sensitivity and specificity were calculated for different durations of rupture. The values obtained are shown in Table 9.8.

The ROC curve obtained by plot at different cutoffs is shown in Figure 9.10. A statistical software found that the area under the curve (AUC) is $C = 0.898$ with its standard error (SE) = 0.029 and 95% CI from 0.841 to 0.956. AUC is explained later in this section.

With $C = 0.898$ out of a possible 1.0, it seems from this ROC that duration since rupture of membrane itself is a good indicator to anticipate Cesarean delivery. The best cutoff that maximizes (sensitivity + specificity) is 9.75 h. At this duration, the sensitivity is 0.74 and specificity is 0.90 (1 − specificity = 0.10).

SIDE NOTE: Although not shown in this example, the Bishop score was not found to be as good an indicator of impending Cesarean as was duration since rupture in this example. Otherwise also, Bishop's score is subjective and nonreproducible because of higher inter- and intraobserver variability.

Example 9.12 illustrates how ROC can be effectively used for medical decisions. The example is illustrative only and should not be construed to mean that duration since rupture can be solely used to anticipate Cesarean. For this, studies in different locales are needed.

ROC curve is useful in (1) finding optimal cutoff point to least misclassify diseased and nondiseased subjects, (2) evaluating the discriminatory ability of a test to correctly pick up diseased and nondiseased subjects, (3) comparing efficacy of two or more medical tests for assessing the same disease, and (4) comparing two or more observers measuring the same test (interobserver variability). The details are as follows.

9.5.1.1 Methods to Find the "Optimal" Threshold Point

ROC curve can help to identify a threshold that gives the highest sum of sensitivity and specificity in situations where sensitivity and specificity are

TABLE 9.8

Sensitivity and (1 – Specificity) for Cesarean Delivery at Different Duration of Rupture of Membrane

Duration (h) Greater Than or Equal to	Sensitivity	1 – Specificity
0.00	1.000	1.000
0.63	1.000	0.976
0.88	1.000	0.969
1.25	1.000	0.890
1.75	1.000	0.866
2.13	1.000	0.819
2.38	1.000	0.811
2.75	1.000	0.780
3.25	1.000	0.717
3.75	1.000	0.709
4.50	1.000	0.646
5.13	0.971	0.583
5.38	0.971	0.575
5.75	0.971	0.551
6.25	0.914	0.378
6.75	0.914	0.346
7.13	0.857	0.291
7.38	0.857	0.283
7.75	0.857	0.276
8.25	0.800	0.189
8.75	0.800	0.181
9.25	0.743	0.110
9.75	0.743	0.102
10.25	0.543	0.039
10.75	0.543	0.031
11.50	0.457	0.024
12.50	0.400	0.008
13.50	0.343	0.000
14.50	0.286	0.000
15.50	0.257	0.000
16.50	0.200	0.000
17.50	0.171	0.000
18.50	0.143	0.000
19.50	0.114	0.000
20.25	0.057	0.000
21.50	0.000	0.000

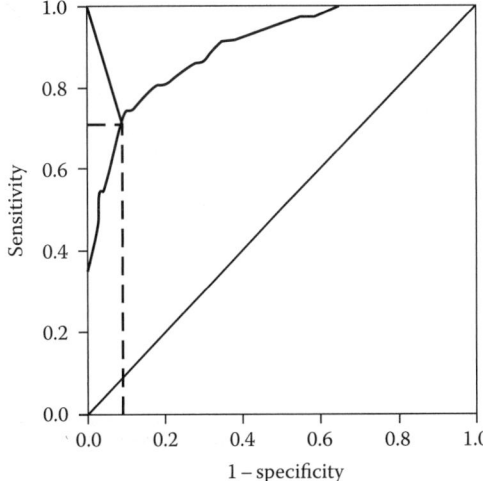

FIGURE 9.10
ROC curve of duration since rupture membrane for Cesarean delivery.

available for a large number of values. In Example 9.12, the values are on continuous scale but ROC can also be obtained when they are in equally spaced categories. The number of such categories must be at least five for ROC curve to be adequate.

Three criteria are used to find optimal threshold point from ROC curve. First two methods give equal weight to sensitivity and specificity and impose no ethical cost and no prevalence constraints. The third criterion considers cost, mainly the financial cost, for correct and false diagnosis, cost of discomfort to person caused by treatment, and cost of further investigation when needed. This method is rarely used in medical literature because it is difficult to estimate the respective costs and prevalence is often difficult to assess. These three criteria are known as points on curve closest to the (0, 1), Youden index, and minimize cost criterion, respectively.

If s_n and s_p denote sensitivity and specificity, respectively, the Euclidean distance between the point (0, 1) and any point on the ROC curve is $d = \sqrt{[(1 - s_n)^2 + (1 - s_p)^2]}$. To obtain the optimal cutoff point to discriminate the disease with nondisease subject, calculate this distance for each observed cutoff point and locate the point where the distance is minimum. Most of the ROC analysis softwares calculate the sensitivity and specificity at all the observed cutoff points allowing you to do this exercise.

The second is Youden index that maximizes the vertical distance from line of equality to the point (x, y) as shown in Figure 9.11. x represents (1 − specificity) and y represents sensitivity. In other words, the Youden index J is the point on the ROC curve that is farthest from line of equality (diagonal line). The main aim of Youden index is to maximize the difference between true positive rate (s_n) and false positive rate $(1 - s_p)$ and

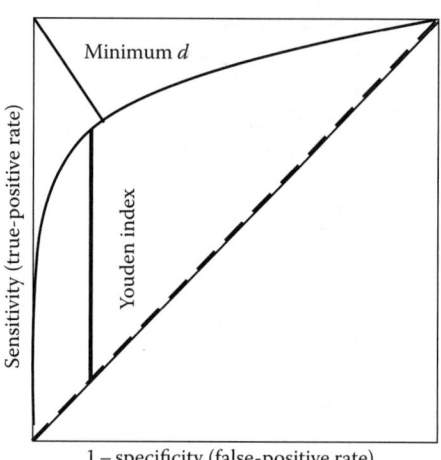

FIGURE 9.11
Finding best cutoff from the ROC curve.

little algebra yields $J = \max(s_n + s_p)$. The value of J can be located by doing a search of plausible values where sum of sensitivity and specificity can be maximum. Youden index is more commonly used criterion because this index reflects the intension to maximize the correct classification rate and is easy to calculate. Many authors advocate this criterion. In Example 9.12, such a threshold is 9.75 h corresponding to sensitivity = 0.74 and specificity = 0.90.

These procedures for finding the optimal threshold are applicable when both sensitivity and specificity are equally important. If they are not, expert judgment may be required to find an appropriate cutoff depending upon whether you are interested in high sensitivity or high specificity.

9.5.1.2 Area under the ROC Curve

Primary utility of ROC curve lies in the AUC. Apart from the inherent validity mentioned earlier, total area under ROC curve is a single index for measuring both the performances of a test. This area is denoted by statistic C. The larger the AUC, the better the overall performance of the medical test to correctly identify diseased and nondiseased subjects. The closer the ROC curve to the left and top border (see Figure 9.12), the larger is the AUC and more valid is the test in terms of sensitivity and specificity. In this figure, test A is better than test B and test B is better than test C. If the test is lousy, for every true positive, as the level (e.g., of duration since rupture) increase, you are likely to encounter a false positive. The ROC curve tends to flatten in this case and you get nearly a diagonal line.

The maximum AUC is 1.0 including the right lower half. The actual area measures the test validity in the sense of its ability to correctly classify those known with and without the disease. Because of inherent variations and

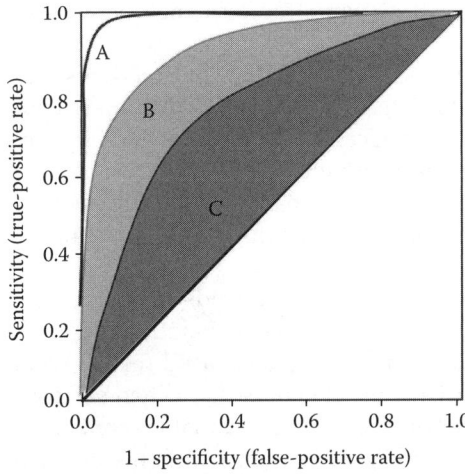

FIGURE 9.12
Three ROC curves with different areas under the curve.

uncertainties in all biological phenomena, no test can be perfect. It is considered excellent if AUC is 0.90 or more, and good if AUC is between 0.80 and 0.89. An area of 0.50 corresponds to the diagonal and indicates that the test is absolutely not helpful. In a rare case, if this area is less than 0.50, conclude that the test is misleading. In this case, reverse the definition of positive and negative, that is, in place of higher values signifying the presence of disease, examine if lower values will correctly pick up the disease.

As mentioned for sensitivity and specificity, validity of ROC and AUC depends on the "gold" really being so. If the gold is suspect, a high AUC does not necessarily mean a good test. Also, for ROC to be valid, the test must not be affected by what gold is. All other constraints mentioned earlier for sensitivity–specificity also apply. AUC fails to balance the differential risks of misdiagnosis and missed diagnosis effectively. Note that these risks cannot come from the present data and should come from the experience of clinicians.

While statistically comparing two tests, the decision regarding which test is better depends on AUC. Thus, this area should be obtained by using an appropriate method. Statistical software provide nonparametric and parametric methods for obtaining the area under ROC curve. The user has to make a choice. The following details may help.

Nonparametric methods are distribution free and the resulting area under the ROC curve is called empirical. First such method uses trapezoidal rule. If sensitivity and specificity are denoted by s_n and s_p, respectively, the trapezoidal rule calculates the area by joining the points $(s_n, 1 - s_p)$ at each interval value of the continuous test and draws a straight line joining the x-axis. This forms several trapezoids, one corresponding to each interval, and their area can be easily calculated and summed. Another nonparametric method uses Mann–Whitney statistics, also known as Wilcoxon rank-sum statistic and the *C*-index for calculating area. Both these nonparametric methods of estimating AUC have been found equivalent [25].

Parametric methods are used when the statistical distribution of test values in diseased and nondiseased is known. **Binormal distribution** is commonly used for this purpose. This is applicable when test values in both diseased and nondiseased follow normal distribution. In this case, the relevant parameters can be easily estimated by the means and variances of test values in diseased and nondiseased subjects. For details of how to obtain the area, see Zhou et al. [26].

The choice of method to calculate AUC essentially depends upon availability of statistical software. Binormal method produces a smooth ROC curve and further statistics can be easily calculated but gives biased results when data are degenerated and when the distribution is bimodal. When software for both parametric and nonparametric methods is available, conclusion should be based on the method that yields greater precision of the estimate of AUC.

Equal AUCs of two tests represent similar overall performance of tests, but this does not necessarily mean that both the curves are identical. They may cross each other. Figure 9.13a has hypothetical ROC curves of two medical tests A and B applied on the same subjects to assess the same disease. Tests A and B have nearly equal area but cross each other. Test A performs better than test B where high sensitivity is required, and test B performs better when high specificity is needed.

In cases where ROC curves for two tests cross each other and in some other situations, the interest may be restricted to specific values of sensitivity or specificity. You may be interested in a test with high sensitivity as for a disease with grave prognosis (cancer). Then, the interest will be

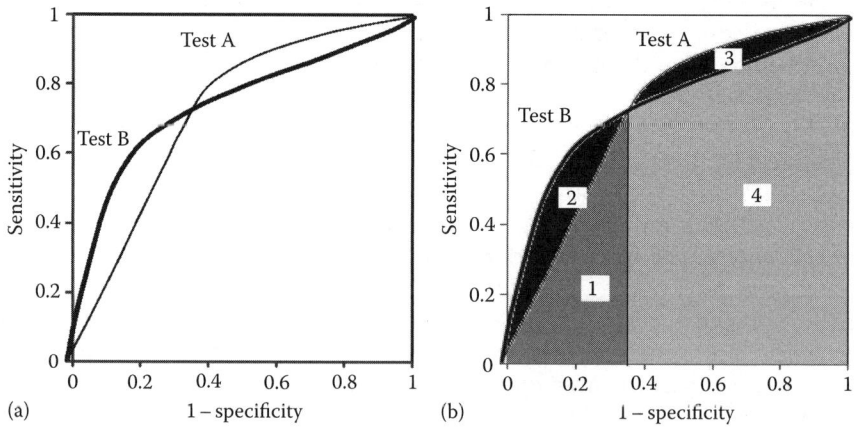

FIGURE 9.13
(a) Two ROC curves crossing each other but with nearly the same area. (b) Illustration of partial area under the ROC curve.

in test A and that too for specificity ≤0.65 or (1 − specificity) >0.35. In that case, the area of interest is 3 + 4 as shown in Figure 9.13b. This is called **partial AUC**. Some software calculate this also and, if you want for easy interpretability, you can standardize it to 1 by considering total area = 1 of rectangle from (1 − specificity) = 0.35.

Variance of AUC can also be obtained by using parametric and nonparametric methods. The formulas are complex. Software will give you this easily. This variance can be used to obtain the confidence interval assuming Gaussian pattern as per the procedure given in a later chapter.

Formulas of sample size for testing hypothesis on comparing AUC with a prespecified value and for comparison of two on the same subjects or different subjects are complex. Refer [26] for details.

There are many more topics for interested reader to explore, such as combining the multiple ROC curve for meta-analysis, ROC analysis to predict more than one alternative, and ROC analysis in the clustered environment, and for tests repeated over time. For these, see Refs. [26,27].

Software illustration: For an illustration of how to obtain an ROC curve with MedCalc software, see Appendix C.

9.5.2 Predictivity-Based ROC Curve

The conventional ROC curve just described considers sensitivity–specificity but not predictivities. The threshold based on this ROC would be valid across the population since sensitivity–specificity is not affected by prevalence. However, the diagnostic efficiency is obtained by the predictivities and not the sensitivity–specificity. Thus, the ROC curve between positive predictivity and (1 − negative predictivity) may be more useful in a local setup as it takes prevalence into account and uses the right kind of indexes.

The relationship between predictivities and sensitivity–specificity can be used to find the criterion that is best to confirm the diagnosis (i.e., maximally increase the positive predictivity) and to exclude the disease (i.e., maximally increase the negative predictivity). For example, instead of using the criterion of at least 250 U/L of total CPK for MI, one can think of using a threshold of 200 or 300 U/L. The procedure then would be to obtain sensitivity–specificity for various thresholds. These possibly can be easily obtained from established cases of infarction and suspected cases established as noninfarction. Substitute these sensitivity and specificity values in the formulas given in Equations 9.28 and 9.29, and calculate predictivities for different prevalence rates. A graph of the type given in Figure 9.14 can be thus obtained.

Figure 9.14 is drawn for thresholds 350, 250, and 150 U/L assuming that (sensitivity, specificity) for these thresholds, respectively, are (0.60, 0.90), (0.80, 0.80), and (0.95, 0.50). Shown are positive predictivity on the upper left side of the diagonal and (1 − negative predictivity) on the lower right side

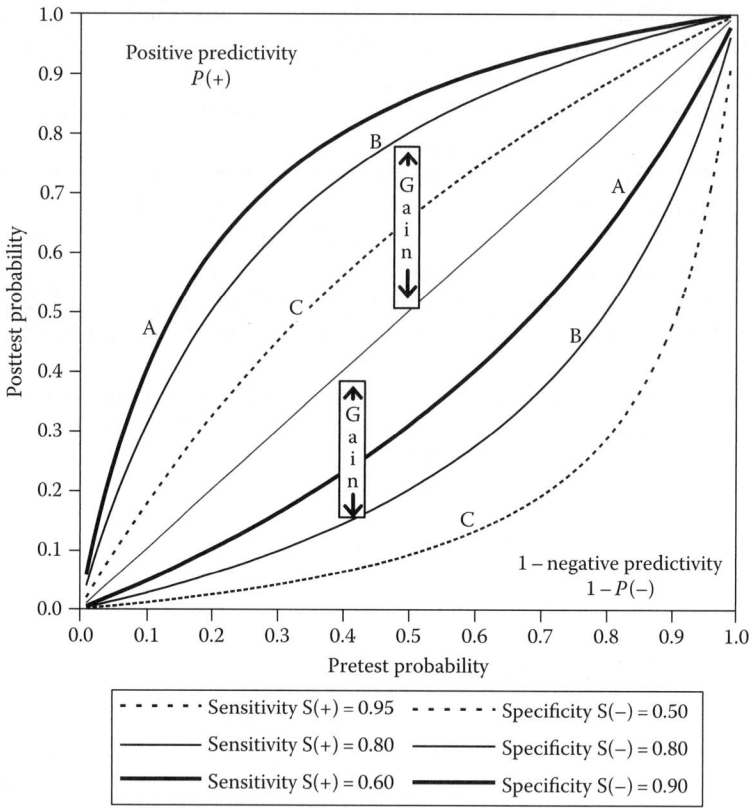

FIGURE 9.14
Illustration of the relationship between pretest and posttest probability.

of the diagonal. Thus, this is a predictivity counterpart of the ROC curve. The curves are marked as A, B, and C for the three (sensitivity, specificity) levels, respectively, corresponding to the three CPK levels under consideration. When sensitivity and specificity are equal, the curves are symmetrical as illustrated by curve B in this figure. Depending on the pretest probability for the patient in hand, which could be either the known prevalence of infarction in these types of cases or the personal probability on the basis of history and signs and symptoms, and the CPK level present, you can immediately obtain the posttest probability (or predictivity) of the presence of infarction with such curves. Gains by curve B relative to the diagonal are illustrated in Figure 9.14 for some specific values. If you estimate that the chance of infarction in a patient with specific signs and symptoms is 60% (pretest probability 0.60) and the CPK level is found to be 150 U/L, then curve C applies and the posttest probability can be read as 0.70. This is a gain of merely 10% over pretest probability. If the CPK level is 350 U/L, then curve A applies and the posttest probability is 90%—a handsome gain of 30% over the pretest probability for the presence of the disease. These gains can be utilized to find a threshold CPK level that is best in the sense of highest gain over a particular pretest probability. Now, note the following:

1. All the discussions of the sensitivity–specificity and predictivities assume that these can be exactly obtained. In practice, these will be based on the study of a sample and are subject to sampling error. Thus, caution should always be exercised. Similar variation is also expected in the prevalence rate. This too will be generally based on a sample. Even when a pretest probability is based on the clinician's belief concerning the presence of disease after taking the history and examination into consideration, it would most likely vary from clinician to clinician. Thus, the values of sensitivity–specificity and predictivities serve only as guidelines and do not have much utility in an absolute sense. The ultimate decision, as always, rests with the attending clinician to give or not give credence to such indicators.

2. There is another reason for clinicians to be judicious in using sensitivity–specificity and predictivities. All these measure probabilities, and probabilities are never absolute—they yield expected results in the long run but may fail in a particular case.

References

1. Cnathingius S, Haglund B, Kramer MS. Difference in late fetal death rate in association with determinants of small for gestational age fetuses: Population based cohort study. *Br Med J* 1998; 316:1483–1487.
2. Ledford AW, Cole TJ. Mathematical models of growth in stature throughout childhood. *Ann Hum Biol* 1998; 25:101–115.
3. WHO. *Child Growth Standards: Length/Height-for-Age, Weight-for-Age, Weight-far-Length, Weight-for-Height, and Body Mass Index-for-Age—Methods and Development.* Geneva, Switzerland: World Health Organization.
4. Freeman N, Power C, Rodgers B. Weight-for-height indices of adiposity: Relationships with height in childhood and early adult life. *Int J Epidemiol* 1995; 24:970–976.
5. CDC. United States: Growth Charts, Center for Disease Control and Prevention, Washington, DC, 2000. www.cdc.gov/growthcharts (last accessed on September 17, 2005).
6. Cole TJ. 3-in-1 weight-monitoring chart. *Lancet* 1997; 349:102–103.
7. Must A, Dallal GE, Dietz WH. Reference data for obesity: 85th and 95th percentiles of body mass index (wt/ht^2) and triceps skinfold thickness. *Am J Clin Nutr* 1991; 53:839–846.
8. Cole TJ, Freeman JV, Preece MA. British 1990 growth reference centiles for weight, height, body mass index and head circumference fitted by maximum penalized likelihood. *Stat Med* 1998; 17:407–429.

9. Preece MA, Baines MJ. A new family of mathematical models describing the human growth curve. *Ann Hum Biol* 1978; 51:1–24.

10. Cole TJ, Bellizzi MC, Flegal KM, Dietz WH. Establishing a standard definition for child overweight and obesity worldwide: International survey. *Br Med J* 2000; 320:1240–1243.

11. Coles JE. *Lung Function Assessment and Application in Medicine.* Oxford, U.K.: Blackwell Scientific Publications, 1993.

12. Avis NE, Smith KW, Hambleton RK, Feldman HA, Selwyn A, Jacobs A. Development of the multidimensional index of life quality: A quality of life measure for cardiovascular disease. *Med Care* 1996; 34:1102–1120.

13. Sprangers MA, Cull A, Groenvold M, Bjordal K, Blazeby J, Aaronson NK. The European Organization for Research and Treatment of Cancer approach to developing questionnaire modules: An update and overview: EORTC Quality of Life Study Group. *Qual Life Res* 1998; 7:291–300.

14. Katz S, Akpom CA. Index of ADL. *Med Care* 1976; 14 (Suppl 55):116–118.

15. Heikkinen E, Waters WE, Brezinski ZJ (Eds.). *The Elderly in Eleven Countries: A Sociomedical Survey.* Copenhagen, Denmark: World Health Organization, Regional Office for Europe, 1983.

16. Buysse DJ, Reynolds CF III, Monk TH, Berman SR, Kupfer DJ. The Pittsburgh Sleep Quality Index: A new instrument for psychiatric practice and research. *Psychiatry Res* 1989; 28:193–213.

17. American Psychiatric Association, DSM-IV. *Diagnostic and Statistical Manual of Mental Disorders*, 4th edn. Arlington, VA: American Psychiatric Association, 1994.

18. Arnesen E, Refsum H, Bonaa KH, Ueland PM, Forde OH, Nordrehaug JE. Serum total homocysteine and coronary heart disease. *Int J Epidemiol* 1995; 24:704–709.

19. Guillaume M, Lapideus L, Beckers F, Lamert A, Bjorntorp P. Cardiovascular risk factors in children from the Belgian province of Luxembourg: The Belgian Luxembourg Study. *Am J Epidemiol* 1996; 144:867–880.

20. Murphy EA. The normal, and the perils of the sylleptic argument. *Perspect Biol Med* 1972; 15:566–582.

21. Aravanis C, Mensink RP, Corcondilas A, Ioanidis P, Feskens EJM, Katan MB. Risk factors for coronary heart disease in middle-aged men in Crete in 1982. *Int J Epidemiol* 1988; 17:779–783.

22. Chatfield C. Confession of a pragmatic statistician. *Statistician* 2002; 51 (Part-1):1–20.

23. Rosner B, Polk BF. Predictive values of routine blood pressure measurements in screening for hypertension. *Am J Epidemiol* 1983; 117:429–442.

24. Sackett DL, Haynes RB, Gyatt GH, Tugwell P. *Clinical Epidemiology: A Basic Science for Clinical Medicine*, 2nd edn. Boston, MA: Little Brown & Company, 1991.

25. Hanley JA, McNeil BJ. The meaning and use of the area under a receiver operating characteristic (ROC) curve. *Radiology* 1982; 143:29–36.

26. Zhou Xh, Obuchowski NA, McClish DK. *Statistical Methods in Diagnostic Medicine.* New York: John Wiley & Sons, Inc., 2002.

27. Kester AD, Buntinx F. Meta analysis of curves. *Med Decis Making* 2000; 20:430–439.

10

Clinimetrics and Evidence-Based Medicine

Indicators discussed in the previous chapter are epidemiological in nature, though applicable to individuals. They do assess particular aspects of health but are not directly used for diagnosis or prognosis. Clinimetrics is a science concerning the development and adequacy of *quantitative measurements*, particularly of composite nature such as scoring systems used for managing patients. The purpose again is to minimize the specter of uncertainty. Thus, clinimetrics is statistical in content.

Evidence-based medicine is the judicious mix of currently available evidence with clinical acumen for making a decision in the best interest of individual patients. Evidence is obtained from the review of literature and other data and thus is mostly empirical. Quite often, it is quantitative rather than qualitative. Hence, the overlap with clinimetrics is substantial.

Both clinimetrics and evidence-based medicine are yet to carve a niche for themselves in the field of patient management. Many of you may argue that the science of medicine is better off without such tools. Indeed, measurements and evidence may fail to match the experience and wisdom that some clinicians already have or would acquire in future. Nonetheless, the trend increasingly is to depend on quantitative assessments than on subjective evaluation, although this is not done at the cost of compromising the best judgment of a clinician.

This chapter: The major components of clinimetrics are indicators, indexes, and scores that are used as aids for diagnosis and in assessing prognosis. As a baseline, the distinction between indicators, indexes, and scores, as well as their advantages and limitations, is explained in Section 10.1. Although a huge list of such statistical tools, possibly running into thousands, can be drawn, Section 10.2 focuses on some specific clinimetric tools, mostly scoring systems in common use, and discusses their merits and demerits and some methods of their development.

Section 10.3 shifts focus to evidence-based medicine with quantitative overtones. This section presents some methods of decision making in medicine based on the risks and benefits of options available for managing a patient and related statistical tools such as the etiology diagram and expert system.

10.1 Indicators, Indexes, and Scores

The terms indicators, indexes, and scores seem to be widely used *interchangeably* for tools of measurement. This text also uses these terms in this generic sense at several places. Technically though, these terms have distinct meanings. Broadly speaking, an indicator is a quantitative univariate measure of specific aspect of health. An index combines two or more indicators, and a score may even incorporate qualitative characteristics such as signs and symptoms. However, there are exceptions. For example, intensity of pain on a visual analogue scale is called score although this is univariate.

10.1.1 Indicators

A factor is a characteristic and an indicator is a tool to measure its level. This distinction works well when the characteristic is graded and it has a level. This is not always true. The site of cancer and the gender of a person are characteristics that cannot be assigned a grade at the individual level. For groups, though, percentage of patients with cancer of different sites and sex ratio in a group of arthritis cases are quantities. These are indicators at the group level.

The term indicator is generally used when the focus is on a *specific* aspect of a characteristic. Severity of disease in a patient can be assessed by the severity of pain, inability to perform essential functions of life, prognostic implications, the chance of death within a week, etc. Each of these is an indicator since each concerns with a particular aspect of severity of disease. An indicator is a univariate assessment and perhaps the most direct measurement of particular aspects of characteristics of interest. Indicators provide the exactitude that you may like to have in your assessment or you may like to have in the resource material you consult to update knowledge.

10.1.1.1 Merits and Demerits of Indicators

Critics may be concerned with the limited utility of indicators because they focus on one particular aspect of health and ignore the other aspects, howsoever related. Case-fatality rate is an indicator of the chance of death but is oblivious to the pain and suffering it brings to the patient and the family. Weight of a person is an indicator that by itself looses importance unless related to height. Glomerular filtration rate (GFR) is better interpreted in the context of creatinine excretion. At the same time, though, the focus of an indicator on a particular aspect is its greatest strength. Case-fatality does measure the most important aspect of prognosis that is important for both patient and the doctor, measuring weight does help when monitored in the same person over a period, and GFR level does give an indication about kidney function. They are good indicators as stand alone, although putting them in the context of other related parameters does help in providing a better interpretation.

10.1.1.2 Choice of Indicators

Quite often, multiple indicators are available for apparently the same characteristic and one may have to make a choice. Triglyceride, high-density lipoprotein (HDL), and low-density lipoprotein (LDL) are indicators of different lipids. Many times these are considered together but if one has to choose, the selection would depend on the relevance for the condition of the patient and the specific aspect of the lipid of interest. For example, triglyceride level can be a good indicator of lipoprotein called natural fats, and LDL level can be a good indicator of risk of developing atherosclerosis. Conversely, in a research setup, you may like to find, say, in coronary events, which of the three is a better marker, and why others are not.

The nutrition level of a child can be assessed anthropometrically by weight, height, and skinfold thickness. Even though these are assessed in relation to age, they are univariate, and thus are indicators in a relaxed sense of the term. Weight-for-height requires two measurements, and hence is bivariate, and technically transgresses to the realm of indexes, discussed in the next section. The choice between weight, height, and skinfold thickness as anthropometric indicators of nutrition level in children would depend on whether one is looking at short- or long-term nutrition or at the adiposity.

Blood hemoglobin (Hb) concentration and hematocrit (Hct) (also called PCV) values assist in evaluating plasma dilution, blood viscosity, and oxygen-carrying capacity. Hemoglobin concentration, in general, is easily affected, whereas Hct is a long-term measure. A related measure is red blood cell (RBC) (also called erythrocyte) count. Each of these is used in specific contexts and you should be able to decide which ones would serve your purpose.

At the community level, events such as child mortality can be measured by neonatal mortality, postneonatal mortality, and infant mortality. The neonatal period can be divided into early (<7 days) and late (7 to <28 days) periods. Although all of them could be used together to provide a holistic picture, a professional may like to concentrate on one indicator that best measures the specific aspect of interest. If the focus is on antenatal care, maternal nutrition, and skilled attendance at birth, perhaps early neonatal mortality rate is the best indicator. If the focus is on breastfeeding, infections, and child nutrition, postneonatal mortality may be better. If the interest is to study the inability to thrive in the face of repeated infections and midterm sequelae of low birth weight, mortality between 1 and 4 years may be better. Indeed, almost all around the world, the interest has shifted to neonatal mortality as the mortality beyond this period has relented.

In summary, different indicators may seem similar but each indicator has a specific application. The focus of application will help you decide which particular indicator to use.

10.1.2 Indexes

Technically, an index is a combination of two or more indicators. A popular example is the body mass index (BMI), which is mathematically derived from weight and height. In this sense, mean corpuscular hemoglobin concentration (MCHC) is an index, which is calculated from Hb and Hct values. Similarly, mean corpuscular hemoglobin (MCH) is an index based on Hb and RBC count, and mean corpuscular volume (MCV) is an index based on Hct and RBC count. A large number of such indexes are used in medicine and health.

10.1.2.1 Some Commonly Used Indexes

Body mass index is a classical example that illustrates how an incompletely looking stand-alone indicator such as weight attains very useful meaning when considered in conjunction with another indicator, namely, height, in this case. An indicator value in the context of another can provide a very different perspective. **Ankle-brachial (pressure) index** is the ratio of systolic BP in the ankle to that of in the arm. Normally, BP at the ankle is the same or slightly higher than brachial pressure so that the normal ratio of these two pressures is 1.0 or 1.1. A ratio much less than 0.95 indicates peripheral arterial disease. A ratio between 0.8 and 0.5 indicates moderate disease and one less than 0.5 indicates severe disease. A ratio greater than 1.3 is also abnormal, suggesting peripheral vascular disease instead of peripheral arterial disease.

Another useful index is the **bispectral index** that has many applications but is mostly used to measure the effect of an anesthetic agent on the hypnotic state of the patient's brain during surgery. Thus, this measures the depth of anesthesia. This index is based on the analysis of electroencephalogram (EEG) signals. The value ranges from 0 for complete EEG silence to 100 in fully awake adults. A value between 40 and 60 is generally recommended for general anesthesia. Sometimes, anesthetic agents are titrated accordingly. Although called an index, an intermediary step in the calculation of bispectral index is a scoring system for grading the hypnotic level. This score, however, is not for a qualitative characteristic. The name is derived from a complex multivariate statistical procedure called bispectral analysis (not discussed in this book), which is used to select the "best" variables for computing the index.

Bispectral index illustrates that no index should be indiscriminately used in varying conditions. It may be a good predictor of movement and hemodynamic response to incision but not good for assessing movement after administration of common drugs such as opioids.

Glycemic index measures the blood glucose increasing potential of a food after consumption. For this, the area under the curve that represents blood glucose level after 2 h is measured for that particular food. This is compared with the same area in the graph for a reference food, namely, glucose sugar in the same quantity. Foods with low glycemic index are considered to contain slowly digested carbohydrates. An index less than 55 is considered low, between 55 and 69 is considered middling, and 70 or more is considered high.

Craniofacial index is the ratio of the breadth of the cranium to the breadth of the face. This is used to assess facial deformities, especially in the newborn. One of its useful articulations is the **craniofacial variability index**. A series of anthropometric measurements of the head and face are taken and each measurement is converted to a Z-score relative to the standard. A high variance of these scores results in pronounced differences in the direction and size of individual anthropometry and indicates craniofacial anomalies.

Occupation stress index measures the burden of job stresses, particularly in relation to cardiovascular risk. This index is mostly based on cognitive ergonomics and brain research rather than on social constructs. For example, an employee may feel handicapped because of limited options available to him for decision making or may find that the information required for decision making is either incomplete or contradictory. The job may involve high risk for a momentary lapse. All these go into making the occupation stress index.

10.1.2.2 Advantages and Limitations of Indexes

The purpose of describing some commonly used indexes is two-fold: (1) for illustrating a cross section of situations where various kinds of indexes are used in health and medicine, and (2) to show that some indexes are simple, others could be quite complex. These two features also express the advantages and limitations of indexes. They allow two or more indicators to be considered together and enhance their utility and sometimes even generate new information by placing them in a proper context.

An index is quantitative and therefore involves calculation. This is a nemesis for some clinicians. Some indexing instruments come ready with a software to perform the calculations and directly provide the results. Bispectral index is automatically calculated by a software. A high-pressure liquid chromatography (HPLC) automatically calculates the peak area of intensity of signals corresponding to concentrates of drug-evoked potentials. Thus, for some indexes, calculations are not much of a problem. Perhaps, a greater problem is their validity and reliability. A large number of indexes are available and many are being devised every year, but studies that provide evidence of their reliability and validity are rare. The most widely used index is the BMI but its utility too is sometimes questioned in comparison with the waist–hip ratio (WHR), which is seen as a better correlate of coronary events. The utility of WHR in cancer and lung disease has not been fully evaluated.

10.1.3 Scores

Medical professionals can be statistically dichotomized as those preferring qualities and subjective assessments using clinical acumen and those who prefer quantities that bring in exactitude. Some of those belonging to the latter group may like to quantify even signs and symptoms of a disease. How to convert qualities into quantities at an individual level? Through the scoring system. Acute Physiology and Chronic Health Evaluation (APACHE) and Apgar scores are everyday examples of such conversions. Pain on a visual analogue scale is an indicator of pain intensity, but the McGill pain score comprises qualities like throbbing pain, shooting pain, and stabbing pain. Scoring systems also reduce a multivariate entity into a univariate quantity, thus increasing comprehension and utility. They also help in reducing some of the epistemic uncertainties that can arise from inadequate realization of how much weight is to be given to various pieces of information.

Scoring systems are gradually gaining importance similar to laboratory and radiological investigations. There was a time when such investigations were considered below the dignity of a clinician, but now hardly any medical decision is taken without recourse to such investigations. A large number of scoring systems on various aspects are being developed: the list already runs into hundreds. These systems have found useful applications in gradation of severity of a disease as they consider several individual assessments together. Scoring systems have also been applied for establishing a diagnosis, particularly for differentiating one condition from another similar-looking condition. Both types of scoring systems—for diagnosis and for severity of a disease—can add to the knowledge with which a medical condition is managed.

10.1.3.1 Scoring System for Diagnosis

Attempts have been made periodically to quantify the field of medicine and developing scoring systems that can help in diagnosis.

Hypothyroidism poses a challenge to physicians at the time of diagnosis. It is not easy to distinguish between hyperthyroidism, euthyroidism, and hypothyroidism on the basis of clinical signs and symptoms. Assays for measuring T_3, T_4, and thyroid-stimulating hormone (TSH) levels help in reaching a diagnosis, but these tests are slightly expensive, and some hospitals in developing countries may not have the facility for such testing. Clinically clear cases of hyper- and hypothyroidism can be treated without taking into account laboratory results, and the tests are ordered only in doubtful cases. This can also reduce the load on laboratory services. See Example 10.1 on a scoring system that can help to understand its utility.

Example 10.1: A Hypothyroid Diagnostic Index

A diagnostic index for hypothyroidism was devised by Billewicz et al. [1], taking into account the frequency of various symptoms and signs. They developed the scoring system excluding four features with zero score (Table 10.1).

The scores are based on the logarithm of the ratio of frequency of presence and absence of symptoms in established hypothyroid and euthyroid cases. According to this scoring system, a clear diagnosis can be made on the basis of signs and symptoms alone if the score is 25 or more (hypothyroid) or −30 or less (euthyroid). Laboratory assistance is required only when the total score is between −29 and +24. This can reduce the load on the laboratory by more than 50%.

SIDE NOTE: This scoring system was later simplified by Zulewski et al. [2]. Although this is a score, the authors liked to call it an index. Many such examples of mixed use of the terms exist in the literature.

TABLE 10.1

Scoring System for Thyroidism

Signs and Symptoms	Score	
	Present	Absent
Physical tiredness	0	+2
Slow cerebration	−3	+2
Diminished sweating	+6	−2
Dry skin	+3	−6
Cold intolerance	+4	−5
Dry hair	−2	+2
Weight increase	+1	−1
Constipation	+2	−1
Hoarseness of voice	+5	−6
Paresthesia	+5	−4
Deafness	+11	0
Slow movements	+4	−3
Coarse skin	+7	−7
Cold skin	+3	−2
Periorbital puffiness	+4	−6
Pulse rate <75/min	+4	−4
Ankle jerk	+15	−6
Total score	≤−30	Euthyroid
	−29 to +24	Doubtful
	≥+25	Hypothyroid

See Example 10.2 for scoring for diagnosis of catheter-related infections. In recent times, a similar scoring system has been developed for the diagnosis for necrotizing soft-tissue infections [3]. This is based on computed tomography findings. A simple 10-point scoring system has been separately proposed for prognostic assessment of acute otitis media in children [4]. A simple scoring has been utilized for diagnosis of chronic lymphocytic leukemia [5].

Not many researchers look for a simple scoring system these days. Instead, a logistic regression equation (see Chapter 17) is used as a scoring system. Examples are scoring systems for repeat biopsies in suspected patients of prostate cancer [6] and for initial treatment failure in suppurative kidney infections [7]. These are some examples of various scoring systems available for diagnostic purposes. Many others are available.

Almost all such scoring systems are based on data from developed countries. Because of nutritional and environmental factors, they may not directly apply to the subjects in developing countries. These scoring systems have to be appropriately modified before using them for patients in such countries.

Such systems should be used only when a valid diagnosis is difficult to establish or when the diagnosis depends on physicians' preferences, their expertise, or results of laboratory or radiological investigations that lack credibility. Such inadequacies in the diagnostic process are more common than are otherwise apparent. A useful strategy is to use these scores as just additional adjunct to the clinical and laboratory evidence and take a decision in a holistic manner.

> **Example 10.2: Scores for Diagnosis of Catheter-Related Infections**
>
> Clinical diagnosis of catheter-related infections is difficult. Lugauer et al. [8] show that it can be done with relative ease with the help of a scoring system comprising rate of rise of body temperature, concomitant shivering, identification of pathogens in blood or catheter tip cultures, improvement in the clinical course after catheter removal, signs of catheter exit site inflammation, and results of diagnostic tests for other possible sources of infection. These criteria were graded using points and weighted according to their specificity. The patients were also diagnosed using the existing complex clinical criteria. The scoring system was in agreement with the clinical diagnosis up to 85% in a group of 65 cases. There were no false-negative cases and 10 were false positive. It turned out that 9 of these 10 were not false positive when additional findings are considered (clinical diagnosis criteria were expanded). Thus, the scoring system appears to be more sensitive than the existing diagnostic criteria, without loss of specificity.

10.1.3.2 Scoring for Gradation of Severity

Prognostic assessment and the management of a patient depend to a large extent on the severity of the disease. There is considerable epistemic uncertainty about how to assess this severity. Different professionals use different methods. For uniformity and exactitude, at times scoring is considered desirable.

The Glasgow Coma Scale [9] is used to grade coma patients by using numeric scale for eye, motor, and verbal response. APACHE is used to assess severity in critically ill hospitalized adults [10]. Various variations of this score are available. Mortality prediction modeling [11] uses systolic BP, level of consciousness, prior cardiopulmonary resuscitation (CPR), age, presence of cancer, and presence of infection to assess chances of survival in patients admitted to intensive care units. Yale observation scale [12] is used to identify serious illness in febrile children. Indrayan smoking index (Chapter 9) measures the lifelong burden of smoking. The Apgar score is used to assess prognosis in a neonate. These are just a few examples. In case you are stuck with a problem of grading the severity of patients, see if a valid scoring system is available. If not, you may like to devise a scoring system yourself.

As already stated, all scoring systems try to convert multiple measurements into a single unified but meaningful index. They transform multivariate data into a univariate score. Sometimes several scoring systems are available for the same condition and it would be difficult to choose the right system. For example, severity in peritonitis cases can be assessed by APACHE score, Peritonitis Severity Score (PSS), Mannheim Peritonitis Index (MPI), Hacettepe score, American Society of Anesthesiologists (ASA) score, etc. Choose a scoring system that looks more appropriate for your patients.

Example 10.3: Risk Score for Predicting Death in Chagas Disease

Chagas disease is an important health problem in Latin America, and cardiac involvement in this disease increases the severity and risk of death. Rassi et al. [13] developed a risk score based on the evaluation of 424 outpatients from a regional Brazilian cohort as given in Table 10.2. The score is obtained as a sum of these points. They divided the score into three groups: low risk, 0–6 points; intermediate risk, 7–11 points; and high risk, 12–20 points. The risk pertains to the risk of death. The 10-year mortality rates for these three groups were 10%, 44%, and 84%, respectively.

TABLE 10.2

Scoring System for Death in Chagas Disease

Risk Factor	Points
New York Heart Association Class III or IV	5
Evidence of cardiomegaly on radiography	5
Left ventricular systolic dysfunction on echocardiography	3
Nonsustained ventricular tachycardia on 24 h Holter monitoring	3
Low QRS voltage on electrocardiography	2
Male sex	2

10.2 Clinimetrics

Metrics is generally identified with any science concerning quantitation and calculations. Many mathematical formulas have unknowingly entered into medical practice back door through computer-based systems, which directly produce results without the user being aware of the back-end calculations. Examples such as the bispectral index, HPLC, and evoked potential have already been cited that directly produce results, without revealing the complex calculation. Clinimetrics goes beyond these calculations and seeks to place the quantities and calculations in a clinical context so that decisions in the interest of the patient can be taken on the basis of measurements after assessing their implications. In a way, this takes away subjectivity from decisions, which might otherwise be difficult to justify logically. You would agree that logic is the cornerstone for medical decisions.

Indicators, indexes, and scores can be considered components of clinimetrics, but the major focus of this gradually developing science is on the methodology so that the tools developed have the requisite applicability. These components of clinimetrics are concerned with the quality of the measuring instruments and look at the process of tool development rather than the tools themselves.

Reliability, validity, and responsiveness of the measuring tools are an integral part of their quality. A measuring tool should also be sufficiently sensitive to detect clinically relevant improvements attributable to therapeutic interventions. Close collaboration between clinicians, biostatisticians, and epidemiologists is required for the development of clinimetrics as a science of consequence.

10.2.1 Method of Scoring

Characteristics that have gradients are relatively easy to score than measurements on the nominal scale. If the gradient is already numeric such as pain on the visual analogue scale, the score is immediately obtained for this variable. The problem arises for ordinal factors that have gradients but no numeric scale is available. Examples are severity of disease, which is categorized as mild, moderate, serious, and critical, and degree of satisfaction categorized as completely dissatisfied to fully satisfied. Characteristics on the nominal scale such as presence or absence of signs and symptoms in any case defy quantitation at the individual level. The method of assigning numerics to ordinal and nominal characteristics should be one that can reduce uncertainties and not introduce additional epistemic uncertainties. A more general method based on logistic regression is discussed later in this section but consider the following for the time being.

10.2.1.1 Method of Scoring for Graded Characteristics

The most simple scoring is linear, for example, 0 for no disease, 1 for mild disease, 2 for moderate disease, 3 for serious disease, and 4 for critical condition. As mentioned in a previous chapter, such a scoring assumes that the difference between mild and no disease is the same as between critical and serious disease. It is legitimate to ask for such a scoring why scores 0, 3, 6, 9, and 12 are not more appropriate, which also are linear, or even why geometric scoring such as 0, 1, 2, 4, and 8 is not better. Very few studies have been carried out to investigate various alternatives, and thus nothing can be stated with confidence. Nevertheless, the 0, 1, 2, 3, and 4 type of scoring remains the most widely used scores because of their simplicity. They go unnoticed. No explanation is generally required when such simple scores are used. Any other scores are expected to be accompanied by justification. A more complex procedure to generate scores is illustrated in Example 10.4.

Example 10.4: Scoring for Motor Components in Multiple Sclerosis

Patient mobility is affected in multiple sclerosis, and various scales are used to assess functionality in the patients. Among those relatively easy to perform are the 10 m timed walk (TMTW) and nine-hole peg test (NHPT) for the right and left hands. Vaney et al. [14] developed a short and graphic ability score (SaGAS) as (2 × TMTW + NHPTright + NHPTleft) after taking the logarithm of these timed values. The authors demonstrated good correlation of SaGAS with established tests such as Multiple Sclerosis Functional Composite, Expanded Disability Status Scale, and Rivermead Mobility Index. Features of SaGAS such as simplicity, intuitiveness, and nonphysician-based measurability were enumerated as advantages for its use in multiple sclerosis patients.

SIDE NOTE: Functionality in multiple sclerosis patients has a gradient and its metric measurement has always been a challenge. The authors developed an easy-to-implement scoring system. Its good correlation with other more complex scoring methods is one way to demonstrate its validity. For a real assessment, comparison with a gold standard is more convincing. Such a standard may not exist in this case; in most cases, the gold standard might be too cumbersome and expensive. Equivalence of the new scoring system with the gold standard when demonstrated would establish its real validity.

10.2.1.2 Method of Scoring for Diagnosis

Scoring might help in situations where diagnosis or differential diagnosis is difficult and requires a lot of expertise that may not be immediately available. See Example 10.1 on the diagnosis of hypothyroidism that uses scores on clinical signs and symptoms. Here, the scoring helps to establish or rule out hypothyroidism in nearly half the cases. Thus, the need of further investigations is reduced to one-half. For another example, see Guo et al. [15]

who provide a scoring system for the diagnosis of Hirschsprung's disease in neonates. Rosen et al. [16] proposed a smaller 5-item index as a diagnostic tool for erectile dysfunction compared with the larger 15-item scoring. Their index uses linear scores for each item.

Signs and symptoms are qualitative and they play a significant role in establishing diagnosis. Converting such qualities to a numerical score has always been a challenge and no widely acceptable method is available yet. Examples 10.5 and 10.6 and their discussion are based on the methods used by some workers who ventured into this area.

Example 10.5: Delphi Method of Scoring

Consider measuring nursing workload in an ICU by a scoring system. Yamase [17] assessed this workload by 88 items relating to (1) number of nurses required, (2) muscular exertion, (3) mental stress, (4) skill, and (5) intensity. A three-round Delphi survey among 20 skilled ICU nurses assigned a consensus four-grade (0–3) score to each of the 73 items after excluding 15 items considered unnecessary. These scores were confirmed by surveying 118 nurses in other ICUs. The "comprehensive nursing intervention score" is the sum of all these individual item scores. The scoring system was confirmed as reflecting true workload by applying it to the daily care of 107 patients.

Example 10.5 illustrates how a simple method can be used to develop a scoring system. Validation of the individual scores by another group of nurses and of the scoring system by using it in actual conditions tends to enhance the confidence in the scoring system.

There are examples of assigning arbitrary scores. Obviously, they provide tentative results.

10.2.1.3 Regression Method for Scoring

A common and acceptable method of assigning scores to individual characteristics is based on the regression coefficients that are estimated using multiple regression method when the outcome is quantitative. For qualitative outcomes, particularly those that are dichotomous, such as the presence or absence of a disease, logistic regression coefficients are used. These regression methods are presented later in this book. The factors surmised to determine the outcome are antecedents whose coefficients in the logistic regression equations are significant. You would soon see that they can be interpreted as the log of odds ratio (OR). Larger the OR, better is the predictive utility of the factor. Thus, the score in proportion of the OR can be assigned to those factors that turn out to be significant predictors. The method is illustrated in Example 10.6. This example also is more on the gradation of a disease than on diagnosis although the outcome is dichotomous: either death or survival.

Example 10.6: Scoring System to Stratify Risk in Unstable Angina

Unstable angina is a complex syndrome prognosticated by a host of factors such as age, hypertension, diabetes, hypercholesterolemia, smoking, previous myocardial infarction (MI), ST segment deviation in ECG, troponin test, etc. Piombo et al. [18] studied a large number of such factors in 473 patients and found four of them to be significant predictors in a multivariate logistic regression of in-hospital occurrence of refractory ischemia, acute MI, or death. These three together formed an unfavorable outcome. The ORs were 4.03, 2.29, 2.21, and 2.0 for ST segment deviation, age $\geq$70 years, previous coronary artery bypass grafting, and positive troponin test ($T \geq 0.1\,mg/mL$), respectively. The scores assigned were 4, 2, 2, and 2 corresponding to the respective ORs. The highest possible score was 10. It was divided into three categories: 0 or 2 for low risk, 4 or 6 for intermediate risk, and 8 or 10 for high risk. Under this scoring system, a score of 1, 3, 5, 7, or 9 is not possible.

The scoring system was validated in another group of 242 patients that provided similar results. Nearly 63% of patients were assigned to the low-risk group, 31% to the intermediary-risk group, and 6% to the high-risk group. The predictive power of the scoring system assessed by the C-statistic (area under the receiver operating characteristic [ROC] curve) was 0.72. The C-statistic is one of the important criteria that determine the validity of a scoring system.

SIDE NOTE: The authors called score categories 0–2, 4–6, and 8–10 as tertiles, which is not a correct use of the term since these categories do not comprise one-third subjects each that tertiles would. Also the predictive power of 72% as measured by the area under the ROC curve is not adequate to inspire confidence. The authors have discussed the limitations of their study, one being the selection of patients for the trial with strict inclusion and exclusion criteria that may have compromised representativeness. In practical application, many patients of unstable angina may not have undergone a previous angiography and this variable would not be available. In addition, serum cardiac markers were not well defined.

Example 10.6 was chosen not because it provides a valid scoring system but because it uses an appropriate method for selection of factors and for assigning them proper score. There are many other examples of this type. Purasiri et al. [19] combined the results of clinical examination, mammogram, ultrasonograph, and fine needle aspiration cytology by assigning them weighted scores using stepwise logistic regression. This combined score performed better than any of the others individually in differentiating malignancy from benign lesions in suspected cases of breast cancer. In this study, the confirmed diagnosis was later available so that the "gold" was present. However, the authors used the term index and not score. Another example is the scoring system developed by Chiu et al. [20] for early detection of oral submucous fibrosis based on a self-administered questionnaire. This had a C-statistic of 0.90. Rassi et al. [13] also assigned scores proportional to the regression coefficients to the independent significant factors for death in Chagas disease. The C-statistic was 0.84.

The method of assigning scores proportional to the regression coefficients is valid only when the values of the predictive factors are standardized to mean zero and variance one before the regression is run. As described later in the chapter on regression, the regression coefficients, without this standardization, are not comparable and are severely affected by the unit of measurement. Since computation is not a limitation these days, regression equation itself can be used as the scoring system without any standardization.

Since regression coefficient after standardization measures the contribution of the factor to the outcome, this method of scoring looks at least face valid as it assigns higher score to the factor that contributes more to the outcome. Also the regression is able to identify the factors that are significant independent contributors and need to be included in the scoring system. Thus, this method has some desirable properties. However, it may fail to provide a valid scoring system in some situations as discussed next.

10.2.2 Validity and Reliability of a Scoring System

Although qualities such as ease of understanding and ease of implementation can be cited, basic statistical qualities of scoring systems are validity and reliability. Of these two, validity is more important and difficult to assess too. If a scoring system were not valid, its good reliability would seldom be useful.

10.2.2.1 Validity of Scoring System

How to assess that a scoring system is providing the right result? First, it should look just about right. It should correspond well to the knowledge of experts. If a scoring system surprises you and the experts, reconsider the elements that cause this surprise. But the most important assessment of validity is against a gold standard. The basic difficulty in assessing this is in identifying and implementing the gold standard against which the validity is checked. If the gold standard is easy to perform, there is no need of a scoring system. In the case of pregnancy-induced hypertension, Thurnau et al. [21] compared the scoring results with clinical manifestation. *If clinical manifestation is to be considered the gold standard and if clinical assessment is relatively easy, nothing additional is gained by the scoring system.* A scoring system is useful only when it really adds to the clinical picture, or when it replaces a complicated procedure. The latter can happen when a final diagnosis is based on consensus of experts or when it emerges later in the course of disease. An explicit advantage is the objectivization that scores introduce, which clinical assessment may lack.

When the results from the gold standard are not available, the worth of a scoring system is assessed using alternatives. One is to see whether the scoring system gives results that are consistent with undisputable outcomes such as death. Rassi et al. [13] reported for their scoring system for predicting death in Chagas disease that patients with low (0–6 points) score had 10% 10-year mortality rate, those with medium (7–11 points) score had 44%, and those with high (12–20 points) score had 84%. This provides an indirect evidence of validity of the scoring. Note in this case that the gold standard is death and there is no way to assess it in advance except by prognostic factors summarized by the total score.

The second aspect of validity is establishing it in a different sample. This, in fact, testifies repeatability. If another sample of similar nature gives similar results, it is safe to conclude that the scoring system is not sample specific and has at least some generalizability. In almost all examples discussed in this section, the **validation sample** is different from the **development sample**, and the results were shown to replicate adequately.

The third type of validation of a scoring system is its comparison with an established and more cumbersome system. This is to check if an easier version provides the same results. Both could be excellent or both could be poor but that is not the issue in this kind of comparison. The comparison is not with a gold standard in this case. Such concurrent validity provides evidence that the easier version can replace the cumbersome procedure. Moreno and Morais [22] compared a 28-question simplified therapeutic intervention scoring system (TISS) with the standard 72-question TISS for nursing workload in intensive care units and came up with the conclusion that the simplified version is just as good. Evans et al. [23] developed a scoring system for identifying BRCA $\frac{1}{2}$ mutation and found that it outperforms existing models.

The statistical performance of a scoring system is generally judged by the C-statistic that measures area under the ROC curve. Recall that an area of 0.5 indicates that the scoring system is not helpful in properly identifying the disease or any health condition. An unattainable area of 1.0 implies a perfect system. The area between 0.70 and 0.79 is considered as satisfactory, between 0.80 and 0.89 as good, and an area of 0.90 or more as excellent. Very few scoring systems, for that matter any diagnostic aid, will attain an area of 0.90 or more.

10.2.2.2 Reliability of a Scoring System

In addition to being valid, any medical assessment tool should also be reliable with respect to repeatability and reproducibility. Reliability may suffer if the wordings are not precise and instructions are not explicit so that there is room for subjective interpretation by the assessor, by the assessee, or by both.

The reliability of a scoring system is assessed interobserver (also called interrater) as well as intraobserver. For both these assessments, intraclass correlation is used. This correlation must be in excess of 0.90 for good reliability and a value between 0.80 and 0.89 is acceptable. Any tool with less than 0.80 is suspect. Use a scoring system with such low intraclass correlation with caution.

Another measure of reliability is the narrow width of the confidence interval (CI). This can be calculated for sensitivity, specificity, predictivities, and the area under the ROC curve.

Example 10.7A: Area under the Curve of a Scoring System to Predict Prostate Cancer

Xu et al. [6] studied 129 patients with suspicion of prostate cancer who underwent transrectal sonography–guided repeat biopsies. They devised a scoring system based on results of multivariate logistic regression. ROC analysis found that the best cutoff is 2.5 at which area under the ROC curve was 0.816. Sensitivity was 76.5% and specificity 74.7%.

SIDE NOTE: Relatively low values of sensitivity–specificity and not particularly high AUC show just about satisfactory validity of this scoring system but not high validity. CIs were not calculated. They would show that the reliability also was not adequate.

Example 10.7B: Reliability of Scores for Severity of Rickets

Severity of nutritional rickets can be assessed by the degree of metaphyseal fraying and cupping, and the proportion of the growth plate affected, based on radiographs of wrists and knees. Thacher et al. [24] evaluated the utility and reproducibility of a 10-point scoring system that progresses in half-point increments from 0 (normal) to 10 points (most severe). They found that interobserver correlation of the score was 0.84 or greater for all observer pairs used by them, and intraobserver correlation was 0.89 or greater for each observer. Thus, there is a fair amount of consistency. The authors conclude that this score should be useful to objectively assess the severity of rickets.

10.3 Evidence-Based Medicine

A large number of unidentified flying objects (UFOs) have been reported as cited, yet the "evidence" of this occurrence is considered weak. Some of these instances provide graphic details that are hard to deny. Physical plausibility requires that stronger evidence must emerge. Such paramount is the role of evidence in science.

Interest may have slowed down now at the beginning of second decade of the twenty-first century, but evidence-based medicine caught a lot of the people's imagination in the early 1990s. This paradigm requires medical decisions to be based on critical appraisal of documented risks and benefits of various aspects of decision-making process, including interventions. Under this paradigm, we strive to critically appraise the best evidence available about diagnostic methods, effectiveness of treatment prognosis, or magnitude and causes of important health problems [25]. The underlying evidence is categorized as level-1 for strong evidence based on properly conducted randomized controlled trials to level-6 for opinions based on experience (see Table 2.2). Note the low priority accorded to the opinion, even if belonging to the experts. Experts too are supposed to back up their opinions by the evidence under this paradigm.

Besides clinical acumen, evidence-based medicine requires expertise in retrieving, collating, and interpreting the evidence available in literature or records. It requires assessment of validity and reliability of diagnostic procedures, efficacy and safety of medical intervention, and the sensitivity of diagnostic markers. Converting these to delineate risks and benefits of different action may be even more daunting, particularly since the available evidence may not be the best you would like to have. Developing such skills is not easy and resources required for instant access to evidence may be woefully inadequate in most medical settings. This might be one reason that the interest in evidence-based medicine has not increased. Epistemic uncertainties due to lack of reliable and valid evidence for many clinical situations may be another bottleneck. Also the evidence that evidence-based medicine works better than opinion-based medicine is slow to surface. The intent is laudable, but the delivery so far has not been very convincing.

Although the evidence is indeed the sheet anchor, patient preferences, expectations, and dilemmas get due consideration. Thus, many times hard evidence does not translate into medical guideline. External evidence may not be applicable to a particular patient.

Evidence is used as a supplement and not as a replacement to clinical judgment. The purpose is to assist in arriving at a decision that is in the best interest of the patient considering the known risks and benefits. The accompanying argument must be convincing and flawless. The key is the balanced approach. The ultimate aim is to reduce the areas of uncertainty and identify medical management steps that have the best likelihood of success. For a critique of evidence-based medicine, see Jenicek [26].

The practice of evidence-based medicine requires several steps such as transforming your information requirements to framing questions, searching the literature and other evidence, evaluating the evidence and assessing its clinical applicability, and realizing its limitations. This text restricts to those aspects that are primarily statistical. They are not the core but are ancillaries in arriving at an evidence-based decision. Also, the discussion is focused on clinical setting but applies equally well to community setting.

10.3.1 Decision Analysis

In many situations, the price paid for a false-positive diagnosis is more than that for a false-negative one, and in some situations it is vice versa. Misdiagnosis of severe schizophrenia, requiring admission to a psychiatric ward, can cause severe strain on the patient, the family, and the medical care system. A false-positive diagnosis is more costly than a false-negative diagnosis in this case. On the other hand, a missed diagnosis of leukemia is much more expensive in terms of loss of years of life than a false-positive diagnosis that can possibly be rectified later on.

The chance of error cannot be eliminated altogether, but efforts can be made to keep both types of errors to a minimum. This is done by using a sufficiently valid test or by combination of tests where feasible. The fact, however, is that errors do occur. The question is what type of error is more affordable considering the monetary cost, pain, and the risks involved. An approach can be evolved for each patient separately to minimize such costs. The following is the most commonly advocated approach.

10.3.1.1 Decision Tree

Two important components of evidence-based medicine are probabilities of various outcomes as available in the literature or record, and value judgment regarding action to be taken at different stages. The probabilities are assessed in terms of prevalence, incidence, risk, sensitivity, specificity, predictivity, etc. They must have an effective interface with clinical acumen so that they are examined in the context of actual condition of a patient. Judgments regarding advising a test or not, treating or not treating, treating by medication or by surgery, discharging from the hospital or not discharging, etc., are subjective assessments based on the experience and knowledge of the physician. The final outcome depends on a judicious mix of these probabilities and judgments. A decision tree helps to visualize various possibilities and helps to act accordingly. Value of a decision tree substantially enhances when utility is assigned to each possible outcome. This utility can be either to the patient, such as 0 for death and 1 for full recovery, or to the society. Thus, a decision tree maps all the pertinent courses of action and their consequences. For details, see Hunink et al. [27].

10.3.2 Other Statistical Tools for Evidence-Based Medicine

Laboratory investigations and imaging are extensively used for helping clinicians to narrow down the spectrum of possible diseases in a patient. In some cases, they provide infallible evidence, but in most cases, the clinician's judgment is still important. Besides the decision tree, there are other statistical tools that can help in controlling some uncertainties arising from subjective interpretation and limitation of knowledge of the individual clinician. Among these, one is the etiology diagram and the other is the expert system. Both thrive on arranging and documenting experts' knowledge in a manner that a person lacking expertise can utilize to reduce opportunities of error. It is presumed that experts' knowledge is based on evidence.

10.3.2.1 Etiology Diagram

Ishikawa diagram, whose name comes from its original proposer [28], tries to present a visual picture of various sources of nonconformities and their relationships. These relationships could be very complex and are represented by, what is also called, a fish-bone diagram or cause-and-effect diagram. A simple version of this could be called an **etiology diagram** of the type given in Figure 10.1 for MI. The postulated independent factors are encircled in this figure. All others are presumed to be consequence of these factors. It is also assumed that no other disease is present.

The process leading to MI is intricate, and the representation in Figure 10.1 is simple. The exact web of causation is perhaps not fully known. Thus, the figure is a model and suggests a hypothesis that postulates various ways that an MI can occur. According to this hypothesis, the factors that independently influence the occurrence of MI are (1) aging, (2) genetic predisposition, (3) personality profile, and (4) lifestyle. All others are dependent on these factors. For example, the diagram shows that obesity is a by-product of genetic predisposition and lifestyle (physical exercise and diet pattern). Obesity, in conjunction with aging, can produce changes in arterial walls, leading to hypertension and atherosclerosis. These, in turn, can cause coronary occlusion, ischemia, and then MI. Even though the diagram is a simple version, it can provide substantial help to a clinician by alerting him to look especially for the underlying factors before assessing the risk or before giving preventive advice.

Some factors mentioned in Figure 10.1 can influence in a reverse direction. For example, hypertension can sometimes lead to changes in the arterial walls in place of age-dependent changes in the walls of arteries leading

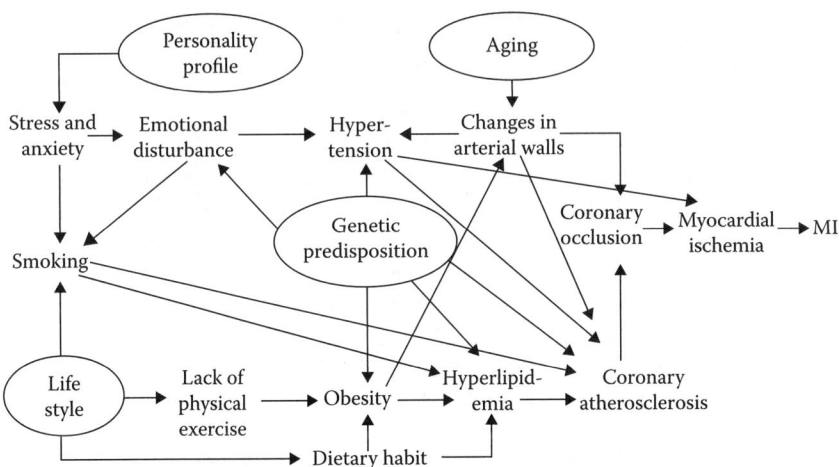

FIGURE 10.1
Suggested etiology diagram for MI.

to hypertension. The attempt here is limited to tracing of MI. For this tracing, such inverse relationships are not important.

10.3.2.2 Expert System

Epistemic uncertainties arise not only because of universal ignorance about some biological processes but also substantially from the failure to judiciously apply the available knowledge. Knowledge base in medicine is enormous, and it is rapidly increasing. It is becoming difficult for a physician to remember and recall everything about a disease at the time of confronting a patient. Errors while prescribing drugs are not uncommon. Computers have tremendous capacity to store information in a systematic manner and to retrieve it selectively as needed at an instant's notice. It can be programmed to take signs and symptoms as inputs and provide prompts on likely diagnoses along with the probability of each. Also laboratory, radiological, and other investigations can be suggested that might help in focusing on a specific diagnosis. Computer alerts on alternative strategies for treatment, as also on the prognostic indicators, can be obtained after a plausible diagnosis is identified. For this, a database is prepared containing various possibilities, and rules are devised to selectively retrieve the information after proper matching (for details, see Waterman [29]). All this is put together in a software and is called an expert system. The success of an expert system largely depends on (1) acquiring wide and valid information; (2) articulated pooling of this information into a knowledge base; (3) devising correct rules for diagnosis, treatment, and prognosis; and (4) adequate programming that uses knowledge base and rules effectively for practical management.

An expert system can have an interface with the introduction of new diagnostic and treatment strategies so that they are automatically incorporated when approved. It can also warn against the possibility of a drug reaction, allergy, or overdose. The function of this system is not more than a reminder on the basis of the inputs, and the decision always remains with the physician. He may or may not agree with the investigation, diagnosis, and the treatment suggestions of the expert system.

An appropriately prepared expert system can be of considerable help in reducing errors in medical decisions. It may also suggest alternatives available for the management of the patient. The advantage of such an expert system is that it is capable of processing a large amount of information without error as per the program or the knowledge given to it by a group of experts after considerable discussion. An individual clinician may not know as much. As always, there are ifs and buts. Medical care is much more than just diagnosis and treatment. Timely detection and timely intervention, adequate care of the patient, proper advice on prognostic implications, etc., are important ingredients. Deficiency in one can cause a chain reaction and upset the whole process. Simultaneous consideration of all such factors tends to make it even more difficult to devise an expert system.

You can see that expert systems are easy to talk about but are extremely difficult to develop. No expert system can be better than the expertise given to it. This implies that the system should be based on the knowledge of real experts. They are rare, and each expert is a specialist of one's own subdiscipline. To put various experts together for developing a comprehensive computer-based system has turned out to be almost an unachievable ideal. For this reason, efforts remain limited to specific diseases. Some are briefly described in Example 10.8.

Irrespective of the expertise of the system, it is seldom able to think as critically as a human mind can. It restricts itself to the structured inputs. It could never be helpful in a situation that has not been thought of at the time of its preparation. Such situations can always arise because of highly individualistic interactions of various biological and environmental factors. An expert clinician can perceive alternative scenarios for a particular condition and can immediately react to an unforeseen observation, but an expert system cannot. Thus, the use of expert systems should be done with sufficient caution. They must be considered only as an aid and not as a guide. The decision at each step has to be entirely that of the attending clinician. He may or may not agree with the suggestion of the expert system. An expert system cannot replace a clinician but can act as a supplementary, perhaps a very useful supplementary, when prepared with a sound knowledge base and correct rules. The nature of help of an expert system is the same as that of laboratory and radiological investigations.

For the reasons just explained, not many good expert systems are available. Most of those in the market are of dubious quality. Thus, use them with care. The purpose of introducing expert systems here is to explain that they can really help in minimizing uncertainties, both in clinical and research setups, when properly developed and judiciously used.

Example 10.8: Some Examples of Expert Systems

Here are the briefs of some expert systems.

1. Differential diagnosis of disorders and diseases manifested by tall stature [30]: For this, the diagnostic criteria were developed by a panel of seven experts. In addition, manuals and textbooks and databases and online resources were also consulted. Linguistic terms were also studied. An interface was made up by a set of slides. The major sources of information were the London Dysmorphology Database and Orphanet. The expert system produces the most probable five diagnostic possibilities and ranks them in order of likelihood depending on the inputs provided by the patient.

2. Automated visual fields: Feldon et al. [31] used visual fields from 189 nonarteritic anterior ischemic optic neuropathy eyes with nonischemic optic neuropathy decompression to develop

a computerized classification system. The expert panel had six neuro-ophthalmologists, who described definitions for visual field pattern defects using 19 visual fields and several levels of severity representing a range of pattern defect types. The expert panel subsequently used 120 visual fields to revise the definitions. These were converted to rule-based computerized classification system. The system was subsequently used to categorize visual field defects for an additional 95 nonarteritic anterior ischemic optic neuropathy. The agreement with the experts was not really high, and further modification was done.

SIDE NOTE: The two expert systems cited earlier are not for complex conditions and are focused on specific conditions. Yet, they illustrate the kind of problems encountered in developing an expert system.

Among others that you may like to review are (1) decision support system to improve clinician's interpretation of abnormal liver function test [32], (2) a three-stage expert system based on support rector machines for thyroid disease diagnosis [33], and (3) medical expert system for the diagnosis of ectopic pregnancy [34].

References

1. Billewicz WZ, Chapman RS, Crooks J et al. Statistical methods applied to the diagnosis of hypothyroidism. *Q J Med* 1969; 38:255–266.
2. Zulewski HK, Muller B, Exer P, Miserez AR, Staub J. Estimation of tissue thyroidism by a new clinical evaluation of patients with various grades of hypothyroidism and controls. *J Clin Endocrinol Metab* 1997; 82:771–776.
3. McGillicuddy EA, Lischuk AW, Schuster KM et al. Development of a computed tomography-based scoring system for necrotizing soft-tissue infections. *J Trauma* 2011; 70:694–699.
4. Casey JR, Block S, Puthoor P, Hedrick J, Almudevar A, Pichichero ME. A simple scoring system to improve clinical assessment of acute otitis media. *Clin Pediatr (Phila)* 2011; 50:623–629.
5. Promsuwicha O, Sontgmuang W, Auewarakul CU. Utilization of a scoring system for diagnosis of chronic lymphocytic leukemia in Thai patients. *J Med Assoc Thai* 2011; 94(Suppl 1):5232–5238.
6. Xu B, Min Z, Cheng G et al. Evaluating possible predictor of prostate cancer to establish a scoring system for repeat biopsies in Chinese men. *J Ultrasound Med* 2011; 30:503–508.
7. Stojadinovic MM, Milovanovic DR, Gajic BS. Scoring system development and validation for initial treatment failure in suppurative kidney infections. *Surg Infect (Larchmt)* 2011; 12:119–125.
8. Lugauer S, Reganfus A, Boswald M et al. A new scoring system for the clinical diagnosis of catheter-related infection. *Infection* 1999; 27(Suppl 1):549–553.

9. Jennnett B, Teasdale G, Braakman R, Minderhoud J, Heiden J, Kurze T. Prognosis of patients with severe head injury. *Neurosurgery* 1979; 4:283–289.

10. Knaus WA, Wagner DP, Draper EA et al. The APACHE III prognostic system: Risk prediction of hospital mortality for critically ill hospitalized adults. *Chest* 1991; 100:1619–1636.

11. Lemeshow S, Teres D, Avrunin JS, Gage RW. Refining intensive care unit outcome predictions by using changing probabilities of mortality. *Crit Care Med* 1988; 16:470–477.

12. McCarthy PL, Sharpe MR, Spiesel SZ et al. Observation scales to identify serious illness in febrile children. *Pediatrics* 1982; 70:802–809.

13. Rassi A Jr, Rassi A, Little WC et al. Development and validation of a risk score for predicting death in Chagas heart disease. *N Engl J Med* 2006; 355:799–808.

14. Vaney C, Vaney S, Wade DT. SaGAS, the Short and Graphic Ability Score: An alternative scoring method for the motor components of the Multiple Sclerosis Functional Composite. *Mult Scler* 2004; 10:231–242.

15. Guo W, Zhang Q, Chen Y, Hou D. Diagnostic scoring system of Hirschsprung's disease in the neonatal period. *Asian J Surg* 2006; 29:176–179.

16. Rosen RC, Cappelleri JC, Smith MD et al. Development and evaluation of an abridged 5-item version of the International Index for Erectile Function (IIEF-5) as a diagnostic tool for erectile dysfunction. *Int J Imp Res* 1999; 11:319–326.

17. Yamase H. Development of a comprehensive scoring system to measure multifaceted nursing workloads in ICU. *Nurs Health Sci* 2003; 5:299–308.

18. Piombo AC, Gagliardi JA, Guelta J et al. A new scoring system to stratify risk in unstable angina. *BMC Cardiovasc Disord* 2003; 3:8.

19. Purasiri P, Abdalla M, Heys SD et al. A novel diagnostic index for use in the breast clinic. *J R Coll Surg Edinb* 1996; 41:30–34.

20. Chiu CJ, Lee WC, Chiang CP, Hahn LJ, Kuo YS, Chen CJ. A scoring system for the early detection of oral submucous fibrosis based on a self-administered questionnaire. *J Public Health Dent* 2002; 62:28–31.

21. Thurnau GR, Dyer A, Depp OR III, Martin AO. The development of a profile scoring system for early identification and severity assessment of pregnancy-induced hypertension. *Am J Obstet Gynecol* 1983; 146:406–416.

22. Moreno R, Morais P. Validation of the simplified therapeutic intervention scoring system on an independent database. *Intensive Care Med* 1997; 23:640–644.

23. Evans DG, Eccles DM, Rahman N et al. A new scoring system for the chances of identifying a BRCA1/2 mutation outperforms existing models including BRCAPRO. *J Med Genet* 2004; 41:474–480.

24. Thacher TD, Fischer PR, Pettifor JM, Lawson JO, Manaster BJ, Reading JC. Radiographic scoring method for the assessment of the severity of nutritional rickets. *J Trop Pediatr* 2000; 46:132–139.

25. Jenicek M. Towards evidence based critical thinking medicine? Uses of best evidence in flawless argumentations. *Med Sci Monit* 2006; 12:RA149–RA153.

26. Jenicek M. Evidence-based medicine: Fifteen years later. Golem the good, the bad, and the ugly in need of a review? *Med Sci Monit* 2006; 12:RA241–RA251.

27. Hunink M, Glasziou P, Siegel J et al. *Medical Decision Making in Health and Medicine: Integrating Evidence and Values.* Cambridge, U.K.: Cambridge University Press, 2002.

28. Montgomery DC. *Introduction to Statistical Quality Control.* New York: John Wiley & Sons, 1985; pp. 149, 351–373.

29. Waterman D. *A Guide to Expert Systems*. New York: Addison Wesley, 1986.
30. Paghava I, Tortladze G, Phagava H, Manjavidze N. An expert system for differential diagnosis of tall stature syndrome. *Georgian Med News* 2006; 131:55–58.
31. Feldon SE, Levin L, Scherer RW et al. Development and validation of a computerized expert system for evaluation of automated visual fields from the Ischemic Optic Neuropathy Decompression Trial. *BMC Ophthalmol* 2006; 6:34.
32. Chevrier R, Jaques D, Lovis C. Architecture of a decision support system to improve clinician's interpretation of abnormal liver function test. *Stud Health Technol Inform* 2011; 169:195–199.
33. Chen HL, Yang B, Wang G, Liu J, Chen YD, Liu DY. A three-stage expert system based on support vector machines for thyroid disease diagnosis. *J Med Syst* 2011; February 1, doi: 10.1007/s10916-011-9655-8.
34. Kitoporntheranunt M, Wiriyasttewong W. Development of medical expert system for the diagnosis of ectopic pregnancy. *J Med Assoc Thai* 2010; 93(Suppl 2):S43–S49.

11

Measurement of Community Health

Chapters 9 and 10 were devoted to the epidemiological measurement of various aspects of health of an individual and to the interpretation of such measurements. However, the term health is used as much for communities as for individuals. A different set of instruments are required to measure the health of a *group* of people. Such a measurement helps to compare the state of health over a period in the same community, among people of different social and ethnic groups, among people residing in different areas, and among people of different biological groups such as of different age, of different gender, of different race, with different health condition, etc. It helps to make statements such as that the state of health of children in Kenya will be better in the year 2012 than in the year 1998, that the health of elders in Japan is better than that in Bulgaria, and that the health of cardiovascular disease patients is better than those of cancer patients.

This chapter: As explained in the previous chapter, the term indicator is generally used for the measurement of a specific aspect of health. At the community level, there are indicators of mortality and there are indicators of morbidity. Section 11.1 describes indicators of mortality such as infant mortality and proportional deaths due to cardiovascular diseases, and Section 11.2 describes measures of morbidity such as incidence and prevalence. In addition to these measures of physical health, there are separate indicators of social and mental health. These are presented in Section 11.3. Efforts are sometimes made to combine many indicators and present a more comprehensive picture. Such composite indices are discussed in Section 11.4.

11.1 Indicators of Mortality

Death is easy to identify in nearly all cases and the date of death is generally available in records. Thus, mortality statistics are considered reliable and used all across the world. A higher rate of mortality is considered an indicator of poor health, although this may not always be so, as explained later. Mortality rates are generally calculated per year.

11.1.1 Crude and Standardized Death Rates

Different types of death rates are computed depending upon groups of interest, for example, groups classified by age and gender. Described in the following are the rates based on overall deaths in a population and later in this section more specific death rates for various groups are discussed.

11.1.1.1 Crude Death Rate

This is the number of deaths in an area in a year per 1000 population counted at midyear, that is,

$$\text{Crude death rate} = \frac{\text{Number of deaths in 1 year}}{\text{Midyear population}} * 1000. \qquad (11.1)$$

It ranged from a low of 1.4 per 1000 population in Qatar to a high of 17.2 in Chad in the year 2008. Thus, this rate varies widely from country to country.

A problem with crude death rate (CDR) is that it disregards the age structure of the population. For this reason, it is called *crude*. If people in an area are predominantly old, a high CDR is not as bad as in an area where the population is predominantly young. Thus, a CDR of 8 per 1000 population in Sweden should not be construed to mean that the health status is nearly the same as in India, where also the CDR is nearly the same. The CDR can be misleading. India has only 8% of the population of age 60 years or more, whereas Sweden has more than 20%. The death rate among old people is high, and therefore the CDR naturally becomes high in a country like Sweden. A valid comparison is obtained when the rate is recomputed by assuming the same age structure in the two countries. This is one form of standardization. The other brings the age-wise mortality pattern to a common base. Both require age-specific death rates (ASDRs).

11.1.1.2 Age-Specific Death Rate

When the numerator and the denominator in Equation 11.1 are restricted to a particular age-group, we get the specific death rate for that age-group. For example, the ASDR for age-group 65–74 years is the deaths that occur in this age-group per 1000 persons of age 65–74 years. Such a rate provides an adequate comparison of the health status in two areas or at two different times for that particular age-group. In the year 2006, the ASDR in the age-group 5–14 years was 1.0 in Peru but only 0.08 in Sweden per 1000 population of that age. The rate in Peru was more than 10 times. This shows the qualitative difference in deaths in these two countries.

11.1.1.3 Standardized Death Rate

Standardization is a method of adjustment. This is generally done for age differentials but can also be done for other factors. The objective is to remove the effect of differential structure of the subgroups in the two populations under comparison. The rates are then brought to a common base, and thus made comparable. Two methods of standardization are in use. Both require assumption of a **standard** or a reference population. This could be real or hypothetical, but in both cases, the standard is arbitrary. In case of age standardization, the standard may have a predefined age structure or a predefined ASDR. These two methods of standardization can give entirely different results.

In the direct method, actual ASDRs in the study population are used on the standard population that has a predefined age structure. Thus,

$$\text{Directly standardized death rate} = \frac{\Sigma_k P_{ks} d_k}{\Sigma_k P_{ks}}; \quad k = 1, 2, \dots, K; \quad (11.2)$$

where

population is divided into K age-groups

P_{ks} is the predefined standard population (in percent or count) in the kth age-group

d_k is the ASDR in the kth age-group of the study population

Thus, the age structure is standardized. When **direct standardization** is done for two populations, any difference between the two can be ascribed to the difference in their ASDRs.

Direct standardization is possible only when ASDRs in the study population are known. If they are not known, the **indirect method** is used. In this method, predefined values of ASDRs in the standard population are used on the actual age structure of the study population. Thus,

$$\text{Indirectly standardized death rate} = \frac{\Sigma_k p_k D_{ks}}{\Sigma_k p_k}; \quad k = 1, 2, \dots, K; \quad (11.3)$$

where

p_k is the actual *study* population (in percent or count) in the kth age-group

D_{ks} is the predefined ASDR in the kth age-group of the standard population

When this is done for two populations, both are then based on the same ASDRs and any difference between the two would be due to their differential age structure.

Example 11.1: Standardized Death Rate for the United States and Venezuela

Shown in Table 11.1 are the age structures of the populations in the United States and Venezuela and their ASDRs [1]. The standard population given is as suggested by the World Health Organization (WHO) [2]. The standard ASDRs in the last column are proposed by me. No widely acceptable standard is available for ASDRs.

Note that the CDR in the United States is nearly twice of the CDR in Venezuela. Also note that the United States has more people in the old age-group. From the formula given in Equation 11.2, the directly standardized death rate

$$\text{for the United States} = \frac{1.7 \times 8.9 + 0.2 \times 17.3 + \cdots + 148.3 \times 0.6}{8.9 + 17.3 + \cdots + 0.6}$$

$$= 5.5 \text{ per 1000 population,}$$

and the directly standardized death rate

$$\text{for Venezuela} = \frac{3.5 \times 8.9 + 0.3 \times 17.3 + \cdots + 140.0 \times 0.6}{8.9 + 17.3 + \cdots + 0.6}$$

$$= 5.8 \text{ per 1000 population.}$$

TABLE 11.1

Population and Death Rates in Different Age-Groups in the United States and Venezuela, and the WHO World Standard

Age-Group (Years)	United States 2002		Venezuela 2002[a]		Standard	
	Population (%)	ASDR per 1000	Population (%)	ASDR per 1000	Population (%)	ASDR per 1000
A	B	C	D	E	F	G
0–4	6.8	1.7	11.1	3.5	8.9	10.0
5–14	14.2	0.2	21.6	0.3	17.3	1.0
15–24	14.1	0.8	19.4	1.8	16.7	0.5
25–34	13.8	1.0	15.6	2.1	15.5	0.5
35–44	15.6	2.0	12.9	2.4	13.7	1.0
45–54	13.9	4.3	9.2	4.5	11.4	5.0
55–64	9.2	9.5	5.5	8.8	8.3	10.0
65–74	6.3	23.1	3.2	21.5	5.2	15.0
75–84	4.4	55.6	1.3	52.4	2.4	35.0
85+	1.6	148.3	0.3	140.0	0.6	100.0
Total	100.0	8.5 CDR	100.0	4.2 CDR	100.0	5.0 CDR

[a] Data from United Nations Statistics Database.

The difference has now almost vanished after standardization.

From the formula given in Equation 11.3, indirectly standardized death rate

$$\text{for the United States} = \frac{10.0 \times 6.8 + 1.0 \times 14.2 + \cdots + 100.0 \times 1.6}{6.8 + 14.2 + \cdots + 1.6}$$

$$= 6.8 \text{ per 1000 population,}$$

and the indirectly standardized death rate

$$\text{for Venezuela} = \frac{10.0 \times 11.1 + 1.0 \times 21.6 + \cdots + 100.0 \times 0.3}{11.1 + 21.6 + \cdots + 0.3}$$

$$= 3.9 \text{ per 1000 population.}$$

This death rate for Venezuela is much lower.

Note the following for Example 11.1:

1. Venezuela has higher population in younger age-groups, and the ASDRs are also higher in younger age-groups.
2. For direct standardization of death rate in the United States, ASDRs in column C are multiplied by the standard population in column F, added, and divided by the total of standard population in column F. The numerator so obtained is the expected number of deaths that would have occurred if the age structure were standard. For direct standardization of death rate in Venezuela, its ASDRs in column E are multiplied by the standard population in column F.
3. For indirect standardization, the population (column B for the United States and column D for Venezuela) is multiplied by the ASDRs in the standard population (column G), added, and divided by the total population in the country, that is, the total of column B for the United States and of column D for Venezuela. The numerator in this case is the expected number of deaths that would have occurred if the ASDRs were the same as in the standard population.
4. Directly standardized DR is less than the CDR in the United States because the standard population is less in the age-groups where the ASDR in the United States is high. When the standard population is more in higher mortality groups as in Venezuela, the directly standardized DR is more than the CDR.
5. Directly standardized rate brings the age structure of the two populations to the same pattern. When this is done, the death rate in the United States becomes nearly the same as that in Venezuela.

6. Indirect standardization has a different effect in this case. The indirectly standardized DR is higher in the United States. The difference between the indirectly standardized rates in the United States and Venezuela is mostly due to the difference in age structure in the two countries.

7. In Example 11.1, the age structure is given in terms of percentages, but the actual population can also be used in the formulas given in Equations 11.2 and 11.3.

8. As in this example, the two methods of standardization can give entirely opposite results. Thus, it is important that the method is correctly chosen in accordance with your objective of doing the standardization. In addition, the interpretation of the standardized rate should be proper for the method used.

The standardized rate depends heavily on the standard chosen. No universal standard is available, and this is arbitrarily chosen. If a desirable structure exists or can be constructed, that can be chosen as the standard. If not, a structure that is of middling type or commonly seen or easy to implement can be chosen as the standard. In column F of Table 11.1 is the WHO World Standard population [2]. This actually extends to age 100 years and over and is given in 5-year intervals. In Table 11.1, an abridged version with 85 years and over as the last interval and age intervals of 10 years is given.

Since the standard is arbitrary anyway, it should be simple. This is not the case with the WHO World Standard population. The ASDRs in column G are my own standards and are simple. For interregional comparison within a country, the age structure of the total country or its ASDRs can be used as the standard.

The direct method seems more appropriate in most cases because this gives the death rate expected for standard age structure. But this method cannot be used when the ASDRs are not known. Also, this method should not be used when ASDRs are unstable and are based on a small number of subjects. Indirect standardization is generally used for disease mortality because of unstable ASDRs. This, when used for small groups instead of the general population, is called the standardized or adjusted mortality rate and leads to a popular measure known as the standardized mortality ratio. This is discussed later in this book. Such an adjustment can be done not only for age but also for any other factor that might influence the mortality pattern.

The illustration in Example 11.1 is for population mortality as it is the most common application of standardized death rate. However, this can also be used for disease-specific mortality. For example, if ASDRs for circulatory diseases are known and if the purpose is to compare two agewise diverse populations, the comparison should be based on standardized rates. For emphasizing the importance of the choice of the standard in this context, it was shown [2] for the year 1995 that standardized death rate due to circulatory disease in males in the United States using the WHO World Standard is 285 per 100,000 population but is 372 when a Scandinavian standard is used. This reflects a large difference of 23% and underscores the heavy dependence of standardized death rate on the standard chosen for this purpose.

11.1.1.4 Comparative Mortality Ratio

The term comparative mortality ratio has not yet achieved the globally accepted meaning. This can be used to express one mortality rate compared to the other when both have the same base:

$$\text{Comparative mortality ratio} = \frac{\text{Death rate-1}}{\text{Death rate-2}} * 100, \qquad (11.4)$$

where the two death rates have some common base.

One popular meaning of comparative mortality ratio is the ratio of expected number of deaths arrived by direct standardization to the actual deaths. In Example 11.1, this comparative mortality ratio for the United States is $100 \times 5.5/8.5 = 65\%$, and for Venezuela it is $100 \times 5.8/4.2 = 138\%$. This again is heavily dependent on the age structure chosen as the standard.

The utility of comparative mortality ratio increases when calculated for standardized rates. This ratio for the United States compared to Venezuela in Example 11.1 is $100 \times 5.5/5.8 = 95\%$. Since both rates refer to the same standard population, this can be interpreted as saying that the death rate in the United States is 95% of that in Venezuela. Although both rates under comparison are standardized for the same age structure, yet the effect of the chosen standard is not eliminated. If another standard age structure is chosen, the ratio can become very different.

The utility of comparative mortality ratio increases further when several populations are compared. For example, if the U.S. age-standardized heart disease death rate in White males is considered 100, the comparative mortality ratio for age-standardized death rate is more than 120 for many health service areas (HSAs) in the mid-southeast and less than 80 for many HSAs in the west. Thus, comparative mortality ratio is one more method to know which segment of population has relatively higher or lower mortality. Strategies for controlling mortality can be accordingly devised.

11.1.2 Specific Mortality Rates

Death rates, whether crude or standardized, are measures of overall mortality. But there are rates for specific groups as well. The ASDR has already been described. Besides this, there are many other indicators that measure mortality in specific groups.

11.1.2.1 Fetal Deaths and Mortality in Children

Almost all fetal and child deaths are preventable. Such mortality is considered a definite indication of lack of health. Several indicators with a focus on different age groups are calculated. Beginning with conception, the age-groups considered important for mortality studies are shown in Figure 11.1.

The rates are generally computed on an annual basis and can be enumerated as follows. The numerator and the denominator are numbers related to the same period although this is not explicitly stated in the following expressions:

$$\text{Abortion rate} = \frac{\text{Abortions}}{\text{Females of reproductive age-group}} * 1000. \tag{11.5}$$

The reproductive age-group for females is 15–49 years:

$$\text{Abortion ratio} = \frac{\text{Abortions}}{\text{Live births}} * 1000, \tag{11.6}$$

$$\text{Stillbirth ratio} = \frac{\text{Stillbirths}}{\text{Live births}} * 1000. \tag{11.7}$$

Generally, pregnancy terminations are considered stillbirths when the fetus or the infant is born dead and weighs at least 500 g or has body length (crown-heel) of at least 20 cm. When these measurements are unavailable, the gestational age should be at least 22 weeks. Such deaths are also called late fetal deaths.

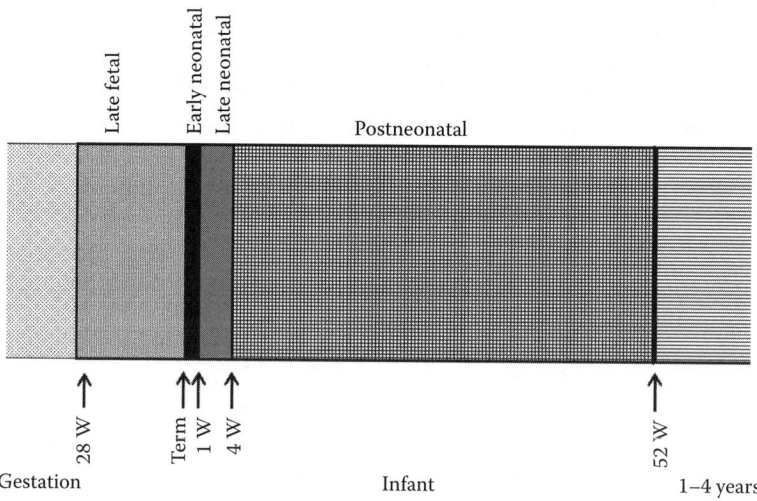

FIGURE 11.1
Childhood age-groups important for study of mortality in a community.

Other rates that take into account fetal life are as follows. These rates are particularly important for developing countries where mortality in early life is high:

$$\text{Stillbirth rate (SBR)} = \frac{\text{Stillbirths}}{\text{Still + live births}} * 1000, \quad (11.8)$$

Perinatal mortality rate (PMR)

$$= \frac{\text{Late fetal deaths + early neonatal deaths}}{\text{Late fetal deaths + live births}} * 1000, \quad (11.9)$$

Perinatal mortality ratio (PM ratio)

$$= \frac{\text{Late fetal deaths + early neonatal deaths}}{\text{Live births}} * 1000. \quad (11.10a)$$

Early neonatal deaths are those occurring during first week of life (<7 days). A more precise definition for perinatal mortality is

$$\text{PM ratio} = \frac{\text{Late fetal deaths + early neonatal deaths weighing} \geq 1000\,\text{g}}{\text{Live births weighing} \geq 1000\,\text{g}} * 1000.$$

$$(11.10b)$$

Even though frequency of occurrence is the essential feature of a rate, note that the preceding text uses the term *rate* when the numerator is part of the denominator, otherwise it should be considered a *ratio*. But this is not true for some rates.

The measures of mortality after a live birth has taken place are the following:

$$\text{Neonatal mortality rate (NMR)} = \frac{\text{Neonatal deaths}}{\text{Live births}} * 1000, \quad (11.11)$$

$$\text{Postneonatal mortality rate} = \frac{\text{Postneonatal deaths}}{\text{Live births}} * 1000. \quad (11.12)$$

The neonatal period is the first 27 days of life and the postneonatal period is 28–364 days. In the case of rate in the formula given in Equation 11.12, the postneonatal deaths can be calculated out of those who survive the neonatal period. Thus, Equation 11.12 is not a rate in the strict sense. Yet, the convention is to call it a rate.

Neonatal mortality is generally determined by the health of the mother and the adequacy of services available at the time of birth. The main causes of neonatal mortality are prematurity and birth asphyxia. Postneonatal mortality is mostly due to infections and undernourishment since birth. International efforts have succeeded in controlling much of the postneonatal mortality, but the neonatal mortality is yet to respond to these efforts. Emphasis is now given to maternal nutrition and skilled attendance at birth in developing countries to control neonatal mortality.

Neonatal and postneonatal periods together form the infantile period. Note that an infant is a child *less* than 1 year of age and a neonate is *less* than 4 weeks of age, which is the same in completed days as just stated. The sum of the neonatal and postneonatal mortality is the infant mortality when their denominator is the same live births:

$$\text{Infant mortality rate (IMR)} = \frac{\text{Deaths of infants}}{\text{Live births}} * 1000. \qquad (11.13)$$

A popular mortality rate for children is the under-5 mortality rate (U5MR). It is calculated as

$$\text{U5MR} = \frac{\text{Deaths of children} < 5\,\text{years}}{\text{Live births}} * 1000. \qquad (11.14)$$

The U5MR is also called the child mortality rate, although this term is sometimes used for mortality in the 1–4 year age-group. These rates are different from the respective ASDRs because the denominator is the number of live births and not the population of that age-group. It is customary to use the term "mortality rate" when the denominator is live births and the term "death rate" when the denominator is population.

The mortality count used in the numerator in the formulas given in Equations 11.13 and 11.14 is not necessarily out of the live births in the denominator. A child born in the month of October may die in the month of January in the next calendar year. The birth and the death of the same child, thus, would be counted in different years. Despite this anomaly, these are called rates and the terms are retained as such because of simplicity in counting. The births are not required to be followed up to count the deaths for these rates. The effect of this anomaly on the rate is minimal because of the averaging-out phenomenon.

Some professionals modify the denominators for calculating some of these rates. For example, for calculating postneonatal mortality rate, they use (live births–neonatal deaths) as the denominator. Although this can be justified, such a modification introduces complexity and compromises the feature of additivity of some of these rates. For this reason, this text uses and advocates simple denominators as stated.

All these childhood mortality rates are considered very sensitive indicators in a global context. Attention is particularly paid to IMR because this is relatively easily understood and is now almost universally available. A large number of background and proximate factors such as education, affluence, nutrition, availability and utilization of health-care facilities, and cultural practices affect IMR. This rate is affected rather quickly by health programs in developing countries, and so is also used to measure the effectiveness of such programs.

Wide differentials exist in these rates from country to country and from one segment of population to another segment within the same country. For example, IMR was 1 per 1000 live births in Luxembourg in the year 2009, while it was 134 in Afghanistan. U5MR was 2 per 1000 live births in San Marino and 239 in Chad in the year 2009.

11.1.2.2 Maternal Mortality

Maternal mortality is measured as follows:

$$\text{Maternal mortality ratio (MMR)} = \frac{\text{Maternal deaths}}{\text{Live births}} * 100,000. \qquad (11.15)$$

The term maternal mortality applies when the cause of death is related to or aggravated by pregnancy or its management. For practical reasons, the WHO definition includes only maternal deaths that occur either during pregnancy or within 42 days after the termination of pregnancy. The cause may be direct, such as hemorrhage, sepsis, eclampsia, and abortion, or indirect, such as heart diseases in pregnancy and hepatitis. However, accidental or incidental deaths are excluded.

The denominator in the formula given in Equation 11.15 is live births, but maternal mortality can also occur at the time of stillbirth or abortion. Although data on stillbirths may be available, it is extremely difficult to count abortions even in the best of conditions. The group at risk comprises women of reproductive age. Because the group at risk is not used as the denominator for the event in the numerator, MMR is called a ratio and not a rate. Where possible, multiple births are counted as one in the denominator.

Maternal deaths in developing countries are now receiving more attention. Many of these countries do not have a sufficiently strong system to monitor every maternal death. Some deaths escape attention. Thus, the reported rate is low compared with the actual rate. WHO, UNFPA, the World Bank, and UNICEF made joint efforts to arrive at more realistic estimates of MMR in all countries. They used surrogates such as proportion of births with different types of birth attendants and general fertility rate to arrive at these estimates [3]. Data on these surrogates are easily available and are more accurate. According to these estimates, the minimum MMR in the year 2008 was nearly 2 in Greece, and the maximum was 1,400 in Afghanistan per 100,000 live births.

11.1.2.3 Adult Mortality

Child morality is a huge concern in some populations, but others who have low child mortality shift focus to adult mortality. Since geriatric mortality does not contribute much to the study of adverse health, adult mortality is concerned with deaths between the age of 15 and 59 years and ignores deaths at age 60 years and beyond. Thus,

$$\text{Adult mortality rate} = \text{Probability of death between 15 and 59 years per 1000 population of this age.} \quad (11.16)$$

This can be obtained through a life table, which is described in Chapter 18. Adult mortality rate arouse passion as it concerns the most productive segment of population and is supposed to be the most healthy period of life.

Adult mortality is generally higher in males than females. This differential is attributed to greater exposure of males to hazards at work, stress, and strain, and also to their biological vulnerability. Adult mortality rate in males in Austria is 102 per 1000 and in females less than half, only 50, in the year 2009. In the United States, these are 134 and 78, respectively. HIV-affected countries have very high adult mortality rate.

11.1.2.4 Other Measures of Mortality

It is useful for health authorities to know the extent of deaths occurring because of various causes. The number one killer in the United States is vehicular accidents, whereas in Indian rural areas it is respiratory diseases. Similar statements can also be made for the age-groups contributing to deaths. Such proportional mortality can be measured in several different ways:

$$\text{Proportional deaths due to cause A} = \frac{\text{Deaths due to cause A}}{\text{Total deaths}} * 100. \quad (11.17a)$$

For age-group 60+ years, proportional deaths

$$= \frac{\text{Deaths in the age-group 60 years and above}}{\text{Total deaths}} * 100. \quad (11.17b)$$

This can be computed for any cause or any age-group. The denominator in both is total deaths. The proportional deaths for any cause are different from the cause-specific death rate:

$$\text{Cause-specific death rate (due to cause A)}$$

$$= \frac{\text{Deaths due to cause A}}{\text{Population}} * 100,000. \quad (11.18)$$

This basically is the same as the CDR but is restricted to a particular cause of death. The sum of specific death rates for all causes would be the same as the CDR. If the rates in two areas are to be compared, then a standardized proportional death rate is compared using the direct or the indirect method described earlier:

$$\text{Case-fatality rate (CFR)} = \frac{\text{Deaths among those affected}}{\text{Number of individuals affected}} * 100. \quad (11.19)$$

CFR represents the severity of a disease to cause death. This rate is typically used in studying acute conditions. CFR is high for diseases such as rabies and tetanus and low for diseases such as typhoid and influenza. Case-fatality can be considered to measure the virulence of a disease. This is related to duration. One-year case-fatality of leukemia is very different from a 5-year case-fatality. For this reason, this term is customarily used for diseases of short duration such as peritonitis.

11.1.3 Death Spectrum

The health-care profession is making every effort to prevent mortality from any cause. Few realize that the probability of death is one—it can only be postponed and not denied. All causes of deaths cannot be eliminated. In fact, various causes tend to compete with one another. Depending upon the biological, environmental, and demographic factors, the causes only change hands. If one does not die of tuberculosis, he may die of cancer or in an accident. If one does not die in infancy, he may die at the age of 100 years. Perhaps, duration of survival is more important than the cause of death.

Over a period, the spectrum of causes of death and age at death, which I am calling death spectrum, has undergone a dramatic transition. Because of various health-promoting steps, infant deaths have substantially decreased and correspondingly deaths due to chronic ailments such as cancers, diabetes, and coronary artery disease have increased. This transition is the direct result of better health and increased longevity.

This raises the question whether some causes of death are more desirable than others. Medical science seems to have completely ignored this issue. The thrust all around is to control all the causes. This simply is not possible. Time has come to debate which causes should, in fact, be promoted for death in *old age* and which should be controlled. Indrayan [4] has emphasized this aspect. In his opinion, more people prefer sudden death in old age instead of a protracted slow death, which necessarily will be painful. There is no condition yet that would bring slow death but would still be not painful. A disease such as Alzheimer's may not cause physical pain but causes an intense psychological trauma. In his opinion, myocardial infarction (MI) could be the most desirable cause of death in old age since it causes sudden death in many cases. But sudden death has negative features also. The person does not get time to meet near and dear ones, to pass on messages, or to settle accounts, etc. Since the concern here is with death in old age, one can counterargue that the person should do all this at the time of reaching old age, say, at 80 years or earlier, and not when death becomes inevitable.

If the contention that MI is the most desirable cause of death in old age is accepted, all research around the world will have to be reoriented. The risk factors for such deaths in old age have to be identified not for controlling death but for nurturing these factors so that the chances of death by this desirable cause increase and correspondingly for other painful causes such as cancer decrease.

Perhaps, a choice should be available to decide to die suddenly or slowly. The choice will be exercised not at the time of death but during lifetime by controlling risk factors of one type and promoting the other types that increase the chance of death by the promoted factors.

All of this is for deaths in old age only. Deaths in young age by any cause, including MI, have to be averted as much as possible. Thus, there is a need to differentiate between risk factors of death in old age from the preferred cause and risk factors of death in young age—the former to be nurtured and the latter to be controlled. This kind of orientation is currently missing from medical research.

11.2 Measures of Morbidity

Morbidity is departure from health. This results in or has a potential to result in at least some restriction in performing normal activities of life. Morbidity could be in terms of disease, injury, burn, handicap, mental depression, insomnia, pain, etc. Sometimes medical measurements outside the normal range, such as diastolic BP ≥ 90 mmHg, are considered morbid whether or not they cause any restriction on activity. Such measurements have the potential to cause disruption later on in life if not immediately. The magnitude of morbidity in a community can be measured by (1) the number or percentage of persons affected, (2) the average number of episodes or spells of sickness, particularly for acute conditions (one person can have more than one episode in a period) per unit of time, (3) the average duration of illness or of restricted activity in the affected persons, and (4) the percentage of patients with different severity.

Morbidity is not as easy to measure as mortality. This is because morbidity can recur but mortality cannot, duration of morbidity varies from seconds to years but mortality is instantaneous, and sometimes morbidity is not well defined.

11.2.1 Prevalence and Incidence

Prevalence is the presence of morbidity, and incidence is its fresh occurrence. Thus, prevalence is computed on the basis of the existing cases and incidence on the basis of the new cases. They can be obtained either by counting the subjects affected or by counting the episodes that have occurred. Prevalence is considered to measure the load on health-care services, and incidence is considered to measure the risk of getting the disease. By its nature, incidence is related to a period such as a week, a month, or a year, but prevalence is related to a point of time. Yet, there are concepts of point prevalence and period prevalence.

11.2.1.1 Point Prevalence

Point prevalence is the number of cases existing at a specific point of time. A survey to identify affected cases generally takes weeks or months, but the count obtained is a point prevalence when the inquiry is with regard to the presence or absence of morbidity at the time of contact or a particular reference time point. This can be obtained by a cross-sectional survey. Point prevalence for the episodes would be the same as for the subjects because at any point of time a person cannot have two episodes of the same illness. In any case, point prevalence is generally obtained for chronic conditions rather than for acute illnesses.

For the purpose of comparison between groups, areas, diseases, etc., it is customary to calculate the prevalence *rate* (per cent, per thousand, or per million persons) at a particular point in time. *It is actually a proportion but is conventionally called a rate.* It does not measure speed of occurrence. Persons counted for the denominator are those who are exposed or at risk for the disease in question. Others are excluded. For example, for smoking, children below the age of 12 years can be excluded. A prevalence rate can be calculated for a specific age, gender, occupation, etc. The prevalence rate estimates the probability of presence of morbidity in a randomly selected person from that group. In Chapter 9, this was also called a pretest probability in the context of predictivities of medical tests although there prevalence is calculated among those suspected. If peptic ulcer is found in 5% cases of hypertension, this is the prevalence rate of peptic ulcer among cases with hypertension.

11.2.1.2 Period Prevalence

For calculating period prevalence, the number of persons affected or the number of episodes of illness present during a specific period such as 1 week or 1 month in a defined population is counted. This includes the number of cases arising before but extending into or through the period as well as those arising during that period. The information sought from the respondent in this case is whether he is suffering from the disease at the time of inquiry or had suffered any time during the last 1 week, 1 month, etc. For measuring period prevalence, generally only a short duration is considered, and it is mostly obtained for acute conditions.

11.2.1.3 Prevalence Rate Ratio

If prevalence of a disease in males is 17% and in females 24%, it could useful to say that the prevalence in females is nearly 1½ times of that in males. Ratio of prevalence in two mutually exclusive groups is called prevalence rate ratio (PRR). Thus,

$$\text{PRR} = \frac{\text{Prevalence rate in one group}}{\text{Prevalence rate in another nonoverlapping group}}.$$

If n_1 is the number of subjects with diabetes and a of them are also hypertensive, the prevalence of hypertension among diabetics is a/n_1. If n_2 is the number of subjects without diabetes and b of them are hypertensives, the prevalence in the group is b/n_2. In this case,

$$\text{PRR} = \frac{a/n_1}{b/n_2}.$$

This is useful to measure how common is the condition in one group relative to the other.

11.2.1.4 Incidence

Incidence is the number of conditions or the number of persons having onset of a condition in a specified period. A condition is considered to have had its onset when it was first noticed. This could be the time when the person first felt sick or injured, or it could be the time when the person (or the family) was first told of a condition that was previously unknown. Thus, a person who has been having diabetes for a long time is considered to have onset at the time when diabetes was first detected.

Incidence rate is calculated per cent, per thousand, per million, etc., as convenient, of exposed subjects, and measures the risk of developing morbidity in a randomly chosen subject from that group. Incidence rate may increase if the reporting or case detection improves. Thus, the increase may not be a real rise. A real rise indicates that the existing strategy to control the disease has not succeeded. Alternative or improved strategies may then be needed. Analysis of differences in incidence in various socioeconomic, biological, and geographical groups may provide useful tips to devise a better strategy for control of the disease. Also, note the following:

1. Incidence rate is necessarily associated with a duration. If the period is 1 year, the rate obtained will also be for that year. Notice that incidence can be obtained only through a follow-up study.

2. Incidence reflects causal factors. This is used to formulate and test hypotheses on the etiology of the disease. Incidence is better interpreted as risk.

3. Incidence rate can also be calculated for spells or episodes instead of the affected persons. This would count a person two or more times if the same person has repeated spells within the reference period. Then, the term **incidence density** is preferred in this case.

11.2.1.5 Concept of Person-Time

It is sometimes not possible to observe each person in a cohort for the same duration. Also, the duration of exposure may vary from subject to subject. For example, persons in a stressful environment for different periods may be observed for incidence of peptic ulcer disease. One person may be under stress for 12 years, the second for 5 years, the third for 8 years, etc. These durations are totaled and called person-years. If the ith person is observed for x_i years, then

$$\text{Total person-years for } n \text{ persons} = \Sigma x_i; \quad i = 1, 2, \ldots, n.$$

Person-years is the most frequently used form of person-time, but this could also be calculated in terms of person-months, person-weeks, etc. For example, in the case of use of oral contraceptives, person-months are used and incidence of a complication or of pregnancy is calculated per 100 person-months of use. The concept of person-time is valid only when initial period is as important or as unimportant as the later period. In the oral contraceptive example, the risk of complication should be the same in first month as in, say, 10th month of use for person-month to be a valid tool. In practice, this may not be so. In general,

$$
\begin{aligned}
&\text{Incidence rate per 100 person-years} \\
&= \frac{\text{New cases occuring in the observed period}}{\text{Person-years observed}} * 100. \quad (11.20)
\end{aligned}
$$

Thus, a uniform follow-up is not needed to calculate incidence rate, although a follow-up is required in any case. However, the incidence must be evenly distributed over different periods. The multiplier in the formula given in Equation 11.20 is not necessarily 100 and can be chosen as per convenience.

Just as in case of prevalence, you can obtain incidence rate ratio for two exclusive groups. If incidence of liver diseases in alcohol users is 5 per 1000 per year and in nonusers 0.5 per 1000 per year, the incidence ratio is 10. However, this has a better and more popular name, the relative risk. This is discussed at length in a later chapter.

Precise incidence or prevalence rates are central to epidemiology of any disease but they are also very difficult to obtain. Full enumeration is typically too expensive, very labor intensive, and almost impossible to achieve for monitoring of disease frequency. Thus, an estimate is obtained by studying a sample. A critical component in disease monitoring is the degree of undercount. One way to improve the estimate of the rates is by correcting them for the level of ascertainment. Comparison of different populations, communities, or countries is accurate when the incidence estimates from a surveillance system or otherwise are adjusted for the degree of under-ascertainment. Among various methods, capture–recapture methodology can be applied to evaluate the degree of undercount and to get a corrected rate.

11.2.1.6 Capture–Recapture Methodology

The capture–recapture methodology was originally devised for obtaining counts of animals in the wild. Consider a sample of 48 deer ($n_1 = 48$) who are captured, marked, and then released. They mix with other deer. Subsequently, a second sample of 35 deer ($n_2 = 35$) is captured from the same area. If 16 ($m = 16$) of them are found to have the mark, an estimate of the total number of deer is obtained by inflating by a factor of 35/16. Thus, an estimate of the total count N is $48 \times 35/16 = 105$. A statistically better estimate in the long run is obtained when 1 is added to each of these numbers and finally 1 is subtracted. Thus,

$$\text{Capture–recapture estimate } N = \frac{(n_1 + 1)(n_2 + 1)}{m + 1} - 1, \qquad (11.21)$$

where
  n_1 is the size of the first sample
  n_2 is the size of the second sample
  m is the number of cases found to appear in both the samples

In health and medicine, the capture–recapture methodology helps to estimate the total number of cases in hard-to-reach population when their incomplete count is available from two *independent* sources. These sources could be hospital records, physicians in private practice, death certificates, or any other such list of cases. The count of duplicate cases, which appear in both lists, can help to substantially improve the estimate of the total cases. Hook and Regal [5] argued that the capture–recapture methodology can improve prevalence estimates even for apparently exhaustive surveys.

**Example 11.2: Capture–Recapture Estimate
of Childhood Diabetes Cases**

The methodology was adopted to assess the prevalence of childhood
diabetes in Madrid, Spain [6]. A population-based registry identified
432 cases through hospital inpatient records. Another source was the
Spanish Diabetes Association, which recorded 138 cases. It was found
on matching that 119 cases were common to the two sources. Thus,
an estimate of the total number of cases of childhood diabetes using
Equation 11.21 is

$$\frac{(432+1)(138+1)}{119+1} - 1 = 501.$$

The results can be converted to prevalence rates as shown in Table 11.2.
     The cases in either sources are undercounts, but the duplicates helped
to come up with an estimate of the total cases as well as an improved
prevalence rate.

Note the following for capture–recapture methodology:

1. The capture–recapture methodology assumes that there are no
   intermediary additions or deletions. If there are any, the estimate
   may have to be revised accordingly.
2. It is necessary that the captured subjects move around freely and are
   homogeneously mixed when released.
3. The two sources should be independent of one another. This is
   not fulfilled in a situation where, for example, some cases seen by
   the first source tend to be referred to the second. In such cases, the
   capture–recapture estimate may be too low.
4. Like all estimates, the estimate of the total count by capture–recapture
   methodology is subject to sampling fluctuation. Methods are available to find the standard error of this estimate and to construct a 95%
   confidence interval. If interested, see Laporte et al. [7].

TABLE 11.2

Capture–Recapture Estimate of Childhood Diabetes in Madrid

	Hospital Records	Diabetes Association	Common	Capture–Recapture Estimate
Number of cases	432	138	119	501
Prevalence rate per 100,000 population (total population 4.4 million)	9.8	3.1	—	11.4
Extent of undercount (as percent of the last column)	14%	72%	—	0%

11.2.2 Duration of Morbidity

Acute conditions such as typhoid, malaria, and diarrhea last a few days, but chronic conditions such as diabetes, malignancy, and epilepsy can go on for years. The duration depends on the course a disease takes. This, in turn, depends on its natural history, severity, adequacy of treatment if any, the subject's own capability to mitigate the disease, etc.

For duration of disease to be correct, it is necessary that the onset and the termination points are properly defined. The starting point could be the day of appearance of the first sign or first symptom, the day of diagnosis, the day of a positive test, the day of reporting, etc. The termination could be in terms of disappearance of complaints, ability to resume normal activities, negative test, discharge from the hospital, etc. Note how different definitions can yield very different durations of a disease.

In the case of acute conditions, a measure of the duration of sickness for a group is its mean or the median. It is also advisable to calculate the standard deviation and include this in the report so that the reader knows the extent of variation.

In the case of chronic conditions, the distribution of duration of sickness may be far from a Gaussian pattern. A longer duration will be more common than a shorter duration. This makes the distribution skewed to the right. If so, the mean may not be a true representative of the central value. The median and 3rd to 97th percentiles are calculated instead. If the disease is severe and end-point is death, as in leukemia and acquired immunodeficiency syndrome (AIDS), the duration of sickness is the same as the duration of survival.

Duration of sickness has more exactitude when it is replaced by duration of restricted activity or duration of disability. In an economic sense, this can be measured in terms of loss of days of normal activity. Average duration of sickness can be computed separately for each disease or for each demographic group, such as age 40–49 years and female sex. The methods described in Chapter 18 for calculating the duration of survival can also be used to analyze duration of morbidity.

11.2.2.1 Prevalence in Relation to Duration of Morbidity

The duration of morbidity does not affect incidence but severely affects prevalence. Prevalence tends to accumulate and becomes higher when the duration is long. Incidence is the inflow; prevalence is the stock. Figure 11.2 illustrates the effect of duration on prevalence when the incidence is the same. The disease on the left side of the figure has shorter duration than the one on the right side. The prevalence in the latter case is higher although the onset in both the cases is the same. For example, on day 6, the point prevalence is 3 cases on the left side and 10 cases on the right side.

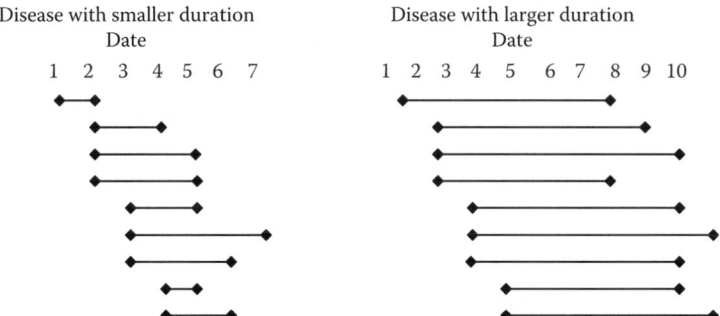

FIGURE 11.2
Effect of duration of morbidity on prevalence.

If treatment reduces the duration of illness, then the prevalence rate would decrease. But if the treatment were such that it prevented death without full recovery, then, paradoxically, the prevalence would increase. On the other hand, if the decrease in duration were sufficiently large, the prevalence would decline even if the incidence increases.

11.2.2.2 Incidence from Prevalence

A follow-up study is relatively more expensive than a cross-sectional survey. Thus, it is easier to obtain the prevalence than the incidence. The duration of sickness can also be generally easily obtained. Because prevalence depends on the incidence and duration, the relationship can be exploited to find the incidence based on prevalence and duration. *If there are no intervening factors*, then

$$\text{Incidence} = \frac{\text{Prevalence}}{\text{Average duration of sickness}}, \tag{11.22}$$

where incidence and duration are in the same time unit. If the annual incidence is to be calculated, then the duration too is to be measured in terms of years. If the average duration of sickness for a particular disease is 15 days, then this is $15/365 = 0.041$ years. If the prevalence rate is 1.2 per 1000 and the duration is 15 days, the incidence rate is $1.2/0.041 = 29$ per 1000 per year. If the prevalence rate is per 1000 persons, then incidence too is a rate per 1000 persons. Incidence can be calculated for specific groups, by age, gender, occupation, and region, by inserting the prevalence and duration for the chosen group in Equation 11.22.

The concept of duration of sickness is applicable to all acute conditions but not so much to chronic conditions. Some not so severe conditions such as hypertension, varicose veins, and lower vision can seldom be fully reversed in a manner that would allow normal life without ongoing treatment.

For such conditions, the duration of disease is anybody's guess. If the condition is more concentrated in the elderly, such as diabetes mellitus, mortality affects the prevalence. Mortality itself may be higher in the group with disease than in the nondiseased group. Such conditions interfere with the parameters of Equation 11.22 and the equation becomes invalid. More elaborate calculations may be required to estimate incidence from the prevalence of such conditions.

There is yet another set of chronic ailments that are reversed in some subjects but not in all subjects. This can happen with any disease but is very common with conditions such as cataract blindness. This can be reversed by surgery, but many affected subjects may not use the option for various reasons. An additional consideration in cataract blindness is that it is a disease of old age, when mortality is high. Also, this is an example of a disease in which the death rate among the diseased is generally higher than among the nondiseased. Equation 11.22 is not applicable for such conditions. Podgor and Leske [8] have given the procedure for estimating incidence from the prevalence of such conditions, although they consider only irreversible diseases.

11.2.2.3 Epidemiologically Consistent Estimates

It is easy to understand that the incidence, prevalence, duration, remission, and case-fatality for any disease are interrelated (Figure 11.3). Also, age at onset, remission, and duration are related to age-specific mortality for that disease. Quite often these are estimated from disparate sources, such as prevalence and age-specific mortality from a cross-sectional survey; incidence from one cohort study; and remission, duration, and case-fatality from another longitudinal study. There is a great likelihood in this case that various rates are not internally consistent. Age-specific mortality may not be the same as expected on the basis of age at onset, duration, and case-fatality, or prevalence may not be the same as expected from the incidence, duration,

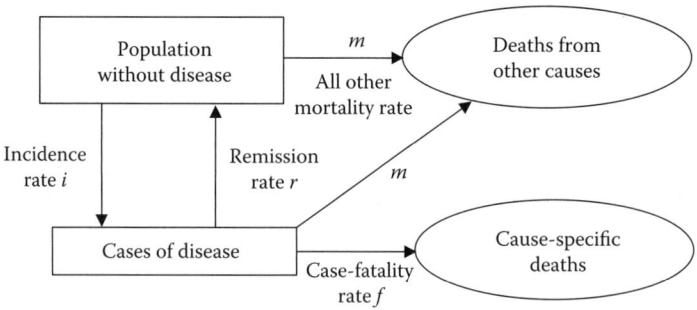

FIGURE 11.3
Interrelations for epidemiological consistency of data. (From Mathers, C.D. et al., *National Burden of Diseases Studies: A Practical Guide*, 2nd edn., Global Programme on Evidence for Health Policy, World Health Organization, Geneva, Switzerland, 2002.)

remission, and mortality estimates. A software package, called DISMOD, is available from the WHO website that can be used to check the internal consistency of these estimates with some limitations. In case they are not internally consistent, more reliable rates should be used to generate consistent estimates of the other rates.

This package can also be used to generate estimates of rates that are not available at all. The generated rate would be epidemiologically consistent but may not be plausible in terms of the knowledge of the experts. In that case, iterations may be needed in terms of reentering a new set of known rates so that the final estimates are not only internally consistent but also plausible.

11.2.3 Morbidity Measures for Acute Conditions

In the case of acute conditions, particularly infections, it is easier to talk in terms of attacks than incidence. The same person can have two or three attacks of diarrhea or of cold in 1 year. Some noncommunicable conditions such as angina and asthma also have the same feature. Thus, the emphasis here is on disease spells rather than on affected persons.

11.2.3.1 Attack Rates

Before measuring attack rates, the following indicator may be useful in some situations:

$$\text{Infection rate} = \frac{\text{Manifest cases} + \text{cases with inapparent infection}}{\text{Exposed subjects}} * 1000. \quad (11.23)$$

Inapparent infection is generally determined by serological examination. The preceding rate is based on the individuals rather than disease spells, is related to a duration of exposure, and measures the number of subjects affected when they are exposed for that duration:

$$\text{Attack rate} = \frac{\text{New spells during a specified time interval}}{\text{Total population at risk during the same interval}} * 100. \quad (11.24)$$

The attack rate is based on disease spells rather than persons. This is generally used when the exposure is for a limited period such as during an epidemic. This can also be calculated per person-year:

$$\text{Secondary attack rate (SAR)} = \frac{\text{New spells within the range of incubation period among those exposed}}{\text{Subjects exposed to the primary cases that can spread the disease}} * 100.$$

$$(11.25)$$

The denominator and the numerator are restricted to the susceptible contacts. This rate is generally used for diseases such as measles and chickenpox that are infective for only a short period. SAR measures the intensity of spread of infection or risk among the susceptible contacts after exposure to an infective case. When the primary case, also called **proband**, is infective for a long period as in tuberculosis, the duration of exposure becomes important. SAR then is computed per 100 person-weeks, person-months, or person-years of exposure.

SAR is a useful measure not only for infectious diseases but also for diseases of unknown etiology such as Hodgkin's disease to find out whether it is communicable. SAR is also useful in evaluating the effectiveness of control measures such as isolation and immunization.

> **Example 11.3: SAR of Type-1 Diabetes in Colorado Families**
>
> Risk of diabetes in nondiabetic siblings of children diagnosed with type-1 diabetes is higher. Steck et al. [9] analyzed the family history of 1586 patients in Colorado with type-1 diabetes diagnosed before 16 years of age and interviewed during 1999–2002. Probands are those who were initially affected and secondary cases are those who appeared later in the family of probands.
>
> SAR by age 20 years in siblings was 4.4%, but it was significantly higher in siblings of probands diagnosed under age 7 years than in those diagnosed later. In the parents too, the SAR by age 40 years was higher when proband was diagnosed under 7 years.
>
> **SIDE NOTE:** The study found that the median age at onset of type-1 diabetes in the probands was 7.1 years and the median duration of diabetes was 3.5 years.

Example 11.3 illustrates the use of the concept of SAR in a very different context. It is not related to the repeated episodes, yet delineates the communicability as explained next. Such usage is not uncommon and it is nice to be familiar with different contexts in which a term can be used.

11.2.3.2 Disease Spectrum

The term "disease spectrum" has at least three usages. This can be illustrated by considering pain as an example. First, one can think of characteristics of pain such as the nature of pain (e.g., throbbing pain), the duration of pain (e.g., intermittent or continuous), the extent of pain (e.g., local or generalized), etc. Second, one can consider the magnitude of pain as mild, moderate, severe, etc., or 7 on a 10-point scale. Prognostic severity, for example, stage of cancer, also comes under this category. A third possible use of disease spectrum can be the description of the progression of infection

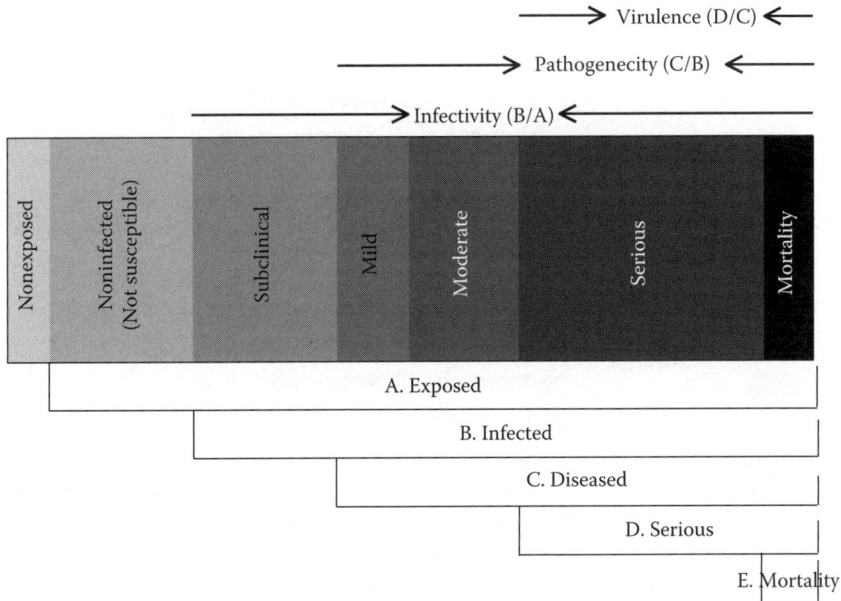

FIGURE 11.4
Disease spectrum.

from susceptibility to virulence. This is illustrated in Figure 11.4. This figure shows as though a large percentage of exposed cases are infected and a substantial portion manifest the disease, but actually these proportions are small in practice.

Susceptibility is proneness to a disease. An effectively immunized person against diphtheria, pertussis, and tetanus (DPT) is not susceptible to these diseases for at least 1 year. A child is not susceptible to MI. Susceptibility is a property of the host. Infectivity is the property of the agent that causes infection. Measles virus is highly infectious for a susceptible person but HIV is not. Infectivity is the actual performance in terms of the percentage infected when exposed. An infection may or may not manifest the disease. The percentage of susceptible persons that get the actual disease after the exposure can be called the pathogenicity of the disease. Out of those diseased, many will have a mild episode and recover easily. The percentage that have the severe form of disease requiring hospitalization and have a risk of death can be called the virulence of the disease. Cholera is highly pathogenic but less virulent, whereas rabies is less pathogenic but more virulent. AIDS is less infective but very pathogenic and highly virulent. At the end of this spectrum is mortality, which is the ultimate for virulence. Mortality can be measured per 1000 population or as percent affected. The latter is called case-fatality, as already defined.

The following indicators can be calculated to delineate the spectrum of a disease:

$$\text{Infectivity} = \frac{\text{Number of subjects infected}}{\text{Number of subjects exposed}} * 100, \qquad (11.26)$$

$$\text{Pathogenicity} = \frac{\text{Number of subjects manifesting the disease}}{\text{Number of subjects infected}} * 100, \quad (11.27)$$

$$\text{Virulence} = \frac{\begin{array}{c}\text{Number of subjects with serious}\\ \text{disease (including mortality)}\end{array}}{\text{Number of subjects with disease}} * 100. \qquad (11.28)$$

The terms pathogenic and virulent are borrowed from the field of infectious diseases but can be used for chronic diseases as well, as long as the meaning is explained. Division of cases into such spectrum can help in choosing treatment strategies and in prognostic assessments.

Disease spectrum is likely to be very different during times of epidemics than in normal times. An **epidemic** is said to have occurred when the incidence is clearly in excess of the usual rate. Although a statistical threshold such as +2SD can be used to assess whether the occurrence is clearly in excess of the usual, practically this is rarely done. If hospitals and health centers find a sudden rise in cases among the patients they attend, an epidemic can be safely presumed. Calling a slight excess as an epidemic and creating a premature alarm is not as bad as being caught unawares for a high occurrence of cases. Thus, even a suspicion of increased incidence should be considered enough for health managers to take preventive steps.

Another related concept is transmissibility. If an infected person is able to infect at least one person on average during the entire period of infectivity, then the infection will sustain or increase. This is called the **reproductive rate of infection**. In countries where hepatitis B infection is on the rise, this rate is more than 1. If reproductive rate is less than 1, expect that the infection will die down or stabilize at a low level.

11.3 Indicators of Social and Mental Health

Social and mental aspects are an integral part of the health spectrum. These are often overlooked or sidetracked because of difficulty in measuring them in a satisfactory manner. However, indirect measurements can be made that serve well as valid surrogates. The common ones in use are described in this section.

11.3.1 Indicators of Social Health

Social health is generally measured in terms of social status comprising education, income, and occupation. These, independently as well as in combination, are believed to have a deep impact on the well-being of the people. Education increases awareness and adds to the productivity of people. Income causes a sense of well-being by allowing access to a wider set of choices. Occupation helps to attain satisfaction of doing something worthwhile for society. These together are sometimes considered **distal measures** of health that affect proximal measures such as diet, smoking, and hygiene. The **proximal measures** affect physiological and pathophysiological processes that lead to disease and infirmity. Genetic factors and environmental factors such as pollution and water supply contribute to this process.

11.3.1.1 Education

At least four different types of indicators are used to assess the level of education in a community. For developing countries, where a substantial segment of the population is illiterate, the literacy rate can be a useful indicator. This is the percentage of literate among people of age 6 years and above. Children less than 6 years are supposed to be neither literate nor illiterate and thus are excluded from calculation. This gives

$$\text{Literacy rate} = \frac{\text{Literates among population of age 6 years and above}}{\text{Total population of age 6 years and above}} * 100. \quad (11.29)$$

When calculated for age 15 years and above, this is called the **adult literacy rate**. In the year 2008, the adult literacy rate was 100% in DPR Korea and 32.7% in Chad. This rate is showing rapid rise all across the world.

The second indicator of the level of education is the average years of schooling in adults:

$$\text{Average years of schooling} = \frac{\begin{array}{c}\text{Sum total of schooling of all}\\ \text{individuals in a population (25+ years)}\end{array}}{\text{Total population (25+ years)}}. \quad (11.30)$$

For maintaining its validity, the age is restricted to 25 years and above for the numerator as well as for the denominator. It is presumed that all schooling will be over in practically all individuals in the population of this age. The years of schooling for each individual in a community are rarely available and thus this average is difficult to compute. As a middle path, the third indicator is

$$\text{Percentage of adults with high school diploma}$$
$$= \frac{\text{Adults with high school diploma}}{\text{Total adults}} * 100. \qquad (11.31)$$

Adults can be defined according to the local conditions. The information on the addition of high school diploma holders each year can be obtained from schools, boards, or councils awarding such a diploma. But keeping track of exits, by way of migration or death, can be difficult. To overcome this problem, the fourth indicator used is

$$\text{Net enrollment ratio}$$
$$= \frac{\text{Children of age (6 – 16) years enrolled in a school}}{\text{Total children of age (6 – 16) years in the population}} * 100. \qquad (11.32)$$

This measures the extent to which children are using educational facilities. In place of 6–16 years, any other age-group can be used. The enrollment figures can be easily obtained from schools, and the age structure of the population is generally available from other sources. Thus, this indicator can be easily computed.

Net enrollment ratio can be computed separately for primary, secondary, and tertiary schools by adjusting the age accordingly. There might be children in schools who are less than 6 years and more than 16 years. When the numerator of Equation 11.32 is relaxed to include such children also, it is called the **gross enrollment ratio**. Since the age-group in the denominator cannot be relaxed, the gross enrollment ratio can exceed 100 as occurring in Australia for primary, secondary, and tertiary levels of education combined [10].

11.3.1.2 Income

The most widely used measure of the level of income of a community is the per capita gross domestic product (GDP), popularly called per capita income. Almost all countries compute this for the nation as a whole and usually for each of their states separately. Further subdivisions may not be easily available.

Income per se in the national currencies is seldom comparable across countries because of differential purchasing power. A Singapore dollar cannot buy as much in Singapore as the Yuan equivalent can in China. The World Bank has done considerable work in this sphere and obtained incomes of various countries in terms of **purchasing power parity in**

international dollars (PPPInt.$). Thus, an average per capita income of nearly 44,000 rupees per annum in 2010 in India was equivalent to Int.$3340. This is for international comparison. While comparing with the United States, an Indian rupee may be worth 10 U.S. cents in terms of purchasing power. Thus, the average per capita income in India in 2010 is equivalent to U.S.$ 4400 in the United States. Note, however, that income is mostly a family trait rather than of an individual because children are not supposed to earn. If the average size of a family is four, the average family income is four times the per capita income.

An indirect way to assess the economic status of a community is by the presence of amenities such as telephones, televisions, computers, and automobiles per family or per 1000 population.

11.3.1.3 Occupation

Ideally speaking, all jobs have dignity and deserve full respect. Practically, though, professionals such as attorneys, doctors, chartered accountants, and consultants are considered to enjoy better social health. Industrial laborers and coal mine workers do not generally enjoy the same status. Thus, there is an inbuilt hierarchy in occupation in most societies. Except for the percentage in different occupations, no worthy measure is available to measure occupational distribution.

11.3.1.4 Socioeconomic Status

Education, income, and occupation can be combined to devise a system of classification of people into various socioeconomic classes.

Table 11.3 suggests a scoring system for the three components and a classification based on the aggregate score. The categories of schooling years and of occupation are absolute in this table and can be probably used anywhere in the world without alteration. But the income level in one country can seldom be compared with that in another country except possibly in terms of the purchasing power parity just mentioned. But this is far too complex to calculate. Thus, a classification of income is on the basis of percentile, which will be specific to the area. There will always be 20% of the population below the 20th percentile and another 20% between the 20th and 40th percentiles, etc. Yet, the classification may be valid for comparing one subpopulation within a country to another subpopulation and to some extent for an international comparison also. Such percentiles are generally available from the distribution of per capita income of the nation.

A large number of common occupations are listed in Table 11.3 but certainly not all. An indication of the score for other occupations not included in this table should be available from the pattern.

The minimum possible aggregate is 0 and the maximum is 15. I suggest a classification as given in the bottom of Table 11.3. The percentage of the

TABLE 11.3

Scoring for Social Classification

Score	Years of Schooling	Income[a]	Occupation
0	Nil (illiterate)	Below poverty line[b]	Unproductive or burden on society (e.g., begging), including unemployed
1	<5	<20th percentile[c]	Unskilled labor
2	5–10	20th–40th percentile	Skilled labor, artisan, small business, student, small farmer, and soldier
3	10–15 including vocational	40th–60th percentile	Clerk, medium business, medium farmer, technician, and salesperson
4	15+ but nonprofessional nontechnical	60th–80th percentile	Teacher, researcher, industrialist, big farmer, big business, government officer, and manager
5	15+ some of which is professional or technical	≥80th percentile	Executive, doctor, attorney, consultant, and engineer

Aggregate Score	Social Class
0–3	V
4–6	IV
7–9	III
10–12	II
13–15	I

[a] In case of a family, calculate per capita income in the family.
[b] Income required to purchase low-cost balanced food to provide 2400 calories.
[c] Excludes those that are below poverty line.

population in different classes would give an indication of the social health of people. Many diseases and other health conditions have been observed to be associated with such social classification of the subjects.

11.3.1.5 Dependency Ratio

A large family living together under one roof with many children is not considered conducive to a healthy environment. Thus, the average size of a family in a population can be considered an inverse indicator of social health. Perhaps, a more sensitive indicator is

Dependency ratio

$$= \frac{\text{Population of age} < 15\,\text{years} + \text{population of age} \geq 65\,\text{years}}{\text{Population of age } 15 - 64\,\text{years}} * 100. \quad (11.33)$$

The child and geriatric populations are considered to be nonworking and dependent on the working-age population. The dependence is not necessarily economic but can be social and psychological. To be exact, the formula given in Equation 11.33 gives the **total dependency ratio**. This can be divided into child dependency and old-age dependency ratios. The total dependency ratio in the year 2004 was 80 in Laos and only 41 in China. It is generally believed that the higher this ratio, the less is the social health. This may be economically true but disregards the wisdom of the elderly as well as the contribution of children's presence to the well-being of the family.

Dependency ratio has obvious economic consequences. But what is not generally appreciated by the people, although the experts realize, is the **demographic bonus** brought about by the transition as a population moves from the high fertility/high child mortality stage to the low fertility/low child mortality stage. A phase comes in this transition when the percentage of adult population swells and the dependency ratio substantially declines. When all such adult population is put to productive work, the economic progress can be fast because they have few people to support, reflecting tremendous improvement in overall health. This is what turned around some countries and is now known as the East Asia miracle.

11.3.1.6 Dietary Indices

Generally, three methods are used to assess nutrient intake. First is multiple 24 h recall, second is 1 week diet record, and the third is food frequency questionnaire (FFQ). FFQ typically asks about diet over the past month, generally in terms of grocery purchased and consumed. This may not be able to provide accurate information on occasional intakes such as in parties and restaurants. When properly adjusted, this method may be epidemiologically more relevant as it provides a long-term perspective of effect that a diet is most likely to have on health parameters.

Some biomarkers such as urinary and plasma measurements are more sensitive to recent intake and their correlation with diet may be exaggerated when assessed with 1 week record. If 1 week record is used, the biomarkers should be assessed remote in time. Natural week-to-week variation in 1 week dietary record lowers the validity of this method. This is true for 24 h recall also. Even if done several times, 24 h recall cannot provide information on dietary intake in the long term. Most dietary studies have interest in food intake also in addition to nutrients. But 24 h recall generally provides more accurate information as the recall lapse is minimal.

You can see that one method has advantage in one setting and another in another setting. Some associations may be better detected by one method than the other. Choose the method that meets your objective most appropriately.

11.3.1.7 Health Inequality

Health inequality among different segments of population has received attention as it affects the rate of improvement in population health. Inequalities definitely work to the disadvantage of the deprived sections but affect all as disparities hamper the generation of human capital and inhibit sustainable improvement. While inequalities are inherent in social fabric, health inequality caused by factors amenable to human intervention is considered unjust.

Level of health in top and bottom income quintiles is a customary measure of health inequality. Ratio of maximum to minimum can be used but this ratio disregards the dispersion. Most popular is **Gini coefficient**, which was originally designed for income disparities:

$$\text{Gini coefficient: } G = \frac{\Sigma_i \Sigma_j |x_i - x_j|}{2n^2 \bar{x}}; \quad i, j = 1, 2, \ldots, n. \qquad (11.34)$$

Note that Gini coefficient is applicable to positive values only. If negative values are present and mean is close to 0, the Gini coefficient may blow up to an unacceptable level. Note also that Gini coefficient is similar to coefficient of variation. Very different distributions can give same value of Gini.

The value of Gini coefficient lies between 0 and 1. A value less than 0.2 can be considered tolerable, between 0.2 and 0.4 middling type, and more than 0.4 high.

Example 11.4: Inequality in Infant Mortality Rates

IMRs in different income quintiles in a developing country are given in Table 11.4. For these data,

$$\text{Gini coefficient} = \frac{\left[|73 - 68| + |73 - 51| + \cdots + |73 - 16| + |68 - 73| + \cdots + |16 - 32|\right]}{(2 \times 5^2 \times 48)}$$

$$= 0.25.$$

TABLE 11.4

IMRs in Income Quintiles in a Developing Country

Quintile	IMR
Lowest quintile	73
Second quintile	68
Third quintile	51
Fourth quintile	32
Top quintile	16
Average ($\bar{x}$)	48

SIDE NOTE: Low value 0.25 in this example despite a ratio of more than 1:4 between IMR in minimum and maximum quintile shows that Gini coefficient is not too sensitive to the inequalities. Now you know why an apparently low value such as 0.4 is considered high. You will soon notice that for other statistical measures, such as correlation, 0.4 is not considered high.

11.3.2 Indicators of Health Resources

Availability of health resources is considered an important determinant of the health status of a community. Among various items comprising resources, the following two groups are important.

11.3.2.1 Health Infrastructure

Availability of hospitals, health centers, dispensaries, and doctors, on the one hand, and availability of food, safe drinking water, and sanitation facilities, on the other, together form the health infrastructure. The second component of this infrastructure (food, safe water, etc.) is not important for developed countries because nearly 100% of the population in those countries have these facilities. But it is very important for developing countries. The following indicators can be used to measure the extent of availability of the health infrastructure:

$$\text{Population served per bed} = \frac{\text{Population}}{\text{Number of beds}}. \qquad (11.35)$$

This includes beds in hospitals, nursing homes, health centers, etc. The population served per bed was 625 in Mexico in the year 2009 and 294 in Canada:

$$\text{Population served per doctor} = \frac{\text{Population}}{\text{Number of medical doctors}}. \qquad (11.36)$$

The denominator of the formula given in Equation 11.36 includes doctors of all systems—allopathy, homeopathy, naturopathy, Ayurvedic, Unani, etc.—when those exist in the society. However, all these should be qualified, registered, and licensed doctors according to the law of the land. Quacks are excluded. The population served per doctor in the year 2010 varied from an average of 300 persons in industrial countries to an average of 2500 in developing countries. The formulas remain the same for other personnel such as nurses, pharmacists, and health workers. These can be inverted to find the **density** per 1000 population:

Per capita availability of food grains per day

$$= \frac{(\text{Total production of food grains}) - (\text{Waste}) - (\text{Exports}) + (\text{Imports}) \text{ in a year}}{\text{Population} \times 365}.$$

$$(11.37)$$

Children do not eat as much as adults but this factor is generally ignored. In the context of calories, this factor is taken care of by converting to **consumption units**. An adult male with average physical activities is considered one unit. The units for the others are as follows: female adults, 0.9; lactating and antenatal woman, 1.1; infant, 0.0; child 1–3 years, 0.4; 3–5 years, 0.5; 5–7 years, 0.6; 7–9 years, 0.7; 9–12 years, 0.8; and 12+ years, 1.0. The age-sex structure of the population in an area is generally available and can be easily converted to consumption units. The calorie consumption per capita (not converted to consumption units) in the year 2010 was 3200 in New Zealand but only 2200 in Bangladesh.

Health services in a developing country are considered accessible when they can be reached by the public on foot or by local means of transport in no more than an hour. **Water supply** is considered safe when it is treated surface water or untreated but uncontaminated water such as from springs, sanitary wells, and protected boreholes. Excreta disposal (**sanitation**) facilities are considered adequate if they can effectively prevent human, animal, and insect contact with excreta. The percentage of the population with such access is an important indicator of health infrastructure. Among other indicators that could be considered for infrastructure is expenditure on health.

11.3.2.2 Health Expenditure

A general belief is that health services can be purchased just like any other commodity. Thus, total expenditure on health is considered a good indicator of health resources. At the national and state levels, this includes the government expenditure as well as the private expenditure. Some specific indicators used for health accounting are as follows:

1. Total expenditure on health as percent of the GDP.

   In the year 2007, the United States spent 15.7% of its GDP on health, which is just about the highest in the world. Australia spent 8.9% and Congo only 2.4%. In a way, this indicates the priority health receives against other items of government expenditure such as military and education, and of private expenditure such as food, clothing and transport.

   The total expenditure can be divided into government and private expenditure, and private into out of pocket and others. There is a debate whether higher percentage out-of-pocket expenditure is good for the country or a lower percentage is better. Perhaps it depends on local conditions.

   The second kind of accounting is done on per capita basis.

2. Per capita total expenditure on health and per capita government expenditure on health.

   For international comparison, these are computed at international dollar rates, which factor the differential pricing in different countries.

In the year 2007, Norway spent 4005 international dollars on health per capita and the United States spent 3317. On the other extreme are countries such as Afghanistan and Madagascar who spent less than 30 international dollars per capita. Since these numbers are in international dollars, they can be compared and it would be safe to conclude that the United States spends more than 100 times per capita on health as some underdeveloped countries do. The main reason is the availability of funds but the other reason is the priority accorded to health.

11.3.3 Indicators of Lack of Mental Health

It is difficult to separate mental health from social and physical health. They affect one another. Yet, for the purpose of separate study of these components, a new set of indicators can be proposed to measure mental health (or rather its lack). This again should be considered surrogate because mental health per se is abstract. Mental health is the sense of well-being rather than the well-being itself. There are people who feel very contented even without full food, adequate clothing, and adequate housing, perhaps even when a limb is lost. There are saints in the Himalayas who are bereft of basic necessities but still seem to be in perfect mental (and physical) health. Widely acceptable indicators are not available for measuring mental health, but the following surrogates can be suggested for measuring its lack.

11.3.3.1 Smoking and Other Addictions

A person may start smoking as a way to derive pleasure, but smoking is widely perceived to indicate a poor state of mental health. Its adverse effects on physical health are well known.

Smoking among youngsters is considered even worse. The percentage of smokers among teenagers can be used as a measure of lack of mental health in a community. Where cigarettes are just about the only mode of smoking, the average number of cigarettes consumed per person per year could be an indicator. In a country like Bangladesh, where the mode is multiple—cigarettes, bidi, hukka—a more comprehensive indicator may be required. Many sections of the population in several countries also consume tobacco orally.

Prevalence of smoking is more in Europe than in the United States. In American adults, cigarette smoking was 37% in the year 1970 and declined to nearly 23% in the year 2002 [11]. It was 21% in the year 2009.

The percentage of adults addicted to drug abuse or alcohol may also serve as an indicator of lack of mental health.

11.3.3.2 *Divorces*

A divorce generally indicates that the choice of spouse was not made with sufficient care. It also often causes trauma to the family, especially to the couple and the children, if any. Thus, the divorce rate per 1000 persons or per 1000 couples can be used as an indicator of lack of mental health. Arguments can certainly be raised against this as well as all other indicators. Some such arguments are discussed at the end of this section.

11.3.3.3 Vehicular Accidents and Crimes

Vehicular accidents are mostly due to lack of mental concentration while driving. Number of accidents per 1000 population or per 1000 vehicles and injuries or deaths due to vehicular accidents seems to be a promising indicator of lack of mental health of a community.

Suicides, homicides, and rapes, per million population per year, may be among the least controversial indicators of mental health. The data on these crimes are relatively more complete in most countries of the world because these form part of the police records. According to a UNDP report, the population victimized of crimes was nearly 8% in Baku (Azerbaijan) and 41% in Kampala (Uganda) in the late 1990s.

11.3.3.4 Other Measures of Lack of Mental Health

Pre- and extramarital sex, lack of family support, perception of inadequate well-being of handicapped persons, stress, dissatisfaction (self-perception), unavailability of leisure time, etc., are among indicators that are also candidates to be considered for measuring lack of mental health. Note the following for indicators of mental health:

1. The preceding list of indicators of social and mental health is by no means comprehensive. For example, indicators of environment and pollution, urbanization, population density, etc., are not included. Depending on the need, suitable indicators can be developed to measure any particular aspect.
2. The utility of an indicator depends on its validity and on availability of the information required to compute the indicator. It should also be readily interpretable and should indicate the action needed to improve the health status.
3. Whereas morbidity and mortality are widely accepted indicators of lack of physical health, the indicators of presence or lack of mental and social health are still in the developmental stage. Advancing arguments in terms of the local situation can negate the role of education and income and of smoking, divorces, and accidents. For example, one might say that divorce is a happy ending of an unsuccessful marriage and thus indicates a better state of well-being compared with a couple living together in constant conjugal stress. Similarly, it can be argued that an infant death is better than lifelong

suffering due to hunger and poverty or that early death is not bad as long as the life lived is productive. After all, the great Indian mathematician Ramanujan lived for just 36 years. It is a question of perception. The listed indicators are those that may be acceptable in most parts of the world.

4. Health is defined as physical, mental, and social well-being, but the latter two components are generally ignored because they are difficult to measure and hotly debated. However, the assessment of health is grossly incomplete without the mental and social aspects. Validity cannot be sacrificed for expediency. If the importance of social and mental components is realized, efforts to develop and standardize relevant indicators will automatically follow.

5. One of the desiderata of a good indicator is that it should be action oriented. Addictions (including smoking), violence and crime, pre- and extramarital sex, and divorces are areas where education and awareness can help to improve the mental health of the population.

6. The WHO has now added a spiritual component to the definition of health. This aspect is even more difficult to measure.

11.4 Composite Indexes of Health

Indicators enumerated in the preceding sections measure specific aspects of health but not health in its holistic sense. Efforts are made periodically to combine various indicators and come up with a comprehensive measure. An index can be defined as a combination of two or more indicators, stated relative to a base or a standard. No index is yet available that measures health in its entirety. Since health in its comprehensive form is also understood as a state of well-being, it can encompass diverse aspects such as income, satisfaction, social relations, and cultural activities. Office of National Statistics of United Kingdom is trying to develop a measure of national well-being. A debate is going on regarding what should this contain. For example, this may have indexes that measure how people feel in control of themselves, make choices, and have a sense of purpose and belonging. The following are two categories of some of the currently more popular indexes of community health.

11.4.1 Indexes of Status of Comprehensive Health

Described first are the positive indicators that measure health of a society in its comprehensive form. Composite indexes of lack of health are described in the next section.

11.4.1.1 Human Development Index

Human development is considered to be an important aspect of well-being and thus of health. UNDP each year computes a human development index (HDI) for each country by using a common base so that they can be realistically compared. This is calculated by combining three components of development, viz., education, income, and life expectancy. The methodology is periodically reviewed and revised. The methodology contained in their 2010 report requires the following:

1. Per capita income in real terms by converting it to PPP Int$: Complex calculations are done for this conversion. As explained earlier, this is needed to restore parity in income in different countries with different price levels.
2. Life expectancy at birth (ELB).
3. Mean years of schooling.
4. Expected years of schooling: This is the number of years of schooling that a child of school entrance age can expect to receive if prevailing patterns of age-specific enrolment rates were to stay the same throughout the child's life.

Each of these four components is converted to an index relative to the observed minimum and maximum value. These are known as goal posts. For per capita income, the minimum estimated by UNDP is PPP Int.$163 and the maximum is PPP Int.$ 108,211, and the logarithm is taken. For ELB, these goal posts are 20 and 83.2 years, and for mean year of schooling and expected years of schooling these are 13.2 and 20.6 years, respectively. Education index is computed by taking geometric mean of these two indexes using a maximum threshold of 0.951. The HDI is the geometric mean of income index, life expectancy index, and education index. Example 11.5 taken from Human Development Report 2010 illustrates the calculations.

Example 11.5: HDI of China for the Year 2010

Per capital income (PPP Int.$) = 7263

ELB (years) = 73.5

Mean years of schooling (years) = 7.5

Expected years of schooling (years) = 11.4

$$\text{Mean years of schooling index} = \frac{7.5 - 0}{13.2 - 0} = 0.568,$$

$$\text{Expected years of schooling index} = \frac{11.4 - 0}{20.6 - 0} = 0.553.$$

Thus,

$$\text{Education index} = \frac{\sqrt{0.568 \times 0.553} - 0}{0.951 - 0} = 0.589,$$

$$\text{Income index} = \frac{\ln(7263) - \ln(163)}{\ln(108,211) - \ln(163)} = 0.584,$$

$$\text{Life expectancy index} = \frac{73.5 - 20}{83.2 - 20} = 0.847.$$

Therefore,

$$\text{HDI} = \sqrt[3]{0.589 \times 0.584 \times 0.847} = 0.663.$$

Indrayan et al. [12] proposed that the index of components of HDI henceforth be computed on the basis of percentage of population reaching a minimum threshold. For income, this threshold could be the minimum required for adequate food, house, and clothing (say PPP Int.$1000 per capita); for ELB, percentage of persons of age 60 years and above; and for education, the percentage of adults (18 years and above) with at least high school education. This modification will make the index and its components readily interpretable as it would indicate how far the country has been able to meet the minimum level.

11.4.1.2 Physical Quality of Life Index

At the individual level, quality of life is mostly the subjective component of well-being. It can be defined as a composite measure of physical, mental, and social well-being as perceived by each individual or by a group of individuals. It includes aspects such as happiness and satisfaction as experienced in life, for example, in health, marriage, finances, education, and creativity. The quality of life can be evaluated by assessing a person's subjective feeling of happiness or unhappiness about various concerns of life. At the individual level, attempts are made to reach one composite index by combining a number of health indicators.

The physical quality of life index (PQLI) at the community level consolidates three indicators: infant mortality, life expectancy at age 1 year, and literacy. For each of these three components, the performance of individual countries or communities is placed on a scale of 0–100, where 0 represents an absolutely defined worst performance and 100 represents an absolutely defined best performance. This is similar to the one explained for HDI. The composite index is calculated by averaging these three indicators, giving equal weight to each of them. The resulting PQLI thus is also scaled 0–100.

11.4.2 Indexes of Health Gap

Health gap can be easily understood as the difference between ideal health and the achieved health. It defines the need and delineates the shortfalls. Among many indices that measure health gap, the ones that look important for the society are burden of disease measured by the disability adjusted life years (DALYs) lost, human poverty index (HPI), and the index of need for health resources.

11.4.2.1 DALYs Lost

Ideal health is that everybody lives for 100 years (even more!) without falling sick for a single day. Practically, this does not look feasible by any stretch of imagination. Perhaps, the next best hope is that the life expectancy becomes as much as the highest actually seen in a population, and this is attained without being sick. This is the basic premise on which the concept of DALYs lost is based.

Japan has the highest life expectancy in the world. Life tables for males and females were prepared by WHO to correspond fairly well with the Japanese experience. Any death contributes to life-years lost according to the remaining life expected as per their life table. Thus, a death even at the age of 90 years means 4 years lost because this is the expectancy at age 90 years in the Japanese model. Deaths of people at different ages add up to, what is called, the **years of life lost (YLLs)**, with a modification as shortly described.

The second component of DALYs is the disability arising from illnesses and impairments. Disease severity is assigned **disability weight** that, in turn, is converted to equivalent years in full health lost, by using a concept such as time trade-off. For details, see WHO's National Burden of Disease Manual [13]. Essentially this means that 1 year of paraplegia counts several times more than 1 year of suffering from hypertension. Death is given weight 1.0, and a state of complete restriction of movements (bed ridden) but no other restrictions (as in the case of fracture) can be given weight 0.6. Lifetime duration of diseases together in terms of equivalent years forms, what is called, **years lost due to disabilities (YLDs)**.

The sum of YLLs and YLDs is called DALYs lost. This is considered a comprehensive measure of health gap or burden of disease. A big advantage of this measure is that it can be calculated separately for each disease or adverse health condition. See Chapter 18 on survival for additional information. WHO has computed DALYs lost for nearly 130 health conditions.

The calculation of DALYs as done by WHO values life around age 25 years much more than at the childhood or old age, that is, nearly 1.4 compared with 1.0 at the age of 10 and 55 years, respectively, and 0.4 at the age of 90 years. Future years lost are discounted for equivalence with the current year.

The concept of DALYs is criticized first because the unavailable information is substantial that has to be estimated by modeling, expert panels, and such other presumptive methods; second because it attaches debatable values to qualities such as severity of disability and age at affliction; and third because it has been linked to resource allocation exercise. Nevertheless, it remains an important measure of burden of disease in a population. It should not be used as a summary measure of comprehensive health because social and many aspects of mental health are not included in this measure.

According to 2004 estimates, an average of 237 DALYs were lost in the world per 1000 population. This implies nearly a quarter of life in full health is lost due to early mortality and various diseases during lifetime. YLLs contributed nearly two-thirds of this loss and YLDs nearly one-third.

If the world average is considered, nearly 48% of DALYs were lost in the year 2004 because of noncommunicable diseases such as neuropsychiatric conditions, cardiovascular disease, and malignancy; nearly 40% because of communicable diseases, nutritional, and perinatal conditions; and nearly 12% because of injuries. The spectrum widely differs from country to country.

11.4.2.2 Human Poverty Index

Whereas HDI is an index of development, HPI is an index of deprivation. This is based on negative indicators such as probability of not surviving to a particular age, illiteracy, and unemployment. The contents of HPI are separate for developing countries (HPI-1) than for developed countries (HPI-2). The formulas are as follows:

$$\text{HPI-1} = \left[\frac{P_1^3 + P_2^3 + P_3^3}{3} \right]^{1/3}, \tag{11.38}$$

where
P_1 is the probability at birth of not surviving to age 40 years (times 100)
P_2 is the adult illiteracy rate (%)
P_3 is the unweighted average of population without access to an improved water source and children underweight for age (%)

$$\text{HP1-2} = \left[\frac{P_1^3 + P_2^3 + P_3^3 + P_4^3}{4} \right]^{1/3}, \tag{11.39}$$

where
P_1 is the probability at birth of not surviving to age 60 years (times 100)
P_2 is the percentage of adults lacking functional literacy skills
P_3 is the percentage of population below income poverty line
P_4 is the rate of long-term unemployment (lasting 12 months or more) (%)

The purpose of cubing the Ps is to deliberately deny overload if any deprivation as measured by the corresponding P is high. All Ps are less than 1 and raising to the third power attenuates its weight.

For the year 2004, HPI-1 was 32.5 for Namibia and 38.7 for Nepal, and HPI-2 for Australia was 12.8. The lower the HPI, the lower is deprivation, and better for the society.

11.4.2.3 Index of Need for Health Resources

Chandra Sekhar et al. [14] investigated more than 40 indicators of different aspects of lack of health for states in India and tried to obtain each as a linear combination of few underlying factors that influence these 40 odd indicators. After factor analysis (Chapter 19), two factors were identified that together accounted for nearly 80% variation among the indicators. These are (1) socioeconomic background and (2) proximate determinants of health. These two factors were combined in proportion to their contribution to the total variation. This gives an index that can be used to assess the need for health resources. The name arises because the indicators used to construct the index are mortality, fertility, accidents, crimes, etc., which measure negative aspects of health. The positive indicators such as literacy and per capita income were assigned negative weights. A high value of the index indicated greater lack of health, and thus more need for health resources in the concerned states.

In the context of a community, the need for resources is also related to the size of the population. The preceding index measures need per person. There are countries having a population less than one-tenth of a million and countries with population more than 1000 million. Countries with larger population obviously need greater resources. Per capita need for resources can be considered to decline marginally as the population increases. It can be subjectively postulated that a country with the largest population needs one-tenth the resources per head of a country with the smallest population. The population of countries that fall in between can be accordingly moderated to assess health needs.

References

1. Hoyert DL, Kung H-C, Smith BL. Deaths: Preliminary data for 2003. *Natl Vital Stat Rep* 2005; 53 (15):7.
2. Ahmad OB, Boschi-Pinto C, Lopez AD, Murray CJL, Lozano R, Inoue M. Age standardization of rates: A new WHO standard. GPE Discussion Paper Series: No. 31. Geneva, Switzerland: World Health Organization, 2001.
3. Trends in maternal mortality: 1990 to 2008: Estimates developed by WHO, UNICEF, UNFPA and the World Bank. Geneva, Switzerland: World Health Organization, 2010.

4. Indrayan A. Can I choose the cause of my death? *Br Med J* 2001; 322:1003.
5. Hook EB, Regal RR. Capture–recapture methods in epidemiology: Methods and limitations. *Epidemiol Rev* 1995; 17:243–264.
6. McCarty DJ, Tull ES, Moy CS, Kwoh CK, LaPorte RE. Ascertainment corrected rates: Applications of capture–recapture methods. *Int J Epidemiol* 1993; 22:559–565.
7. Laporte RE, Tull ES, McCarty D. Monitoring the incidence of myocardial infarctions: Applications of capture–mark–recapture technology. *Int J Epidemiol* 1992; 21:258–262.
8. Podgor MJ, Leske MC. Estimating incidence from age-specific prevalence for irreversible diseases with differential mortality. *Stat Med* 1986; 5:573–578.
9. Steck AK, Barriaga KJ, Emery LM, Fiallo-Scharer RU, Gottlieb PA, Rewers MJ. Secondary attack rate of type 1 diabetes in Colorado families. *Diabetes Care* 2005; 28:296–300.
10. UNDP. *Human Development Report 2007/2008*. New York: UNDP, 2007.
11. Mendez D, Warner KE. Adult cigarette smoking prevalence: Decline as expected (not as desired). *Am J Public Health* 2004; 94:251–252.
12. Indrayan A, Wysocki MJ, Chawla A, Kumar R, Singh N. 3-Decade trend in human development index in India and its major states. *Social Indicators Res* 1999; 96:91–120.
13. Mathers CD, Vas T, Lopez AD, Salomon J, Ezzati M (Eds.). *National Burden of Diseases Studies: A Practical Guide*, 2nd edn. Global Programme on Evidence for Health Policy. Geneva, Switzerland: World Health Organization, 2002.
14. Chandra Sekhar C, Indrayan A, Gupta SM. Development of an index of need for health resources for Indian states using factor analysis. *Int J Epidemiol* 1991; 20:246–250.

12

Confidence Intervals, Principles of Tests of Significance, and Sample Size

Medicine is an empirical science, and decisions are almost invariably based on the experience gained on a sample of subjects. Different samples contain different individuals. Since individuals differ from one another, it is natural that the samples also vary from one another. For example, it is possible in a small therapeutic trial of a new regimen that six patients respond in a first sample of 10, three in a second sample, and seven in a third sample of the same size. Similarly, mean reduction in cholesterol level in a first sample of 15 hypertensive–hypercholesterolemic subjects may be 30 mg/dL, in a second sample 24 mg/dL, and in a third sample 25 mg/dL after the same therapy. Different samples tend to give different results, and this generates considerable uncertainty. One of the main objectives of statistical methods is to study these sampling fluctuations and develop strategies to draw valid conclusions for the target population on the basis of just one sample.

This chapter: Fortunately, statistical methods allow us to draw valid conclusions by assessing the expected magnitude of intersample variability on the basis of just one sample. If this variability is high, greater caution is required in drawing conclusions. A way to study the intersample variability is to find the sampling distribution. This is discussed in Section 12.1.

The objective of sampling is to provide as accurate a picture of the target population as possible with minimal efforts. The inference drawn on the basis of the samples is deemed to be applicable to the target population, especially to statistical parameters such as proportion π and mean μ. Two kinds of inference are generally drawn on these parameters. (1) The sample suggests that a plausible value of a parameter such as π or μ is in the range (a, b). The values of a and b can be determined on the basis of the sample and (a, b) is called the confidence interval (CI). This is described in Section 12.2 for various situations. (2) The information obtained from the sample is used to find whether the population parameter can be a specified value or whether two or more different samples can be considered to have come from the same parent population. This is called the test of statistical significance. The *principles* of such tests are the subject matter of Section 12.3. These two are the main components of statistical inference. The first is referred to as estimation and the second as testing of hypothesis. Most of the methods for the first are described in this chapter, whereas the methods for the second spill into later chapters.

Gaussian pattern is an important prerequisite for many statistical procedures. Graphical methods for assessing Gaussianity are presented in an earlier chapter. Mathematical methods for this are presented in Section 12.4.

CIs have some features that are similar to the tests of significance. Also, statistical significance does not necessarily indicate medical significance. Such issues are addressed in Section 12.5. Sample size determination is intimately dependent on the CI and the test of significance. Some aspects of sample size determination are presented in Section 12.6.

12.1 Sampling Distributions

Samples by themselves are a great source of uncertainty. Yet, sampling is considered a preferred strategy in most situations because of its overwhelming advantages enumerated earlier. Statistical methods help in arriving at a conclusion regarding a population parameter on the basis of just one sample. The basic tool used for this purpose is the sampling distribution.

12.1.1 Basic Concepts

Before discussing sampling distributions, understanding of some basic concepts is necessary.

12.1.1.1 Sampling Error

Sampling error is a term used for variation across samples when repeated samples are taken. The result obtained on the basis of one sample, in all likelihood, will not be the same as that obtained on the basis of another sample from the same population. Thus, *sampling error actually is not an error.* This is just a variation and better understood as **sampling fluctuation**. This is endogenous to the investigation. On the other hand, nonsampling error is indeed an error that can arise from misreporting, misjudgment, misrecording, nonresponse, etc. This is exogenous in nature. As you might have already observed, this book primarily addresses sampling errors, which are core to statistical methods, but nonsampling errors are not ignored either and they also are discussed from time to time.

12.1.1.2 Point Estimate

Depending upon whether the variable is qualitative or quantitative, the basic parameter generally of interest is proportion (or probability) π or mean μ, respectively. In case of anemia, the interest could be in either the proportion π (or percentage) of subjects responding to iron–folic acid supplementation or the mean μ rise in hemoglobin (Hb) or hematocrit (Hct) level. In a sample of 200 subjects, if the proportion responding is $p = 0.30$ (or 30%) or the mean rise in Hb level is $\bar{x} = 0.8\,g/dL$ after supplementation for 100 days, then it is generally expected that the entire target population would also show a similar picture, provided the sample is adequately representative. The sample values p and $\bar{x}$ are considered to be **point estimates** of π and μ, respectively. Such sample values are also called statistics where the usage is in the plural sense. When a large number of samples are actually available, the mean of ps over such samples will approach π and mean of $\bar{x}$s will approach μ. Much of the uncertainty generated by the sampling error can be satisfactorily resolved by studying the behavior of the statistics such as p and $\bar{x}$ from sample to sample. Their intersample variability is measured by their respective standard error (SE).

12.1.1.3 Standard Error of p and $\bar{x}$

As in the case of individuals, the variability in p and $\bar{x}$ from sample to sample is measured in terms of their standard deviation (SD), but it is now called SE. Thus, SE is the measure of variability of estimates from sample to sample. This special term helps to distinguish interindividual variability (measured by SD) from intersample variability (measured by SE). The SE of p measures the variability in proportion p from sample to sample, and the SE of $\bar{x}$ measures the variability in mean $\bar{x}$ from sample to sample. The larger the SE, the greater is the variability and the lesser is the confidence in the sample results. Thus, inverse of SE is also a measure of reliability. You can intuitively realize that in the sample, some values are small and some large so that the sample variance is relatively large. For sample mean though, large values average out with small values and mean tends to be relatively stable across samples—more so as n increases. Thus, SE is always less than SD.

The SE is calculated on the basis of all possible samples of particular size n from the specified target population. However, these samples are not actually drawn. Statistical theory helps to obtain SE on the basis of just one sample, provided it is randomly drawn at least at one stage. This underscores the need to work with random samples. In the case of simple random sampling (SRS), the SEs are computed as follows:

$$\text{SEs: SE}(p) = \sqrt{\frac{\pi(1-\pi)}{n}}, \tag{12.1}$$

and

$$SE(\bar{x}) = \frac{\sigma}{\sqrt{n}}, \tag{12.2}$$

where

n is the size of the sample
π is the proportion in the population
σ is the SD of individual measurements in the population

If measurements of all N subjects in the entire target population are really available, the value of σ is computed with the formula given by Equation 7.11, but the denominator used is N and not $(N - 1)$. In practice, though, the population parameters are seldom known and cannot be calculated (if they are known, there is no need of a sample). They are *estimated* from the corresponding values in the sample. Because the estimator of π is p and of σ is s,

$$\text{Estimated SEs: } \widehat{SE}(p) = \sqrt{\frac{p(1-p)}{n}}, \tag{12.3}$$

and

$$\widehat{SE}(\bar{x}) = \frac{s}{\sqrt{n}}, \tag{12.4}$$

where s now is computed with denominator $(n - 1)$ as shown in Equation 7.11. This adjustment in the denominator helps to achieve a more accurate estimate in the long run (called **unbiased** in statistical parlance). The following comments in this context are useful:

1. $SE(p)$ is maximum when $\pi = 0.5$ and smaller when π is either small or large. But its interpretation requires additional care. For $n = 100$ and $\pi = 0.05$, $SE(p) = \sqrt{0.05 \times 0.95/100} = 0.022$, but for same n, $\pi = 0.25$, $SE(p) = \sqrt{0.25 \times 0.75/100} = 0.043$. In absolute terms, the SE in the first case is nearly half of what it is in the second case. In relative terms, though, the first SE is 44% of π, while the second is only 17% of π. Thus, the first is higher in a relative sense. These are referred to as the **absolute precision** and **relative precision**, respectively, of p. This has serious repercussion on CIs, as shown shortly.

2. Estimate in Equation 12.3 fails when $p = 0$ or $p = 1$ because then $SE = 0$. This case requires a separate consideration.

3. $SE(\bar{x})$ is large when σ is large. That is, if the individuals vary too much from one another, the sample means too will exhibit a large

variation from sample to sample. Conversely, if the individual measurements are nearly alike or homogeneous, the sample mean in one sample will be nearly the same as in the other sample.

4. Both SEs are inversely proportional to the square root of sample size n. They decrease as n increases. This implies that two samples containing 100 subjects each from the same population would not differ as much as two samples containing only 20 subjects each. Larger n thus increases the confidence in the sample results. *It is this feature that propels statisticians to suggest a larger sample.* But this increase can be counterproductive if the cost becomes prohibitive. Thus, a balanced approach is needed.

5. Note that p is the proportion of cases possessing a specific attribute in a group of subjects. For example, one may say that 11% ($p = 0.11$) in a sample of $n = 70$ normal births have a gestation period of less than 265 days. In this case, $SE(p) = \sqrt{(0.11 \times 0.89)/70} = 0.037$ or 3.7%. This result should be stated as "the percentage of births with gestation less than 265 days in a sample of 70 normal births is 11, with SE = 3.7%." This SE measures the expected variation in the percentage from one sample of size 70 to another sample of the same size from the same population.

6. In case of the mean, there is a special need to maintain the distinction between SE and SD. SE is for the aggregated picture obtained from the sample and SD is for individual values. If the average duration of the gestation period in 30 normal births is 280 days with SD = 5 days, then SE is only $5/\sqrt{30} = 0.91$ days. It would be inappropriate to say that the average gestation period is 280 ± 0.91 days because the actual measure of variation of gestation period is SD = 5 days. The correct statement is that the average gestation period is 280 with SD = 5 days. The mean should generally be accompanied by SD and not by SE. Also the usage of the $\pm$ sign should be discouraged for stating SD or SE because the actual variation is more than 1*SD or 1*SE.

7. Formulas for actual SEs are in Equations 12.1 and 12.2. However, π and σ would be rarely known and it would be necessary to use the estimates. This would yield Equations 12.3 and 12.4. The hat ($\wedge$) sign indicates that these are estimates and not actual SEs. But this makes the notation very complex. From now on, for simplicity, this text ignores the hat sign and uses Equations 12.3 and 12.4 as though they are the actual SEs.

12.1.2 Sampling Distribution of p and x̄

As already mentioned, the values of proportion p and mean $\bar{x}$ tend to vary from sample to sample. Just as individual measurements have their distribution pattern such as Gaussian, skewed, J-shaped, or any other, the sample proportion and sample mean have a specific distribution pattern when many samples are available. This is called the sampling distribution. This distribution actually can be very complex, depending on the form of the underlying distribution of the individual values, but it can be approximated by Gaussian under some conditions.

12.1.2.1 Gaussian Conditions

It has been theoretically established that in almost all situations arising in medical practice, the distribution pattern of the sample mean tends to become Gaussian when n is large even when the underlying distribution among individuals is not Gaussian. This statistical result is called the **central limit theorem** (CLT) and is very useful in drawing inferences when n is large. The criteria generally suggested to decide that n is large or not are as follows:

$$\text{For proportion, } n \text{ is large if } np \geq 8 \quad \text{and} \quad n(1-p) \geq 8; \qquad (12.5)$$

$$\text{For means, } n \text{ is large if } n \geq 30 \text{ in most situations.} \qquad (12.6)$$

The criterion given in condition (12.5) could mean very large n if p is really small. If $p = 0.30$, then n should be a minimum of 27, and if $p = 0.02$, then n should be at least 400. Poisson distribution (Chapter 13) can be advised for p less than 0.01, but that too can be approximated by a Gaussian distribution when n is really large. This is already built in the condition (12.5). For example, for $p = 0.0003$, n should be at least 26,667 for Gaussian condition to hold. Such small p can arise when an extremely rare event is being counted, such as epilepsy in children and deaths from accidental falls in home out of total deaths from all causes.

When the variable is qualitative and the interest is in the proportion π, then the underlying distribution is always what is called **Bernoulli**. There is no need to carry out any check about its distribution. This takes value 0 for absence and 1 for presence of the characteristic that p is counting. Proportion p actually is the average of these 0s and 1s for n subject. For sufficiently large n, the assumption of a Gaussian pattern for p will not be far from correct, provided p and $(1 - p)$ each is not zero. For small n, use the binomial method given in Section 13.1.

When the variable is quantitative and the interest is in the mean, the sampling distribution of $\bar{x}$ can almost invariably be approximated as Gaussian for large n regardless of the nature of the underlying distribution. See, for example, Figure 12.1, which shows that the approximate distribution of the sample mean ($n = 30$) is Gaussian although the distribution of homocysteine level among individuals is skewed. (It incidentally also illustrates the earlier contention that the variability in sampling distribution of the mean is much smaller than among individuals.) Thus, there is hardly any need to carry out check about the form of the distribution when n is large. But this check is

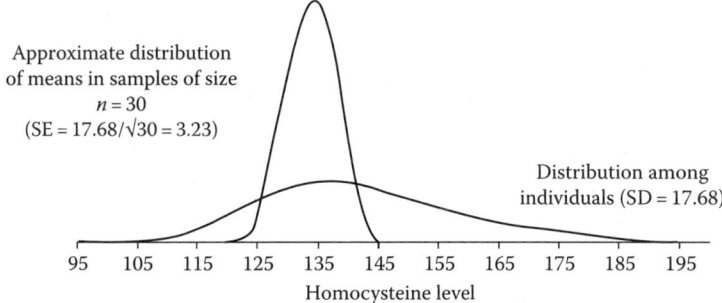

Approximate distribution
of means in samples of size
$n = 30$
$(SE = 17.68/\sqrt{30} = 3.23)$

Distribution among
individuals (SD = 17.68)

95   105   115   125   135   145   155   165   175   185   195
Homocysteine level

FIGURE 12.1
Distribution of homocysteine level and its mean for large n.

TABLE 12.1

Sampling Distribution in Various Situations

Nature of Variable	Parameter of Interest	Sample Size n^a	Underlying Distribution	Sampling Distribution
Qualitative	Proportion (π)	Small	Bernoulli	Binomial
		Large	Bernoulli	Gaussian
Quantitative	Mean (μ)	Small	Gaussian	Gaussian
			Non-Gaussian	Non-Gaussian
		Large	Gaussian	Gaussian
			Non-Gaussian	Gaussian

a Suggested criteria for considering n large are in conditions given by conditions (12.5) and (12.6).

needed when n is small and the variable is quantitative. All this is summarized in Table 12.1.

For the mean, this text considers that Gaussian conditions prevail when n is large irrespective of the form of the underlying distribution. Exceptions exist but those are rare and disregarded in this text. Gaussian conditions also exist for small n if the underlying distribution is approximately Gaussian. Thus, the only **non-Gaussian condition** for the mean is that n is small *and* the underlying distribution is far from Gaussian. Statistical inference methods for non-Gaussian conditions are different from those for Gaussian conditions.

If you know that 60% of patients with sexually transmitted diseases are males, what is the chance that they are only 30% in a random sample of 50 such patients or the chance that they are only 10% in such a sample? Sampling distributions are needed to obtain such probabilities. This chapter discusses such probabilities in the context of the distribution of p and $\bar{x}$. Before that is an explanation of the procedure for obtaining probabilities from a Gaussian distribution. This will also help in understanding probabilities of various types and how they are used for inference.

12.1.3 Obtaining Probabilities from a Gaussian Distribution

When a variable has mean μ, the difference (variable − μ) has mean zero. The SD, however, remains σ as before. This difference is called a **deviate**. When a deviate is divided by its SD, σ, it is called a relative deviate or a standardized deviate. For a variable x, the standardized deviate is $(x − μ)/σ$. This is the same as Z-score defined earlier. The SD of this deviate is one. The standardized deviate is denoted by Z. Thus,

$$\text{Standardized deviate: } Z = \frac{\text{Variable} - \text{Its mean}}{\text{Its SD}}. \tag{12.7}$$

The mean of Z is always 0 and SD always 1. The deviate Z is said to follow a standard Gaussian distribution when the variable itself has a Gaussian pattern. If the variable is x, then

$$Z = \frac{x - μ}{σ}. \tag{12.8}$$

If the variable is $\bar{x}$ or p, the SD is replaced by SE, and

$$Z = \frac{\bar{x} - μ}{σ/\sqrt{n}} \quad \text{or} \quad Z = \frac{p - π}{\sqrt{π(1 - π)/n}}. \tag{12.9}$$

12.1.3.1 Gaussian Probability

The standardized Gaussian deviate can be used to find a probability of any type related to a Gaussian distribution with any mean and any SD. For this, convert the variable to its standard form with Equation 12.7 and then use Table B.1. This table gives the probability to the right of a given value of Z, that is, $P(Z ≥ z)$. (Capital Z is the notation for the standardized deviate and lowercase z for its value in a specific case.) This probability is the same as the corresponding area under the curve. See the shaded area in Figure 12.2.

The probability values given in Table B.1 go up to four decimal places. This kind of accuracy is rarely needed. Also, the values of z for which these are

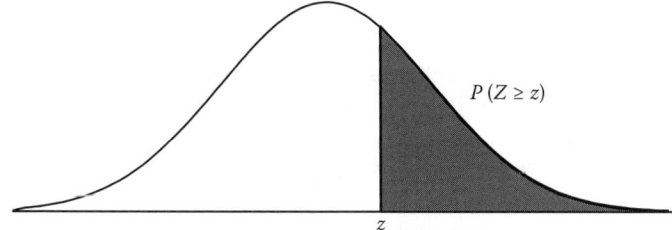

FIGURE 12.2
Probability given in Table B.1.

tabulated go up to two decimal places. The actual value of z in applications can have three or more decimal places. This book approximates z to two decimal places and also *the final* probability calculation to two decimal places. This serves the purpose quite well and redundant accuracy is avoided.

The value of Z such that the probability to its right is a is denoted by z_a, that is, $P(Z \geq z_a) = a$. For example, from Table B.l, $z_{0.025} = 1.96$. This notation is different from the one used in many statistical text but is more convenient. The following example illustrates the calculation and one use of these probabilities.

Example 12.1: Calculating Probabilities Using Gaussian Distribution

You know that heart rate (HR) varies from individual to individual even in healthy subjects. Suppose this follows a Gaussian pattern in a population with mean HR = 72/min and SD = 3/min. (1) What is the probability that a randomly chosen subject from this population has HR 74 or higher? (2) What percentage of people in this population will have HR between 65 and 70 (both inclusive) per minute?

Since mean = 72 and SD = 3, the standardized Gaussian deviate, from Equation 12.7, is

$$Z = \frac{HR - 72}{3}$$

For HR = 74, $z = (74 - 72)/3 = 0.67$. Thus, $P(HR \geq 74) = P(Z \geq 0.67)$. From Table B.1, this probability is 0.2514. In other words, nearly 25% of these healthy subjects are expected to have HR 74 or higher.

For (2), proceed as follows:

$$P(65 \leq HR \leq 70) = P(HR \leq 70) - P(HR < 65)$$

$$= P\left(\frac{HR - 72}{3} \leq \frac{70 - 72}{3}\right) - P\left(\frac{HR - 72}{3} < \frac{65 - 72}{3}\right)$$

$$= P(Z \leq -0.67) - P(Z < -2.33), \text{ see Equation 12.7 for } Z$$

$$= P(Z \geq 0.67) - P(Z > 2.33) \text{ because of the symmetric}$$
$$\text{property of the Gaussian distribution (Figure 12.3)}$$

$$= 0.2514 - 0.0099 \text{ from Table B.1}$$

$$= 0.24.$$

Thus, nearly 24% of these subjects are expected to have HR between 65 and 70.

SIDE NOTE: This answer is far too approximate. The reason is explained in the following paragraph.

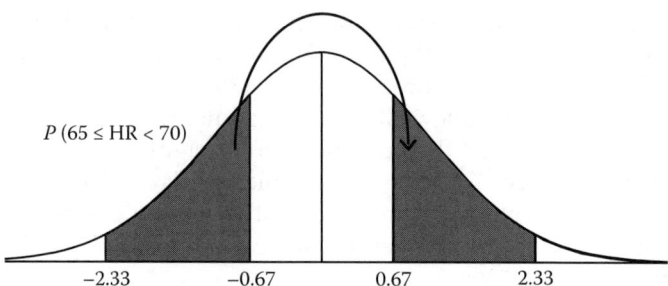

$P\,(65 \le HR < 70)$

-2.33   -0.67   0.67   2.33

FIGURE 12.3
Use of the symmetric property of the Gaussian distribution.

12.1.3.2 Continuity Correction

The Gaussian distribution is meant for continuous variables. For a really continuous variable, $P(Z > 2.33) = P(Z \ge 2.33)$, that is, it does not matter whether or not the equality sign is used. This is what was done in the preceding calculation. Consider the following.

It was mentioned in Chapter 7 that a variable such as HR is measured as discrete but can be considered continuous because of the large number of its possible values. However, in doing so, it is assumed that the rate 70 is a manifestation of values between 69.5 and 70.5. Strictly speaking, when HR is measured, it is not necessarily exactly 70/min. While counting beats, it is possible that they are 70 in 59.7 and 0.3 s remain. In other words, if these are counted for 10 min, the number may reach 704. Thus, a rate of 70.4/min is not impossible. In that sense, it is not wrong to say that the rate 70 really means that it is between 69.5 and 70.5. This is called correction for continuity.

When this is acknowledged, HR between 65 and 70 (both inclusive) is actually HR between 64.5 and 70.5. Thus, to be exact, the probability that HR is between 65 and 70 (both inclusive) in Example 12.1 is actually HR between 64.5 and 70.5. Thus, the probability that HR is between 65 and 70 is

$$P(64.5 \le HR < 70.5) = P(HR < 70.5) - P(HR < 64.5)$$

$$= P\left(\frac{HR - 72}{3} < \frac{70.5 - 72}{3}\right) - P\left(\frac{HR - 72}{3} < \frac{64.5 - 72}{3}\right)$$

$$= P(Z < -0.50) - P(Z < -2.50), \text{ see Equation 12.7 for } Z$$

$$= P(Z > 0.50) - P(Z > 2.50) \text{ because of the symmetry}$$

$$= 0.3085 - 0.0062 \text{ from Table B.1}$$

$$= 0.30.$$

Now, with the correction for continuity, nearly 30% of subjects in this healthy population are expected to have an HR between 65 and 70. This answer is more accurate than the 24% reached earlier without the continuity correction. Note how this correction can affect the probability. The probability of $HR \geq 74$ calculated earlier will also change accordingly.

12.1.3.3 Probabilities Relating to the Mean and the Proportion

The same sort of calculations can be done to find various probabilities for mean $\bar{x}$ and proportion p, provided they follow an approximate Gaussian distribution as per the conditions stated in Table 12.1. The following examples may fix the ideas.

Example 12.2: Calculating Probability Relating to Gaussian Mean

Suppose a sample of size $n = 16$ is randomly chosen from the same healthy population as in Example 12.1. What is the probability that the mean HR of these 16 subjects is 74/min or higher? Since the distribution of HR is given as Gaussian, the sample mean also will be Gaussian despite n not being large. Thus,

$$P(\bar{x} \geq 74) = P\left(\frac{\bar{x} - \mu}{\sigma/\sqrt{n}} \geq \frac{74 - 72}{3/\sqrt{16}} \right)$$

$$= P(Z \geq 2.67), \text{ now see first part of Equation 12.9 for } Z$$

$$= 0.0038 \text{ from Table B.1.}$$

This probability is less than 1%, whereas the probability of individual $HR \geq 74$ is nearly 0.25 (see the first part of Example 12.1). This happens because the SE of $\bar{x}$ is $3/\sqrt{16} = 0.75$, which is substantially less than $SD = 3$. The lower SE indicates that the values of $\bar{x}$ will be very compact around its mean 72 and very few $\bar{x}$s will ever exceed 74/min if the sample size is $n = 16$.

SIDE NOTE: Realize that the mean, not only of HR but of any variable, in any case can be a fraction. In this example, the mean 74 is actually 74.0. Thus, there is no need for any continuity correction when calculating probabilities for the mean.

Example 12.3: Calculating Probability Relating to p Based on Large Sample

Now, take an example of qualitative data where the interest is in proportion instead of mean. Consider an undernourished segment of a population in which it is known that 25% of births are preterm (<36 weeks). Thus, $\pi = 0.25$. In a sample of $n = 60$ births taking place in a random month in this population, what is the chance that the number of preterm births would be less than 10?

Since $n\pi = 15$ in this case, which is more than 8, the Gaussian approximation can be safely used. The probability required is

P(preterm births < 10) = $P(p < 10/60)$, where p is the proportion of preterm births in the sample. Since the mean of p is π and $SE(p) = \sqrt{\pi(1-\pi)/n}$, the probability can be obtained as follows:

$$P(\text{preterm births} < 10) = P\left(\frac{p-\pi}{SE(p)} < \frac{10/60 - 0.25}{\sqrt{0.25 \times 0.75/60}}\right)$$

$$= P(Z < -1.49), \text{ see second part of Equation 12.9 for } Z$$

$$= P(Z > 1.49) \text{ because of the symmetry of } Z$$

$$= 0.0681 \text{ from Table B.1.}$$

Thus, there is nearly a 7% chance that the number of preterm births in this population on a random day would be less than 10 out of 60.

12.1.4 Case of σ Not Known (*t*-Distribution)

For a sampling distribution of mean $\bar{x}$, it is useful to remind what was stated in Table 12.1. If the underlying distribution is Gaussian, the distribution of $\bar{x}$ is Gaussian even for small n. Large n is required only if the underlying distribution is not Gaussian. Now add another rider. The standardized deviate given in Equation 12.7 has SD in its denominator that converts to SE when the variable is mean. This is what was used in Example 12.2. The SE of $\bar{x}$ is $\sigma/\sqrt{n}$. In practice, however, σ is rarely known. It is then replaced by its sample estimates. After this replacement, the deviate in Equation 12.7 for sample statistics such as $\bar{x}$ does not remain Gaussian but follows what is known as **Student *t*-distribution**. A prerequisite is that the underlying distribution of x is Gaussian. One form of t is

$$t = \frac{\bar{x} - \mu}{s/\sqrt{n}}. \tag{12.10}$$

This is exactly the same as Equation 12.9 for $\bar{x}$ when σ is replaced by s. You will see other forms of t as we go along. The t-distribution is very similar to Gaussian but has a larger variance. The t-values for some commonly used probabilities are given in Table B.2. They will be needed for computing CIs and later for hypothesis testing in situations of unknown σ.

It is true that the underlying distribution must be Gaussian for t to be valid, but it is robust to mild departures, even for moderate departure. Second, t remains t whether n is large or small. But for large n, t approximates the Gaussian distribution. There is, however, no need to use this approximation. Use the t-table in the Appendix for large n as much as for small n.

Just as the weight distribution of children of age 3 years is different from that of children of age 4 years, the t-distribution is different for different degrees of freedom (df). This concept is explained for chi-square in Chapter 13. The same is applicable to t. The df depends on the sample size.

12.2 Confidence Intervals

Medical empiricism is all about groups rather than individuals. Suppose a research finds that the positive predictivity of a new diagnostic procedure is 70%. How confident you could be that a similar study on another group of subjects would not give predictivity less than 70%? Since sample values differ from sample to sample, it is useful to find how different the results are likely to be in different samples. However, repeated samples actually are not studied. Methods are available that would provide an interval within which the actual result is likely to lie in repeated samples. This interval can be obtained by using the data of only one sample when randomly drawn. The likelihood of the interval containing the actual value is called confidence level, and the interval is called **confidence interval**. This is also called an **interval estimate**. The basic function of CI is to provide a range that cannot be denied.

Because of profound medical uncertainties, particularly the sampling fluctuations about which I have been talking from time to time in this text, it is never possible to work out an interval with 100% confidence. Generally, a confidence level of 95% is used. A 95% CI has a probability of 0.95 that it will contain the actual value of the parameter. More correctly, the chance is small (5%) that it will not contain the actual value. Statistically, a CI gives a range with a substantial hope that it will include the parameter of interest. The confidence level associated with the interval (say, 90%, 95%, or 99%) gives the percentage of all such possible intervals in repeated samples that will actually include the true value of the parameter. Note that a CI tells us what to expect in the long run. It does not say anything about a particular sample.

This 95% CI gives rise to the **bikini syndrome**. "What it reveals is interesting but what it conceals is vital." Some researchers may argue that they already know about 95% and want to know about the other behind-the-scene 5%. Luckily for biostatistics, rarely in medical research, if ever, 95% will be known a priori. Even when 5% is masked, the revelation of the other 95% as plausible turns out to be a good piece of information.

12.2.1 Confidence Interval for π, μ, and Median: Gaussian Conditions

The proportion of subjects with a specified characteristic such as with a particular sign or symptom, or those responding to a therapy, is just about the most common summary measure used in medical practice. For quantitative measurements, the most common summary measure is mean, such as of urinary creatinine in cases of particular kidney disease and mean forced vital capacity in asthmatic children of age 6–10 years. CI for population proportion π for large n and population mean μ under certain conditions is obtained as follows.

12.2.1.1 Confidence Interval for Proportion π (Large n)

Suppose a new procedure for kidney stones is successful in all 10 cases on which it was tried. Can it be concluded that failure rate would continue to be zero for all such operations in future? Statisticians have worked out that the failure rate can still be 27%. If none failed in a string of 50 operations, statistical methods suggest that the failure rate in the long run may not exceed 6%. For details, see Example 12.9 later in this chapter. This underscores the importance of the size of the trial. Such information is obtained by CIs or confidence bounds for a population proportion π.

When *n* is sufficiently large and *p* not too small, the sample proportion *p* has an approximately Gaussian distribution. For this situation, property-2 of the Gaussian distribution is invoked (Section 9.2.1) to calculate the CI for π. Instead of approximate 2 as multiplier, exact 1.96 is used.

$$95\% \text{ CI for } \pi \text{ (large } n\text{): } p \pm 1.96 * SE(p)$$

or

$$\left(p - 1.96\sqrt{\frac{p(1-p)}{n}}, \quad p + 1.96\sqrt{\frac{p(1-p)}{n}} \right). \tag{12.11}$$

In the long run, if repeated samples were taken, the sample proportion would vary around the proportion π in the target population. Property-2 of the Gaussian distribution says that a distance of 1.96SE on either side of π would contain *p* in 95% of samples. This is transformed to the statement that 1.96SE on either side of *p* is not an unlikely range for the value of the parameter π. Note the reverse direction of this statement.

The limits in CI are sometimes called ±2SE limits as approximation. The lower limit is 2SE less than *p* and the upper limit is 2SE more than *p*. It is customary to use the estimate of SE in place of SE itself because the actual SE would not be known. This can be safely done when *n* is large, but not for small *n*. The multiplier 2 is an approximate value of exact 1.96. For the confidence levels other than 95%, this multiplier will change and can be read from Table B.1 corresponding to the desired confidence.

Example 12.4: CI for Proportion with Poor Prognosis in Bronchiolitis Cases with High Respiration Rate

The management of cases of bronchiolitis in infants may become easier if somehow the course of the disease can be predicted on the basis of the condition at the time of hospital admission. One simple criterion for this could be the respiration rate (RR). Consider an investigation in which cases with RR ≥ 68/min are observed during their stay in the hospital. Suppose in a random block of 80 consecutive cases of bronchiolitis coming to a hospital with RR $\geq$ 68, a total of 51 (64%) are ultimately observed to have had a severe form of the disease, that is, they had a prolonged stay in the hospital, developed some complication, required endotracheal intubation or mechanical ventilation, or died. This 64 is the percentage observed in the present sample. Another sample from the same hospital may give a different percentage. What could be the percentage of cases with a severe form of the disease in the entire population of patients admitted to the hospital with a diagnosis of bronchiolitis and RR $\geq$ 68/min?

The best point estimate according to this sample is 64%. However, this estimate is likely to differ from the actual percentage in the whole population or when another sample is taken. Since $n = 80$ and $p = 0.64$, np and $n(1 - p)$ are large enough for an approximately Gaussian pattern. The ±1.96SE limits in this example are

$$p - 1.96SE(p) = 0.64 - 1.96\sqrt{\frac{0.64 \times 0.36}{80}} = 0.53,$$

and

$$p + 1.96SE(p) = 0.64 + 1.96\sqrt{\frac{0.64 \times 0.36}{80}} = 0.75.$$

Thus, the percentage with a poor prognosis can be anywhere between 53 and 75 in cases of bronchiolitis with RR $\geq$ 68. In other words, there is a chance between 53% and 75% that a case of bronchiolitis with RR $\geq$ 68/min at the time of hospitalization will require special handling.

SIDE NOTE: Suppose that 6% of those with RR $\geq$ 68/min in the preceding example fail to survive. The 95% CI for the proportion dying is

$$\left(0.06 - 1.96\sqrt{\frac{0.06 \times 0.94}{80}}, \quad 0.06 + 1.96\sqrt{\frac{0.06 \times 0.94}{80}} \right),$$

or

$$(0.01, \quad 0.11).$$

Thus, the actual fatality rate could be anywhere between 1% and 11%. This is a wide interval relative to the case fatality of 6% observed in the sample in the sense that the case fatality could, in fact, be nearly double of what was actually observed. Compare it with the (53%, 75%) interval obtained earlier for cases with poor prognosis. This interval is narrow relative to the 64% rate observed in the sample. In general, the CI is narrow relative to p when p is around 0.5, say, between 0.3 and 0.7, and wide relative to p when p is either very low or very high.

Other points to remember are as follows:

1. Where the study group size is small and the proportion of interest too is small, use exact methods based on binomial distribution as described in the next section.

2. CI in Equation 12.11 is valid only when the observed sample proportion p is neither 0 nor 1. Methods for such extreme values are also given in the next section. Also, if p is too small or too large, the CI in Equation 12.11 can yield lower limit less than 0 or upper limit more than 1. This clearly is not possible for π. In such cases, it is customary to keep 1 as upper limit and 0 as lower limit although this amounts to an approximation. Examine whether lower or upper confidence bounds as discussed next are more appropriate in such cases.

3. Strictly speaking, a 95% CI implies probability is 0.95 that such a *random* interval contains the value of the parameter. The value of the parameter is fixed. However, in practical applications, 95% CI is interpreted as though this interval contains the parameter with probability 0.95. This is a loose statement but helps to grasp the essential feature of a CI. Henceforth, this text adopts an easy but not so accurate course and interprets 95% CI on a more practical basis as the interval that contains the parameter with probability 0.95.

Sensitivity, specificity, positive predictivity, and negative predictivity are all proportions, and their CIs are obtained as is usually done for proportions. Opposed to this, parameters such as relative risk (RR) and odds ratio (OR) are ratios and they require a different approach. CIs for RR and OR are presented in Chapter 14.

12.2.1.2 Lower and Upper Bounds for π (Large n)

Instead of two-sided CI, it is sometimes prudent to state the one-sided bound—the lower or the upper bound—which provides information about likely minimum or likely maximum, respectively. For vitamin A deficiency in an area, it is probably wise to say that a *minimum* of 4% children in an area are affected rather than saying somewhere between 3% and 11% are affected.

The ±1.96SE gives limits such that 2.5% largest values of p and 2.5% smallest values of p are excluded. One-sided bound with confidence 95% is such that all 5% smallest values in the case of lower bound and all 5% largest values in the case of upper bound are excluded. For large n, under Gaussian approximation, these are obtained as follows:

$$95\% \text{ lower bound for } \pi: p - 1.645 * SE(p),$$

or

$$p - 1.645 \sqrt{\frac{p(1-p)}{n}}. \tag{12.12}$$

95% upper bound for π: $p + 1.645 * SE(p)$,

or

$$p + 1.645 \sqrt{\frac{p(1-p)}{n}}. \tag{12.13}$$

The multiplier of SE now for bounds is 1.645 in place of 1.96 used for two-sided CI because all the 5% exclusions are in one tail of the Gaussian distribution. This multiplier is obtained from Table B.1. If the required confidence level is other than 95%, this multiplier will change. For example, for one-sided bound for confidence 90%, this multiplier is 1.28. Again, lower bound for π cannot be negative and upper bound cannot exceed one.

12.2.1.3   Confidence Interval for Mean μ (Large n)

Consider a new herbal drug, which is tried on a group of 50 coronary disease patients, and reduced lipoprotein(a) level by an average of 9 mg/dL in 3 months. Because of sampling fluctuation, another group may give a very different reduction or no reduction at all. The CI delineates the limits of the likely values; in fact, specifies values beyond which the average reduction is very unlikely.

The CLT tells us that the distribution of sample mean is nearly always Gaussian for large *n*. The underlying distribution of measurements on individual subjects is then immaterial. The distribution of the duration of survival of patients after detection of leukemia is skewed and far from Gaussian. Yet, when mean survival times are obtained in many samples, each of large size, these means still follow a nearly Gaussian pattern when there are no or few dropouts. Property-2 of the Gaussian distribution can be invoked again to obtain 95% CI for μ as $[\bar{x} - 1.96 SE(\bar{x}), \bar{x} + 1.96 SE(\bar{x})]$. This gives

$$\text{95\% CI for } \mu \text{ (σ known):} \left(\bar{x} - 1.96 * \frac{\sigma}{\sqrt{n}}, \quad \bar{x} + 1.96 * \frac{\sigma}{\sqrt{n}} \right). \tag{12.14}$$

However, σ is rarely known. It is then replaced by the sample SDs. Because of this replacement, as explained earlier, the Gaussian distribution can no longer be used. You need to use Student *t*-distribution instead. Thus,

$$\text{95\% CI for } \mu \text{ (σ not known):} \left(\bar{x} - t_v * \frac{s}{\sqrt{n}}, \quad \bar{x} + t_v * \frac{s}{\sqrt{n}} \right), \tag{12.15}$$

where t_v is the value of t at v df from Table B.2. For 95% CI, this corresponds to the probability column 0.025 in Table B.2 so that the total probability outside $\pm t_v$ is 0.05. In this case, $v = (n - 1)$. The CI in Equation 12.15 is valid only when the underlying distribution is Gaussian, especially for small n. For large n, the CI given in Equation 12.15 can be used even when the underlying distribution is not Gaussian because the distribution of $\bar{x}$ is still approximately Gaussian for such n. In other words, when underlying distribution is Gaussian, use Equation 12.14 for known σ and Equation 12.15 for unknown σ, irrespective of n being small or large. For non-Gaussian distribution, they are valid for large n only. Do not use them for small n when underlying distribution is non-Gaussian.

Example 12.5: CI for Mean Decrease in Diastolic Level

A random sample of 100 hypertensives with mean diastolic BP 102 mmHg is given a new antihypertensive drug for 1 week as a trial. The mean level after the therapy came down to 96 mmHg. The SD of the decrease in these 100 subjects is 5 mmHg. What is the 95% CI for the actual mean decrease? Since $n = 100$, $v = 100 - 1 = 99$. Thus, the 95% CI for the mean decrease is

$$\left(6 - t_{99} \times \frac{5}{\sqrt{100}}, \quad 6 + t_{99} \times \frac{5}{\sqrt{100}} \right).$$

From Table B.2, by interpolation, $t_{99} = 1.99$ in the 0.025 probability column. This means that the range (–1.99, +1.99) contains 95% of values of t. Therefore, the 95% CI is

$$\left(6 - 1.99 \times \frac{5}{\sqrt{100}}, \quad 6 + 1.99 \times \frac{5}{\sqrt{100}} \right) = (5,\ 7)\,\text{mmHg}.$$

There is a 95% chance that the actual mean decrease after 1 week regimen would be between 5 and 7 mmHg. A more appropriate interpretation of such a CI is that there is only 5% chance that the decrease after 1 week of administration would be either less than 5 mmHg or more than 7 mmHg when the study is repeated.

Some intricacies of CI can be explained with the help of Example 12.5:

1. Note that the CI (5, 7) mmHg in Example 12.5 is for the mean decrease. The decrease in 95% of *individual* patients is likely to vary between $(\bar{x} - 1.96s, \bar{x} + 1.96s)$ or between $(6 - 10, 6 + 10)$, that is, $(-4, 16)$ mmHg, provided the underlying distribution is Gaussian. Individual variation is much higher than the variation in means. The minus sign indicates that some patients may show a rise in diastolic BP to the extent of 4 mmHg despite the antihypertensive drug instead of a decline. Care is required to maintain the distinction between SD and SE.

2. It is important to realize that a 95% CI gives limits that contain those values of the parameter which are not unlikely. Thus, the CI is to be interpreted as a useful quantity in the long run. If 100 such trials are done, nearly 95 of them may give a mean decrease between 5 and 7 mmHg. The other five trials can give either a higher or a lower decrease. As emphasized earlier, the value of CI is more in what it does not contain. In this example, the CI suggests that the chance of a mean decrease being either less than 5 mmHg or more than 7 mmHg is very remote.

3. Sometimes, an approximate multiplier 2 is used instead of exact 1.96. Such an approximation is often preferred not only for convenience but also because this slight increase tends to cover the mild departure from the Gaussian pattern. In practice, the distribution is seldom exactly Gaussian. This approximation, however, can create an anomaly in some situations. In Example 12.5, $t_{99} = 1.99$, and this is less than 2. Thus, the CI from Equation 12.15 based on t would be smaller than CI if 2 is used. The fact is that t has a larger variance and it should always give a larger CI relative to the Gaussian CI. The exact value of t does not allow any departure from the Gaussian condition that the approximate multiplier 2 based on Z does.

4. Discussion is restricted to confidence level 95% because that is the most commonly used level. However, the CI can be obtained with any other confidence level. The Gaussian multiplier for large n is approximately 2.58 for 99% confidence, that is, the 99% CI is obtained by ±2.58SE limits. The 90% CI is obtained by ±1.645SE, and the 80% CI by ±1.28SE. The exact value of the multiplier can be obtained from Table B.1.

5. For small n, when the underlying distribution is Gaussian and σ is not known, the value of t_v could be very different from the Gaussian values cited in the preceding paragraph. For example, if $n = 6$, then v is 5 and t_v from Table B.2 for 95% CI is 2.571 and for 99% CI is 4.032. These are very different from Gaussian values 1.96 and 2.58, respectively. For large n, the t-value can be approximated by the Gaussian Z value (see the last row of Table B.2) but, as stated earlier, there is no need to use such an approximation because exact values are available or can be interpolated in this table for any specific df.

6. *Role of sample size in obtaining a narrow CI.* In Example 12.5, the 95% CI for a decrease in diastolic BP level is between 5 and 7 mmHg. This is fairly narrow. If $n = 10$, with no other change then the 95% CI is

$$\left(6 - t_9 \times \frac{5}{\sqrt{10}}, \quad 6 + t_9 \times \frac{5}{\sqrt{10}}\right),$$

or

$$\left(6-2.262\times\frac{5}{\sqrt{10}}, \quad 6+2.262\times\frac{5}{\sqrt{10}}\right),$$

or

$$(2.4, 9.6)\, mmHg.$$

Small n increases the CI in two ways: (1) by increasing the SE and (2) through an increase in t-value. Thus, the CI for $n = 10$ is relatively very wide compared with $n = 100$. A focused conclusion is illusive when n is small. This again underscores the statistical importance of large sample size. Also note that the width of CI depends mostly on n and SE and not on the effect size.

   Proportion and mean (or their functions such as difference) are just about the two most common parameters on which CIs are drawn. There might be isolated examples in which the interest is in CI for the median or for a decile, even σ. The basic methodology to obtain 95% CI is to get the 2.5th and 97.5th percentiles of the distribution of the corresponding "statistic" in the sample. The difficulty is that the distribution of statistics other than the mean, and hence the 2.5th and 97.5th percentiles, is not easy to obtain. The regular statistical packages are not of much help in obtaining such CIs. CI for the median is given later in this chapter. For details of how to obtain CI for other quantiles, see Conover [1]. CI for ratios such as OR and RR are discussed in a later chapter.

12.2.1.4 Confidence Bounds for Mean μ (Large n)

As explained for proportions, the interest sometimes is not in the upper and lower limits but in only one of them. For example, for noninferiority trial, only the lower bound is calculated, and for superiority trials only the upper bound. Equivalence trials use two-sided CI and the entire CI should fall within the prespecified equivalence margin.
   Under Gaussian conditions, for unknown σ,

$$\text{Upper bound for mean } \mu: \bar{x} + t_v * \frac{s}{\sqrt{n}}, \tag{12.16}$$

and

$$\text{Lower bound for mean } \mu: \bar{x} - t_v * \frac{s}{\sqrt{n}}. \tag{12.17}$$

Example 12.6: Upper Bound for Mean Number of Amalgams

Consider a case–control study conducted in Montreal, Canada, on dental amalgam and multiple sclerosis [2]. The report contains detailed results, including that the difference between cases and controls was not statistically significant, but this example is restricted to the number of amalgams in the controls. These controls were chosen at random from the general population. The mean number of amalgams was 8.78 per person and SE = 0.51 in 128 subjects for whom dental information was available. What highest mean number of amalgams per person can be expected in this population?

The value of σ is not known in this case. Therefore, we need to use Student t. Since $n = 128$, df = 128 – 1 = 127. For one-sided 95% confidence, the value of t_{127} from Table B.2 is approximately 1.66. For the one-sided 95% confidence bound, the column now consulted in Table B.2 is for probability 0.05. (For two-sided CI, it was for probability 0.025.) Since SE = 0.51, the 95% upper bound for the mean is 8.78 + 1.66 × 0.51 = 9.63. The observed mean in the sample is 8.78, but it could go up to 9.63 in repeated samples. There is only a remote chance that the mean number of amalgams in repeated samples of size 128 would go beyond 9.63 per person.

SIDE NOTE:

1. Authors in this case chose to give SE instead of SD. This runs counter to the plea that SD should be given instead of SE. They stated this as 8.78 ± 0.51 for mean amalgam. This gives the erroneous impression that the number of amalgams ranged from, say, 8 to 10. The fact is that SD = $0.51 \times \sqrt{128}$ = 5.77, and 95% of the people thus were likely to have amalgams ranging from (mean – 2SD) = 0 (since it cannot be negative) to (mean + 2SD) = 20. The confusion arises partly because of the ± sign and partly because of the use of SE.

2. They actually had 202 eligible controls, but dental information was available for only 128. This can introduce bias because the records are more likely to be available for those who had frequent problems. Thus, the mean 8.78 and upper bound 9.63 may not be applicable to the general population. The actual mean in the general population may be very different.

Gaussian conditions required for validity of the bounds in Equations 12.16 and 12.17 are the same as for the corresponding CIs, namely, (1) if n is small, the underlying distribution must be Gaussian, and (2) if n is large, departure from Gaussianity is admissible—larger the n, greater the admissible departure.

In an unlikely case of known σ, the multiplier should be from the Gaussian table instead of the t-table.

12.2.1.5 CI for Median (Gaussian Distribution)

For sufficiently large n, when the underlying distribution is Gaussian,

$$SE(median) = 1.253 * \frac{\sigma}{\sqrt{n}}. \tag{12.18}$$

You can see that this SE is about 25% larger than the SE for mean. This indicates that sample median is not a precise estimate and should not be used in situations where mean can be used. If your interest remains firm with median because of specific interest in the middle value, when the underlying distribution is Gaussian and n is large, use

$$\text{95\% CI for population median: Sample median} \pm 1.96 * 1.253 * \frac{\sigma}{\sqrt{n}}.$$

If you do not know σ, replace it with sample SD s and replace 1.96 by t-value corresponding to the appropriate df as in case of computing CI for mean. Note, however, that this requires large n and underlying Gaussian. Under these two conditions, you may never use median and may never need CI for median since mean is so appropriate. This CI is given here only for the sake of completeness.

12.2.2 Confidence Interval for Differences (Large *n*)

In many situations, the interest is in the magnitude of the difference in proportions or in means of two groups under study. The types of differences that are of special importance in medicine are between a placebo and a drug, between drug-1 and drug-2, between males and females, etc.

The preceding discussion on CI for μ and π has paved the way to describe a general strategy that works fairly well in most cases when n is large. This is as follows:

1. Find the sample analogue of the population parameter for which CI is needed.
2. Estimate its SE.
3. 95% CI is the (sample analogue $\pm$ 1.96SE).

12.2.2.1 Two Independent Samples

Note the following *estimated* SE for two-sample proportion difference when both samples are large and independent of one another. The subscripts are for the two samples:

$$SE(p_1 - p_2) = \sqrt{\frac{p_1(1-p_1)}{n_1} + \frac{p_2(1-p_2)}{n_2}} \qquad (12.19a)$$

95% CI for difference in two proportions: $(p_1 - p_2) \pm 1.96 * SE(p_1 - p_2)$. (12.20)

In case of two sample means,

$$SE(\bar{x}_1 - \bar{x}_2) = \sqrt{\frac{s_1^2}{n_1} + \frac{s_2^2}{n_2}}. \qquad (12.19b)$$

If variances are nearly equal (Levene test is used for this purpose as described in a later chapter), you can pool the variances of the two samples and get a better estimate. Then

$$\text{SE}(\bar{x}_1 - \bar{x}_2) = s_p \sqrt{\frac{1}{n_1} + \frac{1}{n_2}},$$

where $s_p = \dfrac{(n_1 - 1)s_1^2 + (n_2 - 1)s_2^2}{n_1 + n_2 - 2}$ is the pooled variance. Then

$$\text{CI for difference in two means: } (\bar{x}_1 - \bar{x}_2) \pm t_v * \text{SE}(\bar{x}_1 - \bar{x}_2). \tag{12.21}$$

The df for this t are $v = n_1 + n_2 - 2$.

For unequal variances, Welch t is used. The details are presented in Chapter 15 in the context of test of hypothesis. CI remains same as in Equation 12.21, but the df of t reduce to some extent by using a complex formula stated later.

Example 12.7: CI for Difference in Response to Two Regimens in Peptic Ulcer

In a randomized controlled trial, patients with peptic ulcer were put on two treatment regimens: one based entirely on drugs and the other based on minimal drugs but supplemented by a change in lifestyle. In the first group, 12 of 30 responded after a month, and in the second group, 28 of 50. Find the 95% CI for the difference in proportions in the two groups, and explain its meaning.

In this example, $p_1 = 12/30 = 0.40$ and $p_2 = 28/50 = 0.56$. Since n_1 and n_2 are both sufficiently large, the general procedure as stated earlier can be used. The sample analogue of $(\pi_1 - \pi_2)$ is $(p_1 - p_2)$. From Equation 12.19a,

$$\text{SE}(p_1 - p_2) = \sqrt{\frac{0.40 \times 0.60}{30} + \frac{0.56 \times 0.44}{50}} = 0.1137.$$

The 95% CI for $(\pi_1 - \pi_2)$ is

$$(0.40 - 0.56) - 1.96 \times 0.1137, \ (0.40 - 0.56) + 1.96 \times 0.1137,$$

or

$$(-0.38, \ +0.06)$$

SIDE NOTE: The 95% CI for the difference in the response rate is −0.38 to +0.06, but it also says there is a 5% chance that the difference will be outside these limits. The negative difference indicates that it is not unlikely in the long run that the response rate in the "change in the life-style" group will be even lower than in the exclusive drug group.

Note also that $(\pi_1 - \pi_2)$ in some cases can be interpreted as attributable risk. Inverse of this is called "number needed to treat." This and CI on this are discussed in Chapter 14.

12.2.2.2 Paired Samples

Paired samples in medicine arise primarily in two situations: (1) in a before–after study with no parallel controls where the same subjects are measured twice—first before the stimulus and second after the stimulus and (2) in a strict one-to-one matching where a parallel control group of different subjects is present but each control subject is matched to the corresponding case for nearly all the characteristics that can affect the outcome, except of course the stimulus itself. The control group is not necessarily placebo. It can be active control that has received another stimulus under comparison. For matched controls, mild matching by one or two characteristics such as age and sex generally is not considered adequate for paired analysis—instead such control group is considered independent and analyzed by the methods described in the previous paragraph.

Example 12.5 on diastolic BP before and after a therapy is, in fact, for paired samples. Once the difference between two subjects of each pair is taken, it reverses to one-sample situation and Equations 12.14 and 12.15 apply for CIs and Equations 12.16 and 12.17 for upper and lower bounds, respectively. The mean and SD in these formulas are for the differences in this situation. The restriction of Gaussian condition also applies as before.

Proportions in the case of paired samples are not so straightforward. The responses in this case would be as given in Table 12.2. The counts are of subjects who remain at the same level after the stimulus and those who change.

Consider a sample of $n = 70$ patients with kidney disease. They were dialyzed by the regular machine for 1 week and 49 (70%) showed significant improvement. Next week they were dialyzed by a new machine and 56 (80%) showed significant improvement. Since the patients are the same, these are not independent samples and the formulas earlier stated for difference in proportions are not applicable.

Suppose 40 patients responded well to both the machines. This is a in Table 12.2. Suppose five patients did not respond well to either machine. This is d in Table 12.2. When such numbers are stated, the data are as given in Table 12.3.

TABLE 12.2

Counts for Paired Sample Response

Characteristics before the Stimulus	Characteristics after the Stimulus		
	Present	**Absent**	**Total**
Present	a	b	$a + b$
Absent	c	d	$c + d$
Total	$a + c$	$b + d$	n

TABLE 12.3

Example of Counts in Paired Setup

Significant Improvement with Machine-1	Significant Improvement with Machine-2		Total
	Yes	No	
Yes	40	9	49
No	16	5	21
Total	56	14	70

In a paired setup such as this, the difference in proportions is measured in terms of the ratio

$$d = \frac{|b - c|}{n},$$

with

$$SE(d) = \sqrt{\frac{(b + c)n - (b - c)^2}{n^3}}, \qquad (12.22)$$

where the notations are the same as in Table 12.2. Thus, for large samples,

95% CI for difference in proportions in paired setup : $d \pm 1.96 * SE(d)$. (12.23)

For the dialysis data in Table 12.3, this CI is

$$\frac{|9 - 16|}{70} \pm 1.96 \sqrt{\frac{(9 + 16)70 - (9 - 16)^2}{70^3}},$$

or

$$0.10 \pm 1.96 \times 0.0704,$$

or

$$(-0.04, \quad 0.24).$$

The improvement by machine-2 could be up to 24%, but machine-2 can give 4% less performance than machine-1. This result is not unequivocal for these machines.

12.2.3 Confidence Interval for π, μ, and Median: Non-Gaussian Conditions

This section covers some of the settings that were left out in the previous section. These are (1) CI for π when n is small and (2) CI for μ when n is small *and* the underlying distribution is non-Gaussian. The guidelines for considering *n* large are given in the conditions stated in (12.5) and (12.6). Any other *n* is regarded as small. Sometimes, it is extremely difficult to study a large number of cases. This can happen particularly if the health condition is rare. For a trial on cases of cancer of the colon admitted to a particular hospital, it may be very difficult to get a large number of cases. If the subjects in a study are those with head injuries, a large sample is again very difficult. Statistical methods should be able to provide a valid conclusion for such small samples also. For this, more exact methods are required, which are mathematically complex. Given next are details of some of the easily explainable methods that incidentally are also among the most commonly used.

12.2.3.1 Confidence Interval for π (Small n)

Suppose the interest is in estimating the chance of uterine prolapse in women who come with complaints of micturition disturbance and vaginal discharge. If only $n = 12$ women with such complaints could be examined and three had uterine prolapse, the proportion is $3/12 = 0.25$. Another sample may give another proportion. How does one find limits within which this proportion is not unlikely to lie in all such patients? In this case, since $np = 3$ is small, the Gaussian approximation cannot be used. The CI for π in case of small *n* is obtained by using a binomial distribution. The quantity π in this case is the actual proportion of women with uterine prolapse among the population of women with those complaints. Instead of going through the rigmarole of complex mathematics, presented next is a graphical method for obtaining such a CI when sampling is random. This involves some approximation but is still useful for practical applications.

The 95% CI for π corresponding to different values of the observed sample proportion *p* can be read from Figure 12.4. This is drawn for some specific values of *n*. The upper and lower limits are read off the vertical axis using the pair of curves corresponding to the sample size *n*. The sample proportion *p* is on the horizontal axis. If your *n* is not exactly as shown, visual interpolation can be done to get an approximate CI.

For those interested, the curves in Figure 12.4 are drawn by using the relation between the sum of binomials and the β-function, which, in turn, can be obtained from the *F*-distribution (see Johnson and Kotz [3] for details). This involves some approximation because an exact 95% probability cannot be obtained for discrete distributions such as the binomial. This interval is sometimes referred to as **Clopper–Pearson interval**.

Example 12.8: CI for Percentage of Women with Uterine Prolapse

Consider the example of three women with uterine prolapse out of 12 examined with complaints of micturition disturbance and vaginal discharge. Assume that this is a random sample. In this example, $n = 12$ and $p = 0.25$. Figure 12.4 does not give a CI curve for $n = 12$. However, visual

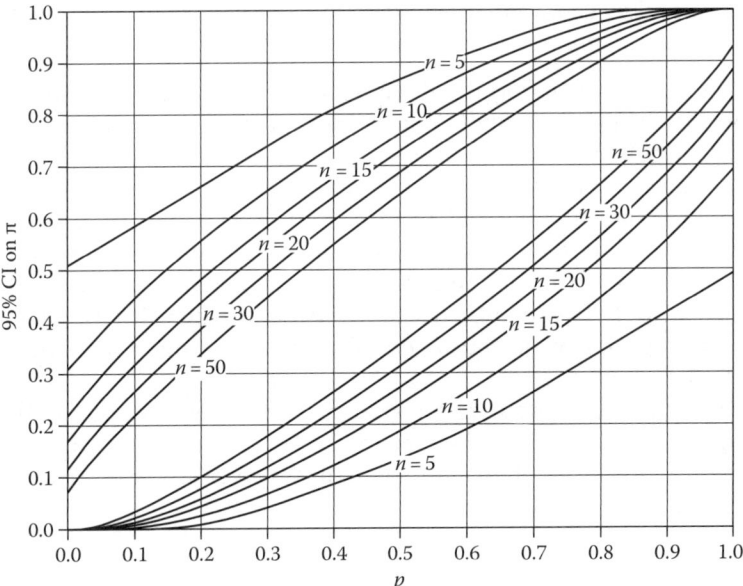

FIGURE 12.4
95% CI for π for different sample sizes.

interpolation for $n = 12$ corresponding to $p = 0.25$ gives (0.06, 0.58) as the CI. Thus, the chance of uterine prolapse in women with those complaints can be anywhere between 6% and 58%. Such a wide interval may not be helpful, but that is what could be obtained on the basis of this small sample.

SIDE NOTE: If $n = 50$, the 95% CI from Figure 12.4 is (0.14, 0.39) approximately. This is much narrower. You may wish to calculate CI for $n = 120$, now using Equation 12.11 because of the large n. For this n, the 95% CI is (0.17, 0.33). Note how the CI narrows as n increases.

Two useful observations can be made on the CIs obtained for different values of n in Example 12.8:

1. As n increases, the CI narrows, but this gain decreases with increasing n. The width of the interval for $n = 12$ is $0.58 - 0.06 = 0.52$, for $n = 50$ is $0.39 - 0.14 = 0.25$, and for $n = 120$ is $0.33 - 0.17 = 0.16$. Increasing n from 12 to 50 reduced the width to less than half, but increasing n from 50 to 120 reduced it to only two-thirds. Thus, the law of diminishing returns is applicable here also.
2. Relative to $p = 0.25$, the first interval for $n = 12$ is highly asymmetric. The lower limit 0.06 is closer to 0.25 than the upper limit 0.58. This asymmetry declines as n increases, and the CI for large n becomes symmetric around the value of p. Also, symmetry increases as p becomes closer to 0.5.

12.2.3.2 Confidence Bound for π When the Success or the Failure Rate in the Sample Is Zero Percent

Consider again a situation in which a surgeon performs the same operation on 10 different patients for kidney stone with complete success without a single complication. Thus, the complication rate is $p = 0$ in this sample. Can it be concluded that the complication rate would continue to be zero for all such operations in future? Or is this just good luck for the 10 patients who happened to be operated on during that period? In statistical language, can $p = 0$ be used as an estimate of π in the SE(p) given in Equation 12.3? The answer obviously is no. In such situations, the true complication rate can be estimated only by obtaining a one-sided confidence bound for π.

The 95% confidence bounds for extreme results for various sample sizes are displayed in Table 12.4. These are again based on the binomial distribution and are more exact than CIs shown in Figure 12.4. These are also called **Clopper–Pearson bounds**.

Example 12.9: Upper Bound for Complications in Surgeries

Consider the previous example of a surgeon with no complication in 10 surgeries for kidney stone. For this surgeon, $p = 0$ and $n = 10$. The upper bound for the true complication rate, corresponding to 95% confidence, from Table 12.4 is 27%. Thus, you could be 95% confident that complication rate in the long run would not exceed 27%; that it could be as high as 27% may have set alarm. The claim of 0% complication rate based on the experience for 10 subjects is not tenable.

SIDE NOTE: If no complication is observed in a series of 50 such surgeries, then the confidence bound corresponding to $n = 50$ from Table 12.4 is only 6%. Note again how important the size of the sample is in determining the bound and in arriving at a focused conclusion.

An exactly similar situation arises when the observed p is 1.0. In this case, for example, for $n = 15$, the lower bound for π corresponding to 95% confidence is 0.81. The limits in Table 12.4 are given for specific values of n. For other values of n, such as $n = 18$, an approximate CI can be obtained by suitably interpolating between the relevant limits.

Although the discussion is for $p = 0$ and $p = 1$, it suggests extra caution in obtaining CI for π when p is extremely small or extremely high. The conventional two-sided CI in this case can go beyond 1 or can start from a negative value. Obviously, such values are impossible for any proportion. In such cases, examine if one-sided bounds serve the purpose instead of two-sided CIs. In many situations, bounds serve the purpose well.

TABLE 12.4

95% Confidence Bounds for Extreme Results

If the Sample Size n Is	If the Sample Percentage Is 0%, the True Percentage Could Be as High As	If the Sample Percentage Is 100%, the True Percentage Could Be as Low As
1	95	5
2	78	22
3	63	37
4	53	47
5	50	50
6	40	60
7	38	62
8	31	69
9	29	71
10	27	73
15	19	81
20	14	86
25	11	89
30	10	90
35	8	92
40	7	93
45	6	94
50	6	94
55	5	95
60	5	95
65	5	95
70	4	96
75	4	96
80	4	96
85	3	97
90	3	97
95	3	97
100	3	97
150	2	98
300	1	99

12.2.3.3 Confidence Interval for Median: Non-Gaussian Conditions

Median becomes relevant when the measurements follow a highly skewed pattern, as often in sick subjects. Median is not a summation type statistic as mean is, and thus CLT does not apply. For non-Gaussian distributions, large n does not help median to attain Gaussianity.

A different method is needed if the underlying distribution is far from Gaussian, particularly if n is small. For example, it is known that the distribution of serum glucose level in diabetics is right skewed and that of the Hb level in anemics is left skewed. (In both these cases, it may be truncated depending on the threshold used for defining diabetes and anemia, respectively.) When no such a priori information about the form of the distribution is available, it is difficult to judge the shape of the distribution from small samples. When the distribution is known or suspected to be far from Gaussian, the methods to be used for small n are **nonparametric** or **distribution-free methods**. These methods do not depend on the exact shape of the underlying distribution.

Nonparametric methods mostly require that the values observed in the sample are arranged in ascending order $X_{[1]}, X_{[2]}, ..., X_{[n]}$. The median is considered the central value in these methods rather than the mean. The 95% CI for the population median (generally denoted by $\tilde{\mu}$) is in terms of the ordered values $X_{[k]}$ and $X_{[n-k+1]}$, where k is largest integer such that the probability in between these two values is at least 0.95. This is obtained by using binomial distribution with $\pi = \frac{1}{2}$. The values of k for different n are given in Table 12.5. With the exception of $n = 17$, the order of values in the table nearly agrees with the following:

$$\text{Lower limit} = \text{Integer part of} \left[\frac{n+1}{2} - 0.9789\sqrt{n} \right],$$

$$\text{Upper limit} = \text{Integer next to} \left[\frac{n+1}{2} + 0.9789\sqrt{n} \right].$$

In fact, these formulas work for $n = 6$ through $n = 283$ except for $n = 17$ and $n = 67$. Thus, you can use these for practically every situation—Gaussian, non-Gaussian, small n, and large $n \leq 283$.

Example 12.10: CI for Median Number of Diarrheal Episodes

The following are the numbers of diarrheal episodes (of at least 3 days duration) during a period of 1 year in 12 children of age 1–2 years:

3	7	12	2	4	3	5	8	1	2	3	4

In this case, $\bar{x} = 4.5$ and $s =$ Eight of 12 observations are below the mean and only 4 are above the mean. This indicates that there is a lack of symmetry and the distribution is unlikely to be Gaussian. In ascending order, the durations are

$$X_{[1]} = 1, \quad X_{[2]} = 2, \quad X_{[3]} = 2, \quad X_{[4]} = 3, \quad X_{[5]} = 3, \quad X_{[6]} = 3,$$

$$X_{[7]} = 4, \quad X_{[8]} = 4, \quad X_{[9]} = 5, \quad X_{[10]} = 7, \quad X_{[11]} = 8, \quad X_{[12]} = 12.$$

TABLE 12.5

Value of k for Different n—95% CI for Median Is $(X_{[k]}, X_{[n-k+1]})$

n	k	95% CI
≤ 5		CI cannot be computed
6	1	$(X_{[1]}, X_{[6]})$
7	1	$(X_{[1]}, X_{[7]})$
8	1	$(X_{[1]}, X_{[8]})$
9	2	$(X_{[2]}, X_{[8]})$
10	2	$(X_{[2]}, X_{[9]})$
11	2	$(X_{[2]}, X_{[10]})$
12	3	$(X_{[3]}, X_{[10]})$
13	3	$(X_{[3]}, X_{[11]})$
14	3	$(X_{[3]}, X_{[12]})$
15	4	$(X_{[4]}, X_{[12]})$
16	4	$(X_{[4]}, X_{[13]})$
17	5	$(X_{[5]}, X_{[13]})$
18	5	$(X_{[5]}, X_{[14]})$
19	5	$(X_{[5]}, X_{[15]})$
20	6	$(X_{[6]}, X_{[15]})$
21	6	$(X_{[6]}, X_{[16]})$
22	6	$(X_{[6]}, X_{[17]})$
23	7	$(X_{[7]}, X_{[17]})$
24	7	$(X_{[7]}, X_{[18]})$
25	8	$(X_{[8]}, X_{[18]})$
26	8	$(X_{[8]}, X_{[19]})$
27	8	$(X_{[8]}, X_{[20]})$
28	9	$(X_{[9]}, X_{[20]})$
29	9	$(X_{[9]}, X_{[21]})$
30+		Order value at $\text{Int}\left[\dfrac{n+1}{2} \pm 0.9789\sqrt{n}\right]$

Sample median = $(X_{[6]} + X_{[7]})/2 = (3 + 4)/2 = 3.5$. From Table 12.5, for $n = 12$, the 95% CI is $(X_{[3]}, X_{[10]})$, that is, in this case, (2, 7). There is a rare chance, less than 5%, that the median number of diarrheal episodes in the child population from which this sample was drawn is less than 2 or more than 7.

SIDE NOTE: When proceeded with the Gaussian pattern, the 95% CI for median from Equation 12.18 would be $3.5 \pm 2.201 \times 1.253 \times 3.12/\sqrt{12}$ or (1.0, 6.0). This Gaussian CI is not very different in this case but sometimes can be very different.

Small sample CIs for median have the following implications:

1. CI obtained by the nonparametric method just described has *at least* a 95% confidence level. The ordered observations rarely allow the level to be exactly 95%. In some cases, it could be as high as 98%. If the CI is narrowed, the confidence level becomes less than 95%.

2. Small sample CIs for differences such as in medians are quite complex. Details are beyond the scope of this book, and there is no useful reference. In the case of paired quantitative data, the difference reduces to the one-sample problem.

3. Small samples can seldom provide precise information. This is illustrated by Example 12.10, where the 95% CI for the median is fairly wide and not able to provide very useful information.

See Table S.1 at the beginning of the book for various methods of obtaining CI in different situations.

12.3 *P*-Values and Statistical Significance

It is legitimate to wonder how statistical tests pervade so much in empirical decisions. Data uncertainty propels use of such methods sometimes to the point of satiation. The medical literature is full of *P*-values. Symbols such as * and ** are sometimes used to show significance. If you are not in research, you may not do tests of statistical significance, but they are frequently talked about regarding what specific observations caused significance, how they are affected or not affected by confounders, what biological explanation is available, etc. For this, it is necessary to understand the basic philosophy behind statistical significance. This section explains the concept. The actual methods appear in Chapters 13 through 15. Realize though that significance in common language is that the effect is big enough to take notice of. In statistics, this means that the evidence against no or specified difference is enough.

12.3.1 What Is Statistical Significance?

Previous section discusses CI as a method to deal with intersample variability. Intersample variability has another type of implication as well.

If the mean reduction in cholesterol level after therapy in a sample of 60 subjects of age 40–49 years is 9 mg/dL and in another sample of 25 subjects of age 50–59 years 13 mg/dL, can it be safely concluded that the average decrease in the two age groups is really different? Or, has this difference just occurred by chance in these samples? A difference apparently seen in average lipoprotein(a) levels in a sample of 30 male and 40 female hypertensives could vanish if the study is done in another group of similar subjects or may be real that would recur in repeated samples. Statistical significance is decided by the probability that the difference will not vanish if repeated samples of the same nature are studied. Again, repeated samples are actually not drawn. Instead, statistical methods are used to evaluate the probability on the basis of just one sample. For this, a hypothesis is setup and tested for significance.

To understand the concept of statistical significance more fully, the following is a brief visit to the methodology followed in all empirical conclusions. This will also help in understanding the concepts of null hypothesis and of *P*-values that are so vital to the concept of statistical significance. I would like to explain them with the help of two everyday analogies that should make these concepts absolutely clear.

12.3.1.1 Court Judgment

The concepts are best understood with the help of an example of a court decision in a crime case. Consider the possibilities mentioned in Table 12.6(a). When a case is presented before a court of law by the prosecution, in most societies the judge is supposed to start with the presumption of innocence. It is up to the prosecution to present evidence against the innocence of the person. The evidence should be enough to change the initial opinion of the judge. Guilt should be proved beyond reasonable doubt. If the evidence is not sufficient, the person is acquitted whether he actually committed the crime or not. Sometimes, the circumstantial evidence is strong and an innocent person is wrongly pronounced guilty. This is considered a very serious error. Special caution is exercised to guard against this type of error even at the cost of acquitting some criminals.

Table 12.6(a) assumes that the decision is either guilty or not guilty. In Scotland, the verdict can also be "guilt not proven." This occurs when the court believes that the accused has probably committed the crime but the evidence is not compelling enough.

TABLE 12.6

Errors in Court Judgment, Diagnosis, and Statistical Decision

(a) Court Setting			(b) Diagnosis			(c) Statistical Decision		
	Assumption of Innocence			Disease Actually Present			Null Hypothesis	
Judgment	True	False	Diagnosis	No	Yes	Statistical Decision	True	False
Pronounced guilty	Serious error	✓	Disease present	Misdiagnosis	✓	Rejected	Type-I error	✓
Pronounced not guilty	✓	Error	Disease absent	✓	Missed diagnosis	Not rejected	✓	Type-II error

12.3.1.2 Errors in Diagnosis

There is no doubt that clinicians these days examine a great deal of evidence before reaching a diagnosis. This evidence is in terms of clinical features, laboratory investigations, images, records, etc. Despite this, errors are not uncommon. In place of a differential diagnosis, let us restrict our discussion to the presence or absence of a specific disease. If the disease is actually not present but wrongly diagnosed as present, this is called **misdiagnosis** (Table 12.6(b)). This is obviously a more serious error than **missed diagnoses** that occur when the disease is present but is missed. Both are errors, but a missed diagnosis is likely to be detected in a subsequent encounter because the patient is likely to come back with complaints. Misdiagnosis can mean a great deal of inconvenience, cost, and side effects to a person who is actually unaffected and an unnecessary load on the health-care system. The same sort of errors can occur in statistical decisions.

12.3.1.3 Null Hypothesis

In the case of statistical decisions also (Table 12.6c), the assumption initially made is that there is no difference between the groups. This is equivalent to the presumption of innocence in the court setting and is called the null hypothesis. The notation used for this is H_0. This is the hypothesis under scrutiny and is sought to be refuted by conducting a study on a sample of subjects just as investigations are done to find evidence against a suspect. Thus, null is something you suspect or do not believe to be true.

The sample observations serve as evidence. Depending upon this evidence, the H_0 is either rejected or not rejected. In an empirical setup, the H_0 is never accepted. The conclusion reached is that the evidence is not enough to reject H_0. This may mean two things: (1) carry out further investigations and collect more evidence and (2) continue to accept the present knowledge as though this investigation was never done. The "truth" remains unchanged in this case. This can be easily explained with the help of a telling example described in the next paragraph.

12.3.1.4 Philosophical Basis of Statistical Tests

Ptolemy, in the second century AD, propounded that the sun revolves round the earth. It remained the "truth" for 14 centuries until Galileo came up with evidence against it in the sixteenth century and established a new truth that says that earth revolves round the sun. Newton's laws of mechanics were accepted until Einstein discussed circumstances in which some of them did not hold. Peptic ulcer was believed to be caused by acidity until sometime ago when *Helicobacter pylori* was found to be a culprit in some cases. As of today, coronary heart disease is not considered caused by any infection but who knows that a rebuttal comes soon.

Empirical strategy is to find evidence against a **hypothesis** that is stated in null form. Similar to a court judgment, if this evidence is sufficient the null hypothesis is rejected, otherwise continues to be considered "truth" by default. Transplanted to the development of a new therapeutic regimen, it seems reasonable to demand evidence against the hypothesis that there is no effect at all. This process is difficult to implement without setting up a null hypothesis.

Since samples by their very nature are uncertain, the conclusion depends on what sort of data is obtained from the subjects. Statistical tests are precisely meant to deal with uncertainties arising from sampling fluctuations. They provide the answer to the question: What is the likelihood of sample values given that null hypothesis is true? If this likelihood is exceedingly small, say, less than 5%, the null is considered implausible and rejected. At the same time, it would be ridiculous to accept a hypothesis whose likelihood of giving the obtained sample is only 15%. Thus, the only conclusion drawn in this case is that the sample fails to provide sufficient evidence to reject the null hypothesis. It is not accepted and the situation reverses back to what it was before that study.

Many would consider it idiosyncratic that the sample values are searched for evidence against a null without recourse to finding what they are for—what they support. The alternative hypothesis that would clarify this dilemma will be described soon, but the process of finding evidence against is widely followed in empirical setup.

As another example, consider the claim of a manufacturer that his drug is superior to the existing angiotensin-converting enzyme (ACE) inhibitors in improving insulin sensitivity in diabetic hypertensives. In a trial of matched cases, improvement was seen in 7 out of 10 patients who were given the new drug compared with 6 out of 10 who were given the existing drug. The methodology for testing the statistical significance of this difference will be described shortly, but you can see that the sample size $n = 10$ in each group is small and the difference is too small to provide confidence to pronounce that new drug is better. The difference could have arisen due to sampling fluctuation. If so, the claim of superiority is not tenable. The manufacturer needs to withdraw the claim forever or until such time that more evidence is available for scrutiny. The medical fraternity is expected to continue with their existing practice and not take cognizance of the claim until the claim is adequately substantiated.

This section started with an example of *difference* between the two groups and explained the concepts in this context. However, the concept of null hypothesis is not restricted to the difference between two groups. It could concern difference between many groups, association or relationship between two or more characteristics or variables, or any other aspect of the problem under investigation. Some of these are discussed in later chapters.

12.3.1.5 Alternative Hypothesis

If a null hypothesis is false, what alternative is true? The alternative hypothesis, denoted by H_1, is the opposite of H_0 that must be true when H_0 is found false. If the null is that a drug is not effective, then the alternative has to be that it is effective. In the preceding example, the claim is that of superiority of the new drug. This is the alternative hypothesis in this case. This is a **one-sided alternative** if inferiority is ruled out. Most often it is not possible to claim that one group is better than the other, and the only claim is that they are different. In the case of peak expiratory flow rate (PEFR) in factory workers exposed to different pollutants, there may not be any a priori reason to assert that they would be affected more by one pollutant than another. Then, the alternative is that the mean PEFRs in workers exposed to different pollutants are unequal. This is called a **two-sided alternative**. The null is that the effects are equal in various groups. One-sided and two-sided H_1 are also sometimes called one-tail and two-tail H_1, although these terms should be used for the corresponding probabilities of Type-I error. This error is explained in the next section.

The one-sided alternative can be considered as saying that one group is better than the other and the two-sided as saying that one group is either better or worse than the other. In clinical trials, the former can be equated to superiority when the margin of superiority is also specified. The latter, in terms of the null hypothesis, signifies clinical equivalence. These two types of alternatives are discussed in the next three chapters where actual methods are presented.

12.3.1.6 One-Sided Alternatives: Which Tail Is Wagging?

One-sided alternative hypothesis can be tricky in some situations. The existing knowledge is that the percentage of hypertensives among diabetic patients is higher than among normal adults. If there is no doubt, there is no point in investigating it further. The study is meaningful if there is a doubt and it is to be confirmed or refuted. If the prevalence of hypertension among normal subjects is known as 20%, the problem can be approached in one of two ways: (1) If in a sample of diabetic patients, the percentage of hypertensives happen to be 22%, can it still be less than 20% in the population of diabetic patients? (2) If a sample of diabetic patients reveals 17% hypertensives, can this be 20% or higher in the population of diabetic patients. The approach will depend on the claim. If the claim is that diabetic subjects have at least 20% hypertensives, the null hypothesis for rejection is H_0: $\pi \geq 0.20$ and the alternative is H_1: $\pi < 0.20$. The sample proportion should be substantially less than 0.20 to provide evidence against the null and in favor of this alternative. If sample proportion is 0.20 or more, there is no evidence against the null, and there is no need to proceed further. If the claim is that diabetic subjects do not have more than 20% hypertensives, the null is H_0: $\pi \leq 0.20$

and the alternative is H_1: $\pi > 0.20$. In this situation, the evidence against the null and in favor of the alternative is a sample proportion $p \gg 0.20$. If sample proportion is less than 0.20, there is no evidence against the null, and no need to proceed further.

12.3.2 Errors, *P*-Values, and Power

As stated earlier, the values observed in the sample serve as evidence against H_0. The sample values are used either to reject or not reject an H_0. But these values are subject to sampling fluctuation and may or may not lead to a correct conclusion. Errors do occur.

12.3.2.1 Type-I Error

It is quite common in court judgment that a real offender is acquitted because of weak evidence. This is not considered a serious error. But the other error also occurs. An innocent is pronounced guilty because there is a strong circumstantial evidence. This is serious.

If there is no real difference between the groups but the data strongly disagree, the true null hypothesis has to be undesirably rejected. A false-positive conclusion is reached. This is serious and called **Type-I error** (Table 12.6) or **alpha error**. Seriousness of this error can be understood from the setup of a trial on a new drug. This type of error occurs when an ineffective drug is declared effective. This gives false assurance. Then, the ineffective drug would be unnecessarily marketed, prescribed, and ingested, and side effects tolerated. Imagine the cost and inconvenience caused by this error. Statistical procedure requires that the probability of this type of error be kept low, generally within 5%.

The probability of Type-I error is called **P-value**. Statistically, *P*-value is the probability of rejecting the true null. It can also be understood as the probability that a true null hypothesis is wrongly discarded. In the case of trials, this could occur if the patients included in the trial by chance happen to exhibit a difference between the groups when actually none exists. Because of interindividual variation, it is not an unlikely scenario. *P*-value helps in increasing or decreasing faith in the results. This gives an idea of how strongly the data contradict the H_0. Loosely speaking, if $P = 0.67$, concluding that a difference exists could be wrong two-thirds of the time, and if $P = 0.03$, the chances are only 3%. When *P*-value is small, it is generally considered safe to conclude that a difference or a relationship is indeed present.

Seriousness of Type-I error requires that a threshold of *P*-value is fixed *in advance* beyond which it would not be tolerated. This is called **significance level** and denoted by α. This is also called **alpha level**. Note the distinction between alpha error mentioned earlier and alpha level stated now. This specifies the criteria of doubt beyond which a null hypothesis is rejected. Statisticians—almost all empirical scientists—generally use a threshold of 5% chance of error.

Omnipresent variations and uncertainties do not allow *P*-value to be zero, and it is many times difficult to reduce this to 1% or 2%. Five percent is an internationally accepted norm for significance level that is rarely breached in medical literature. In situations where 5% chance of error can translate into serious consequences such as death, a lower level can be fixed. To theorists, 5% chance of error may look high. Luckily, yes luckily, humans and animals flicker much more than chemical reactions or electron motion in physics. Thus, 5% chance of error is not high. Medical theories are hard to evolve. All medical professionals are expected to be well informed about the pros and cons of various errors so that they can be judicious in their decisions despite 5% chance of Type-I error.

When *P*-value is less than the level of significance, presence of difference or relationship is concluded. When *P* is more than or equal to, say, 0.05, sampling fluctuation cannot be excluded as a likely explanation of the observed difference. There is a convention to call a result statistically significant when $P < 0.05$ and highly significant when $P < 0.01$. A probability less than 0.01 is considered exceedingly small. The following comments may be helpful:

1. It is important to distinguish between a *P*-value and an α-level. Both measure the probability of Type-I error. The first is obtained for the dataset in hand, whereas the second is fixed in advance. You may say for a problem that not more than 3% chance of Type-I error can be tolerated. Then $\alpha = 0.03$. The calculations later on may reveal that the *P*-value for your dataset is only 0.012, 0.362, or any other number. A decision regarding rejecting or not rejecting a null hypothesis is reached according to the *P*-value being higher or lower than the predetermined α-level.

2. A null hypothesis is generally a statement of the status quo. The alternative hypothesis signifies a change. If it were very costly to make changes from the status quo, you would want to be very sure that the change would be beneficial. The risk of Type-I error is then kept very low, say, less than 1%. This may be needed when, for example, a drug presently in extensive use or an existing lifesaving drug is sought to be replaced. In most medical situations, $\alpha = 0.05$ is considered adequate.

12.3.2.2 Type-II Error

The second type of error is failing to reject H_0 when it is false (see Table 12.6(c)). This occurs when a study fails to detect a real difference. This corresponds to a missed diagnosis as well as to pronouncing a criminal not guilty. The probability of this error is denoted by β. In a clinical trial setup, this is equivalent to declaring a drug ineffective when it actually is effective. When this occurs, society will continue to be without the drug just as it was before the trial. A drug that could possibly provide better relief to scores of patients is denied entry into the market. If the manufacturer believes that the drug is really effective, it will carry out further trials and collect further evidence. Thus, the effect of Type-II error in this case is that the introduction of the drug is delayed but not denied. This is not so much of an error.

A Type-II error is calculated after fixing the level of significance α and for a specific value under the alternative hypothesis. This should be clear from the following discussion on power. Note, however, that statistical procedures are based on chance of Type-I and Type-II errors. Once you have made a decision, you could have made only one error—wrongly reject the null or wrongly not reject the null. And you would not know unless you are divine to know the truth. Also, there is no Type-I or Type-II error if there is no test of hypothesis.

12.3.2.3 Power

The complementary of the probability of Type-II error is called statistical power and is denoted by $(1 - \beta)$. The power of a statistical test is the probability of correctly rejecting H_0 when it is false. In other words, this is the proportion of times that repeated samples of similar nature would give P-values less than α. This is the probability of getting a statistically significant result when a difference exists. Thus, this is the ability to detect a difference. The power of a test is high if it is able to detect a small difference and thus can easily reject H_0. Suppose the mean PEFR in workers in a tire manufacturing industry is 296 L/min and that in workers in a paint varnish industry is 307 L/min. Thus, the mean difference is 11 L/min. This difference seems small relative to the PEFR values. A test with high power is needed to detect this difference and to call it statistically significant. A test with low power will not be able to reject the H_0 of equality in this case and will lead to the conclusion that the difference is likely to have arisen by chance in the samples studied. Power measures the degree of assurance that the specified difference will not be missed even when clouded by high variability in the measurements.

Statistical power becomes an especially important consideration when the investigator does not want to miss a *specified* difference. For example, an antihypertensive drug may be considered useful if it reduces diastolic BP by an average of at least 10 mmHg after use for, say, 1 week. A sufficiently powerful statistical test would be needed to detect this difference with high probability. Thus, the magnitude of $(1 - \beta)$ is an important consideration in this setup. However, one would like the minimum medically important difference (10 mmHg in this case) is chosen on some objective basis. The choice may not be easy in some situations.

The number of subjects in the study that determines the SE is the most important consideration for power when all sources of bias and uncertainties, such as due to lack of knowledge and inadequate design, are in control. The best approach to achieve good power for detecting a minimum medically relevant difference is to increase the number of subjects in the study. Formulas are available that can give this number for different settings (see Section 12.6). Power calculation depends on whether the characteristic under assessment is quantitative or qualitative, the form of its statistical distribution in the target population, the variance across subjects, the minimum difference between groups that can be considered medically relevant, and the chosen level of significance. A power of 0.8 or 0.9 seems to have become the norm for medical studies. For analogy with statistical significance, it is customary to call existence of a medically relevant difference as **medical significance** of results.

There is a trade-off between Type-I and Type-II errors. In court analogy also, eagerness to convict a guilty has direct bearing on innocent being convicted. Type-II error is preferred over Type-I error in court to uphold civil rights if the evidence is not sufficiently convincing. Power has a kind of direct relationship with the level of significance. The lower is the level of significance, the lower is the power. Using 1% level of significance instead of the conventional 5% will make it difficult to reject a false null. Thus, power will be reduced. A relationship between significance level α and power $(1 - \beta)$ can be obtained in terms of a receiver operating characteristic (ROC) curve similar to the one described earlier between sensitivity and (1 – specificity). However, utility of such ROC is limited since α is mostly fixed in advance.

Also, the power is low if variance is high. You can intuitively feel that when values are highly variable from subject to subject, the difference between two or more groups will be difficult to detect. In addition, irrespective of α and variance, a smaller difference is difficult to detect and the corresponding power will be low. Larger the minimum medically relevant difference, more is the power. The basic statistical consideration is sample size. An adequately large sample can override the limitation of low level of significance, high variance, and small medically relevant difference for detection.

12.3.3 General Procedure to Obtain *P*-Value

When the probability of Type-I error, *P*, is less than a low threshold such as 0.05, the null hypothesis is rejected and the result is said to be statistically significant.

The exact criterion for obtaining the *P*-value depends mostly on (1) the nature of the data (qualitative or quantitative), (2) the form of the distribution such as Gaussian or non-Gaussian when the data are quantitative, (3) the number of groups to be compared (two or more than two), (4) the parameter to be compared (it can be the mean, median, correlation coefficient, etc., in case of quantitative data; it is always a proportion π or a ratio in case of qualitative data), (5) the size of the sample (small or large), and (6) the number of variables considered together (one, two, or more). All this is analogous to saying that the criterion for assessing the health of a person depends on the age, gender, purpose, general health or organ focused health, etc. The exact criteria used to test different statistical hypotheses are described in subsequent chapters. The general principle used for obtaining the *P*-value can be described as follows:

Step 1. Set up a null hypothesis and decide whether the alternative is one sided or two sided. Also decide the level of significance α and thus fix the threshold of Type-I error that can be tolerated for the problem in hand.

Step 2. Identify a criterion, such as Student t and chi-square, suitable for the setup in hand. This would depend on items (1) through (6) just enumerated. Some of these criteria for different setups are mentioned in subsequent chapters. These criteria are also called **tests of statistical significance**. The distributional form of these criteria has been obtained and is known.

Step 3. Use sample observations to calculate the value of the criterion *assuming that the null hypothesis is true*.

Step 4. Compare the calculated value with its known distribution, and assess the probability of occurrence of a value of the criterion that is *as extreme as or more extreme toward H_1* than that obtained in Step 3. This probability is the P-value. This is calculated for both the negative and the positive side when the alternative is two sided. Since the comparison is with the distribution under H_0, a probability that is not very low indicates that the sample is not inconsistent with H_0 and thus H_0 cannot be rejected. In this case, sampling fluctuation cannot be ruled out as a likely explanation for the values observed in the sample. A very low probability indicates that the H_0 is very unlikely to give rise to the observed values and deserves to be rejected.

Tables in Appendix B give different values of various criteria for popular thresholds α, particularly 0.05. These are called **critical values**. In fact, what is required is the P-value for each value of the criterion (in Table B.1, this is given for large number of values of Z), but that is difficult to tabulate in most cases. Most statistical softwares give exact P-values associated with the value of the criterion obtained. *In that case, the appendix tables are redundant.*

Step 5. Reject H_0 if P is less than the predetermined level of significance. Generally, $P < 0.05$ is considered low enough to reject H_0. Statistical significance is said to have been achieved when H_0 is rejected. Such a result can be stated in a variety of ways:

The evidence against the null hypothesis is sufficient to reject it.

Sample values are not consistent with the null hypothesis.

Sampling fluctuation is not a likely explanation for the observed difference.

The alternative hypothesis is accepted.

The P-value is less than a predetermined threshold such as 0.05.

The probability of wrongly rejecting H_0 (Type-I error) is very small.

The result has achieved statistical significance.

All these statements are equivalent. They illustrate how a decision is reached despite the presence of uncertainty. When P is more than (or equal to) the predetermined level of significance, the null hypothesis H_0 cannot be rejected. The implications of this are discussed later in this section. But note for the time being that conceding a null hypothesis can be similarly stated in a different ways. Different statements are used at different times in this text and sometimes two or more of these statements are stated together to emphasize the nature of the conclusion reached.

The following example may help to understand the steps and the philosophy behind the statistical tests. Many examples are given in subsequent chapters where procedures for different situations are discussed. The concept of P-values and statistical significance should become very clear once you are through with those chapters.

Example 12.11: Test of Hypothesis for Percentage of Premature Births

Suppose an argument erupts concerning the percentage of births in a community that are premature. Based on everyday experience, a practitioner asserts that 10% of births are premature, neither less nor more. To test this assertion, a random sample of 60 births is systematically observed and 8 of them are found to be premature. This is 13.3%. Can we conclude that the percentage of premature births in the population is not 10%?

Now, follow the steps just outlined.

1. Assertion of 10% premature births in this case is the null hypothesis. This is what is to be tested (and possibly to be refuted). Thus, H_0: $\pi = 0.10$. Since there is no assertion about a particular direction of difference, it could be negative or positive. That is, H_1: $\pi \neq 0.10$. Let the level of significance be fixed at $\alpha = 0.05$. That is, the chance of error of Type-I should be less than 5%.

2a. Quantity of consequence in this case naturally is the proportion, p, actually observed in the sample. For this sample of 60 births, $p = 8/60 = 0.1333$.

2b. Since n is 60 and $np \geq 8$, invoke CLT and assume that p will approximately have a Gaussian distribution.

2c. Standard Gaussian distribution mentioned earlier makes it easy to consult a table of probability. The standard Gaussian deviate given by Equation 12.7 in this case is

$$Z = \frac{p - \pi}{SE(p)}.$$

This can be used as a test criterion.

3. You know that when H_0: $\pi = 0.10$ is true, then from Equation 12.1,

$$SE(p) = \sqrt{\frac{0.10 \times 0.90}{60}} = 0.0387.$$

Thus, for this sample, under H_0,

$$z = \frac{(0.1333 - 0.10)}{0.0387}$$

$$= 0.86.$$

4. The *P*-value is the probability of obtaining this value of z or more extreme toward H_1. Since H_1 is two sided,

$$\text{P-value} = P(Z \le -0.86) + P(Z \ge 0.86) = 0.1949 + 0.1949 = 0.39$$

from Table B.l. This *P*-value is certainly very high in comparison with the conventional threshold 0.05. If H_0 is rejected, the chances of its being wrongly rejected are as much as 0.39 (or 39%). This is too high an error.

5. Since $P \ge 0.05$, H_0 is plausible and the assumption $\pi = 0.10$ cannot be rejected. The sample does not provide sufficient evidence against H_0. This is true for many other values of π as discussed in one of the following paragraphs.

The sample in this example can provide sufficient evidence to reject some other values of π. For the purpose of illustration, change H_0 to $\pi = 0.25$. Under this value of H_0,

$$Z = \frac{0.1333 - 0.25}{\sqrt{0.25 \times 0.75/60}} = -2.08.$$

Now, $P = P(Z \le -2.08) + P(Z \ge 2.08) = 0.0188 + 0.0188 = 0.038$. This is less than 0.05. Thus, this H_0 is not plausible and is rejected. It is concluded that the percentage of premature births is not 25. The percentage (13.33%) observed in the sample is sufficiently different from 25 in a statistical sense but not sufficiently different from 10. Thus, the H_0 that the premature births are 25% can be rejected but not the H_0 that they are 10%.

12.3.3.1 Subtleties of Statistical Significance

Note that H_0 is only "not rejected" but never accepted. This notion can be further enforced by noting that a null hypothesis $\pi = 0.20$ or $\pi = 0.16$ in Example 12.11 would also be not rejected. If you accept, which value of π would you accept? If a conclusion were reached that premature births *are* 10% just because this could not be rejected, then this would be wrong because the values 20% and 16% are also plausible. Whenever statistical significance is not reached, the evidence is not considered in favor of H_0—it is only not sufficiently against it. Samples provide evidence against H_0 and in favor of H_1 but never in favor of H_0 and against H_1. Absence of evidence is not evidence of absence. Not being able to reject H_0 is analogous to pronouncing in a court of law that a person is not proved guilty. This is different from saying that the person is innocent. The other way that this could be understood is that a null hypothesis is conceded but not accepted. Distinction must also be made between "not significant" and "insignificant." Statistical tests are for the former and not for the latter. A statistically not significant difference is not necessarily insignificant.

A statistically significant result does not mean that chance cannot account for the finding: only that such an explanation is highly unlikely. A result that is not statistically significant does not mean that chance is responsible for the result—only that it cannot be excluded as a likely explanation. With statistical inference, the results can seldom, if ever, be absolutely conclusive. The *P*-value never becomes zero. There is always a possibility, however small, that the observed difference in the samples arose by chance alone.

The procedure adopted in Example 12.11 is for the two-sided alternative. In the second part of this example, the H_0: $\pi = 0.25$ is rejected and the conclusion is that the percentage of premature births is either more than 25 or less than 25 but not 25. A one-sided H_1 can also be considered, in which case the *P*-value will change.

You may have noticed that generally a null hypothesis is not setup to say that some *unspecified* difference is present. It is generally stated in the form of "no difference." However, as discussed later, there can be an H_0 that states that the difference between two groups is a specified quantity. For example, a null hypothesis could be that the difference in mean decline in blood glucose level in normal and obese subjects on a particular diet is a minimum of $10\,mg/dL$. The alternative hypothesis in this case would be that the difference in decline is less than $10\,mg/dL$.

If $P = 0.06$ in an experiment, it is not necessary to conclude that the experiment has failed to provide sufficient evidence. In some cases, even $P = 0.08$ may give a medically relevant result. Remember that P is the probability of obtaining the observed data (or the data that are more extreme in favor of H_1) if H_0 is true. It measures the plausibility of the null hypothesis. It should be interpreted in the light of its real meaning instead of depending solely on a rigid cutpoint such as 0.05.

After all the time and effort spent on obtaining accurate measurements through sophisticated and expensive devices in a medical study, it is ironical that the final conclusion is to be based on a *P*-value that is a probability statement and where a threshold such as 0.05 is wholly arbitrary. It is legitimate to ask why 0.05 and why not 0.02. Such a threshold has come to stay by convention rather than for any scientific reason. It seems to be globally accepted and rarely questioned. A practice now gaining ground is to state an exact *P*-value in place of merely saying that it is more than or less than a threshold such as 0.05. Nevertheless, to arrive at a conclusion about whether the difference is present or not, it is necessary to take recourse to a threshold. If $P = 0.03$ in a particular problem, what conclusion should be drawn if a threshold such as 0.05 is missing from the scene? Should the difference be regarded as present or not present? It seems that a cutpoint such as 0.05 is necessary to draw an inference one way or the other.

12.4 Assessing Gaussian Pattern

As you may have noted, the requirement of Gaussian pattern of original values is not rigorous for most sampling distributions. Yet, many times you would want to know whether the pattern is Gaussian or not—for example, to decide whether median would be more appropriate central value or mean. For sampling distribution also, knowledge about the pattern of original values is helpful. For example, there is a tendency for small sample $\bar{x}$ to follow the same kind of distribution as the individual xs. The distribution of duration of labor in childbirth is known to be skewed to the right. That is, a long duration (relative to the mode) is more common than a short duration. The distribution of mean duration in a small sample of, say, eight women is also likely to follow the same pattern, although in attenuated form. In such cases if the pattern is not known, it is worthwhile to investigate.

Some gross methods for assessing Gaussianity were discussed in an earlier chapter. These include (1) studying shape of a histogram or stem-and-leaf plot, (2) inequality among the values of mean, median, and mode, (3) quartile plot or box plot, and (4) proportion-by-probability (P-P) and quantile-by-quantile (Q-Q) plot. The other alternative, not mentioned earlier, is to calculate standardized deviate (see Section 12.1.3) for each observed value, order them from minimum to maximum, and plot them against the corresponding Gaussian probability on a probability paper. This is called normal probability plot. If the distribution of observed value is Gaussian, the plot will be nearly a straight line. You can also check if mean ± 1SD covers nearly two-thirds and mean ± 2SD nearly 95% of the values. The range should be nearly 6SD.

More exact methods are based on calculations that check statistical significance from the postulated pattern such as Gaussian. These are described next.

12.4.1 Significance Tests for Assessing Gaussianity

Although the following tests are discussed in the context of assessing Gaussianity, the methods are general and can be used to assess whether the observed values fall into any specified pattern. Ironically, all these methods require large n in which case the sampling distribution of $\bar{x}$ tends to be Gaussian anyway. The methods would still be useful if your interest is in assessing the distribution of original values rather than of sample mean. A useful method is goodness-of-fit test based on chi-square. This is based on proportions in various class intervals and is presented in Chapter 13 on proportions. Other methods are as follows.

Among several statistical tests for Gaussianity, three most popular are Shapiro–Wilk test, Anderson–Darling test, and Kolmogorov–Smirnov test. All these are mathematically complex that you know are being avoided in this book. Statistical software packages generally have a routine for these tests that you can easily apply. However, it is important that you understand the implications.

Shapiro–Wilk test focuses on lack of symmetry particularly around the mean. This test is not much sensitive to differences present toward the tails of the distribution. Opposed to this, **Anderson–Darling test** emphasizes lack of Gaussian pattern in the tails of the distribution. This test performs poorly if there are many ties in the data. That is, for this test, the values must be truly continuous. **Kolmogorov–Smirnov test** works well for relatively larger n and when mean and SD are known a priori and do not have to be estimated from the data. This also tends to be sensitive near the center of the distribution than at the tails.

Critical value beyond which the hypothesis is rejected in Anderson–Darling test is different when Gaussian pattern is being tested than when another distribution such as lognormal is being tested. Shapiro–Wilk critical value also depends on the distribution under test. But Kolmogorov–Smirnov test is distribution free as the critical values do not depend on whether Gaussianity is being tested or some other form.

May sound strange to some but all these statistical tests cannot confirm Gaussianity although they confirm, with reasonable confidence, lack of it when present. Gaussianity is presumed when its lack is not detected. For reasonable assurance of Gaussianity, equivalence test discussed later in other context can be possibly devised.

12.5 Initial Debate on Statistical Significance

There is a considerable debate about the actual utility of the concept of statistical significance. Two aspects in particular are discussed. The first pertains to the relationship between CI and tests of significance, and which of these is better for inference. This is the subject matter of Section 12.5.1. The second concerns medical significance versus statistical significance. Sometimes, these two significances may be in conflict. A way to resolve this is suggested in Section 12.5.2.

12.5.1 Confidence Interval versus Test of H_0

It may not be immediately evident from the presentation so far that CI and the statistical tests are intimately related. In fact, they have a one-to-one correspondence.

Consider Example 12.11, where 8 births in a random sample of 60 are observed to be premature. The 95% CI for the proportion of premature births, by Equation 12.11, is

$$\frac{8}{60} - 1.96\sqrt{\frac{(8/60)(52/60)}{60}}, \quad \frac{8}{60} + 1.96\sqrt{\frac{(8/60)(52/60)}{60}}$$

$$= (0.047, \quad 0.219).$$

That is, there is a 5% chance that the actual percentage of premature births is less than 4.7 or more than 21.9. Any percentage between 4.7 and 21.9 is not unlikely. In other words, any percentage between 4.7 and 21.9 is plausible, whereas any outside these limits is not. Thus, this sample does not provide sufficient evidence to reject the null at $\alpha = 0.05$ if the hypothesized value of π is between 0.047 and 0.219 and will reject if the hypothesized value is either less than 0.047 or more than 0.219.

12.5.1.1 Equivalence of CI with Test of H_0

A test of hypothesis, which is also called a test of significance, can be alternatively performed as follows. Calculate the CI corresponding to the level of confidence ($1 - \alpha = 0.95$ or any other) you desire. Check whether the value of the parameter under H_0 falls within the CI. If yes, do not reject H_0; otherwise reject it. The level of significance for this test is α—the complement of the confidence level. If the test is to be carried out at $\alpha = 0.02$, then obtain a 98% CI and check whether this contains or does not contain the hypothesized value of the parameter. Conversely, set the value of the test criterion equal to its critical values (upper and lower) at the desired level, and solve for the values of the parameter. That will give the required $100(1 - \alpha)\%$ CI. However, for two-sample situation, if CIs for individual group means overlap, the sample means could still be significantly different. Consider means 10 and 22, each with SE = 4. The 95% CIs are (2.2, 17.8) and (14.2, 29.8) respectively. There is a substantial overlap. In this case, $Z = (22 - 10) / \sqrt{4^2 + 4^2} = 2.12$ and $P < 0.05$. The two means are significantly different despite overlapping CIs.

The first point of the **debate** arises from the equivalence of CI with the corresponding test of significance. If both are equivalent, which one is preferable? Many consider CI a better procedure than the test because CI gives the spectrum of unacceptable values that are outside the interval. Marginal values, both inside and outside CI, can be identified. It is thus possible to exercise more caution when the hypothesized value is on the margin. A CI provides a range of probable values of a population parameter rather than a dichotomy as significant or not significant the way a statistical test does. A test of significance considers only the null value. Medical implications of range of plausible values may be an important element to reach to a valid conclusion.

There is another view that supports CIs. They use inductive logic. From the observed data, you come up with a range that is not unlikely. Thus, CIs can generate new hypotheses and provide new learning. CIs do not depend on null hypothesis which is assumed true for calculating P-values.

In that sense, statistical tests use deductive logic of inference. Deductive logic is easy, direct, objective, and definitive, but it does not expand our knowledge. Inductive knowledge provided by CIs expands horizon and does not limit to preconceived values although it tends to be tentative.

12.5.1.2 Valid Application of Test of Hypothesis

The tests have particularly valid applications in situations in which the present knowledge or a claim is to be refuted. If a decision is to be taken one way or the other on the basis of a statistical result alone, perhaps test of hypothesis provides clear evidence. They also have a special place in comparison of two or more groups where the only objective is to find out whether they are different and the exact effect is not of immediate concern. Some feel lost without a *P*-value. In a regression setup, test of H_0 helps to decide whether to keep a variable or not as a suitable predictor. For further details of the debate about CI versus test of hypothesis, see Gardner and Altman [4].

The scientific community tends to accept new (data-based) findings when their statistical significance is adequately demonstrated. A $P < 0.05$ is generally accepted. Thus, $P = 0.06$ receives the same fate as $P = 0.30$. These two are clearly different and should have different implications. Note that the *P*-value is the probability of Type-I error, which is of rejecting H_0 when true. A 6% chance of error is higher than the threshold 5%, yet is fairly low. A 30% chance of error is very high. Similarly, a *P*-value 0.03 is different from *P*-value 0.001. Using cutpoint 0.05 as is sometimes done for testing of hypothesis masks this difference. Thus, another point of the debate is whether *P*-values, as they are, should be stated or only a cutoff such as 0.05 should be stated. The former is a more exact way of describing the situation, but the latter is simple to understand. Opinion is getting around to stating the exact *P*-values. However, as mentioned earlier, a cutoff is required in any case to make a decision one way or the other without forgetting that *P*-values can provide graded evidence instead of simply binary yes or no.

12.5.2 Medical Significance versus Statistical Significance

The term significant in common parlance is understood to mean noteworthy, important, or weighty as opposed to trivial or paltry. Statistical significance has the same connotation but it can sometimes be at variance with medical significance. A statistically significant result may be of no consequence in the practice of medicine, and a medically significant finding may sometimes fail a test of statistical significance.

Some medical professionals consider statistical methods notorious for discovering significance where there is none and not discovering where one really exists. The difficulty is that statistical methods give high importance to the number of subjects in the sample. If a difference exists in 12 cases of chronic cirrhosis of liver and 12 cases of hepatitis with respect to average aspartate aminotransferase (AST) levels, it can be considered a fluke because of the small size of the groups. This is like weak evidence before the court of law. But if the same difference is exhibited in a study on 170 cases of each type, it is very likely to be real. The *conventional statistical methods of testing hypothesis only tell whether a difference is unlikely or not. They do not say how much.* The difference in average AST levels between cirrhosis and hepatitis cases could be only 3 units/mL, which has no clinical relevance, but it would turn out to be statistically significant if this occurs in large groups of subjects. Medical significance of such a small difference should be separately evaluated using clinical criteria, and it should not depend exclusively on statistical significance. However, note that clinical criteria could be very subjective in many situations.

Example 12.12: Clinically Unimportant Difference Can Be Statistically Significant

Suppose it is known that 70% of cases with sore throat are automatically relieved within a week without treatment because of a self-regulating mechanism in the body. A drug was tried on 800 patients and 584 (73%) were cured in a week's time. Thus, for H_0: $\pi = 0.70$,

$$z = \frac{0.73 - 0.70}{\sqrt{0.70 \times 0.30/800}}$$

$$= 1.85.$$

In this case, suppose that the drug is pretested and there is no stipulation that the drug can make the relief rate even less than 70%. Then, the alternative hypothesis is one sided, H_1: $\pi > 0.70$. For this H_1, a one-tail probability is required. Therefore,

$$P\text{-value} = P(Z \geq 1.85)$$

$$= 0.0322 \text{ from Table B.1.}$$

This probability is very low and certainly small in comparison with the conventional level of significance 0.05. The null hypothesis is extremely unlikely to be true and is rejected. Statistical significance is achieved and the conclusion is reached that the 73% cure rate observed in the sample is really more than 70% seen otherwise. But, is this difference of 3% worth pursuing the drug? Is it medically important to increase the chance of relief from 70% to 73% in case of sore throat by introducing a drug? Perhaps not. Thus, a statistically significant result can be medically not significant.

More examples are given later to strengthen the distinction. One possibility is that a medically significant difference such as 10% is specified and then a statistical test is used to check whether the difference is beyond this threshold. This aspect is discussed in Chapter 13 for proportions and in Chapter 15 for means.

The inverse possibility is not so convincing, but that can also occur. A medically important difference should be statistically significant for it to be acceptable. Consider the same setup as in Example 12.12 where sore throat is assumed to be relieved in a week in 70% of cases without any intervention. Suppose now that the drug is tried in a sample of $n = 50$ patients and 40 ($p = 0.80$) respond in a week's time. Note that np and $n(1 - p)$ are both at least 8, so a Gaussian distribution can still be used. Now, under H_0: $\pi = 0.70$,

$$z = \frac{0.80 - 0.70}{\sqrt{0.70 \times 0.30/50}}$$

$$= 1.54.$$

This gives P-value = $P(Z \geq 1.54) = 0.0618$.

If the probability of Type-I error to be tolerated is less than 0.05, then the H_0 cannot be rejected. Despite 10% improvement shown by the drug in the sample, the sample still cannot be considered to provide sufficient evidence in favor of the alternative H_1: $\pi > 0.70$. Thus, this sample of size 50 does not favor a recommendation of use of the drug for patients with sore throat.

The rise of 10% in efficacy, from 70 to 80, may be considered medically important, and sufficient to pursue the drug. But this rise in this sample of 50 subjects could well have arisen due to chance—due to sampling fluctuation. There is some likelihood that this rise would fail to be reproduced in another sample of 50 subjects or that it would fail to persist in the long run. However, the results could be considered to justify a bigger trial because the P-value is only slightly more than 0.05.

Perhaps, it needs to be emphasized that the difference must be statistically significant for it to be medically relevant. If it is not statistically significant, nobody can be confident that the difference is actually present. Thus, the first step is to assess statistical significance. If not significant and the statistical power is adequate, there is no need to worry about its medical relevance because it is likely to be there by chance. If significant, further statistical testing is used to judge if it reaches a medically relevant threshold. This threshold comes from medical acumen. A value judgment is still required. If in an RCT, 3 deaths occur in 100 subjects in placebo group and none out of 100 in treatment group, the difference is not statistically significant, but would you not try this treatment for your family if everything else has failed? Additional factors such as environment, family condition, and availability of health infrastructure are also considered while taking a final decision regarding the management of a patient. For further debate on medical significance versus statistical significance see Section 15.4.

You may also like to make a distinction between (statistical) inference and the results of a (statistical) test. Inference considers several sources of evidence of which P-value is just one. Background factors, covariates, consequences, etc., are also considered while drawing an inference.

12.6 Sample Size Determination in Some Cases

It is customary to talk about sample size determination in a chapter on sampling. I deferred the discussion till this point in this book because knowledge of the concepts of level of confidence and power is required.

Sample size, *n*, is just about the first thing that comes to mind when planning an investigation. Indeed, it is among the most important considerations that determine the utility of a study. The sample size should be neither too large nor too small. A large *n* could mean a waste of resources. If a study of only 200 subjects can give a reliable answer, why spend resources to study 250? A small sample, on the other hand, may not give evidence one way or the other and the study may fail to achieve its objectives. Thus, this also is a waste of resources. An unduly large sample is unethical in experiments because it means that some subjects are unnecessarily exposed to an intervention whose utility is in doubt. An unduly small sample, too, is unethical because then the subjects are unnecessarily exposed in an experiment that is not going to yield a result one way or the other.

Statistically, larger the sample more is the reliability of the results. Administratively, a large sample is hard to execute as it tends to become a burden on the investigator in some situations. Small samples are not useless altogether. They are valid for a pilot, exploratory, or feasibility study. When honestly done, they do allow us to move forward from knowing less. Small sample can be adequate if the anticipated effect size is large. In most other situations, small samples fail to provide a definite conclusion.

A statistician is sometimes expected to be able to suggest an adequate sample size immediately on being told of the title and the objectives of a study. This is like expecting a physician to prescribe a treatment regimen for pain in the abdomen without going into the details. A physician would need some more information such as the duration and intensity of pain, exact location, palpability or tenderness, any accompanying complaint, sometimes an x-ray image, etc. Similarly, the question about sample size can be answered only after some deliberation on the variability of the observations, precision required, confidence desired, etc. These are explained in the following sections. The procedure to determine the sample size is different for estimation than for testing of hypothesis.

12.6.1 Sample Size Required in Estimation Setup

Estimation of a parameter such as mean and proportion is generally the primary objective of descriptive studies. Although a descriptive study could be a case series but that is rarely based on a random sample. This leaves sample survey as the only relevant format of descriptive studies for calculation of sample size in estimation setup. But estimation is done in analytical studies as well. Following are the details of the concerns that need to be addressed to arrive at a sample size for estimating a population parameter. These details focus on mean and proportion but the same sort of arguments apply to the other parameters.

12.6.1.1 General Considerations in the Estimation Setup

The following may look like commonsense considerations in determining the sample size for estimation:

1. For calculation of sample size, first consideration is the *variable of interest*. In antecedent–outcome setup, the sample size will be based on the outcome variables. In most situations, you will collect data on a large number of variables but, generally, you will measure the outcome primarily by one or two parameters of selected priority variables. If you are measuring blood pressure, body mass index, blood glucose, and cholesterol level in a study on noncommunicable diseases, the sample size can be based on any of them and different variables will give you different answers. You may like to calculate for all of them and choose the maximum as your sample size. Generally, though, sample size calculation is done for a specific target variable of primary interest.

2. Another point of consideration is that your *parameter of interest* is mean, proportion, OR, correlation coefficient, or what. Sample size formula will depend on the choice of such a parameter.

3. What is the *variability* between subjects in the population? Cholesterol level tends to differ widely even among healthy subjects. Thus, a small sample of, say, seven subjects would not be adequate to reveal the full spectrum of cholesterol levels present in the target population. That is, another sample of seven subjects can yield an entirely different picture. Body temperature, on the other hand, is relatively stable in healthy subjects and a small sample can be adequate. If the material to be sampled is as homogeneous as blood in the body, even one drop may be enough to reveal nearly the full picture in a person. The sample size requirement increases as the variance increases. In the case of sample proportion p, the variance depends on the value of π. Thus, the question becomes what proportion of subjects is expected to possess the characteristic under investigation? In relative sense, a large sample is required if π or $(1 - \pi)$ is small.

4. What is the *minimum degree of precision* required? It may be important for a pharmaceutical company to know very precisely the percentage of those with tuberculosis who are becoming drug resistant. This can help the company to assess the exact market for a new antitubercular drug. On the other hand, in surveys on opinions, attitudes, and behaviors, an error in the estimate to the extent of as much as 10% or sometimes even higher may be tolerable. The sample size should evidently be larger when greater precision is required.

In case of proportions, as explained earlier, the precision can be expressed in absolute terms, such as 2%, or in relative sense, such as 10% of p. It is important to realize the difference.

5. What *least confidence* in the estimate would be tolerable? No empirical result is 100% dependable, but the investigator may wish to be sufficiently confident that the result will be replicated in repeated samples. There is always a chance of it being false. The chance of reaching a wrong result can be kept low, generally less than 5%, by including a reasonably large number of subjects in the sample. In some cases, such a large sample may not be feasible because of constraints of time and resources. If a chance of as much as 10% or 20% of being wrong can be tolerated, as in the case of some opinion surveys, then a small sample may be adequate.

6. Are there *any subgroups of interest*? If conclusions are to be drawn for various subgroups, as in stratified sampling in some cases, then each subgroup needs to be adequately represented. The sample size quickly multiplies and can become an enormous number if cross-classifications of a number of factors are under consideration. For only two factors, viz., age at menarche as <12 and ≥12 years, and age at first live birth as <25 and ≥25 years, the four cross-classifications are (a) menarche at <12 years and first live birth at <25 years, (b) menarche at <12 years and first live birth at ≥25 years, (c) menarche at >12 years and first live birth at <25 years, and (d) menarche at ≥12 years and first live birth at ≥25 years. If there are four factors and all are dichotomized, the number of cross-classifications is $2^4 = 16$. Results would be reliable if each of these classifications has an adequate number of subjects.

7. How much, if any, *nonresponse* is expected? Although nonresponse raises questions about the validity of the survey, it also reduces the effective number of subjects that can be utilized to draw conclusions. Thus, a larger sample is planned if it is feared that many subjects may not be available, may not cooperate, or may have to be dropped because of development of, say, undesirable side effects. If the subjects are healthy and an invasive procedure is involved, even as small as a pin prick, the attrition could be large.

8. What *sampling procedure* is to be used? For example, cluster sampling may require double the size or even larger relative to SRS because of the clustering effect. Different procedures have different limitations. Procedures such as stratified and two-stage can require smaller samples to achieve the same precision if the units can be rationally divided into homogeneous groups. This may be difficult to achieve when sufficient information is not available or when the subjects are widely scattered. Then, the sample size could increase. This will depend on the design effect. At the same time, sampling procedure loses much of its

bite when you plan have a huge sample. Such large sample is expected to provide reliable results anyway irrespective of sampling plan.

9. Although ignored in most practical situations, the sample size also depends on the sampling distribution of the summary measure under study. This text assumes Gaussian distribution, which is likely at least for mean and proportion when the sample size is large.

10. Sample size also depends on the number of variables to be *simultaneously* considered. Higher the number of covariates, more is the requirement of bigger sample. Because of complexity, this text discusses only one variable at one-time setup.

11. Sample size also depends on the design of your study. Designs such as matching and repeated measures require different approach than independent samples.

In summary, bigger sample is required if

1. Interindividual variability is higher or expected prevalence is low
2. More confidence or higher precision is required in the result
3. There are subgroups for whom results are to be applied
4. Higher nonresponse is expected
5. Sampling method is other than simple random (in most situations)

12.6.1.2 General Procedure for Determining Size of Sample for Estimation

Let the population parameter under estimation be denoted by τ and its sample estimate by t. Let δ be the difference between the two, that is, $\delta = |\tau - t|$. Suppose the investigator requires that this difference does not exceed a specified limit L in at least $100(1 - \alpha)\%$ of repeated samples. The quantity L is called precision and is also the half-width of the CI. The quantity $(1 - \alpha)$ is the confidence level. If a Gaussian form of distribution can be assumed for sample estimate t, which, in fact, could be so in most cases when the sample size is large, it can be shown that

$$L = z_{\alpha/2} * SE(t), \tag{12.24}$$

where the coefficient $z_{\alpha/2}$ is taken from the standard Gaussian distribution. The table of probability for a Gaussian curve needs to be consulted to find a cutoff $z_{\alpha/2}$ such that the probability between $-z_{\alpha/2}$ and $+z_{\alpha/2}$ is $(1 - \alpha)$. The exact value of $z_{\alpha/2}$ for $\alpha = 0.05$ is 1.96. For $\alpha = 0.10$, $z_{\alpha/2} = 1.645$ and for $\alpha = 0.01$, $z_{\alpha/2} = 2.58$. Thus, for confidence level 90%, $L = 1.645 * SE(t)$; for 95%, $L = 1.96 * SE(t)$; and for 99%, $L = 2.58 * SE(t)$.

Equation 12.24 is basic for calculation of sample size in an estimation setup. $SE(t)$ would almost invariably have n in the denominator, which could then be worked out when other values are known. But a difficulty is that $SE(t)$ would also contain an unknown parameter such as σ, which is to be replaced by its estimate. Where would you get this estimate before the survey is conducted? Either from a previous study or from a pilot study. Note that $SE(t)$ also depends on the sampling method proposed to be followed.

Since $SE(t)$ always contains n, and L is specified, it is possible to obtain n from Equation 12.24. This may have to be inflated to adjust for expected nonresponse. If estimates for various subgroups are required, then this calculation is done separately for each subgroup.

If there are many parameters under estimation, two approaches are available. The first is to calculate the sample size for the most important parameter if that can be identified. The second is to calculate the size for all the parameters and use the one that is largest. The latter would give better-than-required precision of some estimates but each estimate will have *at least* the specified precision. The following example may clarify the procedure to be followed in some cases.

Example 12.13: Sample Size for Estimating Mean Blood Glucose Level

A sample survey is planned to estimate the average level of fasting blood glucose in surviving and apparently healthy females of age 60 years and above of a particular ethnic group. It is desired that the estimate should be within 1.2 mg/dL of the population mean with probability 0.95. A previous study revealed an estimate of SD = 6.3 mg/dL. How big a sample is required if an SRS is to be followed?

In this case, $L = 1.2$. For SRS, $SE(\bar{x}) = \sigma/\sqrt{n}$. If we replace σ by its estimate $s = 6.3$, then Equation 12.24 gives $1.2 = 1.96 \times 6.3/\sqrt{n}$. Or, $n = 105.9$. This is always approximated upward, in this case $n = 106$.

SIDE NOTE: If greater precision (lower L) is required, say, $L = 0.4$, then $n = (1.96 \times 6.3/0.4)^2 = 953$. Note how rapidly the sample size increases when the required precision is enhanced. If the confidence level $(1 - \alpha)$ is raised to 0.99, then, for $L = 1.2$, $1.2 = 2.58 \times 6.3/\sqrt{n}$ and $n = 184$. If the subject-to-subject variability in the population is smaller, say, $s = 2.4$ only, then, for $L = 1.2$ and $\alpha = 0.05$, n is 16.

A bigger sample is required whenever (1) greater precision is required, (2) more confidence is required, or (3) the variability in the population is large. The quantity that affects it most is the precision you desire of the estimate. For double precision (i.e., half L), the sample size required is four times.

If separate estimates for age groups 60–69 and 70+ years are required, and if SD and other values are same for each age group in Example 12.13, then the same sample size is required for subjects of 60–69 years and for those of 70+ years. The total sample size doubles.

For proportions and $(1 - \alpha) = 0.95$, Equation 12.24 gives $L = 1.96 * SE(p)$. If the objective is to estimate the *proportion* of apparently healthy subjects who can be suspected as diabetic on screening using the cutoff 120 mg/dL of fasting blood glucose level and if it is desired that the estimate should not differ by more than one-fifth of the proportion of π in the population, then $L = \pi/5$. This is the relative precision that can be expressed as 20% of π or 0.20π. Since $SE(p) = \sqrt{\pi(1-\pi)/n}$, we need an estimate of π for the sample size calculation. If it is anticipated to be between 0.02 and 0.03, both limits can be used to calculate the sample size. For $\alpha = 0.05$, Equation 12.24 gives

$$\frac{0.02}{5} = 1.96\sqrt{\frac{0.02 \times 0.98}{n}} \quad \text{for } \pi = 0.02, \quad \text{and}$$

$$\frac{0.03}{5} = 1.96\sqrt{\frac{0.03 \times 0.97}{n}} \quad \text{for } \pi = 0.03.$$

The former gives $n = 4706$ and the latter $n = 3106$. The bigger of the two is statistically safe. Thus, a sample of size nearly 4700 is required.

Sometimes it is desirable to do reverse calculations. If resources permit a specific n, Equation 12.24 can be used to find the precision corresponding to that n. For example, if not more than 30 subjects can be studied due to resource limitations in Example 12.13, and if the values of s and the level of confidence are to remain the same, then from Equation 12.24,

$$L = 1.96 \times \frac{6.3}{\sqrt{30}}$$

$$= 2.3\,\text{mg/dL}.$$

Thus, the estimate of mean fasting blood glucose level can be 2.3 mg/dL away from its true value in the population. If this error is acceptable, go ahead with a sample of size 30. If this is too high, reconsider carrying out the study on only 30 subjects. A larger sample may be required.

If L continues to be 1.2 mg/dL as in Example 12.13, then the level of confidence can be calculated corresponding to a given size of sample. For $n = 30$, $1.2 = z_{\alpha/2} \times 6.3/\sqrt{30}$ or $z_{\alpha/2} = 1.04$. From Table B.1, get $\alpha/2 = 0.1492$. Thus, $\alpha = 0.30$ and the confidence level is only 70%. When n is 30, there is only 70% chance that the estimated mean would be less than 1.2 mg/dL away from the actual mean in the population. This chance may be too low to proceed with a study on such a small sample.

12.6.1.3 Formulas for Sample Size Calculation for Estimation in Simple Situations

The sample size formulas for simple situations are given in Table 12.7. These are based on the procedure explained earlier. Further details are omitted. The notations are also explained in the table. The formulas are valid only for situations in which Gaussian approximation is applicable. The parameter values such as π and σ that appear in these formulas have to be replaced by their working estimates as explained earlier. All formulas assume SRS and a univariate setup.

If educated guess of π is not available, you may like to use $\pi = 0.5$. This will give maximum n. In case of sensitivity and specificity, which too are

TABLE 12.7

Sample Size Calculation for Some Estimations (Valid for Large n Only)[a]

Problem	Formula for Computing n	Description of the Notations
a. Population proportion with specified absolute precision	$\dfrac{z_{\alpha/2}^2 \pi(1-\pi)}{L^2}$	π = Anticipated value of the proportion in the population L = Absolute precision required on either side of the proportion
b. Population proportion with specified relative precision	$\dfrac{z_{\alpha/2}^2 \pi(1-\pi)}{(\varepsilon\pi)^2}$	π = Anticipated value of the proportion in the population ε = Relative precision in terms of fraction
c. Population mean with specified precision	$\dfrac{z_{\alpha/2}^2 \sigma^2}{L^2}$	σ = Population SD (can be estimated from a pilot study) L = Specified precision of the estimate on either side of the mean
d. Difference between two population proportions with specified absolute precision—equal n in the two groups	$\dfrac{z_{\alpha/2}^2[\pi_1(1-\pi_1)+\pi_2(1-\pi_2)]}{L^2}$	π_1, π_2 = Anticipated proportions in the two populations L = Absolute precision required on either side of the difference in proportions
e. Difference between means of two populations with specified precision—equal n in the two groups	$\dfrac{z_{\alpha/2}^2(\sigma_1^2 + \sigma_2^2)}{L^2}$	σ_1, σ_2 = Population SD of the two populations (can be estimated from a pilot study) L = Specified precision of the estimated difference on either side of the mean difference

[a] Large n is needed so that the distribution of p can be approximated by Gaussian form. For $\alpha = 0.10$, $z_{\alpha/2} = 1.645$, for $\alpha = 0.05$, $z_{\alpha/2} = 1.96$. For other values, consult Table B.1. The formulas are based on two-sided CI. In case of one-sided confidence bounds, replace $z_{\alpha/2}$ by z_α.

proportions, obtain the sample size as usual for proportions and divide by prevalence rate of disease for sensitivity and by (1 − prevalence) for specificity. If prevalence of disease you are investigating is 12%, divide by 0.12 to obtain sample size for estimating sensitivity and by 0.88 to obtain sample size for estimating specificity. This ensures that the requisite number of disease positive and disease negative, respectively, are available in the sample.

Example 12.14: Sample Size for Estimating the Difference in Two Proportions

To plan a study on the difference in the prevalence of filariasis in agricultural and nonagricultural workers in an endemic area, a pilot study was carried out with 30 workers of each type. The prevalences, respectively, were 33% and 20%. What sample size is needed if the difference is to be estimated within 3 percentage points with 90% confidence?

The calculation requires anticipation that the proportions in the population would be nearly the same as in the pilot study. Thus, $\pi_1 = 0.33$ and $\pi_2 = 0.20$. Confidence $100(1 - \alpha) = 90\%$ or $\alpha = 0.10$. Since $L = 0.03$, from formula (d) in Table 12.7, get

$$n = \frac{z_{0.05}^2 (0.33 \times 0.67 + 0.20 \times 0.80)}{(0.03)^2}$$

$$= \frac{(1.645)^2 \times (0.3811)}{(0.03^2)}$$

$$= 1146.$$

Thus, a sample of nearly 1150 is required in each group of workers.

Example 12.15: Too Large a Sample in a Descriptive Study on Prescribing Pattern of Drugs for Osteoarthritis and Rheumatoid Arthritis

Schnitzer et al. [5] studied prescribing patterns of rofecoxib and celecoxib in 47,935 patients of osteoarthritis (OA) and 10,639 patients of rheumatoid arthritis (RA) in the United States. The most frequently prescribed daily dose of rofecoxib was found to be 25 mg in both OA and RA, but the most frequently prescribed dose of celecoxib was 200 mg in OA and 400 mg in RA. They did not perform any test of statistical significance because such a big sample would have found a medically irrelevant small difference as significant.

Despite such a big sample, the authors rightly remarked that their conclusions are limited by lack of clinical information, inability to ascertain actual use, the potential for selection bias, etc.

SIDE NOTE: Although the authors called this an observational, retrospective cohort study in the title, they also called it as a primarily descriptive study in the text. Such confusion in the nomenclature of the study methodology is frequent in medical literature. The main conclusion too is only on the most common type of prescription. Thus, it is best categorized as a descriptive study.

The formulas provided in Table 12.7 are for estimating mean, proportion, and their difference in two populations. It does not contain formulas for ratios such as OR and RR. These are provided in Chapter 14.

12.6.2 Sample Size for Testing a Hypothesis with Specified Power

The primary aim of analytical studies is to investigate antecedent–outcome relationship. Although other setups are possible, most common is to find whether two groups are different or not with respect to antecedent factor in the case of retrospective studies, and with respect to outcome in the case of prospective studies. Statistically, this leads to testing of hypothesis setups.

The method for determining sample size in the testing-of-hypothesis setup depends on the actual criterion used for testing. Various criteria are discussed later in Chapters 13 through 15. At this stage, it may be stated that the sample size calculation requires the following information in most situations encountered in practice.

12.6.2.1 *General Considerations in a Testing-of-Hypothesis Setup*

Following is a list of general considerations. These are mostly the same as in the estimation setup but there are some changes:

1. Same as in the estimation setup.

2. Same as in the estimation setup.

3. Same as in the estimation setup.

4a. How much minimum *difference* (δ) between the actual value of the parameter and its value under the null hypothesis is *medically important*. This is the test of hypothesis counterpart of "degree of precision" in the estimation setup. A medically important difference is specified under the alternative hypothesis H_1. A small difference is difficult to detect, and a large sample would be required for a small difference to be statistically significant. If the minimum difference to be detected were large, a small sample would be adequate. This is determined on clinical considerations. Do not confuse it with actual effect size, which could be more or could be less.

4b. What is the statistical *power* required? This is the probability of detecting a specified difference and calling it statistically significant. The notation for this is $(1 - \beta)$. Power depends on the magnitude of the difference specified in (4a). If the power is to be 99% for the specified difference, obviously a bigger sample is required than for power 80%. Power and n have a direct relationship for any fixed level of significance.

5a. What *level of significance* is required? In the testing-of-hypothesis setup, the complement of confidence level is the level of significance α. On the analogy stated for CI, a large sample is required if α is to be kept small. If a large α can be tolerated, a relatively small sample would be enough.

5b. Is the test a *one-tail test* or a *two-tail test*? A one-tail test with $\alpha = 0.05$ is equivalent to a two-tail test with $\alpha = 0.10$ in most situations. A high α-level in a two-tail setup is a relatively small α-level in a one-tail setup. This is true for *P*-values also. Thus, one-tail testing requires a smaller sample size then a two-tail set up. If $n = 300$ gives $P = 0.03$ in a two-tail test, the same n would generally give $P = 0.015$ in a one-tail test. For one-tail $P = 0.03$, the required n would be smaller than for a two-tail $P = 0.03$.

6. Same as in the estimation setup.

7. Same as in the estimation setup.

8. Same as in the estimation setup.
9. Same as in the estimation setup.
10. Same as in the estimation setup.
11. Same as in the estimation setup.

In the case of large n, when Gaussian conditions prevail, these considerations lead to an equation of the following type in most situations:

Power
$$= P(Z \geq z_\alpha \text{ when the specified medically important difference is present})$$
$$(12.25)$$

where
    Z is the criterion, assumed Gaussian here
    z_α is the table value of Z corresponding to α level of significance

The sample size n would occur on the right side of Equation 12.25. This can be solved to obtain n when everything else is specified. However, the specific form of Z would depend on the criterion used for testing. That, in turn, depends on considerations (1) through (6) listed in the introductory remarks of Section 12.3.3.

12.6.2.2 Sample Size Formulas for Test of Hypothesis in Simple Situations

Since the solution of Equation 12.25 requires the actual test procedure described in subsequent chapters, the formulas of n are given in Table 12.8 only for some simple situations. Again, these are valid only for large n where Gaussian approximation is applicable and when SRS is used. For sample size for testing hypothesis on OR and RR, see Chapter 14, and for survival analysis, see Chapter 18.

Sometimes, approximate calculations are done for simplicity. For example, for power 80%, $1 - \beta = 0.80$ and $z_\beta = 0.84$. For $\alpha = 0.05$ and two-sided hypothesis, $z_{\alpha/2} = 1.96$. Thus, formula (b) in Table 12.8 becomes

$$n = \frac{(1.96 + 0.84)^2 \sigma^2}{\delta^2} \approx 8 \frac{\sigma^2}{\delta^2}.$$

Just find how many times is σ of the difference to be detected, square and multiply by 8.

This is easy to remember and can be used for 80% power without being too far off.

Example 12.16: Sample Size for Testing Prevalence of Worm Infestation

Suppose it is believed that the prevalence of worm infestation in agricultural workers in an area is around 30%. What sample size should be chosen so

TABLE 12.8

Sample Size Calculation for Some Testing of Hypothesis Situations (Valid for Large n Only)—Two-Sided H_1

Problem	Formula for Computing n	Description of the Notations
a. For a population proportion	$$\left[\dfrac{z_{\alpha/2}\sqrt{\pi_0(1-\pi_0)}+z_\beta\sqrt{\pi_a(1-\pi_a)}}{\delta^2}\right]^2$$	π_0 = Value of π under H_0 π_a = Medically important value of population proportion under H_1, that is $\delta = (\pi_0 - \pi_a)$ is the difference proposed to be detected
b. For a population mean	$$\dfrac{\sigma^2(z_{\alpha/2}+z_\beta)^2}{\delta^2}$$	σ = Population SD (can be estimated from a pilot study) δ = Minimum medically important difference between means under H_0 and H_1 that is proposed to be detected
c. For difference between two population proportions— equal n in the two groups	$$\left[\dfrac{z_{\alpha/2}\sqrt{2\bar\pi(1-\bar\pi)}+z_\beta\sqrt{\pi_1(1-\pi_1)+\pi_2(1-\pi_2)}}{\delta^2}\right]^2$$	π_1, π_2 = Anticipated proportions in the two populations, that is, $\delta = (\pi_1 - \pi_2)$ is the difference proposed to be detected. $\bar\pi = (\pi_1 + \pi_2)/2$, $H_0: \pi_1 = \pi_2$
d. For difference between two population means—equal n in the two groups	$$\dfrac{(\sigma_1^2+\sigma_2^2)(z_{\alpha/2}+z_\beta)^2}{\delta^2}$$	σ_1, σ_2 = Population SD of the two populations (can be estimated from a pilot study) They will be equal in most situations δ = Minimum medically important difference between means under H_1 that is proposed to be detected
e. For independence in matched pairs	$$\left[\dfrac{z_{\alpha/2}\sqrt{\pi_{10}+\pi_{01}}+z_\beta\sqrt{\pi_{10}+\pi_{01}-(\pi_{10}-\pi_{01})^2}}{\delta^2}\right]^2$$	$H_0: \pi_{10} = \pi_{01}$, equivalent to $\pi = \frac{1}{2}$ π_{01} and π_{10} are discordant probabilities $\delta = \pi_{10} - \pi_{01}$ is the difference proposed to be detected

Notes: α is the level of significance and $(1 - \beta)$ is the statistical power. z_α is such that $P(Z \geq z_\alpha) = \alpha$. For $\alpha = 0.05$, $z_{\alpha/2} = 1.96$ and $z_\alpha = 1.645$. For $\beta = 0.10$, $z_\beta = 1.28$; for $\beta = 0.05$, $z_\beta = 1.645$. For other values, consult Table B.1. If the alternative hypothesis is one sided, replace $z_{\alpha/2}$ by z_α.

that this null hypothesis is rejected with probability 0.90 if the actual prevalence is 25% or lower? Keep the probability of Type-I error less than 0.05.

In this example, $\alpha = 0.05$, $1 - \beta = 0.90$, $\pi_0 = 0.30$, and $\pi_a = 0.25$, where π_a is
the value of π under the alternative hypothesis. It is a one-sided test because
the concern is with the lower values only. From formula (a) in Table 12.8, get

$$n = \frac{\left[1.645\sqrt{0.30 \times 0.70} + 1.28\sqrt{0.25 \times 0.75} \right]^2}{(0.30 - 0.25)^2}$$

$$= \frac{(0.7538 + 0.5543)^2}{(0.05)^2}$$

$$= 685.$$

The survey should include 685 workers. If nonresponse is expected, increase
the size accordingly. If the test of significance is to be done separately for
male and female workers, take a sample of 685 males and 685 females.

Example 12.17: Sample Size for Detecting a Minimum Clinically Important Difference

Suppose a difference of at least 10 mg/dL in triglyceride level (TGL) is
considered clinically important. Suppose it is also known that persons
on vegetarian diet have lower average TGL level than persons on nonvegetarian diet. But the actual difference it not known. A group of nonvegetarians and another group of vegetarians are proposed to be investigated.
Previous studies suggest that SDs are 15.7 and 12.5 mg/dL, respectively.
The researcher wishes to detect a difference of 10 mg/dL with probability
0.80. What sample size should be chosen if the level of significance is 0.10?

In this case, $\delta = 10$ mg/dL, $\sigma_1 = 15.7$, and $\sigma_2 = 12.5$ (all are estimates
though). Since vegetarians can have only lower level, H_1 is one-sided.
Now, $\alpha = 0.10$ and $(1 - \beta) = 0.80$ give $z_\alpha = 1.28$ and $z_\beta = 0.84$. With these
values, formula (d) in Table 12.8 gives

$$n = \frac{[(15.7)^2 + (12.5)^2](1.28 + 0.84)^2}{(10)^2}$$

$$= \frac{402.74 \times 4.4944}{100}$$

$$= 19 \text{ when approximated upward.}$$

A sample of 19 vegetarians and 19 nonvegetarians is enough for this
study. This relatively small sample is adequate because the difference to
be detected, 10 mg/dL, is quite large. If this is 5 mg/dL then

$$n = \frac{[(15.7)^2 + (12.5)^2](1.28 + 0.84)^2}{(5)^2}$$

$$= 73.$$

Conversely, if resources permit $n = 100$, then for $\delta = 5$,

$$100 = \frac{[(15.7)^2 + (12.5)^2](1.28 + z_\beta)^2}{(5)^2}.$$

This gives $z_\beta = 1.21$. From Table B.1, $z_\beta = 1.21$ gives $\beta = 0.1131$ and $(1 - \beta) = 0.8869$. Thus, a size of 100 will have a probability of nearly 89% of detecting a difference of $5\,\text{mg/dL}$.

Example 12.18: Sample Size for a Prospective Study on Disability in Work-Related Musculoskeletal Disorders

Turner et al. [6] report a prospective study of workers in the United States who file claims for work-related musculoskeletal disorders. The primary outcome of interest was duration of work disability in 1 year after filing the claim. The purpose was to develop statistical models that could predict the duration of chronic work disability after the initial suffering from the disorder. The required sample size was calculated as 1800 workers for low back injuries and 1200 for workers with carpal tunnel syndrome (CTS). Statistical powers used for these calculations were 0.96 for low back and 0.85 for CTS, and the significance level chosen was $\alpha = 0.05$ (two-tailed) for both.

SIDE NOTE: The example illustrates that the sample size could be quite large for prospective studies when power and significance level are considered. The concept of power is related to the minimum medically relevant difference but the abstract of this article does not mention this difference. A large sample size was feasible in this case because the study is mostly based on administrative database and follow-up interviews were conducted over telephone.

It is an irony that sample size calculations require an estimate of the variance of the observations even before the sample is studied. As advised, this may be taken from a previous study or estimated from a pilot study. Sometimes it may involve guesswork. The sample size so calculated will be approximate. To be safe, it is better to inflate this n slightly to compensate for possible error in the estimate of SD. When SD from previous studies is not available and the range is available instead, a conservative guess for calculation of sample size is SD = range/4, particularly if the underlying distribution is nearly Gaussian. As already mentioned, for proportion π, a conservative estimate is 0.5 that will give higher sample size than any other π.

For sampling methods other than SRS, the SEs of these methods will be used in Equations 12.24 and 12.25. This, for example, in a two-stage sample, can raise questions on balancing the number of primary units for first-stage sampling and the number of subjects to be selected at the second stage. The calculations can become complex and the information required for computing the size can also become difficult to manage. Some details of this are given by Som [7].

When two or more groups are planned to participate in a study, it is ideal to calculate n separately for each group. But this might make the exercise too complex. Generally, the size is calculated for the most important among the study groups and the same n is used for the other groups as well. For many groups in analysis of variance (ANOVA) setup, the significance level for calculation of sample size may be taken as α/H, where H is the number of comparisons. This comes from Bonferroni procedure explained in Chapter 15. For example, for four groups, if all pairwise comparisons are of interest, $H = 6$. In addition, medical and health studies are notorious for nonresponse. As stated earlier, the sample size needs to be inflated if the nonresponse is expected to be high.

12.6.2.3 Nomograms and Tables of Sample Size

Many medical researchers find formulas in Tables 12.7 and 12.8 difficult to adopt. Most statistical softwares incorporate tools to calculate sample size based on your requirement. Two other alternatives are available, although they may not be readily available. One is nomogram. This is a graph containing lines and curves and needs only a ruler to read the sample size that would meet the specifications. Altman [8] has given a nomogram for size of SRS for comparing two groups. Kumar and Indrayan [9] have developed one such nomogram for reading the sample size required in cluster sampling (Figure 8.9). Neter and Wasserman [10] have given nomograms for sample sizes for ANOVA situations where three or more groups are compared for means. If you are weary of the formulas, and the relevant software is not available, try to locate a nomogram for your situation. Nomogram is especially helpful to find sample sizes for several scenarios and choose the scenario considered most suitable. It can be easily used and reused and can be carried in your pocket to the field areas if necessary.

Second alternative is table of sample sizes. Some authors have worked out sample sizes for various situations and tabulated them. One such compilation is by Lwanga and Lemeshow [11] and the other is by Machin et al. [12]. A'Hern [13] has provided sample size tables for phase II traials on small samples. These tables are based on exact binomial distribution. Sample size required for the specified values can be directly read from such tables. An interpolation may be required when the desired exact specification is not present in the table.

12.6.2.4 Thumb Rules

Thumb rules lack scientific basis and many scientists dislike them. When no baseline information for computation of sample size is available and the constraints do not permit pilot study either, the following thumb rules can be used.

A large-sized medical trial should include nearly 300 subjects in *each* group, a midsized trial nearly 100 in each group, and a small-sized trial at least 30 in each group. The last can be used for postgraduate theses where the time and resources are limited. Bigger study is multicentric with these numbers in each center. Same norms can be used for a retrospective or case–control study. However, in the case of a prospective study, the number to be followed up should be such that at least 30 persons are finally available with the outcome of interest in *each group*. This applies to field trials also. In this case, an extremely large group may be needed to yield a rare outcome such as hepatitis B infection in at least 30 subjects after administration of a protective vaccine. To calculate exact numbers, use the formulas given in different tables in this book. A study of these formulas would indicate that a laboratory experiment on animals can be done on a smaller number because of fairly standard and controlled conditions in a laboratory that minimize variations.

For a descriptive study that seeks to find normal levels in healthy subjects, the thumb rule is to include at least 200 subjects in each group for which norms are required. For pathological levels in patients, the group-size could be smaller. Depending upon the targeted reliability of the results, exact sample-size requirement can be calculated in each case. However, all these may have to be modified because of feasibility considerations when the resources and time are limited. Such limitations obviously compromise the reliability.

12.6.2.5 Power Analysis

Good researchers determine sample size on the basis of power required to detect a specified minimum difference and try to do study on the size so determined. Because of some exigencies, sometimes it becomes difficult to study as many subjects. This reduces the power. Sometimes, a study is done on specific *n* without basing it on power. The formulas given in Table 12.8 can be used to inversely calculate power for a specific *n*, when other pieces of information are available. This is illustrated in Example 12.17. Nomograms and other tables can also be used. This process, sometimes referred to as power analysis, is becoming a requirement these days as awareness has increased. In case you are doing a study yourself or consulting other studies to update your knowledge, examine if the power of the study is adequate to inspire confidence.

There is a considerable overlap in the literature about the term power analysis. It is used for calculation of power for given *n* as well as for calculation of *n* for specified power. Both calculations require prior specifications based on literature or previous experience. Real utility of power analysis is in designing a study and not so much in interpreting nonsignificance of the result once obtained. Nonetheless, midcourse power calculation can be done on the basis of the observed data and can tell you how much more *n* you need. Post hoc calculation can tell you what might have gone wrong. Use this experience for planning a better study next time.

There is another exception, though. If the study throws up very different p or SD than assumed earlier for calculation of sample size, and the effect size is found statistically not significant, recalculation of power with new p or SD is justified and can give you better leads about the strength of your result. This can lead to sample size reestimation as described in the next section.

12.6.3 Sample Size in Adaptive Clinical Trials

Clinical trials are costly and difficult to repeat. There are several other problems. Among them, statistical problems relate to missing data due to drop outs, partial compliance, and multiplicity of end-points. Design of the trials tends to focus on primary end-point but also keep secondary end-points in mind. Clinical trial must be planned for adequate sample size so that there is no wastage in a trial that would not give clear evidence one way or the other, nor should it be unduly large if clear evidence could emerge from a relatively small trial. Unnecessary large trial would mean longer period of trial and more cost in management, in addition to the cost of patients and delayed approval of the regimen.

You now know that clinical trial is an analytical study involving test of hypothesis. The sample size calculation requires specification of level of significance and power but most of all mean and SD (or proportion) and the minimum clinically relevant gain you wish to detect. These are not easy to specify. For example, minimum clinically relevant difference varies according to perception. In addition, for low prevalent disease, even a difference of 10% in mortality may not be worth investigating since the trial needs huge resources. Wherever you can, use the formulas in Table 12.8 to calculate the sample size for the primary end-point of your choice. You may like to calculate this for secondary end-points as well and choose the maximum as your sample size.

Almost always a trial is planned on the basis of some previous study that may not be on similar population, may be based on small sample that gives unreliable estimates, may have followed different methodology, may be on relatively homogenous subjects, may have been on a regimen similar to the present one but not exactly same, etc. If you find a study exactly same as you plan, you probably would not do the new study at all. Statistically, the effect size estimate and SD you used for calculation of sample size may be different from what you actually get in your trial. Thus, the execution of the trial on planned lines may not produce expected results. It is wise to inflate the sample size by atleast 10% to cover such uncertainties.

While minor variations can be taken in to stride, major variations can completely derail the trial. This applies to all sample size calculations but clinical trials are much more sensitive for reasons already stated. Thus, an opportunity for a trial to correct itself while on course is considered welcome, provided the final results remain valid and reliable. One such opportunity is to stop the trial at interim appraisals and the other is sample size reestimation.

12.6.3.1 Stopping Rules in Case of Early Evidence of Success or of Failure: Lan–deMets Procedure

An ongoing trial can be stopped for two reasons. One is that evidence appears of the adequate efficacy of the regimen and the other is that the efficacy is found too low to go any further. First is stopping for efficacy and the second is stopping for futility. Futility has special appeal for phase II trial because the primary aim of this phase is to provide proof of concept in the sense that the regimen has minimum efficacy. Futility will save it from going through the expensive phase III. Stopping for efficacy has application to both phases II and III but has special appeal to phase III as it can save substantial time and resources. For these reasons, stopping rules are relatively relaxed for phase II trials as they allow early stopping for futility if the concept is not found sufficiently rigorous. Phase III trials do not want to stop early unless the evidence of efficacy is firm.

1. *Stopping for futility*: Statistical changes in an ongoing trial are done in a manner that the prefixed level of significance and power remain unaltered. You know that power is associated with the effect size you want to detect. The procedure to decide futility becomes intricate in view of these constraints. A software is required. Among many procedures, the following, called Simon method, is relatively simple to understand [14].

   Consider first a single-stage design with only one arm with $\pi_0 = 0.20$ and $\pi_a = 0.40$. This means that you are looking for at least 20% improvement in efficacy over existing 20%. For power $(1 - \beta) = 0.90$ and $\alpha = 0.05$, the sample size formula gives $n = 49.08 \approx 50$. Calculations of the type shown for power will tell that you need at least 16 positive responses out of 50 (31%) in order to reject H_0: $\pi_0 \leq 0.20$. If power $= 0.90$ for alternative $\pi_a = 0.40$, any response <16 out of 50 says that the efficacy is extremely unlikely to reach to 0.40. Thus, it is futile to continue the trial—for example, to go from phase II to phase III. Stop the trial if the number of positive responses is 15 or less.

   Single-stage design promises a large n upfront and this can expose such a large number of needy patients to an unproven regimen which may also have side effects. A two-stage sequential design can partially save from these vagaries. Stopping rules for futility for two-stage design can also be worked out by following a similar procedure with the help of a software, and the method can be extended to multistage sequential design and multiarm trial.

As an illustration, consider a two-stage design with the same parameters, namely, $\pi_0 = 0.20$, $\pi_a = 0.40$, power $(1 - \beta) = 0.90$, and $\alpha = 0.05$. The software can be asked to find a suitable n_1 for interim look at the data and propose a final n. The algorithm that minimizes the expected requirement of total n gives $n_1 = 19$ and $n_2 = 35$ for a total of $n = 54$ in this case [14]. Plan a study of this size. This is more than worked out earlier for a single-stage trial for the same power. However, there is a chance that you stop early in this case. The software will also tell you that if the number of positive responses ≤ 4 in the first phase out of $n_1 = 19$, it is futile to continue as the efficacy of 0.40 is extremely unlikely to accomplish at completion when the number of responses are so few at first stage.

A computer may take substantial time to arrive at these numbers as the computations are enormous. You can see how complicated this procedure could become if there are two arms (case and control) and many stages. For such rules for some other values of π_0 and π_a, see Simon [15].

2. *Stopping for efficacy:* This requires that the level of significance α is judiciously apportioned at each appraisal such that the total Type-I error does not exceed α. A simple though inadequate approach is to spend α in K equal parts if K appraisals are planned including the final analysis of all the data. This means that the hypothesis at first appraisal is tested at α/K level of significance, second at $2\alpha/K$ level of significance, and the last at α level of significance. More generally, in this case, level of significance at kth appraisal is $\alpha(\tau_k) = \tau_k \alpha$, where τ_k is the proportion covered at kth appraisal. If $\alpha = 0.05$ and $K = 2$ (only one midterm appraisal), use $\alpha = 0.025$ level of significance to test the null at midterm stage. This implies for two-tail Gaussian Z-test that the critical value will be ± 2.24 (from Gaussian table for $\alpha = 0.025$) in place of the conventional ± 1.96. This, in fact, is more stringent than apparently looks because at this stage the sample size available is also one-half of what would be eventually available. If your results meet this criterion, reject the null of equality of the test and control group. Conclude that you have enough evidence in favor of the alternative hypothesis, and stop the trial.

Statistically, equal-α spending procedure just stated is too liberal than what it should be to control Type-I error and would relatively easily reject the null. It sounds reasonable to have a procedure that is even more stringent at initial stages. For equally spaced group sequential designs (Chapter 6), one such popular procedure is due to O'Brien–Fleming. An improvement over this is due to Lan–deMets, which is flexible and accommodates unequally spaced appraisals. This can be used for equally spaced sequential designs as well. This preserves the overall Type-I error regardless of timing of the appraisals but makes it difficult to stop the trial early unless there is a strong evidence of the desired efficacy. This also imposes a small penalty at the end for interim looks and can be stated as follows:

$$\text{Lan--deMets } \alpha\text{-spending function: } \alpha(\tau_k) = 2\left[1 - \Phi\left(\frac{z_{\alpha/2}}{\sqrt{\tau_k}}\right)\right], \qquad (12.26)$$

where
  Φ is the cumulative Gaussian probability
  τ_k is the proportion of the information available at kth appraisal

To some, this may look like a complex expression, but that really is not so. One-half of the subjects completing one-half trial corresponds to $\tau_k = 0.25$. For two-sided Gaussian test at 0.05 significance level, $z_{\alpha/2} = 1.96$ and $z_{\alpha/2}/\sqrt{\tau_k} = 1.96/\sqrt{0.25} = 3.92$ at this τ_k. At this value, $(1 - \Phi) < 0.00005$ and $\alpha(\tau_k) < 0.0001$. Thus, critical value 3.92 for rejecting the null corresponds to P-value less than 0.0001 at this stage. This critical value assumes that the trend of results later in the trial would be on the same pattern as accrued so far.

A suitable computer program is needed to find the Lan–deMets critical values for specified appraisals. For example, if there are three appraisals at 50%, 80%, and 100%, the critical values for one-tail $\alpha = 0.025$ are given in Table 12.9. The total of adjusted P-values is 0.025 and the critical value at last appraisal at completion of the trial is 2.0302 in place of the usual 1.96. There is a slight penalty for conducting previous two appraisals.

No matter which method you use, there would be a bias involved in early stopping. If you plan $n = 400$ subjects in a trial and stop it after analysis of data on 50 subjects since the data tell you that strict significance has reached, the question arises whether these 50 are as representative of the population as 400 would have been. If the first 50 subjects are not random, further bias is apparent. For details, see Chow et al. [16].

12.6.3.2 Sample Size Reestimation in Adaptive Designs

Sample size can be reestimated on the basis of accumulated data from an ongoing trial to make it more likely that the trial provides convincing evidence. Appraisal of the data may indicate to you that the originally planned

TABLE 12.9

Lan–deMets α-Spending Function for Unequally-Spaced $K = 3$ Appraisals

Appraisal No. k	$\tau_k(\%)$	Critical Value	Nominal P-Value from Gaussian Table (One-Tail Test)	Adjusted P-Value
1	20	2.9626	0.0015	0.0015
2	50	2.2682	0.0117	0.0107
3	100	2.0302	0.0212	0.0128
			Total	0.0250

clinically relevant difference for detection can be reduced or that the SD is larger than stipulated. To compensate, sample size can be increased to retain adequate power. For example, if you initially used SD = 50 and your data give you SD = 70, the required sample size may nearly double to detect the same gain. This calculation presumes that the data so far collected are a better indication of what might occur eventually compared with the data of any of the previous studies. Thus, sample size reestimation is an exercise to increase the chance of reaching to a valid conclusion.

When sample size reestimation is planned, it is prudent to start with a relatively small sample and ask for increase if the interim analysis indicates that the originally planned sample size was not adequate. The trial can be stopped at such appraisal when convincing evidence one way or the other is available, but sample size reestimation is rarely done to decrease the total sample size.

Some researchers question the validity of such a design as it starts with a small sample. This may be considered to indicate that enough planning was not done. Others consider this a valid and efficient strategy since such reestimation is planned in advance at design stage and can save the trial from failure. Adaptive designs that build in the feature of revision while on course always have budgetary problems since the sample size cannot be decided in advance. Total size depends on what results are obtained at interim analysis.

Expectedly, the method is complex and a software would be needed. Again, many algorithms are available for this evolving strategy. The following is an illustration using Lin and Shih's procedure with the same specifications as before, namely, $\pi_0 = 0.20$, $\pi_a = 0.40$, power $(1 - \beta) = 0.90$, and $\alpha = 0.05$ for a two-stage adaptive design with a single arm [14]. This illustration is for deciding futility. Compared with stopping rules, this algorithm of sample size reestimation requires additional specifications from the user, namely, the high and low target response rate and power for first and second stages. Suppose now that the high target response continues to be the same $\pi_2 = 0.40$ although the notation now is π_2 instead of π_a. If the low target response rate is $\pi_1 = 0.35$ and $\beta_1 = 0.20$ and $\beta_2 = 0.10$, a software gives the following, where x is the number of positive responses:

Sample size for first stage – $n_1 = 23$.

If $x \leq 5$ out of 23, decide futility.

If $x = 6$, continue the trial for additional $n_2 = 22$ subjects for a total of $n = 45$ subjects.

Decide futility if $x \leq 12$ out of 45.

If $x \geq 7$ continue for additional $n_2 = 51$ subjects for a total of $n = 74$ subjects.

Decide futility if $x \leq 20$ out of 74.

Note how the total sample size changes in this adaptive design depending on the number of positive responses at first stage.

Another useful application of adaptive strategy is in modifying the sample size after you find that your initial estimate of π or of SD was either too high or too low. I will illustrate this for SD. The sample size for comparing case and control group, from formula (d) in Table 12.8 for equal σ, is $n = 2Q^2 s_0^2 / \delta^2$ per group, where $Q = (z_{\alpha/2} + z_\beta)$ and s_0^2 is the initial estimate of σ^2. In a blinded trial, you would not be able to get groupwise estimate but will get one common estimate. Suppose you decide to do a review after observing first n_0 subjects per group. If these $2n_0$ subjects give you an estimate s_1^2, the revised sample size is $n_{rev} = 2Q^2 s_1^2 / \delta^2$. This cannot be less than n_0 as those many have already been observed, but it can still be less than the originally planned n if $s_1^2 < s_0^2$.

This simple procedure has obvious flaws. First, the estimate based on $2n_0$ subjects is biased, and second, interim look at the data affects Type-I error and power. Corrections for bias have been suggested but it is for controlling Type-I error and achieving specified power. The procedure becomes complex. Help of a software is needed. Such softwares are gradually coming up in the market as the procedure itself is also evolving.

References

1. Conover WJ. *Practical Nonparametric Statistics*, 3rd edn. New York: John Wiley & Sons, 1999, pp. 143–144.
2. Bangsi D, Ghadirian P, Ducic S et al. Dental amalgam and multiple sclerosis: A case-control study in Montreal, Canada. *Int J Epidemiol* 1998; 27:667–671.
3. Johnson NL, Kotz S. *Discrete Distributions*. New York: John Wiley & Sons, 1969, p. 59.
4. Gardner MJ, Altman DG. Confidence intervals rather than *P*-values: Estimation rather than hypothesis testing. *Br Med J* 1986; 292:746–750.
5. Schnitzer TJ, Kong SX, Mitchell JH et al. An observational, retrospective cohort study of dosing pattern of rofecoxib and celecoxib in the treatment of arthritis. *Clin Ther* 2003; 25:3162–3172.
6. Turner JA, Franklin G, Fulton-Kehoe D et al. Prediction of chronic disability in work-related musculoskeletal disorders: A prospective population based study. *BMC Musculoskelet Disord* 2004; 5:14.
7. Som RK. *Practical Sampling Techniques*, 2nd edn. New York: Marcel Dekker, 1996.
8. Altman DG. *Practical Statistics for Medical Research*. London, U.K.: Chapman and Hall, 1991, p. 456.
9. Kumar R, Indrayan A. A nomogram for single-stage cluster sample surveys in a community for estimation of a prevalence rate. *Int J Epidemiol* 2002; 31:463–467.
10. Neter J, Wasserman W. *Applied Linear Statistical Models: Regression, Analysis of Variance and Experimental Designs*. Homewood, IL: Richard D. Irwin, Inc., 1974, pp. 827–828.
11. Lwanga SK, Lemeshow S. *Sample Size Determination in Health Studies: A Practical Manual*. Geneva, Switzerland: World Health Organization, 1991.

12. Machin D, Campbell MJ, Tan S-B, Tan S-H. *Sample Size Tables for Clinical Studies*, 3rd edn. Chichester, UK.: Wiley Blackwell, 2008.
13. A'Hern RP. Sample size tables for exact single-stage phase II designs. *Stat Med* 2001; 20:859–866.
14. Groulx A, Moon K, Chung SC. Using SAS® to determine sample sizes for traditional 2-stage and adaptive 2-stage phase II cancer clinical trial designs. Paper 188-2007. *SAS Global Forum* 2007.
15. Simon R. Optimal two-stage designs for phase II clinical trials. *Cont Clin Trials* 1989; 10:1–10.
16. Chow S-C, Wang H, Shao J. *Sample Size Calculations in Clinical Research*, 2nd edn. Boca Raton, FL: Chapman & Hall, 2007.

13

Inference from Proportions

Statistical inferential methods are different for qualitative variables than for quantitative variables. This is analogous to different methods of assessment of cardiovascular diseases than for gastrointestinal tract diseases. Those variables can be considered qualitative that are concerned with proportions of subjects in different categories. A variable such as birth weight is, in fact, quantitative but is treated as qualitative when the interest is in knowing the percentage of newborns with weight <2500, 2500–3499, and ≥3500 g. The actual scale can be metric but the variable becomes qualitative when such categories are formed. There is a rider though. The number of such categories must be small for it to be qualitative. If, instead of three broad categories, the birth weight is divided into a large number of 100 g categories such as <2000, 2000–2099, 2100–2199, etc., the interest would rarely be in a proportion of births in these categories. It would then be in parameters such as mean or median birth weight. The methods to draw inferences from mean are presented in Chapter 15.

Although 100 g categories of birth weight also give rise to categorical data just as the broad categories do, the statistical methods of analysis of data in these two setups are different. This text therefore avoids using the term "analysis of categorical data" that many other texts do. For the methods described in this chapter, categories are not a prerequisite either. A metric variable such as parity with values 0, 1, 2, and 3+ has no conventional kind of categories (except the last with parity 3+) but is covered by the methods described in this chapter as long as the interest is in the proportion of subjects with different parity. As you will shortly see, the statistical methods for inference from proportions are distribution free, generally called nonparametric. There is no requirement that the values follow a particular distribution such as a Gaussian distribution. Outliers also do not affect these methods.

The statistical inference primarily discussed in this chapter is the test of significance. However, the corresponding confidence interval (CI) is also explained when not covered in the previous chapter. As already outlined for testing of hypothesis in Chapter 12, the procedure is to set up a null hypothesis and check whether the sample provides enough evidence against it for its rejection. A large number of statistical tests are available for different situations even when restricted to proportions. It is not possible to describe all of them in this text. The attempt is to include those methods that are commonly used in health and medicine.

This chapter: Section 13.1 is devoted to a situation in which only one qualitative variable is under consideration. The interest in this case would be in finding out whether or not the population proportion is a specific value or whether or not a specified pattern exists in proportions in different polytomous categories. Section 13.2 describes the association between two dichotomous variables. This includes equivalence tests. The methods for two dichotomous variables are extended to polytomous situations in Section 13.3. Section 13.4 describes the analysis of three or more qualitative variables considered together in a contingency table setup, including log-linear models.

13.1 One Qualitative Variable

The proportion of obese among people with hypertension, the proportion of persons who are sick by various diseases at a particular point of time, and the proportion with different grades of disability among those surviving head injury are examples of proportion based on one qualitative variable. The variables in these examples are obesity, sickness, and disability, respectively. It is binary (obese or nonobese) in the first and polytomous (sickness of various types and disability into mild, moderate, serious, and critical) in the second and third cases. This section restricts discussion to the pattern in different categories. There is no comparison group. If comparison with another group is the objective, the problem can be stated as that of association, which is discussed later in this chapter and further in Chapter 17.

13.1.1 Dichotomous Categories: Binomial Distribution

Consider a simple example of 5-year survival among patients with cervical cancer. There is no noncancer or any other group for comparison. Two kinds of statistical question can arise in this case. (1) If the proportion surviving for 5 years among patients with cervical cancer is known to be 30%, what is the chance that at least six will survive for at least 5 years in a random sample of 10 patients? (2) If the number surviving in a random sample of 20 patients is only 4, can the survival rate in the long run still be 30%? You will shortly see that these two questions are, in fact, two sides of the same coin. The answer to these questions depends on what is called a binomial distribution.

13.1.1.1 Binomial Distribution

For simplicity, let us call the event of interest as a "success." Denote its probability by π. In the preceding example, the event of interest is survival for 5 years and $\pi = 0.3$. It can be shown by the law of multiplication that the probability of x successes in n independent trials is

$$\text{Binomial probability: } P(x) = {}^{n}C_{x}\,\pi^{x}(1-\pi)^{n-x}, \tag{13.1}$$

where

$$
{}^{n}C_{x} = \frac{n!}{x!(n-x)!} \quad \text{and} \quad x! = x(x-1)(x-2)\ldots 3\cdot 2\cdot 1
$$

$$
\left(\text{e.g., } {}^{5}C_{3} = \frac{5!}{3!(5-3)!} = \frac{5\times 4\times 3!}{3!\times(2\times 1)} = 10 \right),
$$

and π is the probability of success in one trial. In this example, a trial corresponds to one patient. Patients behave independently in the sense that survival of one patient does not affect the chance of the survival of another. Also, the chance of survival for each patient should be the same. This would be so when the patients are homogeneous with respect to prognostic factors. When these conditions are fulfilled, the probability in Equation 13.1 can be used to answer the two questions proposed earlier.

Example 13.1a: Binomial Probability

In question (1), $n = 10$ and $\pi = 0.3$. The success in this case is survival for at least 5 years. You need to find $P(x \geq 6)$. Because not more than 10 successes are possible in 10 patients and the successes are mutually exclusive, by law of addition,

$$P(x \geq 6) = P(x = 6) + P(x = 7) + \cdots + P(x = 10)$$

$$= {}^{10}C_{6}(0.3)^{6}(0.7)^{4} + {}^{10}C_{7}(0.3)^{7}(0.7)^{3} + \cdots + {}^{10}C_{10}(0.3)^{10}(0.7)^{0},$$

from Equation 13.1

$$= 0.0368 + 0.0090 + 0.0014 + 0.0001 + 0.0000$$

$$= 0.047.$$

Thus, the chance that at least six will survive for at least 5 years in a sample of 10 patients is only 4.7%. You may not have expected it to be so low.

Example 13.1b: Binomial Probability for Extreme Values

As explained in Chapter 12, question (2) is more appropriately answered by obtaining the probability of $x = 4$ or a more extreme value. Such a value in this case is $x \leq 4$. Because of mutually exclusive values of x, for $\pi = 0.3$ and $n = 20$,

$$P(x \leq 4) = P(x = 0) + P(x = 1) + P(x = 2) + P(x = 3) + P(x = 4)$$

$$= {}^{20}C_0(0.3)^0(0.7)^{20} + {}^{20}C_1(0.3)^1(0.7)^{19} + {}^{20}C_2(0.3)^2(0.7)^{18}$$

$$+ {}^{20}C_3(0.3)^3(0.7)^{17} + {}^{20}C_4(0.3)^4(0.7)^{16}$$

$$= 0.0008 + 0.0068 + 0.0278 + 0.0716 + 0.1304$$

$$= 0.238. \tag{13.2}$$

This probability is fairly high. Thus, it is not unlikely that the survival rate in the long run is 30% even when only 4 survive out of 20.

Whereas the first example is a simple exercise in calculating binomial probability, the second, in fact, is an exercise in statistical testing of a hypothesis. The null hypothesis in Example 13.1b is H_0: $\pi = 0.3$ and the probability in Equation 13.2 is the P-value when H_0 is true. Since the P-value is not less than 0.05, the H_0 cannot be rejected. Hence, the conclusion that the survival rate could be 30%.

Consider the following interesting example of the implication of **independent trials**, which is an assumption implicit in the binomial expression. If there is a couple with three female children, the probability of the fourth birth being female is still as much ($\pi = 1/2$) as the probability of a male birth. The probability for each birth is the same and is not affected by what happened in the past. This process has no memory and each birth is independent for gender of the child. This condition is fulfilled in a large number of practical situations.

Suppose for some reason the interest is in finding the probability of at least two survivals out of $n = 10$. In notation, this is $P(x \geq 2)$. Calculating all the probabilities for $x = 2, 3, \ldots, 10$ would take a lot of time and efforts, particularly when done by a hand-held calculator. In this situation, use your knowledge that $P(x \geq 2) = 1 - P(x \leq 1)$. Now, this is to be computed only for $x = 0$ and $x = 1$. Similarly, for example, $P(x \leq 8) = 1 - P(x \geq 9)$.

13.1.1.2 Large n: Gaussian Approximation to Binomial

The calculation of the probability given in Equation 13.1 can become complex when n is large and when it is to be computed for several different values of x.

When the condition (12.5) given in Chapter 12 is satisfied, which is likely when n is large, the binomial distribution can be approximated by a Gaussian distribution. This approximation arises again from central limit theorem as the binomial x also is a summation type of variable—this time sum of 1s and 0s for "success" and "failure," respectively. For Gaussian approximation of binomial, the following are needed:

Mean of a binomial variable $x = n\pi$ and $SD = \sqrt{n\pi(1-\pi)}$.

Thus, for large n, from Equation 12.7,

$$Z = \frac{x - n\pi}{\sqrt{n\pi(1-\pi)}} \tag{13.3}$$

is a standard Gaussian variate. When n is large, this can be used to answer the same types of questions as posed earlier. This is illustrated in Examples 13.2a and 13.2b.

Example 13.2a: Binomial Probability for Large n

If the proportion surviving for at least 3 years among cases of cancer of the cervix is 60%, what is the chance that at least 40 will survive for 3 years or more in a random sample of 50 such patients?

For $n = 50$ and $\pi = 0.60$, mean number of survivors, $n\pi = 50 \times 0.60 = 30$, and $SD = \sqrt{n\pi(1-\pi)} = \sqrt{50 \times 0.60 \times 0.40} = 3.464$.

Since $n\pi \geq 8$ and $n(1 - \pi)$ also ≥ 8, the Gaussian approximation can be used. With continuity correction, $P(x \geq 40) = P(x \geq 39.5)$. Thus,

$$P(x \geq 40) = P\left(\frac{x - \text{mean}}{SD} \geq \frac{39.5 - 30}{3.464}\right)$$

$$= P(Z \geq 2.74), \quad \text{where } Z \text{ is as in Equation 13.3}$$

$$= 0.0031 \text{ from Table B.1.}$$

This low probability indicates that there is practically no chance that 40 or more patients will survive for at least 3 years in a sample of 50 when the survival rate is 60%. This might seem unbelievable but it is true.

SIDE NOTE: The probability just calculated is the approximate probability based on a Gaussian distribution. The exact binomial probability by Equation 13.1 for $x \geq 40$, when $n = 50$ and $\pi = 0.60$, is 0.0008.

Example 13.2b: Binomial Probability for Large *n* for Extreme Values

If the percentage surviving in a random sample of 100 patients is 20, could the survival rate in the long run in such patients still be 30%?

Now, H_0: $\pi = 0.30$. Under this H_0, mean number of survivors in samples of size 100, $n\pi = 100 \times 0.30 = 30$, and SD = $\sqrt{n\pi(1-\pi)} = \sqrt{100 \times 0.30 \times 0.70} = 4.58$.

Note in this case that if the survival rate cannot be 30% since it is only 20% in the sample, it certainly cannot be more than 30%. Thus, the alternative hypothesis is one sided, namely, H_1: $\pi < 0.30$. Since 20% of 100 is 20, you need to compute $P(x \le 20)$. Since H_1 is left tailed (less-than type), the *P*-value to be obtained is also left tailed. With the continuity correction,

$$P(x \le 20) = P\left(\frac{x - \text{mean}}{\text{SD}} \le \frac{20.5 - 30}{4.58}\right)$$

$$= P(Z \le -2.07)$$

$$= P(Z \ge 2.07) \text{ because of symmetry of } Z$$

$$= 0.0192 \text{ from Table B.1.}$$

This *P*-value is exceedingly small, certainly less than the conventional level of significance $\alpha = 0.05$. Thus, the null hypothesis is not likely to be true. The evidence against H_0 is sufficient and this cannot be conceded. It is exceedingly unlikely that the survival rate in the long run would be 30% when 20 survive in a sample of 100.

13.1.2 Poisson Distribution

You may have noted that binomial *x counts* the number of successes out of *n*. If *n* becomes extremely large and the probability of success becomes extremely small, ultimately in the limit we get a Poisson distribution. Applicationwise, if migraine occurs three times in a year on average in established cases, the probability that it will occur six or more times in a year in a random case can be obtained by Poisson distribution. This distribution is given by

$$\text{Poisson distribution: } P(x) = \frac{e^{-\mu}\mu^x}{x!}; \quad x = 0, 1, 2, \ldots;$$

where μ is the mean. In migraine example,

$$P(x \ge 6) = 1 - P(x \le 5) \text{ by complementary rule}$$

$$= 1 - \left(\frac{e^{-3}3^0}{0!} + \frac{e^{-3}3^1}{1!} + \frac{e^{-3}3^2}{2!} + \frac{e^{-3}3^3}{3!} + \frac{e^{-3}3^4}{4!} + \frac{e^{-3}3^5}{5!}\right)$$

$$= 1 - e^{-3}\left(1 + 3 + \frac{9}{2} + \frac{27}{6} + \frac{81}{24} + \frac{243}{120}\right)$$

$$= 1 - 0.0498 \times 18.4$$

$$= 0.084.$$

Thus, the chance is nearly 8.4% that a random migraine patient from this "population" will have six or more attacks during one year. Poisson tables are available in statistical books to give you these probabilities. Statistical packages in any case have provision to give you exact Poisson probabilities. The value of μ, being the average, does not have to be an integer. In most situations, this will be in decimals.

Mean μ of Poisson is estimated by sample mean $\bar{x}$ as usual and the variance of Poisson is also μ so that its estimate is also $\bar{x}$. Equality of mean and variance is considered a defining property of Poisson. In any set of data on rate where mean and variance are nearly equal, you should especially look for Poisson distribution. Examples of Poisson variable are

1. Number of deaths occurring in a cardiac hospital per day
2. Number of myocardial infarction (MI) cases arriving in a hospital per day
3. Number of measles cases occurring in a city per year
4. Number of handicap persons per 1000 population

In most practical situations where Poisson is applicable, the mean would be small, possibly less than 1. If it gets bigger, say, more than 10, the Poisson too tends to behave in the same manner as Gaussian. For example, if $\mu = 20$, then for Poisson $\sigma^2 = 20$ and by Gaussian approximation the chance that this number is 15 or less in a specific case is

$$P(x \leq 15) = P(x < 15.5) \text{ with continuity correction}$$

$$= P\left(z < \frac{15.5 - 20}{\sqrt{20}}\right) \text{ using Equation 12.8}$$

$$= P(z < -1.01)$$

$$= 0.1562.$$

Exact answer with Poisson tables is 0.157. Gaussian approximation is not far off.

Despite a clear and useful application of Poisson to rare events, use of this distribution in medicine has been infrequent. I do not discuss this distribution further in this book.

13.1.3 Polytomous Categories (Large *n*): Goodness-of-Fit Test

Now, extend the method to a situation where the variable has several categories. For example, these categories could be none, mild, moderate, and serious forms of anemia; calendar month of occurrence of sudden infant death syndrome (SIDS) in a year; or cirrhosis, hepatitis, and malignant forms of liver disease. The first two are on an ordinal scale and the last is on a nominal scale. The first has 4 categories, the second 12, and the last 3 categories.

Let the interest be in finding whether the subjects in the target population follow a prespecified pattern. Examples are (1) whether SIDS occurs twice as often in winter as in summer months; (2) whether the patterns of none, mild, moderate, and serious hypertension in cases of MI are 10%, 20%, 40%, and 30%, respectively; and (3) whether the proportions of full, partial, and no recovery within 2 years after surgery in cases of breast cancer are 40%, 50%, and 10%, respectively. Such a problem is known as a problem of goodness of fit because the interest is in finding whether or not the pattern observed in the sample fits the specified pattern well. The procedure is best explained with the help of an example.

Example 13.3a: Blood Group Pattern of AIDS Cases

The blood group of a random sample of 150 patients with acquired immunodeficiency syndrome (AIDS) is investigated to examine the possibility of a preponderance of a particular blood group in AIDS cases. If there were no preponderance, the profile would be the same as in the general population. Suppose this is 6:5:8:1 for blood groups O, A, B, and AB, respectively. The sample observations are as shown in Table 13.1. If the cases are in the same ratio as in the general population, the number of cases with blood group O should have been $6 \times 150/20 = 45$. The observed number is 57. In view of this difference, does this pattern in AIDS cases really conform to that in the general population? How to draw a conclusion such that the chance of Type-I error does not exceed 0.05? A solution is provided in the following.

Since the categories are mutually exclusive and exhaustive, Table 13.1 is a one-way contingency table. This particular sample happens to have AIDS patients with this pattern, but another sample may very well show a different pattern, possibly exactly the same as in the general population. What is the chance that this sample has indeed arisen from the population with blood group pattern in the ratio 6:5:8:1?

Denote the population proportions in the four blood groups by π_1, π_2, π_3, and π_4, respectively. The null hypothesis in this case is

$$H_0: \pi_1 = \frac{6}{20} = 0.30, \quad \pi_2 = \frac{5}{20} = 0.25, \quad \pi_3 = \frac{8}{20} = 0.40,$$

$$\text{and} \quad \pi_4 = \frac{1}{20} = 0.05. \tag{13.4}$$

TABLE 13.1

Blood Group Pattern in a Hypothetical Sample of AIDS Cases

Blood Group	O	A	B	AB	Total
Number of AIDS patients	57	36	51	6	150

These πs are in the same ratio as in the general population and they add up to 1.00 since the categories are mutually exclusive and exhaustive. The interest is in testing whether the sample provides enough evidence against this H_0. The alternative hypothesis H_1 is that the pattern is any other than specified in the null hypothesis stated in Equation 13.4. Now proceed as follows.

13.1.3.1 Chi-Square and Its Explanation

Denote the observed frequencies in the four groups in the sample by O_1, O_2, O_3, and O_4, respectively. That is, $O_1 = 57$, $O_2 = 36$, $O_3 = 51$, and $O_4 = 6$. If H_0 is really true, then expected frequencies, denoted by Es, would be in the ratio specified in the hypothesis (13.4), that is, $E_k = n\pi_k$ ($k = 1, 2, 3, 4$). For $n = 150$, $E_1 = 150 \times 0.3 = 45$. Similarly, $E_2 = 37.5$, $E_3 = 60$, and $E_4 = 7.5$. A large difference between Os and Es would suggest that the observed pattern is different from that stipulated in the null hypothesis. This would be evidence against H_0 and in favor of H_1. It seems that the examination of the differences $(O_k - E_k)$ for different k would be helpful. Since the total of the expected frequencies has to be the same 150 as that of the observed frequencies, it is imperative that some of these differences are negative and some positive. The sum $\Sigma(O_k - E_k)$ would be always 0. As in the case of deviations $(x_i - \bar{x})$ for calculating SD, the square of these differences gets rid of the negative sign. This gives $(O_k - E_k)^2$. The magnitude of these squares is the key to the plausibility of H_0. But a difference of 1.5 over the expected 7.5 in blood group AB has a different meaning than the same difference over the expected 37.5 in blood group A. The former difference is one-fifth of the corresponding expected frequency, whereas the latter is not even one-twentieth. Thus, the squared differences should be viewed in relation to the expected frequencies. The quantity $[(O_k - E_k)^2/E_k]$ becomes relatively free of the differentials in the magnitude of the expected frequencies in different groups and helps to give nearly equal weight to the groups. In place of taking the average of these quantities, this time obtain the sum $\Sigma[(O_k - E_k)^2/E_k]$. This quantity is based entirely on frequencies and thus is unit free. This obviates the need to take square root as was done at the time of calculating SD. To indicate that the quantity is a square, the sum is called chi-square (χ^2). This is the test criterion in this case as per Step 2 laid down in Section 12.3.3. Thus,

$$\text{Chi-square (one-way table): } \chi^2 = \sum \frac{(O_k - E_k)^2}{E_k}; \quad k = 1, 2, \ldots, K; \quad (13.5)$$

where K is the total number of cells in the contingency table. In Example 13.3, $K = 4$.

Note that Es are obtained assuming that H_0 is true. Thus, the value of χ^2 in the formula given in Equation 13.5 is under H_0. When H_0 is true, the difference between O_k and E_k, that is, $(O_k - E_k)$, should be small and consequently the value of χ^2 should also be small. In other words, a large value of χ^2 is unlikely if H_0 is true. If the sample gives a large χ^2, it provides evidence against H_0.

As noted in the preceding chapter, P-value in this case is the probability of occurrence of the value of the criterion as extreme as or more extreme than obtained for the sample data. This requires distribution of the criterion under H_0. Such a distribution of χ^2 is known and the critical values are given in Table B.3 for specific values of α. For P to be less than α, the calculated value of χ^2 must be more than or equal to the critical value. The exact shape of the distribution varies according to what is called the degrees of freedom (df). The df in turn depends mostly on the number of cells K. A different distribution of χ^2 for different df is analogous to a different distribution of diastolic BP in different age groups. The concept of df is explained next.

13.1.3.2 Degrees of Freedom

In Table 13.1, four categories of blood group are listed, namely, O, A, B, and AB. However, the frequency in only three of them can be freely chosen, the fourth is automatically determined by the total. If the frequencies chosen for O, A, and AB are 70, 20, and 10, respectively, then the frequency in group B has to be 50 because the total is 150. If the frequencies chosen for O, A, and B are 60, 30, and 20, respectively, then the frequency in the group AB has to be 40. Thus, there is freedom to choose only three out of four cells. This is called the df. For K cells in a one-way contingency table, when the sample values have no restriction other than that the total is fixed, the df $= K - 1$.

Example 13.3b: Blood Group Pattern of AIDS Cases (continued)

In the case of the data in Table 13.1, the preceding discussion yields df $= K - 1 = 4 - 1 = 3$. Calculations are presented in Table 13.2, which show $\chi^2 = 4.91$. If a computer software is used, it will automatically compare the calculated value of χ^2 with its known distribution for 3 df and give $P = 0.178$. Otherwise, because 4.91 is less than the critical value 7.815 for 3 df in Table B.3, the P-value is more than 0.05. Thus, the value 4.91 of χ^2 obtained for these data is not all that unlikely when H_0 is true. That is, the frequencies observed in different blood groups in the example are not very inconsistent with H_0 stated in Equation 13.4. The sample values do not provide sufficient

TABLE 13.2

Calculation of Chi-Square for the Null in Example 13.3

	Blood Group				
	O	A	B	AB	Total
Observed frequency (O_k)	57	36	51	6	150
Expected frequency under H_0 (E_k)	45.0	37.5	60.0	7.5	150.0
$O_k - E_k$	12.0	−1.5	−9.0	−1.5	0
$(O_k - E_k)^2 / E_k$	3.20	0.06	1.35	0.30	$4.91 = \chi^2$

evidence against H_0 and it cannot be rejected. A preponderance of any blood group in cases of AIDS cannot be concluded on the basis of this sample.

SIDE NOTE: This inference is drawn despite an apparently clear excess of blood group O (57 subjects vs. an expected 45) in this sample of AIDS cases. This is because such a frequency *pattern* is not very unlikely to occur when the sample comes from the general population where the blood group ratios are as given in the hypothesis (13.4). Further analysis is given a little later.

13.1.3.3 Cautions in Using Chi-Square

Chi-square does not require the frequency pattern to be Gaussian nor does it require any other specific pattern. Thus, *chi-square is a distribution-free procedure.* The following cautions are advised:

1. The use of chi-square for categorical data is well established but χ^2 itself is a continuous variable. Theoretically, it is based on an approximation. This approximation works fine when the expected frequency in any cell is not less than 5. When the number of categories K is large, there can be a small relaxation. A rule of thumb is as follows. Not more than one-fifth of categories (i.e., $K/5$ cells) should have $E_k < 5$ and almost none should be less than 1. When small frequencies are expected in many cells under H_0, either because of a small sample or because of very small π in some cells, an exact multinomial test should be used. (See the Section 13.1.5 for this test.)

2. It is necessary to realize that chi-square is calculated from the actual frequencies in the cells. Percentages cannot be used.

3. Chi-square test is basically a two-tailed test. Significance in this case implies only presence of some difference from H_0 and it can seldom be labeled positive or negative. If the observed frequency is less than the expected frequency in one cell, it has to be more in one or more of the other cells because the total for both the observed and the expected frequencies is the same. In Example 13.3, the observed frequency is more for blood group O but less for the other three blood groups. Thus, the alternative hypothesis H_1 is two-sided. Some other tests discussed later can have a one-sided alternative.

4. The χ^2 criterion is the sum $\Sigma[(O_k - E_k)^2/E_k]$. This would be large even if one particular difference $(O_k - E_k)$ is large. Thus, rejecting H_0 tells us only that there is at least one cell where the observed frequency is substantially different from the expected under the null hypothesis. It does not say where. On the other hand, if a large difference is present in only one cell, this can be masked by small differences in the other cells. This is what might be happening for blood group O in Example 13.3. For a more focused inference, further analysis may be helpful.

13.1.3.4 Further Analysis: Partitioning of Table

Consider Example 13.3 again for further analyzing the data. Examination of the data in this example reveals that the observed frequency in blood group O is very much higher than expected from the pattern in the general population. But other differences are not as large. To check that this really is so, check whether the pattern in blood groups A, B, and AB is nearly the same as expected, and then check the difference in blood group O. The corresponding null hypotheses are

$$H_{01}: \text{A, B, and AB are in the ratio 5:8:1,}$$

and

$$H_{02}: \pi_1 = \frac{6}{20} = 0.3, \quad \pi_2 + \pi_3 + \pi_4 = \frac{14}{20} = 0.7.$$

The former ratio is the same as in Example 13.3 and the latter combines A, B, and AB. Not stating H_{01} in terms of π is deliberate because that might give rise to confusion. The sum of πs in all contingency tables should be 1 and therefore the same πs as in hypothesis (13.4) cannot be used for the three cells covered by H_{01}. For these two null hypotheses, calculations for χ^2 are shown in Table 13.3.

The division of the earlier four-cell table into two tables as shown is called **partitioning**. The first partition gives $\chi_I^2 = 0.38$. This has $3 - 1 = 2\,df$. From Table B.3, $P > 0.10$ when $\chi^2 = 0.38$, and from a software package, $P = 0.83$. Since it is more than 0.05, H_{01} cannot be rejected. The evidence is not sufficient to

TABLE 13.3

Partitioning of Table 13.1 and Calculation of Partitioned Chi-Square

I	Blood Group			
	A	B	AB	Total
O_k	36	51	6	93
E_k	33.2	53.1	6.6	93
$(O_k - E_k)^2 / E_k$	0.23	0.09	0.06	$0.38 = \chi_I^2$

II	Blood Group		
	O	Others	Total
O_k	57	93	150
E_k	45.0	105.0	150
$(O_k - E_k)^2 / E_k$	3.20	1.37	$4.57 = \chi_{II}^2$

conclude that the pattern of blood groups A, B, and AB in AIDS cases is not the same as in the general population.

Part II of the table has only two cells, so χ^2_{II} has only 1 df. For $\chi^2 = 4.57$, from Table B.3, $P < 0.05$, and from a software package, $P = 0.033$. This is statistically significant. It can be concluded that the pattern in part II is not the same as in the general population without much chance of error. Since the grouping now is O and others, it can be safely concluded that blood group O is *more* common (note that the observed frequency 57 is more than the expected 45) in AIDS cases. Nothing specific can be said about the other three groups.

The conclusion reached after partitioning is different from that reached earlier when all the cells were considered together. This is because the lack of difference in A, B, and AB groups masked the difference in the O group. Partitioning helped to uncover this difference. Note the following:

1. When n is large, the values of χ^2_I and χ^2_{II} based on the partitioned table should add *approximately* to the overall χ^2 based on all the cells. In this example, $\chi^2_I + \chi^2_{II} = 0.38 + 4.57 = 4.95$, which is only slightly different from 4.91 obtained earlier when all four cells were considered together. For this reason, this is also called **partitioning of chi-square**.

2. The data in Table 13.1 are hypothetical. This is used only to illustrate the methodology of the goodness-of-fit test. Whether or not AIDS has any association with any blood group is anybody's guess at this stage of our knowledge.

3. Any conclusion about association reached by chi-square should be interpreted as casual and not causal unless a series of many other conditions is met. Details are given in Chapter 21.

13.1.4 Goodness of Fit to Assess Gaussianity

One very useful application of goodness-of-fit chi-square is in assessing Gaussianity pattern of a distribution, for that matter any specified distribution. It requires comparing the observed frequencies in different class intervals with the expected under postulated pattern. Example 13.4 illustrates the method.

Example 13.4: Whether Cholesterol Level Distribution in Table 8.1 Can Be Considered to Follow a Gaussian Pattern?

Observed frequencies O_k are already given in Table 8.1 and they are repeated in Table 13.4. To obtain expected frequencies under Gaussian pattern, you need mean and SD. Suppose these are known a priori as mean $\mu = 270 \, \text{mg/dL}$ and SD $\sigma = 30 \, \text{mg/dL}$. Thus, the null hypothesis

TABLE 13.4

Observed Frequencies and Expected Frequencies under the Specified
Gaussian Pattern

Cholesterol Level (mg/dL)	−199	200–239	240–259	260–279	280–299	300–319	320–339	340–399	Total
Observed frequency (O_k)	3	13	16	17	24	6	0	3	82
Expected frequency (E_k)	0.88	12.20	17.38	21.21	17.38	9.04	3.08	0.80[a]	81.97

[a] See comment regarding such small frequencies.

in this case is that the distribution of cholesterol level is Gaussian with
mean = 270 mg/dL and SD = 30 mg/dL. Then, for example,

$$P(\text{cholesterol level between 200 and 240})$$

$$= P(200 < x < 240)$$

$$= P\left(\frac{200 - 270}{30} < \frac{x - \mu}{\sigma} < \frac{240 - 270}{30}\right)$$

$$= P(-2.33 < Z < -1.0)$$

$$= P(Z > 1.0) - P(Z > 2.33) \text{ by symmetry}$$

$$= 0.1587 - 0.0099$$

$$= 0.1488.$$

Expected frequency in this interval = $0.1488 \times n = 0.1488 \times 82 = 12.20$.
For all other intervals, expected frequencies under postulated Gaussian
pattern can be similarly obtained. Thus, you get the last row in Table 13.4.

Details of calculations are omitted but for the data in Table 13.4 Equation
13.5 gives $\chi^2 = 18.76$. In this case, cholesterol level is in eight categories so that
$K = 8$. The table value of chi-square for $K - 1 = 8 - 1 = 7$ df is 14.07 for $\alpha = 0.05$.
Since the calculated value is larger, $P < 0.05$, and you can safely reject the null
hypothesis and conclude that the observed frequencies in different cholesterol
level categories do not follow Gaussian distribution with mean $\mu = 270$ mg/dL
and SD $\sigma = 30$ mg/dL.

In this example, mean and SD are given and df = $K - 1$. One df is lost
because of the restriction that the totals of observed and expected frequencies must be the same. In many situations, mean and SD are not known and
have to be estimated from the sample values. Using sample mean imposes
a restriction and removes 1 df. Similarly, using sample SD imposes another
constraint and removes one more df. Thus, in that case, df = $K - 3$.

There is a limitation in using chi-square in Example 13.4. One of the conditions for validity of chi-square is that at least four-fifths of expected frequencies should be 5 or more and none should be less than 1. In this example, the expected frequency in the first and the last intervals is less than 1 and one more frequency is less than 5. The procedure in this constrained situation should be to merge first two categories and last two categories as <239 and 320–399 mg/dL, respectively. You may like to recompute chi-square with this change and see how it affects the result.

13.1.5 Polytomous Categories (Small *n*): Exact Multinomial Test

The procedures mentioned so far are approximate and applicable only for large *n*. Also, the expected frequency in at least four-fifths of the cells should be 5 or more. If not, then exact methods are required.

13.2.5.1 Goodness of Fit in Small Samples

For small *n*, the hypothesis that cells follow a specified pattern is tested by computing probabilities by a multinomial test. Multinomial probability can be calculated as

Multinomial probability:

$$P = \frac{n!}{O_1! O_2! \cdots O_K!} * \pi_1^{O_1} \pi_2^{O_2} \ldots \pi_K^{O_K}; \quad \Sigma O_k = n \quad \text{and} \quad \Sigma \pi_k = 1; \quad (13.6)$$

where the πs are the probabilities under H_0. The condition $\Sigma \pi_k = 1$ ensures that the categories are mutually exclusive and exhaustive. The P-value is obtained after summation of *P* in Equation 13.6 over the configurations of the cell frequencies that are as observed in the sample or more extreme favoring H_1. Manual computation can become too complex even for moderate *n*. It is advisable to use a software for calculating this probability.

Example 13.5: Multinomial Probability for Angina Attacks

Suppose a regimen for control of angina pectoris is considered effective if at least 60% of patients on this regimen do not have any attack in 1 year of follow-up and not more than 10% have two or more attacks. Thus, the desired ratio is as follows:

Number of angina attacks	0	1	2+
Desired percentage of patients	60	30	10

These define the null hypothesis—H_0: $\pi_1 = 0.60$, $\pi_2 = 0.30$, $\pi_3 = 0.10$.
A higher percentage of patients with a lower number of attacks is even better. A lower percentage in these categories is H_1. The null hypothesis

TABLE 13.5

Configurations Adverse to H_0 and Favoring H_1 in Example 13.4

No. of Angina Attacks	Notation	Configurations Favoring H_1 (as or More Extreme than the Observed)																	
		1	2	3	4	5	6	7	8	9	10	11	12	13	14	15	16	17	18
0	O_1	2	2	2	2	2	1	1	1	1	1	1	0	0	0	0	0	0	0
1	O_2	3	0	1	2	4	0	1	2	3	4	5	0	1	2	3	4	5	6
2+	O_3	1	4	3	2	0	5	4	3	2	1	0	6	5	4	3	2	1	0

in this case is for the effectiveness of the regimen, whereas generally it is stated for ineffectiveness. The regimen under trial is lifestyle changes such as yoga, dietary changes, and physical exercise. After excluding other causes, only six eligible volunteers could be followed up for 1 year. The data obtained are as follows:

Number of angina attacks	0	1	2+
Observed number of patients	2	3	1

The observed ratio is loaded more toward a higher number of attacks than postulated under H_0. Is the regimen ineffective according to the criterion?

In this case, $O_1 = 2$, $O_2 = 3$, and $O_3 = 1$. The configurations adverse to H_0 and favoring H_1, beginning from the observed, are shown in Table 13.5. This includes all configurations adverse to H_0 that have two or less patients with no attack. (It can be debated whether configuration (2, 2, 2) is adverse to (2, 3, 1). In the first case, two patients have one attack and another two have two or more attacks. In the second case, three patients have one attack and one patient has two or more attacks.) There are 18 such configurations. From Equation 13.6, the probability of observing the first configuration under H_0 is

$$P_1(O_1 = 2, O_2 = 3, O_3 = 1) = \frac{6!}{2!3!1!}(0.60)^2(0.30)^3(0.10)^1$$

$$= 0.058.$$

Similar probabilities can be calculated for the other configurations. But there is no need to do so here because P_1 itself is more than 0.05. The sum of the probabilities for these 18 configurations is going to be higher in any case. Since this P-value is not sufficiently small (P_1 itself is more than 0.05), the null hypothesis cannot be rejected. The evidence is not sufficient to call the regimen ineffective in controlling angina attacks.

Example 13.4 once again highlights the negative nature of statistical conclusions. The null hypothesis is not accepted but could not be rejected. Conceding H_0 should not be construed to mean that the regimen *is* effective in controlling angina attacks in a stipulated pattern. The only conclusion is that the sample observations fail to provide sufficient evidence against it. A sane advice in this situation is to draw a conclusion that nothing can be confidently stated about the ineffectiveness of the regimen. Note that this study is on volunteers. Thus, the results are applicable only to the class of volunteers that are represented in the sample. The results are not applicable to the general class of patients.

You may have noticed that with only $n = 6$ subjects and just $K = 3$ categories in Example 13.4, the number of configurations is already large. If, for example, $n = 12$ and $K = 4$, the number of configurations, even those adverse to H_0, may become enormous. This is the reason for the advice to use a software to calculate this probability. The calculations in this example are given only to enhance your understanding of the underlying procedure.

The categories in Example 13.4 are metric. But the method of computing probabilities considers them nominal. The only use made of the metric scale of categories is in identifying the configurations adverse to H_0.

13.2 Proportions in 2 × 2 Tables

Among the problems now discussed is whether the proportion of subjects having a particular characteristic is the same in one group as in another. This section is restricted to the dichotomous variable, that is, a characteristic present or absent, just as in the case of a binomial distribution. However, now introduce a second dichotomous variable that identifies the group. Group-I and group-II can be with and without disease, with disease-A and with disease-B, male and female, young and old, or any other such groups. The setup is essentially bivariate. Both variables have two categories. Such a 2 × 2 table can also arise in a variety of other situations, as you will shortly see. The kind of contingency table arising in such cases is shown in Table 13.6. Inside parentheses in each cell in the table are the corresponding probabilities. This is also known as a **fourfold table**.

TABLE 13.6

General Structure of a 2 × 2 Contingency Table

Variable-2 (Outcome)	Variable-1 (Antecedent)		
	Present	Absent	Total
Present	$O_{11}(\pi_{11})$	$O_{12}(\pi_{12})$	$O_{1\bullet}(\pi_{1\bullet})$
Absent	$O_{21}(\pi_{21})$	$O_{22}(\pi_{22})$	$O_{2\bullet}(\pi_{2\bullet})$
Total	$O_{\bullet1}(\pi_{\bullet1})$	$O_{\bullet2}(\pi_{\bullet2})$	n

13.2.1 Structure of 2 × 2 Table in Different Types of Study

Table 13.6 is stated in the classical antecedent–outcome format. Three situations are possible as discussed in Chapter 2.

13.2.1.1 Structure in Prospective Study

Because the investigation is from antecedent to outcome in a prospective study, the column totals $O_{\bullet 1}$ and $O_{\bullet 2}$ are fixed in advance. They can also be denoted by n_1 and n_2, respectively. These are the numbers of exposed and nonexposed subjects followed up for appearance of outcome. The row totals $O_{1 \bullet}$ and $O_{2 \bullet}$ become known only after the investigation is over. The relevant hypothesis in this case is H_0: $\pi_{11} = \pi_{12}$. This states that the incidence rate in the two groups is same. In this case, $\pi_{11} + \pi_{21} = 1$ and $\pi_{12} + \pi_{22} = 1$. Thus, H_0 is equivalent to $\pi_{21} = \pi_{22}$.

13.2.1.2 Structure in Retrospective Study

The direction of the investigation in a retrospective study is from outcome to antecedent. Thus, the row totals $O_{1 \bullet}$ and $O_{2 \bullet}$, say, with and without disease, are fixed in advance and the column totals $O_{\bullet 1}$ and $O_{\bullet 2}$ are obtained through the study. The fixed row totals can also be denoted by n_1 and n_2. The null hypothesis now is that the rate of presence of antecedent in those with a positive outcome is the same as in those with a negative outcome, that is, H_0: $\pi_{11} = \pi_{21}$. In this case, $\pi_{11} + \pi_{12} = 1$ and $\pi_{21} + \pi_{22} = 1$. The H_0 implies $\pi_{12} = \pi_{22}$ also.

13.2.1.3 Structure in Cross-Sectional Study

In this case, n subjects are simultaneously cross-classified by the antecedent and outcome. Neither the column totals nor the row totals are fixed in advance and both become known only after study of the subjects is over. In this case, $\pi_{11} + \pi_{12} + \pi_{21} + \pi_{22} = 1$. According to the law of multiplication of probabilities, the antecedent and outcome are independent if and only if H_0: $\pi_{rc} = \pi_{r \bullet} * \pi_{\bullet c}$ ($r, c = 1, 2$) holds, where $\pi_{r \bullet} = \pi_{r1} + \pi_{r2}$ and $\pi_{\bullet c} = \pi_{1c} + \pi_{2c}$.

The H_0 in the prospective and retrospective setup is called **hypothesis of homogeneity** (column homogeneity and row homogeneity, respectively), and the H_0 in the cross-sectional setup is called **hypothesis of independence**. All these situations can be viewed as subjects divided by two qualitative characteristics with the objective to investigate if one characteristic has any association with the other—whether one is occurring more commonly with the other than expected by chance.

13.2.2 Two Independent Samples (Large *n*): Chi-Square Test and Proportion Test

Even though the null hypotheses in the three situations just enumerated are different and consequently the interpretation is different, it can be shown that the test criterion is the same for all of them. Under any of the three H_0s, the expected frequency in the (r,c)th cell, when the samples are independent, is given by

$$E_{rc} = (O_{r\bullet} * O_{\bullet c})/n; \quad r,c = 1, 2. \tag{13.7}$$

Consider, for example, a study where H_0: $\pi_{11} = \pi_{21}$. This hypothesis implies, in the context of Table 13.6, that the proportion of subjects with antecedent in the two outcome categories should be the same. Each of these proportions would be the same as in the two categories combined. A similar statement should also be true for subjects without antecedent. This gives

$$\left(\frac{E_{11}}{O_{1\bullet}}\right) = \left(\frac{E_{21}}{O_{2\bullet}}\right) = \left(\frac{O_{\bullet 1}}{n}\right)$$

and

$$\left(\frac{E_{12}}{O_{1\bullet}}\right) = \left(\frac{E_{22}}{O_{2\bullet}}\right) = \left(\frac{O_{\bullet 2}}{n}\right)$$

or

$$E_{11} = \frac{O_{1\bullet}O_{\bullet 1}}{n}, \quad E_{21} = \frac{O_{2\bullet}O_{\bullet 1}}{n}, \quad E_{12} = \frac{O_{1\bullet}O_{\bullet 2}}{n}, \quad E_{22} = \frac{O_{2\bullet}O_{\bullet 2}}{n}.$$

These are exactly the same as stated in the formula given in Equation 13.7. The other two hypotheses can also be shown to lead to the same formula.

13.2.2.1 Chi-Square Test

Now, the test criterion analogous to criterion in Equation 13.5 is

$$\text{Chi-square for } 2\times 2 \text{ table: } \chi^2 = \Sigma_{rc} \frac{(O_{rc} - E_{rc})^2}{E_{rc}}; \quad r,c = 1,2 \tag{13.8}$$

The justification is the same as for the criterion in Equation 13.5 and the applicability also requires each expected cell frequency to be at least 5 as before. In a 2×2 table, df = 1. There is freedom to choose the frequency arbitrarily in only one cell. The others are automatically decided because the row and column totals are considered fixed. The test procedure is to calculate χ^2 and find the probability P of obtaining this or a higher value. A small value of P, as before, is evidence against H_0. If the P-value is sufficiently small, less than the predetermined level of significance, reject H_0, otherwise not.

TABLE 13.7

Anemia and Parity in a Cross-Sectional Study of 100 Women

Anemia	Observed in the Survey			Expected under H_0	
	Parity ≤ 2	Parity ≥ 3	Total	Parity ≤ 2	Parity ≥ 3
Present	14	16	30	18	12
Absent	46	24	70	42	28
Total	60	40	100	60	40

Example 13.6: Relation between Anemia and Parity Status

Let the interest be in finding whether or not the prevalence of anemia in women is related to their parity status. Parity status is divided as two or less and three or more. Suppose the observed prevalence in a *cross-sectional* survey of 100 randomly selected women from a specified population is as given in Table 13.7. In these data, 14 × 100/60 = 23% of women of parity ≤ 2 have anemia versus 16 × 100/40 = 40% women of parity ≥ 3. Is this association really present in the population?

The null hypothesis in this cross-sectional study would be of independence. That is, H_0: $\pi_{rc} = \pi_{r\bullet} * \pi_{\bullet c}$. If parity status has nothing to do with anemia status, then the ratio of anemics in both parity groups would be the same. That is, 30% of 60 women with parity ≤ 2 and 30% of 40 women with parity ≥ 3 should be anemic. Such expected frequencies under H_0 are also given in Table 13.7. These can be verified to follow the formula given in Equation 13.7. Now, from criterion set in Equation 13.8,

$$\chi^2 = \frac{(14-18)^2}{18} + \frac{(16-12)^2}{12} + \frac{(46-42)^2}{42} + \frac{(24-28)^2}{28} = 3.17.$$

Only one cell frequency in a 2 × 2 table can be freely determined. If the number of nonanemics in the group with parity ≥ 3 is 18, then the number of anemics in this group has to be 22, that of anemics of parity ≤ 2 has to be 8 and that of nonanemics in this group 52. Only that will keep the row and column totals unaltered.

A computer-based statistical package gives $P(\chi^2 \geq 3.17) = 0.075$. Otherwise, from Table B.3, for 1 df, $P < 0.05$ only when $\chi^2 > 3.841$.

Since χ^2 is less, the *P*-value is not sufficiently small in this example. Thus, the H_0 cannot be rejected. The evidence in this sample of 100 women is not sufficient to conclude that the prevalence of anemia in women is related to parity status. The initial assumption (H_0) of no relation can be conceded in the absence of sufficient evidence against it.

13.2.2.2 Yates Correction for Continuity

As remarked earlier, for a 2 × 2 table also, the frequencies in a contingency table give rise to a discrete distribution whereas chi-square is a continuous variable.

The approximation works fine for large n, but this can sometimes be improved in a 2 × 2 table by Yates correction for continuity:

$$\text{Continuity corrected } \chi^2 = \Sigma_{rc}\left[\frac{(|O_{rc} - E_{rc}| - 0.5)^2}{E_{rc}}\right],$$

where $|a|$ means the absolute value or modulus of a. For the data in Table 13.6, corrected $\chi^2 = 2.43$. This is substantially less than 3.17 obtained earlier without correction. Now $P = 0.119$. Yates correction will give a lower value of χ^2 and consequently a higher value of P. This improves the approximation in some cases but can make the test overly conservative in other cases. Clear guidelines are not available regarding situations in which this is helpful. I avoid this correction in two independent samples setup but use it in matched pair setup where it is less controversial.

13.2.2.3 Z-Test for Proportions

The other way to test hypothesis of homogeneity or of independence for large n in a 2 × 2 table is by comparing proportions by Z-test. This is given by the Gaussian test for comparing proportions as

$$Z = \frac{|p_1 - p_2|}{\sqrt{p(1-p)(1/n_1 + 1/n_2)}}, \tag{13.9}$$

where, for a retrospective study, $p_1 = O_{11}/n_1$, $p_2 = O_{21}/n_2$, and $p = O_{\bullet 1}/n$. In fact, $n_1 = O_{1\bullet}$ and $n_2 = O_{2\bullet}$, but notations n_1 and n_2 are easy to understand in this case. The denominator is the estimated SE of the numerator in the case of independent samples. The value of Z is referred to a Gaussian distribution and a two-sided P-value is obtained. For the data in Table 13.7,

$$Z = \frac{|14/30 - 46/70|}{\sqrt{60/100(1 - 60/100)(1/30 + 1/70)}} = 1.78.$$

Corresponding to this value of Z, from Table B.1, $P = 2 \times 0.0375 = 0.075$ when both negative and positive sides are considered. For large n, this should give the same P-value as obtained by the chi-square criterion in Equation 13.8. In fact, there is a theoretical relationship saying that $Z^2 = \chi^2$ with 1 df. This is called the equivalence of the 1 df χ^2 to Z. An advantage of the criterion in Equation 13.9 is that H_0 can be directly tested against the *one-sided alternative*. Then, the P-value is obtained for one tail only. This would be twice the two-tail probability if the one-tail probability is not directly available. In case of chi-square in Equation 13.8, the P-value obtained would be two tailed.

Most software packages require only the entry of data in a specific format and a command to compute chi-square. P-values with and without Yates correction will be automatically obtained. Many packages issue a warning in case n or expected cell frequencies are small or automatically compute Fisher exact test in this case (see Section 13.2.4). The purpose of providing formulas is to explain the underlying principles. Many users these days would not actually use these formulas and would not manually calculate the value of χ^2 or Z or, for that matter, any other given in this book.

13.2.2.4 Detecting a Medically Important Difference in Proportions

As noted in Chapter 12, a small difference between groups can become statistically significant when the sample size is large. If the cure rate after 1 month of a particular new therapy is 40% in a sample of subjects versus 30% with the existing therapy, the difference is statistically significant if the number of subjects in each group is 123 or more. (This sample size can be obtained by using the method of Section 12.6.2.) But this small difference may not be worth the trouble of switching over to the new therapy if it is relatively more difficult to implement. A difference of even 1% can be statistically significant for a sufficiently large sample but very few clinicians if at all would change their practice for 1% gain. Thus, medical significance of the difference is always a potent consideration.

In view of the importance of medical significance of the result vis-à-vis statistical significance, I discuss this aspect in detail in Chapter 15. For the time being, the concern is with a method for ensuring that a medically important difference is not missed when present. It is for the medical profession to specify the minimum difference that would be considered medically important. The biostatistician has little role.

If a difference of more than 20% is considered of some consequence, then H_0: $(\pi_1 - \pi_2) = 0.20$. Conventional null is $(\pi_1 - \pi_2) = 0$ (nil), but $(\pi_1 - \pi_2) = 0.20$ is equally valid null despite not being nil. If this is rejected, the alternative hypothesis H_1: $(\pi_1 - \pi_2) > 0.20$ is accepted. If the difference in sample proportions is 0.20 or less, there is no scope for H_1 to be true. There is no way, then, that H_0 can be rejected in favor of the alternative that says that the difference is more. There is no need to carry out the test of statistical significance in that case. The question of testing arises only when the sample difference is more than 0.20 and the intention is to find whether it can still be 0.20 (or less) in the target population. This type of argument applies to most one-sided tests. In some situations, the argument can be reverse. The sample difference is less and you want to know if it still can be a higher value in the population.

For two independent samples of large size, the criterion to test H_0: $(\pi_1 - \pi_2) = \pi_0$ is only slightly different from the criterion in Equation 13.9. This is

$$Z = \frac{(p_1 - p_2) - \pi_0}{\sqrt{(p_1 q_1 / n_1) + (p_2 q_2 / n_2)}}, \tag{13.10}$$

where the denominator is the estimated SE of $(p_1 - p_2)$ when π_1 and π_2 are different. The criterion in Equation 13.9 is based on the null hypothesis that $\pi_1 = \pi_2$, which is no longer true in this new setup. In this example, $p_1 = 0.40$, $p_2 = 0.30$, and $\pi_0 = 0.20$. The sample difference is less than 0.20. Thus, H_0 cannot be rejected in favor of H_1: $(\pi_1 - \pi_2) > 0.20$. However, for the purpose of illustration, consider H_1: $(\pi_1 - \pi_2) < 0.20$. If $n_1 = n_2 = 50$, the value of the criterion in Equation 13.10 is

$$Z = \frac{(0.40 - 0.30) - 0.20}{\sqrt{(0.40 \times 0.60/50) + (0.30 \times 0.70/50)}} = -1.05.$$

For H_1: $(\pi_1 - \pi_2) < 0.20$, smaller values of Z would favor H_1. So we need to find $P(Z \le -1.05)$. Note again that H_1 and the P-value both have same direction, in this case, the less-than type. From Table B.1, this is 0.1469. This P-value is not small and H_0 cannot be rejected. The chance that the difference in proportions in the target population is 0.20 is not small. As already explained, this does not mean that the difference is 0.20.

Note that in this example, the therapy group is labeled as the first group and the control group as the second group. Then, $(\pi_1 - \pi_2)$ was expected to be $+0.20$ under H_0. If the labels were reversed, H_0 would specify this difference to be -0.20.

13.2.2.5 Crossover Design with Binary Response (Large n)

As stated in Chapter 5, crossover is an effective strategy for minimizing the effect of interindividual variation in case there is no carryover effect of the drug and the disease bounces back to the original level when the drug is discontinued. A patient is given one drug for a specified duration, then the other drug with a washout period in between.

Sometimes, the sequence of administration makes a difference in the outcome. That is, patients who receive drug-B and then drug-A (BA sequence) may give different results than those who receive the AB sequence. To study the effect of different sequences, the patients are divided into two random groups. One is given the BA sequence and the other the AB sequence. The two groups contain different set of individuals, thus the groups are independent. For further explanation and an example on quantitative data, see Section 15.1.3.

The concern in this section is with crossover experiments in which the response or the outcome is binary. This is a yes/no, present/absent, or relieved/not relieved type of response. In such cases, the number of subjects with different responses in a crossover trial can be listed as in Table 13.8.

TABLE 13.8

Format of Binary Response in a Crossover Trial

	Response[a]				
Group (Sequence)	(0, 0)	(0, 1)	(1, 0)	(1, 1)	Total
Group-I—sequence AB	a_1	a_2	a_3	a_4	n_1
Group-II—sequence BA	b_1	b_2	b_3	b_4	n_2

[a] 0, no relief; 1, relief.

TABLE 13.9

Relevant Numbers for Analyzing
Crossover Design

		Response	
		(0, 1)	(1, 0)
①	Group-I—sequence AB	a_2	a_3
	Group-II—sequence BA	b_2	b_3

In this table, for example, b_2 is the number of subjects who were not relieved by drug-B but were relieved by drug-A when the sequence of administration was B followed by A. Other notations also have similar meaning.

The analysis of data from crossover trials is not fully standardized. One of the methods is as follows. For finding statistical significance in this case, the concordant pairs in the first and last columns are ignored. Since the response with drug-A is the same as with drug-B in these two types of pairs, they cannot help in deciding which drug is better. The decision is based on the discordant pairs in the second and third columns. The frequencies in these two middle columns are analyzed as a usual 2 × 2 table. For large n, the chi-square criterion stated in Equation 13.8 is computed and the inference drawn as per the procedure already explained. The relevant numbers for this setup are given in Table 13.9.

This table will indicate whether the response with one treatment is different from the response with the other treatment. If $(a_2 + b_3)$ is large relative to $(a_3 + b_2)$, then drug-B is more effective. If $(a_3 + b_2)$ is larger, then drug-A is more effective. The prerequisite for validity of this test, however, is that there is no effect of different sequence of drug administration and no carryover effect.

It is helpful to investigate whether sequencing (A precedes B and B precedes A) alters the response. This can be done by making another 2 × 2 table (Table 13.10) that gives counts of relieved subjects in the two groups. Again, only discordant pairs are counted.

Chi-square is computed and an inference drawn as usual. If sequence effect in Table 13.10 is significant, crossover is not a suitable strategy. Redo the trial with some other design. If interaction between sequencing and response is to be statistically tested, prepare a table as Table 13.11 and calculate chi-square as usual.

TABLE 13.10

Numbers for Testing Sequencing Effect in a Crossover Trial

		Relieved with A	Relieved with B
②	Group-I—sequence AB	a_3	a_2
	Group-II—sequence BA	b_2	b_3

TABLE 13.11

Numbers for Testing Interaction
in Crossover Trial

	Response	
	(0, 0)	(1, 1)
③ Group-I—sequence AB	a_1	a_4
Group-II—sequence BA	b_1	b_4

These three procedures are stated in reverse order. In practice, do ③, then ②, and then ①. If the presence of interaction is detected by ③, it is not worthwhile to do ② or ①. In this case, find out why the interaction is occurring and do the trial again after taking steps to remove the likelihood of interaction.

No example is given here but the method is illustrated shortly while describing this procedure for small n. Further details about crossover trials appear in Chapter 15 in the context of quantitative data.

13.2.3 Equivalence Tests

Various kinds of equivalences are explained in Chapter 6 in the context of clinical trials, but the ideas are general and apply to many other setups. The thrust here is to test whether the two groups are essentially equivalent with respect to a particular end-point and the difference if any is of no medical consequence. Superiority and noninferiority are also discussed.

13.2.3.1 Superiority Equivalence and Noninferiority

One regimen can be considered medically equivalent to the other regimen when the difference between the two does not exceed by more than a pre-specified medically unimportant difference. Group-I is superior to group-II when the response in group-I is higher by at least the specified margin, and group-I is noninferior to group-II when the response is lower by not more than the specified margin. This is illustrated in Figure 13.1.

Comments on the left side on this figure assume that the same margin Δ can be used for noninferiority, equivalence, and superiority and also assume that higher values are in favor of the test regimen. A trial gives evidence of superiority of the test regimen when the lower limit of the CI for (μ_{test} − μ_{ref}) exceeds $+\Delta$, gives evidence of noninferiority when upper limit of CI is more than $-\Delta$, and of equivalence when the entire CI is within $-\Delta$ and $+\Delta$ (Figure 13.1).

You can easily see that the superiority or noninferiority is tested by specifying the medically relevant difference and conducting a one-tailed test. However, equivalence within a specified margin needs a little more consideration.

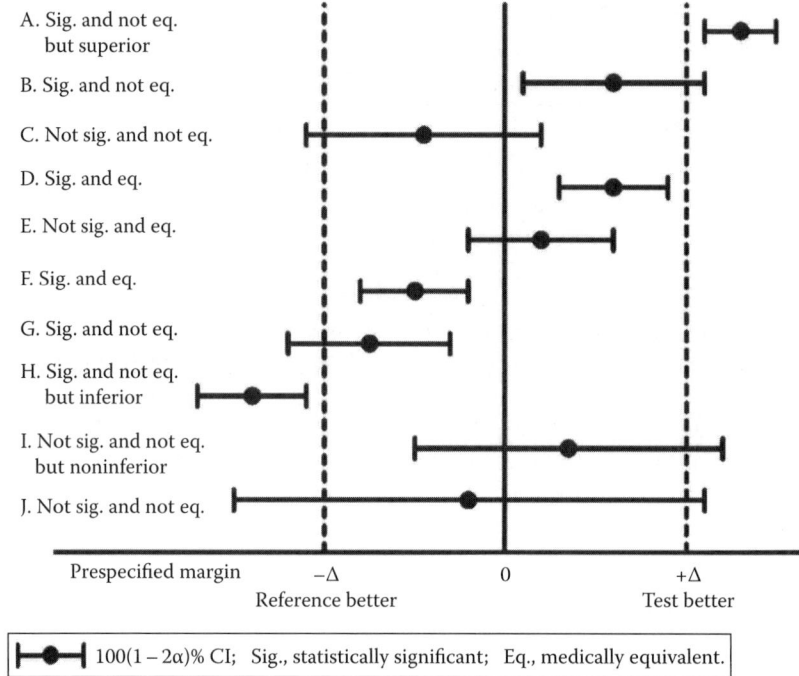

FIGURE 13.1
Statistical significance and medical equivalence.

13.2.3.2 Equivalence

For equivalence, the margin is specified for either side, generally saying that the difference does not exceed $\varepsilon\pi$. If proportion of success in one group is $\pi = 0.50$ and 4% of π on either side is considered to arise due to natural factors, then $\varepsilon = 0.04$ and $\varepsilon\pi = 0.04 \times 0.50 = 0.02$. The limits of tolerance for equivalence thus are 0.48–0.52. This is for therapeutic or clinical equivalence and not for bioequivalence.

In equivalence testing, the null hypothesis has a reverse nature: the difference is a priori specified Δ or more. Rejecting this null would mean equivalence, and the burden of rejecting the null is on the data as before. Type-I error in this case is concluding that the difference is less when, in fact, it is Δ or more, and Type-II error is concluding that the difference is at least Δ when, in fact, it is less. These are kind of inverse of the conventional setups.

An approach to test equivalence is to perform two one-sided tests (TOSTs). But easier procedure is to obtain CI for the difference between two groups at $100(1 - 2\alpha)\%$ confidence level instead of the usual $100(1 - \alpha)$ level. Since the difference on either side is admissible in equivalence setup, limits for confidence level $100(1 - 2\alpha)\%$ provide lower and upper bounds at $100(1 - \alpha)\%$ confidence. If this CI contains zero, the difference is not statistically significant at 2α% level of significance. If this interval is completely contained in $-\Delta$ to $+\Delta$, equivalence is concluded. Any difference less than Δ is considered trivial and within medical indifference. In exceptional cases, the difference for lower bound can be different than for the upper bound. Thus, Δ_1 and Δ_2 can be separately specified.

Under this procedure, the difference between two proportions may be statistically significant, yet the two could be equivalent in terms of the difference not exceeding the medical tolerance. The reverse can also happen. Two proportions may not be statistically significantly different, yet not equivalent. In Figure 13.1 are the $100(1 - 2\alpha)$% CIs for difference in 10 different experiments. Experiments A and H provide unambiguous results—A has statistical significance and superiority and H has statistical significance and inferiority. In experiment E, the interval contains zero and the entire interval is within the limits of medical indifference. Thus, the difference is neither statistically significant nor medically significant. In experiments D and F, the interval does not contain zero so that the difference is statistically significant but the interval again lies entirely within the limits of medical indifference, thus being medically equivalent. These are the examples that illustrate that statistical significance does not imply medical significance. In experiments I and J, the CIs overlap with the limits of equivalence. In these two cases, since the CI is relatively large, a restudy with a larger sample is advisable.

Main difficulty in equivalence tests is not the statistical procedure but specification of tolerance Δ. This is primarily a medical decision where statistics plays a secondary role. This limit depends on how much latitude can be given without compromising patient management and what variation is expected due to unforeseen and unknown factors such as in obtaining exact measurements. By its very nature, this has to be fairly small and trivial that does not alter the patient management. This is what makes the two regimens clinically indistinguishable. Identifying such a limit could be a challenge in some situations and can force you to be inexact. See also my comments on determining noninferiority margin later in this section.

Typically, tests for establishing equivalence need a larger n than tests for establishing difference. This is because clinically unimportant Δ chosen for equivalence is generally much smaller than the clinically important difference in the usual comparative trials.

There are other ifs and buts attached to equivalence: (1) equivalence can be achieved when both treatments are ineffective—thus the reference under comparison must have established efficacy; (2) unless special care is taken, equivalence in efficacy is oblivious of difference in patient compliance and losses that can be more with one regimen than the other; (3) if equivalence is found for a particular dose, it may not carry over to the other doses; and (4) if both regimens have high efficacy, the difference could be masked and equivalence can be fallacious.

Equivalence trials are affected much more than the usual comparative trials by violation of the protocol. Equivalence should be concluded only when both the per-protocol analysis and intention-to-treat analysis (discussed later) reach to the same conclusion.

Example 13.7: Equivalence in Vaccine Coverage

Disparity in vaccine coverage in different population groups could be a threat to public health. Barber et al. [1] examined the National Immunization Survey (NIS) for the year 2000 for coverage of various vaccinations in Whites, Blacks, Hispanics, and Asians in the United States. The results for Blacks and Hispanics in comparison to Whites for some vaccinations are given in Table 13.12. This table also contains 95% CIs for differences and 95% CIs for equivalences. Since five of the six CIs for differences do not contain zero, the difference in coverage is statistically significant in these cases. Only the difference in MMR vaccine coverage in Hispanics is not significant from that of Whites. This is given in bold letters in the table.

SIDE NOTE: The authors considered coverage within ±5% from Whites as medically equivalent. The last two columns of the table show that all coverages, except of MMR in Blacks, are medically equivalent to those in Whites. If the equivalence criterion is more strict at ±3%, only the coverage of DPT3 and MMR in Hispanics is equivalent to that in Whites.

13.2.3.3 Determining Noninferiority Margin

Noninferiority trials are done to find if the test regimen can be as good as the one with established efficacy. The procedure for noninferiority testing is essentially the same as for equivalence trials. But, in this case, the lower limit of 95% CI should be more than $-\Delta$ and upper limit can be any value (experiment I in Figure 13.1). Interpret noninferiority with caution. It does not mean that the regimen is not inferior—only that it is not worse by more than the prespecified clinically unimportant margin. That is, the loss of efficacy has no clinical relevance. Such a regimen may have other benefits such

TABLE 13.12

Vaccination Coverage in Different Ethnic Groups in the United States (2000)

Vaccination	NIS Coverage			95% CI for Difference		95% CI for Equivalence	
	Whites	Blacks	Hispanics	Blacks	Hispanics	Blacks	Hispanics
DPT3[a]	95.0	92.1	93.3	1.2, 4.6	0.4, 3.0	**1.5, 4.3**	**0.6, 2.8**
MMR[b]	91.5	87.9	90.0	1.7, 5.5	−0.1, 3.1	2.0, 5.2	**0.2, 2.8**
Hep B[c]	91.4	89.2	88.2	0.3, 4.0	1.5, 1.9	**0.7, 3.7**	**1.8, 4.6**

[a] DPT3: Three or more doses of diphtheria, pertussis, and tetanus vaccine.
[b] MMR: One or more dose of measles–mumps–rubella vaccine.
[c] Hep B: Three or more doses of hepatitis B vaccine.

as convenience, cost, and safety. There are other implications too. If clinical unimportant margin is 2%, a regimen with 86% efficacy is noninferior to one with 88% efficacy, and one with 84% efficacy is noninferior to the one with 86% efficacy. Thus, the standard can progressively slip down, and it is important to be judicious in choosing Δ, especially for noninferiority trials.

Noninferiority margin is specified in advance and must always be justified for the specific regimen and disease you are considering. This margin does not depend on the size of the trial nor on statistical power of the study. It primarily depends on natural variability in the difference between efficacies of test and reference regimen that can arise from trial to trial. It also depends what sacrifice in efficacy you can make in exchange of lower cost or less side effects that the regimen under test may have. Review of literature or experience might suggest what difference is clinically irrelevant. The following may also help in deciding the noninferiority margin:

1. Survey the practitioners who deal with that disease and find the range they consider unimportant considering other advantages of the regimen under investigation. This will take care of cost, convenience, acceptability, etc.

2. If there are many regimens that are interchangeably used at present for treating the same condition, the difference in their efficacies as reported in the literature can give fairly good idea of what can be clinically unimportant margin.

3. Consider natural variation you expect due to unavoidable errors in measurement and other assessments.

4. If there is a definite safety advantage, larger Δ can be chosen.

If the outcome of interest is death, it could be ethically difficult to specify clinically unimportant deaths. You may like to plan a superiority trial in this case, possibly with relaxed significance level.

13.2.4 Two Independent Samples (Small *n*): Fisher Exact Test

The usual chi-square fails to oblige when *n* is small. An exact method is needed in this case. For a 2 × 2 table, the null hypothesis can be tested by Fisher exact test. A similar strategy is needed to analyze crossover designs with small *n*.

13.3.4.1 Fisher Exact Test

The test applicable to a 2 × 2 table in case of small *n* is called Fisher exact test, also sometimes called Fisher–Irwin test. In terms of notations presented in Table 13.6, the probability in this test is computed as follows:

$$\text{Fisher exact test for } 2 \times 2 \text{ table: } P = \sum \frac{O_1.!O_2.!O._1!O._2!}{n!O_{11}!O_{12}!O_{21}!O_{22}!}, \qquad (13.11)$$

where the sum is over all the configurations in a 2×2 table that are observed as or more extreme in favor of H_1, without altering the marginal totals. This can be easily calculated manually for small n (say, $n < 10$) but can become difficult for larger n. Most statistical software packages would give the exact P-value for small n using Equation 13.11. Reject H_0 if $P < 0.05$; otherwise, be content with the assertion made in H_0.

The probability in Equation 13.11 gives the one-tail P-value. This is typical for Fisher exact test. If a two-tail value is needed, proceed as follows: (1) Double the one-tail P-value if any (row or column) marginal totals are equal. Equal totals imply symmetry and this allows such doubling. (2) Calculate a separate P-value from Equation 13.11 for each tail if the marginal totals are not equal.

Fisher exact test assumes fixed marginal totals. This restriction is valid for a situation where, for example, you have six persons with disease and six without disease, and the test is built up in a manner that it is constrained to give six positive and six negative results. You can see that this is an unnatural restriction and would rarely hold in practical situations. This restriction makes the test overly conservative (i.e., fails to reject H_0 where it should). To overcome this problem, another test called **Bernard test** is advocated, which does not require fixed margins. This test is computationally difficult and is not popular yet. If you happen to work with statistical software that computes Bernard test, prefer this. For comparison between Fisher exact and Bernard test, see Mehta and Senchaudhuri [2].

13.2.4.2 Crossover Design (Small n)

For small cell frequencies in a crossover design, the statistical significance of the difference between treatments A and B can be investigated by using Fisher exact test on the discordant pairs. The concordant pairs are ignored. This is shown in Example 13.8.

> **Example 13.8: Crossover Trial for Urinary Problems in Enlarged Prostate**
>
> A new drug-A for relief from urinary problems in subjects with enlarged prostates was compared with an existing drug-B. Each was given for 1 month to 50 subjects in sequence BA and to another 50 subjects in sequence AB. There was a washout period in between as needed in a crossover design. The results are shown in Table 13.13.
>
> For the purpose of comparison, the first and the last columns are ignored. The frequencies in the discordant cells are small and so Fisher test would be used. The given configuration and more extremes are as follows:

1	7	8		0	8	8
5	2	7		6	1	7
6	9	15		6	9	15

TABLE 13.13

Crossover Trial with Small Frequencies in Some Cells

	Response[a]				
	(0, 0)	(0, 1)	(1, 0)	(1, 1)	Total
Group-I—sequence AB	2	1	7	40	50
Group-II—sequence BA	7	5	2	36	50

[a] 0, no relief; 1, relief.

No other extreme configuration is possible because marginal totals have to remain the same, and one cell frequency is already zero in the second configuration. Therefore, from Equation 13.11,

$$P = \frac{8!7!6!9!}{15!1!7!5!2!} + \frac{8!7!6!9!}{15!0!8!6!1!}$$

$$= 0.0336 + 0.0014$$

$$= 0.035.$$

This is less than 0.05. Since it is a one-tail probability, the conclusion too would be one sided. In this case, there are seven subjects in sequence AB that had relief from drug-A but not from drug-B and five subjects with relief from A but not from B in sequence BA. Of a total of 15 discordant pairs, 12 favor drug-A and only 3 favor drug-B. A small P-value shows that this is statistically significant. Thus, the conclusion is that the new drug-A is significantly better than the existing drug-B. For such one-sided conclusions, one-sided α is used.

13.2.5 Proportions in Matched Pairs: McNemar Test (Large *n*) and Exact Test (Small *n*)

The procedures described in the previous sections are valid only when the two groups of subjects are independent. Independence is lost when there is one-to-one matching or pairing. Matching is a frequently adopted mechanism in medical studies. Pairing also occurs when the same group of subjects is observed before therapy and after therapy. A matched pair is considered one unit, and the contingency table contains the number of units or pairs with both elements positive, the first element positive and the second negative, the first negative and the second positive, and both negative. These can be denoted, respectively, by *a*, *b*, *c*, and *d* as in Table 13.14. The table is arranged for a prospective study. In all, there are *n* matched pairs in this table. The cell frequencies are the numbers of the pairs. For example, *b* is the number of pairs in which the exposed partner develops the disease and the nonexposed partner does not develop the disease.

TABLE 13.14

Matched Pairs with Dichotomous Antecedent and Dichotomous Outcome: Prospective Study

Partner-2 Antecedent Present (Exposed or Experiment)	Partner-1 Antecedent not Present (not Exposed or Control)		Total
	Positive Outcome (Disease+)	Negative Outcome (Disease−)	
Positive outcome (disease+)	a	b	$a + b$
Negative outcome (disease−)	c	d	$c + d$
Total	$a + c$	$b + d$	$n = a + b + c + d$

13.2.5.1 Large n: McNemar Test

A very popular criterion in case of matched pairs is as follows:

$$\text{McNemar test: } \chi_M^2 = \frac{(|b - c| - 1)^2}{b + c}, \tag{13.12}$$

where b and c are as given in Table 13.14. For large n, this is referred to a chi-square distribution with 1 df for obtaining the P-value. The restriction of no expected cell frequency less than 5 applies to b and c. The subtraction of one in the numerator of the formula given in Equation 13.12 represents a continuity correction similar to Yates correction mentioned earlier. Note again that the concordant pairs a and d do not contribute to the decision. It is based solely on the number of discordant pairs of the two types. Significance would mean that the discordance is not symmetric and the pairs do not match with respect to the outcome. The numbers b and c can be high but if they are equal, the test will not give significance.

Example 13.9: Matched Pairs for a Trial on Common Cold Therapy

To evaluate the role of a therapy in relieving common cold within a week, suppose 50 cases underwent the therapy and another group of 50 cases served as controls. The experimental and control cases were matched one to one for age, gender, and body mass index (BMI) so that these do not act as confounders. Matching on BMI was done to largely rule out nutritional status as a confounder. The results obtained are summarized in Table 13.15. There are 22 pairs in which both types of subjects—with therapy and without therapy—felt relieved in 1 week's time. In 15 pairs, the subject with therapy felt relieved but the subject without therapy did not feel so. The frequencies in the second row can be similarly explained. In this table, $b = 15$ and $c = 5$. Therefore,

$$\text{McNemar } \chi_M^2 = \frac{(|15 - 5| - 1)^2}{15 + 5} = 4.05.$$

TABLE 13.15

Trial for Therapy for Common Cold: Matched Pairs

With Therapy (Experimental Group)	Without Therapy (Control Group)		
	Relieved Within 1 Week	Not Relieved Within 1 Week	Total
Relieved within 1 week	22	15	37
Not relieved within 1 week	5	8	13
Total	27	23	50

A software package gives $P = 0.044$, otherwise from Table B.3, $P < 0.05$ for 1 df. The null hypothesis in this case is that the therapy has no effect. But the likelihood of this being true is extremely small—less than 5%. Thus, reject H_0 and conclude that the therapy is helpful in relieving common cold within 1 week. Note that the number of those relieved by therapy (15 subjects) is much more than those relieved without therapy (5 subjects) among the discordant pairs.

13.2.5.2 Small n: Exact Test (Matched Pairs)

As in most cases, McNemar test in Equation 13.12 also ceases to follow the chi-square distribution when n is small. The test is then done with the help of a binomial distribution.

If there is no association between antecedent and outcome, the number of discordant pairs b should be nearly the same as the number of discordant pairs c. Each of these $(b + c)$ pairs should be divided with equal probability in these two groups. The probability is equal when it is $\frac{1}{2}$ for either type of discordant pairs; that is, $\pi = \frac{1}{2}$ or $1 - \pi = \frac{1}{2}$. Thus, by binomial probability (Equation 13.1), the P-value under H_0 is

$$\text{Exact test for matched pairs: } P = \sum_{x=b}^{(b+c)} {}^{(b+c)}C_x \left(\frac{1}{2}\right)^{(b+c)}. \tag{13.13}$$

This is the probability of obtaining b or more discordant pairs when H_0 of no association is true. The H_1 in this case is that a positive outcome is more in those with an antecedent present. Thus, b should be large if H_1 is true and c should be small. The probability in Equation 13.13 is of the configurations as much or more extreme favoring H_1 when actually there is no association. If this is small, say, less than 0.05, reject H_0.

Example 13.10: Exact Test for Matched Pairs

Consider the same setup as in Example 13.8; however, the number of pairs is small and suppose they are distributed in a 2×2 table as follows:

5   3

1   6

There are a total of 15 pairs in this example and a total of $(b + c) = 4$ discordant pairs. Out of these 4, therapy has been effective in $b = 3$ pairs. This is the number in favor of H_1. In addition, $b = 4$ would be in favor of H_1. Statistically, it cannot exceed 4 because $(b + c) = 4$. If the therapy actually is not effective, the probability of obtaining $b = 3$ or $b = 4$ from the formula given in Equation 13.13 is

$$P = \sum_{x=3}^{4} {}^{4}C_x \left(\frac{1}{2}\right)^4$$

$$= {}^{4}C_3 \left(\frac{1}{2}\right)^4 + {}^{4}C_4 \left(\frac{1}{2}\right)^4$$

$$= 0.31.$$

This is very high relative to the conventional threshold of 0.05. The observed configuration could well have arisen from sampling fluctuation when there is actually no association. Thus, H_0 of no association cannot be rejected. It cannot be concluded that the therapy is more effective in relieving common cold within a week.

13.2.5.3 Comparison of Two Tests for Sensitivity and Specificity: Paired Setup

The procedure of matched pairs can be used to compare sensitivities and specificities of two medical tests when both the tests are performed on the same group of people. For comparison of sensitivities, the performance of the two tests on true positives is compared, and for comparison of specificities, the performance on true negatives is compared. When both sensitivities and specificities are compared on the same set of data, the level of significance needs to be adjusted so that the total Type-I error does not exceed the prefixed threshold.

Example 13.11: Comparison of Sensitivities and Specificities of Two Tests on the Same Group of Subjects

Suppose 200 suspected cases of some disease were thoroughly investigated and 80 were finally found to have the lesion of interest, and the remaining 120 were not found to have that lesion. These 80 confirmed cases with the lesion and 120 confirmed cases without the lesion were tested for the presence of the lesion by computed tomography (CT) scan and ultrasound. The results obtained are given in Table 13.16. The results of both the tests for 80 cases with lesion and 120 cases without lesion are also given in Table 13.16.

Comparison of sensitivities will be based on cases with lesion in Table 13.16(c). This gives

$$\text{McNemar } \chi^2 = \frac{(|5 - 15| - 1)^2}{5 + 15}$$

$$= 4.05.$$

TABLE 13.16

Results of CT Scan and Ultrasound on the Same Group of Patients

(a) All Cases				(b) All Cases			
	CT Scan				Ultrasound		
Lesion	+	−	Total	Lesion	+	−	Total
+	70	10	80	+	60	20	80
−	20	100	120	−	25	95	120
Total	90	110	200	Total	85	115	200

(c) Cases with Lesion				(d) Cases without Lesion			
	CT Scan				CT Scan		
Ultrasound	+	−	Total	Ultrasound	+	−	Total
+	55	5	60	+	18	7	25
−	15	5	20	−	2	93	95
Total	70	10	80	Total	20	100	120

This is significant ($P < 0.05$). Thus, sensitivities of the two tests are significantly different. Table 13.16(a) gives a sensitivity of $70/80 = 0.875$ for the CT scan and Table 13.16(b) gives sensitivity of $60/80 = 0.75$ for ultrasound. Based on the McNemar test, this difference of 12.5% is statistically significant.

For comparing specificities, from Table 13.16(d),

$$\text{McNemar } \chi^2 = \frac{(|7-2|-1)^2}{7+2}$$

$$= 1.78.$$

This is not significant ($P > 0.05$). The specificity of the CT scan is $100/120 = 0.833$ and of ultrasound is $95/120 = 0.792$. Based on the McNemar test, this difference of 4.1% is not significant. Thus, the tests have different sensitivity but not specificity.

Can we make a conclusion regarding sensitivity and specificity together? Since both are based on the same sample of subjects in Table 13.16, Type-I error can be controlled to less than 0.05 by using the Bonferroni procedure (Chapter 15) when each is tested at the 0.025 level. You may like to confirm that at this level sensitivities too are not significantly different. Thus, the evidence is not enough to conclude that the CT scan and ultrasound differ for sensitivity *and* specificity.

13.3 Analysis of $R \times C$ Tables (Large n)

The preceding section is on analysis of tables with two rows and two columns. This means that both the characteristics (or variables) are dichotomous. But a large number of variables are not dichotomous. Blood group has four categories. Subjects may be categorized for smoking as those who have never smoked, ex-smokers, mild smokers, and heavy smokers. All polytomous categories can be dichotomized, such as blood group into B and non-B groups and smoking into yes and no, but quite often such dichotomy fails to serve the purpose. In some situations, such as in assessing trend in proportion of sexually mature girls (with regard to, say, breast stage) of age 14 years with different grades of anemia, it is better to have as many grades of anemia as is feasible. Thus, $R \times C$ contingency tables are not uncommon. Here R is the number of rows and C is the number of columns. For this section, either or both of them are more than 2.

The inference generally needed from such tables is whether the two variables are associated. If an association is found, further analysis can be done to measure the degree of association, to ascertain the presence of trend, if any, or to find which particular cell or cells in the contingency table are contributing to the relationship. The method for finding the presence or absence of association is basically the same for $R \times C$ tables as for 2×2 tables, although some generalization is needed. This section first discusses $2 \times C$ (or $R \times 2$) tables and then goes on to discuss $R \times C$ tables. The concern in this section is with a setup where both variables are qualitative. If one of them is quantitative, examine whether methods such as logistic regression (Chapter 17) can be used. As in the case of 2×2 tables, presumption throughout this section is that n is large and the frequency in at least 80% of cells is at least 5. For small frequencies, Fisher exact test can be extended to $R \times C$ tables. For more on this, see Mehta and Patel [3].

13.3.1 One Dichotomous and the Other Polytomous Variable ($2 \times C$ Table)

Consider the following example to fix ideas. This is based on the data in Table 13.17.

Example 13.12: Enlarged Prostate after Different Dosages of Dioxin in Mice

Dioxins are by-products of combustion and other processes. They persist at lower levels virtually everywhere—in air, water, and soil. They are known to be disrupters that can off-balance the endocrine system. Children are particularly vulnerable. This can affect behavior, immune function, neurological development, and gender development. In an experiment, a group of 100 mouse fetuses are randomly divided into four equal groups and are exposed to none, low, medium, and heavy doses of a dioxin-like chemical.

TABLE 13.17

Dioxin Dosage and Enlarged Prostate in Mice

Enlarged Prostate	Dosage				Total
	None	Low	Medium	Heavy	
Yes	2	5	6	8	21
No	21	20	19	12	72
Total	23	25	25	20	93

These dosages are in relative terms, but even a heavy dose would be a microdose. This experiment is along the lines reported by vom Saal et al. [4]. The numbers of fetuses that developed enlarged prostate are given in Table 13.17. Some fetuses were lost and could not be observed.

The null hypothesis for rejection is that there is no effect of differential dose on the incidence of enlarged prostate. If this H_0 is true, the enlarged prostate would be divided among the dose groups according to the total number of fetuses in each group. This means that 21 enlarged prostates should be in the ratio 23:25:25:20. This total is 93. The expected frequencies under H_0 therefore are $21 \times 23/93, 21 \times 25/93, 21 \times 25/93$, and $21 \times 20/93$, or 5.19, 5.65, 5.65, and 4.52, respectively. If the expected frequency in the rth row and the cth column is denoted by E_{rc}, it should be clear that

$$E_{rc} = O_{r\bullet} * O_{\bullet c}/n; \quad r = 1, 2, \ldots, R; \quad c = 1, 2, \ldots, C; \qquad (13.14)$$

where

$O_{r\bullet}$ is the total of the rth row
$O_{\bullet c}$ is the total of the cth column
n is the grand total

In this example, $R = 2, C = 4, O_{1\bullet} = 21, O_{2\bullet} = 72, O_{\bullet 1} = 23, O_{\bullet 2} = 25, O_{\bullet 3} = 25, O_{\bullet 4} = 20$, and $n = 93$. From the formula given in Equation 13.14, $E_{11} = 5.19, E_{12} = 5.65, E_{13} = 5.65$, and $E_{14} = 4.52$. These are the same as obtained before. Also, $E_{21} = 17.81, E_{22} = 19.35, E_{23} = 19.35$, and $E_{24} = 15.48$. Now, a test criterion is required.

13.3.1.1 Test Criterion

Following the argument similar to that for the criterion in Equation 13.5, the test criterion now is

$$\text{Chi-square for } R \times C \text{ table: } \chi^2 = \sum_{rc} \frac{(O_{rc} - E_{rc})^2}{E_{rc}}; \quad r = 1, 2, \ldots, R; \, c = 1, 2, \ldots, C;$$

$$(13.15)$$

where E_{rc} is as given in Equation 13.14. The chi-square formula given in Equation 13.15 is the same as the formula given in Equation 13.8 for 2×2 tables but now R and C can be more than 2. The present section is restricted to the case in which $R = 2$ but $C > 2$, but that is not a restriction for the criterion in Equation 13.15. A large value of χ^2 would indicate that the observed frequencies are very different from those expected under the null hypothesis and thus would provide evidence against the null. For it to be statistically significant, the χ^2 value should be as large as is very unlikely to occur under H_0. As usual, it is considered very unlikely when the chances are less than, say, 0.05. Statistical packages would readily give this P-value, so a conclusion can be immediately drawn. Otherwise, consult Table B.3 for the critical value of chi-square at the desired significance level. This would again depend on the df. On the lines discussed earlier, it can be seen for an $R \times C$ table that df $= (R - 1)(C - 1)$. In Table 13.17, only $(2 - 1)(4 - 1) = 3$ cells out of 8 can be freely chosen because the row totals and column totals cannot be disturbed.

For the data in Table 13.17, by the formula given in Equation 13.15,

$$\chi^2 = \frac{(2-5.19)^2}{5.19} + \frac{(5-5.65)^2}{5.65} + \cdots + \frac{(12-15.48)^2}{15.48} = 6.13.$$

Since df $= 3$ for this 2×4 table, it is clear from Table B.3 that the probability of a chi-square value of 6.13 or higher under H_0 is more than 0.05. A statistical package gives $P = 0.105$. This shows that the chance of H_0 being true is not sufficiently small. The plausibility of H_0 is not adequately ruled out. Thus, H_0 cannot be rejected. The conclusion is that the dose level of the chemical does not significantly affect the proportion with enlarged prostate in these data.

The chi-square criterion in Equation 13.15 considers each dose level in Example 13.11 on a nominal scale and is oblivious of its ordinal character. If the gradient or the trend is the concern, proceed as follows.

13.3.1.2 Trend in Proportions in Ordinal Categories

It is easy to see in Example 13.11 that the proportion with enlarged prostate increases with dose of dioxin. But that is the observation in this sample. Is there a substantial likelihood that the trend will persist in repeated samples?

A study of trend in proportions is relatively simple when the ordinal categories can be assigned a valid score. For the data in Table 13.17, these scores for dosages could be 0, 1, 2, and 3 for none, low, medium, and heavy doses, respectively. As mentioned in a previous chapter, such linear scores have an inbuilt assumption that the difference between the effect of the no-dose and low-dose categories is the same as that between the heavy- and medium-dose categories, the effect of a heavy dose is three times that of a mild dose, etc. These scores are considered usual metric quantities amenable to algebraic manipulations. This may not be exactly true in many situations. Yet, such scores seem to work reasonably well as an approximation in most practical situations.

One can think of nonlinear scores. Condom use among sexually transmitted disease patients can be categorized as never, sometimes, often, and almost always. Spouse infection percentage may follow a trend in this case

TABLE 13.18

Dioxin-Treated Fetal Mice with Enlarged Prostate
and Some Calculations for Chi-Square for Trend

	\multicolumn{5}{c}{Dosage Score (x_k)}				
	0	**1**	**2**	**3**	**Total**
Number of mice (n_k)	23	25	25	20	93
With enlarged prostate (O_{1k})	2	5	6	8	21
Proportion with enlarged prostate ($p_k = O_{1k}/n_k$)	0.09	0.20	0.24	0.40	0.2258
$O_{1k}x_k$	0	5	12	24	41
n_kx_k	0	25	50	60	135
$n_kx_k^2$	0	25	100	180	305

depending on the frequency of condom use. The scores to the regularity of condom use can be given as 0 for never, 1 for sometimes, 3 for often, and 6 for always. A scoring that adequately expresses the intensity of categories is not easy to devise, but some methods of determining scores were discussed in Chapter 10. Results would differ depending upon what scores are used.

The data in Table 13.17 are rewritten in Table 13.18. Some additional calculations are shown that are needed to compute chi-square for the trend in proportions. The proportion with enlarged prostate in this example steadily rises from 0.09 for no dosage of a dioxin-like chemical to 0.40 for a heavy dosage. Among various methods available to find the statistical significance of this trend, a simple method is to calculate

$$\text{Chi-square for trend: } \chi^2_{\text{trend}} = \frac{(\Sigma O_{1k}x_k - O_{1\bullet}\bar{x})^2}{p(1-p)(\Sigma n_kx_k^2 - n\bar{x}^2)}; \quad k = 1, 2, \dots, K; \quad (13.16)$$

where $O_{1\bullet} = \Sigma O_{1k}$, $n = \Sigma n_k$, $p = O_{1\bullet}/n$, and $\bar{x} = \Sigma n_kx_k/n$. The notation x_k is for the dosage score. The criterion in Equation 13.16 follows a chi-square distribution with only 1 df. A test can be performed as usual by finding the P-value. Sometimes, this is called **Cochran test for linear trend**.

In the case of Table 13.18, $n = 93$, $O_{1\bullet} = 21$, $\bar{x} = 135/93 = 1.4516$, $p = 21/93 = 0.2258$, $\Sigma O_{1k}x_k = 41$, and $\Sigma n_kx_k^2 = 305$. Thus,

$$\chi^2_{\text{trend}} = \frac{(41 - 21 \times 1.4516)^2}{0.2258 \times 0.7742 \, (305 - 93 \times 1.4516^2)}$$

$$= \frac{110.5947}{19.0610}$$

$$= 5.80.$$

From Table B.3, for this value of chi-square at 1 df, $P < 0.05$. A computer package gives $P = 0.016$. Reject the null hypothesis of no trend and conclude that a trend in proportions is present. This is at variance to the conclusion of no difference arrived earlier. Note the following:

1. When the scores are equally spaced as in Table 13.18 ($x_k = 0, 1, 2$, and 3), the chi-square in the criterion in Equation 13.16 tests linearity. Thus, the test performed is for assessing *linear trend*. With only four categories in Table 13.18, a curve may not be a good idea to investigate. But the presence of a curve can also be tested by suitably modifying the scores. In this example, these could be changed to 0, 1, 4, and 9 (square of x_k) or any other plausible values for investigating a trend other than linear.

2. The overall chi-square value for the data in Table 13.17 is 6.13. The difference between this and the value of criterion in Equation 13.16 is the chi-square for deviation from the trend, that is,

$$\chi^2_{\text{deviation},(K-2)\text{df}} = \chi^2_{\text{overall}(K-1)\text{df}} - \chi^2_{\text{trend},1\text{df}}, \qquad (13.17)$$

where the left-hand side is the chi-square for deviation from the trend. A large value of this chi-square would indicate that a trend other than that studied is significant. In that case, other kinds of trend can be tried. For the data in Table 13.17,

$$\chi^2_{\text{deviation},2\text{df}} = 6.13 - 5.80 = 0.33.$$

This is not significant ($P > 0.05$). The trend, in this case linear, seems adequate and there is no statistical need to study any other kind of trend.

3. Again, this procedure is valid for large n only. In fact, all n_ks should be reasonably large. Also, many p_ks should not be close to 0 or close to 1.

4. Here is a situation where the chi-square for trend is significant although the overall chi-square is not. This can happen because the test for trend has greater power to detect trend than the overall chi-square test has.

5. It is important to realize that although the chi-square for trend and the general chi-square test the same null hypothesis, namely, equal proportions in different categories, they are designed to detect different types of alternatives. The χ^2_{trend} is specially designed for the alternative that a trend exists in proportions, whereas the usual chi-square is an overall test for any type of association.

6. Ordinal data are sometimes perceived to arise from categorical metric data. For investigating association between maternal drinking and congenital malformations, Grauband and Knor [5] categorized average number of alcohol drinks per day as 0, <1.0, 1.0–2.9, 3.0–5.9, ≥6.0. Since the categories are metric, midpoints seem most adequate

scores for this kind of analysis. You should try to resist the temptation to use equally spaced scores such as 0, 1, 2, 3, and 4 in such cases. You can see that the drink categories do not follow this linear pattern.

13.3.1.3 Dichotomy in Repeated Measures: Cochran Q Test (Large n)

McNemar test for 2 × 2 tables can be extended to T repeated measures with dichotomous outcome. It is now called Cochran Q and given by

$$\text{Cochran } Q = (T-1)\frac{T\Sigma_t P_t^2 - P^2}{TP - \Sigma_i S_i^2},\tag{13.18}$$

where

T is the number of repetitions
P_t is the number of positives at tth ($t = 1, 2, ..., T$) repetition
$P = \Sigma_t P_t$ is the total number of positives
S_i is the number of positives for ith ($i = 1, 2, ..., n$) subject in T repetitions

Example 13.13 illustrates the method. You may never need to use Equation 13.18 since the software will give Q-value. For large n, this follows a chi-square distribution with ($T - 1$) df. Thus, P-value can be obtained. Exact distribution of Q is also known, which is valid for small n also. A good statistical package will give you exact P-value as well.

Example 13.13: Self-Perceived Satisfaction with Health at Four Time Points of BPH Patients

Thirty patients of BPH whose disease was under control were assessed for their perceived health at monthly intervals for 4 months. The data obtained are shown in Table 13.19.

A software package gives $Q = 10.862$ and exact P-value = 0.012.

For those who want to see what is going on underneath, note for this data that $T = 4$, $P_1 = 9$, $P_2 = 17$, $P_3 = 18$, and $P_4 = 21$. Thus, $P = 9 + 17 + 18 + 21 = 65$. Also $S_1 = 3$, $S_2 = 1$, $S_3 = 3,...$, $S_{30} = 2$ as given in the last column of the table. Substituting these values gives

$$Q = (4-1)\frac{4\times(9^2 + 17^2 + 18^2 + 21^2) - 65^2}{4\times65 - (3^2 + 1^2 + \cdots + 2^2)} = 3\times\frac{4\times1135 - 4225}{260 - 173} = 10.86.$$

This is the same as just obtained by software package. At $T - 1 = 4 - 1 = 3$ df, critical value of chi-square is 7.815 at $\alpha = 0.05$. Since the calculated value is higher, $P < 0.05$ and the result is statistically significant. Conclude that self-perceived satisfaction in these people is different at different time points.

For $T = 2$, Cochran Q reduces to McNemar χ^2 as it should. If you find Q statistically significant and want to find where actually this difference is, do McNemar for pairs of interest using Bonferroni alpha. Bonferroni procedure is explained in Chapter 15 for multiple comparisons. If there

TABLE 13.19

Self-Perceived Satisfaction with Health of BPH Patients at Monthly Intervals (0 = Not Satisfied, 1 = Satisfied)

Subject (i)	Time-1	Time-2	Time-3	Time-4	S_i
1	0	1	1	1	3
2	0	1	0	0	1
3	1	0	1	1	3
4	1	1	1	1	4
5	0	0	0	1	1
6	0	1	1	1	3
7	1	0	0	0	1
8	0	0	1	0	1
9	0	1	0	0	1
10	1	1	1	1	4
11	0	1	1	1	3
12	0	1	1	0	2
13	0	0	1	1	2
14	1	1	0	0	2
15	0	1	0	1	2
16	0	0	1	1	2
17	0	0	0	1	1
18	0	0	1	1	2
19	1	1	0	1	3
20	1	1	1	0	3
21	0	0	0	1	1
22	0	1	1	1	3
23	0	0	1	1	2
24	1	1	1	1	4
25	0	1	1	1	3
26	0	0	0	1	1
27	0	1	1	0	2
28	0	0	0	0	0
29	1	1	0	1	3
30	0	0	1	1	2
Sum P_t	9	17	18	21	

are a total of K pairwise comparisons of interest, Bonferroni procedure requires that you use α/K for each comparison instead of α. This keeps the total Type-I error less than α.

13.3.2 Two Polytomous Variables

Now, extend the method to $R \times C$ tables where both R and C are more than 2. The objective is to find whether one qualitative variable is related to or affected by the other. For example, the interest may be in the relationship of extent of smoking (none, mild, moderate, and heavy) to social classification of people. Or, it may be on the influence of nutritional status (good, fair, and poor) of pregnant women on maternal complications. Both variables are polytomous in these examples. The last three columns of Table 7.1 on the gender and age distribution of subjects coming to a cataract clinic is a 5 × 2 table.

13.3.2.1 Chi-Square Test for Large n

The basic method for finding whether the two qualitative variables with multiple categories are associated continues to be the same chi-square as given in Equation 13.15 provided the condition of large n is met.

> **Example 13.14: Association of Age at Death in SIDS with Calendar Month of Death**
>
> Douglas et al. [6] investigated the seasonality of SIDS by age at death in the United Kingdom. A total of 13,990 such deaths were observed. They provided data for each calendar month of death and each month of age. A summary is given in Table 13.20.
>
> This table has $R = 4$ rows and $C = 3$ columns excluding totals. The total n is very large, and no expected cell frequency under H_0 would be less than 5. Thus, this is a very appropriate case for applying the chi-square test. Statistically, this is a cross-sectional study in which 13,990 sudden deaths are divided by age at death and calendar month of death. The null hypothesis is that age at death of infants was not associated with the calendar month of death. The expected frequency under this H_0 can be obtained for each cell by the formula given in Equation 13.14 and the chi-square value by the formula in Equation 13.15. A software package obtained $\chi^2 = 40.46$ for these data. This has $(4 - 1)(3 - 1) = 6\,df$. The package gives $P = 0.000$ (it can be stated as $P < 0.001$). This is far less than 0.05. Thus, H_0 was rejected and it was concluded that age at death was indeed associated with the calendar month of death. A perusal of the data indicates that the deaths were proportionately more in January to April for those who died after five months of age. Biological implication of this result is not clear.

As noted earlier, the explanation of the expected frequency in Equation 13.14 and of χ^2 in Equation 13.15 for $R \times C$ tables is valid for retrospective and

TABLE 13.20

Age and Calendar Month of Sudden Infant Deaths

Age at Death (Months)	Calendar Month of Death			Total
	January–April	May–August	September–December	
<2	831	490	745	2,066
2–4	3,163	1,833	3,022	8,018
5–8	1,457	750	1,104	3,311
9–12	283	140	172	595
Total	5,734	3,213	5,043	13,990

prospective studies as much as for cross-sectional studies. Thus, the method remains the same for the three types of design. However, the nature of the null hypothesis, and hence of the conclusion, changes. As explained for 2 × 2 tables, in the case of a cross-sectional study, the null hypothesis is of independence of the two variables. If this is rejected, an association is concluded. In the case of retrospective and prospective studies, the null hypothesis is of a similar pattern of proportions in different groups. This is the hypothesis of homogeneity.

The chi-square in the formula given in Equation 13.15 treats all categories as nominal. If they are ordinal, an analysis for trend, similar to the one given in the previous section, can be done. Metric categories can be dealt with similarly. For details, see Agresti [7].

For completeness, it should be reiterated for $R \times C$ tables that the method of chi-square is again suitable only for large n. Strictly speaking, no expected cell frequency E_{rc} should be less than 5. If there are a large number of cells, then possibly some relaxation can be made, but not more than 20% of expected cell frequencies should be less than 5 and none should be very small, say, less than 1, as stated earlier. When the numbers are really small, you may have to collapse adjacent rows or columns or both. In an extreme case, the table may have to be collapsed to 2 × 2 and use Fisher exact test. Collapsing should be done in a manner that biological relevance is not lost.

The chi-square mentioned earlier evaluates only whether an association is present or not. As in Example 13.3 on the association of AIDS with blood group, the association may be present in specific cells but the number of subjects in the other cells can mask it. This can be unmasked by a suitable partitioning of the table on the lines done for the data in that example. For partitioning of $R \times C$ tables, see Agresti [8].

Another point of interest may be the degree of association between two qualitative variables. This is discussed in Chapter 17.

13.3.2.2 Matched Pairs: I × I Table

As another extension of McNemar setup, now consider disease severity in each patient before and after a therapy. The degree of severity can have many categories—for simplicity assume only as none, mild, moderate, and serious. Or, you can have two raters for assessing the severity before therapy only. In this case, on both the occasions, there are $I = 4$ categories instead of 2 in classical McNemar setup. This is a square table. The null of interest is $\pi_{ii'} = \pi_{i'i}$ for all i, i'. This null says, for example, that the probability of mild–severe combination before–after therapy is the same as the probability of severe–mild combination. For this reason, this is called hypothesis of symmetry. For this, use the following **McNemar–Bowker test:**

$$\text{McNemar–Bowker test for symmetry: } \chi^2 = \sum_{i<i'} \frac{(O_{ii'} - O_{i'i})^2}{O_{ii'} + O_{i'i}}.$$

This is referred to chi-square with $I(I - 1)/2\, df$ for large n.

Example 13.15: Type of Cataract in Left and Right Eyes

The following are the data on type of cataract seen in a random sample of patients attending a particular eye hospital:

Left Eye	Right Eye			Total
	Nuclear	Cortical	Subcapsular	
Nuclear	18	11	6	35
Cortical	3	15	7	25
Subcapsular	10	9	16	35
Total	31	35	29	95

The null hypothesis in this case is that discordance in type of cataract between left and right eyes is the same as between right and left eyes. In other words, the side of the eye does not matter.

If you look at the data, the diagonal frequencies are maximum. This suggests that one eye has tendency to have same type of cataract as the other eye. But, if they are to be different types, when left eye has nuclear cataract, right eye has cortical cataract in 11 subjects. Against this, when left eye has cortical, right eye has nuclear cataract in three subjects. This large difference suggests that nuclear cataract in left eye may have some association with cortical cataract in right eye. Whether the patterns on the whole are really different in the two eyes?

For these data, McNemar–Bowker test gives

$$\chi^2 = \frac{(11-3)^2}{11+3} + \frac{(6-10)^2}{6+10} + \frac{(7-9)^2}{7+9}$$

$$= 4.5714 + 1.0000 + 0.2500 = 5.82.$$

Since $I = 3$ in this case, the df $= 3(3-1)/2 = 3$. The table value of chi-square at 3 df is 7.815 at 5% level. Since the calculated value is less, you cannot reject the null at 5% level of significance. Despite big difference between 11 and 3, when other cells are considered, the data are not enough to conclude that the pattern of type of cataract in left eye is different from that in the right eye.

13.4 Three-Way Tables

The methods described so far are restricted to one-way and two-way contingency tables. A three-way contingency table arises when the classification of the subjects is done with respect to three variables. Thus, this is a genuine multivariate categorical data setup. The variables could be either on a nominal or on an ordinal scale. If any of them is on a metric scale, then it should be either discrete, with a small number of values, or continuous, divided into a small number of broad categories for the methods of this section. This means that a variable such as aspartate aminotransferase (AST), formerly called serum glutamic-oxaloacetic transaminase (SGOT), should have categories such as (in U/L) 0–49, 50–99, and 100–199, at least for constructing a contingency table, and not small categories like 0–9, 10–19, 20–29, etc. In any case, the categories should be mutually exclusive and exhaustive. The interest must be in proportions and not the mean.

Table 7.1 is an example of a three-way table where the variables are visual acuity (VA), age group, and gender. This is reproduced here as Table 13.21 so that you do not have to flip back to Table 7.1. Besides row and column, the third dimension is called layer. The numbers of rows, columns, and layers can be denoted by R, C, and L, respectively. In Table 13.21, $R = 5$, $C = 3$, and $L = 2$. I have chosen to consider VA categories as main columns and gender as a subclassification (layer). You may wish to switch the labels, and that would be equally valid. The order of this table is $5 \times 3 \times 2$. The body of the table has 30 cells. The totals are not included in this count.

Denote the probability in the (r, c, l)th cell by π_{rcl} ($r = 1, 2, ..., R; c = 1, 2, ..., C; l = 1, 2, ..., L$), and the observed frequency in the cell by O_{rcl}. The methods described in the present section are valid for large n and need $E_{rcl} \geq 5$ in at least 80% of the cells as before, where E is the expected cell frequency under H_0. One method for obtaining Es for three-way tables will be described shortly. The case for small n is too complex and is beyond the scope of this text. In any case, it is seldom prudent to classify a small number of subjects simultaneously by three characteristics. Each of these characteristics will have a minimum of two categories, and thus, there are at least eight cells in a three-way contingency table.

The analysis of three-way contingency tables described in this section is not necessarily a comparison of three groups. One group of subjects can be cross-classified by levels of two variables. For example, Table 13.21 can be considered to cross-classify subjects in three VA groups by age and gender.

TABLE 13.21

Distribution of 1000 Subjects Coming to a Cataract Clinic by Age, Gender, and VA in the Worse Eye

Age Group (Years)	VA ≥ 6/60			6/60 < VA ≤ 1/60			VA < 1/60			Total		
	M	F	P	M	F	P	M	F	P	M	F	P
–49	11	8	19	37	32	69	12	10	22	60	50	110
50–59	18	21	39	69	73	142	13	16	29	100	110	210
60–69	25	21	46	183	142	325	42	47	89	250	210	460
70–79	10	11	21	54	44	98	26	25	51	90	80	170
80+	3	4	7	9	14	23	8	12	20	20	30	50
Total	67	65	132	352	305	657	101	110	211	520	480	1000

M, male; F, female; P, person.

13.4.1 Assessment of Association in Three-Way Tables

As in the case of two-way tables, the null hypothesis in a three-way table could be that of homogeneity of different types or of independence, depending upon individual variables being factors or responses. In Table 13.21, if a sample of 520 males and 480 females was chosen and their age and VA elicited, then gender is a factor and the other two are responses. In this case, H_0 could be that age and VA have the same pattern of relationship in males as in females. If a sample of 1000 subjects is chosen and they are cross-classified by age, gender, and VA, then all three are responses, and H_0 is that of independence of the variables. Luckily, no matter what type H_0 is, the calculation proceeds along the same lines. Only the interpretation differs. The calculation is in terms of the usual $\chi^2 = \Sigma(O - E)^2/E$ although the calculations of Es are not so straightforward in this case. These are illustrated in the following example.

Consider a survey carried out in Brazil among female family planning clients to study the profile of women who approve sterilization [9]. The profile was studied in terms of age of the women, the number of living children (LC), and the age of the youngest child. The data obtained on 1250 approvers are given in Table 13.22.

TABLE 13.22

Women Approving Sterilization in a Survey in Brazil

| | Observed Frequencies | | | |
| | Age of the Youngest Child (Years) | | | |
Age of Woman and LC[a]	−1	1–4	5+	Total
Age ≤ 30 years				
LC < 3	37	95	22	154
LC = 3	63	58	12	133
LC > 3	77	47	3	127
Total-1	177	200	37	**414**
Age > 30 years				
LC < 3	40	91	176	307
LC = 3	57	65	79	201
LC > 3	136	105	87	328
Total-2	233	261	342	**836**
All ages				
LC < 3	77	186	198	**461**
LC = 3	120	123	91	**334**
LC > 3	213	152	90	**455**
Total	**410**	**461**	**379**	**1250**

Source: Adapted from Lassner, K.J. et al., *Stud. Fam. Plann.*, 17, 188, 1986, Table 3. With permission of the Population Council.

[a] LC, number of living children.

This is a cross-sectional study and all the three variables are responses. The appropriate H_0 is thus of independence. This H_0 says that sterilization approvers are not concentrated in any specific combination of the categories of age, LC, and age of the youngest child.

Consider LC to be in rows so that $R = 3$, age of the youngest child in columns so that $C = 3$, and the age of woman in layers so that $L = 2$. The corresponding subscripts are r, c, and l, respectively.

The bottom part of the table is a two-way table with age collapsed. Total-1 and Total-2 give another two-way table with LC collapsed. The last row is a one-way table with age of women and LC collapsed.

Following a procedure similar to the one used for two-way tables, the expected frequencies for some cells under H_0 are calculated next. These are obtained by multiplication of the corresponding row, column, and layer totals divided by the grand total.

$$E_{111} = 1250 \times \frac{461}{1250} \times \frac{410}{1250} \times \frac{414}{1250} = 50.08 \text{ (observed} = 37),$$

$$E_{211} = 1250 \times \frac{334}{1250} \times \frac{410}{1250} \times \frac{414}{1250} = 36.28 \text{ (observed} = 63),$$

$$E_{311} = 1250 \times \frac{455}{1250} \times \frac{410}{1250} \times \frac{414}{1250} = 49.43 \text{ (observed} = 77), \text{ and}$$

$$E_{112} = 1250 \times \frac{461}{1250} \times \frac{410}{1250} \times \frac{836}{1250} = 101.13 \text{ (observed} = 40).$$

These frequencies are for the first four cells (excluding Total-1) of the first column in the table. The numerators are the corresponding marginal totals. All other expected frequencies can be calculated in a similar manner. These and their totals are arranged in Table 13.23. The totals are in bold face in these two tables, which should match.

Note that the totals given in bold italics match the corresponding observed totals. The calculation of χ^2 is done as usual with the following formula:

$$\text{Three-way table: } \chi^2 = \sum_{rcl} \frac{(O_{rcl} - E_{rcl})^2}{E_{rcl}}; \quad r = 1, 2, \ldots, R;$$

$$c = 1, 2, \ldots, C; \quad l = 1, 2, \ldots, L. \tag{13.19}$$

Under H_0, this follows a chi-square distribution with $(R - 1)(C - 1)(L - 1)$ df. In this example, df = $(3 - 1)(3 - 1)(2 - 1) = 4$ and software gives $\chi^2 = 277.47$. Compare it with the value of $\chi^2 = 9.488$ at 4 df for $\alpha = 0.05$ in the Appendix Table B.3. The calculated value is much larger, indicating that $P \ll 0.05$. Thus, the hypothesis of independence is rejected. Conclude that at least two of the three variables are associated among the family planning approvers.

TABLE 13.23

Expected Frequencies of Women in the Example When They Are Not Affected by Age of Woman, LC, and Age of the Youngest Child

Age of Woman and LC	Expected Frequencies			
	Age of the Youngest Child (Years)			
	−1	1–4	5+	Total
Age ≤ 30 years				
LC < 3	50.08	56.31	46.29	152.68
LC = 3	36.28	40.80	33.54	110.62
LC > 3	49.43	55.58	45.69	150.70
Total-1	135.79	152.69	125.52	*414*
Age ≥ 30 years				
LC < 3	101.13	113.71	93.48	308.32
LC = 3	73.27	82.38	67.73	223.38
LC > 3	99.81	112.23	92.26	304.30
Total-2	274.21	308.32	253.47	*836*
All ages				
LC < 3	151.21	170.02	139.77	*461*
LC = 3	109.55	123.18	101.27	*334*
LC > 3	149.24	167.81	137.95	*455*
Total	*410*	*461*	*379*	*1250*

The detailed calculations are not shown. The purpose is to illustrate the *type* of calculations required in a three-way setup. Analysis of such tables of higher dimensions is indeed complex. However, this should not be a worrying factor as long as you can draw valid conclusions. Statistical software easily do such calculations.

Chi-square in the formula given by Equation 13.19 is an overall test. If significant, it indicates only that an association is present *somewhere*. To find exactly where this association is, one approach is partitioning on the lines illustrated earlier for a one-way table. Further discussion is beyond the scope of this book. The second and more versatile approach is using log-linear models as described next. When one of the variables can be considered dependent as in the case of prospective and retrospective studies, use logistic regression (Chapter 17). This can handle a large number of variables together.

13.4.2 Log-Linear Models

As explained next, the logarithm of expected frequencies in a contingency table can be expressed in terms of additive factors relating to the variables under study. Hence, the name "log-linear" for these models. These are useful only when no variable is considered dependent on the other or others. No distinction is made between outcome and antecedent. Thus, this model is best suited for data obtained from cross-sectional studies. The dependent variable in these models is the number of subjects in a cell of the contingency table. The objective is to find whether the categories of different variables, individually or jointly, are especially contributing to determining the cell frequency.

13.4.2.1 Log-Linear Model for Two-Way Tables

Log-linear models are easily understood with the help of two-way tables. Be it hypothesis of independence or of homogeneity, the expected frequency under H_0 in case of two-way tables is

$$E_{rc} = \frac{(O_{r.}O_{.c})}{n}; \quad r = 1, 2, \ldots, R; \quad c = 1, 2, \ldots, C.$$

This gives

$$\ln(E_{rc}) = -\ln(n) + \ln(O_{r.}) + \ln(O_{.c})$$

$$= \mu + \alpha_r + \beta_c, \tag{13.20}$$

where $\mu = -\ln(n)$, $\alpha_r = \ln(O_{r.})$, and $\beta_c = \ln(O_{.c})$. These are interpreted as the general mean and the main effects of rows and columns, respectively. They can be redefined to satisfy conditions such as $\Sigma_r \alpha_r = 0$ and $\Sigma_c \beta_c = 0$. Such conditions sometimes help to make the quantities α_r and β_c more interpretable. The model in Equation 13.20 describes the logarithm of expected cell frequency as a linear combination of an overall effect, the effect of the rth category of the first variable, and the effect of the cth category of the second variable. The observed frequency O_{rc} will not be the same as in Equation 13.20 but will be something else. Let the difference be denoted by θ_{rc}. Thus, in general,

$$\text{Log-linear model (two-way): } \ln(O_{rc}) = \mu + \alpha_r + \beta_c + \theta_{rc}. \tag{13.21}$$

The component θ_{rc} measures interaction. $\theta_{rc} = 0$ for all (r, c) implies that the observed frequencies conform to the hypothesis of independence or homogeneity. The higher the value of θ_{rc}, the stronger is the interaction or dependence. The test of the hypothesis of no interaction ($\theta_{rc} = 0$ for all r, c) is the usual χ^2. In the case of log-linear models, however, it is more convenient to use another criterion called G^2. For a two-way table, this is defined as

$$G^2 = 2\Sigma_{rc} O_{rc} \ln\left(\frac{O_{rc}}{E_{rc}}\right). \tag{13.22}$$

This also follows a chi-square distribution with $(R - 1)(C - 1)$ df when at least four-fifths $E_{rc} \geq 5$ and n is large. In other words, χ^2 and G^2 are equivalent for large n.

13.4.2.2 Log-Linear Model for Three-Way Tables

The preceding discussion for two-way tables was just to explain the concept of log-linear models. These models are really effective for three-way or higher-way tables. An explanation similar to that for a two-way table can also be given for a three-way table. In this case,

Log-linear model (three way): $\ln(O_{rcl}) = \mu + \alpha_r + \beta_c + \gamma_l + \theta_{rc} + \omega_{rl} + \delta_{cl} + \phi_{rcl}$;

$$r = 1, 2, \ldots, R; \quad c = 1, 2, \ldots, C; \quad l = 1, 2, \ldots, L; \tag{13.23}$$

where
  μ is the general mean
  α_r, β_c, and γ_l are the main effects
  θ_{rc}, ω_{rl}, and δ_{cl} are the two-factor interactions
  ϕ_{rcl} is the three-factor interaction

Thus, it contains all possible interactions for a three-way table and is called a **saturated model**. This will always be an exact fit. The estimates of the parameters of the model depend on the hypothesis of interest. The adequacy of fit under the null, or its lack, is tested by G^2 similar to the one in the formula given in Equation 13.22. The primary interest in log-linear models is to test significance of interaction terms because those indicate dependence. In the previous example, a null hypothesis could be that LC among sterilization approvers is independent of age of the youngest child. In terms of notation, this transforms to H_0: $\omega_{rl} = 0$. The less the number of interaction terms, the more is the parsimony of the model. In a log-linear model, your interest usually will not be in testing that the main effects are zero.

The procedure is to calculate the expected cell frequencies under H_0 and obtain G^2 by a three-way analogue of the formula given in Equation 13.22. Use the chi-square distribution to check whether or not $P < 0.05$. If it is, reject H_0 and conclude that $\omega_{rl} \neq 0$ or that LC and age are related in Table 13.22. The category actually dominating is identified by further analysis. Interested readers may consult Knoke and Burke [10] for details of this kind of analysis. The model can be easily extended to four or more variables although the interpretation becomes increasingly complex as the number of variables increases.

Different statistical software packages give different types of output, but everything would start making sense after some experience. A model is generally specified by leaving out the interaction or the main effect under test from the model given in Equation 13.23 and *P*-value corresponding to the value of G^2 obtained from the data. A series of models may have to be fitted to come to a focused conclusion. The software may also help in identifying particular cells that are most divergent, causing the significance.

Among many uses of log-linear models, one is to evaluate the net association between two variables after removing the effect of the others. Tiensuwan et al. [11] used log-linear models to conclude that the site of cancer in both males and females is related to marital status, diagnostic evidence, and treatment. This just illustrates the use of log-linear models but does not demonstrate the utility. The following remarks may be helpful to understand implications of log-linear models:

1. Simultaneous consideration of three or more factors raises the question of mutual independence, joint independence, and conditional independence. For details, see Le [12].

2. Log-linear models can be adjusted for structural zeros in cells of the contingency tables. These are not zero frequencies by chance in the sample but are present by design. In such a case, fit a model to the subset of cells that remain.

3. Just as in the case of classical $\chi^2 = \Sigma[(O - E)^2/E]$, log-linear models disregard the order, if any, in the categories of the variables. Each variable is considered to be on a nominal scale. In the data in Table 13.21, the age of the youngest child is actually ordinal in the sense that −1, 1–4, and 5+ years must come in this order but the log-linear model considers them as though 1–4 can be shifted after 5+. The same is true for the other two variables in this example. If an ordinal scale is present and is to be given due consideration, different models are required. See Hilderbrand et al. [13] and Agresti [7] for these models. Ordinal and metric categories allow investigation of the presence or absence of a trend or a gradient in proportion as well as its nature similar to the one discussed earlier for 2 × C tables.

4. One way to assess goodness of log-linear models is to compute standardized deviate $Z = (O - E)/\sqrt{E}$ for each cell, where O is the observed frequency and E is the expected frequency under the model. When the cell frequency is large, this Z follows an approximately Gaussian pattern. If almost all values of Z are between −2 and +2, the fit can be considered good. When the value of Z is large for one or more cells, those cells can be considered to contribute to the association. If $|Z| > 2$, the contribution is statistically significant. If there are many such cells, the value of each $|Z|$ should be more than 3. Raising the cutoff from 2 to 3 is based approximately on Bonferroni considerations, which are explained in Chapter 15.

Example 13.16: Log-Linear Models for Sterilization Approver Data

Pursue the sterilization approver data in Table 13.22. Under the log-linear model of complete independence (all interaction terms zero in the model in Equation 13.23), the expected cell frequencies are the same as in Table 13.23. Thus, the value of chi-square is also the same, namely, $\chi^2 = 277.47$. A log-linear model can be used to test the significance of individual variables. When this is done with the help of a software package, the following values are obtained:

	G^2	*P*-value
LC	25.59	0.000
Child's age	8.18	0.017
Woman's age	145.31	0.000

A *P*-value 0.000 means that it is less than 0.0005. Thus, all the three values are significant.

The details are omitted, but the log-linear procedure can also be used to obtain a model that explains the cell frequencies reasonably well. This, for the data in Table 13.22, is

$$\ln(E_{rcl}) = \mu + \alpha_r + \beta_c + \gamma_1 + \theta_{rc} + \omega_{rl} + \delta_{cl}. \tag{13.24}$$

Much like in ANOVA, there could be constraints on these parameters such as $\Sigma_r \alpha_r = 0$, $\Sigma_c \beta_c = 0$, etc. These conditions reduce the number of independent parameters. This number is $1 + (R - 1) + (C - 1) + (L - 1) + (R - 1)(C - 1) + (R - 1)(L - 1) + (C - 1)(L - 1)$. For R = 3, C = 3, and L = 2 in Table 13.21, the number of independent parameters is

$$1 + (3-1) + (3-1) + (2-1) + (3-1)(3-1) + (3-1)(2-1) + (3-1)(2-1)$$
$$= 1 + 2 + 2 + 1 + 4 + 2 + 2 = 14.$$

For the model in Equation 13.24, a software package gives $G^2 = 2.76$ and $P = 0.60$, indicating that expected frequencies from the model are not significantly different from the observed frequencies. The model says that the cell frequencies are affected not only by the category of each variable but also by their interaction with each other. Only the three-variable interaction ϕ_{rcl} is not statistically significant. You may wish to fit the model yourself with the help of a software package and check whether you really obtain this model. The parameters of the model can also be estimated with the help of a computer package.

A large variety of methods have been discussed in this chapter. It would be helpful to get a bird's eye view of these methods along with the situation in which each is applicable. This is given in **Table S.2** at the beginning of the book.

References

1. Barber LE, Luman ET, McCauley MM, Chu SY. Assessing equivalence: An alternative to the use of difference tests for measuring disparities in vaccination coverage. *Am J Epidemiol* 2002; 156:1056–1061.
2. Mehta CR, Senchaudhuri P. Conditional versus unconditional exact test for comparing two binomial. 2003. www.cytel.com/papers/twobinomials.pdf (last accessed on October 12, 2011).
3. Mehta CR, Patel NR. A network algorithm for performing Fisher's exact test in $(r \times c)$ contingency tables. *J Am Stat Assoc* 1983; 78:427–434.
4. vom Saal FS, Timms BG, Montano MM et al. Prostate enlargement in mice due to fetal exposure to low doses of estradiol or diethylstilbestrol and opposite effects at high doses. *Proc Natl Acad Sci USA* 1997; 94:2056–2061.
5. Grauband BI, Knor EL. Choice of column scores for testing independence in ordered $2 \times K$ contingency tables. *Biometrics* 1987; 43:471–476.
6. Douglas AS, Helms PJ, Jolliffe IT. Seasonality of sudden infant death syndrome (SIDS) by age at death. *Acta Paediatr* 1998; 87:1033–1038.
7. Agresti A. *Analysis of Ordinal Categorical Data.* New York: John Wiley & Sons, 1984.
8. Agresti A. *Categorical Data Analysis.* New York: John Wiley & Sons, 1990.
9. Lassner KJ, Janowitz B, Rodrigues CMB. Sterilization approval and follow-through in Brazil. *Stud Fam Plann* 1986; 17:188–198.
10. Knoke D, Burke PJ. *Log–Linear Models (Quantitative Application in Social Sciences).* New York: Sage Publications, 2004.
11. Tiensuwan M, Yimprayoon P, Lenbury Y. Application of log-linear models to cancer patients: A case study of data from the National Cancer Institute. *Southeast Asian J Trop Med Public Health* 2005; 36:1283–1291.
12. Le CT. *Applied Categorical Data Analysis and Translational Research*, 2nd edn. New York: John Wiley & Sons, 2009.
13. Hilderbrand DK, Laing JD, Rosenthal H. *Analysis of Ordinal Data (Quantitative Application in the Social Sciences).* New York: Sage Publications, 2005.

14

Relative Risk and Odds Ratio

Relative risk (RR) and odds ratio (OR) are extensively used in medicine and health for assessing the chance of outcome or antecedent in one group relative to the other. The concepts have been found extremely useful in describing the importance of the presence of a characteristic in altering risk. In view of the wide applicability of RR and OR, this chapter is exclusively devoted to these concepts. Statistically, this is an exercise in comparing proportions in two groups. You may find many similarities between the methods described here and the methods described in the preceding chapter for 2×2 tables. RR and OR, when considered alone, do emanate from 2×2 tables, and thus similarities are apparent. For this reason, a frequent reference to Chapter 13 is made in this chapter.

This chapter: Risk, hazard, and odds are terms that seem to convey the same meaning—chance—to most people. The statistical distinction between them is explained in Section 14.1. This section also explains RR and attributable risk (AR) and describes the methods to obtain confidence interval (CI) and test of hypothesis on these in the case of independent samples and matched pairs. Section 14.2 describes the methods to calculate OR, CI, and test of hypothesis on OR in different situations. This includes stratified analysis and the Mantel–Haenszel (M-H) procedure for OR. Formulas for calculating sample size for estimation of and test of hypothesis on OR and RR in different situations are given in Section 14.3.

14.1 Relative and Attributable Risks (Large *n*)

An extremely useful and common application of comparison of two proportions is in the study of RR and AR. Before describing these, it is important to understand some terms.

14.1.1 Risk, Hazard, and Odds

Medical literature commonly uses the terms risk and odds interchangeably. Actually they do not mean the same. Hazard is not so commonly used, but statistically this term also has a very different meaning.

14.1.1.1 Risk

In common parlance, risk is the likelihood of an adverse outcome under given conditions. In medicine, it is generally the chance that a person without a disease will develop the disease in a defined period. It can be any other event or outcome of interest such as accidental injury or vision becoming <6/60. Statistically, the term is generic and not restricted to adverse outcomes. It could be risk of survival, risk of reduction in side effects, or risk of conception. Risk could be 1 in 1000 or 0.05 or 0.20 but cannot exceed one. It is a decimal number but often expressed as percentage. Although the implication is for future events, the calculation is based on previous experience.

The denominator for calculating risk is the number of persons exposed to the risk and numerator is that part of the denominator that develops the outcome. For example, 10-year risk of coronary heart disease (CHD) may be 9% in a population of age 40–79 years. Such an estimate can be obtained only after follow-up of subjects for a 10-year period. This is based on past experience but would be used on future subjects. The population at risk for this calculation is not what was in the middle of the period but what was in the beginning of the period because all are exposed to that risk. A measure of risk is the incidence rate, where also the denominator is the population at risk.

14.1.1.2 Hazard

Risk cannot exceed one but hazard has no such restriction. The hazard of death is exceedingly high in a plane crash than at any other time. Hundreds die within a few hours after the accident. Hazard can be understood as the instantaneous rate of death—for that matter of any event of interest. It is sometimes called the force of mortality for deaths, force of morbidity for ill health, and force of event occurrence in general. It can also be understood as the intensity with which events occur at a *point of time*. Hazard can increase steeply in situations such as a calamity or an epidemic. If constant over a period, hazard can be obtained by dividing the number of events per unit population by the exposure period.

14.1.1.3 Odds

Whereas the term risk is used for the outcome, the term odds is used for the antecedent. If 67% of all subjects with hypertension are obese in a particular segment of population, the odds of obesity are said to be 67:33 or 2 to 1 in that population. That is, a subject with hypertension is twice as likely to be obese as being nonobese. This is a quirky measure but has been found extremely useful as explained in this chapter.

14.1.1.4 Ratios of Risks and Odds

Risks, hazards, and odds are important measures by themselves but their importance increases manifold when they are expressed as a ratio in one group relative to another. Hazard ratio is explained in a later chapter. This chapter gives details of RR and OR as well as of AR, which is the difference in the risks in two groups.

Under certain conditions, RR and OR can measure the strength of relationship between an exposure and an outcome. The terms are used differentially in the context of prospective and retrospective studies. This chapter contains some details of the analysis of data from such studies when the data can be represented by a 2 × 2 table. RR, AR, and OR inherently compare one group with the other. If there are more than two groups, the comparison can be made between two groups at a time. Then, several RR, AR, and OR values are computed. The present discussion is restricted to large n. The case of small n can become too complex and is outside the scope of the present book. In any case, the concepts of risk and odds are mostly relevant to large n only.

14.1.2 Relative Risk

The concept of risk, and thus of RR and AR, is applicable only to prospective studies. The analogous concept of OR is applicable to retrospective and cross-sectional studies. This section describes RR and considers two situations: (1) If there is a study of two groups that are not matched or not paired, then these can be considered independent samples, and (2) if they are matched or paired one to one, then these are matched pairs. Both situations were explained in a preceding chapter.

14.1.2.1 RR in Independent Samples

RR is the ratio of the risk of developing an outcome such as disease (D) in those with an antecedent factor (A) compared with those without this factor. This obviously requires a prospective study, giving rise to data as presented in Table 14.1. The notations a, b, c, d for observed frequencies are more convenient than Os used in the preceding chapter. For totals, Os are used. πs are the group probabilities in the population. The dot in subscript indicates total for that particular subscript. The antecedent factor would generally be an exposure believed to cause the disease. The term risk here has the same meaning as incidence rate per unit.

TABLE 14.1

Structure of a Study for RR: Independent Samples

	Antecedent		
Outcome	Present	Absent	Total
Present	$a(\pi_{11})$	$b(\pi_{12})$	$O_1.(\pi_1.)$
Absent	$c(\pi_{21})$	$d(\pi_{22})$	$O_2.(\pi_2.)$
Total	$n_1(1)$	$n_2(1)$	

In terms of probabilities,

$$RR = \frac{P(D+/A+)}{P(D+/A-)}, \tag{14.1}$$

where

D is for disease

A is for antecedent

+ sign is for presence, similar to that used earlier

− sign is for absence, similar to that used earlier

In terms of the notation of Table 14.1, RR = π_{11}/π_{12} subject to the condition that $\pi_{11} + \pi_{21} = 1$ and $\pi_{12} + \pi_{22} = 1$. It measures the degree of association of outcome with the antecedent factor. If the incidence of lung cancer among heavy smokers is 6% and among nonsmokers 0.5%, then RR = 6, that is, heavy smokers have 12 times as much risk of developing lung cancer as non-smokers. The value of RR in the target population can be estimated from a sample of subjects:

$$\text{Estimate of RR: } \widehat{RR} = \frac{a/(a+c)}{b/(b+d)} = \frac{p_1}{p_2}, \tag{14.2a}$$

where the notations are same as in Table 14.1. None of the cell frequencies should be very small, say, less than 5. Otherwise the estimate becomes unstable. If any cell frequency is 0 or very small, say, less than 1, then a modified estimate of RR is

$$\widehat{RR}_{mod} = \frac{(a+0.5)/(a+c)}{(b+0.5)/(b+d)}. \tag{14.2b}$$

These and several other expressions in this section are estimates, but I *will now ignore the hat (^) sign for simplicity.* Even in the literature, an indication that these in fact are estimates is rare.

Example 14.1: RR of Anemia in Lower Parity Women

Consider 120 women of parity ≤ 2 and 120 women of higher parity who are subsequently assessed for anemia. This example is similar to Example 13.6, but now the study design is *prospective* instead of cross-sectional. Parity in this case is the antecedent and anemia is the outcome. If the incidences of anemia in the two groups are the same 23% and 40% as in Example 13.6, then $a = 28$ and $b = 48$. These give

$$RR = \frac{28/120}{48/120} = 0.58.$$

Thus, the risk of anemia in lower parity women is nearly half of that in higher parity women.

The following comments are helpful in clarifying certain issues regarding RR:

1. RR = 1 implies independence, that is, the risk in exposed subjects is the same as in nonexposed subjects. It might seem paradoxical to some of you that correlation = 1 means perfect relationship but RR = 1 means no relationship. RR > 1 means a higher risk in the exposed subjects and RR < 1 means a lower risk. RR < 1 can be interpreted as a protective effect in place of risk. RR cannot be negative but can be 100 or more.

2. RR alone is rarely enough to draw a valid conclusion about a cause–effect type of relationship. Other possible explanations have to be ruled out. A discussion on this aspect is in Chapter 21.

3. It is preferable to keep the adverse category of antecedent in the first column of a 2×2 contingency table and the adverse outcome in the first row. The interpretation is then easy. However, in the case of the data in Example 13.6, which is the same as used in Example 14.1 with the numbers changed, the adverse antecedent category, in the sense that it is more prone to anemia, is parity ≥ 3. But this is in the second column. The interpretation of RR is in terms of the first row/first column. Thus, RR < 1 in this case means less risk of anemia for parity ≤ 2. This is equivalent to saying that the risk of anemia is more for parity ≥ 3. This risk can be obtained as a reciprocal of the earlier risk. In Example 14.1, $1/0.58 = 1.72$. Thus, the risk of anemia for parity ≥ 3 is 1.72 times the risk for parity ≤ 2.

4. In Example 14.1, anemia is not incident, that is, it is not a new occurrence. This is a prevalent condition. Thus, the concept of RR is also applicable to prevalence. Some caution is advised when interpreting RR based on prevalences. It tells about the chance of presence of disease instead of the risk of developing the disease.

5. A term sometimes used for RR is **risk ratio**.

6. Note that an RR does not tell what the actual risks were. For example, the same RR can be obtained as 27/9, 4.5/1.5, 0.045/0.015, etc.

Actual risks are entirely different in these cases but RR = 3 in all of them. Risk may decrease over time without an alteration in the RR. This will happen when the risk in the comparison group also declines in the same proportion.

7. Context can be important for interpretation of RR. If you find diabetes doubles the risk of stroke, you would be careful without worrying about baseline. If risk of vehicular death in regular conditions is 1/10,000 and in bad weather conditions 2/10,000, you may still take it lightly despite double risk.

8. Assessment of RR is considered important to discuss the consequences with the patient and in preparing a management strategy. Remember, however, one great limitation of RR is that if the risk in the unexposed group is low, the RR can be a high value. If the risk in the unexposed group is only 2% and in the exposed group 70%, the RR is 35. On the other hand, if the risk in the unexposed group is high, say, 60%, RR cannot exceed 100/60 = 1.67. It cannot exceed 2 if risk in unexposed is 0.5 or more. Thus, RR heavily depends on the risk in the unexposed group. In view of this limitation, interpret an RR with caution.

9. For RR to be a valid measure, it is necessary that the two groups under comparison are identical except for the presence of risk factor in one and its absence in the other.

14.1.2.2 Confidence Interval for RR (Independent Samples)

It would not be wrong to surmise that all users of statistical methods use software to do the calculations. Nearly all standard statistical software do have a provision to calculate the RR and the CI for RR corresponding to the prefixed confidence level. Nevertheless, the following details may be helpful in understanding the underlying procedure.

Suppose there are two populations in which the probabilities that an individual shows an outcome of interest are π_1 and π_2, respectively. Suppose also that a random sample of size n_1 from the first population has a subjects showing the outcome (and a proportion $p_1 = a/n_1$), while the corresponding values for an independent sample from the second population are n_2, b, and $p_2 = b/n_2$, respectively. Then, the estimated RR in Equation 14.2a is the simple ratio of these proportions. Thus, lnRR (natural logarithm of RR) is a linear combination of the frequencies. The central limit theorem works not just for mean but for any linear combination. Because of this, lnRR has a Gaussian distribution for large n. It is thus easy to use lnRR for inference when n is large. It is known for large samples that

$$\text{SE}(\ln \text{RR}) = \sqrt{\frac{1}{a} - \frac{1}{n_1} + \frac{1}{b} - \frac{1}{n_2}}.$$

(14.3)

Therefore,

$$95\% \text{ CI for RR: } \exp[\ln RR \pm 1.96 * SE(\ln RR)], \qquad (14.4)$$

where exp is the exponent to the Naperian base e. This is the inverse of the logarithm. As you can see, the CI is first obtained for lnRR and then the exponent is taken to convert it to RR. The CI for lnRR will be symmetric but this interval will not be symmetric to RR when exp is taken. Equations 14.3 and 14.4 can also be used to obtain one-sided bound by the procedure already explained in Chapter 12.

Note, however, that the CI 0.125–0.5 when estimated RR is 0.25 has the same implication as CI 2–8 when estimated RR is 4. The width looks larger in the later case but both are one-half to two times of the estimated RR. Width of CIs for ratios is tricky. Ratios make better sense when examined on log scale.

Example 14.2: CI for RR of Respiratory Tract Illness in Boys Born to Younger Women

Martinez et al. [1] reported a prospective study on wheezing lower respiratory tract illness (LRI) during the first year of life of 500 boys and an almost equal number of girls. They were enrolled in Tucson, Arizona. One among the objectives was to explore maternal age as risk factor for wheezing LRI. The data for the boys are given in Table 14.2.

The outcome is LRI. Rows and columns are switched in this table compared with Table 14.1. The estimated RR of wheezing LRI in boys with mother's age <26 years compared with mother's age ≥26 years is

$$RR = \frac{48/165}{65/335} = 1.50.$$

Thus, boys born to younger women were 1.5 times likely to get LRI during their first year of life than boys born to older women.

Also, lnRR = ln(1.50) = 0.4055,
and

$$SE(\ln RR) = \sqrt{\frac{1}{48} - \frac{1}{165} + \frac{1}{65} - \frac{1}{335}} \quad \text{from the formula given in Equation 14.3}$$

$$= 0.1648.$$

TABLE 14.2

Maternal Age and LRI in Infant Boys

Maternal Age (Years)	LRI		
	Yes	No	Total
<26	48	117	165
≥26	65	270	335

Therefore, 95% CI for RR from the formula given in Equation 14.4 is

$$\exp(0.4055 \pm 1.96 \times 0.1648)$$

or

$$\left(e^{0.082},\ e^{0.729}\right) \quad \text{or} \quad (1.09,\ 2.07).$$

It can be stated with 95% confidence that this interval contains the true RR in the population, provided the sample subjects can be considered random representatives.

14.1.2.3 Test of Hypothesis on RR (Independent Samples)

As for CI, the test of hypothesis on RR too can be directly carried out with the help of any standard statistical software. However, the following details may help to grasp some of the intricacies involved in this procedure.

Equivalence of CI with the test of hypothesis procedure was explained in a previous chapter. Because the 95% CI in Example 14.2 does not include RR = 1, the H_0: RR = 1 can be safely rejected at significance level $\alpha = 0.05$. It can be concluded that the risk of wheezing LRI in the first year of life was indeed different in boys born to younger women than to those born to older women. To directly test H_0, calculate

$$Z = \frac{\ln RR}{SE(\ln RR)}, \tag{14.5}$$

and reject if the corresponding two-tail P-value is less than 0.05. The second procedure for testing H_0: RR = 1 against H_1: RR $\neq$ 1 is by the usual chi-square explained for 2 × 2 tables and illustrated in Example 14.3.

Example 14.3: Test for Hypothesis on RR Using Chi-Square

For the data in Table 14.2, the expected frequencies under H_0 can be calculated as usual and chi-square can be calculated by the formula given in Equation 13.8. This gives

$$\chi^2 = \frac{(48 - 37.29)^2}{37.29} + \frac{(117 - 127.71)^2}{127.71} + \frac{(65 - 75.71)^2}{75.71} + \frac{(270 - 259.29)^2}{259.29} = 5.93.$$

From Table B.3, the critical value of chi-square at 1 df is 3.841 for $P = 0.05$. Since the calculated value is higher, it follows that $P < 0.05$. A statistical software package gives $P = 0.015$. The null hypothesis is rejected. Conclude that RR $\neq$ 1.

SIDE NOTE: Chi-square is a two-tailed test and so the conclusion too is two sided. However, it is the same as reached earlier. In the case of CI also, a two-sided CI was used and not a one-sided confidence bound. If there is an a priori reason to ensure that RR would be more than 1, then H_1: RR > 1. It may then be prudent to use the formula given in Equation 14.5 and refer it to the Gaussian distribution table to get the one-sided P-value. In this example,

$$Z = \frac{0.4055}{0.1648} = 2.46.$$

From Table B.1, $P(Z \geq 2.46) = 0.0069$. This P-value is not only less than 0.05 but also less than 0.01. The RR can be said to be *highly* significantly more than RR = 1. There is only an exceedingly small chance that RR = 1 will give frequencies as extreme as observed in this sample.

Example 14.4: Prevalence Ratios as Surrogate for Risk Ratios

Brenner et al. [2] reported adjusted prevalence ratios for *Helicobacter pylori* infection among persons who consumed up to 10, 10–20, and 20+ g of alcohol per day compared with nondrinkers as 0.93, 0.82, and 0.71, respectively, in a German population of adults. Prevalence of the infection decreased as the alcohol consumption increased. This relationship became stronger when recent drinkers were excluded. These results are based on prevalence ratios and not on risk ratios.

SIDE NOTE: The findings support the hypothesis that moderate alcohol consumption may facilitate spontaneous elimination of *H. pylori* infection among German adults.

If the alternative hypothesis is H_1: RR < 1, the question of a test arises only if the observed RR is less than 1. In this case lnRR would be negative and the relevant P-value will be $P(Z \leq z)$ and this will be based again on the formula given in Equation 14.5.

14.1.2.4 RR in the Case of Matched Pairs

A general situation for matched pairs with regard to antecedent and outcome in a prospective study is shown in Table 13.14. In this case, RR is estimated as follows:

$$\text{Relative risk (matched pairs): } RR_M = \frac{a+b}{a+c}, \qquad (14.6)$$

where the notations are same as in Table 13.14. The meaning of a, b, c, d is different here than in Table 14.1. The numerator of Equation 14.6 is the number of subjects developing the disease among those exposed and the denominator is the number of subjects developing the disease among those nonexposed. For the data on common cold in Table 13.15, $RR_M = (22 + 15)/(22 + 5) = 1.37$, that is, in this sample of 50 pairs, the estimated chance of relief within 1 week is 1.37 times in the therapy group than that of the nontherapy group.

The exact expression for the SE of RR_M is complex. However, it is known that the OR for matched pairs, OR_M, approaches RR_M for rare outcomes. Thus, the same CI and test criterion can be used for RR_M as for OR_M in most practical situations (see Section 14.2.2).

14.1.3 Attributable Risk

AR (or absolute risk reduction [ARR]) is the difference in the risk among the exposed and the nonexposed subjects. If the risk of lung cancer among smokers is 7.6% and in nonsmokers of same age–gender is 1.2%, the risk attributable to smoking is the difference, that is, 6.4%. This can also be understood as the risk difference. For this to be valid, again, it is necessary that the groups are similar with respect to all factors except the antecedent under review, that is, no other factor is present to alter the risk.

14.1.3.1 AR in Independent Samples

In terms of probabilities in Table 14.1, for independent samples,

$$AR = \pi_{11} - \pi_{12}, \tag{14.7}$$

provided $\pi_{11} + \pi_{21} = 1$ and $\pi_{12} + \pi_{22} = 1$. This is estimated as

$$AR = \frac{a}{n_1} - \frac{b}{n_2} = p_1 - p_2. \tag{14.8}$$

For the data on anemia and parity in Table 13.7,

$$AR = \frac{14}{60} - \frac{16}{40} = -0.17.$$

This is $(0.17 \times 100)/(16/40) = 42.5\%$ of the risk in the higher parity group. The risk of anemia in parity ≤ 2 is 42.5% *lower* than in parity ≥ 3.

AR is the difference between risks in two groups and a CI for such a difference for large n can be obtained by the usual method for difference in proportions. The test of hypothesis H_0: $AR = 0$ is equivalent to H_0: $\pi_{11} = \pi_{12}$ as well as to H_0: $RR = 1$. The former can be tested by the method of Section 13.2.2, which includes a chi-square test in Equation 13.8 (although the notations are different) and the Z-test in Equation 13.9. The hypothesis that AR is a specified quantity π_0 can also be tested by the Z-test in Equation 13.10. The H_0 on

RR is also tested either by Z in Equation 13.9 or by the usual chi-square in Equation 13.8 for 2×2 tables. All these procedures are equivalent for large n.

AR measures the expected reduction in risk if the exposure factor is eliminated. Thus, this has public health importance. A slight modification of AR is the **AR fraction**. This is calculated as a proportion of the incidence in the exposed group, that is,

$$\text{AR fraction} = \frac{\pi_{11} - \pi_{12}}{\pi_{11}} = \left(1 - \frac{1}{\text{RR}}\right). \tag{14.9}$$

In short, this is called the **attributable fraction**. It is the proportion contributed by the exposure to the elevated risk, and thus measures the preventability of the disease from eliminating a particular factor. The other name for the same quantity is **etiologic fraction**. When multiplied by 100, this becomes AR percent. This is 42.5% in the example on parity and anemia just cited. If you conclude, for example, that hypertension is responsible for 20% of the risk of myocardial infarction among adult males in your area, then AR fraction = 0.20.

RR and AR can sometimes lead to very different conclusions. Consider the data in Table 14.3, which are from the famous Doll and Hill [3] study of British doctors. It compared the mortality from lung cancer and cardiovascular disease in nonsmokers and heavy smokers (>25 cigarettes per day) from 1951 to 1961.

The RR for lung cancer was high, that is, 32.43. This indicates a very strong association of lung cancer deaths with heavy smoking and underscores the importance of smoking in the etiology of lung cancer. This association of cardiovascular disease death with heavy smoking was mild, only 1.36. But the AR in the two cases was nearly the same. During 1951–1961, elimination of heavy smoking among British male doctors would have reduced the cause-specific mortality for lung cancer almost as much in absolute terms as for cardiovascular disease.

Also note that the risk of death by lung cancer in nonsmokers is low—only 0.07 per 1000—causing RR to appear so high, as remarked earlier.

TABLE 14.3

Comparison of RR and AR of Death due to Lung Cancer and Cardiovascular Diseases in Heavy Smokers among British Male Physicians

| | Annual Death Rate/1000 | | | |
| | | Heavy | | |
Cause of Death	Nonsmokers	Smokers	RR	AR
Lung cancer	0.07	2.27	32.43	2.20
Cardiovascular disease	7.32	9.93	1.36	2.61

Against this, the cardiovascular disease mortality is high even among non-smokers so that the RR is not so high.

Since AR and RR are functions of p_1 and p_2 only, p_1 and p_2 can be uniquely determined when AR and RR are known. Simple algebra yields

$$p_1 = \frac{AR \times RR}{RR - 1} \quad \text{and} \quad p_2 = \frac{AR}{RR - 1}.$$

14.1.3.2 AR in Matched Pairs

AR in case of matched pairs (large n) is estimated as

$$AR_M = \frac{|b - c|}{n}, \tag{14.10}$$

where b and c are same as in Table 13.14.

The test of hypothesis H_0: $AR_M = 0$ again is equivalent to H_0: $RR_M = 1$. For matched pairs, this can be tested by McNemar χ^2_M as in the formula given in Equation 13.12 when n is large. For CI, however, SE is needed. For matched pairs, it is given by

$$SE(AR_M) = \sqrt{\frac{(b + c)n - (b - c)^2}{n^3}}. \tag{14.11}$$

Thus,

$$95\% \text{ CI for } AR_M: AR_M \pm 1.96 * SE(AR_M), \tag{14.12}$$

where $SE(AR_M)$ is as in the formula given in Equation 14.11. Most of these formulas are the same as stated in Chapter 12 for difference in proportions in matched pair setup.

Example 14.5: CI for AR in Common Cold

For the data on common cold in Example 13.9, $AR_M = (15 - 5)/50 = 0.20$. Thus, the therapy seems to increase the 1 week relief rate by 20%. At least this is the finding from these 50 pairs. Also,

$$SE(AR_M) = \sqrt{\frac{(15 + 5)50 - (15 - 5)^2}{50^3}} = 0.08485.$$

Thus, the 95% CI for AR_M is

$$(0.20 - 1.96 \times 0.08485, 0.20 + 1.96 \times 0.08485)$$

or

$$(0.03, 0.37).$$

The AR is 20% in this sample but the CI says that the actual AR in the corresponding population could be as low as 3% or as high as 37%.

SIDE NOTE: Since the CI does not contain zero, H_0: AR = 0 can be rejected at the 5% level. If McNemar criterion as in Equation 13.12 is used to test this hypothesis, then

$$\chi_M^2 = \frac{\left(|15 - 5| - 1\right)^2}{15 + 5} = 4.05.$$

This is more than the critical value 3.841 of chi-square at 1 df. Thus, H_0: AR = 0 is rejected by this method also. The excess relief with the therapy is statistically significant.

Note the following regarding RR and AR:

1. As remarked earlier, risk in a statistical sense does not necessarily refer to an adverse outcome. In the data in Table 13.15, the risk is for relief within 1 week, which is a positive feature. Earlier, the term protective effect was used for such a factor.

2. This section assumes large n, but the binomial methods similar to those mentioned in Section 13.1.1 can be used to find CI and to test H_0 on RR or AR when n is small.

3. All the RR-related parameters can be estimated from retrospective or case–control studies by replacing RR by OR as an approximation under certain conditions. As explained shortly, OR is a good approximation to RR in most situations. This obviates the need to carry out expensive prospective studies. However, *AR does not have this feature.*

4. Another advantage with RR opposed to AR is that RR often comes close to multiplying individual RRs when two independent factors act jointly in concert. This does not happen with AR.

5. The same AR such as 4% between 60% and 64% risks has an entirely different implication than, say, between 2% and 6%.

6. RR generally represents the magnitude of the association and provides information that can be used in making a judgment on causality. Once a causality inference is drawn, AR assumes importance in delineating the public health importance of the exposure.

7. Many times AR gives more valid information to the health managers than the RR. This happens particularly when the disease is rare. Suppose oral cancer has a risk of 0.0002 in non-tobacco-chewing

persons but has a risk of 0.015 in those who chewed tobacco for 10 years or more. Thus, the RR = 0.015/0.0002 = 75 whereas AR = 0.015 − 0.0002 = 0.0148 (less than 1.5% only). Changing habits of tobacco chewing in this setup will not make much of a difference to the overall incidence of oral cancer; RR = 75 notwithstanding. As shown next, this is all the more true if only a small percentage, such as 2%, of the population chews tobacco.

14.1.3.3 Number Needed to Treat

The number needed to treat (NNT) is conversion of the ARR to a useful and readily understood quantity. ARR is the same as AR but we retain ARR as this is more popular for NNT. NNT = 1/ARR, and it measures the average number of patients that would need to be treated to prevent one additional adverse outcome or for one additional success. For the data on common cold in Table 13.15, ARR = 0.20 as worked out in Example 14.5, and NNT = 1/0.20 = 5. Thus, on average, five cases of common cold need to be treated to provide relief within 1 week to one case more than the control that has no therapy.

The concept of NNT has gained popularity because of its simplicity. It is especially useful in comparing the efficacy of two treatment regimens with binary outcome. For example, the NNT to cure one patient of atherosclerosis by diet–exercise regimen may be 7 and by drug-X may be 5. Then, drug-X therapy would be considered more effective. This number also measures absolute efficacy in the sense that five patients are required to be treated by drug-X in this example to cure one patient; and thus this drug may not be considered a sufficiently effective strategy. Thus, NNT is a useful aid to clinical decision process. Clinicians understand and appreciate NNT much better than probabilities. However, NNT does not work for one patient.

Also, if 12 patients of atherosclerosis needed to be treated to prevent one death and 8 cases of diabetes to prevent one death, you have some idea of where to put better part of your resources, particularly if the financial and social cost of treatment of these types of cases is known. Also, due consideration should be given to the factors such as age, severity of disease, and duration of sickness while interpreting NNT.

Even more caution is required when a regimen is to be used for mass adoption on the basis of NNT. Heller [4] gives example of thrombolysis with NNT = 7 for prevention of one adverse outcome in secondary prevention after nonhemorrhagic stroke, and aspirin with NNT = 33. But only 4% of the stroke patients are eligible for thrombolysis due to time window and other factors, whereas 70% are eligible for aspirin. You can see that the total benefit of a policy of using aspirin is nearly four times of that of using thrombolysis.

Distributional properties of NNT are not fully known and a makeshift procedure is adopted to get CI. This is easily implemented when the difference in efficacy of two treatments under comparison is statistically significant.

In this case, first step is to calculate CI for ARR using the method given in Section 14.1.3 for AR. Call this (ARR$_L$ to ARR$_U$). Then, CI for NNT is (1/ARR$_U$ to 1/ARR$_L$). Note that L and U have switched their position. If the CI for ARR is from 0.04 to 0.15, the CI for ARR is from 1/0.15 = 6.67 to 1/0.04 = 25. For not significant ARR, consider the following.

Generally, ARR would be small in a trial. If efficacy of a drug-A is 70% and of drug-B is 75%, $(p_1 - p_2) = 0.05$. This has the same interpretation as ARR. Since $1/(p_1 - p_2) = 20$, NNT = 20 in this case. That is, 20 subjects are needed to be treated by drug-B to get one extra success. The CI for $(p_1 - p_2)$ in this case can include negative values also such as −0.02 to +0.12. This interval includes zero where NNT is infinity. This causes problems. Zero within this CI indicates that ARR is not statistically significant. If nonsignificance is disregarded, the reciprocals give −50 to +18 as CI for NNT. This interval does not even include the original point estimate NNT = 20. Thus, this method of obtaining CI for NNT fails in this case. NNT is one example where sensible CI is obtained only when ARR is statistically significant.

For independent samples, NNT can also be obtained in terms of RR instead of AR:

$$\text{NNT} = \frac{1}{\text{AR}} = \frac{1}{R_2 - R_1} = \frac{1}{R_1(\text{RR} - 1)} = \frac{1}{p(\text{RR} - 1)},$$

where p is the proportion affected in the control group. Note that p and R_1 are the same.

Just as RR can be approximated by OR in special cases, so can NNT. Then, it is possible to calculate NNT through case–control studies as well. In most situations, however, an immaculately carried out RCT is needed to correctly calculate NNT.

14.1.3.4 Relative Risk Reduction

For measuring the impact of intervention, another measure used is risk reduction. The ARR is the same as AR, whereas the RR reduction is AR as percentage of the risk in the exposed group. This is the same as earlier called attributable fraction. So many terms for the same quantity can be confusing. A standardized terminology is yet to emerge. In the example on oral cancer, RR reduction is 0.0148 × 100/0.015 = 98.7%. This percentage of risk *among tobacco chewers* would be reduced if tobacco chewing were eliminated. Do not overrate it by looking at such a high percentage and realize that this percentage is out of 0.015 risk among the chewers. The importance of RR reduction cannot be assessed without knowledge of the risk in the exposed group. An RR reduction of 25% could be from risk 4% to 3% or from risk 80% to 60%. For something like a vaccine, RR reduction measures protective efficacy.

14.1.3.5 Population Attributable Risk

In view of the public health importance of AR, it is sometimes of interest to estimate the excess rate of disease attributable to the exposure in the *total population* under study. This excess is called the population attributable risk (PAR) and is calculated as

PAR = Incidence rate in the total population (including exposed subjects)
    − Incidence rate in the nonexposed subjects.   (14.13)

This can also be called population risk difference. The first component of Equation 14.13 can be estimated only when the sample contains the exposed and the nonexposed subjects in the same proportion as in the total population. Otherwise, an extraneous estimate would be needed. Note that PAR is the rate of disease in *the population* minus the rate in the unexposed group. This is different from AR because the population comprises both the exposed and the nonexposed groups of people. In fact PAR = AR $*$ p, where p is the prevalence of exposure. This measures the impact of eliminating the exposure from the entire population.

As with the AR fraction, the PAR fraction can be calculated as proportion of the rate of disease in the population. This measures the proportion of disease in the population that is attributable to the exposure and is the proportion of incidence that could be eliminated if the exposures were eliminated.

PAR fraction (or population attributable fraction) can be directly obtained from RR as follows when the proportion of persons with the given risk factor is known:

$$\text{PAR fraction} = \frac{p(\text{RR}-1)}{p(\text{RR}-1)+1},$$ (14.14)

where p is the proportion of persons having the given risk factor. This is the same as the prevalence of exposure.

Whenever RR can be approximated by OR, PAR can be estimated on the basis of case–control data. For a method of doing so, see Bruzzi et al. [5].

Example 14.6: PAR and PAR Fraction for Oral Cancer in Tobacco Chewers

Pursue the example just cited where oral cancer has a 10-year risk of 0.0002 among nonchewers of tobacco and 0.015 among the chewers. If only 2% are chewers as before, there are 2,000 chewers in 100,000 population and 98,000 nonchewers. At the risk presumed earlier, 20 among nonchewers and 30 among chewers will develop oral cancer in a 10-year period. These two together amount to 50 per 100,000 or 0.0005 as the risk in the total population. Thus, the PAR = 0.0005 – 0.0002 = 0.0003. If chewing of tobacco is completely eliminated from the population through educational campaigns or otherwise, 3 cases per 10,000 population would be saved at the end of 10-year period. A health administrator may not view this as a substantial improvement considering the cost involved in the educational campaign and considering the fact that 2 cases per 10,000 would in any case occur among nonchewers.

SIDE NOTE: The result may look dramatically different if PAR *fraction* is calculated. Using the formula given in Equation 14.14, since RR = 75 in this example,

$$\text{PAR fraction} = \frac{0.02(75-1)}{0.02(75-1)+1} = 0.60.$$

This means 60% of risk of oral cancer in the population is attributable to chewing of tobacco. This is the same as PAR = 0.0003 out of population risk of 0.0005.

14.2 Odds Ratio

The term OR is used in the context of retrospective or case–control studies. The comparison in such studies is between the frequency of occurrence or presence of an antecedent among the cases relative to the controls. Ideally, all other possible factors are appropriately matched so that they do not influence the result. If there are some that are not matched, the statistical analysis is geared to minimize the influence of these factors on the result. For this, methods such as that of logistic regression (Chapter 17) are used.

In betting, it is stated, for example, that the odds of winning are 1:3. This means that a loss is three times more likely than a win. Similarly, in case–control studies, the odds are the frequency of presence of antecedent relative to its absence. This is calculated separately for the cases and the controls. The ratio of these two odds is called the OR. Just as RR does in the case of prospective studies, OR measures the strength of relationship between an exposure and an outcome in case–control studies. The relationship may or may not be of cause–effect type.

14.2.1 OR in Two Independent Samples

In case–control studies for OR, the data take the form as given in Table 14.4. Note how it is different from Table 14.1, which is for RR in prospective study. In this case, $\pi_{12} = 1 - \pi_{11}$ and $\pi_{22} = 1 - \pi_{21}$.

TABLE 14.4

Structure of Study for OR: Independent
Samples

Outcome	Antecedent		
	Present	**Absent**	**Total**
Present (cases)	$a(\pi_{11})$	$b(\pi_{12})$	$n_1(1)$
Absent	$c(\pi_{21})$	$d(\pi_{22})$	$n_2(1)$
Total	$O._1(\pi._1)$	$O._2(\pi._2)$	n

The odds of the presence of an antecedent among cases are $\pi_{11}/(1 - \pi_{11})$ and among controls are $\pi_{21}/(1 - \pi_{21})$. Thus, the OR is

$$\text{OR} = \frac{\pi_{11}/(1-\pi_{11})}{\pi_{21}/(1-\pi_{21})} = \frac{\pi_{11}\pi_{22}}{\pi_{12}\pi_{21}}. \tag{14.15}$$

If estimate of π_{11} is p_1 and of π_{21} is p_2, OR is estimated as

$$\text{OR} = \frac{p_1/(1-p_1)}{p_2/(1-p_2)} = \frac{ad}{bc}, \tag{14.16a}$$

where a, b, c, d are as in Table 14.4. (Note that these p_1 and p_2 are different from p_1 and p_2 in RR.) Since the numerator is the product of the elements in the leading diagonal in Table 14.4 and the denominator is that of the elements in the other diagonal, OR is also sometimes called the **cross-product ratio**.

Equation 14.16a loses definition when any cell frequency is zero, and becomes inflated if any observed frequency is exceedingly small. In that case,

$$\text{Modified estimate of OR: OR}_{\text{mod}} = \frac{(a+0.5)(d+0.5)}{(b+0.5)(c+0.5)}. \tag{14.16b}$$

The interpretation of OR is similar to that of RR. An OR = 2 means that the presence of the antecedent is twice as common among the cases as in the controls. It can be shown that the OR approximates the RR fairly well when the outcome of interest is rare, say, less than 5% in the target population. Most outcomes of medical interest are rare. If the outcome is not rare, an OR more than 1 leads to an overestimate of RR and less than 1 leads to an underestimate. This can be serious if the risk or the probability of outcome is more than 20%.

Example 14.7: OR of Lower Parity in Anemia

Consider the data in Table 14.5 on parity status of 210 anemic and 140 nonanemic women. Anemic and nonanemic women were assessed for parity so that the design now is retrospective:

$$\text{OR} = \frac{98 \times 48}{92 \times 112} = 0.46.$$

TABLE 14.5

Parity Status in Anemic and Nonanemic Women

Anemia	Parity ≤ 2	Parity ≥ 3	Total
Present	98	112	210
Absent	92	48	140

The likelihood of parity ≤2 among anemics is less than one-half that among nonanemics.

SIDE NOTE: Anemia in women is not rare, particularly in developing countries. Thus, in this case, OR may fail to approximate RR. If it is really rare, then the conclusion could be that parity ≤2 has less than one-half the risk of anemia compared to the risk in parity ≥3. Or, the risk of anemia in women with parity ≥3 is more than twice that for women with parity ≤2.

Some features of OR are as follows:

1. The sample OR is always a good estimate of the population OR irrespective of the disease being rare in the population.

2. If, instead of odds of exposure in the control group, odds of exposure in the population is used in calculating the OR (i.e., when the control group is representative of the population and not selected from the noncases), then OR is exactly equal to RR. There is no need for the disease to be rare in this case. The format of the study is no longer the classical case–control type because the control group also contains diseased subjects.

3. In case–control studies, the response refers to presence or absence of an antecedent characteristic. For this, it could be inappropriate to use the term incidence. Nor does the term risk seem appropriate to indicate the presence of an antecedent characteristic. The term odds is used, which appropriately describes the situation.

4. Although OR = 0.5 and OR = 2 are half and double of OR = 1, respectively, their average is not 1. OR is a ratio and logarithm removes this discrepancy—lnORs can be averaged as usual but if you need to average ORs, use geometric mean.

5. RR can lead to a very different conclusion depending on whether a positive outcome or a negative outcome is being measured. OR is not affected by such a consideration. For instance, consider the data in Table 14.6.

   The estimated RR of death within 1 year among breast cancer versus lung cancer cases in these data is $(2/50)/(1/100) = 4$. If it is considered a retrospective study, the same data lead to the OR $(2/48)/(1/99) \approx 4$.

TABLE 14.6

Deaths in Breast Cancer and Lung Cancer Cases

	Deaths within 1 Year	Survival after 1 Year	Total
Breast cancer	2	48	50
Lung cancer	1	99	100

Thus, both measures indicate nearly four times the risk of dying among breast cancer patients compared with lung cancer patients.

Now use the same data, this time with respect to survival instead of death. The RR of survival is (48/50)/(99/100) = 0.97, while the OR of survival is (48/2)/(99/1) = 0.24. 1/RR = 1.03 and 1/OR ≈ 4. The RR results are different depending on whether the study is summarized with respect to death or with respect to survival. For the OR, it really does not matter. Breast cancer has about four times the odds of death within 1 year compared with lung cancer or about one-fourth the odds of survival compared with lung cancer. Both convey the same result in case of OR.

6. The OR from a cross-sectional study is computed the same way as that from a retrospective study. However, *it is necessary that the sample is a true reflection of the proportion of subjects with and without outcome as well as of the proportion with and without antecedent.* Thus, for valid results, a cross-sectional sample must be a truly random sample from the target population.

14.2.1.1 CI for OR (Independent Samples)

Similar to RR, OR also is a ratio and its natural logarithm (ln) becomes a linear function. The distribution of OR can be shown to be highly skewed but lnOR has a nearly Gaussian pattern for large n. It has been established for large n that the estimated

$$\text{SE}(\ln \text{OR}) = \sqrt{\frac{1}{a} + \frac{1}{b} + \frac{1}{c} + \frac{1}{d}}, \qquad (14.17)$$

where a, b, c, d are as in Table 14.4. If any of these is zero or small, add ½ to each of these in the denominators in the formula given in Equation 14.17. For large n,

$$95\% \text{ CI for OR: } \exp[\ln \text{OR} \pm 1.96 * \text{SE} (\ln \text{OR})], \qquad (14.18)$$

where exp is the exponent on Naperian base e. This is exactly similar to the procedure followed earlier for RR although SE is now different.

14.2.1.2 Test of Hypothesis on OR (Independent Samples)

The H_0 in this case almost invariably is that OR = 1. This says that the presence of the antecedent is as common in cases as in controls. Since OR = RR if the outcome is rare, H_0: OR = 1 also says that the presence or absence of the antecedent does not influence the outcome. A simple statement that takes care of both directions of the relationship is that there is no association under H_0 between antecedent and outcome. The alternative could be one-sided H_1: OR < 1 or H_1: OR > 1 or could be two-sided H_1: OR ≠ 1. The latter is applicable when there is no priori assurance that the relationship could be one sided. The two-sided hypothesis is tested by classical chi-square as stated in Section 13.2. For the one-sided alternative, use Z = ln OR/SE(ln OR) and refer it to a Gaussian distribution to obtain the P-value.

Example 14.8: Association of HIV with Extramarital Contacts

Suppose that 80 promiscuous males who were recently found to be HIV positive and 160 promiscuous controls (two controls per case) of the same age and socioeconomic–cultural–ethnic background are asked about their sexual behavior in terms of average frequency of extramarital contacts per month during the last 2 years. The data obtained are in Table 14.7.

The column totals in this case are 89 and 151, respectively, and $n = 240$. Thus, for chi-square, with notations of the previous chapter,

$$E_{11} = \frac{80 \times 89}{240} = 29.67, \quad E_{12} = \frac{80 \times 151}{240} = 50.33,$$

$$E_{21} = \frac{160 \times 89}{240} = 59.33, \quad E_{22} = \frac{160 \times 151}{240} = 100.67.$$

Therefore,

$$\chi^2 = \frac{(8.33)^2}{29.67} + \frac{(8.33)^2}{50.33} + \frac{(8.33)^2}{59.33} + \frac{(8.33)^2}{100.67} = 5.58.$$

A computer package gives for 1 df $P(\chi^2 \geq 5.58) = 0.018$. If a package is not available, consult Table B.3 for 1 df and get $P < 0.05$. The P-value is exceedingly small. There is an exceedingly small chance that H_0: OR = 1

TABLE 14.7

Extramarital Contacts in HIV+ Subjects and Controls

Group	Average Extramarital Contacts/Month		Total
	One or More	Less than One	
HIV+	38	42	80
Controls	51	109	160

is true. Reject H_0 and conclude that OR $\neq$ 1. HIV positivity is associated with one or more average extramarital contacts per month in promiscuous males. In this case, OR = $(38 \times 109)/(51 \times 42)$ = 1.93. This shows that the odds of one or more extramarital contacts per month are nearly twice as much in HIV-positive males as in controls. Since HIV positivity even in promiscuous males (the target population in this example) is rare, OR can be interpreted as by RR and it can be inferred that an average of one or more extramarital contacts per month doubled the *risk* of HIV compared with less than one average number of extramarital contacts per month. Also,

$$SE(\ln OR) = \sqrt{\frac{1}{38} + \frac{1}{42} + \frac{1}{51} + \frac{1}{109}} = 0.2809.$$

The sample size n is sufficiently large in this case and a Gaussian distribution of lnOR is expected. Thus, the 95% CI for OR is

$$\exp[\ln(1.93) \pm 1.96 \cdot 0.2809],$$

or

$$\left(e^{0.107}, e^{1.208}\right) \quad \text{or} \quad (1.11, 3.35).$$

The OR in this sample is 1.93 but it can well range from 1.11 to 3.35 in the target population. If only the one-sided (lower) bound with confidence 95% is required, this is $\exp[\ln(1.93) - 1.645 \times 0.2809] = 1.22$. It can be stated with 95% confidence that the OR in the population is not less than 1.22.

SIDE NOTE: Extramarital contacts per month in promiscuous males is a quantitative variable. The dichotomy in this example is on the basis of average extramarital contacts per month during the 2-year period preceding the date of inquiry. Categorization of this into one or more and less than one is arbitrary. Another cutoff point can give a different result. The other way to compare HIV positive subjects with controls is by computing the mean number of such contacts per month per person in the two groups. Comparison of such means is generally done with the help of Student *t*-test as described in Chapter 15.

Compare SE(lnOR) in Equation 14.17 with SE(lnRR) in Equation 14.3 and note that SE(lnOR) is larger. As a consequence, the CI for OR would be larger than for RR. This statistical loss can be compensated by conducting a case–control study on a larger number of subjects within the cost of a prospective study. A strategy sometimes advocated to get a narrower CI for OR is to enhance the number of controls by including multiple controls for each case. Inclusion of a large number of controls does not generally involve much cost. Multiple controls are discussed in a subsequent paragraph.

A very useful method for comparing the cases with controls with respect to one or more dichotomous antecedents is logistic regression. The OR arises naturally in this kind of model. This is discussed in Chapter 17. This model also gives an adjusted OR when more than one factor is under consideration.

14.2.2 OR in Matched Pairs

Consider Table 14.8 on matched pairs. This is similar to Table 13.14 but now uses A, B, C, D as notation for cell frequencies in place of a, b, c, d to distinguish the case–control setup from the prospective setup. Note also that the labeling of the cells has now changed.

The total number of pairs is $A + B + C + D$. In this table, A is the number of pairs with both case and control subjects found exposed, and D is the number of pairs with both found nonexposed. These two together are the concordant pairs. The OR is computed on the basis of the discordant pairs: B is the number of pairs in which the case partner is exposed but the control partner is nonexposed, and C is the number of pairs in which the case partner is nonexposed but the control partner is exposed. In case of a positive association between exposure and disease, clearly B should be more than C:

$$\text{Odds ratio (matched pairs): } OR_M = \frac{B}{C}. \tag{14.19}$$

14.2.2.1 Confidence Interval for OR (Matched Pairs)

The distribution of $\ln OR_M$ is nearly Gaussian for large n. This also implies that no B or C is small, say, less than 5. For large B and C,

$$SE(\ln OR_M) = \sqrt{\frac{1}{B} + \frac{1}{C}}. \tag{14.20}$$

The 95% CI for log of OR, as usual, is $\ln OR_M \pm 1.96 SE(\ln OR_M)$. In this case this becomes

$$\ln \frac{B}{C} \pm 1.96 * \sqrt{\frac{1}{B} + \frac{1}{C}}.$$

TABLE 14.8

Matched Pairs in a Case–Control Study with Dichotomous Antecedent

	Controls (Partner-1)	
Cases (Partner-2)	**Antecedent Present (Exposed)**	**Antecedent not Present (Nonexposed)**
Antecedent present (exposed)	A	B
Antecedent not present (nonexposed)	C	D

Take the exponential of the limits and get

$$\text{95\% CI for OR in matched pairs: } \frac{B}{C} e^{\pm 1.96\sqrt{(1/B)+(1/C)}}. \tag{14.21}$$

While the CI for $\ln OR_M$ is symmetric, it is not symmetric for OR itself.

14.2.2.2 Test of Hypothesis on OR (Matched Pairs)

The relevant null hypothesis in this case is H_0: $OR_M = 1$. To test this against a one-sided alternative H_1: $OR_M > 1$ or $OR_M < 1$, calculate

$$Z = \frac{B-C}{\sqrt{B+C}}. \tag{14.22}$$

For large n, refer it to the usual Gaussian distribution to find out whether the P-value is sufficiently small. For a two-tailed test, it may be easier to calculate

$$\text{McNemar } \chi^2_M = \frac{\left(|B-C|-1\right)^2}{B+C} \tag{14.23}$$

and refer it to chi-square with 1 df. This incorporates correction for continuity.

Example 14.9: Birth Defects and Sex of the Children

Consider a case–control study of births with multiple malformations. The malformations considered are cleft lip, cleft palate, anal atresia, heart defects, hypospadias, etc. They are considered multiple when at least two are present. The controls were one-to-one matched for birth order, maternal age, socioeconomic status, and the place of delivery. The objective is to find any excess of one gender over the other in such births. Suppose the data obtained are as shown in Table 14.9.

In these data, $B = 60$ and $C = 50$. Thus,

$$OR_M = \frac{60}{50} = 1.20.$$

TABLE 14.9

Sex of Children with and without Multiple Malformations

Cases	Matched Control		Total
	Male	**Female**	
Male	70	60	130
Female	50	40	90
Total	120	100	220

The odds are 1.2 times in these subjects that the malformed child is male and not female. Now,

$$SE(\ln OR_M) = \sqrt{\frac{1}{60} + \frac{1}{50}} = 0.1915.$$

Consider the following three procedures:

a. From the formula given in Equation 14.21, 95% CI for the odds ratio is $(1.2 \times e^{-1.96 \times 0.1915}, 1.2 \times e^{1.96 \times 0.1915})$ or $(0.82, 1.75)$.

Note that this interval contains $OR_M = 1$, which shows that this null hypothesis cannot be rejected.

b. Otherwise, if the null hypothesis of $OR_M = 1$ is to be tested by the criterion in Equation 14.22, then $Z = \dfrac{60 - 50}{\sqrt{60 + 50}} = 0.95.$

This gives $P = 0.3422$ for a two-sided alternative (double the one-tail probability in Table B.1).

c. Also, McNemar chi-square in Equation 14.23 in this case is

$$\chi^2_M = \frac{(|60 - 50| - 1)^2}{60 + 50} = 0.74.$$

This again gives $P > 0.05$.

None of the three procedures is able to reject H_0: $OR_M = 1$. They are equivalent for large n. Thus, these data do not allow the conclusion that either gender is more prone to multiple malformations at birth.

There are several points that you can note in Example 14.9: (a) The CI is not symmetric to $OR_M = 1.2$. It is skewed to the right because $1.75 - 1.20 = 0.55$ is more than $1.20 - 0.82 = 0.38$. (b) The antecedent characteristic is not conventional exposure but is gender. (c) Multiple malformations are rare and thus case–control is a suitable design. Several confounders that could have influenced the outcome, namely, birth order, maternal age, socioeconomic status, and place of delivery, are controlled by matching. Perhaps, there are no more confounders except those in the epistemic domain. Thus, any difference between male and female children can be ascribed to their gender and, of course, sampling fluctuations.

The results obtained by three methods for H_0: $OR_M = 1$ are consistent in Example 14.9 as all three fail to reject this null. But you might see some variation in other situations. This variation tends to vanish as B and C increase. Use McNemar test (Equation 14.23) if the alternative is two sided and not the CI because the CI uses a greater degree of approximation.

14.2.2.3 Multiple Controls

Because the SE depends on the number of subjects in the study, a useful strategy could be to include more control subjects. In most practical situations, controls are easily available, less cumbersome to investigate, and more cooperative.

Thus, their number can be increased without a corresponding increase in the cost. In that case, in place of one-to-one matching, C:1 matching can be done, that is, C controls per case. This increase in the number of controls reduces the SE and increases the power of the study to detect an association. However, the returns are progressively diminishing. In general, as the ratio of controls to cases increases beyond 4:1, the additional gain in statistical power may be small compared with the cost involved. However, in a study of asthma severity and first prescription of inhaled fenoterol as opposed to inhaled salbutamol, Blais et al. [6] used up to 30 controls per case in one group. Such a large number of controls is unnecessary in my opinion.

Multiple controls can also be used as an effective strategy to reduce the number of cases when they are difficult to enlist or are very costly. But the reduction in the number of cases is not substantial. It can be shown, for example, that a study with 100 cases and 100 controls has the same statistical power as a study with 75 cases and 150 controls, or one with 62 cases and 252 controls, or one with 55 cases and 550 controls. Note that as the number of cases decreases, the total number of subjects (cases + controls) steeply increases. Therefore, apply this strategy only where it is really helpful.

The analysis of data from a study with multiple controls proceeds in a slightly different manner. For details, see Selvin [7].

14.3 Stratified Analysis, Sample Size, and Meta Analysis

When a qualitative confounding factor is suspected to be working behind the scene and spoiling the picture, a clear picture is obtained by preparing a separate contingency table for each level of the confounder. This is called stratification. The analysis is done by using the M-H procedure. The second issue for this section is determination of sample size for reliable inference on RR and OR. The third, again disjointed, is meta-analysis. This can be used for any type of effect but is explained here with an example of OR.

14.3.1 Mantel–Haenszel Procedure

Consider recurrence of eclampsia in second or subsequent pregnancies when it once occurred at primigravida stage. It is widely suspected that occurrence at the time of first pregnancy is related to a positive family history of eclampsia, and so is the recurrence. But the risk of recurrence could increase if male and female family members have hypertension. This could be a confounding factor. To examine if it really is, the association between recurrence of eclampsia and family history should be examined separately in the group with (1) no family history of hypertension, (2) only female members with history of hypertension,

(3) only male members with history of hypertension, and (4) both male and female members with history of hypertension. Thus, the data on recurrence of eclampsia and history of eclampsia are stratified into four groups. If the association in each of these groups is the same, it can be concluded that family history of hypertension has no role. Otherwise confounding is suspected. If any other factor is suspected to confound then that also should be examined.

Stratified analyses are considered fundamental to causal thinking. When confounders exist, overall results are not valid and stratum-specific results should be reported. In the absence of confounding, unadjusted associations may be reported. For adjustment, methods such as the M-H chi-square could be used to get a combined chi-square. In a situation where confounding is present, this procedure helps to get a pooled measure of association while controlling for confounding.

14.3.1.1 Pooled Relative Risk

Add a subscript k to the observed frequencies in 2 × 2 Table 14.1. Let this subscript index the $k = 1, 2, ..., K$ stratified tables. For prospective studies, the adjusted M-H pooled RR is

$$\text{Pooled RR: } RR_{MH} = \frac{\Sigma_k \left[a_k(b_k + d_k)/n_k \right]}{\Sigma_k \left[b_k(a_k + c_k)/n_k \right]}, \tag{14.24}$$

where

a_k, b_k, c_k, d_k are the frequencies in the kth 2 × 2 table such as in Table 14.1
n_k is the total number of subjects in the kth stratum, that is, $n_k = (a_k + b_k + c_k + d_k)$

In case any cell frequency is zero, 0.5 is added to all cell frequencies for computing pooled RR. The sum is over the values of k. But pooled RR works only when RRs are uniform across strata.

Example 14.10: Pooled RR after Adjustment for Sex Differentials

Consider two strata (males and females) and frequencies as in Table 14.10. The results for this table are as follows:

$$RR_1 \text{ for males} = \frac{3/100}{13/450} = 1.04 \quad \text{from the first two columns of Table 14.10.}$$

$$RR_2 \text{ for females} = \frac{97/400}{12/50} = 1.01 \quad \text{from the middle two columns of Table 14.10.}$$

TABLE 14.10

Cell Frequencies in Two Strata and Combined

| | Males | | Females | | Combined | |
| | Antecedent (Exposure) | | Antecedent (Exposure) | | Antecedent (Exposure) | |
Outcome	Present	Absent	Present	Absent	Present	Absent
Present	3	13	97	12	100	25
Absent	97	437	303	38	400	475
Total	100	450	400	50	500	500

These two are not much different.

$$RR_c \text{ for combined} = \frac{100/500}{25/500} = 4.00 \quad \text{from the last two columns of Table 14.10.}$$

All these are calculated from the usual formula given previously.

Note how the RR for the combined is so different from each of the strata. This has mainly occurred because in males only 100 are exposed out of 550 but in females 400 are exposed out of 450. This is a situation where the M-H procedure is very effective. In this example, by the formula given in Equation 14.24,

$$RR_{MH} = \frac{(3 \times 450/550) + (97 \times 50/450)}{(13 \times 100/550) + (12 \times 400/450)} = \frac{13.2323}{13.0303} = 1.02.$$

The M-H procedure restores the right value, which was masked in the combined data. This is adjusted for confounding effect of sex.

14.3.1.2 *Pooled Odds Ratio and Chi-Square*

For retrospective studies, the M-H adjusted

$$\text{Pooled OR: OR}_{MH} = \frac{\Sigma_k(a_k d_k/n_k)}{\Sigma_k(b_k c_k/n_k)}. \tag{14.25}$$

For a simple explanation and an example, see Le [8]. As for RR, this pooling is admissible when the ORs are homogenous in different strata. Breslow–Day test is used to check this homogeneity.

For cross-sectional studies, the corresponding pooled chi-square is given by

$$\text{M-H chi-square} = \frac{\left[\Sigma\{(a_k d_k - b_k c_k)/n_k\}\right]^2}{\Sigma\left[(a_k + b_k)(c_k + d_k)(a_k + c_k)(b_k + d_k)/\{(n_k - 1)n_k^2\}\right]}. \tag{14.26}$$

This has 1 df. It can be shown with some algebra that the numerator of Equation 14.26 is the square of the (observed – expected) frequencies, and the denominator is the variance of these squared differences.

One difficulty with M-H procedure is that only one confounder can be conveniently studied at one time. If there are many, this procedure fails. Also the confounder must be categorical. If it is continuous, categorize this first into meaningful groups and then use M-H test.

Example 14.11: Inadequate Stratified Analysis for Assessing Neonatal Deaths Following HBV

All deaths under 29 days of age in a birth cohort of more than 350,000 live births from 1993 to 1998 in California were examined with the objective to find if reduced neonatal deaths are associated with hepatitis B vaccine (HBV) at birth [9]. The data obtained can be reconstructed as given in Table 14.11.

Although absolute numbers are not mentioned, it is stated that 67% received HBV at birth. Against this, only 5% of total deaths occurred in this group. This difference is statistically significant ($P < 0.01$). Thus, apparently it looked that HBV at birth is substantially associated with reduced neonatal mortality. This conclusion is acceptable, provided the unvaccinated group was not underprivileged or undernourished. However, the authors performed a stratified analysis and divided deaths by causes classified as "expected" and "unexpected." They found no significant difference ($P = 0.6$) between 31% unexpected deaths in the vaccinated group and 35% in the unvaccinated group. The conclusion of the authors is that no relationship could be established between HBV and neonatal deaths. This conclusion is the reverse of what was obtained on the basis of total deaths that ignores the cause of death.

The lessons learned from Example 14.10 are the following:

1. A neonate has been defined as a child of age under 29 days. The standard definition is under 28 days.
2. The term "association" between HBV at birth and neonatal mortality is correct usage since the cause–effect type of relationship is not implied.

TABLE 14.11

Deaths due to HBV Infection at Birth

HBV at Birth	Expected Deaths	Unexpected Deaths	Total Deaths	Total Births
Vaccinated	50 (69%)	22 (31%)	72 (100%)	67%
	(6%)	(5%)	(5%)	
Unvaccinated	839 (65%)	452 (35%)	1291 (100%)	33%
	(94%)	(95%)	(95%)	
Total	889 (65%)	474 (35%)	1363 (100%)	100%
	(100%)	(100%)	(100%)	

3. The conclusion depends on what causes are called "unexpected."

4. The initial conclusion is based on 67% immunized but only 5% of total deaths occurring in this group. Stratification of causes of deaths into expected and unexpected has been done to find if unexpected deaths are same in the immunized group as in the unimmunized group. A careful look at the table in the example shows that both expected and unexpected deaths are nearly 5% in the vaccinated group and the remaining 95% in the unvaccinated group. The vaccinated group has substantially lower deaths of both the types. Thus, the conclusion lacks justification. This is an example of a statistical fallacy that so commonly creeps into medical literature, intentionally or unintentionally, and can remain unnoticed for a long time.

Table S.3 at the beginning of the book provides a snapshot of the test of hypothesis methods discussed in this chapter.

14.3.2 Sample Size Requirement for Statistical Inference on RR and OR

You may also review Section 12.6 on sample size determination. A large number of formulas are given in Tables 12.7 and 12.8 for calculating n in different situations. But they do not include n for estimation and test of hypothesis for RR or OR. These are now given in Table 14.12. As explained earlier, if the objective is estimation, then it is necessary to specify the confidence level $(1 - \alpha)$ and the precision desired on either side of RR or OR. This is generally specified as a proportion of RR or OR. The anticipated RR or OR or the anticipated probability of disease (or of exposure) in the respective groups should also be specified. If the objective is to test a hypothesis, then the specification of the relative precision is not required and the significance level is required in place of the confidence level. Specification of the statistical power $(1 - \beta)$ is needed for a testing-of-hypothesis setup. For this the difference to be detected must be specified.

The formulas in Table 14.12 require the anticipated probabilities in the two groups. If only one of these can be reasonably anticipated, then the anticipated value of RR (or of OR) would be needed. In that case, the other probability can be calculated by using the definition of RR or of OR. The notation now is π_1 and π_0 for disease probability in the exposed and unexposed group, respectively, and exposure probabilities by π_1' and π_0' in the case and control groups. In terms of these probabilities, you know that RR $= \pi_1/\pi_0$ and OR $= [\pi_1'/(1-\pi_1')]/[\pi_0'/(1-\pi_0')]$.

An astute reader would note that formula (b) in Table 14.12 for test of hypothesis on RR uses $\bar{\pi} = (\pi_1 + \pi_0)/2$ but such an average is not used in formula (d) for OR. Instead, only π_0' is used. This is the exposure rate among the controls. Quite often, this rate is reliably known and there is no need to

TABLE 14.12

Sample Size Required for Two-Sided Inference on RR, OR, and AR

Problem	Formula for Computing n (Large n)	Explanation of the Notations
RR		
a. Estimating RR	$\dfrac{z_{\alpha/2}^2}{[\ln(1-\varepsilon)]^2}\left[\dfrac{1-\pi_1}{\pi_1}+\dfrac{1-\pi_0}{\pi_0}\right]$	π_1 = anticipated probability of disease among the exposed subjects π_0 = anticipated probability of disease among the nonexposed subjects ε = relative precision in terms of proportion of RR $\bar{\pi}=\dfrac{(\pi_1+\pi_0)}{2}$
b. Hypothesis testing for RR = 1 Equal exposed and unexposed groups	$\dfrac{1}{(\pi_1-\pi_0)^2}\left[z_{\alpha/2}\sqrt{2\bar{\pi}(1-\bar{\pi})}+z_{\beta}\sqrt{\pi_1(1-\pi_1)+\pi_0(1-\pi_0)}\,\right]^2$ Or, $\dfrac{[z_{\alpha/2}\sqrt{2\bar{\pi}(1-\bar{\pi})}+z_{\beta}\sqrt{\pi_0(1+r)-\pi_0^2(1+r^2)}\,]^2}{[\pi_0(1-r)]^2}$	π_1, π_0 as above Medically relevant least RR is $r=\pi_1/\pi_0$
OR		
c. Estimating OR	$\dfrac{z_{\alpha/2}^2}{[\ln(1-\varepsilon)]^2}\left[\dfrac{1}{\pi_1'(1-\pi_1')}+\dfrac{1}{\pi_0'(1-\pi_0')}\right]$	π_1' = anticipated probability of exposure among the cases π_0' = anticipated probability of exposure among the controls ε = relative precision in terms of proportion of OR π_1', π_0' as above
d. Hypothesis testing for OR = 1 One control per case	$\dfrac{1}{(\pi_1'-\pi_0')^2}\left[z_{\alpha/2}\sqrt{2\pi_0'(1-\pi_0')}+z_{\beta}\sqrt{\pi_1'(1-\pi_1')+\pi_0'(1-\pi_0')}\,\right]^2$	Medically relevant least OR is $[\pi_1'/(1-\pi_1')]/[\pi_0'/(1-\pi_0')]$
C controls per case	$\dfrac{1}{(\pi_1'-\pi_0')^2}\left[z_{\alpha/2}\sqrt{(1+1/C)\pi'(1-\pi')}+z_{\beta}\sqrt{\pi_1'(1-\pi_1')+\pi_0'(1-\pi_0')}\,\right]^2$	$\pi'=\dfrac{\pi_1'+C\pi_0'}{C+1}$; π_1', π_0' as above In terms of OR, $\pi_1'=\dfrac{\pi_0'(\text{OR})}{1+\pi_0'(\text{OR}-1)}$

(continued)

TABLE 14.12 (continued)

Sample Size Required for Two-Sided Inference on RR, OR, and AR

Problem	Formula for Computing n (Large n)	Explanation of the Notations
Incidence rate		
e. Estimation of an incidence rate	$\dfrac{z_{\alpha/2}^2}{\varepsilon^2}$	ε = specific relative precision as fraction of the anticipated incidence rate
f. Hypothesis testing for incidence rate	$\dfrac{(z_{\alpha/2}\pi_0 + z_\beta\pi_a)^2}{(\pi_0 - \pi_a)^2}$	π_0 = incidence rate (in terms of proportion) under the null hypothesis π_a = incidence rate in terms of proportion under the alternative hypothesis (difference to be detected is $\pi_0 - \pi_a$)
g. Hypothesis testing for difference in incidence rates per year (**attributable risk**)	$\dfrac{\left[z_{\alpha/2}\sqrt{2f(\pi)} + z_\beta\sqrt{f(\pi_1) + f(\pi_0)}\right]^2}{(\pi_1 - \pi_0)^2}$	$\pi = \dfrac{(\pi_1 + \pi_0)}{2}$ $f(\pi) = \pi^3 T/(\pi T - 1 + e^{-\pi T})$ if the duration of the study (in years) is fixed as T (censored observations) $f(\pi) = \pi^2$, if T is not fixed For $f(\pi_1)$ and $f(\pi_0)$, replace π by π_1 and π_0 in these expressions

Source: Lwanga, S.K. and Lemeshow, S., *Sample Size Determination in Health Studies: A Practical Manual*, World Health Organization, Geneva, Switzerland, 1991.

Note: Two independent samples and same n for each group. α is the significance level and $(1 - \beta)$ is the statistical power, z_a is such that $P(Z \geq z_a) = a$; for $\alpha = 0.05$, $z_{\alpha/2} = 1.96$; for $\beta = 0.05$, $z_\beta = 1.645$; for $\beta = 0.10$, $z_\beta = 1.28$; for $\beta = 0.20$, $z_\beta = 0.84$.

use the average. If you are not fully confident about π_0', the first part in formula (d) should become $\sqrt{2\bar{\pi}(1-\bar{\pi})}$ just as in formula (b).

The formulas in Table 14.12 are restricted to the following: (a) two independent samples (not matched pairs), and (b) a large n so that the Gaussian pattern is applicable. Relative precision ε is on either side of RR or OR, and the alternative hypothesis is RR $\neq 1$ or OR $\neq 1$. Both these inferences are two sided. For one-sided alternative, replace $z_{\alpha/2}$ by z_α. The formulas for matched pairs can be worked out similarly but are too complex for this book.

Example 14.12: Sample Size for Clinical Trial on Control of Nausea

Consider a clinical trial in which a new regimen is compared with an existing regimen for their role in control of nausea while treating depression. The objective is to estimate the RR of nausea. The anticipated values are as follows:

Proportion with nausea in new regimen $\pi_1 = 0.40$
Proportion with nausea in existing regimen $\pi_0 = 0.50$
Relative precision desired, $\varepsilon = 0.10$ (i.e., estimated RR should be within 10% of its actual value)
Confidence level 95%, $\alpha = 0.05$

Thus, $z_{\alpha/2} = 1.96$, and, from formula (a) in Table 14.12,

$$n = \frac{(1.96)^2}{[\ln(1-0.10)]^2}\left[\frac{1-0.40}{0.40}+\frac{1-0.50}{0.50}\right] = 865,$$

with upward approximation.

This is the number of subjects required for each of the regimens. If the relative precision is relaxed to $\varepsilon = 0.20$, then $n = 193$. The sample size requirement declines steeply when the precision requirement is relaxed.

Example 14.13: Sample Size for Case–Control Study on Thyroid Status

Consider the plan of a case–control study (one control per case) of people with beta-thalassemia major to evaluate their thyroid status. Thyroid status is divided into only normal and abnormal categories but under very stringent criteria. The objective is to assess whether there is any association between thyroid status and beta-thalassemia. Thus, this is a testing-of-hypothesis setup. Suppose the anticipated values are as follows:

Proportion with abnormal thyroid among cases, $\pi_1' = 0.25$
Proportion with abnormal thyroid among controls, $\pi_0' = 0.10$
[Thus, anticipated OR is $(0.25/0.75)/(0.10/0.90) = 3$.]
Significance level $\alpha = 0.05$
Power $(1-\beta) = 0.80$

Since $z_{\alpha/2} = 1.96$ and $z_\beta = 0.84$, we get from formula (d) in Table 14.12 (one control per case).

$$n = \frac{1}{(0.25 - 0.10)^2} \times \left(1.96\sqrt{2 \times 0.10 \times 0.90} + 0.84\sqrt{0.25 \times 0.75 + 0.10 \times 0.90} \right)^2$$

$$= \frac{1}{0.0225}(0.8316 + 0.4425)^2 = 73,$$

with upward approximation.

Only 73 cases and 73 controls are required for this study.

14.3.3 Meta-Analysis

Meta-analysis is pooling varying results of various studies on the same parameter after a systematic review. For this, studies meeting prespecified quality criteria are selected after a comprehensive search of literature. A particular relevant parameter such as OR, RR, or mean difference is chosen and its value with CI is extracted from each selected study.

Forest plot provides graphical summary view of the varying results obtained in different studies. An example is in Figure 14.1 where ORs of probiotics in prevention of antibiotic-associated diarrhea found in different

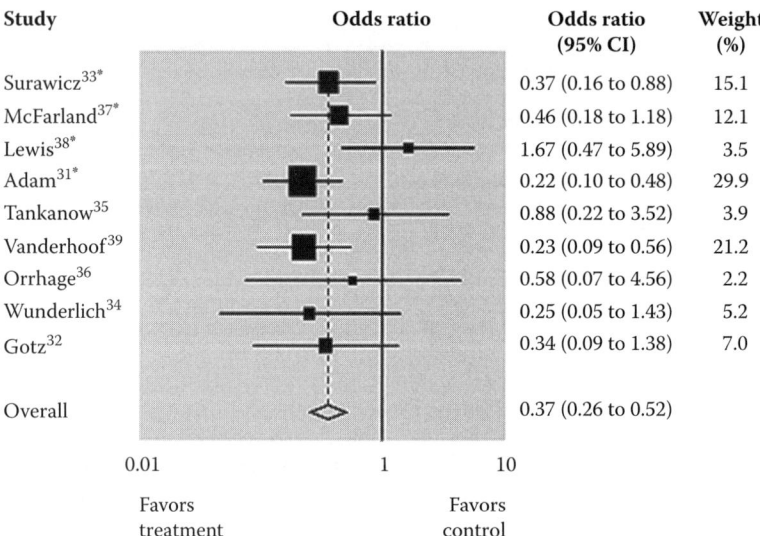

FIGURE 14.1
Plot of the log of ORs for the proportion of patients free of diarrhea in treatment groups compared with control groups. (Reproduced from D'Souza, A.L. et al., *Br. Med. J.*, 324, 1361, 2002. With permission.)

studies are shown. In the left column, you can have study identifier such as name of the first author. Black square is located at the value of OR (or any other parameter of interest) scaled on *x*-axis. For a ratio such as OR and RR, log scale is used as in this figure. Horizontal lines are the 95% CIs. Actual values are stated in a column on right side. For pooling the results, the average is calculated but all studies are not given equal weight. Generally, studies with larger sample size or with smaller SE get more weight. You may use any other criterion. The area of the black square represents this weight—this also is mentioned in another column on the right side in this figure.

Pooled result is shown by a diamond touching the *x*-axis. The width of this diamond is the pooled CI. This CI is based on combined sample size and thus more reliable. If this touches or crosses the line of no effect (OR = 1 in this case), the pooled conclusion is that the effect is not statistically significant. In the case of OR, diamond toward less than one indicated decreased likelihood of the presence of antecedent of interest and toward more than one indicates increased likelihood.

In this figure, the diamond is on the left side of OR = 1 and it does not touch (or cross) the vertical line of OR = 1. Thus, decreased diarrhea is significantly associated with more probiotics. In other words, probiotics have protective effect.

Results based on small sample size or with high SE in different studies will obviously spread across a broad range of values. If you plot ORs in three studies with small sample size each, they are likely to be far apart from one another compared with ORs in other three studies with large sample size each. If you are reviewing a large number of studies—some of small size and some of large size—and plot OR on horizontal axis and sample size on vertical axis, the plot generally will be as shown in Figure 14.2. This is called **funnel plot** because of its resemblance with inverted funnel.

In place of OR, you can have any other effect size such as RR and difference in means or proportions. On the vertical axis, you can have inverse of the SE instead of sample size. An asymmetric shape of the funnel plot raises

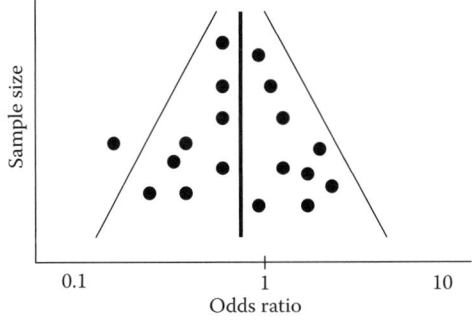

FIGURE 14.2
Typical funnel plot.

suspicion over the results of meta-analysis since the selected studies may suffer from publication bias, favoring either higher or lower effect size. It also suggests the possibility of a systematic bias in smaller studies. Check if most of smaller studies tend to give larger (or smaller) effect size compared to larger studies. If so, the bias is evident and the results of meta-analysis would be invalid. When biased studies are not included in meta-analysis, heterogeneity among results of various studies does not cause much of a problem. Your final CI would depict this. Sometimes you would want to know the extent of heterogeneity. For this an index denoted by I^2 is calculated. This measures the percentage variance attributed to between-study variation; $I^2 = 0.25$ is low, 0.50 is middling, and 0.75 is high. If this is high, you may like to identify studies causing this and exclude them from your meta-analysis.

In place of aggregate results of studies, the emphasis now is on individual participant data. Since almost all studies around the world have data on individual subjects in electronic form, they can be easily pooled in a collaborative effort that would provide a direct estimate based on a large number of subjects. Thompson et al. [10] have provided this kind of analysis based on 154,211 participants in 31 studies on hazard ratio for CHD per 1 g/L higher baseline fibrinogen. Care is needed while pooling because the participants within each study form a cluster with shared similarities, and clustering is factored into pooling.

References

1. Martinez FD, Wright AL, Holberg CJ, Morgan WJ, Taussig LM. Maternal age as a risk factor for wheezing lower respiratory illnesses in the first year of life. *Am J Epidemiol* 1992; 136:1258–1268.
2. Brenner H, Berg G, Lappus N, Kliebsch U, Bode G, Boeing H. Alcohol consumption and *Helicobacter pylori* infection: Results from the German National Health and Nutrition Survey. *Epidemiology* 1999; 10:214–218.
3. Doll R, Hill AB. Mortality in relation to smoking: Ten years' observations of British doctors. *Br Med J* 1964; 1:1399–1410.
4. Heller RF. Development of modern epidemiology: Clinical epidemiology. In *The Development of Modern Epidemiology: Personal Reports from Those Who Were There* (Eds. Holland WW, Olsen J, du V Florey C). Oxford, U.K.: Oxford University Press, 2007, p. 269.
5. Bruzzi P, Green SB, Byar DP, Brinton LA, Schairer C. Estimating the population attributable risk for multiple risk factors using case-control data. *Am J Epidemiol* 1985; 122:904–914.
6. Blais L, Ernst P, Suissa S. Confounding by indication and channeling over time: The risks of β_2-agonists. *Am J Epidemiol* 1996; 144:1161–1169.
7. Selvin S. *Statistical Analysis of Epidemiological Data*, 3rd edn. Oxford, U.K.: Oxford University Press, 2004.

8. Le CT. *Applied Categorical Data Analysis and Translational Research*, 2nd edn. New York: John Wiley & Sons, 2009.
9. Eriksen EM, Perlman JA, Miller A et al. Lack of association between hepatitis B birth immunization and neonatal death: A population based study from the vaccine safety datalink project. *Pediatr Infect Dis* 2004; 23:656–662.
10. Thompson S, Kaptoge S, White I et al. Statistical methods for the time-to-event analysis of individual participant data from multiple epidemiological studies. *Int J Epidemiol* 2010; 39:1345–1359.

15

Inference from Means

This chapter focuses on quantitative data. Blood pressure (BP), body mass index (BMI), parity, and pain score are examples of measurements that give rise to quantitative data. As noted earlier, some of these are discrete, that is, they can take one of only a small number of possible values, whereas others are continuous with theoretically infinite number of possible values within the specified range.

The summary measure under scrutiny in case of quantitative data is mostly mean. Mean can be calculated even for categorical data on a metric scale and is valid as long as the categories are many and have small width. Sampling distribution of mean depends on the pattern of distribution of values in the target population. Thus, forms such as the Gaussian distribution are especially important in drawing inferences from sample means.

Recollect that mean is an appropriate summary measure for quantitative data in most situations, but is not so in some. For example, when outliers are present, mean can be a highly distorted value. Another criticism mounted against mean is that an average patient does not exist. This is indeed true, but empirical evidence suggests that a large number of patients revolve around the average. Empirical conclusion is based on groups of subjects and applied to individual patients. The presumption is that individuals *mostly* behave as the group average suggests. Although each individual patient is managed on a personal basis, this requires guidelines, and these guidelines are obtained by group studies.

As always, because of interindividual variability and sampling fluctuation, the mean of any variable would differ from sample to sample. If the mean thyroxine (T_4) level in a sample of male thalassemic children is 0.65 ng/dL and in a sample of female children 0.56 ng/dL, can it be concluded that the T_4 level is affected by the gender? How can one be confident that another sample of male and female children will not give equal means? If these children are divided into several groups by growth pattern (normal, slightly retarded, moderately retarded, and severely retarded) and a difference in thyroid function tests is observed, can it be confidently stated that this difference would persist in repeated samples? Or, can it be concluded that this difference is genuinely present in such subjects in the target population and is not a chance occurrence in the sample?

Specifically, the objective could be to know (1) whether the mean of the target population from which the sample is drawn has a specified value; (2) when there are two groups for comparison such as test and control, whether the respective population means are different or differ by more than some specified medically relevant difference; and (3) when there are three or more groups, which specific group or groups are really different from others with respect to their means or whether they follow a particular pattern. Thus, a host of questions arise pertaining to the uncertainties inherent in sample means. Most of these can be satisfactorily answered by application of statistical methods discussed in this chapter.

This chapter: This chapter is divided into four sections. Section 15.1 is on comparing the mean of a group with a prespecified value and on comparing means of two groups. These comparisons are done with the help of Student *t*-test under Gaussian conditions. Section 15.2 is on comparison of means in three or more groups. The popular method of analysis of variance (ANOVA) is discussed in this section. The methods of these two sections are generally applicable to large *n* because this allows relaxation of the condition of Gaussian form of the distribution of the underlying variable. When *n* is small, these methods can be used only if the underlying distribution is Gaussian. Nonparametric methods, applicable to small *n* from a non-Gaussian distribution, are discussed in Section 15.3. The debate on statistical significance initiated in Chapter 12 is further discussed in Section 15.4. This may provide more insight into what statistical significance is all about.

15.1 Comparison of Means in One and Two Groups (Gaussian Conditions): Student *t*-Test

Two different problems are discussed in this section. First is comparing the mean in a sample of subjects with a prespecified value. Second is comparing the mean in one sample with that in another sample. These two samples could be paired or may represent two different groups. Gaussian conditions are presumed.

15.1.1 Comparison with a Prespecified Mean

Let the interest be in finding whether patients with chronic diarrhea have the same average hemoglobin (Hb) level as normally seen in healthy subjects in the area. Suppose the normal level of Hb is 14.6 g/dL. This is assumed to be known and fixed for the present example. Since chronic diarrhea can only decrease the Hb level and not increase it, it is a one-tail situation. In an unlikely event of sample mean being >14.6 g/dL, the evidence is immediate that Hb level does not decrease in chronic diarrhea patients, and there is no need to proceed further. A higher mean can occur by chance in the sample.

Suppose further that a random sample of 10 patients with chronic diarrhea is investigated and the average Hb level is found to be 13.8 g/dL. Thus, the sample mean is lower than normal. Since lower level is not uncommon even in healthy subjects, a lower mean can occur if the sample happens to include such subjects, and lower mean may not be necessarily due to diarrhea. If another sample of 10 patients is studied, the average could well be 14.8 g/dL. Can it be concluded with reasonable confidence on the basis of the sample with mean 13.8 g/dL that patients with chronic diarrhea indeed have a lower Hb level on average?

There is only one sample in this example, and the comparison is with the known average in the healthy subjects. It is a one-sample problem, although the comparison is of two means—one observed in the sample and the other known for the healthy population.

15.1.1.1 Student t-Test for One Sample

The null hypothesis in the preceding example is H_0: $\mu = 14.6$ g/dL. Since the possibility of a higher average Hb level in patients with chronic diarrhea is ruled out, the alternative hypothesis is one sided. That is, H_1: $\mu < 14.6$ g/dL. If H_0 is rejected, then H_1 is considered true.

According to the steps described in Section 12.3.3, there is a need to choose an appropriate criterion to test the hypothesis. The value of this criterion is then calculated assuming that H_0 is true. Then the probability of the observed or a more extreme value is obtained. This is the P-value. If this probability is very small, H_0 is considered not plausible and rejected. The conclusion reached is that H_1 is true. If the P-value is not sufficiently small, say, not less than 0.05, the null hypothesis is conceded. As mentioned before, this does not imply that H_0 is accepted. It is just that it cannot be rejected on the basis of that sample because sampling fluctuation is not adequately ruled out as a likely explanation.

Heuristically, the answer depends on the magnitude of difference between the sample mean and the known mean of the healthy subjects. In the preceding example, this difference is $13.8 - 14.6 = -0.8$ g/dL. This magnitude is assessed relative to the expected variation in means from sample to sample. This variation is measured by the SE of the mean, $\sigma/\sqrt{n}$. In a rare case when σ is known, the criterion as in Equation 12.9 is

$$\text{Gaussian test: } Z = \frac{\bar{x} - \mu_0}{\sigma/\sqrt{n}}.$$

This follows Gaussian distribution, provided the underlying distribution of xs is also Gaussian. Value of this criterion is compared with its value in Table B.1 to find if P-value is sufficiently small.

In practice, the SD σ would be rarely known and is replaced by its estimate s. Thus, the criterion for this setup is

$$\text{Student } t\text{-test (one sample): } t_{n-1} = \frac{\bar{x} - \mu_0}{s/\sqrt{n}}, \tag{15.1}$$

where μ_0 is the value of the mean under H_0. This criterion is valid under mild conditions. One of the conditions is that the observations are independent. In the context of this example, this means that the Hb level in one subject should not have any influence on the level in other subjects. This is clearly satisfied in this example, but examples are given later where independence is violated. The other condition is that distribution of xs themselves is Gaussian. You now know that this can be relaxed if sample size is large.

The higher the value of t, the greater is the chance that the sample has not come from a population with mean μ_0. In that case, the decision to reject H_0 would have less chance of error. When this probability of Type-I error is low, say, less than 0.05, H_0 can be safely rejected. As in the case of chi-square, the exact P-value for a particular value of t is provided by most of the standard statistical packages. Alternatively, consult Table B.2 to check whether P is less or not less than the threshold, such as 0.10, 0.05, or 0.01. As mentioned earlier, the distribution of t depends on degrees of freedom (df) (on the analogy that the distribution of BP depends on age). For t in the formula given in Equation 15.1, df $= (n - 1)$. This is specified as a subscript of t.

Example 15.1: Significance of Decrease in Hb Level in Chronic Diarrhea

Consider again the example of the Hb level in chronic diarrhea patients. Suppose that in a random sample of size $n = 10$, the levels in g/dL are as follows:

11.5	12.2	14.9	14.0	15.4	13.8	15.0	11.2	16.1	13.9

These give mean $\bar{x} = 13.8\,\text{g/dL}$ and SD $s = 1.67\,\text{g/dL}$.

The hypothesis under test is that the average Hb level in the patients with chronic diarrhea is the same $14.6\,\text{g/dL}$ that is normal in healthy subjects. Thus, H_0: $\mu = 14.6\,\text{g/dL}$. The alternative as already explained is H_1: $\mu < 14.6\,\text{g/dL}$. In this case, under H_0,

$$t_9 = \frac{13.8 - 14.6}{1.67/\sqrt{10}} = -1.51.$$

A statistical package gives $P(t < -1.51) = 0.0827$. This is the probability of getting the sample mean this much or more extreme in favor of H_1. Since this H_1 is one-sided, the probability required is also one-tailed. If an exact P-value is not available from a computer package, Table B.2 gives $P > 0.05$ for 9 df when t is 1.51. The distribution of t is also symmetric just as is the Gaussian distribution. Thus, $P(t < -a) = P(t > a)$. In this case, P is more than 0.05 because 1.51 is less than the critical value 1.833 in the table. Thus, the chance is more than 5% that H_0 is true. Therefore, this H_0 cannot be rejected. The difference between the sample mean 13.8 g/dL and the population-normal 14.6 g/dL is not statistically significant. This sample does not provide sufficient evidence to conclude that the mean Hb level in chronic diarrhea patients is less than normal.

The conclusion in Example 15.1 is partly the result of the high variability in the Hb level in the sample patients. Whereas it was only 11.2 g/dL in one patient, it was 16.1 g/dL in another. Widely scattered values gave a high value of sample SD s and led to the expectation of high intersample variability. Consequently, it became difficult to say anything definite about the lower mean Hb in the patients.

15.1.2 Difference in Means in Two Samples

Now, let us consider a situation where two samples are available. These could be from two groups, such as males and females, patients suffering from disease-A and disease-B, of age 20–39 years and age 40+ years, or could be from one group only, which is measured before and after a treatment. The latter is called a **paired samples** setup. The general form of the criterion in a two-sample setup is

Student t (two-sample)

$$= \frac{\text{Sample mean difference} - \text{Population mean difference under } H_0}{\text{Estimated SE of difference in sample means}}. \qquad (15.2)$$

This takes different form in paired setup than in independent samples setup.

15.1.2.1 Paired Samples Setup

The procedure used to calculate the SE of the difference in case of paired samples is different from that for unpaired samples. Observations are said to be paired when they are obtained twice for the same subject, such as BP before and after treatment or erythrocyte sedimentation rate (ESR) measured by two methods in the same group of subjects. They are also considered paired when the subjects in the two groups are one-to-one matched such as in some case–control studies. In the case of paired samples, obtain the difference between the pairs as $d_i = (x_{2i} - x_{1i})$, $i = 1, 2, \ldots, n$, where n is the number of pairs. Calculate the SD of these ds as usual by

$$s_d = \sqrt{\frac{\Sigma(d - \bar{d})^2}{n-1}}.$$

The null hypothesis of interest for rejection in this case is H_0: $\mu_1 = \mu_2$ (or $\mu_1 - \mu_2 = 0$), where μ_1 is the mean of one population and μ_2 is the mean of the other population. As already stated, these could be mean diastolic BP before and after treatment, mean ESR obtained by two methods, or mean of any other such population. Under this H_0, the criterion, from the formula given in Equation 15.2, is

$$\text{Student } t_{n-1} \text{ (paired samples)} = \frac{\bar{d}}{s_d/\sqrt{n}}. \tag{15.3}$$

This basically is the same as the one-sample formula in Equation 15.1 for H_0: $\mu_d = \mu_1 - \mu_2 = 0$. After the differences are obtained, the paired sample problem reduces to a one-sample problem for these differences with the null hypothesis that the mean difference is zero. When Student t in Equation 15.3 is significant, the point estimate of the difference $(\mu_1 - \mu_2)$ is $(\bar{x}_1 - \bar{x}_2)$. In the case of measurements before and after treatment, this is the estimate of the treatment effect.

Example 15.2: Paired-t for Mean Albumin Level in Dengue

Following are the serum albumin levels (g/dL) of six randomly chosen patients with dengue hemorrhagic fever before and after treatment. The null hypothesis is that the mean after is the same as the mean before, that is, the treatment of dengue fever (this treatment is mostly symptomatic) has not altered the albumin level.

Before treatment	4.8	4.1	5.3	3.9	4.5	3.8
After treatment	5.2	4.9	5.2	4.8	4.6	4.4
In this case, difference, d_i:	0.4	0.8	−0.1	0.9	0.1	0.6

Mean differences, $\bar{d} = 0.45$, and SD of differences, $s_d = 0.3937$. Thus, under H_0: $\mu_1 = \mu_2$,

$$t_5 = \frac{0.45}{0.3937/\sqrt{6}} = 2.80.$$

Since there is no assertion in this case that the albumin level after the treatment will increase or decrease, the alternative hypothesis is H_1: $\mu_1 \neq \mu_2$. For this H_1, a two-tailed probability $P(|t| > 2.80)$ is needed. For $(n - 1) = 5\,df$, this is $P = 0.038$ from statistical software. Also, since 2.80 is more than the critical value 2.571 of t at $5\,df$ in Table B.2 for one-tail probability 0.025, the two-tail $P < 0.05$. Because the probability of a Type-I error is sufficiently small, reject H_0 and conclude that the mean albumin level after the treatment is different from the mean before the treatment.

15.1.2.2 Unpaired (Independent) Samples Setup

Paired observations are highly desirable in the situation of Example 15.2, but it is possible that the albumin investigations could not be done before the treatment started. Suppose that a separate group of six similar but randomly drawn patients was investigated for albumin level before the treatment. Thus, there is a total of 12 patients—6 in each group. The two groups are now independent. In such cases, the first step is to check that the variances in the two groups are not widely different. Generally, a ratio $s_1^2/s_2^2 < 3$ is considered adequate if each n is around 10 or 15. If n is smaller, even $s_1^2/s_2^2 < 4$ can be tolerated. If n is 30 or more, a ratio of 3 may be too high. The conventional statistical test for H_0: $\sigma_1^2 = \sigma_2^2$ is $F = s_1^2/s_2^2$ with higher SD in the numerator. However, it is necessary for this test that the underlying distribution is Gaussian, at least approximately. Prefer Levene test [1] for equality of variances because it is valid even in case of departure from the Gaussian pattern. Student t-test is applicable even when the underlying distribution is very different from Gaussian, provided n is sufficiently large. Levene test is also applicable in this situation. Now there are two possibilities.

1. *Population variances are equal*: In this case, the samples can be combined to obtain a more reliable estimate of the variance. This is given by

$$\text{Pooled variance: } s_p^2 = \frac{(n_1 - 1)s_1^2 + (n_2 - 1)s_2^2}{(n_1 - 1) + (n_2 - 1)}, \qquad (15.4)$$

   where
   s_1 is the SD in the first sample and s_2 in the second sample
   n_1 is the size of the first sample and n_2 of the second sample

Now, calculate

$$\text{SE(mean difference)} = s_p \sqrt{\frac{1}{n_1} + \frac{1}{n_2}}. \qquad (15.5)$$

From the formula given in Equation 15.2, the criterion for testing H_0: $\mu_1 = \mu_2$ is

$$\text{Student } t \text{ (two independent samples): } t_{n_1 + n_2 - 2} = \frac{\bar{x}_1 - \bar{x}_2}{\sqrt{s_p^2(1/n_1 + 1/n_2)}}, \qquad (15.6a)$$

where $\bar{x}_1$ and $\bar{x}_2$ are the respective sample means. The df for t in the formula given in Equation 15.6a is $(n_1 + n_2 - 2)$.

2. *Population variances are unequal*: You would be rarely interested in equality of variances per se. Statistically, it is this kind of nuisance parameter that we have to deal with anyway. Realize though that when means are very different, there is a good likelihood that variances would also differ. Group with higher mean may have higher variance too. In that sense, equality of variances becomes a major issue. If the population variances are known to be unequal or if the sample variances are very different from one another, then the samples cannot be pooled and the formula given in Equation 15.4 is not valid. Now calculate

Separate variance Student t (two independent samples):

$$t_v = \frac{\bar{x}_1 - \bar{x}_2}{\sqrt{s_1^2/n_1 + s_2^2/n_2}}, \tag{15.6b}$$

where the df

$$v = \frac{(s_1^2/n_1 + s_2^2/n_2)^2}{(s_1^2/n_1)^2/(n_1 - 1) + (s_2^2/n_2)^2/(n_2 - 1)}.$$

The case of unequal variances in two samples is known as the Beherens–Fisher problem in statistical literature. This is now popularly known as **Welch test**. Note that this reduces to the usual Student t in case of equal n and equal variances.

Example 15.3 Unpaired-t for Albumin Level in Dengue

Consider the same data as in Example 15.2, but now the observations are for 12 different patients in place of pairs of observations on 6 patients. The patients measured before treatment are not the same as those measured after the treatment. In this case,

Mean albumin level in before treatment group, $\bar{x}_1 = 4.40\,\text{g/dL}$
SD in this group, $s_1 = 0.5797\,\text{g/dL}$
Mean albumin level in after treatment group, $\bar{x}_2 = 4.85\,\text{g/dL}$
SD in this group, $s_2 = 0.3209\,\text{g/dL}$

In this example, $n_1 = 6$ and $n_2 = 6$ so that df $= n_1 + n_2 - 2 = 10$. For such small samples, the SDs do not differ too much and we can pool them. Thus,

$$s_p^2 = \frac{5 \times 0.5797^2 + 5 \times 0.3209^2}{10} = 0.2195.$$

Therefore,

$$t_{10} = \frac{4.40 - 4.85}{\sqrt{0.2195(1/6 + 1/6)}} = -1.66.$$

The alternative hypothesis continues to be two sided. The probability of getting these sample values or more extreme in favor of H_1 when H_0 is true is $P(|t| > 1.66)$. From a statistical package, this P-value is 0.1272, and from Table B.2, $P > 0.10$ because 1.66 is less than the critical value 1.812 for two-tailed $P = 0.10$ (corresponding to one-tail $P = 0.05$ in Table B.2) for 10 df. This P-value is large. Thus, the null hypothesis of equality of means cannot be rejected. The evidence is not strong enough to conclude that the mean albumin level after the treatment is any different from the mean before the treatment.

15.1.2.3 Some Features of Student t

Examples 15.2 and 15.3 very aptly illustrate the following features of Student t:

1. The t-test is based on magnitude of difference and its variance but ignores the fact that five of the six patients in the paired setup in Example 15.2 showed some rise. If the interest is in the *proportion* of subjects showing a rise and not in the magnitude of the rise, then use the methods of Chapter 13.

2. The same difference is statistically significant in the paired setup in Example 15.2 but not significant in the unpaired setup in Example 15.3. This occurred because the difference in the paired setup is fairly consistent, ranging from -0.1 to $+0.9\,g/dL$. Each patient served as their own control. In the unpaired setup, the inter-individual variation is large. A paired setup may be a good strategy in many situations. Earlier advice of matching of controls with cases is for this reason. Such matching simulates pairing and reduces the effect of at least one major source of uncertainty.

3. The size of the sample is 6 in the paired setup and a total of 12 in the unpaired setup. Both are small. Recall that Student t is valid only when means follow a Gaussian pattern. When n is large, this pattern is nearly always Gaussian due to the central limit theorem, whether or not the underlying distribution of the individual measurements is Gaussian. Where n is small (say, less than 30), the t-test is valid only if the underlying distribution is Gaussian. This is the assumption made in these examples. If the underlying distribution is far from Gaussian and n is small, other methods such as nonparametric (e.g., Wilcoxon) tests are used. Some of these methods are presented in Section 15.3.

4. In Example 15.3, one should examine whether the assumption $\sigma_1^2 = \sigma_2^2$ is violated. The ratio s_1^2/s_2^2 in this example is $0.5797^2/0.3209^2 = 3.26$. For a sample of size 6 each, this may be within the tolerance limit. Many statistical software packages would automatically test this. In case of violation, the P-value is obtained by using Welch t as given in Equation 15.6b in place of the pooled variance t given

in Equation 15.6a. You may wish to calculate this and examine whether a different *P*-value is obtained. Some statistical softwares would also do this automatically or would provide *P*-values based on both types of *t*. However, caution is required when using a separate variance estimate. When the variances are different, it is clear that the two populations are different. Then, equality of means may not be of much consequence. Even if means are equal, the distribution pattern is different. If it is known, for example, that BMI is much more variable in women than in men, equality of their means would rarely help. Thus, use of a separate variance estimate in a two-sample *t* is rare.

5. As mentioned in Chapter 8, in some cases, for example, in estimation of antibody titers, the conventional (arithmetic) mean is not applicable and the geometric mean (GM) is used to measure the central value. The logarithm of GM has the same features as the usual arithmetic mean. Thus, Student *t*-test and other mean-based tests can be carried out on GMs after taking the logarithm of the values. However, in this case, the conclusions will be applicable to log values and not to the values themselves. You should carefully examine whether or not these results can be extended to the original values. In many cases, such extension does not cause any problem.

6. All statistical tests give valid conclusions for the groups and not for individuals. The *t*-test is for *average* values in the groups. Individuals can behave in an unpredictable way even when the difference in means is statistically significant. Thus, use caution in applying results to individual subjects. Clinical features of a person may completely override the mean-based conclusions.

7. It is often mentioned in examples in this text that the samples are random. Statistical inference is not valid for nonrandom samples.

8. The fundamental requirement for Student *t*-test is that the sample values are independent of each other. In Example 15.3, the albumin level in one patient is not going to affect the level in any other patient. Thus, the values are independent. In Example 15.2, there is no such independence because of paired setup, but once the difference is obtained, these differences would be independent. Paired *t*-test is based on the differences.

Consider BP measured for two or more subjects belonging to the same family. Familial aggregation is well known, and thus values belonging to the members of the same family are not independent. Student *t*-test cannot be used for such values unless family effect is first removed. Ingenious methods may be needed for removing this effect. Most practical situations do not have this constraint and Student *t* can be safely used.

15.1.2.4 Effect of Unequal n

Student *t*-test for two independent samples does not have any restriction on n_1 and n_2—they can be equal or unequal. However, equal *n*s are preferred because of two reasons: (1) When a total of 2*n* subjects are available, their equal division among the groups maximizes the power to detect a specified difference, and (2) two-sample *t* is not robust to $\sigma_1^2 \neq \sigma_2^2$ unless $n_1 = n_2$. If smaller sample has larger variance, the problem is aggravated. To reiterate what was stated earlier, if you have prior knowledge or find from the data that $\sigma_1^2 \neq \sigma_2^2$, further testing for equality of means may not be relevant since the distributions are difference anyway owing to different SDs.

Although equal *n*s are desirable, in many medical situations, this may not be a prudent allocation. In clinical trials, many times controls are easy to investigate and more than one control per case could be a good strategy as discussed in an earlier chapter.

15.1.2.5 Difference-in-Differences Approach

A popular method with social scientists, now making its way into medical sciences, is to test significance of difference in differences. This is used when paired observations are available from two independent groups. Suppose the test group is measured before and after the treatment as well as a control group before and after a placebo. If the corresponding population means are μ_{1T} (before treatment), μ_{2T} (after treatment), μ_{1C} (before placebo in the control), μ_{2C} (after placebo in the control), then the actual treatment effect is

$$\text{Difference in differences: } (\mu_{2T} - \mu_{1T}) - (\mu_{2C} - \mu_{1C}),$$

assuming that after values are higher. Statistical significance of this can be tested by unpaired *t* after obtaining the differences in the two groups. If this difference in differences is found significant, the estimate of the treatment effect is obtained by substituting the corresponding sample means.

15.1.3 Analysis of Crossover Designs

As mentioned earlier, crossover could be a very efficient strategy for trials on regimen that provide temporary relief. Hypertension, thalassemia, migraine, and kidney diseases requiring repeated dialysis are examples of conditions that are relieved temporarily by the currently available regimens. The crossover design economizes on subjects because the same subject is used for trials on two regimens. However, each subject should be available twice as long as for a routine trial. Comparison is within subjects and therefore more precise.

In a crossover design, one group receives regimen A then B (AB sequence) and the other group receives regimen B then A (BA sequence). This sequence itself can cause differential effect. This can particularly happen when the intervening period is favorable to one regimen than the other. For example, when lisinopril is given first and then losartan after a 2-week washout gap to nondiabetic hypertensive patients for insulin sensitivity, the first drug may still be in the system. This may not happen when losartan is given first. Crossover design is not appropriate for such regimens that have significant sequence effects.

The primary objective of a crossover trial is to test the significance of the difference in effects of the regimens. The method described earlier in Chapter 13 is for qualitative (binary) data. Now, this section presents the method for quantitative data. Primary difficulty in analyzing data from crossover designs arises from duplicity of factors. If there are three patients, two periods and two treatments, a crossover could yield only six observations instead of $3 \times 2 \times 2 = 12$ in a conventional design. Two factors determine the third. If patient #1 gets AB sequence, the observation for this patient in period-2 must be under treatment-B. All this is explained with the help of an example in the following paragraphs.

Consider a trial on $n = 16$ chronic obstructive pulmonary disease (COPD) patients who were randomly divided into two equal groups of size 8. The first group received treatment-A and then treatment-B, and the second group received treatment-B and then treatment-A. Abbreviate them as trA and trB. An adequate washout period was provided before switching the treatment so that there was no carryover effect. The response variable is forced expiratory volume in one second (FEV_1). The data obtained are in Table 15.1.

15.1.3.1 Test for Group Effect

In the case of crossover trials, the groups identify the sequence and the group effect is the same as the sequence effect. If sequence does not affect the values, the mean difference between trA and trB should be the same in AB group as in BA group. A significant effect means that trA has different effect when in period-1 then when in period-2. Thus, this is also called

TABLE 15.1

FEV_1 in COPD Patients in a Crossover Trial

					Group-1—AB Sequence				
Subject No.		**1**	**2**	**3**	**4**	**5**	**6**	**7**	**8**
$FEV_1(L/s)$									
Period-1	trA	1.28	1.26	1.60	1.45	1.32	1.20	1.18	1.31
Period-2	trB	1.25	1.27	1.47	1.38	1.31	1.18	1.20	1.27
					Group-2—BA Sequence				
Subject No.		**9**	**10**	**11**	**12**	**13**	**14**	**15**	**16**
$FEV_1(L/s)$									
Period-1	trB	1.27	1.49	1.05	1.38	1.43	1.31	1.25	1.20
Period-2	trA	1.30	1.57	1.17	1.36	1.49	1.38	1.45	1.20

treatment–period interaction. Note for crossover that group effect = sequence effect = treatment * period interaction. In this example,

trA – trB Values								
Group-1 (AB)	+0.03	−0.01	+0.13	+0.07	+0.01	+0.02	−0.02	+0.04
Group-2 (BA)	+0.03	+0.08	+0.12	−0.02	+0.06	+0.07	+0.20	+0.00

Since the patients in group-1 are different from patients in group-2, equality of means in these groups can be tested by the two-sample *t*-test. In this case, for these differences,

$$\bar{x}_1 = 0.03375 \qquad \bar{x}_2 = 0.06750$$
$$s_1^2 = 0.0023125 \qquad s_2^2 = 0.0048786 \qquad s_p^2 = 0.0035955$$

Thus,

$$t_{14} = \frac{0.03375 - 0.06750}{\sqrt{0.0035955(1/8 + 1/8)}} = -1.126.$$

This is not statistically significant ($P > 0.05$). Thus, the evidence is not enough for presence of sequence effect. The practical implication is that there is no sequence effect although this can be ascertained only when the sample is large and the power is high. If a sequence effect is present, the reasons should be ascertained and the trial done again after eliminating those causes. In most practical situations where a crossover trial is used, the sequence of administering the drugs does not make much difference. The real possibility is that of a carryover effect that can also make a dent in the sequence effect.

15.1.3.2 Test for Carryover Effect

If a positive carryover effect were present, the performance of a regimen in period-2 would be better than its performance in period-1. Thus, the presence of a carryover effect can be assessed by comparing the performance of each regimen in the two periods. In the preceding example, this is obtained by comparing trA values in period-1 with trA values in period-2. Similarly for trB. It is possible that only one of the regimens has a long-term effect so that carryover is present for that regimen and the other has no such effect. Two two-sample *t*-tests would decide whether one or both have a carryover effect. In this example, these are as follows:

trA			
Period-1	Mean = 1.325	SD = 0.1386	Pooled variance
Period-2	Mean = 1.365	SD = 0.1387	0.019224

$$t_{14} = \frac{1.365 - 1.325}{\sqrt{0.019224(1/8 + 1/8)}} = 0.577.$$

trB			
Period-1	Mean = 1.298	SD = 0.1391	Pooled variance
Period-2	Mean = 1.291	SD = 0.0952	0.014206

$$t_{14} = \frac{1.291 - 1.298}{\sqrt{0.014206(1/8 + 1/8)}} = -0.117.$$

P-values associated with these values of *t* show that the carryover effect is not statistically significant for any of the treatments in this example.

15.1.3.3 Test for Treatment Effect

The two tests just mentioned are preliminaries. The primary purpose of the trial, of course, is to find whether one treatment is better than the other. This can be done only when the sequence (or the group) effect is not significant.

a. *If there is no carryover effect*, the procedure is as follows. Consider two groups together as one because the sequence is not important and there is no carryover effect. Calculate differences trA − trB and use paired *t*-test given in Equation 15.3 on the joint sample. In this example, these differences are

+0.03	−0.01	+0.13	+0.07	+0.01	+0.02	−0.02	+0.04
+0.03	+0.08	+0.12	−0.02	+0.06	+0.07	+0.20	0.00

These give mean = 0.0506 and s_d = 0.06049. Thus,

$$t_{15} = \frac{0.0506}{0.06049/\sqrt{16}} = 3.346.$$

From Table B.2, for 15 df, this gives $P < 0.01$. Thus, statistically the treatment difference is highly significant.

b. *If carryover effect is present*, crossover is not a good strategy. You may then increase the washout period and ensure that no carryover effect is present. If the data from a crossover trial are already available and a carryover effect is found to be significant, then analyze as usual by Student *t* after ignoring the second period. Thus, half of the data (and efforts) will become redundant and the advantages of the crossover design lost.

Remember the following for crossover designs:

1. It is easy to say that a washout period will eliminate a carryover effect. In fact, it can rarely be dismissed on a priori grounds. A psychological effect may persist even in case of blinding. Thus, a crossover design should be used only after there is fair assurance that a carryover effect is practically absent.

2. The test for a carryover effect is based on variation between subjects and thus has less power. A small but real effect may not be detected unless a big trial with a large number of subjects is done. Then, the advantage of the economy of subjects in a crossover design may be lost.

3. Details are given later, but carrying out so many statistical tests on the same set of data increases the total chance of Type-I error. To keep this under control to, say, less than 5%, you should carry out the three *t*-tests at the 2% level each.

4. The preceding is an easy method based on Student *t*-test. This is not so elegant. Quantitative data from crossover trials can be analyzed more meticulously by using the ANOVA method discussed later (although not for crossover trials). For ANOVA-based analysis of crossover trials, see Everitt [2]. Neither methodology is fully standardized.

5. The procedure given in this section assumes fixed treatment effect. That is, it is not allowed to vary from subject to subject.

15.1.4 Analysis of Data of Up-and-Down Trials

The details of the up-and-down method of clinical trials are given in Chapter 6. This method can be used when the response is of the yes/no type. For this kind of trial, fix a dose step that you think would be appropriate escalation or reduction and also guess a median dose with your clinical acumen or past experience. Give this anticipated median dose to an eligible patient and assess if it is effective. If effective, step down the dose by the prior fixed step, if not effective, step up the dose for the second patient, and so on. Generally, a trial on 10–15 patients is considered adequate. An essential prerequisite for this kind of trial is that such variation in dose should not be unethical.

The analysis of data from an up-and-down trial is in terms of estimating least effective dose and obtaining a confidence interval (CI). Generally, the least effective dose is defined as the same as median effective dose. The procedure for obtaining the median effective dose is as follows:

Step 1. Calculate the logarithm for all the doses you tried from minimum to maximum. Find the mean of the differences of successive log doses. Denote this mean difference by $\bar{d}$.

Step 2. Separately list doses that were effective in different patients. Exclude the ineffective ones. Calculate the mean of logarithm of these doses and SD of these logarithms. Denote them by $\bar{y}$ and s_y, respectively.

Step 3. Median effective dose $= \exp(\bar{y} - 1/2 * \bar{d})$.

Calculation of CI requires a constant G, which is obtained from a figure given by Dixon and Massey [3]. See this reference for the method to obtain the CI.

Example 15.4: Up-and-Down Method for Estimating Minimum Effective Dose of a Local Anesthetic Agent

There is considerable uncertainty regarding minimum dose requirement of local anesthetics administered intraspinally for surgery. Sell et al. [4] conducted a study to estimate minimum effective dose of levobupivacaine and ropivacaine in hip replacement surgery.

Forty-one patients were randomly allocated to one of the two local anesthetic groups in a double-blind manner. The authors used the up-and-down strategy, determining the initial dose on previous experience. The minimum effective dose was defined as the median effective dose. The up-and-down method found minimum effective dose of levobupivacaine as 11.7 mg (95% CI, 11.1–12.4) and that of ropivacaine 12.8 mg (95% CI, 12.2–13.4). The authors concluded that these doses are smaller than those reported earlier for single-shot anesthesia.

15.2 Comparison of Means in Three or More Groups (Gaussian Conditions): ANOVA *F*-Test

Statistical methods for evaluating significance of difference in means in three or more groups remain conceptually simple but become mathematically complex. As always in this text, I avoid complex mathematical expressions and concentrate on explanations that may help in understanding the basic concepts, in being more judicious in choosing an appropriate method for a particular set of data, in realizing the limitations of the methods, and in interpreting the results properly.

Student *t*-test of the previous section is valid for almost any underlying distribution if n is large but requires a Gaussian pattern if n is small. A similar condition also applies to the methods of comparison of means in three or more groups. The test criterion now used is called *F*. As in the case of Student *t*, this test also requires that the variance in different groups is nearly the same. This property is called **homoscedasticity**. The third, and more important, prerequisite for validity of *F* is independence of observations. Serial measurements, taken over a period of time on the same unit, generally lack independence because a measurement depends on its value on the previous occasions. A different set of methods, called hierarchical or repeated measures and akin to a paired-*t*, is generally applied for analyzing such data. Random sampling is required, as always, for validity of conclusions from the *F*-test.

The generic method used for comparing means in three or more groups is called ANOVA. The name comes from the fact that the total variance in all the groups combined is partitioned into components such as within-groups variance and between-groups variance. Between-groups variance is the systematic variation occurring due to group differentials. For example, ANOVA may reveal that 60% variation in P3 amplitude in healthy adults is due to genetic differentials, 10% due to age differentials, and the remaining 30% due to other factors. Such residual left after the extraction of the factor effects is considered a random component arising due to intrinsic biological variability between individuals. This is the within-groups variance. If genuine group differentials are present, the between-groups variance should be large relative to the within-groups variance. Thus, the ratio of these two components of variance can be used as a criterion to find whether the group means are different. Details for common setups are given next.

15.2.1 One-Way ANOVA

Consider a study in which plasma amino acid (PAA) ratio for lysine is calculated in healthy children and in children with undernourishment of grades I, II, and III. This ratio is the difference in PAA concentration in blood before and after the meal, expressed as a percentage of the amino acid requirement. There are four groups in this study. The setup is called one way because no further classification of subjects, say, by age or gender, is sought in this case. Groups define the factor: in this case, grade of undernourishment. The response is a quantitative variable. This is a ratio in this example but can still be considered to follow a Gaussian pattern within each group. When other factors are properly controlled, the difference in PAA ratio among subjects would be due to either the degree of undernourishment or the intrinsic inter-individual variation in the subjects in different groups. The former is the between-groups variation and the latter the within-groups variation. These variations are illustrated in Figure 15.1 for a variable observed for four different groups. Group means are denoted by $\bar{y}_{.j}$ ($j = 1, 2, 3, 4$) and are represented by a circle. Within-groups variation is the difference between individual values and their respective group mean. This is summed over the groups. The overall mean is denoted by $\bar{y}_{..}$ and is represented by a line in this figure. Between-groups variation is the difference between group means and the overall mean.

Note that part of the within-groups variation in the preceding example can be due to factors such as heredity, age, gender, height, and weight of the children. But these are assumed under control and disregarded in this setup. All within-groups variation is considered intrinsic and random. In an ANOVA setup, this is generally called **residual or error variance**. This is also called mean square error and popularly written as MSE. The term error does not connote any mistake but stands only for random component, as for sampling error.

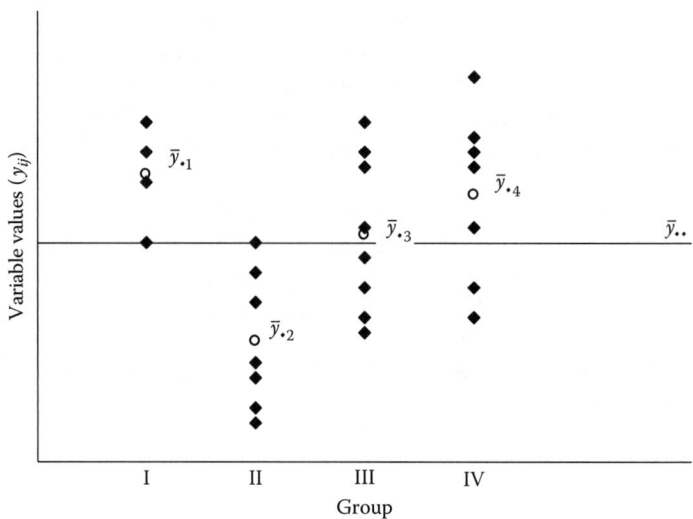

FIGURE 15.1
Graphical display of within-groups and between-groups variances.

If group differences were not really present, the between-groups variance (call it mean square between groups and write it as MSB) and the within-groups variance would both arise from intrinsic variation alone and will be nearly equal. The ratio of these two, with between-groups variance in the numerator, is the criterion F (for this reason, this is also called **variance ratio**). A value of F substantially more than one implies that between-groups variation is large relative to within-groups variation. This is an indication that the groups are indeed different with respect to the mean of the variable under study.

15.2.1.1 Procedure to Test H₀

It was assumed earlier that all the groups have the same variance. This is a prerequisite for validity of the usual ANOVA procedure. It is also stipulated that the pattern of distribution of the response variable is the same. This could be of almost any form if the number of subjects in each group is large. One possibility is shown in Figure 15.2. The distribution in all four groups in this figure is identical except for a shift in location. Groups B and C are close to one another, but C and D are far apart. Only the means differ and other features of the distribution are exactly the same. If n is small, the ANOVA procedure is valid only when this distribution is Gaussian. Thus, skewed distributions of the type shown in Figure 15.2 are admissible for ANOVA only when n is large.

The null hypothesis in the case of one-way ANOVA is

$$H_0 : \mu_1 = \mu_2 = \cdots = \mu_J.$$ (15.7)

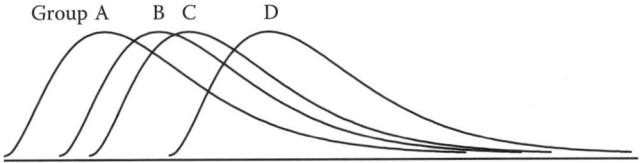

FIGURE 15.2
Distribution in four groups differing only in mean.

This says that there are J groups and the means in all groups are the same. This hypothesis, in conjunction with the conditions mentioned in the preceding paragraph, implies that the response variable has the same distribution in different groups. The alternative hypothesis is that at least one mean is different.

When H_0 stated in Equation 15.7 is true, MSB is also an estimate of the population variance σ^2. MSE is an estimate of σ^2 whether or not H_0 is true. The ratio

$$F = \frac{\text{MSB}}{\text{MSE}} \qquad (15.8)$$

is expected to be one under H_0. When $F \leq 1$, it is surely an indication that group means can be equal. If group means are different, MSB would be large and $F \gg 1$ (substantially more than 1). Just like χ^2 and t, the distribution of F under H_0 is known. In place of a single df, the exact shape of F depends on a pair of df, namely, $(J - 1)$, $J(n - 1)$ in case of the one-way ANOVA. The first corresponds to the numerator of the criterion given in Equation 15.8 and the second to the denominator. In general, these dfs are denoted by (v_1, v_2), respectively. The probability P of wrongly rejecting H_0 can be obtained corresponding to the value of F calculated from the data. If $P < 0.05$, the evidence can be considered sufficient to reject H_0. Standard statistical packages provides a exact P-value so that a decision can be made immediately. Cutoff values of F for $\alpha = 0.05$ and different dfs are given in Table B.4. These can be used if you do not have access to a statistical package.

The following comments provide further information on one-way ANOVA:

1. The primary purpose of ANOVA is to test the null hypothesis of equality of means. However, as shortly explained, an estimate of the effect size of the groups can also be obtained as a by-product of this procedure.

2. ANOVA was basically developed for evaluating data arising from experiments. As explained in Chapters 5 and 6, this requires random allocation of the subjects to different groups and then exposing these groups to various interventions. In the PAA example, the groups are preexisting and there is no intervention. The design is

prospective because the undernourishment groups define the antecedent and the PAA ratio is investigated as an outcome. The subjects are not randomly allocated to the different undernourishment groups. But it is important to ensure that the subjects in each group adequately represent their group and that there is no other factor, except the undernourishment grade, that separates the groups. Such precautions in selection of subjects are necessary for a valid conclusion from ANOVA.

3. Although the ANOVA method has been described here for comparing means in three or more groups, the method is equally valid for $J = 2$ groups. In fact, there is a mathematical relationship that says that two-tailed t^2 for two independent samples of size n is the same as F with $(1, 2n - 2)$ df. Note that the df of t for two independent samples, each of size n, is $(2n - 2)$. Both methods lead to the same conclusion, but F is a two-tailed procedure whereas t can also be used for a one-sided alternative. This flexibility is not available with the F-test.

Example 15.5: One-Way ANOVA for the Effect of Various Drugs on REM Sleep Time in Rats

Tokunaga et al. [5] conducted a study for effects of some h(1)-antagonists on the sleep–wake cycle in sleep-disturbed rats. Among the response variables was rapid eye movement (REM) sleep time. Take a similar example of a sample of 20 rats, homogeneous for genetic stock, age, gender, etc., who are randomly divided into $J = 4$ groups of $n = 5$ rats each. Let one group be the control and the others receive drug-A (diphenhydramine), drug-B (chlorpheniramine), and drug-C (cyproheptadine). The REM sleep time was recorded for each rat from 10:00 to 16:00 h. Suppose the data shown in Table 15.2 are obtained.

These data show a large difference in the mean sleep time in various groups. Another experiment on a new sample of rats may or may not give similar results. The likelihood of getting nearly equal means in the long run is extremely remote, as will be clear shortly.

To avoid the burden of calculations, given in the following are values obtained from a statistical package for these data.

TABLE 15.2

REM Sleep Time in Sleep-Disturbed Rats with Different Drugs

Drug	REM Sleep Time (min)					Mean (min)
0 (control)	88.6	73.2	91.4	68.0	75.2	79.28
A	63.0	53.9	69.2	50.1	71.5	61.54
B	44.9	59.5	40.2	56.3	38.7	47.92
C	31.0	39.6	45.3	25.2	22.7	32.76

TABLE 15.3

ANOVA Table for the Data in Table 15.2

Source of Variation	df	Sum of Squares	Mean Squares	F
Drug	3	5882.4	1960.8	21.09
Error	16	1487.4	93.0	
Total	19	7369.8		

$$\text{SST} = 7369.8, \ \text{SSB} = 5882.4, \ \text{and SSE} = 1487.4, \tag{15.9}$$

where
SST is the total sum of squares
SSB is between-groups sum of squares
SSE is error (within-groups) sum of squares

Mathematical expressions of these are being avoided due to complexity. However, note that SST = SSB + SSE in one-way setup.

The df for SSB are $(4 - 1) = 3$ and for SSE are $4(5 - 1) = 16$. Mean squares are obtained by dividing these sums of squares by the corresponding dfs. These values, as well as F, are conventionally shown in the form of a table, popularly called an **ANOVA table**. This is given in Table 15.3 for this example. For these data,

$$F = \frac{5882.3/3}{1487.4/16} = 21.09.$$

The null hypothesis is that there is no effect of dose on REM sleep time. This is the same as saying that means in all four groups are equal.

The df of this F are (3, 16). A statistical package gives $P < 0.001$ under H_0 for $F = 21.09$ or higher. That is, there is less than one in a thousand chance that equal means in groups will give a value of F this large. Therefore, these data must have come from populations with unequal means. In other words, another experiment on the same kind of animals is extremely unlikely to give equal means. The evidence is overwhelming against the null and it is rejected. The conclusion is that different drugs do affect the REM sleep time differentials.

SIDE NOTE: In this example, mean REM sleep time in different drugs not only differs but also follows a trend. Had the groups represented increasing dose levels, it would have implied decline in sleep time as the dose level is increased. The conventional ANOVA just illustrated allows conclusion of different means in different groups but not of any trend. Evaluation of trend in means is discussed in Chapter 16.

In Example 15.5, there are only five rats in each group. Statistically, this is an extremely small sample. The reason that such a small sample can still provide a reliable result is that the laboratory conditions can be standardized and most factors contributing to uncertainty can be controlled. The rats can be chosen to be homogeneous, as in this experiment, so that intrinsic factors such as genetic makeup, age, and gender do not influence the outcome. Random allocation tends to average out any effect of other factors that are not considered in choosing the animals. The influence of body weight, if any, is taken care of by adjusting dose for body weight. Thus, interindividual variation within groups is minimal. On the other hand, the variation between groups is very large in this case, as is evident from the large difference between the means. This provided clinching evidence in favor of the alternative hypothesis.

Among cautions in using ANOVA are the following: (1) The ANOVA is based on means. Any mean-based procedure is severely disturbed when outliers are present. Thus, ensure before using ANOVA that there are no outliers in your data. If there are, examine whether they can be excluded without affecting the conclusion. (2) A problem in comparison of three or more groups by criterion F is that its significance indicates only that a difference exists. It does not tell exactly which group or groups are different. Further analysis, called multiple comparisons, is required to identify the groups that have different means. This is discussed later in this chapter. (3) The small numbers of subjects in different groups in Example 15.5 should put you on alert regarding the pattern of distribution of the measurements and of their means. It should be Gaussian. The validity of other assumptions that underlie an ANOVA F-test should also be checked according to the guidelines given next. (4) When no significant difference is found across groups, there is a tendency to locate a group or even subgroup that exhibits benefit. This post hoc analysis is alright as long as it is exploratory in nature. For conclusion, a new study should be conducted on that group or subgroup.

15.2.1.2 Checking the Validity of the Assumptions of ANOVA

When the assumptions are violated, the P-values are suspect. Important assumptions for the validity of ANOVA are (1) Gaussian pattern, (2) homoscedasticity, and (3) independence. These are actually checked for the residuals and not for the original observations. Residuals are the remainders left after the factor effects are subtracted from the observed values. These are explained in detail in Chapter 16 in the context of regression.

Of the three assumptions for the validity of ANOVA, the assumption of a **Gaussian pattern** is not a strong requirement. The ANOVA F-test is quite robust to minor departures from a Gaussian pattern. A gross violation, if present, can be detected by the methods given in Section 9.2.1. A crude but easy method is to confirm that neither (maximum − mean)/SD nor (mean − minimum)/SD is less than 2 or more than 4. The normal practice is to continue to use F unless there is extraneous evidence that the observations follow a non-Gaussian pattern. When the pattern is really far from Gaussian, it is advisable to use nonparametric methods of Section 15.3.

This is particularly important when n is small. However, always check for outliers. These tend to throw away the entire results out of gear. You should routinely check the data and decide whether to keep an outlier or not. Genuine outliers cannot be disregarded.

If the distributions are non-Gaussian but similar in shape such as positively skewed for all groups, ANOVA may still be a valid method for large n.

Homoscedasticity is the equality of variances in different groups. This requirement should be fairly met. Check this graphically by box-plot (Figure 8.8b) for different groups. Varying height of the boxes would indicate different variances. Statistically, the variation must be substantial for violation of homoscedasticity. Generally, the largest variance should be no more than four times the smallest. The conventional statistical test for checking homoscedasticity is Bartlett test. This is heavily dependent on Gaussian pattern of the distribution of y. For this reason, many softwares now use Levene test as mentioned for two-sample t-test. This is based on median and thus more robust to departure from Gaussianity. Transformation of the data, such as logarithm $(\ln y)$, square (y^2), and square root $(\sqrt{y})$, is tried in cases where violation occurs. Experience suggests that a transformation can be found that converts a grossly non-Gaussian distribution to an approximately Gaussian pattern and, at the same time, stabilizes the variance across groups. If the distribution pattern is already Gaussian and the test reveals that the variances in different groups are significantly different, then also the F-test should not be used. In fact, as stated earlier, there may not be much reason in doing the test for equality of means when the variances are found different. It is rare that the variance of a variable is different from group to group but the mean is the same. The nature of the variable and measuring techniques should ensure roughly comparable variances. Difference in variances is itself evidence that the populations are different. However, in the rare case in which the interest persists in equality of means despite different variances, try transformation of the data as suggested before. But this should not disturb the Gaussian pattern too much.

Except when the data are highly skewed or when most group sizes are less than 10, **Welch test** performs well for testing equality of means in unequal variances and unequal ns situation. Also called Welch–Satterthwaite test, this is an extension of the Welch test for two groups described earlier. Group sizes between 6 and 10 are your call. When group sizes are really small such as $n < 6$ for most groups, unequal ns and unequal variances, use **Brown–Forsythe test**. This uses median in its calculation instead of mean—thus, less sensitive to departures from assumptions. The mathematical procedure for both is complex for this book. The names are given so that you can make an appropriate choice while running a statistical package.

The assumption of **independence** of residuals is the most serious requirement for the validity of the ANOVA F. This is violated particularly in cases in which serial observations are taken and the value of an observation depends on what it was at the preceding time. This is called auto- or serial correlation.

Although this happens in almost all repeated measures where, for example, a patient is measured repeatedly after a surgery, these serial measurements are class apart as discussed in Section 15.1.2 next. The problem arises when, for example, an infant mortality rate (IMR) series from 1980 to 2007 is studied for different socioeconomic groups. Successive residuals, after the time factor is properly accounted for, may be correlated because of other factors that also change but are not accounted for. Independence of residuals can be checked by the Durbin–Watson test [6]. Most statistical packages routinely perform this test and give a *P*-value. If $P < 0.05$, reanalyze after controlling the factors that may be causing serial correlation. A strategy such as that of working with differences of successive values might be adopted in some cases. If $P \geq 0.05$ for Durbin–Watson test, the *F*-test can be used as usual to test differences among means in different groups.

15.2.2 Two-Way ANOVA

The fundamentals of ANOVA may be clear from the details given in the previous section. Now proceed to a slightly more complex situation. Consider a clinical trial in which three doses (including a placebo) of a drug are given to a group of male and female subjects to assess the rise in hematocrit (Hct) level. It is suspected that the effective dose may be different for males than for females. This differential response is called **interaction**. In this example, the interaction is probably between drug dose and gender. In some cases, evaluation of such interaction between factors could be important to draw valid conclusions. The details are given a little later in this section. Consider the design aspect first.

15.2.2.1 Two-Factor Design

There are two factors in the trial just mentioned: dose of the drug and gender of the subject. The response of interest is a quantitative variable, namely, the percentage rise in Hct level. The objective of the trial is to find the effect of dose, gender, and interaction between these two on the response. Such a setup with two factors is called a two-way ANOVA situation. Note that there are three dose groups of male subjects in this trial and another three dose groups of female subjects. The researcher may wish to have $n = 10$ subjects in each of these six groups, making a total of 60 subjects. To minimize the role of other factors causing variation, these 60 subjects should be as homogeneous as possible with respect to all the characteristics that might influence the response; for example, they may be of the same age group and of normal build (say, BMI between 20 and 25). They may, of course, not be suffering from any ailment that may alter the response. Once 30 male and 30 female eligible subjects meeting the inclusion and exclusion criteria are identified, they need to be randomly allocated 10 each to the three dose levels. Such allocation increases the confidence in asserting that any difference that then occurs is mostly, if not exclusively, due to the factors under study, namely, the dose of the drug and the gender of the subjects in this example. A post hoc analysis can be done to check that other influencing factors, such as age and BMI, are indeed almost equally distributed in the six groups under study. Informed consent and other requirements of ethics, in any case, must be met. Blinding may be required to eliminate possible bias of the subjects, of the observers, and even of the data analyst.

15.2.2.2 Hypotheses and Their Test in Two-Way ANOVA

The null hypotheses that can be tested in a two-way ANOVA situation are as follows:

1. Levels of factor-1 have no effect on the mean response; that is, each level of factor-1 has the same response on average. This translates into

$$H_{0a}: \mu_{1\bullet} = \mu_{2\bullet} = \cdots = \mu_{J\bullet}, \tag{15.10}$$

   where J is the number of levels of factor-1.
2. Levels of factor-2 have no effect on the mean response. That is,

$$H_{0b}: \mu_{\bullet 1} = \mu_{\bullet 2} = \cdots = \mu_{\bullet K}, \tag{15.11}$$

   where K is the number of levels of factor-2.
3. There is no interaction between factor-1 and factor-2. This is explained with the help of an example in a short while.

Again, I do not want to burden you with the mathematical expressions. As with Equation 15.9, the total sum of squares in a two-way ANOVA is broken into the sum of squares due to factor-1, factor-2, interaction, and the residual (also called error). These are divided by the respective df to get mean squares. As in the case of one-way ANOVA, each of these mean squares is an independent estimate of the same variance σ^2 of the response y when the corresponding H_0 is true. Mean square due to error is an estimate of σ^2 even when H_0 is false. Other mean squares are compared with MSE and the criterion F, as in Equation 15.8, is calculated separately for each of the two factors and for their interaction. The P-value is obtained as usual corresponding to the calculated value of F. The pair of df for different Fs are

1. For factor-1: $(J-1)$, $JK(n-1)$
2. For factor-2: $(K-1)$, $JK(n-1)$
3. For interaction between factor-1 and factor-2: $(J-1)*(K-1)$, $JK(n-1)$

where n (≥ 2) is the number of subjects with the jth level of factor-1 and the kth level of factor-2 ($j = 1, 2, \ldots, J$; $k = 1, 2, \ldots, K$). This assumes the same number

TABLE 15.4

Average Birth Weight of Children Born to Women with Different Amount and Duration of Smoking

Duration of Smoking in Pregnancy	Amount of Smoking			All
	Mild	Moderate	Heavy	
−18 weeks	3.45 ($n = 15$)	3.42 ($n = 12$)	3.43 ($n = 7$)	3.44 ($n = 34$)
18–31 weeks	3.38 ($n = 8$)	3.40 ($n = 10$)	3.39 ($n = 6$)	3.39 ($n = 24$)
32+ weeks	3.35 ($n = 25$)	3.30 ($n = 23$)	3.18 ($n = 29$)	3.25 ($n = 17$)
All	3.41 ($n = 28$)	3.40 ($n = 25$)	3.32 ($n = 22$)	3.38 ($n = 75$)

Note: Entries are average birth weight in kg.

of subjects for each combination. Separate decisions for factor-1, factor-2, and the interaction are made regarding their statistical significance. When interaction is not significant, the factors are called **additive**. If interaction is significant, the inference for any factor cannot be drawn in isolation. It has to be in conjunction with the level of the other factor.

In the example of a drug for improving Hct level, n could be 10 in each group in a one-way setup. However, a two-factor trial can be done with only $n = 1$ subject in each group. In this case, however, interaction cannot be *easily* evaluated. For simplicity, the preceding example had the same number of subjects in each group. If the situation so demands, you can plan a trial or an experiment with unequal n in different groups. This is called an **unbalanced design** and illustrated in Table 15.4 for birth weight of children born to women with different levels and duration of smoking. See Example 15.6 for more details.

The analysis of an unbalanced design is slightly more complex although the concepts remain the same. For example, the error df would be $\Sigma(n_{ij} - 1)$ where n_{ij} (≥ 2) is the number of subjects for ith level of factor-1 and jth level of factor-2. When using a statistical software package, be careful in selecting an appropriate routine if the data are unbalanced. My advice, though, is to plan and conduct a balanced design (equal n for each group) wherever feasible so that such complications are avoided.

Example 15.6: Two-Way ANOVA for Effect of Smoking on Birth Weight

A study was done by Wang et al. [7] on the effect of maternal smoking on birth weight in the United States. For this example, vary this study and assume that some women, who are otherwise habitual smokers, give up smoking off and on but not completely during pregnancy. Nonsmokers are excluded in this example. Suppose that such women are asked at the time of their last antenatal visit just before birth about the duration of smoking (factor-1) and the amount of smoking (factor-2). At the end of the pregnancy, the former is categorized as <18, 18–31, and ≥32 weeks.

These are the three levels of factor-1. The amount of smoking is categorized as mild (1–9 cigarettes/day), moderate (10–19 cigarettes/day), and heavy (20+ cigarettes/day). The days in this calculation are only those when at least some smoking was done and the amount of smoking is average per smoking day. These are the three levels of factor-2. The outcome of interest (response variable) is birth weight of the children born to these women.

Some possible confounders in this case are race, age, nutrition, education, and parity. All these can affect birth weight. Suppose these factors could be satisfactorily matched in different groups. Thus, the effect of such extraneous factors is minimal. Let the mean birth weight in different groups be as given in Table 15.4.

There are, for instance, 15 women who smoked an average of one to nine cigarettes per day for a total of less than 18 weeks during the entire pregnancy. Average birth weight of their babies was 3.45 kg. There are eight women who smoked on average of one to nine cigarettes per day for a total of 18–31 weeks and the average birth weight of their babies was 3.38 kg, and so on. Because of unequal n, the design is unbalanced.

The question to be answered is whether the difference in mean birth weight in different groups is really present in this population of antenatal women or is it just a chance occurrence in this sample. This can be examined in the following three ways:

1. Differences in mean birth weight in the last row of Table 15.4 for different amounts of smoking are significant or not.
2. Differences in mean birth weight in the last column of Table 15.4 for different durations of smoking are significant or not.
3. Difference in mean birth weight, for instance, in those mildly smoking for 32+ weeks and those heavily smoking for 37+ weeks compared with the difference in those with similar smoking for −18 weeks are significant or not. This is the interaction.

The calculations are done on the original values of birth weight in 75 children. Suppose a software package reveals $P > 0.05$ for F when calculated for amount of smoking (factor-2) and $P < 0.01$ for F when calculated for duration of smoking (factor-1). The first is not statistically significant but the second is. The conclusion then is that this sample of women does not provide sufficient evidence to conclude that the amount of smoking makes a difference in birth weight, but the duration of smoking in the pregnancy does make a difference. Let $P < 0.05$ for F for interaction. This indicates that an interaction between amount of smoking and duration of smoking is present and implies that the effect of duration of smoking on birth weight is not uniform in the three categories of amount of smoking. This can also be easily seen in Figure 15.3, which is drawn from the data in Table 15.4. If the lines for the three duration groups are parallel, an interaction is said to be absent. In this example, the lines are flatter for −18 and 18–31 weeks but shows an unusual dip for smoking heavily for 32+ weeks. This indicates that an interaction is present. Smoking heavily for 32+ weeks steeply reduced the mean birth weight compared with, for example, heavy smoking for 18–31 weeks or moderate smoking for 32+ weeks.

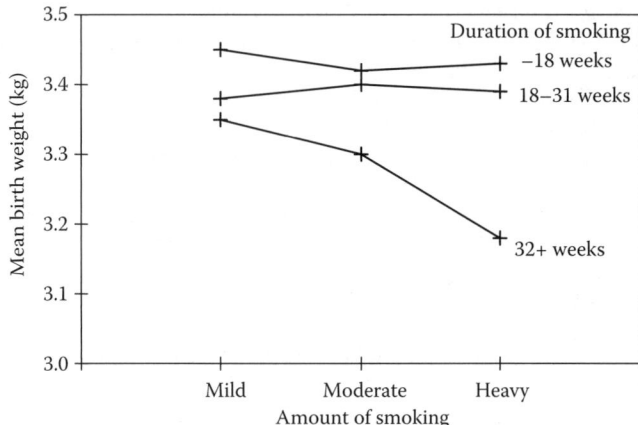

FIGURE 15.3
Interaction between duration and amount of smoking.

The results of ANOVA in Example 15.6 are valid only if the women are randomly chosen from their respective groups. This is apparently missing in the example. The stipulation is that the women included in the study adequately represent the "population" of such antenatal women with respect to influence of amount and duration of smoking on birth weight.

A study like this is useful only if the results can be considered valid for use on other contemporary or future antenatals. For this reason, I have taken the liberty to state the results in the present tense. Many research workers would question such extrapolation. In my opinion, however, the study loses much relevance when its implications for the present and immediate future are denied. At the same time, though, it is important to realize that generalization is valid only to that segment of the population that was adequately represented in the samples.

The statistical design of the study in Example 15.6 is basically prospective because it moves from antecedent (smoking) to outcome (birth weight). ANOVA considers all categories nominal. In Example 15.6, factor-1 and factor-2 both are ordinal. This feature is overlooked in this analysis.

15.2.2.3 Main Effect and Interaction (Effect)

ANOVA is primarily designed to find the statistical significance of the differences between group means, but it can also be used to estimate the average effect of various levels of the factors, called main effects, and the magnitude of different interaction effects. In Example 15.6, the main effect of smoking for 32+ weeks is estimated by the difference of the mean in this category from the overall mean. This is $3.25 - 3.38 = -0.13$ kg. That is, smoking for 32+ weeks in pregnancy reduced birth weight on average by 130 g. (Average birth weight 3.38 kg is among those who smoked since the study is based on smokers only.) Similar main effects can be estimated for the other categories of duration of smoking. They can be calculated for different categories of the amount of smoking as well, but that is inadvisable in this case because these differences are not statistically significant.

Main effects and interactions are easy to explain with the help of notations. As before, let the levels of factor-1 be identified by subscript j ($j = 1, 2, ..., J$), of factor-2 by subscript k ($k = 1, 2, ..., K$), and subjects within each group by subscript i ($i = 1, 2, ..., n$). In Example 15.6, $J = 3$ and $K = 3$ but n in different groups is not equal. For the purpose of explaining the concepts, it is useful to keep the notation relatively simple and assume that all groups have the same number n of subjects. The response of the ith subject belonging to the jth level of factor-1 and the kth level of factor-2 is denoted by y_{ijk}. In Example 15.6, if birth weight of the child born to the fourth woman in the group comprising those smoking mildly (first group of amount of smoking) for 18–31 weeks (second group of duration of smoking) is 3.49 kg, then $y_{421} = 3.49$ kg. With these notations, the mean for the jth level of factor-1 is $\bar{y}_{.j.}$, for the kth level of factor-2 is $\bar{y}_{..k}$, and the overall mean of all nJK observations is $\bar{y}_{...}$. The actual two-way ANOVA model for the corresponding population is in terms of mean as follows:

$$\text{Mean of } y_{ijk} = \mu + \alpha_j + \beta_k + \theta_{jk} + \varepsilon_{ijk},$$

where
  μ is the overall mean
  α_j is the main effect of jth level of factor-1
  β_k is the main effect of kth level of factor-2
  θ_{jk} is the interaction between jth level of factor-1 and kth level of factor-2
  ε_{ijk} is the residual

Now

Estimated main effect of jth level of factor-1: $\alpha_j = (\bar{y}_{.j.} - \bar{y}_{...})$; $j = 1, 2, ..., J$;

$$(15.12)$$

and

Estimated main effect of kth level of factor-2: $\beta_k = (\bar{y}_{..k} - \bar{y}_{...})$; $k = 1, 2, ..., K$.

$$(15.13)$$

In Example 15.6, the main effect of smoking for less than 18 weeks is $3.44 - 3.38 = +0.06$ kg and of moderate smoking is $3.40 - 3.38 = +0.02$ kg. The positive effect is not surprising since these are the effects of those categories on the birth weight *compared with the overall mean* of all categories that include the heavy and long-duration smokers. It is easy to show by some algebra that $\Sigma_j \alpha_j = 0$ and $\Sigma_j \beta_j = 0$. Consequently, only $(J - 1)$ αs and $(K - 1)$ βs are independently determined. One each is automatically determined by these conditions.

The interaction is obtained separately for each combination of the levels of the two factors. This is the excess mean after adjustment for the main effects of the concerned level of factor-1 and factor-2. Thus, the estimated interaction effect between the jth level of factor-l and kth level of factor-2 is

$$\theta_{jk} = (\bar{y}_{\bullet jk} - \bar{y}_{\bullet\bullet\bullet}) - (\bar{y}_{\bullet j\bullet} - \bar{y}_{\bullet\bullet\bullet}) - (\bar{y}_{\bullet\bullet k} - \bar{y}_{\bullet\bullet\bullet})$$

$$= (\bar{y}_{\bullet jk} - \bar{y}_{\bullet j\bullet} - \bar{y}_{\bullet\bullet k} + \bar{y}_{\bullet\bullet\bullet}), \tag{15.14}$$

where $\bar{y}_{\bullet jk}$ is the mean of n subjects in the (j, k)th group. For example, the estimate of the interaction effect between moderate smoking and smoking for 32+ weeks in Table 15.4 is $3.30 - 3.25 - 3.40 + 3.38 = +0.03\,\text{kg}$. Thus, this combination of duration and amount of smoking increases birth weight by $30\,\text{g}$ on average in this sample *relative to the means in respective categories*. Note, again, that the overall mean is based on all women including those who are heavy and long-duration smokers. Mathematically, Equation 15.14 implies $\Sigma_j \theta_{jk} = 0$ and $\Sigma_k \theta_{jk} = 0$.

This example also illustrates that if the interaction is significant, you should condition your conclusion about effect of one factor on the level of the other factor. When the interaction is not significant, the focus is on main effects. In that case, factor-1 levels should not be compared within factor-2 levels because factor-1 effects are not significantly different in this example. Only the average is adequate.

The Equations 15.12 through 15.14 define the main effects and the interactions relative to the overall mean. This is the most commonly used definition. However, if you so wish, you can define them relative to a base value. In Example 15.6, the base could be the category with least smoking where the mean birth weight is $3.45\,\text{kg}$. In that case, for example, the estimated main effect of heavy smoking is $3.32 - 3.45 = -0.13\,\text{kg}$. That is, heavy smoking reduces birth weight by $0.13\,\text{kg}$ on average compared with the mild smokers for less than 18 weeks. A similar interpretation can be given for the other main effects and interactions when measured from a base value.

Note the following for the ANOVA procedure:

1. As stated for one-way ANOVA, criterion F provides an overall test about difference being present *somewhere*. Multiple comparisons procedures as described in the next section can be used after ANOVA to find the group or groups that are different from others.

2. Extension of ANOVA to three or more factors is straightforward with main effects and interactions similar to those in two-way ANOVA.

However, there would now be several two-factor interactions, three-factor interactions, etc. The details are beyond the scope of this book. Interested reader may consult Doncaster and Davey [8].

3. It is desirable to test higher order interactions before lower order ones because it is difficult to attach meaning to the lower order inter-actions when higher order interactions are present.

As always, lack of statistical significance of interaction does not mean inter-action is absent. It only means that this could not be detected from the avail-able data. This might be more so in case of ANOVA since the sample size is generally planned to detect main effects. For detecting interaction, higher sample size is required.

15.2.2.4 Type-I, Type-II, and Type-III Sums of Squares

When the number of factors is two or more, various sums of squares (ss) can be calculated in at least three different ways. Each has relevance in a particu-lar setup.

1. *Type-I ss*: For a two-way ANOVA, Type-I ss measures effect α of factor-1 after adjusting for μ (overall mean), effect β (of factor-2) after adjusting for α and μ, and interaction θ after adjusting for β, α, and μ. Thus, this assesses effect of parameters in the order they appear in the model.

2. *Type-II ss*: This is used for assessing effect of those parameters that do not include it. Interaction includes both α and β. Thus, Type-II ss measures effect α after adjusting for β and μ but not θ.

3. *Type-III ss*: This assesses effect of parameters after adjusting for all other terms in the model. Thus, this measures effect α after adjust-ing for θ, β, and μ and effect β after adjusting for θ, α, and μ. This is the same as Type-II ss if there is no interaction in the model.

In most situations, Type-III ss is computed and it serves the purpose well.

15.2.3 Repeated Measures

In many medical situations, as in the case of administering an anesthetic agent, it is necessary to monitor a subject by repeatedly observing vital signs such as heart rate (HR) and BP at specified intervals. Analysis of such repeated mea-sures requires special methods. Some of these are described in this section.

Repeated measures have potential to provide unnecessary satisfaction because of apparently large volume of data. If BP of 10 subjects is measured hourly 12 times during day time for diurnal variation for a week, you will have a total of 840 data points for systolic and 840 for diastolic level.

This can mislead you to believe that you have a large sample of values. Actually, there are only $n = 10$ subjects.

15.2.3.1 Random Effects versus Fixed Effects

Repeated values could be heavily correlated. When there is only one factor under study, such as doses of a drug, and repeated measures are taken at fixed points of time for each subject, such data can be analyzed by two-way ANOVA considering subjects as factor-2 and $n = 1$. This is despite the basic difference that in two-way design, allocation within blocks (subjects in this case) is randomized. The assumption in this case is that the interest is only in comparison of means in the groups and not the time trend of the response. This, however, violates one basic premise. The conventional ANOVA considers levels of the factors fixed and of specific interest to the problem. Both amount and duration of smoking are **fixed effect** factors in Example 15.6. In the case of fixed effects, no inference can be drawn about any other level except the ones under study. In contrast, for random effects, the inference would be valid for the entire population from which the factor levels are randomly drawn. Subjects are almost invariably a random sample from a defined population. When such subjects are considered factor levels as suggested for repeated measures, these give rise to what is called a **random effect**. The method of calculating F is different when random effects are present. This is sometimes referred to as **variance components analysis**. For details, see Kenward [9]. This reference also discusses analysis of more intricate designs.

Special caution is required in the case of repeated measures. ANOVA for repeated measures requires not only uniform variance of each repeated measure but also uniform covariance (or correlation) between each pair of repeated measures. Small differences do not matter if n is large, but the differences could be unduly large in some situations. Measurements close in time may be more highly correlated than those widely separated in time. Heart rates after 1 and 5 min after anesthesia could be highly correlated and heart rates after 1 and 30 min poorly correlated.

15.2.3.2 Sphericity and Huynh–Feldt Correction

Equality of covariances together with equality of variances within a matrix is known as compound symmetry. A slightly relaxed condition is that the differences between all pairs of times are independent (covariance = 0) and have same variances. This is called **sphericity** and is the actual required condition. This is tested by **Mauchly test**. The procedure for this also is too complex for the level of this book. Again, the name is provided to help you understand the computer output. Realize though that compound symmetry assures sphericity but sphericity can occur in some other situations also.

Mauchly test for sphericity works well for Gaussian data and reasonably well for minor departures from Gaussianity but not when departures are gross and dataset is small. When sphericity is violated, too many hypotheses are falsely rejected. When in doubt, safe bet is to use **Huynh–Feldt correction** to dfs. This correction can be used directly without worrying about Mauchly test results. This correction is explained next.

Many statistical packages provide results with Huynh–Feldt correction and also the corresponding P-value. This is also known simply as Huynh correction. F remains same but the correction factor reduces both the dfs of F by a factor called epsilon. You may find other types of epsilon also in your software output but Huynh–Feldt epsilon is widely accepted. In a rare case, this correction factor can exceed one. In that case, one is used and no correction to df is done.

Preceding discussion is for one group of subjects only, which is measured at different time points. You may have two or more groups such as different dose groups or subjects with different forms of a disease. Each subject in each group can be measured at several time points to study the trend of quantitative outcome such as various enzyme levels. The objective in this case is to find if the average time trend in one group is the same as in other groups. This is the same as assessing absence of interaction group * time. If this interaction is significant, conclude that the average time trend in at least one group is different from the trend in other groups. If the number of subjects in each group is same n, this interaction will have $(K - 1)(T - 1)$ df and will be checked against residual with $K(n - 1)(T - 1)$ df, where K is the number of groups and T is the number of time points of repeated measures.

Besides sphericity, which is for variances–covariances among differences in values at different points of time within each group, you need to worry about homogeneity also of covariance matrices across groups when groups are two or more. For this, **Box M test** is used.

It is often desirable to repeat an experiment three or four times on different groups of similar subjects. Such **replication** helps to examine the consistency of results. Replication itself can be considered a factor in ANOVA. This would be a random effect factor because the targets of inference are not those particular replications but all such replications. If there is only one substantive factor of interest, such as dose levels, the data on replications can again be analyzed by two-way ANOVA with replications as a random effect factor. A difference between replications, if found significant, would raise questions about the validity of the conditions in which the experiment was conducted or the homogeneity of subjects in different replications. If between-replication differences were not significant, conclusions on the differences between levels of factor-1 would have substantially more confidence.

Complexities seem to never end with repeated measures. What happens if the design is unbalanced (unequal ns in different groups) and covariance matrices across groups are also unequal? For this situation, Welch test is recommended in place of F. This is similar to the one you saw for two-sample situation. But the n in smallest group must be at least $3 * (T - 1)$ for testing main effects and at least $5 * (T - 1)$ for testing interaction where T is the number of time points for repeated measures. Quite often in repeated measures, the interest is in group differences in time trend and n must be sufficiently large for testing interaction.

15.2.3.3 Repeated Measures versus Two-Way ANOVA

In repeated measures, successive measurements are fixed but in two-way design allocation within blocks are randomized. Thus, serial correlation is present in repeated measures but not in two-way design. Yet, two-way ANOVA is adequate for repeated measures design (one group of n subjects, repeated T times) subject to the condition that $(x_t - x_{t'})$ have same population variance for every pair of occasions, earlier called sphericity. (Subscript t is for time point.) This is achieved when all x_ts have same variance, and covariances between all pairs of x_t and $x_{t'}$ are equal.

Another alternative is to consider repeated observations on each subject in multivariate setup and do multivariate analysis of variance (MANOVA). This works well for balanced design. Further details are in Chapter 19 on multivariate analysis. If you find MANOVA output difficult to decipher, do univariate repeated measures ANOVA and use Huynh–Feldt correction if Mauchly test is significant.

Software illustration: A software illustration of running a repeated measures ANOVA is in Appendix C.

15.2.3.4 Area under the Concentration Curve

In case of quantitative outcome such as pain score and CD4 count measured at several fixed points of time (e.g., preoperative, perioperative, 5 min after end of surgery, 15 min after surgery, etc.; before drug, one day after drug, 5 days after drug, etc.), the outcome can be plotted against time as in Figure 21.4 in Chapter 21. This is called concentration curve where the term "concentration" come from pharmacokinetic studies that measure concentration of drug in the body at different points of time. Area under the concentration (AUC) curve is sometime used as a measure of overall performance of the intervention. This AUC is different from AUC you saw in the context of receiving operating characteristic (ROC) curve in a previous chapter although both are areas under the curve.

AUC curve for each subject is measured by linear trapezoidal rule. This has minor error since all possible points are infinite but works well in practical situations. The area can be averaged over subjects provided the follow up for each subject is the same. In practice, the follow-up varies from subject to subject. Some patients recover soon and discharged early so that the follow-up truncates. Some may die early and some may survive long enough. When the length of follow-up varies from subject to subject, a time averaged AUC curve is obtained. This has obvious limitation, particularly because it assumes that each period of time has same implication. This still fails to account for voluntary dropouts or those taken out because of complication. It is inadvisable to use AUC curve in this situation. AUC curve as performance measure is best suited when all subjects have uniform follow-up.

Utility of AUC curve is in comparing performance of two or more interventions, one of which can be placebo. This can be done by ANOVA with the usual restrictions. If quantitative outcome is positive such as arterial BP at different points of time after anesthesia-1 versus anesthesia-2, valid conclusion can be reached when AUC curve of one is lower than the other at all time points. If the curves crisscross, the interpretation can be difficult. There is also distinct possibility in this case that the two areas are nearly equal even when the outcome at each time point is very different. This fallacy is discussed further in Chapter 21.

15.2.4 Multiple Comparisons: Bonferroni, Tukey, and Dunnett Tests

Once overall significance is indicated by the F-test, the next step is to identify the groups that are different from one or more of the others. This requires several comparisons. In case of pairwise comparisons, for example, if there are four groups, the comparisons are group-1 with group-2, 1 with 3, 1 with 4, 2 with 3, 2 with 4, and 3 with 4. There are a total of six comparisons and are called multiple comparisons. You now know that means of two groups are generally compared by Student t-test. However, repeated application of this test at, say, 5% level of significance on the same data blows up the total probability of Type-I error to an unacceptable level. If there are 15 tests on the same data, each done at the 5% level, then the overall (experiment-wise) Type-I error could be as high as $1 - (1 - 0.05)^{15} = 0.54$. Compare this with the desired 0.05.

Type-I error allowed for each individual comparison is called **comparison-wise error rate**, whereas the total error for all the comparisons together is called **experiment-wise error rate**. To keep the probability of Type-I error within a specified limit such as 0.05, many procedures for multiple comparisons are available. Each of these is generally known by the name of the scientist who first proposed it. Among them are Bonferroni, Tukey, Scheffe, Fisher's LSD, Newman–Keul, Duncan, and Dunnett. The Bonferroni and Tukey procedures are commonly used in medical and health literature and are also the most suitable ones. Dunnett test is used specifically when each group is to be compared with the control only. These are described next. For other procedures, see Klockars and Sax [10].

15.2.4.1 Bonferroni Procedure

This is the simplest method to ensure that the probability α of Type-I error does not exceed the desired level. Under this procedure, each comparison is done by using Student t-test, but a difference is considered significant only if the corresponding *P*-value is less than α/H, where *H* is the number of comparisons. If there are four groups and *all* pairwise comparisons are required, then *H* = 6. Then, a difference would be considered significant at the 5% level if $P < 0.05/6$, that is, if $P < 0.0083$.

The Bonferroni procedure is conservative in the sense that the actual probability of Type-I error will be much less than α. This means that there is an additional chance that some differences are actually significant but pronounced not significant. This is not a major limitation in an empirical sense for any statistical test as earlier explained. The advantage of the Bonferroni procedure is that *H* can be only as much as the number of comparisons of interest. If there are four groups and the interest is only in comparing group-1 with group-2, 2 with 3, and 3 with 4 (and not, e.g., in comparing group-l with group-3), then *H* = 3. Small *H* improves the efficiency of this procedure.

15.2.4.2 Tukey Test

This is most suitable when the interest is in *all* pairwise comparisons. The procedure works in a slightly different manner. Corresponding to the df(J,v), where *J* is the number of groups to be compared and *v* is the df associated with MSE in the ANOVA table, a value *Q* is obtained from what is called the **Studentized range distribution** for a specified α such as 0.05. This distribution is based on the maximum difference in means. This is the difference between the largest mean and the smallest mean. Values of *Q* are given in Table B.5 for different numbers of groups and different error df when α is 0.05. Use this *Q* value and calculate

$$\text{Tukey test: } D = Q\sqrt{\frac{\text{MSE}}{N/J}}, \tag{15.15}$$

where
   N is the total number of subjects
   MSE is the mean square error in the ANOVA table for those data

Any pairwise difference exceeding *D* in group means is considered statistically significant. This is called the Tukey test and keeps the experiment-wise probability of Type-1 error limited to the specified α-level.

Example 15.7: Tukey Test for Birth Weight in Smoking Data

Consider the data in Example 15.6 on effect of maternal smoking on child birth weight. In this case, only the effect of duration of smoking is statistically significant. To find which duration or durations are making a significant impact, compare mean birth weights for different durations. Thus, all pairwise comparisons are needed and a Tukey test would be appropriate. The number of levels of this factor is *J* = 3. Also, the total number of subjects in this study is *N* = 75. The ANOVA table obtained from a computer package revealed MSE = 0.022 and the error df $v = 66$. From Table B.5, the value of Q at $\alpha = 0.05$ for (3, 66) df is approximately 3.39. Thus,

$$D = 3.39 \sqrt{\frac{0.022}{75/3}} = 0.10.$$

This is the critical difference that should be present between means for it to be statistically significant at $\alpha = 0.05$. The mean birth weights (in kg) in the three duration-of-smoking categories are 3.44, 3.39, and 3.25, respectively. The difference in mean birth weight between –18 and 18–31 weeks does not exceed the critical value but exceeds it between other durations. Thus, the real culprit in this example is smoking for 32+ weeks, which significantly lowers the mean birth weight.

15.2.4.3 Dunnett Test

Many times the interest in experiments is in comparing each group with a particular reference group, and comparison between other groups is not required. Quite often the reference group is control. For example, you may like to compare maternal serum copper concentration in normal pregnancies with pathological conditions such as threatened abortion, blighted ovum, and pyelonephritis. In such a situation, Tukey test and Bonferroni test are not efficient in correctly controlling Type-I error. Instead Dunnett test is used:

$$\text{Dunnett test: } t_d = \frac{\bar{x}_k - \bar{x}_0}{\sqrt{2\text{MSE}/n_h}}, \qquad (15.16)$$

where
$\bar{x}_k$ is the mean of the kth group under comparison ($k = 1, 2, ..., K$)
$\bar{x}_0$ is the mean of the reference group

Thus, under this notation, there are total of $(K + 1)$ groups in the experiment including the reference group. Here, n_h is the harmonic mean of n_0 and n_k, the respective sizes of the reference and the comparison group. Harmonic mean arises from $1/n_0$ and $1/n_k$ that you saw in the denominator of t in Equation 15.6a.

The value of Dunnett t_d is compared with its distribution at $(n - K)$ df. You may find table of t_d in some statistical books; else good software will provide P-value directly.

In such comparisons, the result of reference group is expected to be much more reliable since all the comparisons are with this group. Statistically, this means that the reference group should have larger n, generally $n_0 \geq n_k \sqrt{K}$ for all k. If not, the reference group values should have less variation from person to person than in other groups under comparison.

15.2.4.4 Intricacies of Multiple Comparisons

Following comments explain some of the intricacies of multiple comparisons:

1. Multiple comparisons are generally done only after an F-test reveals significance. In any case, ANOVA is needed to compute MSE. This MSE is required to calculate D from the formula given in Equation 15.15 or t_d from formula in Equation 15.16. Thus, there is no point in sidetracking F and directly calculating D or t_d. On the other hand, the Bonferroni procedure can be used without an F-test if its limitation in terms of altered α is adequately realized.

2. Many statistical packages perform the Tukey test or other multiple comparison procedures at the specified level of significance. They also indicate which groups, if any, are significantly different from others.

3. Multiple comparisons, particularly the Tukey test, may sometimes give results at variance with the results of the F-test. It is possible for the F-test to be significant but none of the pairwise comparisons significant. Conversely, the F-test may not show significance but comparison for a specific pair may still be significant. This happens because both require a Gaussian pattern, but the practical data seldom follow exact Gaussian pattern. The F- and Tukey tests behave differently for departure from a Gaussian pattern. The problem may arise more frequently for small n and then for large n because large n is insulation against violation of a Gaussian pattern in most cases.

4. The foregoing procedures are valid only for preplanned comparisons. Sometimes the idea of testing statistical significance between two or more groups arises after seeing the observed data (this is sometimes called **data snooping**). Although multiple comparisons are often described as post hoc tests, the tests indicated by data are also post hoc of a different type. Test done after looking at the data do distort the chances of error. For procedures applicable to tests suggested by the data, see Klockars and Sax [10].

5. Multiple comparisons procedures are also helpful in **ranking** of groups with regard to the value of mean. For example, you may find that group-3 has highest mean, group-1 and group-4 have nearly same but lower mean, and group-2 lowest mean.

Software illustration: See Appendix C for a simple software illustration of one-way ANOVA and use of Tukey test for multiple comparisons.

15.3 Non-Gaussian Conditions: Nonparametric Tests for Location

Continue with the setup in which the response is quantitative, practically continuous, and the interest is in comparing two or more groups. The difference now, however, is that n is small and the underlying distribution of the response variable does not follow a Gaussian pattern. Non-Gaussian distribution can arise when, for example, studying duration of labor at the time of childbirth in undernourished women, blood glucose level in diabetics, or Hb level in anemics. The distribution pattern of these variables in such restricted class of subjects can be highly skewed. Nonparametric methods, also called distribution-free methods, are needed for this setup. One popular nonparametric method already discussed is **Kolmogorov–Smirnov test** for large n, which is used to test whether or not a set of data follows a specific stipulated pattern, such as Gaussian. Nonparametric methods are especially suitable when outliers are present, which are genuine in the dataset (not artifacts) and cannot be ignored. Outliers do not affect these methods. In any case, nonparametric are the methods of choice when the underlying distribution is far from Gaussian and n is small. When Gaussian conditions are present, the performance of nonparametric methods is not as good as those of parametric methods such as Student t-test and the ANOVA F-test.

The term nonparametric method implies a method that is not for any specific parameter. Student t-test, for example, is a parametric method because it is concerned with a parameter, namely the mean. In the case of nonparametric methods the hypothesis is concerned with the pattern of the distribution as in a goodness-of-fit test or with some characteristic of the distribution of the variable such as randomness and trend. More commonly, though, the interest would be in the location of the distribution without specifying the parameter. This is illustrated in Figure 15.2, where the distributions are identical except for location. A location shift of the entire distribution is clear and it is not necessary to talk of mean or median or any such parameter in this case.

Distribution-free methods are those that are based on functions of sample observations whose distribution does not depend on the form of the underlying distribution in the population from which the sample was drawn. You have already seen chi-square that is distribution free. Chi-square remains valid for any categorical data whether the underlying distribution of the variable is Gaussian or not.

Although nonparametric methods and distribution-free methods are not synonymous, most nonparametric tests are also distribution free. Both categories of methods are generally considered as nonparametric methods.

These methods derive their strength from transforming the values to ranks. Thus, they are applicable to ordinal data as well.

Nonparametric methods are not as developed as parametric methods. Nor they are as efficient under Gaussian conditions as are parametric methods. The details given in the following are further restricted to the commonly used methods. If you have data more suitable for a nonparametric method for which such a method is not readily available, the recommended procedure is to use the parametric method on the actual data as well as on the rank transformed data, which makes it nonparametric [11]. If the two methods give nearly identical result, you are done. If not, take a closer look at the data for outliers or for highly skewed distribution and use nonparametric method.

15.3.1 Comparison of Two Groups: Wilcoxon Tests

As in the case of Student *t*-test, the comparison of two groups can be done in two types of situations: paired and unpaired. Nonparametric methods for these situations are described next.

15.3.1.1 Case I: Paired Data

As explained earlier, paired data arise when the same subject is measured twice, such as before the treatment and after the treatment. They also arise in case of one-to-one matching such as in some case–control studies. Once the difference between two observations of each pair is known, the problem reduces to that of a one-sample setup.

A **sign test** is one option in the case of paired data. The only information it utilizes is the direction of difference, negative or positive, within pairs. Under the null hypothesis that there is no difference, a negative sign is as likely as positive sign. Thus, H_0: $\pi = \frac{1}{2}$, where π is the probability of, say, a positive sign. The test can be carried out by calculating the binomial probability given in Equation 13.1 under this H_0.

Let (x_i, y_i) be the *i*th pair out of n. Exclude the pairs with **ties** $x_i = y_i$. This may leave n' pairs, where $n' \leq n$. If the distribution of x is suspected to be shifted to the right (that is, xs are generally larger than ys) then the difference $d_i = (x_i - y_i)$ should be more commonly positive than negative. Thus, in this case, H_1: $P(+) > P(-)$, while H_0: $P(+) = P(-)$. If there are r positive signs out of n in the sample, then the probability of this (or more extreme in favor of H_1) happening under H_0 is

$$P = \sum_{x=r}^{n'} {}^{n'}C_x \left(\frac{1}{2}\right)^{n'}. \tag{15.17a}$$

This is the P-value for the right-sided H_1. For left-sided H_1: $P(+) < P(-)$,

$$P = \sum_{x=0}^{r} {}^{n'}C_x \left(\frac{1}{2}\right)^{n'}. \tag{15.17b}$$

And for two-sided H_1: $P(+) \neq P(-)$

$$P = \sum_{x=0}^{r} {}^{n'}C_x \left(\frac{1}{2}\right)^{n'} + \sum_{x=n'-r}^{n'} {}^{n'}C_x \left(\frac{1}{2}\right)^{n'}. \tag{15.17c}$$

Reject H_0 in favor of H_1 if the P-value calculated is less than the predetermined α, such as 0.05. The P-values in Equations 15.17a–c are valid for small n' (as well as for large n'). If n' is large, these can be approximated by the corresponding Gaussian probability as explained for binomial probability in Chapter 13. In this case, n' is considered large if it is 20 or more.

The sign test is considered deficient because it ignores the magnitude of difference. But there are situations where only sign is important and the magnitude can be ignored. This can happen, for example, in a behavioral problem where a judgment can easily be made about "greater than" or "less than" between pairs of performances but not about the magnitude of difference. In an iron supplementation program for antenatal women, the interest may be in responders who show an increase in Hb level in excess of, say, 0.5 g/dL and not in the actual amount of increase. Note, however, that such considerations convert the quantitative data to qualities, and thus the power of the procedure to detect a difference may be compromised. This can be compensated by increasing n.

If the magnitude as well as the direction of the differences is important, a more powerful test is the **Wilcoxon signed-rank test**. The method for this test is as follows:

Step 1. Let there be n pairs and the observed value for the ith pair (x_i, y_i), $i = 1, 2, \ldots, n$. Calculate the difference $d_i = (x_i - y_i)$ in the variable values for each pair. For the purpose of illustration, consider x_i as the value before the treatment and y_i the value after the treatment.

Step 2. Ignore the + or – sign of d_i and assign rank starting from 1 to the smallest $|d_i|$ to n' to the largest $|d_i|$, where n' is the number of pairs with nonzero difference. The pairs with zero difference are omitted. Thus, $n' \leq n$. If two or more $|d_i|$ are equal (ties), they are each assigned the average rank of the ranks they would have received individually if ties in the data had not occurred.

Step 3. Reaffix the + or – sign of the difference to the respective ranks. That is, indicate which ranks arose from negative d_is and which from positive d_is.

Step 4.   Calculate the Wilcoxon test criterion as the sum of the positive ranks. That is,

Wilcoxon signed–rank test :
$$W_S = \text{ sum of the ranks with positive sign} \qquad (15.18a)$$

The minimum possible value of W_S is zero when no rank has a positive sign. The maximum occurs when all ranks are positive. If there was no difference between the groups, the criterion W_S would take on a value close to its mean $n'(n' + 1)/4$. A big difference from this mean would be evidence against the null hypothesis of no difference between before and after measurements.

Step 5A. Find *P*-value corresponding to the calculated value of W_S. Reject H_0 if P is less than the predetermined significance level. Tables of *P*-values for different values of W_S are available in the literature [12]. Described here is a procedure based on critical values of W_S corresponding to $\alpha = 0.05$. For this α and $n' \leq 19$, reject H_0 if

a. $W_S \geq w_r$, if H_1 is that before measurements are *higher* than after measurements (right-sided H_1)

b. $W_S \leq w_l$, if H_1 is that before measurements are *lower* than after measurements (left-sided H_1)

c. Either $W_S \leq w_1$ or $W_S \geq w_2$, if H_1 is that before measurements are *different* from after measurements (two-sided H_1)

The critical values w_r, w_l, or w_1 and w_2 for $\alpha = 0.05$ for n' from 5 to 19 are given in Table 15.5. The one-sided critical value for $n' = 5$ is 0 or 15. These are the minimum and the maximum possible values with $n' = 5$. Thus, the decision to reject H_0 in favor of one-sided H_1 can be reached in this case only if all five differences are positive (or all negative). When $n' \leq 4$, no difference can be statistically significant at the 5% level when H_1 is one sided. If H_1 is two sided, no difference can be statistically significant by this test at the 5% level for $n' \leq 5$.

Step 5B. For $n' \geq 30$, it is safe to use Student *t*-test for means. If n' is between 20 and 29, use the Gaussian approximation to the Wilcoxon test. This is given by

$$Z = \frac{W_S - \mu_{W_S}}{\sigma_{W_S}}, \qquad (15.18b)$$

where
μ_{W_S} is the mean value of W_S, given by $\mu_{W_S} = n'(n' + 1)/4$
σ_{W_S} is the SD of W_S, given by $\sigma_{W_S} = \sqrt{n'(n' + 1)(2n' + 1)/24}$
A one-tailed or two-tailed test can be carried out as usual with this Z

TABLE 15.5

Critical Values of Wilcoxon Signed-Rank Test W_S
for Matched Pairs ($\alpha = 0.05$)

n'	Right-Sided H_1 w_r	Left-Sided H_1 w_1	Two-Sided H_1 w_1	w_2
≤4	a	a	b	
5	15	0	b	
6	19	2	0	21
7	25	3	2	26
8	31	5	3	33
9	37	8	5	40
10	45	10	8	47
11	53	13	10	56
12	61	17	13	65
13	70	21	17	74
14	80	25	21	84
15	90	30	25	95
16	101	35	29	107
17	112	41	34	119
18	124	47	40	131
19	137	53	46	144

Source: Adapted from Wilcoxon, F. and Wilcox, R.A., *Some Rapid Approximate Statistical Procedures*, Lederle Laboratories, Pearl River, New York, 1964, Table 2.

[a] No difference can be statistically significant by this test at $\alpha = 0.05$ (one-tail) when $n' \leq 4$.

[b] No difference can be statistically significant by this test at $\alpha = 0.05$ (two-tail) when $n' \leq 5$.

A good statistical package will do all these steps for you. It will give the P-value that you can interpret to draw conclusions. The following comments may be helpful:

1. A procedure equivalent to the Wilcoxon test is the **Mann–Whitney test**. In fact, they can be shown to be algebraically identical.

2. The Wilcoxon test is less powerful than Student t-test if the underlying distribution is Gaussian. Thus, it should not be used when the data follow Gaussian pattern. Also, the sign test is less powerful than the Wilcoxon test because it ignores the magnitude of differences.

3. The Wilcoxon signed-rank test and, for that matter, all the tests discussed in this section are based on ranks. These are discrete and not continuous. For this reason, it is rarely possible to specify a critical value for exact $\alpha = 0.05$. The critical values given in Table 15.5 (as well as those in similar tables) actually have $\alpha \leq 0.05$. In some cases, this could be substantially less. But that is the nearest one could reach if α is to be 0.05.

TABLE 15.6

Duration of Labor in Obese and Nonobese Women

	Duration of Labor (h)						
Obese	18	15	17	20	14	12	18
Nonobese	17	15	18	18	11	10	14

TABLE 15.7

Calculation for Wilcoxon Signed-Rank Test for Data in Table 15.6

d_i	1	0	−1	2	3	2	4
Rank of $\lvert d_i \rvert$	1.5	—	1.5	3.5	5	3.5	6
Signed ranks	+1.5	—	−1.5	+3.5	+5	+3.5	+6

Example 15.8: Wilcoxon Signed-Rank Test for Prolonged Labor in Obesity

Suppose the role of obesity in prolonged labor is examined in pregnant women of age more than 35 years with persistent occipito-posterior presentation. Seven obese (say, BMI ≥ 30) and seven nonobese women with such presentation, one-to-one matched for age, parity, Hb level, etc., are included in the study. The data obtained are in Table 15.6. Is the evidence sufficient to conclude that obese women have longer labor?

The calculations are shown in Table 15.7. For positive ranks,

$$W_S = 1.5 + 3.5 + 5 + 3.5 + 6$$

$$= 19.5.$$

Since the pair with $d_i = 0$ is ignored, $n' = 6$ in this example. Note how the ranks have been assigned to the ties. Two pairs have the same absolute difference, namely, 1 h. They would have received ranks 1 and 2 but now receive rank 1.5 each because of the tie. The case with ranks 3 and 4, where the difference is 2 h, is similar.

The alternative hypothesis here is one sided (H_1: Obese women have a longer duration of labor than nonobese) because there is no stipulation for obese women to have a shorter duration of labor. Since the calculated value of W_S is more than the cutoff 19 in Table 15.5 for $n' = 6$, reject H_0 of no difference at $\alpha = 0.05$. Conclude that obese women do indeed have a longer duration of labor.

15.3.1.2 Case II: Independent Samples

The counterpart of the Wilcoxon test for two independent samples is called the **Wilcoxon rank-sum test**.

Suppose there are n_1 observations in the first sample and n_2 in the second sample. The Wilcoxon rank-sum test can be used to test whether the locations of the distributions from which these samples have been drawn are the same. For this, the two samples are combined and these $n(= n_1 + n_2)$ observations are assigned ranks from lowest to highest. Ties, if any, are given average ranks as before. For convenience, assume that the labels are such that $n_1 \leq n_2$. Now,

Wilcoxon rank-sum test : W_R = sum of the ranks assigned to the
$\qquad\qquad\qquad n_1$ observations in the first sample. (15.19)

Critical values of W_R for $\alpha = 0.05$ under the null hypothesis of no difference between the two groups are given in Table 15.8 for one-sided and two-sided H_1.

Reject H_0 if the calculated value of W_R is equal to or beyond the critical value for your n_1 and n_2. The table is only for values of (n_1, n_2) between 4 and 9. As in the case of the Wilcoxon signed-rank test, the rank-sum test for two independent samples will not give any statistical significance at the 5% level if both n_1 and n_2 are less than or equal to 3 (also see the footnote in Table 15.8).

When any of these sample sizes is 10 or more, use the following Gaussian approximation:

$$Z = \frac{W_R - \mu_{W_R}}{\sigma_{W_R}}, \qquad\qquad (15.20)$$

TABLE 15.8

Lower and Upper Critical Values of Wilcoxon Rank-Sum Test W_R for (n_1, n_2) between 4 and 9

n_2	α One Tailed	α Two Tailed	n_1 4	5	6	7	8	9
4	0.05	0.10	11, 25					
	0.025	0.05	10, 26					
5	0.05	0.10	12, 28	19, 36				
	0.025	0.05	11, 29	17, 38				
6	0.05	0.10	13, 31	20, 40	28, 50			
	0.025	0.05	12, 32	18, 42	26, 52			
7	0.05	0.10	14, 34	21, 44	29, 55	39, 66		
	0.025	0.05	13, 35	20, 45	27, 57	36, 69		
8	0.05	0.10	15, 37	23, 47	31, 59	41, 71	51, 85	
	0.025	0.05	14, 38	21, 49	29, 61	38, 74	49, 87	
9	0.05	0.10	16, 40	24, 51	33, 63	43, 76	54, 90	66, 105
	0.025	0.05	14, 42	22, 53	31, 65	40, 79	51, 93	62, 109

Source: Adapted from Wilcoxon, F. and Wilcox, R.A., *Some Rapid Approximate Statistical Procedures*, Lederle Laboratories, Pearl River, New York, 1964, Table 1.

Note: No difference can be statistically significant by this test at $\alpha = 0.05$ (two-tail) or $\alpha = 0.025$ (one-tail) when both n_1 and $n_2 \leq 3$. However, significance at this level can be achieved if $n_1 = 2$ and $n_2 \geq 8$, or $n_1 = 3$ and $n_2 \geq 5$. Critical values for these situations are not given in this table. If needed, see Hollander and Wolfe [12].

where

$$\mu_{W_R} = \frac{n_1(n+1)}{2},$$

and

$$\sigma_{W_R} = \sqrt{\frac{n_1 n_2(n+1)}{12}}, \quad n_1 \le n_2 \quad \text{and} \quad n = n_1 + n_2.$$

The critical values in Table 15.8 are used in a manner similar to those illustrated in Example 15.8 for Table 15.5. For the right-sided alternative, use the upper value and for the left-sided alternative use the lower value. For a two-sided alternative, use both values as shown. An $\alpha = 0.025$ for a one-tail test is equivalent to $\alpha = 0.05$ for a two-tail test because the Wilcoxon criterion is symmetric.

Note that both Wilcoxon tests, and for that matter any test based on ranks, are based on the order of the values and not the actual values. If one value is 10.1 and the next higher is 18.2, this receives the same rank even if it is 10.2. Thus, quantities lose half their relevance. For this reason, the Wilcoxon and all other rank-based tests are sometimes not considered adequate. Many nonparametric enthusiasts consider this a strength of the nonparametric test. In any case, they are the best procedures available so far for small samples when the underlying distribution is far from Gaussian. The basic premise of these tests is that they are *nonparametric*. They compare the distribution as a whole for location and not any particular parameter. But they can be considered to compare medians. Some books describe them as procedures for comparing medians.

Wilcoxon rank-sum test basically is a permutation test and is valid only when all the rearrangements of the data are equally likely under the null. This condition is met if and only if the groups differ in only location and nothing else. For example, if variances differ, the test ceases to be a test for difference in location. It mixes the variance differences also.

If each n_1 and n_2 is 30 or more, there is no need to use the criterion given by Equation 15.20. A Student t-test for means can be safely used for such large sample sizes in practically every situation.

15.3.2 Comparison of Three or More Groups: Kruskal–Wallis Test

Consider now one-way layout for which the procedure under the usual Gaussian conditions is ANOVA. You now know that ANOVA should not be used when a small sample size is available from a non-Gaussian distribution.

Consider an example of the cholesterol level in isolated diastolic hypertensives, isolated systolic hypertensives, clear hypertensives, and controls.

These groups are defined later in Example 15.9. All subjects are adult females of medium build and the groups are matched for age. They all belong to the same socioethnic group. Thus, the factors that may affect cholesterol level are controlled to a large extent. It is expected that the *pattern* of distribution of cholesterol level in different hypertension groups would be the same but not Gaussian. It is suspected that one or more groups may have measurements higher or lower than the others. The objective is to find whether or not the location differences between groups are statistically significant. Because the underlying distribution is not Gaussian, and if, in addition, the number of subjects in different groups is small, the conventional ANOVA cannot be used. The nonparametric Kruskal–Wallis test is the right method for such a setup.

Denote number of groups by J, each containing n subjects. For simplicity, let the number of subjects be the same in each group, but that is not a prerequisite. Rank all nJ observations jointly from smallest to largest. Denote the rank of the ith subject ($i = 1, 2, ..., n$) in the jth group ($j = 1, 2, ..., J$) by R_{ij}. Let the sum of the ranks of the observation in the jth group be denoted by $R_{\cdot j}$; that is, $R_{\cdot j} = \Sigma_i R_{ij}$. If there is no difference in the location of the groups, then $R_{\cdot 1}, R_{\cdot 2}, ..., R_{\cdot J}$ should be nearly equal and their variance nearly equal to zero. The following criterion exploits this premise:

$$\text{Kruskal–Wallis (K-W) test: } H = \frac{12}{nJ(nJ+1)} \frac{1}{n} \Sigma_j R_{\cdot j}^2 - 3(nJ+1). \qquad (15.21)$$

When ties occur and the observations are assigned average ranks, the criterion given by Equation 15.21 changes slightly. For details, see Hollander and Wolfe [12]. Standard statistical packages automatically take care of such contingencies.

If the null hypothesis of equality of groups were true, the value of H would be small. The distribution of H under H_0 is known and the P-value corresponding to the calculated value of H for a set of data can be obtained. Again, it is better to leave this part to a statistical software package. This will give the P-value. If you are interested in consulting a table of probabilities of H, see Table 15.9. This is restricted to $\alpha = 0.05$ and $J = 3$ groups. The table gives critical values of H for varying ns (group sizes can be unequal) but none exceeds 5. The ns are labeled such that $n_1 \geq n_2 \geq n_3$. Note from Table 15.9 that the differences between the groups cannot be statistically significant by the Kruskal–Wallis test unless $(n_1 + n_2 + n_3) \geq 7$ and at least two groups have two or more subjects. Your software may give you exact P-values for higher J and higher n than given in this table. If not, for four or more groups or any $n_j \geq 6$, H in Equation 15.21 can be approximated by chi-square with $(J - 1)$ df. In that case, reject H_0 if the calculated value of H exceeds the critical value of chi-square (Table B.3) at the desired significance level.

TABLE 15.9

Critical Values of Kruskal–Wallis Test H ($\alpha = 0.05$)

Sample Sizes[a]				Sample Sizes[a]			
n_1	n_2	n_3	Critical Value	n_1	n_2	n_3	Critical Value
2	1	1	b	5	1	1	b
2	2	1	b	5	2	1	5.00
2	2	2	b	5	2	2	5.16
3	1	1	b	5	3	1	4.96
3	2	1	b	5	3	2	5.25
3	2	2	4.71	5	3	3	5.65
3	3	1	5.14	5	4	1	4.99
3	3	2	5.36	5	4	2	5.27
3	3	3	5.60	5	4	3	5.66
4	1	1	b	5	4	4	5.66
4	2	1	b	5	5	1	5.13
4	2	2	5.33	5	5	2	5.34
4	3	1	5.21	5	5	3	5.71
4	3	2	5.44	5	5	4	5.67
4	3	3	5.79	5	5	5	5.78
4	4	1	4.97				
4	4	2	5.45				
4	4	3	5.60				
4	4	4	5.69				

Source: Adapted from Kruskal, W.H. and Wallis, W.A., *J. Am. Stat. Assoc.*, 47, 583, 1952; errata 48, 907, 1953. With permission. Copyright 1952 by the American Statistical Association. All rights reserved.

[a] n_1, n_2, and n_3 are labeled such that $n_1 \geq n_2 \geq n_3$.

[b] No difference can be statistically significant by this test at $\alpha = 0.05$ in these cases.

Example 15.9: Kruskal–Wallis Test for Cholesterol Level in Different Types of Hypertension

The cholesterol level in females with different types of hypertension and controls are given in Table 15.10. Definition of various types of hypertension is also given. Isolated systolic and isolated diastolic hypertensions are no longer considered benign conditions—they are now recognized as cardiovascular risk factors. Also given in parenthesis in this table are the joint ranks. The last column is the sum of the ranks ($R_{\cdot j}$) for the group.

In this case, $J = 4$ and $n = 5$. Using the sum of the ranks in the groups (last column), from the criterion given by Equation 15.21,

$$H = \frac{12}{5 \times 4(5 \times 4 + 1)} \frac{1}{5}(34.5^2 + 46^2 + 57.5^2 + 72^2) - 3(5 \times 4 + 1)$$

$$= 4.4.$$

TABLE 15.10

Cholesterol Level in Women with Different Types of Hypertension

Hypertension Group	Total Plasma Cholesterol Level (mg/dL)[a]					Sum of Ranks
No hypertension—control	221	207	248	195	219	
(DBP < 90, SBP < 140)	(8)	(2.5)	(16)	(1)	(7)	(34.5)
Isolated diastolic hypertension	217	258	225	215	228	
(DPB ≥ 95, SBP < 140)	(5)	(17)	(9)	(4)	(11)	(46)
Isolated systolic hypertension	262	227	207	245	230	
(DBP < 90, SBP ≥ 150)	(18)	(10)	(2.5)	(15)	(12)	(57.5)
Clear hypertension	218	238	265	269	240	
(DBP ≥ 90, SBP ≥ 140)	(6)	(13)	(19)	(20)	(14)	(72)

The hypertension categories are not exhaustive: For example, a subject with DBP = 92 and SBP = 138 mmHg would not be in this study.

DBP, diastolic blood pressure (mmHg); SBP, systolic blood pressure (mmHg).

[a] Rank of the value in the parentheses.

Since the number of groups is four, you can use the chi-square approximation. Table B.3 gives $P(H \geq 4.4) > 0.05$ for 3 df. A computer package gives $P = 0.2203$. This is not small, and therefore the null hypothesis of equality of locations of the groups cannot be rejected. Note that this conclusion is reached despite major differences in the value of $R_{\cdot j}$ in the four groups. The Kruskal–Wallis test reveals that those differences in this example may have arisen from sampling fluctuation when the groups actually have the same location.

15.3.3 Two-Way Layout: Friedman Test

In Example 15.9 on the cholesterol level in different types or hypertension, suppose the interest is in simultaneously considering obesity also. Let there be another study of the cholesterol level in thin, normal, and obese subjects belonging to each of the four hypertension groups. Now, this is a two-way layout. Factor-1 is hypertension status and factor-2 is obesity. The former has $J = 4$ levels as before and the latter has $K = 3$ levels. For simplicity, take the case of only one subject in each group, that is, $n = 1$. The test for effect of hypertension status and obesity on cholesterol level can be carried out by the Friedman test when the underlying distribution is non-Gaussian and $n = 1$. This test is primarily for correlated values such as at different points of time but can be used in a two-way setup since values at different levels of the same factor could be correlated. The procedure for testing for differences between levels of factor-1 in a two-way layout can be stated as follows:

Step 1. For each kth level of factor-2, rank J observations belonging to J levels of factor-1 from 1 to J in order of magnitude. Denote the rank of the observation for the jth level of factor-1 and kth level of factor-2 by R_{jk} ($j = 1, 2, \ldots, J; k = 1, 2, \ldots, K$).

Step 2. Calculate $R_{j\bullet} = \Sigma_k R_{jk}$. This is the sum of the ranks received by K subjects, one each in K levels of factor-2. If H_0 of no difference between levels of factor-1 is true, $R_{j\bullet}$ would be nearly same for all j ($j = 1$, $2, \ldots, J$). This leads to the criterion S due to Friedman as specified in the next step.

Step 3. Calculate

$$\text{Friedman test for factor-1: } S_1 = \frac{12}{JK(J+1)} \Sigma_j R_{j\bullet}^2 - 3K(J+1). \qquad (15.22a)$$

Step 4. If S_1 is large, the probability of H_0 of equality of levels of factor-1 being true is small. Reject H_0 if the P-value is less than the predetermined significance level. Computer packages may give the P-value right away. Otherwise, use Table 15.11. This table provides critical values of S_1 for $K = 3$ ($J = 3$–13), $K = 4$ ($J = 3$–8), and $K = 5$ ($J = 3$–5) for $\alpha = 0.05$. If K or J is large, S_1 has an approximate chi-square distribution with ($J - 1$) df. In that case, use the chi-square table to obtain the P-value.

TABLE 15.11

Critical Values of Friedman Test (S_1)
($\alpha = 0.05$)

J[a]	$K = 3$	$K = 4$	$K = 5$
3	6.00	7.40	8.53
4	6.50	7.80	8.80
5	6.40	7.80	8.96
6	7.00	7.60	
7	7.14	7.80	
8	6.25	7.65	
9	6.22		
10	6.20		
11	6.55		
12	6.50		
13	6.62		

Source: Hollander, M. and Wolfe, D.A., *Nonparametric Statistical Methods*, Table A.15, 1973. Copyright Wiley-VCH Verlag GmbH & Co. KGaA. Adapted with permission.

[a] For critical values of S_2 (for testing factor-2), the first column is K and the values tabulated are for $J = 3, 4$, and 5.

The preceding procedure is for testing differences in the levels of the factor-1. For testing differences in the levels of the other factor, J and K switch their position.

$$\text{Friedman test for factor-2: } S_2 = \frac{12}{JK(K+1)} \Sigma_k R_{\cdot k}^2 - 3J(K+1). \quad (15.22b)$$

Again, for large J or K, S_2 has an approximate chi-square distribution with $(K-1)$df. If you are using Table 15.11, switch J and K before you read the critical value of S_2; see the explanation given in the table.

Example 15.10: Friedman Test for Effect of Obesity and Hypertension on Cholesterol Level

Consider the data in Table 15.12 on total plasma cholesterol level (in mg/dL) in 12 subjects belonging to different hypertension groups. In parentheses are ranks within each hypertension category. In this example, $J = 4$ and $K = 3$. For levels of obesity (factor-2),

$$S_2 = \frac{12}{4 \times 3(3+1)}(5^2 + 9^2 + 10^2) - 3 \times 4(3+1)$$

$$= 3.5.$$

For this value of S_2, when $J = 4$ and $K = 3$, $P = 0.273$ from a statistical package. This is more than 0.05. Also, the critical value in Table 15.11 for reversed $J = 4$ and $K = 3$ is 7.40 for $\alpha = 0.05$. The value of S_2 has not been able to reach this threshold. This again tells that $P \geq 0.05$. Thus, the evidence is not enough to conclude that obesity affects cholesterol level in these subjects. You may wish to calculate S_1 for hypertension groups. This is $S_1 = 3.40$ and corresponds to $P = 0.446$. Thus, there is no sufficient evidence for difference in cholesterol levels in different hypertension groups either.

TABLE 15.12

Cholesterol Level (mg/dL) in Persons with Different Obesities and Hypertension

| Hypertension Group (Factor-1) | Obesity (Factor-2) | | |
	Thin	Normal	Obese
No hypertension	248	233	263
	(2)	(1)	(3)
Isolated diastolic hypertension	247	249	258
	(1)	(2)	(3)
Isolated systolic hypertension	261	275	267
	(1)	(3)	(2)
Clear hypertension	225	290	285
	(1)	(3)	(2)
Total of ranks ($R_{\cdot k}$)	(5)	(9)	(10)

The following comments for Kruskal–Wallis test and Friedman test may be useful:

1. In case of the Kruskal–Wallis and Friedman tests, notice that the chi-square approximation is used for fairly small sample sizes. This is quite an approximation. Such an approximation becomes necessary mostly because the exact distribution of these criteria, particularly Friedman's, is very difficult to obtain. In case you have a software that gives you exact probabilities, there is no need to use chi-square approximation.

2. Kruskal–Wallis and Friedman criteria arise naturally when the usual ANOVA F is calculated for ranks instead of the actual values. For $K = 2$, Friedman reduces to a two-sided sign test.

3. This section, in fact, the whole of this chapter, is restricted to the testing of hypothesis situation. An overall view of these methods is given in **Table S.4** at the beginning of the book. The CIs for the mean and for the difference in two means under Gaussian conditions and for some situations were presented in Chapter 12. For other nonparametric situations, obtain CI by the methods given by Conover [13]. He also provides advice for experimental designs for which no nonparametric test exists. That book also discusses other nonparametric procedures such as for multiple comparisons and regression.

This chapter is restricted to tests for difference in locations such as means in different groups. A large variety of other inferences can be drawn on means. One that is very common is trend and the relationship among means. This is discussed in Chapter 16.

15.4 When Significant Is not Significant

A debate was initiated in Section 12.5.2 on the real meaning of statistical significance. It primarily emphasized that a statistically significant result may not have any medical significance. Let us carry this debate further in this section at the risk of some duplication.

15.4.1 Nature of Statistical Significance

As already emphasized, nearly all information in health and medicine is empirical in nature. It is gathered from samples time to time. Besides all other sources mentioned in Chapter 1, the samples are a big source of uncertainty by themselves. Samples tend to differ from one another. For instance, there is no reason why 10-year survival rates of cases of breast carcinoma in two groups of women, the first born on odd dates of any month and the second on even dates, should differ. But there is a high likelihood that the rates would be different in a *sample* of women of this type. To recapitulate, this sampling fluctuation depends primarily on two considerations: (1) the sample size n and (2) the intrinsic interindividual variability in the subjects. The former is fully under control of the investigator. The latter is not under human control, but an appropriate design can still minimize its influence on medical decisions. The sources of uncertainty other than those intrinsic in the subjects, such as observer variation and inadvertent measurement errors, are controlled by adopting a suitable method of data collection.

The purpose of mentioning this again is to repeat that sample size plays a dominant role in statistical inference. As demonstrated earlier, larger n could substantially reduce the SE. This helps to increase the reliability of the results when based on the average kind of summary measures. A narrow CI is then obtained, which can help in drawing a focused conclusion. Inference from a test-of-hypothesis procedure can be drawn with less chance of error when n is larger. However, a side effect of a large n is that a very small difference can become statistically significant. This difference may or may not be medically relevant.

Besides the sample size, which can sometimes cause problems, the level of statistical significance can also create confusion. Statistical significance is said to have reached when the probability of Type-I error is very low. Most use 0.05 as the threshold but sometimes 0.10 or 0.01 is also used. Although $P < 0.01$ implies $P < 0.05$, $P < 0.10$ does not imply $P < 0.05$. If $P = 0.08$, the result would be statistically significant at $\alpha = 0.10$ but not at $\alpha = 0.05$. This shows that caution is needed in drawing conclusions from statistical significance. Some of these are explained next.

1. *Whether or not a statistically significant result has any medical significance.* You may wish to revisit Example 12.12, in which the relief rate from sore throat within a week by a drug is 73% as opposed to 70% generally seen in subjects not receiving any drug. The sample size was $n = 800$. Because of such large n, the difference is statistically significant ($P < 0.05$). This example highlights two other aspects of the nature of statistical inference. (1) Statistical significance only means that probability of no difference in the target population is extremely small. It does not say how much difference is present. It cannot be concluded in this example that the difference is 3%. If n is really very large, even a difference of 1% would become statistically significant. (2) Even if it is assumed that the drug really increases the relief rate from 70% to 73%, the question of medical relevance is whether this rise of 3% makes it worth taking the drug. There is a cost involved, not just the price of the drug but also efforts in procuring it and inconvenience of ingesting it. Also there is always a possibility of side effects. Thus, it is important that physicians specify the minimum difference the drug must make for it to be acceptable as a better treatment modality. If this is, say, an increase of minimum 10% in relief rate, obviously the sample of 800 in

this example does not pass the drug. In fact, if the observed difference in the sample is 15%, it may still fail the test of minimum 10%.

2. *Whether or not a plausible medical reason is available for the observed difference.* In the preceding example, it can probably be safely concluded that the increase in relief rate is due to the effect of the drug. Sometimes the difference is difficult to explain. Consider a random sample of 24 male and 15 female patients with leukemia. Suppose four males and seven females survive for 5 years. The difference in their survival rate is statistically significant because P is less than 0.05 for one-sided H_1. No worthwhile reason may be available for this difference in their survival rate. When the level of significance is 5%, there is a 1 in 20 chance that false significance is obtained. On the other hand, there might be hitherto unknown factors that could account for such a difference, and the difference could be real. For example, an inborn resistance in females, which leads to their greater longevity, may be an explanation. Statistical significance without proper medical explanation is rarely useful. However, such an explanation may not be immediately available in some situations and may emerge in future.

3. *Whether or not the P-value obtained is sufficiently small.* The convention is to use a threshold of 0.05 to label a P-value small or large, but this is not uniformly applicable in all cases. In cases where the consequence of accepting H_1 can be grave, as in the case of accepting a drug with major side effects, a smaller α level such as 0.01 should be used. On the other hand, an inflated threshold such as 0.10 can be used in behavioral research (e.g., the opinion on optimum size of sibship reported by males and females or by people of lower and upper socioeconomic status).

A result may be statistically significant at the 10% level but not at the 5% level. This may be misused to convey the result in either way as illustrated in the following example.

Example 15.11: *P*-Value between 5% and 10% for pH Rise by a Drug

Table 15.13 gives the rise in blood pH concentrations in 18 patients with acid peptic disease after treatment for 1 month by a new drug. Ten patients receiving placebo were also investigated for rise in their pH concentration after 1 month. In these data,

pH rise in the drug group: Mean $\bar{x}_1 = 1.219$; SD $s_1 = 1.152$;
pH rise in the control group: Mean $\bar{x}_2 = 0.585$; SD $s_2 = 0.788$.

The sample size is small and the data should be examined for violation of a Gaussian pattern. This was done by the procedures mentioned in an earlier chapter separately for the data in the two groups. The details are not given here, but no serious violation was observed. The null hypothesis of equality of variances also could not be rejected by Levene test. Thus, you can go ahead with the pooled variance Student t-test.

TABLE 15.13

Rise in Blood pH in Patients with Acid
Peptic Disease after a Treatment

pH Rise in the Drug Group					
0.87	1.58	0.53	−1.95	1.88	2.51
0.08	2.89	2.03	1.81	1.32	0.32
0.09	0.71	1.29	1.67	2.53	1.78

pH Rise in the Control Group				
−0.41	−0.38	1.58	0.54	1.21
1.09	1.37	−0.01	−0.28	1.14

The pooled variance

$$s^2 = \frac{(18-1)(1.152)^2 + (10-1)(0.788)^2}{18+10-2}$$

$$= 1.0827.$$

In this case, df = 18 + 10 − 2 = 26. Thus, under H_0: $\mu_1 = \mu_2$,

$$t_{26} = \frac{1.219 - 0.585}{\sqrt{1.0827(1/18 + 1/10)}}$$

$$= 1.54.$$

The alternative hypothesis in this case is one sided (H_1: $\mu_1 > \mu_2$) if the possibility of lower pH in the drug group is excluded. Since this is right-sided H_1, the required P-value is $P(t_{26} > 1.54)$. A statistical package gives this as 0.0675. Thus, the P-value is less than 0.10 but more than 0.05. If the threshold 0.10 is used, it can be claimed that the new drug does increase the blood concentration of pH in cases of acid peptic disease. This claim is not tenable at $\alpha = 0.05$. The pharmaceutical literature on the drug may claim that the "drug is effective ($P < 0.10$)." This statement is true but provides a different perception than saying that the drug was not proved effective in raising pH level at $\alpha = 0.05$. Thus, care is needed in judging the statement made in the literature as well as in making a statement.

4. *Whether or not statistical tests are used for several variables for the same group of subjects.* The procedures mentioned in this chapter are applicable to only one variable at a time. If you measure arterial blood gases HCO_3, PCO_2, and PO_2 in asthma patients and observe statistically significant ($P < 0.05$) alterations in all three gases individually, then composite conclusion jointly for all three of them should not be drawn. Multivariate methods are required when the variables are to be considered together. The results obtained in multivariate setups are not necessarily the same as those obtained by multiple tests on individual variables.

Cupples et al. [14] have illustrated the problem of multiple testing. In their data, individual tests on baseline levels of systolic BP, cholesterol, blood sugar, relative weight, age, lung capacity, uric acid, and ventricular rate revealed statistically significant ($P < 0.05$) differences between those who developed coronary heart disease (CHD) in the long run and those who did not. Only the Hb level was not different ($P = 0.97$) and perhaps cigarette smoking ($P = 0.061$). This analysis of individual variables ignores the possible correlation among some of these variables, for example, between age and lung capacity, and between systolic BP and cholesterol level. This can make the difference in some variables unnecessarily significant when it actually is a by-product of the difference in variables that are correlated. The second problem with this analysis is that the actual probability of Type-I error is much higher. There are 10 variables in this analysis and a separate test has been used for each variable at the 5% level. Thus, total $\alpha = 1 - (1 - 0.05)^{10} = 0.40$. When multivariate analysis (such as the discriminant of Chapter 19) is done, blood sugar, lung capacity, uric acid, and ventricular rate fail to be significant contributors to the long-range development of CHD. On the other hand, cigarettes smoked then become significant, which was not in univariate analysis. This example illustrates for multiple variables that you have to be on guard not only when you draw conclusions regarding statistical significance in your data but also when using the results of other studies.

5. *Whether or not multiple tests are used on the same variable.* In a study on referral behavior, suppose data are collected from 20 physicians on the number of patients referred on various days of the week in a 10-week period. The hypothesis under test is whether the mean number of referrals on Fridays is different from those on other weekdays. Let the difference between Friday and other weekdays not be statistically significant ($P \geq 0.05$) for 19 physicians and significant ($P < 0.05$) for one particular physician. Can it be concluded that this physician has behavior different from the others? Recollect that $P < 0.05$ implies that there is less than 5% chance of the difference really being present in the target population. But this also means that nearly 1 in 20 samples can give $P < 0.05$ even when H_0 is true. That is, rejecting H_0 in one case out of 20 can be in error. On the basis of chance alone, 1 out of 20 samples can lead to a wrong conclusion. Thus, not much value can be attached to one physician yielding $P < 0.05$ out of 20 physicians, particularly if his average referrals are not much different.

The problem of multiple comparisons is already discussed earlier in this chapter. When ignored, this also can result in statistical significance for some comparisons when, in fact, no significance is present.

6. *Whether or not the size of the sample is adequate.* Freiman et al. [15] reexamined 71 negative trials to determine whether a sufficiently large sample was studied. Negative trials are those that report that a difference is not statistically significant. They concluded that 50 of these trials had a greater than 10% risk (Type-II error) of missing a true 50% therapeutic improvement. This happened because the size of the sample was not sufficiently large. Dimmick et al. [16] reported similar findings for surgical trials. The sample size must be adequate to inspire confidence that a medically relevant difference would not go unnoticed. The following example explains this problem. This example is for proportions and not for the means.

Example 15.12: Limitation of Sample Size in Negative Trials

Table 15.14 contains results of a trial in which patients receiving a regular tranquilizer were randomly assigned to continued conventional management and a tranquilizer support group. The null hypothesis is that the two groups are similar. Under this H_0, the expected frequency in each cell is $15 \times 15/30 = 7.5$. Since no expected frequency is less than 5, chi-square can be safely applied. This gives $\chi^2 = 3.33$ and $P > 0.05$ at $1\,df$. Note that the number of patients who stopped taking the tranquilizer in the support group is twice that in the conventional group. Yet the difference is not statistically significant. There is a clear case of a trial on a larger n. If the same type of result was obtained with $n = 30$ in each group, then the difference would be statistically significant.

SIDE NOTE: You may wish to do this as an exercise and examine whether significance is achieved with $n = 30$ in each group and all numbers doubled.

Example 15.12 also brings in the question of distinction between significant, real, and important. As already demonstrated, a very large n can make a medically unimportant difference statistically very significant. A statistically significant difference is *very likely* to be real although there is a small chance that it is not real. On the other hand, a real difference may not be statistically significant if

TABLE 15.14

Duration of Taking Tranquilizer in Support Group and Conventional Group

	Tranquilizer Support Group	Conventional Management Group
Still taking tranquilizer after 16 weeks	5	10
Stopped taking tranquilizer by 16 weeks	10	5
Total	15	15

n is small. Similarly, a large and medically important difference can also be statistically not significant if n is not sufficiently large. A real but small difference such as 3 mg/dL in average total plasma cholesterol between treatment responses in males and females can be medically unimportant or of no prognostic consequence.

7. *Whether or not averages are masking the differences present in a large section of the subjects.* Averages can be deceptive, and there is always a need to be cautious when interpreting them. As mentioned earlier, very small and very large values together could produce a middling kind of mean. Similarly, large variation between individuals can mask the difference between two or more groups. This is illustrated in the Example 15.13.

With all this, it should be clear that a regimen can be abandoned if sufficiently powered study tells you that the effect is not statistically significant. At the same time, do not forget that P-value alone is rarely enough to draw a valid conclusion. Previous knowledge, biological plausibility, and your intuition that would also incorporate epistemic gaps must remain the guiding factors. A valid conclusion is reached when all these are considered together wherein statistical evidence plays a role though not as dominant as made out by some statisticians.

Example 15.13: Difference Masked by Means Is Revealed by Proportions

Consider the data in Table 15.13 on pH rise after a drug in cases with acid peptic disease. Four of 10 controls exhibited a decline while only 1 of 18 cases with acid peptic disease had a decreased pH value. Thus, Table 15.15 is obtained.

The frequencies expected under the null hypothesis of no association are small for the cells in the second column. Thus, Fisher exact test is needed. This gives $P = 0.041$ for one-sided H_1. This is sufficiently small for H_0 to be rejected at the 5% level. The conclusion now is clearly in favor of the drug. This is different from the one obtained earlier in Example 15.11 on the basis of comparison of means. Which one should you believe? The answer depends on the objective of the study. If any rise, small or large, is more relevant than the magnitude, then the method based on Fisher exact test is more valid. If the magnitude of rise is important, then the test based on means is more valid.

TABLE 15.15

Rise and Decline in pH in Cases and Controls in Table 15.13

	Rise in pH	Decline in pH	Total
Cases	17	1	18
Controls	6	4	10

15.4.2 Testing for Presence of Medically Important Difference in Means

The null hypotheses discussed so far are generally for no difference. When this H_0 is rejected, the only conclusion reached is that a difference is present. No statement can be made about the magnitude of difference. The difference could be very small with no clinical implication or could be large enough to be medically important. As mentioned for proportions, this uncertainty is tackled by setting up an H_0 that specifies the magnitude of difference. Consider the following examples:

1. When can a new antihypertensive drug be considered clinically effective in postsurgical cases?
   a. Average decrease in diastolic BP by at least 2, 5, 8, or 10 mmHg? Or,
   b. Achieving a threshold diastolic BP such as 90 mmHg in a large percentage of subjects—60%, 70%, or 90% of subjects?

2. An iron supplementation program in female adolescents is organized. This raises the mean Hb level from 13.6 to 13.8 g/dL after intake for 30 days. Is this gain of 0.2 g/dL in the average sufficient to justify the program? What gain can be considered enough to justify the expenditure and efforts in running the program—0.5, 1 g/dL, or more?

3. The prognostic severity of bronchiolitis in children can be assessed either by respiration rate (RR) alone or by using a bronchiolitis score (BS) comprising RR, general appearance, grunting, wheezing, etc. Suppose the former can correctly predict severity in 65% of cases and the latter when considered together in 69% of cases. BS is obviously more complex to implement. Is it worth the trouble to use BS in place of RR for a gain of nearly 4%? What percentage gain in predictivity warrants adopting a more complex BS instead of simple RR—5%, 10%, 15%, or higher?

4. Suppose the normal intraocular pressure (IOP) is 15.8 (SD = 2.5) mmHg in healthy subjects when measured by applanation tonometry. In glaucoma, it is elevated. In a group of 60 patients with primary open-angle glaucoma, suppose the average was 22.7 (SD = 4.5) mmHg. After treatment with a new beta-adrenergic blocker, it came down to 19.5 (SD = 3.7) mmHg. This reduction is statistically significant, but the reduced level in the treatment group is still higher than the healthy level. Is this reduction still clinically important? What kind of difference from the normal level of 15.8 mmHg is clinically tolerable—½ mmHg, 2 mmHg, or higher?

In all such problems, the medical profession needs to decide the minimum acceptable or tolerable difference that justifies the intervention under consideration. The specification could be either in terms of proportion or in terms of mean. In cases 1b and 3 mentioned earlier, the question is related to proportions, and in 1a, 2, and 4, it is related to means. The methods to detect medically important differences in proportions were described in Section 13.2.2. The corresponding method for means is given next.

15.4.2.1 Detecting Specified Difference in Mean

Let the minimum acceptable (or tolerable) difference be μ_0. Then $H_0: \mu_1 - \mu_2 = \mu_0$ and $H_1: \mu_1 - \mu_2 > \mu_0$. The test criterion for independent samples (pooled variance setup) is

$$t_{n_1+n_2-2} = \frac{\bar{x}_1 - \bar{x}_2 - \mu_0}{s_p\sqrt{1/n_1 + 1/n_2}}. \tag{15.23}$$

This is same as the criterion given in Equation 15.6a except for extra μ_0 in the numerator. Similarly, the criterion for a paired setup is the same as in Equation 15.3 except for the difference μ_0 in the numerator. For right-sided H_1, find the P-value in the right tail. This would be the probability that t is *more* than the value obtained for the sample. Reject the null hypothesis when the P-value so obtained is less than the predetermined level of significance.

The conditions for application of the criterion given in Equation 15.23 are the same as for the two-sample t-test. Gaussian conditions are required. These include large n if the underlying distribution is non-Gaussian. If n is small and the underlying distribution is non-Gaussian, use nonparametric methods.

Example 15.14: Detecting Minimum Reduction in Homocysteine Level

It is now well known that an increased level of homocysteine is a risk factor for myocardial infarction and stroke. It has been observed in some populations that vitamin B converts homocysteine into glutathione, which is a beneficial substance, and reduces homocysteine level. To check whether this also happens in nutritionally deficient subjects, a trial was conducted in 40 adults with Hb < 10 g/dL. They were given vitamin B tablets for 3 weeks and their homocysteine level before and after was measured. Vitamin B has almost no known side effect, but the cost of administering vitamin B to all nutritionally deficient adults in a poor country with a great deal of undernourishment can be enormous. The program managers would consider this a good program if the homocysteine level were reduced by at least 3 μmol on average. In this sample of 40 persons, the average reduction is 3.7 μmol with SD = 1.1 μmol. Can the mean reduction in the target population still be 3 μmol or less?

This is a paired setup. In this case, the medically important difference is a minimum 3 μmol. Thus, $H_0: \mu_1 - \mu_2 = 3$ μmol and $H_1: \mu_1 - \mu_2 > 3$ μmol:

$$t_{39} = \frac{3.7 - 3}{1.1/\sqrt{40}}$$

$$= 4.02.$$

From Table B.2, $P(t > 4.02) < 0.01$. This is extremely small. The null hypothesis is rejected. The conclusion reached is that the mean reduction in homocysteine level is more than $3\,\mu$mol. Evidence from this sample is sufficient to conclude that running this program will provide the minimum targeted benefit.

15.4.2.2 Equivalence Tests for Means

You are now in a position to learn further about equivalence tests. The primary aim of these tests is to disprove a null hypothesis that two means, or any other summary measure, differ by a clinically important amount. This is reverse to what has been discussed so far. Equivalence tests are designed to demonstrate that no important difference exists between a new regime and the current regimen. They can also be used to demonstrate stability of a regimen over time, equivalence of two routes of dosage, and equipotency.

Equivalence can be demonstrated in the form of either at least as good as the present standard, or neither better nor worse than the present standard. The former in fact is noninferiority and the latter is indeed equivalence. Since only the outcome is compared and not the course of disease, this is clinical or therapeutic equivalence and not bioequivalence.

As explained for proportions in Chapter 13, the concern in equivalence tests is to investigate if means in two groups do not differ by more than a prespecified medically insignificant margin Δ. The null hypothesis in this case is that the difference is Δ or more, and the burden of providing evidence against this null is on the sample. If the evidence is not sufficient, the null is conceded and the conclusion is that the two groups are not equivalent on average.

The procedure is essentially the same as described earlier for equivalence of proportions. Construct a $100(1 - 2\alpha)\%$ CI and see if it is wholly contained in the interval $-\Delta$ to $+\Delta$. *Irrespective of statistical significance,* if such CI is within $(-\Delta, +\Delta)$, the two groups are equivalent. Otherwise not. Refresh your memory and see Figure 13.1 again. If the CI overlaps, and the lower or upper limit lies outside the prespecified zone of clinical indifference, the null hypothesis of nonequivalence cannot be rejected. In this case, examine whether increasing the sample size can help to decide one way or the other.

15.4.3 Power and Level of Significance

The concepts of power and level of significance were introduced in Chapter 12 but I would like to go deeper into these concepts and explain them further

now that the actual tests have been discussed. Statistical significance is intimately dependent on these two concepts. Consider the following data on
birth weight in a control group and a group with individual educational
campaign regarding proper diet:

Control group (n_1 = 75): Average birth weight $\bar{x}_1$ = 3.38 kg, SD of birth
weight s_1 = 0.19 kg

Campaign group (n_2 = 75): Average birth weight $\bar{x}_2$ = 3.58 kg, SD of birth
weight s_2 = 0.17 kg.

An individual-based educational campaign requires intensive inputs.
Suppose it is considered worthy of the effort only if it can make a difference
of at least 100 g in the average birth weight. What is the power for detecting
this kind of difference when the significance level is α = 0.05?

There is no stipulation in this case that the average birth weight in the campaign group could be lower. Thus, the alternative hypothesis is one-sided,
H_1: $\mu_1 - \mu_2 < 0$. Note that label 1 here is for the control group and label 2 for
the campaign group. At significance level α = 0.05, the null hypothesis would
be rejected in favor of this H_1 when

$$\frac{\bar{x}_1 - \bar{x}_2}{\text{SE of } (\bar{x}_1 - \bar{x}_2)} < -1.645. \tag{15.24}$$

The left side of inequality (15.24) is the usual statistic Z when H_0: $\mu_1 - \mu_2 = 0$
is true. The right side is the value of Z from the Gaussian table for left-tailed
α = 0.05. The Gaussian form is applicable in this case because n is sufficiently
large. When σ is not known, the actual test applicable is the two-sample
Student t, but for df = $n_1 + n_2 - 2 = 148$ (which is very large), the critical value
would still be –1.645. Table B.1 can also be used for such a large df. When
H_0: $\mu_1 - \mu_2 = 0$ is true, the probability of (15.24) is α = 0.05. The actual Z, as is
evident from Equation 12.7 in Chapter 12, is

$$Z = \frac{(\bar{x}_1 - \bar{x}_2) - (\mu_1 - \mu_2)}{\text{SE of } (\bar{x}_1 - \bar{x}_2)}, \tag{15.25}$$

and it is only under H_0 that it reduces to the left side of the inequality (15.24). Power is the probability of Z in Equation 15.25 less than –1.645
when $(\mu_1 - \mu_2) < 0$. But this probability can be obtained only if the difference
$(\mu_1 - \mu_2)$ is exactly specified. In this example $(\mu_1 - \mu_2) = -100$ g (or –0.1 kg). This
specifies the magnitude of difference between the null and the alternative
hypothesis. The power depends on this difference. The larger the difference,
the greater is the power.

In this example, $s_p = \sqrt{(74 \times 0.19^2 + 74 \times 0.17^2)/148} = 0.18028$. Thus, SE of $(\bar{x}_1 - \bar{x}_2) = 0.18028\sqrt{1/75 + 1/75} = 0.0294$. Now,

$$\text{Power for } (\mu_1 - \mu_2 = -0.1) = P\left(\frac{\bar{x}_1 - \bar{x}_2}{\text{SE of } (\bar{x}_1 - \bar{x}_2)} < -1.645 / (\mu_1 - \mu_2 = -0.1) \right)$$

$$= P\left(\frac{(\bar{x}_1 - \bar{x}_2) - (\mu_1 - \mu_2)}{\text{SE of } (\bar{x}_1 - \bar{x}_2)} < -1.645 - \frac{-0.1}{\text{SE of } (\bar{x}_1 - \bar{x}_2)} \right)$$

$$= P\left[Z < -1.645 - \frac{(-0.1)}{0.0294} \right],$$

where Z is obtained from the Equation 15.25. This gives

$$\text{Power for } (\mu_1 - \mu_2 = -0.1) = P(Z < 1.76)$$

$$= 1 - P(Z \geq 1.76)$$

$$= 0.96 \text{ from Table B.1.}$$

This probability is very high. This study on 75 subjects each is almost certain to detect a difference of 100 g in average birth weight if present.

Now, for the purpose of illustration, calculate the power for detecting a difference of 50 g:

$$\text{Power for } (\mu_1 - \mu_2 = -0.05) = P\left[Z < -1.645 - \frac{(-0.05)}{0.0294} \right]$$

$$= P(Z < 0.06)$$

$$= 1 - P(Z \geq 0.06)$$

$$= 0.52 \text{ from Table B.1.}$$

This probability is not so high. The chance of detecting an average difference of 50 g in birth weight in this case is nearly one-half. If the campaign is really able to improve birth weight by an average of 50 g, this study on 75 subjects in each group is not very likely to detect it. In other words, such a difference can actually be present but can easily turn out to be statistically not significant because samples of this size do not have sufficient power to detect such a small difference.

If the significance level is relaxed to $\alpha = 0.10$, then H_0 would be rejected when

$$\frac{\bar{x}_1 - \bar{x}_2}{\text{SE of } (\bar{x}_1 - \bar{x}_2)} < -1.28,$$

where the right-hand side is the value of Z corresponding to left-tail $\alpha = 0.10$. For this α,

$$\text{Power for } (\mu_1 - \mu_2 = -50\,\text{g or} - 0.05\,\text{kg}) = P\left[Z < -1.28 - \frac{(-0.05)}{0.0294} \right]$$

$$= P(Z < 0.42)$$

$$= 1 - P(Z \geq 0.42)$$

$$= 0.66 \text{ from Table B.1.}$$

Note how the power increases when the significance level is increased.

To see the effect of sample size, increase n_1 and n_2 to 250 each. With these ns, get $s_p = 0.19$ and SE = 0.01699. For $\alpha = 0.10$,

$$\text{Power for } (\mu_1 - \mu_2 = -0.05\,\text{kg}) = P\left[Z < -1.28 - \frac{(-0.05)}{0.01699} \right]$$

$$= P(Z < 1.66)$$

$$= 1 - P(Z \geq 1.66)$$

$$= 0.95 \text{ from Table B.1.}$$

These results are summarized in Table 15.16.

The following conclusions can now be drawn:

1. Power is directly related to the magnitude of difference to be detected. It decreases if the difference to be detected is small. In other words, a difference may be actually present but it is more difficult to call a small difference statistically significant. This may be appealing to your intuition also.

TABLE 15.16

Illustration of Power Variation in Different Situations

$\mu_1 - \mu_2$(g)	α	n_1	n_2	Power
−100	0.05	75	75	0.96
−50	0.05	75	75	0.52
−50	0.10	75	75	0.66
−50	0.05	250	250	0.95

2. Power increases as the sample size increases. A small difference is more likely to be statistically significant when n is large—a fact that has been emphasized over and over again.

3. Power increases if a higher probability of Type-I error can be tolerated. Note that increase in the level of significance is a negative feature whereas an increase in power is a positive feature of a statistical test. These should be balanced.

15.4.3.1 Balancing Type-I and Type-II Error

Since power is the complement of Type-II error, it should be clear that increasing one type of error decreases the other type and vice versa. The two can rarely be simultaneously kept low for fixed sample size.

As explained earlier, Type-I error is like misdiagnosis or like punishing an innocent person and thus more serious than Type-II error. Can Type-II error be more hazardous than Type-I error? Yes, in some rare situations. Suppose H_0 is a drug that is not too toxic and H_1 that it is too toxic. Type-II error in this case can be serious. In the context of drug trials, Type-I error corresponds to an ineffective drug allowed to be marketed and Type-II error corresponds to an effective drug denied entry into the market. In this sense, Type-II error may have more serious repercussions if the drug is for a dreaded disease such as AIDS or cancer. If a drug can increase the chance of 5-year survival for such a disease by, say, 10%, it may be worth putting it in the market provided the side effects are not serious. Not much Type-II error can be tolerated in such cases. This error can be decreased (or power increased) by increasing the sample size but also by tolerating an increased Type-I error. But if the side effects were serious, a rethinking would be necessary. Thus, the two errors should be balanced keeping their consequences in mind.

References

1. Gamst G, Meyers LS, Guarino AJ. *Analysis of Variance Designs: A Conceptual and Computational Approach with SPSS and SAS.* Cambridge, U.K.: Cambridge University Press, 2008.
2. Everitt BS. *Statistical Method for Medical Investigation,* 2nd edn. London, U.K.: Edwin Arnold, 1994, pp. 77–91.
3. Dixon WJ, Massey FJ Jr. *Introduction to Statistical Analysis,* 2nd edn. New York: McGraw Hill, 1957, p. 324.
4. Sell A, Olkkola KT, Jalonen J, Aantaa R. Minimum effective local anaesthetic dose of isobaric levobupivacaine and ropivacaine administered via a spinal catheter for hip replacement surgery. *Br J Anaesth* 2005; 94:239–242.

5. Tokunaga S, Takeda Y, Shinomiya K, Hirase M, Kamei C. Effect of some h(1)-antagonists on the sleep-wake cycle in sleep-disturbed rats. *J Pharmacol Sci* 2007; 103:201–206.

6. Neter J, Wasserman W, Kutner MH. *Applied Linear Regression Models: Regression, Analysis of Variance, and Experimental Designs.* Homewood, IL: Richard D Irwin, 1983.

7. Wang X, Tager IB, van Vunakis H, Speizer FE, Hanrahan JP. Maternal smoking during pregnancy, urine cotinine concentrations, and birth outcomes: A prospective cohort study. *Int J Epidemiol* 1997; 26:978–988.

8. Doncaster P, Davey A. *Analysis of Variance and Covariance: How to Choose and Construct Models for the Life Sciences.* Cambridge, U.K.: Cambridge University Press, 2007.

9. Kenward MG. *Analysis of Repeated Measurements.* Oxford, U.K.: Oxford University Press, 1998.

10. Klockars AJ, Sax G. *Multiple Comparisons (Quantitative Applications in the Social Sciences).* New York: Sage Publications, 2005.

11. Conover WJ. *Practical Nonparametric Statistics,* 3rd edn. New York: Wiley, 1999, p. 419.

12. Hollander M, Wolfe, DA. *Nonparametric Statistical Methods,* 2nd edn. New York: Wiley-Interscience, 1999.

13. Conover WJ. *Practical Nonparametric Statistics,* 3rd edn. New York: John Wiley & Sons, 1999.

14. Cupples LA, Heeren T, Schatzkin A, Colton T. Multiple testing of hypotheses in comparing two groups. *Ann Intern Med* 1984; 100:122–129.

15. Freiman JA, Chalmers TC, Smith H Jr, Kuebler RR. The importance of beta, the type II error and sample size in the design and interpretation of the randomized control trial: Survey of 71 "negative" trials. *N Engl J Med* 1978; 299:690–694.

16. Dimmick JB, Diener-West M, Lipsett PA. Negative results of randomized clinical trials published in the surgical literature: Equivalency or error? *Arch Surg* 2001; 136:796–800.

16

Relationships: Quantitative Data

Relationships are inherent in medicine and health. Eosinophil count normally decreases from birth to adulthood, and the risk of lung cancer increases with the amount and duration of smoking. Total leukocyte count (TLC) and hemoglobin (Hb) levels decrease up to the age of 6–8 years and show some increase thereafter in healthy subjects. Height velocity is at its peak during adolescence. Blood pressure (BP) levels have a positive relationship with waist–hip ratio (WHR), and the incidence of dental caries in children depends negatively on the socioeconomic grade of their family. Such relationships are everyday occurrences in the practice of medicine and health, so much so that they are sometimes taken for granted.

There are two primary purposes of studying such relationships: (1) to be able to predict, with some degree of precision, the value of one with the help of the others, and (2) to help understand the underlying mechanisms in such a relationship, particularly by studying the effect of alteration in the value of one variable when other conditions do not change. The latter can provide useful clues about what determines the outcome and how much influence each variable has.

It is known that height and weight are correlated with age in children but visual acuity is independent of age in this group. Hb level, serum ferritin, and mean corpuscular volume (MCV) are related to one another but each is unrelated to WHR. Risk of breast cancer may be related to dietary intake but unrelated to Hb level. But the presence or absence of relationships is not absolute. Some relationship show up sometimes but vanish at other times. Variation and uncertainties play a prominent role in the presence or absence of such relationships. Some individuals or groups defy the nature and strength of relationships seen in other subjects. Thus, care is needed in their use and interpretation. As always, the attempt is to identify the trend and separate it out from fluctuations. Relationship is just another term for such a trend.

What do we statistically mean by relationship? Consider the example of cholesterol level affected by obesity and hypertension. Cholesterol level is a continuous variable in this case. The degree of obesity can be measured on a continuous scale by body mass index (BMI), but it can also be divided into four groups, namely, thin, normal, overweight, and obese. Then, the quantification of BMI is lost although the gradient remains. That cholesterol is affected (or not affected) by obesity is a perfect conclusion in this case, but such differences in means in various groups of subjects do not generally come under the term statistical relationship. Relation in statistical sense is the *nature or form of relationship*. Thus, this is an expression of quantitative change in one variable per unit change in the other. In the cholesterol level/obesity example, a relationship would be able to indicate how much increase in cholesterol level is expected when BMI increases from, say, 25 to 26. Such a form of relationship is called **regression**. In general, this is expressed as

$$\hat{y} = f(x_1, x_2, \ldots, x_K), \tag{16.1}$$

where

$\hat{y}$ is the predicted value of the outcome or response of interest, called dependent

f is some function that specifies the actual form

$x_1, x_2, \ldots, x_K$ are variables that may predict the variable y, called independents

For the methods of this chapter, each value of y must be independent of other values of y. For example, they cannot belong to different sites of the same body.

K (≥ 2) variables on the right side of Equation 16.1 make it a **multivariable setup**. As explained next, this allows to study net effect of each and joint effect of two or more independent variables on the dependent. While such multivariable setup seems like the only feasible method to assess joint effect, net effect of each individual variable can be studied by stratified analysis also. If you suspect that BMI and sex affect cholesterol level, you can stratify by sex and study effect of BMI separately for males and females. If BMI is in four categories, you need a total of eight categories of subjects. If hypertension is also added as a factor and it is in three categories (e.g., normotension, borderline, and clear hypertension), you need data in a total of 24 categories. This multiplies fast as you add more factors. On top of this, stratified analysis requires adequate sample size in each such category. Also stratified analysis is not feasible if any of these factors is on continuous scale. Multivariable setup obviates these problems.

As explained later, such a relationship in health and medicine is never exact and the right-hand side of Equation 16.1 provides only an estimated or predicted value of y. For this reason, this is denoted by $\hat{y}$. In the example just cited, cholesterol level can be denoted by y and obesity and hypertension status by x_1 and x_2, respectively. In general, there can be many x variables. The interest actually may be in change in y for change in one particular x_k but other xs that may influence y are kept in the regression so that the **net effect** or **independent effect** of x_k can be studied. This is also called the **adjusted effect**. Other xs are called **concomitant** or **confounding** variables in this setup. For example, in a study on change in triglyceride (TG) levels with age in children, the possible concomitant variables are BMI, fat intake, insulin level, physical activity, etc. In some situations, you may be interested in the net effect of each of these xs instead of any particular x_k.

Three kinds of questions generally arise in the context of relationship: (1) the exact form of the relationship, if there is any. This is required to answer the question, "How much does one variable change for a specified change in one or more of the others?" (2) The strength or the degree of relationship. This answers "How accurately can y be predicted from the knowledge of the xs?" (3) The relationship is causal or incidental. The first of these, in particular, requires a somewhat detailed examination although this text tries to address all these questions.

This chapter: Relationships are studied and expressed in several different forms depending upon the nature of y and the xs. But there are some general features. These are described in Section 16.1. They provide the basis for the discussion in the subsequent sections.

You know now that a variable is considered quantitative when it is measured on a metric scale (continuous or discrete) and is not divided into broad categories (fine categories can still be considered quantitative). Variables on nominal and ordinal scales as well as those that have broad categories on a metric scale are considered qualitative. The method for obtaining relationships is different for qualitative variables than for quantitative variables.

Although qualitative measures have precedence in the scheme of presentation in this book, the classical form of regression is obtained when y and all of the xs are quantitative. This serves as an easy portal to explain some basics, and these basics are discussed first in Section 16.2. Some special issues such as confidence band, multilevel regression, and analysis of covariance (ANCOVA) are discussed in Section 16.3.

The second question regarding the strength of relationship is discussed in Section 16.4 for quantitative variables. The third, whether the relationship is causal or incidental, is discussed later in Chapter 21.

There is another type of relationship frequently encountered in the practice of medicine and health. This is between two or more measurements of the same variable obtained on the same group of subjects. HbA2 measured by high-performance liquid chromatography (HPLC) and by the cellulose acetate elution technique and the interpretation of the same x-ray plates by two radiologists are examples of this kind of relationship. The problem under study in most such cases is that of *agreement* between two or more observers, methods, laboratories, etc. If the agreement is good, one that is less expensive or more convenient can be adopted. Methods of assessing such agreements in quantitative variables are presented in Section 16.5.

A summary of methods for assessing relationship between different types of y and xs is provided at the beginning of the book as a ready reference (**Table S.5**).

16.1 Some General Features of a Regression Setup

Even though the specific type of regression depends upon whether y and the set of xs are quantitative or qualitative, some features are common to all setups. These features are described in this section, mostly assuming that y is quantitative but they are also generally applicable to qualitative y.

16.1.1 Dependent and Independent Variables

The general form of regression is given in Equation 16.1. This tries to express a variable y in terms of variables $x_1, x_2, ..., x_K$. Whether or not this relationship is adequate is another question and is discussed later. For the present, note the following. The variable y, whose estimated value appears on the left side of Equation 16.1, is called the **dependent,** outcome, or the **response variable**. The variables $x_1, x_2, ..., x_K$, which appear on the right side, are called **independent or explanatory variables** or input variables, sometimes as **covariates**. These are also sometimes called **determinants or predictors of** y. In general, $x_1, x_2, ..., x_K$ are the **regressors**. These sometimes define the intervention or specify what can be possibly manipulated. Note there are at least seven different names for the same set of variables.

In a survey, the independent variables are the underlying variables and the dependent variable is the target variable that is to be ultimately studied. In a clinical setup, they may be prognostic or status variables. In an experiment or a trial, the independent variables define the initial state and the dependent variable measures the subsequent or the final state. The *independents are considered fixed and known in a regression setup*, although they may actually not be fixed. Only y is considered stochastic. Regression Equation 16.1 is interpreted as the value of y for given values of xs.

Independent variables are the antecedents and the dependent is the outcome. There may be situations where it is difficult to distinguish between independent and dependent variables. For example, among gender and blood group, which is the dependent variable? In general, independents are those that come first in time and are easy to manipulate. If the postulated dependence does not appeal to your conscience, try reversing the order and see if it makes more sense.

An equation such as BMI = weight/height^2 is not a regression equation. BMI does depend on weight and height but BMI is not stochastic; it is not subject to any variation for given weight and height. Similarly, the equation for aortic valve area (AVA) in square centimeters,

$$AVA = \pi * \left(\frac{LVOT}{2}\right)^2 * \left(\frac{SVTI}{MVTI}\right),$$

where LVOT is the left ventricular outflow tract diameter (cm), SVTI is the subvascular velocity time integral (cm), and MVTI is the maximum velocity time integral across the valve (cm), is mathematical and not statistical since the dependent is not subject to any variation. This index is used to assess the severity of aortic stenosis. Normal range is 3.0–4.0 cm², any value less than 1.5 indicates moderate stenosis and a value less than 0.75 indicates serious stenosis.

16.1.1.1 Simple, Multiple, and Multivariate Regression

If the interest is in studying how systolic level of blood pressure (sysBP) is affected by age in healthy people, then there is only one independent (age) and one dependent (sysBP) variable. The form of relationship between these two is called **simple regression**. If the interest is in studying how sysBP is affected by age, BMI, and socioeconomic status together, the relationship becomes **multiple regression**. In this setup, the number of independents is more than one. If the number of dependents is more than one, then it is in the domain of **multivariate regression**. This is briefly discussed in Chapter 19. Erroneously in literature, sometimes multiple regression is called multivariate regression. Multiple regression can be called multi*variable* regression but not multi*variate* regression.

Irrespective of being simple, multiple, or multivariate, regressions are only an expression of the nature of a statistical relationship as revealed by a set of data. They merely indicate how the dependent variable generally behaves when the independent variables change in value. This relationship is not necessarily causal. It could be entirely incidental.

16.1.2 Linear, Curvilinear, and Nonlinear Regressions

Although most regressions used in practice are linear, the utility of curvilinear and nonlinear regressions in health and medicine is being increasingly realized. An understanding of the distinction between these types of regressions is important.

16.1.2.1 Linear Regression

The relationship $\hat{y} = f(x_1, x_2, \ldots, x_K)$ may take any of several different forms. One that is relatively simple to comprehend and most commonly studied is the **linear** form. For K regressors, this is expressed as

$$\hat{y} = b_0 + b_1 x_1 + b_2 x_2 + \cdots + b_K x_K. \tag{16.2}$$

This is linear in the coefficients $b_1, b_2, \ldots, b_K$. Geometrically, this equation defines a hyperplane in $(K + 1)$-dimensional space and reduces to a line when $K = 1$. When there is only one independent variable x_1, the equation is $\hat{y} = b_0 + b_1 x_1$, which depicts a line. It is more convenient to write it as $\hat{y} = a + bx$. This is the equation for simple linear regression.

Equation 16.2 defines multiple linear regression with K independents. The constants $b_1, b_2, ..., b_K$ in Equation 16.2 are called **regression coefficients** and b_0 is called the **intercept**. These, in fact, are estimates of actual regression coefficients in the target population. The corresponding regression parameters in the population are denoted by $\beta_0, \beta_1, \beta_2, ..., \beta_K$, respectively. Equation 16.2 is the sample version. The interpretation of these coefficients is explained later in this chapter. For the present, consider the following two examples. The hat ($\hat{\ }$) sign is ignored in most of the following discussion.

Example 16.1: Simple Linear Regression for Body Surface Area

Body surface area (BSA) determination is useful in several applications related to the body's metabolism such as ventilation, fluid requirements, extracorporeal circulation, and drug dosages. Current [1] found that this could be adequately estimated by using weight (Wt) in well-proportioned infants and children of weight between 3 and 30 kg. The regression equation obtained is

$$BSA = 1321 + 0.3433(Wt), \tag{16.3}$$

where
  BSA is in square centimeters
  Wt is in grams

The left-hand side is, in fact, an estimate, although the hat ($\hat{\ }$) sign is removed for simplicity. Here, $b_0 = 1321$ and $b_1 = 0.3433$. A simplified version of this is BSA = (Wt + 4)/30, where BSA now is in square meters and Wt in kilograms. The author found the equation very useful in acute operating room management of infants and children for the application of pump gas flows and fresh gas flows in noncritical cases.

Example 16.2: Multiple Linear Regression for Length of Children

Marquis et al. [2] studied length (in cm) of 15-month-old (L15m) Peruvian toddlers living in a shantytown of Lima. This length was studied in relation to length at 12 months (L12m), time interval (TI) in months between 12 and 15 month measurements (as per the authors), breast-feedings (BFs) per day between 12 and 15 months of age, and its interaction with a low diet/high diarrhea (LDHD) combination if present. The following regression equation was obtained:

$$L15m = 74.674 + 0.976(L12m) + 0.860(TI)$$
$$+ 0.043(BF) - 0.157(BF * LDHD), \tag{16.4}$$

where the last term represents interaction. The authors also studied other regressors but Equation 16.4 includes only those that are relevant for illustration. Of these, BF was not significant (P > 0.3). The associated signs show that L12m, TI, and BF had a positive contribution to L15m but the interaction BF * LDHD had a negative contribution. That is, when LDHD is present with a high number of BFs, the length of the child is less. This was part of an investigation to show that increased BF did not lead to poor growth but poor health led to increased BF.

Linear regression is an overused model in health and medical setups. Whereas the relationship between pulse rate and body temperature is indeed linear (Figure 16.1a), this may not be so in other situations. In Figure 16.1b, the trend looks like a decline in cholesterol with increased intake of green leafy vegetables just because of the first two points in the scatter. If these two points are ignored, you can see that the other points are randomly scattered, with no visible trend. Thus, you should critically examine statistically obtained relationships. Even more daunting is the relationship between forced vital capacity (FVC) and age from 5 to 70 years (Figure 16.1c). FVC increases up to the age of 35–40 years and declines thereafter. If you try to fit a linear regression of FVC on age 5–70 years, it will have nearly zero slope, indicating that as age increases, FVC remains constant. This absurd result is due to a completely inappropriate application of linear regression. The scatter plot clearly indicates that a curve should be fitted to these data instead of a line. This comes under the rubrics of curvilinear regression.

16.1.2.2 Curvilinear Regression

Linear regression in most cases is a simplified version of the actual relationship, which is generally more complex and nonlinear. The linear version gives satisfactory results when the dependent and explanatory variables have a straight relationship, that is, where a unit change in the level of the explanatory variable *over its entire range* in the target group is accompanied

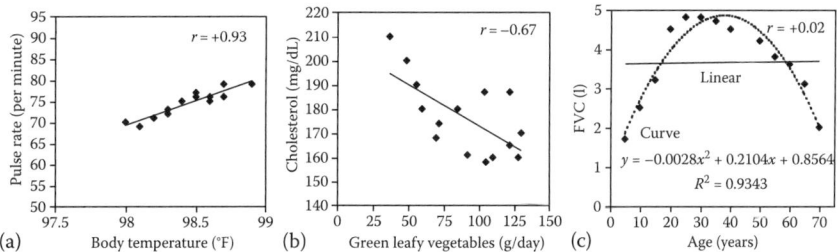

FIGURE 16.1
(a) Good linear regression of pulse rate on body temperature, (b) linear regression of cholesterol level on intake of green leafy vegetables due to two specific points, and (c) inappropriate linear regression of FVC on age.

by nearly the same change in the level of the dependent variable. A straight line can then depict the trend over that range. The relationship of TLC and Hb level with age in healthy subjects is nonlinear over the age range 0–20 years because these two decline up to age 6–8 years (physiological anemia) and then increase. A good example of a nonlinear relationship is that between diastolic BP and age in female adults in a general urban population in India [3]. (The regression for males was linear.) The regression obtained for females is

$$\hat{y} = 58.18 + 1.668t - 0.0453t^2 + 0.000439t^3, \tag{16.5}$$

where
$\hat{y}$ denotes estimated diastolic BP in mmHg
t is age in years up to 60 years

The shape is as shown in Figure 16.2. This is a curve but can be considered linear in $x_1 = t$, $x_2 = t^2$, and $x_3 = t^3$. Each power of t can be regarded as a separate regressor. The regression thus is intrinsically linear. For this reason, such a regression is called curvilinear. Equation 16.5 is a polynomial of degree 3. All polynomial regressions are curvilinear.

The exponential growth curve $y = ae^{bx}$, seen in multiplication of organisms when not interrupted as in a laboratory (Figure 16.3a), is apparently nonlinear but is intrinsically linear. In this case, $\ln(y) = \ln(a) + bx$, which takes the form $z = b_0 + bx$ where $z = \ln(y)$ and $b_0 = \ln(a)$. This is then a straight line but is between $\ln(y)$ and x. There are several other forms of the equation that are intrinsically linear. An example of a curve depicted by $xy = a + bx$ is given later. When x is fixed, this reduces to $y = a(1/x) + b$, which is linear in y and $(1/x)$.

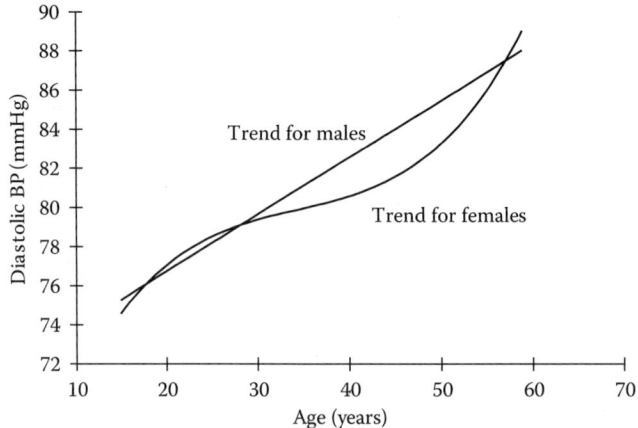

FIGURE 16.2
Diastolic BP in relation to age in adult males and females.

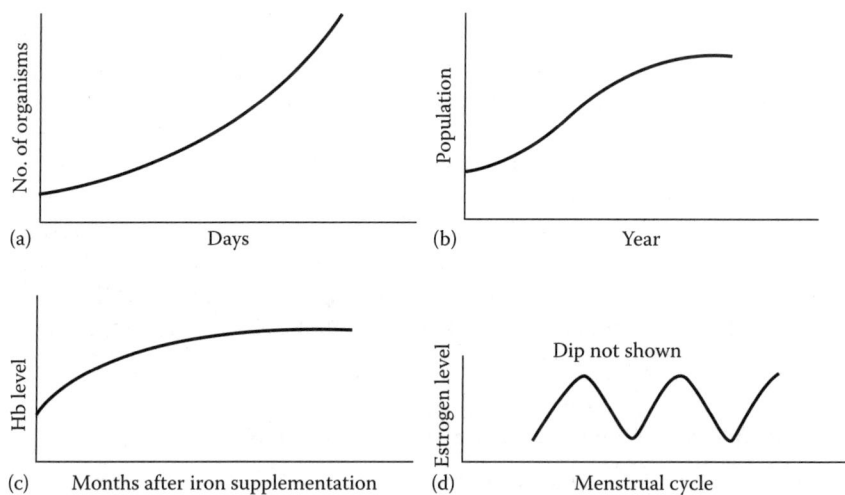

FIGURE 16.3
Some examples of nonlinear relationship: (a) Multiplication of organisms over time (exponential curve), (b) population growth curve, (c) rise in Hb level after iron supplementation (asymptotic curve), and (d) estrogen level in menstrual cycle (cyclic curve).

16.1.2.3 Nonlinear Relationships

Certain relationships are genuinely nonlinear in parameters and cannot be reduced to a linear form. A popular example is the relationship between height and weight in children represented by a growth chart. But this can be easily approximated by a polynomial. If so, it becomes curvilinear. The other example is the equation $y = a/(1 + be^{ct})$, which represents a population growth model where t is the time or the year (Figure 16.3b). There are three parameters, a, b, and c, in this model. No transformation of variables y and t can reduce this equation to a linear form. Thus, this is a genuine nonlinear relation. An example of another type of nonlinear relationship is that between increase in Hb level in anemic women (say, less than $8\,g/dL$) and duration of iron supplementation. The initial increase is high but soon levels off (Figure 16.3c). This can be represented by $y = a - be^{-cx}$, where x is the duration of supplementation and y is the Hb level. This is called asymptotic regression. Another example of a genuine nonlinear relation is a cyclic trend such as that of estrogen level with menstruation phases in women of reproductive age (Figure 16.3d). This level actually exhibits a dip at the time of ovulation, but this is not shown in the figure for simplicity. The level of the drug in the blood over a period of 24 h in an 8 hourly regimen would show a similar cyclic trend.

The discussion in this chapter is restricted to relationships that can be satisfactorily expressed by a linear equation either with or without transformation. For a discussion of nonlinear regression, see Draper and Smith [4].

16.1.2.4 Regression through Origin

When $b_0 = 0$ in Equation 16.2, the regression is said to be through origin. This kind of regression implies that the mean of y is zero when all xs are zero.

Most interesting and most commonly occurring regression through origin is in case of simple linear regression where $y = a + bx$ takes the form $y = bx$. That is, the intercept $a = 0$. You can easily see that $y = 0$ for $x = 0$ and this line passes through coordinates (0, 0)—called origin—when plotted as graph. This is used when y has tendency to be proportional to x. For example, under normal conditions, the Hct level (%) is nearly three times the Hb level in mg/dL. If Hb 10 mg/dL, Hct = 30%. However, this is not a mathematical relationship—variations occur.

> **Example 16.3: Glomerular Filtration Rate in Children**
>
> Glomerular filtration rate (GFR) in children is estimated by the formula
>
> $$\text{GFR (mL/min/1.73 m}^2) = k * \frac{\text{Height in cms}}{\text{Plasma creatinine concentration in mg/dL}}.$$
>
> This is a regression through origin. The value of k varies from population to population depending primarily on the nutrition status of children. For example, Schwartz et al. [5] found $k = 0.55$ in Canadian children of age 1–13 years.
>
> **SIDE NOTE:** The gold method for measuring GFR is 99mTcDTPA renal scan, which is cumbersome for routine clinical use. Length and plasma creatinine levels can be easily measured. Thus, the formula, if accurate, is very useful for adjusting dosages of drugs excreted by kidney and for detecting changes in renal function.

The predictor variable in Example 16.3 on GFR is a ratio of two measurements. This does not invalidate regression so long as both are premeasured. The GFR obtained by this equation is for known height of the child and known creatinine level and therefore known ratio of these two. The estimated GFR may be very different from the actual obtained by DTPA scan even when the value of coefficient k is estimated from data on representative set of children. In some cases, error found is to the extent of ±20%. Thus, the formula does not provide as precise estimate of GFR as is made out to be. This can happen with any regression and you should exercise due care in adopting such equation for clinical use.

I later on discuss adequacy of fit that may indicate when to rely or not rely on a regression equation for prediction purposes.

16.1.3 Concept of Residuals

Consider the simple regression given in Equation 16.3. If this equation holds good, it tells what BSA to expect in children of different weights. The estimated BSA of a child of 20 kg is 8187 cm² and of a child of 25 kg is 9904 cm². Actual values would most likely differ from these predicted values. If the actual BSA of a child of 20 kg is found to be 8093 cm², then the difference from the expected is −94 cm². If another child of 20 kg has BSA 8214 cm², then the difference is +27 cm². Such differences between the expected measurements ($\hat{y}$) and the actual measurements (y) are called **residuals** and are denoted by e. Thus, $e = y − \hat{y}$. The residual for the first child in this example is −94 cm², and for the second child, it is +27 cm². In general, for the subjects giving rise to the regression, some residuals will always be positive and some negative in such a manner that their sum, and hence mean, is always zero.

The magnitude of residuals is crucial in judging the adequacy of a regression function. If it provides a good fit to the data, then the residuals will be mostly small. Because their mean is zero, small residuals amount to a small standard error (SE). Conversely, a small SE of residuals is an indication that the model obtained is a good fit to the data.

16.1.4 General Method of Fitting a Regression

To obtain a regression (this is also called fitting a regression), the data required are the values of y corresponding to different values of the xs. For one set of values of xs, one or many values of y can be observed. For example, for Equation 16.3, you may have three children of the same weight 20 kg. Their BSAs are likely to be different. In this case, for one $x = 20$, there are three values of y, one corresponding to each child. But y should be available for several distinct values (at least four for the regression to be meaningful) of x. The values of the x variables can be *deliberately* chosen to serve the purpose.

The total number of individuals, as usual, is denoted by n. The number of regressors, K, should preferably be considerably less than the number of subjects n (i.e., $K \ll n$), and the same K regressors must be available for each subject in any analysis. The exact method of fitting varies according to qualitative or quantitative y, but the basic steps are common. For qualitative y, the dependent actually is proportion of subjects with specific quality.

Step 1. Identify the dependent variable of interest. It should ideally be an outcome or response of the regressors. In the case of risk of lung cancer and smoking, smoking does not depend on the risk but risk depends on smoking. Thus, risk is y and smoking is x. In some cases, the direction of dependence may not be clear. Between cholesterol level and WHR, there could be a debate regarding which is a precursor of the other. Abnormal histology of the liver can precede cholecystitis or can follow cholecystitis. Both are admissible. In such cases, either could be investigated for its dependence on the other. Regression in these cases is just an expression of relationship and not of dependence.

Step 2. Identify the set of regressor variables. This ideally should consist of all those that are known or suspected to influence the response y. It may not always be feasible to study all possible variables and some selection may be necessary. Choose those that influence more than the others and those that are directly related, in place of those that cause indirect influence. These should preferably be unrelated to one another, at least not intimately related. If they are intimately related, **multicollinearity** is said to exist and can adversely affect the validity of the regression. Regression coefficients shall remain the same but their joint effect is unreliable. These coefficients and their SEs will show large change if only few values are slightly altered. Regression equation will make more sense if you retain only one of the highly correlated variables.

If systolic BP is among the regressors for, say, serum glucose level, then inclusion of diastolic BP among the regressors may not serve much purpose. This is because systolic and diastolic levels are highly correlated. If body fat can be calculated from skinfold at triceps and thigh, there is no use in keeping all three as regressors. Keeping body fat as a regressor is enough. An association or correlation exceeding 0.8 on a scale of 0–1 is considered enough to exclude the variable, particularly when the form of relationship to be examined is linear.

Choice of regressors also depends on the purpose of running regression. As stated earlier, one purpose could be to examine how the regressions affect the outcome. This is called **explanatory regression**. In this regression, explanation you get from one set of variables may be different from what you get from another set of variables. Such differing explanations may not be biologically plausible. In this setup, the choice of variables is crucial and requires deep thinking. Second purpose could be prediction. In **predictive regression**, any set of variables that predicts close to the observed values is good. This set of variables may or may not have much biological meaning. That does not matter for prediction.

In some cases, you may also like to include interaction between variables. This is done by considering product of variables as a new variable. This would be as good a regressor as any other.

For convenience and expedience, when needed, only one variable can be considered as a regressor at a time. The regression also remains valid in this case provided the residuals are not large. Large residuals tend to invalidate the regression even when many regressors are present. Adequacy of regression depends not so much on the number of regressors as on the magnitude of residuals.

Step 3. Decide the general form of relationship that you want to investigate, that is, decide whether you want to restrict to linear regression, curvilinear regression, or a nonlinear form of regression. The best guide for this is the scatter plot, particularly where there is only one regressor, and both y and x are quantitative. If there are many regressors, then many plots would be required, one corresponding to each regressor. The basic function of regression analysis is to provide the best estimate of the regression coefficients once the form of regression is specified. However, several forms can be explored and the best can be objectively identified as per the methods given later. Remember though that *all regressions are compromises. They never depict the data exactly.*

Steps 1, 2, and 3 together specify what is called **the regression model**.

Step 4. Estimate the regression coefficients. This is done in such a manner that the fitted line or the curve passes closest to the points in the scatter plot. The mathematical method generally employed to obtain such regression coefficients is called the **method of maximum likelihood**. This method finds the values of b_0, b_1, ..., b_K that make our observed sample most likely to happen. The other method commonly used is called the **method of least squares**. For this reason, the whole method of regression with quantitative dependent discussed in this chapter is sometimes referred to as **ordinary least squares**. This finds a regression that is closest (least sum of the squared distances) to the points in the scatter diagram. Both the methods tend to produce an equation that *regresses* toward the mean of y across different values of x. These two methods in many practical situations lead to the same estimates. Many software packages are available to perform the calculations required to estimate the regression coefficients. A feature of estimation of the regression coefficients is that some residuals are positive and others negative such that their sum is always zero as already noted.

Step 5. Test the goodness of fit of the regression model. The method for this test depends on whether y is quantitative or qualitative.

Step 6. The overall regression model may or may not be a good fit, but it is also important that individual regressors are tested separately for their significance. The method for testing this is also different for qualitative y than for quantitative y. The regressors found statistically significant are considered to contribute to the relationship and are retained in the model. The "not significant" ones can be dropped without substantially affecting the utility of the model. Such deletions help to enhance the simplicity of the model.

Step 7. Calculate residuals and check for validity of assumptions of the regression model. These assumptions are stronger for quantitative y than for qualitative y. One of them is homoscedasticity, that is, uniform variance of y for different values of the regressors. A transformation such as the logarithm, square, and square root may be sometimes needed to meet this requirement. The other way to overcome this is to modify the method of estimation from least squares to *weighted least squares*. Also for testing or for computing confidence interval (CI), the residuals must follow nearly a Gaussian pattern. If the residuals are skewed, other distributions such as Gamma, Poisson, or negative binomial are tried. These methods are not discussed in this text. A scatter plot of residuals versus each x is often helpful in assessing the validity of assumptions and in suggesting improvements to the model.

Sometimes a relation exists but the regression fails to detect it. The success depends mostly on the adequate specification of the model in steps 1 through 3. Thus, failure to find a relationship through regression does not necessarily mean that no relationship exists.

As mentioned earlier, one of the important functions of regression is to provide an estimate or to predict the value of y for a given set of xs. There is a fine distinction between an estimated value and a predicted value. The former term is generally used for the *mean* of y for a group of individuals with a specific value of $(x_1, x_2, ..., x_K)$, and the latter is used for the value of y for a single individual. The former has less variance than the latter, although both have the same value. Because the estimated and predicted values are the same, these terms are interchangeably used in this text.

16.2 Linear Regression Models

The setup under consideration in this section is the dependent variable (y) as well as all the regressors $(x$s$)$ measured on a metric scale, that is, they are all quantitative. The quantity of interest is the mean of y for each specific set of value of xs. Thus, y should ideally be continuous but can be categorized into class intervals of small width so that a mean based on midpoints can be legitimately calculated. There is no such restriction on the xs as long as they are measured on a metric scale. The linear relationship in this setup takes the form

$$\hat{y} = b_0 + b_1 x_1 + b_2 x_2 + \cdots + b_K x_K. \tag{16.6}$$

Regression, particularly in the case of quantitative variables, is a statement of *trend in averages* of individual values. That this is doubly statistical would be clear if examined from the viewpoint of residuals.

Conceptually, regression requires observation of several values of y for each value of x. For example, for regression of diastolic BP (y) on age (x), you should have, say, 10 individuals of age 20 years, 15 of age 21 years, 6 of age 25 years, 12 of age 28 years, etc. These should be random samples from the population of subjects of different ages. Essentially, *mean* diastolic BP for each age is used to obtain the regression. Individual values would differ from the respective means. One component of each residual arises from this difference. In practice, however, there could be only one value of y for one value of x. This does not affect the interpretation of the residuals.

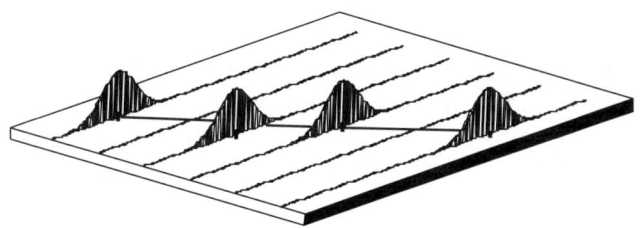

FIGURE 16.4
Conceptual framework of distribution of *y* for each *x* in a regression setup.

Regression seeks a trend in different means of *y*, one each for each specific value of *x*. This is illustrated in Figure 16.4 in the case in which the distribution of *y* for each *x* is Gaussian. There are only four values of *x* in this figure. Regression is the best fit to the means. This may not pass through all the means even when population means are considered and can definitely happen in the case of sample means. For regression of diastolic BP on age, supposing the average diastolic BP in a sample of males of age 30 years is 79.5 mmHg, in males of age 40 years is 82.7, in males of age 50 years is 85.2, etc., then the *trend* is DiasBP = 71.0 + 0.3 (age). The level expected from the trend could differ from the observed means. The second component of residuals arises from the difference of the trend from the observed means.

In summary, residuals arise due to (1) difference of individual values from their respective means and (2) difference of observed means from the trend. Regression, therefore, should be interpreted with caution. Despite such constraints, regressions do lead to useful conclusions in many cases, and the technique is extensively used in the medical literature.

Because regression seeks to identify the trend of the mean of *y* over different values of the *x*s, it is sometimes called **trend analysis or trend assessment**. These terms are particularly appropriate when the *x* variable measures time such as chronological years (e.g., trend of infant mortality rate from 1970 to 2010 in a country) or age (e.g., trend of BP levels over age).

The general method of fitting regression for quantitative variables is the same as that already outlined. The steps are (a) identification of the dependent and (b) regressor variables, (c) specification of the form of regression (linear, curvilinear, or nonlinear) that you want to investigate, (d) estimation of the regression coefficients, (e) test of the goodness of fit of the model, (f) test of the significance of individual regression coefficients, and (g) examination of the residuals for validity of the assumptions. Steps (a) through (c) are fairly general and depend on the problem. They have been already explained in Section 16.1.4. Step (d) is quite mathematical that could be left to a software package provided that a standard software is used. Basic details of steps (e) through (g) are given next. These three together decide whether the fit is adequate.

16.2.1 Adequacy of a Regression—1

You would often want to do more with correlative data than merely draw a line or a curve. You should be able to draw inference for the population for which the sample of subjects was chosen. For this, the regression must be adequate. Several aspects of fitting are considered to examine the adequacy of a regression.

16.2.1.1 Goodness of Fit and η^2

Overall statistical significance of a regression model is checked by an F-test similar to the one described for ANOVA. This significance indicates only that the model is not useless. It may or may not be a sufficiently good fit. A regression model is considered a good fit if the residuals $e = (y - \hat{y})$ are small. Since residuals fluctuate around zero in any case, small residuals would necessarily yield a small sum of squares Σe^2. This is the residual sum of squares, popularly called the sum of squares due to error (SSE). Its magnitude, when compared with the total sum of squares, $\text{SST} = \Sigma(y - \bar{y})^2$, provides a measure of "lack of fit" of the regression and (SST – SSE) is called regression sum of squares (RegSS). These have associated degrees of freedom (dfs) as in the case of ANOVA. RegSS measures how much of the total sum of squares the regression has been able to account for. This is the "goodness of fit" and is denoted by η^2, that is,

$$\eta^2 = \frac{\text{RegSS}}{\text{SST}}$$

$$= 1 - \frac{\text{SSE}}{\text{SST}}. \tag{16.7}$$

The quantity η^2 is interpreted as the proportion sum of squares of y explained by the regression and called **coefficient of determination**. The larger the η^2, the better is the fit. If the residual sum of squares is as small as, say, 10% of the total sum of squares then $\eta^2 = 0.90$. The fit, then, is said to account for 90% variation in y. A fit with such high η^2 should be adequate in most cases, especially if n is large.

16.2.1.2 Multiple Correlation in Linear Regression

The notation η^2 for coefficient of determination is general and applies to all regression setups. However, in case of multiple or simple linear regression, this is denoted by R^2. R is called coefficient of multiple correlation. Most researchers present results in terms of R^2 and not η^2. In case of simple linear regression with one independent variable, $R^2 = r^2$, where r is the notation for correlation coefficient.

You will soon see that high R^2 for small n is not much helpful because SEs are still high and the reliability of the estimates is low. As illustrated later, high R^2 can occur in a structurally inappropriate model. This can also occur in case of multicollinearity where many individual regression coefficients are not statistically significant.

R^2 is also used to compare one model with another. If R^2 for one model is 0.63 and for the other model 0.76, then the model with higher R^2 is considered a better fit. Increasingly better fit can be obtained by progressively adding new regressors. But a large number of regressors makes the model difficult to interpret. Thus, *a model is considered good when it contains a small number of regressors but gives a sufficiently large value of R^2*. A value more than 0.70 is generally considered desirable, between 0.80 and 0.89 good, and 0.90 or more as excellent. $R^2 < 0.50$ can also be useful in rare cases where almost nothing is known. The success in achieving this depends on the proper choice of regressors and on the right specification of the model. In health and medicine, it is many times difficult to obtain a high value of R^2. The appropriate regressors are not known for some situations, and they have to be assumed on the present knowledge or on available data. Both these may be inadequate.

Since R^2 continues to improve as more regressors are added, a realistic assessment is made when adjusted for the number of regressors. This is done by dividing by the respective dfs. Thus,

$$\text{Adjusted } R^2 = 1 - \frac{\text{SSE}/\text{df}_e}{\text{SST}/\text{df}_t},$$

where
  df_e is the df for SSE
  df_t is for SST

In terms of usual R^2 this is

$$\text{Adjusted } R^2 = 1 - \frac{(1-R^2)(n-1)}{n-K-1},$$

where
  n is the sample size
  K is the number of regressors

This adjustment would be substantial if n is small. Since this is adjusted for a number of independent variables, adjusted R^2 can be used to compare two or more regressions with different number of regressors.

A statistical procedure, similar to the F-test for ANOVA, is available to test the statistical significance of R^2 as well as of addition to R^2 made by any additional variable. Most statistical softwares do this easily. It is thus possible to specify a large number of regressors and ask the computer program to include only those that contribute significantly. Methods such as forward selection, backward elimination, or stepwise are used for this purpose. These methods tend to automatically exclude variables that have high multicollinearity with other regressors.

16.2.1.3 Stepwise Procedure

When the number of possible explanatory variables is large, the computer program can be asked to identify and include only the statistically significant variables in the regression model. This is one of those situations where testing of hypothesis strategy opposed to CI is preferred. Many algorithms can do this but the first three of the following are common:

1. **Forward selection**: In this algorithm, the variable with the highest statistical significance (i.e., with the least P-value) is entered first, and then those that are significant in diminishing order are entered sequentially. The variables that do not add significantly to R^2 do not find a place in the model.

2. **Backward elimination**: In this algorithm, all explanatory variables are included in the first step and then deleted one at a time starting from the least significant. The procedure stops when further deletion significantly reduces R^2.

3. **Stepwise**: In both the procedures mentioned earlier, a variable once entered continues to remain in the model and a variable once deleted remains excluded. The third algorithm, called stepwise, reexamines the utility of variables already entered and deletes those that become insignificant after addition of one of the other explanatory variables, or enters one of the already deleted variables. The procedure stops when neither deletion nor addition significantly alters R^2. This procedure is generally started from no variable in the model.

   Software illustration: An example on stepwise method is in Appendix C illustrating use of SPSS software.

4. **Best subset**: All possible combinations of predictors are tried and the one with lowest P-value for R^2 is chosen.

My advice is to use these methods to get a feeling of what predictors may be important. Practically, solution based on real biological worth, if extraneously available, may prove better than purely numerical optimization procedures such as these. You may want to ensure that clinically important variables are not excluded.

These are purely statistical procedures and require alpha-to-enter and alpha-to-remove a variable. The software used may have default values—check that they are adequate for your regression. Generally, these alphas are fixed at 0.10 or 0.15 each; the conventional value 0.05 becomes too restrictive in this situation.

None of these algorithms may yield the "best" model. They may lead to different models. It is a good idea to examine several possible models and choose on the basis of interpretability, parsimony, and convenience in obtaining the data. Do not accept result of stepwise method without critically examining the final model. There are simulation examples where the stepwise procedure at different simulations on the same set of data selected different covariates.

Increasing the number of variables increases the validity in terms of closer prediction on average but can reduce reliability due to compromised df. The variance of the prediction can increase in some cases. Scientists would stake their reputations on the model they propose. If another scientist sampled new data from the same target population and failed to find nonzero coefficients for the variables in the model, or much less than the proposed, few would pay attention to this scientist in future. The best guide, in fact, is not R^2 but the biological utility of the model in doing what it is supposed to do. Even though a regression model is developed after going through so many mathematical steps, its biological plausibility still requires to be explained.

A method that might help in keeping biological plausibility intact in some situations is **backward stepping**. Under this procedure, first obtain the regression with all the explanatory variables in the model and then make intelligent choices yourself for identifying the less useful variables, starting from the least useful. For making this choice, you may like to examine the plots of data and use your prior knowledge about the biological role various factors play in determining the value of the dependent y.

16.2.1.4 Statistical Significance of Individual Regression Coefficients

When all specified regressors are forced into the model and not selected by any of the procedures just mentioned, then some of these may not be statistically significant and can be dropped without significantly affecting the efficiency of the model. In other words, those regression coefficients could be nearly zero. The test of H_0 that a particular regression coefficient is zero or not is done by using Student t-test under some general conditions. Most software packages directly give P-values for each regression coefficient. Statistical significance is a reasonable assurance that the regression coefficient is not zero and that the corresponding regressor is indeed making some contributions in determining the value of the dependent variable.

A computer output for regression would generally also provide the P-value for the intercept b_0. But this needs to be interpreted with added caution. This P-value corresponds to H_0: Intercept = 0. This is plausible only when y is expected to be zero for $x = 0$. This is not plausible, for example, when regression is between age and weight. Birth weight is the weight at age = 0 but is not expected to be zero. A zero intercept will not be of any interest except in few special cases as discussed with regression through origin.

Although not explicitly mentioned, it may be clear by this time that the additional contribution of each regressor to the prediction of y in multiple regression setup can also be assessed by the addition it makes to the RegSS. Its gross individual contribution is obtained by RegSS as percentage of SST in a simple regression with this regressor as the only predictor. Some of this contribution may be coming through other predictors that might be influencing this regressor. The net contribution is obtained by the difference between RegSS with and without this regressor.

If one particular regressor is contributing the major part of RegSS, perhaps this itself is a good predictor and simple regression with this regressor may serve the purpose almost equally well—other regressors can be ignored. If not, you have a genuine multiple regression setup.

In terms of R^2, the contribution of any regressor must be at least a few percent for it to be of any consequence and this addition to R^2 must also be statistically significant.

16.2.2 Adequacy of a Regression—2

Other aspects of adequacy of a regression are as follows.

16.2.2.1 Validity of Assumptions

The basic requirements for validity of the F- and t-tests mentioned earlier in the context of regression on quantitative variables are that the residuals (1) have a Gaussian pattern, (2) are statistically independent of one another in probability sense, and (3) have the same variance for each value of x. These are the same as already discussed for ANOVA. If the model is adequate, residuals would arise mostly due to chance, and these are known to follow a Gaussian pattern for practically any continuous y. A scatter plot of residuals would then show concentration toward zero and lesser values away from zero, evenly scattered both on the negative and positive side of the scatter plot.

Independence of residuals can be violated in a case such as a time series where the value of y depends serially on its value on the previous occasion. Such **serial correlation** can occur because the time effect by itself may not be able to take care of other factors that may be simultaneously changing, such as population size. These spill into the residuals and cause serial correlation. In this case, the residual plot may track snakily or follow some other *pattern*. Inclusion of factors causing serial correlation in addition to time in the model may make residuals free of such correlation. Among several other methods suggested to remove the effect of serial correlation, the simplest is to take the difference $(y_i - y_{i-1})$ and consider this difference as dependent. Some softwares use the **Durbin–Watson test** to check the significance of serial correlation. For time-independent data, such independence remains fairly secure. Ignoring serial correlation among residuals may cause a spurious trend and underestimate the uncertainties in the estimate of the SE. When serial correlation is given due consideration, the actual SE will turn out higher.

Uniformity of variance, which is the same as **homoscedasticity** mentioned in the previous chapter on ANOVA, can be checked by a scatter plot of residuals for different values of x. A random pattern is an indication of uniformity of variance. If the spread of residuals is found to follow a discernible pattern, a transformation such as the logarithm, inverse, square, or square root may help. Calculation for regression coefficients and for their significance will have to be done all over again after such a transformation.

With some experience, you can reasonably judge the validity of these assumptions by examining the process of generation of the data and by looking at the patterns in the data. The transformation, if needed, can be incorporated into the model itself after step 3 in the sequence given in Section 16.1.4.

Any special or identifiable pattern in residuals can be in the variability, in the values themselves, or in both. Two of these patterns are illustrated in Figure 16.5a and b. In the first case, the residuals have U-shape. This shows that the model is inadequate. Incorporation of the x^2 term in this case may help. In the second case (Figure 16.5b), the residuals again are almost equally divided on either side of zero but the variability increases as x increases. This is called fanning and indicates lack of homoscedasticity. A log or square root transformation of y may help achieve uniformity of variance over x in this case.

An example where all the common assumptions of a regression model are violated arises when studying coronary artery calcification (CAC). The amount of CAC helps in predicting future risk of coronary artery disease (CAD) morbidity and mortality. The distribution of CAC would contain a high relative frequency of zeros and would be highly skewed. Even logarithmic transformation [ln(CAC+ 1)] does not help. (To get an interpretable

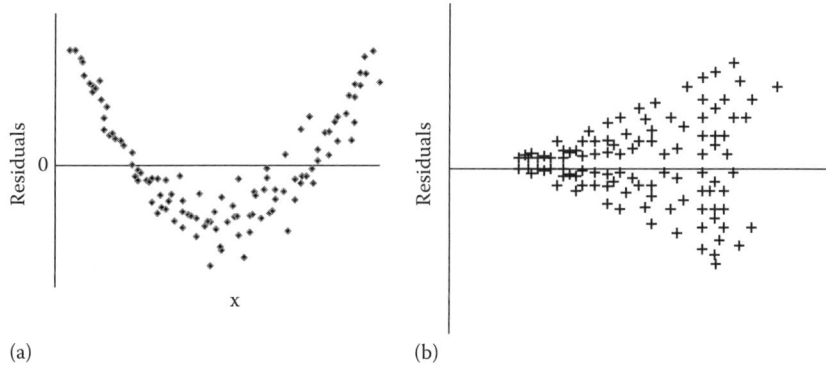

(a)   (b)

FIGURE 16.5
Some special forms of residuals: (a) U-shaped residuals and (b) fanning of residuals.

logarithm, 1 is added, since log of zero is −∞.) Also, the variance of the quantity of CAC increases with age. If it is measured at multiple locations in the same subject, the independence is also lost. This will continue to be so for residuals too. In a situation like this, a complex model, namely, a generalized linear mixed model [6], may be appropriate.

Gaussian pattern, independence, and homoscedasticity are requirements for a valid test of statistical significance but not for estimating the regression coefficients. These estimates can still be obtained by the method of least squares and interpreted in the usual way. But the coefficients so obtained lack credibility if the assumptions are violated. For example, they may not be unbiased or may have inflated variance that makes them less reliable. Also, the practical utility of the model is greatly diminished if one is not able to test its statistical significance, and this can happen if the assumptions are violated.

16.2.2.2 Choice of Form of Regression

Despite all the advancements, statistical softwares do not have intelligence yet to choose an appropriate form of regression. The user must specify the broad shape. The software finds only the best coefficients that would make the specified shape closest to the scatter. The shape of regression is decided by the following considerations:

1. Start from the simple linear regression with the most prominent predictor as the only regressor. Simple linear is easy to understand and easy to explain. Check if (a) the value of R^2 is reasonably high, (b) the regression is close to the scatter, and (c) the relationship is biologically explainable. If you are satisfied on all these three counts, you are done. These three remain your cardinal indicators for judging the adequacy of any regression.

2. If simple linear regression is inadequate, examine the scatter closely for any specific pattern. It could be parabolic, cyclical, exponential, or any other. Such patterns can be fitted by using terms such as square, cube, reciprocal, and logarithm of the independent variable. For cyclical factors such as for estrogen level in women of reproductive age, try sine function. If needed, take help of a statistician to identify the form of regression for investigation.

   An example on quadratic regression is in Appendix C that illustrates use of SPSS software for this kind of fitting.

3. If the outcome measurement looks appreciably affected by two or more factors, consider multiple linear regression. Include only those factors as regressors that can really affect the outcome and are relatively independent of each other (no multicollinearity). If there are many, use stepwise procedure to select few significant ones. Do not forget to consider interaction—for this, use the product (multiplication) of the variables that you think can interact.

4. If multiple linear regression also does not give satisfactory results in terms of the three cardinal indicators listed earlier, you can try multiple nonlinear regression. But this becomes too complex to explain and the parsimony is lost. Weigh the cost–benefit. Sometimes it is worth spending time and resources on investigating complex relationship. Perhaps, many relationships are complex and many researchers probably waste resources on exploring simple relationships where the relationship is complex.

5. If that also fails, conclude that either the choice of regressors was inappropriate or that enough is not known about the factors responsible for that outcome. In this situation, first step is to do basic research and find factors that could be rightfully conjectured as explanatory of the outcome.

Note also that choosing a form of regression implies that you know that this form is appropriate. Quite often you will find that you do not know mechanics of how predictors affect the outcome. Various forms are tried in the hope that one of these would fit the data and then a biological explanation is sought for that kind of relationship. If you think you know the likely from of relationship as indicated by biological processes, try to obtain a regression of this form. Either way, epistemic uncertainties can be prominent because both are based on the existing knowledge—and the existing knowledge can be inadequate.

Uncertainties never end. No matter what form of regression you choose, part of variation will always remain unexplained. Good researchers tighten the control on this unexplained part so that medical decisions are more valid and reliable.

Example 16.4: Regression of GFR Values on Creatinine in CRF Cases

The following are the plasma levels of creatinine and total GFR in 15 cases of chronic renal failure (CRF). GFR is measured in milliliters per minute but can also be expressed as a percentage of normal (100%). Their relationship can be useful, with creatinine level serving as a clinical index of GFR in CRF cases.

GFR (%)	19	6	7	8	72	67	13	33	48
Creatinine (mg/dL)	10.5	23.7	20.1	18.4	2.5	2.8	14.3	7.6	3.0

GFR (%)	57	9	10	18	22	62
Creatinine (mg/dL)	2.5	14.7	12.6	12.0	8.1	2.1

The scatter plot is shown in Figure 16.6. Creatinine level is dependent in this case on GFR, but for creatinine level to serve as a clinical index of GFR, the statistically dependent (y) variable is GFR because that is what is to be predicted.

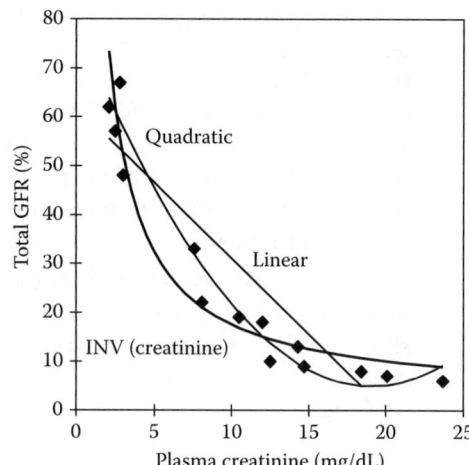

FIGURE 16.6
Scatter plot and different regressions of
GFR on plasma level of creatinine.

Computer processing gives the following result for linear regression:

$$GFR = 62.06 - 3.10 \, (\text{Creatinine})$$

or

$$\hat{y} = 62.06 - 3.10x, \tag{16.8}$$

where
  $\hat{y}$ is an estimate of GFR
  x is the plasma creatinine level

The overall F-test gives $P < 0.01$, which means that the model does help in explaining the relationship between these two variables. The value of R^2 is 0.81, which is fairly high. But the line is far from the plotted points (Figure 16.6). The residual plot (Figure 16.7a) shows that the residuals are large and follow a trend. This indicates that there is room

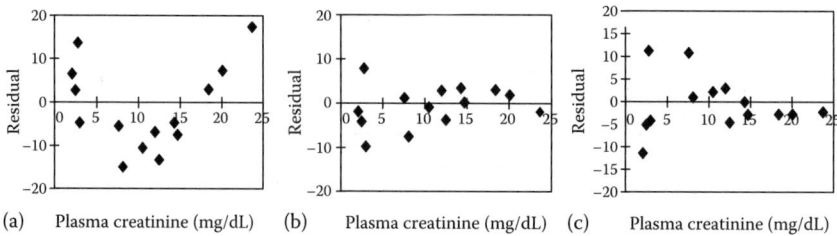

FIGURE 16.7
Residual plots of different regressions in Figure 16.6 (a) residuals of linear regression, (b) residuals of quadratic regression, and (c) residuals of regression on inverse of x.

for improvement. By incorporating the square term, we obtain the following quadratic regression:

$$\hat{y} = 79.43 - 7.82x + 0.205x^2.$$

Now $R^2 = 0.95$, which is extremely good. The plot of this curve is also shown in Figure 16.6 and the residuals in Figure 16.7b.

A model with such a high R^2 would usually be acceptable and there would be no need to go any further. However, the model should also be medically plausible. The quadratic regression is close to the scatter but shows an increasing GFR for some creatinine levels at the upper end. This trend is not acceptable because higher GFR is not associated with a higher creatinine level. You must search for a plausible alternative.

Knowing that the trend should be a decreasing GFR with an increasing creatinine level and that it tends to stabilize at both the upper and lower end points, and considering the shape suggested by the scatter plot, you might like to fit what is mathematically called a hyperbola. This requires that the independent variable should be $1/x$. This is the inverse of the creatinine level. Then the following regression is obtained:

$$\hat{y} = 2.69 + 148.59\left(\frac{1}{x}\right). \tag{16.9}$$

The value of R^2 is now 0.93. This is slightly less than the 0.95 obtained for the quadratic equation, yet the regression (Equation 16.9) is preferable because of its plausibility. The curve obtained is shown as dark thick curve as INC (Creatinine) in Figure 16.6 and the corresponding residuals in Figure 16.7c.

The following are the lessons from Example 16.4:

1. The example illustrates that a statistically superior model (with higher R^2) is not always the best. The model should also be consistent with the knowledge regarding such relationships.

2. The inverse transformation of x to $1/x$ makes it look like nonlinear, but the model (Equation 16.9) is intrinsically linear because it is linear in $1/x$.

3. The example also illustrates that the scatter plot, the plot of the regression model, and the residual plot are useful aids in formulating the model and assessing its adequacy. These plots should be considered in addition to R^2 and P-value for assessing the adequacy of a model.

4. The residual plots in Figure 16.7b as well as in Figure 16.7c indicate that the variance is larger at the low levels of creatinine and smaller

at the higher levels. Thus, the requirement of homoscedasticity seems to be violated. The number of observations, $n = 15$, is too small for a firm conclusion, but a variance-stabilizing transformation cannot be ruled out. You may try to fit a regression of $\sqrt{y}$ on $(1/x)$ and examine the improvement, if any.

5. The regression of y on x is not the same as that of x on y. Example 16.4 considers GFR as dependent because it is to be predicted. Physiologically, though, the creatinine level depends on GFR. The regression of creatinine level on GFR is creatinine = 18.16 − 0.26(GFR). This is different from what is obtained by bringing creatinine to the left side of Equation 16.8. Thus, *a regression equation should not be used as a simple mathematical equation.*

   Another example, more telling, is the curvilinear quadratic regression between FVC and age shown in Figure 16.1c, where $R^2 = 0.93$. Such high R^2 may convince anyone that this is an extremely good fit—thus is an adequate representation of the relationship between FVC and age. Pause for a moment and realize that the peak FVC according to the regression is at around age 37 years. Against this, a careful look at the scatter shows that the actual peak FVC is at around 25 years—and biological knowledge suggests that this is closer to reality rather than 37 years revealed by the regression. Thus, the regression has failed us again as did quadratic regression in Example 16.4 on GFR. It shows again that R^2 is merely a statistical number—it must be supported by biologically plausible relationship to be really useful.

6. All models, including regression models, can be overfit for a particular set of data. A value of R^2 in excess of 0.90, as in Example 16.4, can be an artifact arising in a particular sample. Calculating this again for another sample can be instructive.

16.2.2.3 Outliers and Missing Values

A large value of a residual for any particular y is a definite indication that an outlier is present in the data, but that cannot be relied upon as a sole indicator. An outlier would most likely affect the entire regression, and thus many residuals could be moderately large. They may escape attention. Careful examination of residuals is required to find whether one or more outliers are present. An outlier may be a genuinely high or low value or may be due to data entry error or inappropriate selection. If the number of genuine outliers is large, this may indicate that some relevant factor is missing from the predictors. Some softwares have an option to flag outliers, which makes their spotting easier.

Statistical methods can be annoying as they heavily depend on averages and patterns. Averages and patterns are disturbed whenever outlying values are present. They have disproportionate effect—just one outlier can destabilize the results. See Chapter 21 for an example. There is another example in that chapter that illustrates how trend provided by regression can be fallacious in some cases. In view of their negative effect on validity of results, outliers are deleted from the data before analysis.

The other problem is missing values. They can have exacerbated effect since their pattern is generally very different from the available values. Imputation for missing values can be done as discussed in a later chapter. For the time being, consider deletion.

For analysis purposes, two kinds of deletions are available for missing values. For univariate analysis, values of the concerned independent and dependent are omitted in case anyone is missing. This is called *pairwise deletion*. For multivariable analysis, the entire record for the subject with any missing value is deleted. This is called *casewise deletion*. If y is dependent and x_1, x_2, x_3, and x_4 are independent variables, and if x_2 is missing for subject number 46, x_2 and y are deleted for this subject under pairwise deletion, and all x_1, x_2, x_3, x_4, and y are deleted under casewise deletion. The same applies to deletion of outliers.

Such deletions, if not few, can severely affect the results. As already mentioned, missing values and outliers may follow a distinct pattern and that pattern will not be accounted for in results based on such deletions. Second, deletions reduce the sample size and thus compromise the reliability of results. If there are many outliers or missing values, you should think of this as a pilot study and use it for learning lessons for planning another study where such values are infrequent. This could be very expensive, though.

Software illustration: Appendix C contains a software illustration for fitting a curvilinear regression.

16.2.3 Interpretation of the Regression Coefficients

Coefficients in the case of regression for quantitative variables have a very simple interpretation. In the regression equation $\hat{y} = b_0 + b_1x_1 + b_2x_2 + \cdots + b_Kx_K$ if x_2, for example, increases by unity then $\hat{y}$ increases by b_2 when the other xs remain the same. Thus, b_k is the independent contribution of one unit of x_k when the other xs remain constant. (This in a way also explains why xs should not be strongly correlated. If they are, it is difficult to imagine that only one varies without a change in the others.) These coefficients are "adjusted" for other xs in the model. When a simple regression is obtained between y and x_k alone, the regression coefficient b_k then obtained is called unadjusted. Adjusted b_k can be more than unadjusted b_k, less than unadjusted, or may not change.

Each regression coefficient measures the *average* contribution of the corresponding variable. The actual contribution in a subject can vary. The attempt again is to manage variations and consequent uncertainties so that a valid conclusion can still be drawn.

Example 16.5: Regression of Birth Weight on Parental Weights

Consider regression of birth weight of full-term healthy babies on the weight of the father and mother. Suppose the regression equation obtained on the basis of a random sample is

$$BW = 2.65 + 0.008\,(MW) + 0.004\,(FW); \quad 55 \le MW < 80; \quad 60 \le FW < 90;$$

$$(16.10)$$

where
  BW is birth weight
  MW is mother's weight
  FW is father's weight

All weights are in kilograms. Since there are two regressors, graphically a surface is obtained in place of a line. The response surface of Equation 16.10 is shown in Figure 16.8. Note that the regression is based on the mother's whose weight ranges from 55 to 80 kg and the father's whose weight ranges from 60 to 90 kg. Thus, this regression is valid for these weight ranges only, although slight extrapolation on either side may be admissible.

SIDE NOTE: It can be seen from Equation 16.10 that for every kilogram weight of the mother, the birth weight increases on average by 8 g, and for every kilogram weight of the father, birth weight increases by 4 g. The influence of the father's weight on birth weight is one-half of that of the mother. When the mother's weight is 60 kg and the father's weight 70 kg, the predicted birth weight is $2.65 + 0.008 \times 60 + 0.004 \times 70 = 3.41$ kg. Note that this regression applies only to full-term healthy babies. Even in this

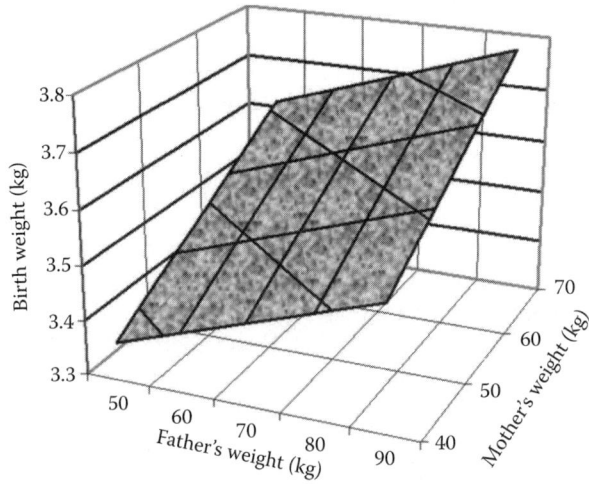

FIGURE 16.8
Response surface for child's birth weight for different weights of the mother and father (Equation 16.10).

group, it is known that the birth weight actually does not depend exclusively on the parental weight. Inclusion of other influential factors in the regression may improve the prediction.

As noted earlier, the prediction in Example 16.5 is for the *average* weight of babies with specific weights of parents. The actual weight in individual births could be different. This difference would be due partly to deviation from the average and partly to the influence of factors other than parental weight.

16.2.3.1 Standardized Coefficients

It is sometimes tempting to compare the contribution of one variable to the prediction of y with the contribution of another variable. In Example 16.5, clearly the weight of the mother has greater contribution, as the corresponding regression coefficient is larger. This comparison is valid in this case because both are weights in the same unit (kg) and the variance of the two may not differ very much. Just think what would happen if mother's weight were in kilograms but the father's weight in grams. The regression coefficient for the father would become 4.0 and would look 500 times that of the mother. If you want to compare the contribution of fat intake in grams to cholesterol level versus that of energy intake in kilo calories, the two will have to be brought to a comparative scale by standardizing. This is done by subtracting mean and dividing by SD, just as is done for Z-score, from the value of each regressor *before* running the regression. The regression then obtained will have standardized coefficients. These coefficients are comparable across various variables despite different units. If x_1 is cost in dollars and x_2 is TG level in mg/dL, the standardized coefficients can still be compared to find which regressor is contributing more to the dependent variable. Such comparison is not possible with unstandardized coefficients. Note, however, that standardized coefficients are not applicable to qualitative categories.

16.2.3.2 Other Implications of Regression Models

The following comments regarding regression clarify certain implications:

1. The predicted value of y will be the best that the regression model can predict, but it may still be far from reality. The practical utility of the model depends on the proper choice of regressors. As mentioned earlier, these should be such that they include all those that influence the outcome y. The model in Equation 16.10 may fail to predict birth weight correctly because many important variables such as nutritional status of the mother (other than weight), dietary intake during pregnancy, and overall health of the mother such as diabetes and hypertension have been ignored. In addition, a linear model is not necessarily the most appropriate.

2. Regression coefficients are the parameters of the regression model. Not only they are estimated with the help of the data and tested for their statistical significance but also a CI on them can be obtained as also on the predicted y. This CI may give a better idea of how much each regressor is contributing to the variability in y. A brief is given in Section 16.3 but note for now that a small SE of $\hat{y}$, a narrow CI, and a high R^2, all indicate that the model is good. In fact, these are interrelated and provide measures of the precision of the model as would be shortly explained.

3. In general, a and b in the regression $y = a + bx$ would be different from b_0 and b_1 in the regression $y = b_0 + b_1x_1 + b_2x_2$. The former regression does not consider x_2 and the latter does so. Because the coefficients measure the contribution of the regressors, it follows that the contribution of one specific variable x to the dependent y can change depending upon how many or what other regressors are present in the model. Similarly, RegSS with three independent variables is not the sum of three RegSS with individual variables.

4. A simple linear regression $y = a + bx$ is the most common form of regression. However, this is valid only when the dependent variable tends to rise or fall with a uniform rate with increasing value of x. This regression is depicted by a line with intercept a, and a coefficient b called the **slope of the line**.

5. Regression models, in fact all models, silently assume that there are no measurement errors. However, these errors remain fact of life despite instrumentation. These errors can disguise an actual relationship particularly if they are systematic such as low values are recorded high and high values recorded low (i.e., positive error for low values and negative error for high values).

6. All models, including regression models, generally work well on the data from which they are derived. The real test of the adequacy of the model is on a new set of data from the same target population. Such external validation is an integral part of the process of development of a model. A model can seldom be recommended for use unless it is externally validated, preferably on a variety of datasets.

7. A regression model can be validly used to predict the level of the response variable only for values of explanatory variables that fall within the range actually used to derive the regression equation. Perhaps, a slight extrapolation, say, not exceeding one-tenth of the range of x actually observed, is admissible. For example, regression Equation 16.3 on BSA and weight cannot be used for predicting the BSA of a child of weight 40 kg because it is valid only for weights between 3 and 30 kg. Also note that this particular regression is valid only for well-proportioned infants and children. For example, this is not applicable to stunted children or obese children.

8. Even within the range observed, the regression could take different shapes for low, middle, and high values. Before using a regression for inferential purposes, check that the shape of the relationship for different values within the range is adequate relative to the scatter of the data.

9. Regression can be severely distorted by a few outliers and also by mixing up of two or more distinct groups of subjects. This aspect is discussed in Chapter 21.

10. Sample size calculation in the case of logistic and classical regression involves complicated formulas. Hsieh et al. [7] suggested a simple method for such a calculation.

16.3 Some Issues in Linear Regression

Regression methods have been found useful in a variety of situations and have been studied extensively. Among many issues that can be discussed, two seem to be important. The first relates to SE of the estimates and CIs, and the second to modifications of the regression method to meet certain exigencies.

16.3.1 Confidence Interval, Confidence Band, and Tests

As all sample estimates, estimates of regression coefficients too are subject to sampling fluctuation. They also have an SE, which provides a measure of their precision.

16.3.1.1 SEs and CIs for the Regression

Most standard statistical software packages readily provide various SEs corresponding to the data entered for regression. Each regression coefficient $b_0, b_1, b_2, \ldots, b_K$ will have its own SE and the regression-estimated $\hat{y}$ also has an SE. Under the usual conditions, particularly when n is large, these estimates follow a Gaussian pattern. Thus, a CI can be constructed and a test of hypothesis can be easily performed for the corresponding $\beta_0, \beta_1, \beta_2, \ldots, \beta_K$. Standard statistical software will readily provide these CIs and the corresponding P-values. The expression for SEs in case of multiple regression are too clumsy. For those interested, the estimated SEs for simple linear regression $y = a + bx$ are as follows:

$$SE(a) = \sqrt{MSE\left(\frac{1}{n} + \frac{\bar{x}^2}{\Sigma(x - \bar{x})^2}\right)} \quad \text{and} \quad SE(b) = \sqrt{\frac{MSE}{\Sigma(x - \bar{x})^2}},$$

where MSE = SSE/$(n-2)$. If β is the slope parameter in the population, under Gaussian conditions,

$$CI \text{ for } \beta : b \pm t_v * SE(b),$$

where dfs for t are $v = (n-2)$. An example in the next paragraph illustrates these CIs as well as confidence band and polynomial regression, which are discussed next.

$$To \text{ test } H_0: \beta = \beta_0, \quad t = \frac{b - \beta_0}{\sqrt{MSE/\Sigma(x - \bar{x})^2}} \quad \text{with } (n-2) \text{ df.}$$

These are for simple linear regression. Similar procedure is used to test significance of various regression coefficients in multiple linear regression. A statistical software does this easily.

16.3.1.2 Confidence Band for Simple Linear Regression

The SE of $\hat{y}$ is different when it is for the mean of y, earlier called as the *estimated value* of y, and when it is for individual value of $\hat{y}$, earlier called as the *predicted value* of y. Prediction of individual values has a much larger SE. These SEs have complex forms for multiple regression but they can be easily stated for a simple linear regression. Sometimes the software does not provide these SEs and you may have to calculate these yourself. Under certain general conditions, for simple linear regression of y on x

$$SE \text{ (estimated mean of } y_x) = \sqrt{MSE\left[\frac{1}{n} + \frac{(x - \bar{x})^2}{\Sigma(x - \bar{x})^2}\right]}, \quad (16.11)$$

and

$$SE \text{ (predicted individual value of } y_x) = \sqrt{MSE\left[1 + \frac{1}{n} + \frac{(x - \bar{x})^2}{\Sigma(x - \bar{x})^2}\right]}, \quad (16.12)$$

where y_x is the value of y at given x. MSE is always generated by the regression software. If the dependent of interest is duration of analgesia (y) induced by different doses of a drug (x) and $n = 15$, MSE = 6.50, $\bar{x} = 11.2$ μg, and $\Sigma(x - \bar{x})^2 = 18.07$, then for $x = 8$, SE(estimated mean of y at $x = 8$) = $\sqrt{[6.50\{1/15 + (8 - 11.2)^2/18.07\}]} = 2.03$ by Equation 16.11; the SE(predicted individual value of y at $x = 8$) = $\sqrt{[6.50\{1 + 1/15 + (8 - 11.2)^2/18.07\}]} = 3.26$ by Equation 16.12. The SE of the predicted value is much higher. Prediction of individual value of y is always less precise than the prediction of mean of y.

The SE of estimated mean in the formula given in Equation 16.11 is used to generate, what is called, a confidence band for the regression line. Although used for individual value predictions, the regression line actually pertains to means. Since xs are considered known, everything is fixed in this SE. As x moves away from its mean, the SE increases because of the term $(x - \bar{x})^2$. This provides a band of the type shown in Figure 16.9.

Under Gaussian conditions, the 95% confidence band is obtained by calculating the limits $\hat{y} \pm t_{n-2}SE(\hat{y})$ for different values of x in the given range, where the value of t corresponds to required confidence as usual. The confidence band specifies the *area* within which the regression line can lie—better phrased as outside which the regression is extremely unlikely to lie. For example, it could be as steep as shown by the dashed line in Figure 16.9 or as flat as shown by the dotted line. Only statisticians seem to know the wide variation that can occur in the estimation of y from a regression line. Many medical professionals use the line obtained from their data as the true line without realizing the variation around it. In case you are reviewing such literature or running a regression yourself, beware of these variations that could be enormous in some situations. Figure 16.10 has such a band for a quadratic regression.

As the number of regressors increase, the width of CI for estimated y decreases in most situations because of reduced SSE and MSE. However, in some situations, width increases because of loss of df. Loss of df can cause the MSE to increase and consequently increase in the width of CI. An adequate sample size becomes doubly important in case of multiple regression. Generally, it is advocated for continuous regressors that a minimum of 10 observations are needed per regressor for multiple regression to provide reliable estimates. This means that if there were $K = 6$ regressors, n must be minimum 60. More the merrier. My take, however, is that the requirement

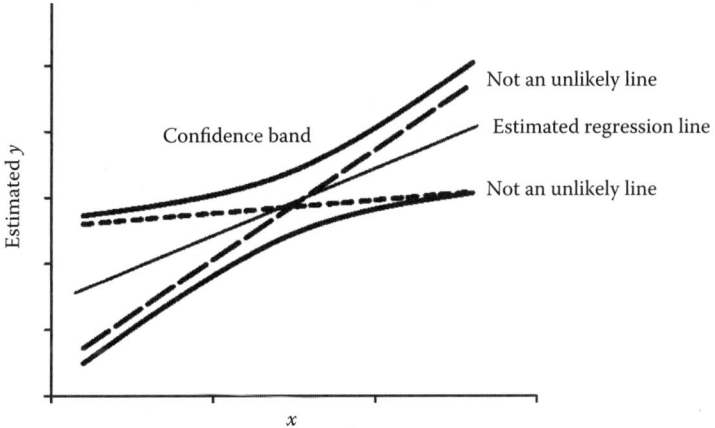

FIGURE 16.9
Confidence band for the regression line.

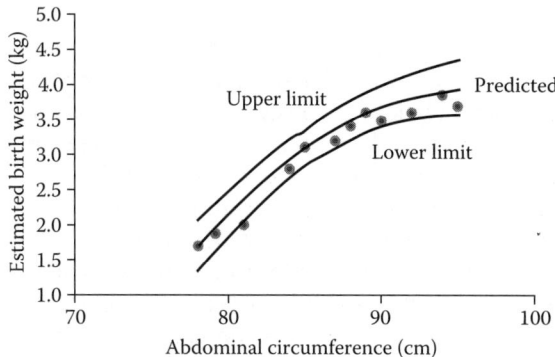

FIGURE 16.10
Confidence band for a quadratic regression.

of n progressively reduces as K increases. If $K = 10$, perhaps $n = 80$ would be enough in place of 100. If n is not as large as you would like, the strategy generally used is to test significance of each individual regressor in simple linear regression and include those in multiple regression that turn out statistically significant in a simple regression. This strategy has obvious flaws since each is being tested in stand-alone setup, whereas the objective of multiple regression is to consider them jointly. Thus, it is desirable to have sufficiently large n wherever the number of regressors under consideration is large.

16.3.1.3 Equality of Two Regression Lines

Quite often the objective is to compare trend in two groups such as treatment and control groups or males and females. Most statistical softwares have still not been given intelligence to do this comparison directly. You will be required to specify the right model or use a modified procedure. Among various ways that this comparison can be done, the following illustrates one for comparing simple linear regressions.

Define an **indicator variable** z, and assign it a value zero for group-1 and value one for group-2. The variable has no meaning except to indicate the group. For this reason, this is also called a **dummy variable**. Remind yourself that in the case of regression, xs are considered fixed and they can have any appropriate value. Thus, such a regressor variable is admissible. Now fit the following regression model to the data. The data should have values of x and y as before, as well as the values of z as just assigned:

$$y = b_0 + b_1 x + b_2 z + b_3 xz. \tag{16.13}$$

You can easily see that for group-1 since $z = 0$, the model is

$$y = b_0 + b_1 x, \tag{16.14}$$

and for group-2 since $z = 1$, the model is

$$y = b_0 + b_1 x + b_2 + b_3 x$$

$$= a + bx, \tag{16.15}$$

where $a = (b_0 + b_2)$ and $b = (b_1 + b_3)$.

Compare Equation 16.14 with Equation 16.15 and note that the corresponding parameters $\beta_3 = 0$ implies same slope but different intercepts, that is, the lines in the two groups are parallel; $\beta_2 = 0$ implies same intercept in the two groups but different slopes; and $\beta_2 = 0$ and $\beta_3 = 0$ together imply that the regression lines in the two groups are identical.

The test of these hypotheses can be done as usual for regression coefficients. The only additional requirement is that the variance in the two groups is same. If the two lines are found not significantly different when n_1 and n_2 are sufficiently large to provide enough power, the data from two groups can be pooled and a unified regression can be fitted that would have better reliability because of increased n. For this, pool the data and run the regression again. The procedure can be easily extended to compare more than two lines.

16.3.1.4 Difference-in-Differences Approach with Regression

Indicator variables are sometimes useful in including one or more qualitative variables in the regression. If you want to obtain different regression equation for males and females by running a single regression, just include an appropriate indicator variable with (0, 1) values for males and females, respectively. One such application is in difference-in-differences approach that was introduced in Chapter 15 using the unpaired Student t-test in the context of before–after values in treatment and control groups. The regression approach to the same problem is as follows. Consider the regression

$$y = b_0 + b_1 x_1 + b_2 x_2 + b_3 x_1 * x_2,$$

where
$x_1 = 0$ for control group and $x_1 = 1$ for treatment group
$x_2 = 0$ for before values and $x_2 = 1$ for after values

These are the indicator variables. Plug these values and obtain the following:

Before values: treatment group $y = b_0 + b_1$; control group $y = b_0$; difference $= b_1$.

After values: treatment group $y = b_0 + b_1 + b_2 + b_3$; control group $y = b_0 + b_2$; difference $= b_1 + b_3$.

Difference-in-differences $= (b_1 + b_3) - b_1 = b_3$.

Thus, b_3 is an estimate of the actual treatment effect compared with the controls. This is the regression coefficient of the product $x_1 * x_2$, which is now called **interaction**. If the interaction is significant, the treatment effect obtained by the difference-in-differences approach is significant.

16.3.2 Some Variations of Regression

Any method, when widely used, is looked at very critically and the inadequacies noted. Thus, several modifications have been suggested for the regression to meet specific requirements. Among those used sometimes in health and medicine are ridge regression and multilevel regression. A more widely used method is ANCOVA. All these are intricate and beyond the scope of this book. They are briefly explained in this section so that you know where to use these methods.

When the variables depend bidirectionally—y depends on x and x depends on y—such as health depends on exercise and exercise depends on health, a **two-stage regression** procedure is used. The details are complex for the level of this book. Those interested may refer to Faries et al. [8].

16.3.2.1 Ridge Regression

Consider predicting body fat with the help of skinfold thickness at triceps, mid-arm, and thigh. These three are highly correlated. As noted earlier, regression coefficients tend to have an increased variance when multicollinearity exists among the regressors. The regression model loses its sheen and its utility is compromised because of the instability of the estimates. Ridge regression is one of the modifications that can remedy this problem. Instead of using the method of least squares, a method called ridge is used that produces slightly biased estimates of the regression parameters but substantially reduced SEs. For more details, see Dodge and Jureckova [9]. Some statistical softwares provide an option to run this kind of regression.

16.3.2.2 Multilevel Regression

The method of regression presented so far works well when the regressors are considered fixed. However, this requirement may not be met in some situations. Consider the duration of stay of patients in critical care in small, medium, and large hospitals. The suspicion is that this duration of stay varies according to the size of hospital that determines the facilities, care available, and the confidence of the patients. In addition, of course the duration will depend on the condition of the patient at the time of admission that can be assessed by APACHE score. Let the size of the hospital be defined by the number of beds: small if the number of beds is less than 100, medium if the number of beds is between 100 and 399, and large if the number of beds is 400 or more. The exact number of beds is not under consideration.

A sample of $n_1 = 15$ patients from a small hospital, $n_2 = 30$ patients from a medium hospital, and $n_3 = 50$ from a large hospital are chosen from critical care. In this example, hospital effect is fixed but the effect of the APACHE score is random as the sample of patients is random. The linear regression model in this situation will be of the following type:

$$y_{ij} = a_i + b_i x_{1i} + c_i x_{2ij} + d_i x_{1i} * x_{2ij}, \tag{16.16}$$

where

y_{ij} is the duration of hospital stay of jth patient of the ith hospital ($j = 1, 2, ..., n_i; i = 1, 2, 3$)

x_{1i} is the size of the ith hospital ($i = 1$ is small, $i = 2$ is medium, $i = 3$ is large)

x_{2ij} is the APACHE score of jth patient of the ith hospital ($j = 1, 2, ..., n_i; i = 1, 2, 3$)

$a_i, b_i, c_i,$ and d_i together define the intercepts and slopes for the hospitals of different sizes

This model will give three different regressions, one for each size of the hospital and called multilevel regression. This example is for two levels (size within hospital and subjects within size) and can be extended for more level. Multilevel regression is also called **mixed regression** as well as **hierarchical regression**. This kind of regression is also used in the analysis of longitudinal data. Suppose after a surgery each patient is recorded for pulse pressure every 5 min till such time the patient stabilizes. The first patient becomes stable in 40 min so that he is recorded for $n_1 = 8$ time points, the second in 30 min so that $n_2 = 6$, etc. If the patients are considered fixed (not a random sample), the situation is the same as mentioned earlier. If they are a random sample, a slight modification of the model will be required.

The basic premise in multilevel analysis in the context of this example is that the duration of hospital stay tends to follow different patterns depending on the size of the hospital and that both hospital size and the APACHE score should be considered together. Depending on the data, the regressions may look as shown in Figure 16.11. The lines have different intercepts and different slopes. The usual line obtained after ignoring the hospital size, which assumes that all durations and APACHE scores are from the same target population, is also given. Note how this composite line hides the differences present in the lines for different hospital sizes.

16.3.2.3 Regression Splines

I alerted you earlier regarding fitting one regression equation to the entire range of x-values that can lead to fallacious results when the pattern of y-values is different between one interval of xs than between another interval of xs. For example, if you fit one regression for risk of mortality over BMI values from 18 to 36 kg/m², you may find it U-shaped with least mortality

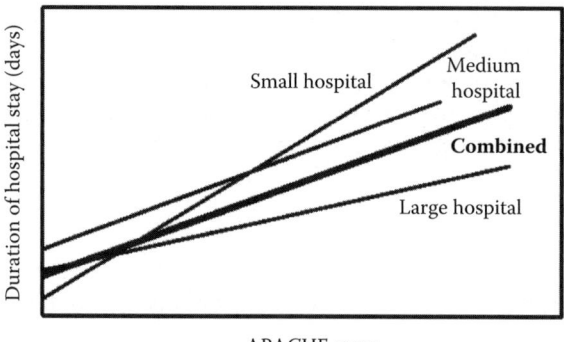

FIGURE 16.11
Multilevel regression.

around BMI = 24 and high at both ends—steeper hike for BMI ≥ 34. However, if you divide BMI into three segments as BMI < 23, 23 ≤ BMI < 27 and BMI ≥ 27, you may find that risk of mortality is nearly flat in the middle range. This finding is lost when one regression equation is obtained for the entire range even when the regression is a curve.

Regression splines fit different regression equations to different segments of x-values. They produce a kink or knot at the specified values of x. Research is going on how to find the values of x where a knot is needed and how many knots are needed. At present they need to be specified by the user and this can become arbitrary. The method of splines can be extended to many regressors, and different curvilinear fitting to different segments of values of xs can be done.

16.3.2.4 Analysis of Covariance

You may have noticed in both the discussions of regression in the present chapter and of ANOVA in the preceding chapter that the dependent variable is quantitative. However, in the case of regression, the independents too are quantitative whereas in the case of ANOVA they are qualitative. What should one do if the independent set contains both quantitative and qualitative factors and the dependent set continues to be quantitative as before? The answer is ANCOVA. You can see that it sits between regression and ANOVA. Some of it is already explained in this chapter such as for comparing regression lines for two groups where groups are coded as (0, 1) with the help of an indicator variable, and in multilevel regression where, for example, small, medium, and large hospitals are given values 1, 2, and 3, respectively. ANCOVA is a generalized procedure and is mathematically intricate. Since software could be essential for performing ANCOVA, the emphasis here will be on explaining the underlying principles and its application.

Consider determinants of BMI of adults of age 20–49 years. This is the dependent variable y and is calculated exactly as weight/height². Among several determinants that can be considered, the important ones are gender (male or female); physical activity as none, mild, moderate, and heavy; and fat and carbohydrate intake in grams per day. Note that these variables affect each other and any conclusion based on any one variable in isolation may be misleading. Thus, all the factors should be considered together. If the primary interest is in knowing the relationship between BMI and physical activity and gender, it needs to be **adjusted** for the accompanying differentials in dietary intake. In this situation, dietary intake of fat and carbohydrate are the covariates. This is done by ANCOVA. It is similar to ANOVA of BMI on gender and physical activity on residuals after removing the effect of dietary differentials.

For running ANCOVA, the quantitative covariates remain as they are but the qualitative variables are assigned values (0, 1) through the indicator variable. In the BMI example, if x_1 is the fat intake and x_2 the carbohydrate intake, then $x_3 = 0$ for males and $x_3 = 1$ for females. For physical activity since there are four categories, the indicator variables can be defined as follows:

For no physical activity:	$x_4 = 0, x_5 = 0, x_6 = 0$
For mild physical activity:	$x_4 = 1, x_5 = 0, x_6 = 0$
For moderate physical activity:	$x_4 = 0, x_5 = 1, x_6 = 0$
For heavy physical activity:	$x_4 = 0, x_5 = 0, x_6 = 1$

The number of indicator variables is one less than the number of categories. This takes care of all the categories and no category is left out in the cold. Now, a regression is run with $x_1, x_2, x_3, x_4, x_5,$ and x_6 as the regressors. The corresponding regression coefficients measure the contribution of each variable and each category, and this contribution is now adjusted for the other variables in the model.

You can see that ANCOVA as explained earlier is another statistical method to reduce the specter of uncertainty. It adjusts the comparison across groups for imbalances in the covariates and enhances the precision by accounting for these imbalances. The devil is in the details. For example, ANCOVA is valid when slopes of regression lines are same for different groups formed by categories of qualitative variables. To check this homogeneity, absence of interaction between covariates and qualitative variables is tested. Thus, do ANCOVA with abundant caution.

Software illustration: A software illustration on ANCOVA using SPSS appears in Appendix C.

Sometimes, the groups formed by qualitative variables differ at baseline, which confounds the results. ANCOVA can be used to find the net differences among groups by using baseline values as a covariate. This would adjust results for differentials at the baseline. One such application is in the following example.

> **Example 16.6: ANCOVA Can Give Substantially
> Improved Conclusions in Some Situations**
>
> For an outcome such as pain score, the baseline values are important and should be adjusted. Vickers [10] reanalyzed original data from four trials of acupuncture for pain relief. Publication of these trials used statistical methods that did not adjust for baseline pain scores. Vickers used ANCOVA in reanalysis to adjust for baseline differences. In two trials, ANCOVA did not change the conclusion that there was no difference between acupuncture and control groups, perhaps because the baseline levels were not much different. In the third trial, the evidence of the effect of acupuncture slightly strengthened. In the fourth trial, ANCOVA reversed the earlier negative result and showed that acupuncture was effective. Vickers concluded that ANCOVA could be more efficient in some situation than unadjusted analysis.

16.3.2.5 Some Generalizations

After discussing so many regression type models, would it not be illuminating to see that they weave into a unified theory? The rubric is called **general linear models** (GLM). Different types of analysis with quantitative dependent (ANOVA, regression, ANCOVA) can be combined under this umbrella term. Standard statistical packages have a program to run GLM. Only that it requires a little more expertise—in defining the variables including the indicator (dummy) variables, in identifying the comparisons of interest, in looking at the interdependence among independent variables, etc.

This is mentioned just for reference of those who may be looking for GLM procedure in this book in the context of what they see in the software. There is another term called generalized linear model, also called GLM, and is discussed in Chapter 17.

16.4 Measuring the Strength of Quantitative Relationship

It is known that the relationship of smoking with cotinine level is strong, mild with BP, and practically nil with intraocular pressure. Statistically speaking, a relationship is strong if the scatter plot follows an exact pattern (other than parallel to the *x*-axis) and weak if the points are widely scattered. Different types of measures of the strength of relationship are computed depending upon the nature of the variables involved. This chapter considers situations where both variables are quantitative. The relationship with the qualitative dependent variable is discussed in the next chapter.

16.4.1 Product–Moment and Related Correlations

While regression expresses the nature of relationship, correlation measures the degree of relationship. Both the variables in the classical setup of correlation are quantitative. The following paragraphs contain an explanation of some of the different types of correlations, their interpretation, and the test of their significance.

16.4.1.1 Multiple Correlation

As stated in Section 16.2.1 in the context of goodness of fit,

$$\eta^2 = 1 - \frac{\text{SSE}}{\text{SST}}$$

measures the proportion of total variation explained by the regressors and called the **coefficient of determination** irrespective of the relation being linear or otherwise. If $\eta^2 = 0.64$, this means that 64% of total variation in y is due to the factors considered in the model. The positive square root, η, exists but has no particular name when the relationship is nonlinear. It is called the coefficient of **multiple correlation** when the relationship is linear and the number of regressors is more than one. This is denoted by R as mentioned earlier. No meaning can be attached to the direction of correlation when the number of regressors is more than one. Thus, the sign of R is always considered positive. This is the correlation between y and its values predicted by the regression. Higher value of R indicates that the predicted values are close to the observed values thus regression is good. However, R by itself is rarely used. Instead R^2 is used whose meaning is already explained.

16.4.1.2 Product–Moment Correlation

When the relationship is linear and when there is only one regressor x, it can be algebraically shown that R^2 is the square of

$$\text{Correlation coefficient: } r = \frac{\Sigma(x - \bar{x})(y - \bar{y})}{\sqrt{\Sigma(x - \bar{x})^2}\sqrt{\Sigma(y - \bar{y})^2}}. \tag{16.17}$$

The notation now used is r in place of R. Lowercase r is used to clarify that only a linear relationship is being considered and there is a total of only two variables under consideration, one being dependent and the other is independent. Uppercase R has no such restriction on the number of independent variables. The quantity r^2 continues to have the same interpretation as R^2 except that it now has only one regressor. The value of r can be negative or positive according to the details given later. As there are many types of correlations, the formula given in Equation 16.17 is sometimes specified as the **product–moment coefficient of correlation**. The name arises from the numerator, which is covariance, as we explain shortly, but is also called product–moment. This is also known as **Pearsonian correlation**. The value of r does not change if you add, subtract, multiply, or divide each value of x or each value of y or both by any positive constant. This property is called invariance under change of origin and scale and is sometimes handy in reducing calculations.

In the case of regression, y is observed for certain *fixed* values of the variable x. Against this, the general presumption in correlation is that both x and y are stochastic. However, r remains valid even if only one of them is stochastic. The observed values of x and y should cover a fair range of values for which inference is desired.

16.4.1.3 Covariance

The real operational quantity in the formula given in Equation 16.17 is its numerator, called covariance when divided by n. This is the average of the products of $(x - \bar{x})$ and $(y - \bar{y})$, which are respective deviations from the mean. An explanation for covariance is as follows.

Consider the scatter diagrams in Figure 16.12 between age (x) and FVC (y). Figure 16.12a is for the age group 10–24 years, Figure 16.12b for 25–39 years, and Figure 16.12c for 40–79 years. Figure 16.12d gives the joint scatter of all the age groups combined. Quadrants are drawn by lines corresponding to the mean of x and mean of y in each case. In Figure 16.12a, FVC is increasing as age increases from 10 to 24 years. Both x and y are moving in the same direction. This is an indication of positive correlation between x and y. Most points are in the first and the third quadrants. In the first quadrant, both $(x - \bar{x})$ and $(y - \bar{y})$ are positive, so their product is also positive. In the third quadrant, both $(x - \bar{x})$ and $(y - \bar{y})$ are negative, so their product again is positive. Thus, the *covariance* would be a positive quantity, signifying a positive relationship.

In Figure 16.12b, FVC is neither increasing nor decreasing when age is between 25 and 39 years. The points are almost equally scattered in the four quadrants. In the second quadrant, $(x - \bar{x})$ is negative but $(y - \bar{y})$ is positive. In the fourth quadrant, $(x - \bar{x})$ is positive but $(y - \bar{y})$ is negative. In both cases, the product $(x - \bar{x})(y - \bar{y})$ is negative. The sum of negative products in the second and the fourth quadrants would mostly cancel out with the sum of positive products in the first and third quadrants, thus giving a final sum nearly equal to zero. Thus, the covariance between x and y in Figure 16.12b is nearly zero. In Figure 16.12c, FVC decreases as age increases from 40 to 79 years. This inverse relationship yields a negative correlation. Most points are in either quadrant II or quadrant IV. They give negative products. The covariance in this case is negative. The illustration in all three cases indicates that covariance does measure the direction of relationship.

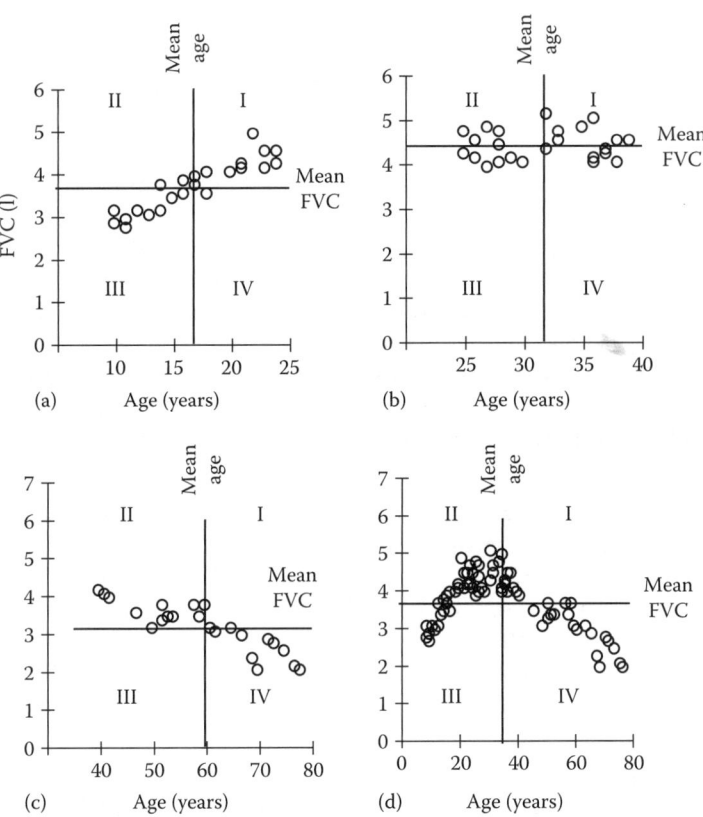

FIGURE 16.12
Scatter of FVC by age: (a) for the age group 10–24 years, (b) for the age group 25–39 years, (c) for the age group 40–79 years, and (d) combined for all the age groups.

The magnitude of covariance depends on the magnitude of $(x - \bar{x})$ and $(y - \bar{y})$, that is, on the variation in x and y. The variation is measured by $\sqrt{\Sigma(x-\bar{x})^2/n}$ and $\sqrt{\Sigma(y-\bar{y})^2/n}$. These are the respective SDs. To make the covariance independent of such variation, it is divided by the product of SDs. This finally gives the formula given in Equation 16.17. The correlation coefficient thus obtained also becomes dimensionless and very easy to interpret.

Example 16.7: Correlation between TG Level and WHR

Consider the TG and WHR data in Table 3.1. The correlation coefficient, when computed on the basis of all 100 subjects in the entire population, is denoted by ρ. This, for these data, is given by $\rho = 0.67$. For the asterisk marked $n = 16$ subjects in the sample (Table 3.1), this is $r = 0.69$.

Now note the following for a correlation coefficient:

1. The correlation coefficient (Equation 16.17) is a pure number without any unit and ranges from −1 to +1. A value close to zero indicates that the two variables are linearly uncorrelated. That is, a change in the value of one is not accompanied by any linear change in the other. The scatter can then be a horizontal line with no slope. An example is cholesterol level and Hb level, which are generally uncorrelated. A value close to +1 indicates a strong positive relationship. Body temperature and heart rate have a strong positive correlation. A value close to −1 indicates a strong negative correlation. Homocysteine and transthyretin have a strong negative correlation among vegetarians. A perfect $r = \pm 1$ would mean that the scatter of y with x forms an exact straight line. Such extreme correlation would seldom occur in health or medicine.

2. Generally speaking, a correlation greater than 0.9 in absolute value can be considered strong, between 0.9 and 0.6 moderate, between 0.6 and 0.4 weak, and between 0.4 and 0 almost nonexistent. Thus, the correlation between TG and WHR in Example 16.7 is moderate. An illustration of scatter plots in weak and strong correlation is shown in Figure 16.13. Note, for example, that $r = 0.3$ means $r^2 = 0.09$. That is, for this r, only 9% of the total variation in y is attributable to the linear variation in x.

3. The magnitude of the correlation coefficient becomes unusually high when wide range of values (very high and very low) of x and y are observed and low when only a small range of values is observed. Thus, it is not desirable to deliberately restrict the observations of x and y to a narrow range or observe only extreme values. The distribution of observed values of at least one of them should be fairly spread out.

4. The product–moment correlation r measures only the *linear* component of the relationship, and this line must have some slope

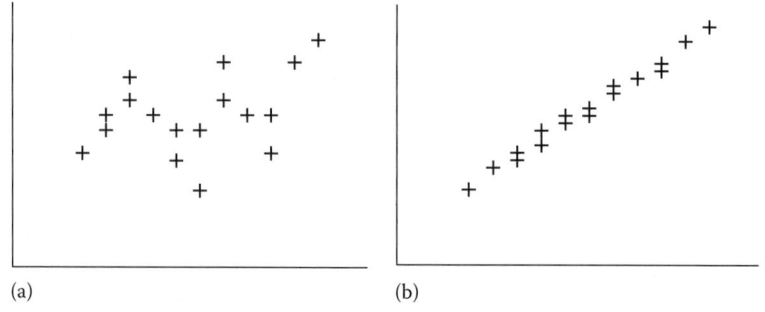

(a)   (b)

FIGURE 16.13
Scatter plot in the case of (a) weak and (b) strong correlations.

(small or big, negative or positive). If all age groups in Figure 16.12a through c are combined, you get Figure 16.12d. This scatter is nearly evenly spread in all four quadrants. Now the negative products $(x - \bar{x})(y - \bar{y})$ will mostly cancel out with positive products. Thus, the covariance will be nearly equal to zero. When all ages between 10 and 80 years are considered together, these data will give a small value of the correlation coefficient (also see Figure 16.1c). This should not be construed to mean that there is no relationship between age and FVC. The relationship, in fact, is strong and can be expressed by a nice parabolic curve, although not by a straight line. The product–moment coefficient of correlation does not measure nonlinear relationships. Thus, a low value of the correlation coefficient needs to be interpreted with caution. Two variables can be heavily interdependent but may not appear correlated by this measure. This may sound like a paradox but is clear when the concept of correlation is correctly understood as concerned only with a linear form of relationship. The right criterion for measurement of a nonlinear relationship is the coefficient of multiple correlation R^2 when a curvilinear form is considered and η^2 when nonlinear form is considered.

5. Sometimes weak correlations too can be medically important. One such example is the correlation between prostate volume and urethral resistance parameters in patients with prostatism [11]. This correlation was found less than 0.4 and indicated that urethral resistance is determined mostly by factors other than the volume of the prostate. This is a significant finding. Another example of weak correlation is that between age and FVC mentioned earlier. The *linear* relationship, as measured by the correlation coefficient, is weak, but they actually have a high degree of relationship of a curvilinear form. A third example is correlation of systolic BP of persons with the level seen in their children. This correlation is weak in many populations but is definite. Such a correlation indicates that parents do have an influence on the level of systolic BP of their offspring, although the influence is minor.

6. The birth rate in Delhi is high in the months of August and September as a consequence of increased sexual activity at the onset of winter in the months of November and December. But August and September also happen to be the months with high pollen density in the Boston area. The correlation between pollen density in Boston and birth rate in Delhi over a 12 month period may turn out to be high. But this correlation has no meaning. A correlation arising because of intervention of a third, irrelevant, variable is called a **spurious or nonsense correlation**. Also, not all explainable correlations can be considered to indicate a cause–effect type of relationship.

16.4.1.4 Statistical Significance of r

A correlation coefficient r is called statistically significant when the probability of it being zero in the population (H_0: $\rho = 0$) is less than 0.05 or any other such predetermined level of significance. This is tested by referring the criterion

$$\text{Student test for correlation: } t = \frac{r\sqrt{n-2}}{\sqrt{1-r^2}} \qquad (16.18)$$

to Student t-distribution with $(n - 2)$ df. The test is valid provided at least one of the variables follows a Gaussian pattern. If this condition is severely violated, it is advisable to use rank correlation, explained in the next section.

A distinction must be made between a strong correlation and a statistically significant correlation. When n is sufficiently large, even $r = 0.1$ ($r^2 = 0.01$) can be statistically significant. If $n = 400$, then

$$t = \frac{0.1\sqrt{400-2}}{\sqrt{1-0.1^2}} = 2.0.$$

This gives $P < 0.05$ and indicates statistical significance. Similarly, a relatively high correlation, say, $r = 0.6$, can be statistically not significant if n is small, say, $n = 5$. In this case,

$$t = \frac{0.6\sqrt{5-2}}{\sqrt{1-0.6^2}} = 1.30,$$

and $P > 0.10$ even for a one-tail test. Statistical significance merely confirms that there is some sort of linear relationship with a nonzero slope. It does not say how strong this relationship is.

A distinction must also be made between the regression coefficient (b in equation $y = a + bx$) and the correlation coefficient r. The regression coefficient defines the slope of the regression line. The value of r could be one even if the slope is exceedingly small. This will happen if y consistently increases by the same small amount for each unit increase in x. Also b can be enormously large yet r may not be significant.

16.4.1.5 Intraclass Correlation

Siblings within a family are likely to be more similar to one another than siblings belonging to different families. This phenomenon is known as **familial aggregation**. The intraclass correlation coefficient is the measure of the degree of consistency or conformity between members belonging to the same subgroup. This is especially suited to study (linear) correlation between twins or other multiple births or between measurements of two eyes or two limbs of the same individual. One useful application of intraclass correlation is in assessing agreement between different observers and different methods when used on the same set of subjects. This application is discussed in the next section where the method of computation is also described.

16.4.1.6 Serial Correlation

Particularly in time-dependent observations, the value at time t is generally related to the value at previous time $(t - 1)$. If pain score of a patient is high 5 min after analgesia compared with the other patients, it is likely to be relatively high at 10 min also. Correlation with a previous value is called **serial correlation** of lag 1. In some cases, the interest might be in serial correlation of lag 2 or even a higher lag. In case of seasonal pattern such as for diarrhea and respiratory infection, when monthly observations are taken, the level in November this year may be related to the level in the same month a year ago. This is a lag of 12 months.

   This kind of correlation was mentioned earlier while talking of independence of residuals as a requirement for valid regression. Serial correlations are the main concern in **time-series** data. For this reason, they are analyzed using autoregressive models or moving average models, or a mixture of both called autoregressive moving average (ARMA) models. This text does not discuss methods for time-series analysis.

16.4.2 Rank Correlation

In some cases, x and y clearly increase together but the relationship is not necessarily linear. This is called a monotonic relationship. The dependence of height on the age of children is an example of such a relationship. This is not linear but definitely exists. The relation of visual acuity with age after 60 years may be monotonically decreasing with a sharp drop after the age of 80 years. The linearizing effect of the product–moment correlation coefficient in such cases amounts to an oversimplification of the relationship. An alternative is the rank correlation. The best known measure of rank correlation is Spearman correlation coefficient.

16.4.2.1 Spearman Rho

For this correlation, the values of x and y are separately ranked from 1 to n in increasing order of magnitude. Ordinary product–moment correlation coefficient is computed between ranks of x and ranks of y. This simplifies to the following:

$$\text{Spearman rank correlation coefficient: } r_S = 1 - \frac{6\Sigma d^2}{n(n^2 - 1)}, \qquad (16.19)$$

where d = (rank of y − rank of x) and where the summation is over all pairs of observations. When computed for all subjects in the population, it is denoted by ρ (rho).

One big advantage of rank correlation is that it attenuates the effect of outliers. As shown later in Chapter 21, only one outlier can substantially distort the value of the product–moment correlation coefficient. The value of rank correlation is not so much affected because a high value is converted just to the next rank.

Example 16.8: Spearman's Rank Correlation between Height and Weight

Given in the following are the height and weight of eight children:

Weight (kg)—y	10	14	26	6	14	18	32	4	
Height (cm)—x	65	92	127	63	95	96	128	55	
Rank (y)	3	4.5	7	2	4.5	6	8	1	
Rank (x)	3	4	7	2	5	6	8	1	
Difference (d)	0	+0.5	0	0	−0.5	0	0	0	$\Sigma d^2 = 0.50$

$$\text{Spearman rank correlation coefficient, } r_s = 1 - \frac{6 \times 0.50}{8 \times 63} = 0.994.$$

Example 16.8 illustrates the following:

1. Equal ranks are assigned to the tied observations. In this example, the second and fifth child have the same weight. They would have received ranks 4 and 5 but now received a midrank of 4.5 each because of the tie.

2. The ordinary product–moment correlation coefficient between weight and height in this example is 0.967. The rank correlation is slightly higher. This would have been exactly +1 if the weight of the second child were 13 kg and the height the same 92 cm. The rank correlation can sometime overrate the strength of the relationship because it partially disregards the actual magnitude of x and y. This is clear from the explanation given next.

3. The value of rank correlation would not change if, for example, the height of the third child were 114 cm instead of 127 cm. This is because the rank is not changed by such an alteration in this case. Rank correlation, thus, is not fully sensitive to the exact values of the variables.

This is a nonparametric procedure and the test of its significance does not require a Gaussian pattern of either x or y. This correlation is thus preferable when the distribution pattern of both x and y is far from Gaussian.

For samples of 10 or fewer pairs, the minimum values of r_s for different significance levels are given in Table 16.1. For $n \geq 11$, the Gaussian pattern

TABLE 16.1

Minimum Value of r_s for Different n (≤ 10) that Is
Statistically Significant for Different α Levels (Two-Tailed)

Sample Size (n)	Minimum Value of r_S			
	$\alpha = 0.10$	$\alpha = 0.05$	$\alpha = 0.02$	$\alpha = 0.01$
1, 2, or 3	None	None	None	None
4	1.000	None	None	None
5	0.900	1.000	1.000	None
6	0.771	0.886	0.943	1.000
7	0.714	0.786	0.892	0.929
8	0.643	0.738	0.810	0.857
9	0.600	0.683	0.783	0.817
10	0.564	0.648	0.733	0.781

Source: Adapted from Snedecor, G.W. and Cochran, W.G., *Statistical Methods*, 8th edn., Iowa State University Press, Ames, IA, 478, 1980, Table A11(ii). With permission.

Note: For $n \geq 11$, use Student t-test.

holds reasonably well and the Student t-test (Equation 16.18) can be used with $(n - 2)$ df.

Different methods for measuring the strength of relationship between two qualitative or quantitative variables are listed in **Table S.6** at the beginning of the book.

16.5 Assessment of Quantitative Agreement

Medical science is growing at a rapid rate. New instruments are invented and new methods are discovered that measure anatomical and physiological parameters with better accuracy and precision, and at a lower cost. Emphasis is on simple, noninvasive, safer methods that require smaller sampling volumes and can help in continuous monitoring of patients when required. Acceptance of any new method depends on a convincing demonstration that it is nearly as good, if not better, as the established method.

Irrespective of what is being measured, it is highly unlikely that the new method would give exactly the same reading in each case as the old method even if they are equivalent. Some differences would necessarily arise. How do you decide that the new method is interchangeable with the old? The problem is described as one of agreement. This is different from evaluating which method is better. The assessment of "better" is done with reference to a gold standard. Assessment of agreement does not require any such standard.

The term agreement is used in several different contexts. The following discussion is restricted to a setup where a pair of observations (x, y) is obtained by measuring the same characteristic on the same subject by two different methods, by two different observers, by two different laboratories, at two anatomical sites, etc. The measurement could be qualitative or quantitative. The method of assessing agreement in these two cases is different. This section is on agreement in quantitative measurements. Agreement in qualitative measurements is discussed in the next chapter.

16.5.1 Agreement in Quantitative Measurements

The problem of agreement in quantitative measurement can arise in at least five different types of situations: (1) comparison of self-reported values with the instrument-measured values, for example, urine frequency and bladder capacity by patient questionnaire and frequency–volume chart; (2) comparison of measurements at two different sites, for example, paracetamol concentration in saliva with that in serum; (3) comparison of two methods, for example, bolus and infusion methods of estimating hepatic blood flow in patients with liver disease; (4) comparison of two observers, for example, duration of electroconvulsive fits reported by two psychiatrists on the same group of patients or of two laboratories when, for example, aliquots of the same sample are sent to two different laboratories for analysis; and (5) intraobserver consistency, for example, measurement of anterior chamber depth of an eye segment two times by the same observer using the same method to evaluate reliability of the method.

In the first four cases, the objective is to find whether a simple, safe, less expensive procedure can replace an existing procedure. In the last case, it is evaluation of the reliability of the method.

16.5.1.1 Statistical Formulation of the Problem

The statistical problem in all these cases is to check whether or not a $y = x$ type of relationship exists in individual subjects. This looks like a regression setup $y = a + bx$ with $a = 0$ and $b = 1$, but that really is not so. The difference is that, in regression, the relationship is between x and the average of y. In an agreement setup, the concern is with individual values and not with averages. Nor should agreement be confused with high correlation. Correlation would be nearly one if there is a systematic bias and nearly same difference occurs in every subject. Example 16.9 illustrates distinction between $y = x$ regression and agreement.

> **Example 16.9: Very Different Values but Regression Is $y = x$**
>
> Following are Hb values reported by two laboratories for the same blood samples:
>
> | Laboratory I (x) | 11.3 | 12.0 | 13.9 | 12.8 | 11.3 | 12.0 | 13.9 | 12.8 |
> |-------------------|------|------|------|------|------|------|------|------|
> | Laboratory II (y) | 11.5 | 12.4 | 14.2 | 13.2 | 11.1 | 11.6 | 13.6 | 12.4 |
>
> $\bar{x} = 12.5, \bar{y} = 12.5, r = 0.945, \hat{y} = x$, that is, $b = 1$, and $a = 0$.

The two laboratories have same mean for these eight samples and a very high correlation (0.945). The intercept is zero and slope is 1.00. Yet, there is no agreement in any of the subjects. The difference or error ranges from 0.2 to 0.4 g/dL. This is substantial in the context of the present-day technology. Thus, equality of means, a high degree of correlation, and regression $y = x$ are not enough to conclude agreement. Special methods are required.

SIDE NOTE: If you notice carefully, first four values of x in this example are the same as the last four values. The first four values of y are higher and the last four values are lower by same margin. Thus, for each x, $\bar{y} = x$ giving rise to the regression $\hat{y} = x$. In this particular case, the correlation coefficient also is nearly one.

16.5.2 Approaches for Measuring Quantitative Agreement

Quantitative agreement in individual values can be measured either by limits of disagreement or by intraclass correlation. The details are as follows.

16.5.2.1 *Limits of Disagreement Approach*

In this method, the differences $d = (x - y)$ in the values obtained by the two methods or observers under comparison are examined. If these differences are randomly distributed around zero and none of the differences is large, the agreement is considered good. A graphical approach is to plot ds versus $(x + y)/2$. A flat line around zero is indicative of good agreement. Depending upon which is labeled x and which is y, an upward trend indicates that x is generally more than y, and a downward trend that y is more than x.

A common sense approach is to consider agreement as reasonably good if, say, 95% of these differences fall within the prespecified clinically tolerable range and the other 5% also not too far. Statistically, note that when the two methods or two observers are measuring the same variable, then the difference d is mostly the measurement error. Such errors are known to follow a Gaussian distribution. Thus, the distribution of d in most cases would be Gaussian. Then, the limits $\bar{d} \pm 1.96 s_d$ are likely to cover differences in nearly 95% of subjects where s_d is the SD of the differences. The literature describes them as the limits of agreement. *They are actually limits of disagreement.* If these limits are within clinical tolerance in the sense that a difference of that magnitude does not alter the management of the subjects, then one method can be replaced by the other. The mean difference $\bar{d}$ is the bias between the two sets of measurements and s_d measures the magnitude of random error. For further details, see Bland and Altman [12].

Enough has already been said about the limitation of correlation. If things are not clear enough, consider the following example. Suppose a method consistently gives a level 0.5 mg/dL higher than the other method. The correlation coefficient between these two methods would be prefect 1.0. Correlation fails to detect systematic bias. This also highlights the limitation of limits of disagreement approach. The difference between measurements by two methods in this case is always +0.5 mg/dL—thus the SD of difference is zero. The limits of disagreement in this case are (+0.5, +0.5). This, in fact, is just one value and not limits. A naïve argument could be that these "limits" are within clinical tolerance and thus the agreement is good. To detect this kind of fallacy, plot the differences against the mean of paired values. This trend can immediately reveal this kind of systematic bias.

Example 16.10: Limits of Disagreement between Pulse Oximetry and Korotkoff Readings

Consider the following data [13] on systolic BP readings derived from the plethysmographic waveform of a pulse oximeter. This method could be useful in a pulseless disease such as Takayasu syndrome. The readings were obtained at the (1) disappearance of the waveform on the pulse oximeter on gradual inflation of the cuff and at the (2) reappearance on gradual deflation. In addition, BP was measured in a conventional manner by monitoring the Korotkoff sounds. The study was done on 100 healthy volunteers. The readings at disappearance of the waveform were observed to be generally higher and at reappearance generally lower. Thus, the average (AVRG) of the two is considered a suitable value for investigating the agreement with the Korotkoff readings. The results are as follows. The scatter and the line of equality are shown in Figure 16.14.

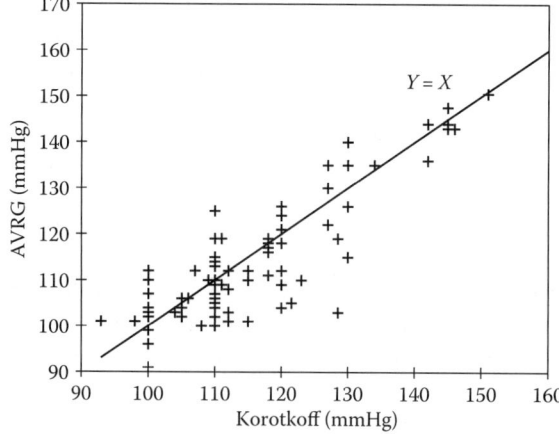

FIGURE 16.14
Scatter of the pulse oximeter and Korotkoff readings (AVRG = average of readings at disappearance and reappearance of waveform).

	AVRG	Korotkoff
Mean systolic BP (mmHg)	115.1	115.5
SD (mmHg)	13.4	13.2
Mean difference (mmHg)	−0.4	
P-value for paired t	>0.50	
Correlation coefficient (r)	0.87	
SD of difference, s_d (mmHg)	6.7	
Limits of disagreement (mmHg)	(−13.5,12.7)	
Intraclass correlation coefficient (r_I) (formula given in next section)	0.87	

Despite the means being nearly equal and r very high, the limits of disagreement show that a difference of nearly 13 mmHg can arise between the two readings on either side (average of pulse oximetry readings can give either less or more than the Korotkoff readings). These limits are further subject to sampling fluctuation, and the actual difference in individual cases can be higher. Now it is for the clinician to decide whether a difference of such magnitude is tolerable. If it is, then the agreement can be considered good and pulse oximetry readings can be used as a substitute for Korotkoff readings, otherwise they should not be. Thus, the final decision is clinical rather than statistical when this procedure is used.

16.5.2.2 Intraclass Correlation as a Measure of Agreement

As mentioned earlier, intraclass correlation is the strength of a linear relationship between subjects belonging to the same class or the same subgroup or the same family. In the agreement setup, the two measurements obtained on the same subject by two observers or two methods is a subgroup. If they agree, the intraclass correlation will be high. This method of assessing an agreement was advocated by Lee et al. [14].

The computation of the intraclass correlation coefficient is slightly different from that of the product–moment correlation coefficient in the formula given in Equation 16.17. No subscript was assigned to x and y in this formula because it was clear in that case that the summation is over the subjects. In the agreement setup, the interest is in the correlation between two measurements obtained on the same subject and the subscripts are needed:

Intraclass correlation coefficient (a pair of readings):

$$r_I = \frac{2\Sigma_i(x_{i1} - \bar{x})(x_{i2} - \bar{x})}{\Sigma_i(x_{i1} - \bar{x})^2 + \Sigma_i(x_{i2} - \bar{x})^2},$$ (16.20)

where
x_{i1} is the measurement on the ith subject ($i = 1, 2, ..., n$) when obtained by the first method or the first observer
x_{i2} is the measurement on the same subject by the second method or the second observer
$\bar{x}$ is the overall mean of all $2n$ observations

Note the difference in the denominator compared with the formula given in Equation 16.17 of product–moment correlation.

This was calculated for the systolic BP data described in Example 16.10 and was found to be $r_I = 0.87$. A correlation more than 0.75 is generally considered enough to conclude good agreement. Thus, in this case, the conclusion on the basis of the intraclass correlation is that the average of readings at disappearance and appearance of the waveform in pulse oximetry agrees fairly well with the Korotkoff readings. This may not look consistent with the limits of disagreement that showed a difference up to 13 mmHg between the two methods. The two approaches of assessing agreement can sometimes lead to different conclusions.

16.5.2.3 Relative Merits of the Two Methods

Indrayan and Chawla [15] studied the merits and demerits of the two approaches in detail. The following are their conclusions on the comparative features of the two methods:

1. The intraclass correlation coefficient does not depend on the subjective assessment of any clinician. Thus, it is better to base the conclusion on this correlation when the clinicians disagree on the tolerable magnitude of differences between two methods (or two observers). And clinicians seldom agree on such issues.

2. The 0.75 threshold to label an intraclass correlation high or low is arbitrary, although generally acceptable. Thus, there is a subjective element in this approach also.

3. Intraclass correlation is unit free, easy to communicate, and interpretable on a scale of zero to one as "no agreement" to "perfect agreement." This facility is not available in the limits of disagreement approach.

4. A distinct advantage of the limits of disagreement approach is its ability to delineate the magnitude of individual differences. It also provides separate estimates of bias ($\bar{d}$) and random error (s_d). This bias measures the constant differences between the two measurements and random error is the variation around this bias. Also, this approach is simple and does not need much calculation.

5. The limits of disagreement can be evaluated only when the comparison is between two measurements. The intraclass correlation, on the other hand, is fairly general and can be used for comparing more than two methods or more than two observers. The formula given in Equation 16.20 then changes and takes the following form.

Intraclass correlation coefficient (several readings):

$$r_I = \frac{\Sigma_i \Sigma_{j \neq k}(x_{ij} - \bar{x})(x_{ik} - \bar{x})}{(M-1)\Sigma_i \Sigma_j(x_{ij} - \bar{x})^2}; \quad i = 1, 2, \ldots, n; \quad j, k = 1, 2, \ldots, M; \quad (16.21)$$

where
n is the number of subjects
M is the number of observers or the number of methods to be compared
mean $\bar{x}$ is calculated on the basis of all Mn observations

A review of the literature suggests that researchers prefer the limits of disagreement approach to the intraclass correlation coefficient approach. A cautious approach is to use both and come to a firm conclusion if both give the same result. If they are in conflict, defer a decision and carry out further studies.

The following comments might help in better appreciation of the procedure to assess quantitative agreement:

1. As mentioned earlier, the limits of disagreement $\bar{d} \pm 1.96s_d$ themselves are subject to sampling fluctuation. A second sample of subjects may give different limits. Methods are available to find an upper bound to these limits. For details, see Bland and Altman [13]. They call them limits of agreement, whereas I prefer to call them limits of disagreement.

2. The intraclass correlation coefficient too is subject to sampling fluctuation. For assessing agreement, the relevant quantity is the lower bound of r_I. This can be obtained by the method described by Indrayan and Chawla [16]. Their method for computing the intraclass correlation coefficient is based on ANOVA, but that gives the same result as obtained by the formula given in Equation 16.19.

3. Though not specifically mentioned, the intraclass correlation approach assumes that the methods or observers under comparison are *randomly* chosen from a population of methods or observers. This is not true when comparing methods because they cannot be considered randomly chosen. Thus, the intraclass correlation approach lacks justification in this case. However, when comparing observers or laboratories, the assumption of a random selection may have some validity. If observers or laboratories agree, a generalized conclusion about consistency or reliability across them can be drawn.

4. Intraclass correlation is also used to measure reliability of a method of measurement as discussed briefly in Chapter 19.

5. Both these approaches are applicable when both the methods could be in error. As mentioned earlier, these methods are not appropriate to compare with a gold which gives a fixed target value for each subject. For agreement with gold, see Lin et al. [16].

16.5.2.4 *Alternative Simple Approach to Agreement Assessment*

The limits of disagreement approach just described is based on average difference and has the limitation applicable to all averages. For example, this approach does not work if the bias or error is proportional. Fasting blood glucose level varies from 60 to 300 mg/dL or more. Five percent of 60 is 3 and of 300 is 15. Limits of disagreement approach considers them different and ignores that both are 5% and proportionately same. Also, if one difference is 10 and the other is 2, not necessarily proportional, limits of disagreement consider only the average. Individual differences tend to be overlooked. A few unusually large differences distort average and are not properly accounted except by disproportional inflation of SD.

To account for small and big individual differences as well as proportional bias, it may be prudent to set up a clinical limit that can be tolerated for individual differences without affecting the management of the condition. Such limits are required anyway for limits of disagreement approach also, albeit for average. These clinical limits of indifference can be absolute or in terms of percentage. If not more than a prespecified, say, 4% of individual differences are beyond these limits in a large sample, you can be safe in assuming adequate agreement. This does not require any calculation of mean and SD. You may like to add a condition such as none of the differences should be more than two times the limit of indifference. Any big difference, howsoever isolated, raises alarm. A plot of y versus x can track that the differences are systematic or random.

Example 16.11: Agreement between Two Methods of Measuring Fasting Blood Glucose Level

Consider the following data on fasting blood sugar level in 10 blood samples:

Method-1 (x)	86	172	75	244	97	218	132	168	118	130
Method-2 (y)	90	180	73	256	97	228	138	172	116	132
$d = x - y$	−4	−8	+2	−12	0	−10	−6	−4	+2	−2
5% of x	4.30	8.60	3.75	12.20	4.85	10.90	6.60	8.40	5.90	6.50

Suppose method-1 is the current standard although this can also be in error. Method-2 is desperately cheap and gives instant results. Suppose also that clinicians are willing to accept 5% error in view of distinct advantages of method-2. Note that this indifference is in percentage and not an absolute value.

None of the differences exceed the clinical limit of indifference in this sample. Thus, method-2 can be considered in agreement with method-1 although a larger sample is required to be confident. However, most differences are negative, indicating that method-2 generally provides lower values. The average difference is 4.2 mg/dL in absolute terms and nearly 3% of y in relative terms. This suggests the correction factor for bias. If you decide to subtract 3% of the level obtained by method-2, you can reach very close to the value obtained by method-1 in most cases. Do this as an exercise and verify yourself.

Now forget about 5% tolerance, and note that some differences are small and some are quite large in Example 16.11. Since $s_d = 2.80$ in this case, the Bland and Altman limits of disagreement are

$$-4.2 \pm 2 \times 2.80 \quad \text{or} \quad -9.8 \text{ to } +1.4.$$

These limits may look too wide and beyond clinical tolerance, particularly on the negative side. These limits do not allow larger error for larger values that proportionate considerations would allow. Also, these are based on average and do not adequately consider individual differences. If one out of 20 values shows a big difference, this can distort the mean and inflate the SD and provide unrealistic limits of disagreement. The alternative approach suggested earlier can be geared to allow not more than 5% individual differences beyond tolerance limit and you can impose an additional condition that none should exceed, say, by 10% of the base value. Since based on individual differences and not average, this alternative approach may be more appealing too.

References

1. Current JD. A linear Equation for estimating the body surface area in infants and children. *Internet J Anesthesiol* 1998; 2(2). http://www.ispub.com/journals/IJA/Vol2N2/bsa.htm. Published April 1, 1998; last updated April 1, 1998.
2. Marquis GS, Habicht J, Lanata CF, Black RE, Rasmussen KM. Association of breastfeeding and stunting in Peruvian toddlers: An example of reverse causality. *Int J Epidemiol* 1997; 26:349–356.
3. Indrayan A, Srivastava RN, Bagchi SC. Age regression of blood pressure in an urban population of age 15–59 years. *Indian J Med Res* 1972; 60:966–972.
4. Draper NR, Smith H. *Applied Regression Analysis*, 3rd edn. New York: John Wiley & Sons, 1998.
5. Schwartz SJ, Haycock GB, Edelmann CM, Spitzer A. A simple estimate of glomerular filtration rate in children derived from body length and plasma creatinine. *Pediatrics* 1976; 58:259–263.
6. Kim K, Timm N. *Univariate and Multivariate Linear Models: Theory and Applications with SAS*, 2nd edn. New York: Chapman & Hall, 2006.
7. Hsieh FY, Bloch DA, Larsen MD. A simple method of sample size calculation for linear and logistic regression. *Stat Med* 1998; 17:1623–1634.
8. Faries D, Leon AC, Haro JM, Obenchain RL. *Analysis of Observational Health Care Data Using SAS*. Cary, NC: SAS Publishing, 2010.

9. Dodge Y, Jureckova J. *Adaptive Regression*. New York: Springer, 2000, pp. 17–19.
10. Vickers AJ. Statistical reanalysis of four recent randomized trials of acupuncture for pain using analysis of covariance. *Clin J Pain* 2004; 20:319–323.
11. Bosch JL, Kranse R, van Mastrigt R, Schroder FH. Reasons for the weak correlation between prostate volume and urethral resistance parameters in patients with prostatism. *J Urol* 1995; 153:689–693.
12. Bland JM, Altman DG. Statistical methods for assessing agreement between two methods of clinical measurement. *Lancet* 1986; i:307–310.
13. Chawla R, Kumarvel V, Girdhar KK, Sethi AK, Indrayan A, Bhattacharya A. Can pulse oximetry be used to measure systolic blood pressure? *Anesth Analg* 1992; 74:196–200.
14. Lee J, Koh D, Ong CN. Statistical evaluation of agreement between two methods for measuring a quantitative variable. *Comput Biol Med* 1989; 19:61–70.
15. Indrayan A, Chawla R. Clinical agreement in quantitative measurements. *Natl Med J India* 1994; 7:229–234.
16. Lin L, Hedayat AS, Sinha B, Yang M. Statistical methods in assessing agreement: Models, issues and tools. *J Am Stat Assoc* 2002; 7:257–270.

17

Relationships: Qualitative Dependent

The regression setup discussed in the preceding chapter requires that the dependent and generally the set of independents also are quantitative. This chapter considers the setup where the dependent is qualitative. The independents can be qualitative or quantitative or a mixture of both. The interest is primarily in the form or nature of relationship and secondarily in the strength of relationship.

As a reminder, qualitative variables are those that are nominal, ordinal, or metric divided into broad categories that can be ordinalized. But the focus of this chapter is on nominal-dependent variables, particularly when these are dichotomous. However, a brief is given for polytomous as well as for ordinal dependents. The dependent of interest is neither the value nor the mean but the probability or the proportion of subjects with a particular characteristic. The probability necessarily lies between 0 and 1, and this restriction disqualifies the usual quantitative regression method of Chapter 16. The statistical method that meets this requirement is the popular logistic regression.

The methods discussed in this chapter are basically distribution free (nonparametric) in the sense that they do not require any specific distribution of the underlying variable values.

This chapter: The basics of logistic regression including its meaning are explained in Section 17.1. This is done for dichotomous-dependent variables. Logistic coefficients, which are similar to the regression coefficients of the preceding chapter, have an extremely useful interpretation as relative risk (RR) or odds ratio (OR). This is described in Section 17.2. This includes the confidence interval (CI) and test of significance for individual coefficients. Issues such as matched data that require conditional logistic, polytomous-dependent, and ordinal categories are briefly addressed in Section 17.3. Section 17.4 describes the Cox hazard regression model and classification and regression trees that have overlap with the quantitative regression method. The measures of strength of relationship among qualitative variables are presented in Section 17.5. This also includes methods of assessing agreement among qualitative data.

17.1 Binary Dependent: Logistic Regression (Large *n*)

Logistic regression is typically suitable in situations where the response or the dependent variable is dichotomously observed, that is, the response is binary. For example, the occurrence of prostate cancer is dichotomous (yes or no) and may be investigated to depend on, say, vasectomy status (yes or no) and dietary habits as the primary variables of interest. Possible confounders such as smoking and age at vasectomy can be other explanatory variables in the model. Hypothyroidism versus euthyroidism can be investigated for dependence upon the presence or absence of signs or symptoms such as lethargy, constipation, cold intolerance, or hoarseness of voice and upon serum levels of thyroid-stimulating hormone (TSH), triiodothyronine (T_3), and thyroxine (T_4). A binary variable is also obtained when a continuous variable is dichotomized, such as diastolic blood pressure (BP) <90 and $\geq$90 mmHg.

In all such dichotomous situations, the dependent variable y is given a value 0 for negative response or 1 for positive response, and nothing in between. Statistically, the dependent is the proportion or probability of subjects providing a positive response. As before, the probability of positive response is $P(y = 1)$ and is denoted by π for population. This is estimated by proportion in the sample, denoted by p.

17.1.1 Meaning of a Logistic Model

The regression discussed in Chapter 16 for quantitative dependent is sometimes referred to in the literature as **ordinary least squares (OLS)**. However, I refer to this as quantitative regression in this text to distinguish it from the logistic regression since logistic is for qualitative dependent.

No restriction is imposed on dependent y in the usual quantitative regression but now for logistic the dependent, since probability, has to be between 0 and 1. Probabilities cannot go in a straight line for ever. If the probability increases with value of a predictor, such as probability of death increasing with increasing severity of disease, the relationship generally takes the form of an S-shaped curve (Figure 17.1). This shape is natural because of restriction of (0, 1) on π.

Note from this figure that very small values of x make little contribution to the probability and so do the very large values. Increase in value of x toward the middle steeply increases the probability in a pretty straight forward fashion.

The transformation,

$$\text{Logit: } \lambda = \ln \frac{\pi}{1 - \pi} \qquad (17.1)$$

removes the (0, 1) restriction since λ can now be between $-\infty$ for $\pi = 0$ to $+\infty$ for $\pi = 1$. It is nice to note that $\lambda = 0$ for $\pi = \frac{1}{2}$. Thus, λ is statistically more suitable for studying the relationship with the predictors. This is called the **logit**

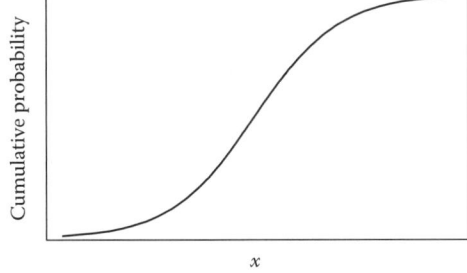

FIGURE 17.1
S-shape of relationship of cumulative probability with the value of the predictor x.

(logistic integral transformation) of π. The functional form of relationship between λ in the formula given in Equation 17.1 and its possible predictors is called a logistic model. This takes the form

$$\hat{\lambda} = b_0 + b_1 x_1 + b_2 x_2 + \cdots + b_K x_K \tag{17.2}$$

for K predictors, where $\hat{\lambda}$ is the estimate of λ. This is also called the logistic regression equation and is similar to Equation 16.2 for quantitative regression. Computer programs easily provide the values of $b_0, b_1, b_2, \ldots, b_K$, which are the estimates of the corresponding parameters $\beta_0, \beta_1, \beta_2, \ldots, \beta_K$, respectively. These are now called **logistic coefficients**. The estimates are obtained by using an iterative reweighted least square method. This is too complex for this book. You may like to leave that to the wisdom of professional statisticians or possibly to a reputed software.

There are several advantages of logistic regression compared to the quantitative regression:

1. There is no need to worry about Gaussian pattern, even for testing of hypothesis. It is a nonparametric procedure.
2. Homoscedasticity is not a requirement.
3. $\pi/(1 - \pi)$ represents the odds for positive response in the subjects, which is a useful statistical quantity. Thus, note that logit in Equation 17.1 is actually log of odds. Also note that earlier use of the term "odd" was for antecedents but now this is for outcome. This is the vagary yet to be properly resolved.
4. Logistic regression does not require *linear* relationship between the dependent π and the predictors. It handles nonlinear effects without explicitly adding nonlinear terms in the model. However, logistic regression does require linear relationship between the predictors and the log of odds of the dependent, as you would realize from the formula given in Equation 17.2. It has been observed that the results lend themselves to easy and useful interpretation when the binary dependent is transformed to logit.

However, independence of various values of the dependent variable is still needed. This is fulfilled when this is assessed for different subjects where response of one does not affect the response of another subject. Nevertheless, it may be violated in some situations as in placental abruption or placenta previa in repeated pregnancies in the same set of women. These conditions are known to recur with greater frequency in subsequent pregnancies. Thus, exercise due caution in such cases.

Although the real objective is to estimate the probability π on the basis of a set of predictors, λ is predicted instead. The logistic relationship recognizes that the probability π is always between 0 and 1. It is easily seen from the formula given in Equation 17.1 that this probability is given by

$$\pi = \frac{1}{1 + e^{-\lambda}}. \tag{17.3}$$

In practice, π is replaced by its estimate p in logit function (Equation 17.1) and correspondingly in the formula given in Equation 17.3 also where λ is replaced by $\hat{\lambda}$.

A useful application of logistic regression is in case–control studies. Such studies have a retrospective design, but for the purpose of the logistic model, case–control studies are considered a binary response of the subjects. The case group may comprise those who have the disease and the control group those who do not have the disease. This is discussed in more detail later in this chapter.

As in the case of the usual quantitative regression, the predictor could be one variable x, providing a simple logistic setup, or a set of many predictors, providing a multiple logistic setup. In the literature, multiple logistic is sometimes described as multivariate logistic but that is not an adequate description. You will note many such similarities between the logistic regression and the usual quantitative regression. For example, interaction can be considered by including the product of the values of the concerned predictors. Multicollinearity affects reliability of the estimates in the case of logistic also. Outliers can throw this out of gear.

Example 17.1: Logistic Regression for Benign Prostate Hyperplasia on Its Risk Factors

Consider a case–control study of adult males to assess the relative importance of various risk factors of benign prostate hyperplasia (BPH). For convenience of illustration, restrict risk factors to only three: x_1 for age, categorized as 50–59, 60–69, or 70+ years; x_2 for self-reported sexual activity, categorized as mild, moderate, or heavy; and x_3 for diet, categorized as vegetarian or nonvegetarian. These are coded as (0, 1, 2), (0, 1, 2), and (0, 1), respectively. This kind of coding for polytomous variables has limitations, described later. The controls in this study are non-BPH males from the same social milieu. Let there be 30 subjects in each group.

The data are entered into a computer and the logistic regression is obtained. Suppose the estimated value of the intercept (sometimes called constant) is −2.65 and the regression coefficients are +0.50, +0.78, and +0.22 for x_1, x_2, and x_3, respectively. Thus, the following logistic regression is obtained:

$$\hat{\lambda} = -2.65 + 0.50x_1 + 0.78x_2 + 0.22x_3.$$

For a person with highest risk factors, $x_1 = 2$ (age 70+ years), $x_2 = 2$ (heavy sexual activity), and $x_3 = 1$ (nonvegetarian diet). Thus, for this person

$$\hat{\lambda} = -2.65 + (0.50 \times 2) + (0.78 \times 2) + (0.22 \times 1) = 0.13.$$

Since $\hat{\lambda} = \ln[p/(1 − p)]$, the estimate of the odds for this person is $p/(1 − p) = e^{0.13} = 1.1388$. For a person with lowest risk, $x_1 = 0$, $x_2 = 0$, and $x_3 = 0$, λ is −2.65. This gives odds $e^{-2.65} = 0.0707$. Therefore, the OR is $1.1388/0.0707 = 16$. If prevalence of BPH is low, the risk of BPH for the first person with all the three risk factors is nearly 16 compared with a person with least risk (of age 50–59 years, with mild sexual activity, and on vegetarian diet).

SIDE NOTE: Also, the estimated probability of BPH for the first person, using the formula given in Equation 17.3, is $p = 1/(1 + e^{-0.13}) = 0.53$ and for the second person is $p = 1/(1 + e^{+2.65}) = 0.066$. The ratio of these probabilities is $0.53/0.066 = 8$. This is not the same as the OR. It should be clear that OR is not the same as probability ratio.

17.1.2 Assessing Overall Adequacy of a Logistic Regression

Before using any model for inferential purpose, you should convince yourself about the adequacy of the model. Biological plausibility is always a prime consideration and should be assessed separately. Statistical adequacy means whether the model is able to explain the data adequately. Among several methods available to assess the statistical adequacy of a logistic model, the following are widely used.

17.1.2.1 Log Likelihood

Although several criteria such as pseudo R^2, generalized R^2, likelihood ratio, and Wald statistic are available to check statistical adequacy of a logistic model, log likelihood seems to have better appeal and wider applicability. This is explained in the following. For other criteria, see Hosmer and Lemeshow [1].

The likelihood, L, is the probability of obtaining the values observed in the sample when the model is correct. A probability is necessarily a small number, that is, less than 1 and your high school math tells you that the logarithm of a number less than 1 is negative. Thus $-\ln L$ is positive. It is helpful to use $-2\ln L$ instead of $-\ln L$ because the distribution of $-2\ln L$ for large n has been found to follow chi-square under H_0 in fairly general conditions. This is called the **deviance**. The H_0 in the case of logistic is that there is no relationship between the dependent and the explanatory variables. If this H_0 is true, then all coefficients are zero and

$$\hat{\lambda} = b_0'. \tag{17.4}$$

Compare this with the fitted model $\hat{\lambda} = b_0 + b_1 x_1 + b_2 x_2 + \cdots + b_K x_K$.

Denote likelihood for the reduced model (Equation 17.4) by L_0 and for the fitted model (Equation 17.2) by L_1. The value of the deviance $-2\ln L_0$ for the reduced model would invariably be more than $-2\ln L_1$ for the fitted model. The difference between these two also follows a chi-square distribution with K degrees of freedom. This is called the model chi-square. Most standard statistical software packages give the value of $-2\ln L$ for the model under consideration. In the BPH example, if $-2\ln L_0 = 83.18$ for the model represented in Equation 17.4 and $-2\ln L_1 = 61.01$ for the model in Equation 17.2 (with $K = 3$ predictors), then model chi-square = $83.18 - 61.01 = 22.17$. At $K = 3\,\mathrm{df}$, this is highly significant ($P < 0.001$). This shows that x_1, x_2, and x_3 (together) in this example are useful in predicting BPH positivity.

The following comments explain some of the implications:

1. If the model is a perfect fit, then the likelihood is 1 and $-2\ln L = 0$. The higher the deviance, the less adequate is the model. Note that since probability $L < 1$, $\ln L$ is always negative and $-2\ln L$ is always positive.

2. One measure of the adequacy of a logistic model (Equation 17.2) is the extent of decrease in the value of $-2\ln L$ relative to the value for the reduced model (Equation 17.4). This can be calculated as follows:

$$\text{Contribution of the model: } C = \frac{(-2\ln L_0) - (-2\ln L_1)}{-2\ln L_0} \tag{17.5}$$

where
  L_0 corresponds to the reduced model (Equation 17.4)
  L_1 corresponds to the fitted model (Equation 17.2) as before

The role of $-2\ln L$ is similar to that of R^2 but has a negative meaning. In the case of R^2, larger value is better, but in the case of $-2\ln L$, smaller value is better. Also, $-2\ln L$ does not fall between 0 and 1 as R^2 does. For the purpose of C in the formula given in Equation 17.5, $\ln L$ can be used if your software provides log likelihood instead of $-2\ln L$.

A similar measure is Nagelkerke pseudo R^2. This also is based on ratio of likelihoods, and ranges from 0 to 1 just as is R^2 for quantitative regression.

3. All models can be improved by adding more explanatory variables. This will invariably decrease the value of $-2\ln L$. But this decrease may or may not be statistically significant. To test this, the decrease in $-2\ln L$ is again referred to chi-square to obtain a P-value. A nonsignificant

decrease indicates that adding those predictors is not helpful. Similarly, one or more predictors can be dropped in search of a more parsimonious model. A nonsignificant increase in deviance justifies dropping because the model still works just as well without those predictors.

Also, as always in a regression setup, increase in number of parameters decreases $-2\ln L$ and improves fit in absolute terms, but sometimes the dfs increase relatively faster and statistical significance declines.

4. Beware of outliers. These can vitiate any model and logistic is no exception. Parameter estimates generally incorporate all values but are affected disproportionately by such extreme values. In the process, the whole model can shift its position. As a result, the likelihood L may decrease and the deviance may increase. At the time of scrutiny of data, try to locate outliers and take a conscientious decision to include or exclude some or all of them from the analysis. This should not be done after you have seen the results.

5. All these considerations may fail if the sample size is not adequate. The general advice is to have at least 10 subjects for each predictor variable. If there are six predictors in the model, data on a minimum of 60 subjects must be available. The method of logistic regression is indeed greedy of sample size. To be safe, it is sometimes argued that the frequency of the rarest outcome (0 or 1) should be at least 10 times the number of parameters. If total n is 1000 but only 36 are 1 s (e.g., deaths) and 964 are 0 s, you should probably not have more than three parameters in your model. Of course, larger n is better, but my experience suggests this rule of 10 for the outcomes is too conservative. I will not mind more parameters in this case so long as the number of subjects in any subcategory does not fall below 5.

6. The logistic method for small n in case of categorical dependent is too complex for inclusion in this book. Interested readers may see Mehta and Patel [2] for an exact logistic regression valid for any n, including small n.

When the number of possible explanatory variables is large, the computer program can be asked to identify and include only the significant variables in the logistic model. This can be done by one of the following three algorithms: (1) forward selection, (2) backward elimination, and (3) stepwise algorithms. These are the same as listed for quantitative regression in Chapter 16. The only difference is that instead of R^2, now $-2\ln L$ is used as the criterion. However, these algorithms are not as well defined for logistic regression as they are for quantitative regression. None of the algorithms may yield the best model. The three may lead to three different models. It is a good idea to examine several possible models and choose on the basis of interpretability, parsimony, and convenience in obtaining the data.

17.1.2.2 Classification Accuracy

Another way to assess the adequacy of a logistic model is by finding what percentage of subjects is correctly classified. For this, the model is used on the same set of data and an estimated probability p is obtained for each subject. A subject is predicted to belong to the control if $p < 0.5$ and to the case if $p > 0.5$. Any other cutoff point can be used if that is more justified, otherwise 0.5 seems like a reasonable cutoff. The terms case and control used here are in a very general sense for the dichotomous group—a case could be either a subject with disease or one with exposure. The probability p is calculated to several decimal places and $p = 0.5$ is unlikely to occur. If it does, the subject is not classified into either group. The actual status of all the subjects is already known. The predicted grouping is compared with the observed grouping and a table similar to Table 17.1 is obtained.

Consider Example 17.1 with 30 non-BPH and 30 BPH subjects as shown in Table 17.1. When the probability (Equation 17.3) is estimated for these subjects, let us suppose 23 out of 30 controls have $p < 0.5$ and 22 out of 30 cases have $p > 0.5$. This information can be used to calculate both positive and negative predictivity type of measures. Mostly the two are considered together. In Table 17.1, a total of 45 out of 60 subjects (75%) were correctly classified by the model. This is not high and there is a room for improvement. This can be done either by including more regressors in the model or by considering an alternative set of variables as regressors that can really predict BPH status. The higher the correct classification, the better is the model. Generally, a correct classification between 80% and 89% is considered good and 90% or more is considered excellent.

One difficulty with this procedure is that the middling probability 0.51 is treated the same way as the high probability 0.97. Also, a subject with probability 0.49 goes to another group than a subject with probability 0.51 despite the minimal difference. The classification is strictly with respect to cutoff point 0.5 and ignores the magnitude of the difference from 0.5.

17.1.2.3 Hosmer–Lemeshow Test

Some statistical softwares provide an option to use the Hosmer–Lemeshow test as a further evidence of the adequacy of model (or the lack of it). This is

TABLE 17.1

Classification Table for the Subjects

| Observed Group | Predicted by Logistic Model | | | Percent Correct |
	Control	Case	Total	
Control	23	7	30	23/30 = 76.7
Case	8	22	30	22/30 = 73.3
Overall				45/60 = 75.0

obtained by dividing the subjects into categories and using the usual chi-square goodness-of-fit test. For this, divide the predictor set of x values into some rationally looking arbitrary number of groups and find the observed frequency of cases and controls in these groups. Corresponding to these groups of xs, find the expected frequencies based on the fitted logistic model. Compare them by the usual chi-square. This chi-square will have 2 df less than the number of groups you formed. Alternatively, divide predicted probabilities into, say, 10 decile groups and compare the observed number of responses in these groups with the predicted number of responses using cells defined by the groups. The predicted are those obtained by the fitted logistic model. For 10 groups, this chi-square will have 8 df. A software will do all this for you. The validity condition of chi-square applies, namely, that not many expected cell frequencies should be less than 5 and none should be close to 0. This implies that n should be really large. This test has been lately criticized for arbitraries that creeps while defining the groups and for being insensitive to deviations in individual cases.

In Hosmer–Lemeshow test, the null hypothesis is that the model is an adequate fit and the alternative is that it is not. When $P < 0.05$, the null is conceded but that does not necessarily mean that the model is good—the only implication is that the evidence against it is not sufficient.

17.2 Inference from Logistic Coefficients

Logistic coefficients have proved to be useful tools in drawing inference regarding the role of various factors in predicting the probability of occurrence of an event. This depends on careful interpretation of logistic coefficients and on the CI and test of hypothesis on these coefficients.

17.2.1 Interpretation of the Logistic Coefficients

The coefficients $b_1, b_2, \ldots, b_K$ in the logistic regression formula given in Equation 17.2 have an extremely useful interpretation in terms of ORs. Their meaning in case of a dichotomous predictor is slightly different from the meaning in case of a polytomous and continuous predictor.

17.2.1.1 Dichotomous Predictors

Consider a simple logistic equation with only one predictor, x_1. Then, $\hat{\lambda} = b_0 + b_1 x_1$. Since $p/(1 - p) = e^{\hat{\lambda}}$

$$\frac{p}{1-p} = e^{b_0 + b_1 x_1}.$$

Say, x_1 is a *binary* predictor with value 1 when present and value 0 when absent. Refer to these as exposure present and exposure absent. Since $p/(1-p)$ is odds for positive response,

$$\text{Odds for exposed subjects } (x_1 = 1) = e^{b_0+b_1} \quad \text{and}$$
$$\text{odds for unexposed subjects } (x_1 = 0) = e^{b_0}.$$

Thus,

$$\text{OR} = \frac{e^{b_0+b_1}}{e^{b_0}} = e^{b_1}.$$

It is now clear that b_1 is the log of the OR corresponding to the variable x_1 when x_1 is binary. If x_1 is *continuous*,

$$\text{odds}(x_1 = a+1) = e^{b_0+b_1a+b_1} \quad \text{and} \quad \text{odds}(x_1 = a) = e^{b_0+b_1a}$$

Thus,

$$\text{OR} = \frac{e^{b_0+b_1a+b_1}}{e^{b_0+b_1a}} = e^{b_1}.$$

Again b_1 is the log of the OR. Whether x_1 is binary or continuous, b_1 is the log of OR when x_1 is increased by 1. In general, when several predictors are present in the logistic model, e^{b_1} is an *independent* contribution of one unit of x_1 to the OR when other xs remain the same. Then, this is also called **adjusted odds ratio**.

In Example 17.1, for a person with $x_1 = 0$, $x_2 = 0$, and $x_3 = 0$, odds = $e^{-2.65}$ = 0.0707. If the person is nonvegetarian ($x_3 = 1$) with other xs same, for $x_1 = 0$, $x_2 = 0$, and $x_3 = 1$, odds = $e^{-2.65+0.22}$ = 0.0880. The OR for BPH for a nonvegetarian against a vegetarian is 0.0880/0.0707 = 1.25. This means that a nonvegetarian diet increases odds by 25%. This is the adjusted OR and measures independent contribution of x_3 when x_1 and x_2 are in the model. Again, since $\ln(1.25) = 0.22$, the logistic coefficient ($b_3 = 0.22$ in this case) is the logarithm of the OR for x_3. When logistic coefficients are available, the corresponding OR relative to a reference can be immediately computed. It is this property of the logistic regression that has made this method so popular. The category coded as $x = 0$ serves as reference and e^b is the ratio of odds for category $x = 1$ relative to the reference category. For example, if $b = 1.3$ for a positive family history in case of breast cancer, OR = $e^{1.3}$ = 3.67. This means that odds of breast cancer in those with a positive family history are 3.67 times the odds in those without a positive family history. Thus, e^b is the predicted change in odds of the dependent for a unit increase in the corresponding predictor. OR less than 1 for any x corresponds to a decrease in odds and

OR more than 1 corresponds to increase. An OR close to 1 indicates that the change in the predictor does not affect the response, and thus the predictor has no predictive utility.

Logistic regression fails if any predictor perfectly predicts the response. Theoretically, this situation is unconceivable in medicine, since medicine is so much riddled with uncertainties. But practically this can happen in a sample, particularly if it is small. In the aforementioned example, it is possible that nearly all persons *in the sample* with BPH happen to be nonvegetarians and nearly all persons without BPH are vegetarians. There may be perfect or near-perfect correspondence between the predictor values and the response values. You may get extremely high OR—running into thousands or even higher—for this kind of predictor. This can also happen when the number of one kind of responses (either positive or negative) is extremely small. The only insulation against this aberration is using a large sample so that a cross section of subjects are represented in the sample and such perfect correspondence does not occur.

In the case of the usual quantitative regression, comparison across predictors regarding their effect is valid only after the regressors are standardized to have mean 0 and variance 1. In the case of logistic, the relative importance of various predictors can be assessed by using OR as the criterion. OR is scale invariant and there is no need to use standardized predictors for such comparison. Remember, however, that absolute difference between OR = 0.1 and OR = 0.2 looks minor but the second is double of the first.

The preceding explanation is in terms of OR because this is how a logistic regression is widely understood. However, in case of prospective studies, the interpretation is in terms of RR and not OR. Also, realize that π ranges between (0, 1), OR between (0, ∞), and logit between ($-\infty$, $+\infty$).

17.2.1.2 Polytomous and Continuous Predictors

There is no restriction in logistic regression for any predictor to be binary. It can also be polytomous or continuous. Interpretation of logistic coefficients for predictors with three or more categories depends on the coding system adopted. If a particular predictor is polytomous ordinal such as disease severity categorized as none, mild, moderate, serious, or critical, many types of coding can be done. The simplest can be 0, 1, 2, 3, and 4 when the number of categories is five. This kind of coding quantifies predictor values and works like a score. In this case, the corresponding e^b is the factor by which OR multiplies when the category moves one up in terms of this score. If $e^b = 1.15$, this means that the odds for category $x = 1$ (mild disease) are 1.15 times the odds for category with $x = 0$ (no disease), and the odds for category $x = 2$ (moderate disease) are 1.15 times the odds for category $x = 1$ and 1.15 $\times$ 1.15 = 1.32 times the odds for $x = 0$ (no disease).

This interpretation of the logistic coefficient is valid for ordinal x. For real nominal categories, suitable contrasts of interest should be defined. They are generally defined in terms of the difference of one group from one or more of the others. If the objective is to examine 5-year survival with site of malignancy, you can have one particular site, say, lung, as the reference category and compare this with oral, prostate, esophagus, etc., by forming contrasts such as (lung–oral), (lung–prostate), etc. Each of these contrasts would appear in the predictor set of the logistic model. If you consider it more appropriate, you can have "prostate cancer" as the reference category. In this case, all other sites will be compared with this cancer. This requires coding accordingly. The advantage with these kinds of contrasts is that each contrast will have its own logistic coefficient, and the coefficient for one contrast can differ from that of another contrast. Thus, differential effects of the categories on the response, if present, will emerge. It is possible to find out if one particular category, say, category-2, is significantly higher contributor to the response than category-3 or category-1. The serial coding 0, 1, 2, 3, and 4 mentioned in the preceding paragraph does not have this feature.

For interpretability, it is necessary that each category of each predictor is explicitly defined. For example, the "others" category that includes leftovers as in a kitchen sink is not admissible. In cases where reference category is used for all comparisons, conclusions are more reliable when the number of subjects in the reference category is reasonably large.

If a predictor is continuous such as age in years, a logistic coefficient of 0.15 would imply OR $= e^{0.15} = 1.16$, indicating that each year of increase in age increases OR by a factor of 1.16. A 10-year increase in age would increase OR by a factor of $(1.16)^{10} = 4.41$. You can say that a 10-year increase in age increases OR by 341%. In this case, the OR may look only slightly more than 1.0, but it can translate into an enormous effect.

Logistic regression is linear and assumes that the effect of increase from, say, 5 to 10 in value of x is the same on logit values as of increase from 70 to 75. This may not be the case with many clinical variables. For example, BP rise from 120 to 130 mmHg may not have same effect on outcome as increase from 170 to 180 mmHg. Linearity of logit values should be checked for any continuous x. However you can have another x, say $x_2 = x^2$.

17.2.2 Confidence Interval and Test of Hypothesis on Logistic Coefficients

When all of the specified explanatory variables are forced into the model and not selected by any of the algorithms stated earlier, there is a need to test the statistical significance of each logistic coefficient. For this, one method is to refer the difference between $-2\ln L$ without and with the variable in question to χ^2 with 1 df. The other method is to refer $Z = b/SE(b)$ to the Gaussian distribution. The SE(b) has a complex expression but its value is routinely provided by statistical software. Both chi-square and Gaussian Z require large n. Most statistical software perform the Gaussian test. The square of this is called the Wald statistic, which is referenced to chi-square with 1 df for testing significance. Consider an elaborate example that illustrates some of these facets of logistic regression.

**Example 17.2: Multiple Logistic for Weaning
a Baby from Breast-Feeding**

Let us pursue Example 16.2 further. Among a group of Peruvian children studied by Marquis et al. [3], 127 were still breast-feeding at the
end of their 13th month of age. These children were thus at *risk* of being
weaned. The objective was to find support for the contention that poor
health of the child determined maternal breast-feeding practice and
maternal breast-feeding is not a precursor to poor health. The following
variables were investigated for their role in determining the weaning:
(1) Z-score of 12-month weight-for-age (WA), (2) complementary food
intake (FOOD) between 9 and 12 months of age in terms of number of
foods (out of 27 groups), and (3) change (CH) in diarrheal incidence from
9–12 to 12–15 months of age. As these are to be considered precursors for
weaning during the 14th month, consider CH as 9–11 and 11–13 months
for our example. The response variable is the probability of weaning
during the 14th month of age. All variables were centered on the mean.
The results obtained by logistic regression are summarized in Table 17.2.
The table contains the estimated logistic coefficients, their SE values, and
the two-tail P-value for assessing their statistical significance. All these
easily come from a statistical software package.

The aim of a logistic model is to predict $\lambda = \ln[\pi/(1 - \pi)]$, and thereby π,
by a linear combination of the predictor variables. In this example, λ is
the logarithm of odds of weaning. The logistic equation obtained is

$$\ln\frac{p}{1-p} = -3.303 - 0.024(WA) + 0.093(FOOD) + 0.542(CH)$$

$$+ 0.046(WA \times FOOD) + 0.336(WA \times CH) - 0.143(CH \times FOOD)$$

$$- 0.202(WA \times CH \times FOOD).$$

A coefficient b close to 0 implies that the corresponding explanatory
variable is not a good predictor. Whether it is significantly different

TABLE 17.2

Results of Logistic Regression of Weaning on Health
of the Children

Variable	b	SE(b)	P-Value
Constant	−3.303	0.644	0.000
12 month Z-score—WA	−0.024	0.446	0.958
9–12 months—FOOD	0.093	0.267	0.733
Diarrheal change—CH	0.542	0.322	0.095
Interaction 1—WA × FOOD	0.046	0.211	0.833
Interaction 2—WA × CH	0.336	0.179	0.064
Interaction 3—FOOD × CH	−0.143	0.144	0.323
Three-factor interaction— WA × FOOD × CH	−0.202	0.096	0.038

from 0 can be tested by referring [b/SE(b)] to the standard Gaussian or its square to the Wald statistic. The P-values thus obtained are given in the last column of the table. Diarrheal change (CH) is statistically significant at $\alpha = 0.10$ but not at the conventional $\alpha = 0.05$. The only explanatory variable that shows significance ($P < 0.05$) in this case is in the last row. This is the interaction between WA, FOOD, and CH. The coefficient b is negative for this interaction, which indicates that logarithm of odds of weaning *declines* as WA, FOOD, and CH change *together* such that WA × FOOD × CH increases. When log-odds decreases, the corresponding probability also decreases. The authors of this investigation explained that probability of weaning decreases especially when low FOOD and low WA are accompanied by increase in diarrheal incidence. These three together indicate poor health of the child. Thus, the inference is that a child who is breast-fed until 13 months of age is less likely to be weaned during the 14th month if its health condition is poor. Hence the conclusion that health condition determines maternal breast-feeding practice.

Note the following points for Example 17.2: (1) The dependent variable is the probability of weaning during the 14th month of age. The observation is in terms of either weaned or not weaned. This is binary as generally required for the application of the logistic model. (2) All the explanatory variables in this case are quantitative and none is categorical. In general, some or all of them can be categorical in a logistic regression. (3) The interest was not just in main effects of the variables but also in their **interactions**. These are studied by including simple multiplication of the levels of the predictor into the model as shown in the last four rows of Table 17.2. Interaction in logistic is departure from multiplicativity, while in quantitative regression, it is departure from additivity.

As mentioned earlier, the logistic coefficient b_k in the model specified in Equation 17.2 quantifies the contribution of the explanatory variable x_k in determining the response. If x_k changes from a to ($a + 1$), the estimate of $\hat{\lambda}$ changes by b_k. In Example 17.2, a unit increase in diarrheal incidence increases the log-odds of weaning by an average of 0.542 and the odds by a factor of $e^{0.542} = 1.72$. This measures the independent contribution of the variable and is already adjusted for the other variables in the model.

Example 17.2 also illustrates the method to test significance of individual coefficients. If there are many predictors as in this example, an alpha level of 0.05 for each individual predictor may be too high because all tests are based on the same data. Exercise caution and see if you would like to use a Bonferroni type of procedure to adjust the alpha level instead of the one used in Table 17.2.

Ninety-five percent CI for each logistic coefficient is obtained as $b \pm 1.96$SE(b) using Gaussian approximation. This is converted to OR by computing its exponents. This procedure is valid only for large n. When such a CI is obtained, there is no need to perform the test of hypothesis $H_0: \beta = 0$ on individual coefficients since CI for any predictor containing zero indicates that the predictor is not significantly affecting the odds of the dependent variable when other predictors are fixed. Thus, that factor is not a useful predictor. When there are a large number of predictors, CIs on individual coefficients may not be very useful for any *joint* conclusion because joint probability may be very different.

Software illustration: Appendix C contains an example of use of SPSS software for running a logistic regression and explains the meaning of various outputs.

17.3 Issues in Logistic Regression

Even though quantitative regression is extensively used, the application of logistic regression is even more widespread in health and medicine. This has applications ranging from clinical trials when the interest is in the probability of outcome (relief or no relief) as the dependent variable to epidemiological prospective studies that may follow up subjects for any outcome of interest and retrospective studies that investigate the presence or absence of risk factors for known outcomes. Many of these studies use one-to-one matched controls. Matching imposes constraints that need to be tackled differently. Also, polytomous nominal and ordinal categories of dependent variables introduce intricacies. The details are beyond the scope of this book. The following is a brief description that should be adequate to alert you about the intricacies involved.

17.3.1 Conditional Logistic for Matched Data

Case–control studies are commonly done on subjects that are one-to-one matched for background characteristics such as age-group, sex, and BMI. Such matching produces a high degree of correlation. For proportions, McNemar test was described earlier for matched data but that cannot be used here because here there is a set of predictors whose relationship with the outcome is under study. One-to-one matching gives rise to n strata corresponding to n matched pairs, each comprising only two individuals. This requires generation of $(n - 1)$ dummy variables. Each of these will have a corresponding logistic coefficient. The number of parameters for estimation becomes too many for n pairs of subjects. Increasing the sample size does not help in this case because the number of dummy variables also correspondingly increases. Each stratum will have a parameter dedicated to the stratum that statisticians call **nuisance parameter** because it does not help but hurts the interpretation.

A conditional logistic method eliminates this nuisance parameter, which is the intercept b_0 in this case, and thus simplifies the estimation method and protects against potential bias that arises in the regular unconditional logistic model when used for matched data. Only the parameters corresponding to the predictors are estimated. You must specify "no intercept" for a conditional model so that a logistic model is fitted with no b_0. In conditional logistic, matching variables are not included in the model, and the entire pair is discarded when information on one part of the pair is missing.

Statistical software packages generally allow one or more than one matched control per case. You must ensure that a suitable package and the correct option in the package is used for analyzing matched data.

17.3.2 Polytomous Dependent

In many practical situations, the dependent has more than two categories. Examples are diagnosis of liver disease such as hepatitis, cirrhosis, and malignancy, which can be investigated for dependence on signs or symptoms and enzyme levels; and smoking status (never smoker, former smoker, current smoker) for its dependence on family history, stress level, and peer pressure. If there are four categories of the dependent variable, then $\pi_1 + \pi_2 + \pi_3 + \pi_4 = 1$. The odds otherwise are expressed as $\pi/(1 - \pi)$, so the option in the polytomous logistic is to use either $\pi_1/(1 - \pi_1)$, which in this case is the same as $\pi_1/(\pi_2 + \pi_3 + \pi_4)$, or to use π_1/π_2, π_1/π_3, etc. This illustration is for π_1 but instead any other π can be used in the numerator. The method and the interpretation would depend on whether the dependent is nominal or ordinal.

17.3.2.1 Nominal Categories: Multinomial Logistic

You now know that nominal categories cannot be ordered and each category must be studied as stand alone. Logistic regression is straightforward if a category is to be compared with the rest. The probability π is for the category of interest and $(1 - \pi)$ is for the rest. In the case of liver diseases, compare hepatitis with (cirrhosis + malignancy), cirrhosis with (hepatitis + malignancy), and malignancy with (hepatitis + cirrhosis). A joint model can be made for studying all three comparisons together that will reduce the number of parameters, but the interpretation of the logistic coefficients becomes complex. Calculations too become more complex, although an appropriate software package can help.

Methods are also available to compare one category with the other without the restriction of two probabilities adding to one. Thus, category-1 can be compared with category-3, and category-2 with category-3. In this case, it is helpful to consider one particular category as reference just as is done for predictors. For the comparisons just cited, since both comparisons are with category-3, this is serving as reference. In the example on liver disease, any one can be considered the reference. If cirrhosis were the reference, hepatitis and malignancy would be compared with cirrhosis. There will be two logistics, one for hepatitis and one for malignancy.

Suppose gender is a predictor coded as 0 for males and 1 for females. Put the value of gender as 0 in both logistics mentioned in the preceding paragraph and let other predictors remain the same. For gender = 0 if logistic coefficient for hepatitis is 0.6 and for malignancy 0.3, the ratio is 0.6/0.3 = 2.0. You can say for males the odds of hepatitis are twice as much as for malignancy. As cautioned earlier, this, however, does not imply that the probability is also two times.

Linear discriminant functions can be used in the case of polytomous-dependent variables to classify, with minimum error, a subject into one of the several possible categories. Each discriminant function, expressed as a combination of many independent variables, defines a line (or a plane) that divides the total number of subjects into groups where the subjects are most likely to belong. This also estimates the probability that a subject with given xs belongs to one of the given categories. However, discriminant function generally requires multivariate Gaussian distribution for these categories together, which may be difficult to ensure. Also this is mostly used to find linear boundaries. For details, see Chapter 19.

17.3.2.2 Ordinal Categories

Several possibilities exist in case of ordinal dependent categories: (1) Compare each category with each of the others. If the dependent is severity of illness categorized as none, mild, moderate, and serious, the possible comparisons are mild, moderate, and serious each with none; moderate and serious each with mild; and serious with moderate. This would require six different logistic runs. (2) Compare each category with the preceding or succeeding category—mild with none, moderate with mild, and serious with moderate. (3) Compare each with the combination of the others. This means comparing none with (mild + moderate + serious), mild with (none + moderate + serious), etc. This is the easiest but may not be sensible in some cases. (4) Compare each category with the combination of the preceding or succeeding categories. This means comparing mild with none, moderate with (none + mild), and serious with (none + mild + moderate). Similarly, comparison with the succeeding categories can also be done if it is considered more meaningful. (5) Compare each with a reference. If none is considered the reference, compare mild with none, moderate with none, and serious with none category.

Perhaps, some other comparisons can also be thought of. In all these cases, it is possible again to prepare a joint model, though this may make the model too complex for interpretation. Such possibilities should be considered only for a small number of categories of the dependent variable. If the number of ordinal categories is large, say, more than four, consider running the usual quantitative regression instead of logistic regression.

**Example 17.3: Ordinal Logistic Regression
for Insurance Coverage of Children**

A large number of children in the United States are not covered by health insurance and many children are not continuously covered. Such unstable coverage can lead to inadequate health-care utilization and poor child health outcomes. The insurance coverage has three ordinal categories: uninterrupted coverage through the reference year, gaps in the coverage in the year, and no coverage in the year. Satchell and Pati [4] studied the dependence of such coverage on several predictors as given in Table 17.3. A separate logistic was run for uninsured versus continuous coverage subjects and subjects with gaps in coverage versus continuous coverage. Thus, continuous coverage was the reference category in this analysis.

Most results in the two ordinal analyses are similar. They indicate that (1) children with chronic conditions were as likely as other children to have gaps in coverage or be uninsured, (2) Hispanic children were most likely to have insurance gaps or be uninsured, and (3) children from poor and near-poor families were 4–5 times (OR between 4 and 5: not stated in Table 17.3) likely to have lapsed coverage than children of high-income families.

TABLE 17.3

Results of Ordinal Logistic Regression of Insurance Gaps in Children

Predictors	Uninsured vs. Continuous Coverage	Gaps vs. Continuous Coverage
Age	Significant	Significant
Gender (male)	Significant	Significant
Race/ethnicity		
Non-Hispanic White	Reference	Reference
Black	Not significant	Not significant
Hispanic	Significant	Significant
Other	Not significant	Not significant
Poverty		
Poor	Significant	Significant
Near poor	Significant	Significant
Middle income	Significant	Significant
High income	Reference	Reference
Receives TANF	Significant	Not significant
Maternal education		
Less than high school	Significant	Significant
High school	Not significant	Significant
More than high school	Reference	Reference

TANF: Temporary Assistance for Needy Families.

The only real difference in the two analyses is with regard to contribution of receiving Temporary Assistance for Needy Families (TANF) and maternal education. TANF is a significant contributor to the odds of no insurance versus continuous coverage but not to the odds of gaps in coverage. Maternal education, when high school, had reverse effect.

17.4 Some Models for Qualitative Data and Generalizations

Qualitative data are more common in health and medicine than quantitative data. Perhaps, they are more meaningful in medical practice. But statisticians have a fancy for quantities. Thus, statistical methods are more developed for quantitative data than for qualitative data. Nevertheless, some models are available for qualitative data as well. These models have overlap with quantitative methods. Two such methods are presented here.

17.4.1 Cox Regression for Hazards

In a disaster such as an earthquake, thousands of people may die in a few hours. The intensity of death or force of mortality in persons exposed to such disasters is extremely high. Three days are less than 1% of a year, and if 800 people die in 3 days, the rate is $800/(3/365) = 97,333$ persons/year! As explained earlier, such a force of mortality at a particular *instant* is called hazard. Note that it is very different from the probability of death. Probability cannot exceed 1, but hazard can. Probability of death competes among different causes (because the sum total of these probabilities from different causes has to be 1) but hazards do not. They are absolute. The concept of hazard can be applied to all health conditions that change from one state to another in the course of time.

While the hazard evidently depends on time, it may also depend on several other factors. In a clinical setup, hazard of serious side effect may depend on the characteristics of the person such as age, gender, and nutritional status, as well as on the type of regimen, type of domiciliary care, alertness, competence of the attending physician, etc. When two or more groups are available, such as an experimental and a control group, these groups can be compared with respect to the hazard rate. For such a comparison, since many factors are involved, one of which is time, it is sometimes helpful to obtain the hazard as a function of various variables. One such model is

$$\text{Cox regression: } \hat{\lambda}(t) = \hat{\lambda}_0(t)e^{b_1 x_1 + b_2 x_2 + \cdots + b_K x_K}, \tag{17.6}$$

where
$\hat{\lambda}(t)$ is the estimated hazard at time t for given values of the covariates $x_1, x_2, \ldots, x_K$
$b_1, b_2, \ldots, b_K$ are the estimates of the corresponding regression coefficients

These coefficients measure OR or RR on a logarithmic scale as in a logistic model and are treated nearly the same way. Positive value of any b_k indicates that higher values of x_k increase the hazard or indicate worse prognosis. A negative value indicates that the corresponding covariate reduces hazard. The first component on the right-hand side of Equation 17.6, $\hat{\lambda}_0(t)$, is the estimate of the time-dependent baseline hazard that is present in any case even when all xs are zero. This is considered same for all subjects. In some cases, this can be understood as the hazard of death at time t for a healthy subject (no risk factor) because of the time factor that in any case works on all life forms. The hazard ratio $\hat{\lambda}/\hat{\lambda}_0$ can be interpreted as relative death rate when hazard of death is under study. The covariates $x_1, x_2, ..., x_K$ could be continuous such as age, polytomous such as type of treatment, or dichotomous such as gender and, of course, would vary from person to person. The difference $[\ln \lambda(t) - \ln \lambda_0(t)]$ is the effect of the covariates.

Because of the presence of time-dependent $\lambda_0(t)$ in the Cox regression, the usual method of maximum likelihood cannot be used to compute bs. Special methods are needed. For details, see Kleinbaum and Klein [5]. The following comments contain some useful information about Cox regression:

1. The term hazard is generic and not restricted to death. It can be used for any other event of interest such as appearance or reappearance of symptoms or even for a favorable event such as discharge from the hospital, cessation of smoking, or resumption of daily activities.

2. In clinical studies, the variables $x_1, x_2, ..., x_K$ may contain not only the personal characteristics of the patient such as age, gender, and nutritional status but also the treatment indicators such as dosage of drug, type of treatment, and kind of care provided. These may change over time. For example, drug dosage may be heavy in the beginning and moderate later on. Such flexibility is not available in the usual quantitative regression. However, when any x is time dependent, the estimation of the bs becomes complex.

3. Consider a simple situation where there is only one x in the Cox regression (Equation 17.6) such that $x = 0$ for standard treatment and $x = 1$ for experimental treatment. The Cox model sometimes assumes that the difference between the logarithms of hazards in the two treatment groups is constant over the follow-up period. This has been seen to be true in many applications. This makes it a *proportional* hazards model. Further explanation is given in Chapter 18 on survival analysis where Cox model is frequently used.

4. The Cox regression also assumes that the covariates affect the hazard in a multiplicative manner. This means that when two factors are simultaneously present, the hazard multiplies instead of adding. Multiplicability amounts to additivity in logarithm terms. This condition, too, is generally satisfied in many situations.

5. In the proportional hazards model, the predictors $x_1, x_2, ..., x_K$ should be time invariant such as sex and race. But the Cox regression is most useful when their value changes with time. While using statistical software yourself, or trying to understand the results of someone else's, ensure that the right package for time-dependent covariates has been used when such covariates are present. If a time-invariant model is used for time-dependent covariates, the results can be misleading.

6. The Cox model has extensive use in modeling survival. This is discussed in Chapter 18.

Example 17.4: Cox Regression for Lung Cancer Mortality and Carbon Black Exposure

Morfeld et al. [6] used Cox modeling to analyze lung cancer mortality in 1528 German carbon black workers. The covariates were cumulative and mean carbon black exposure, duration of work, smoking, and age at hire. Note that some of these are time dependent.

A total of 50 lung cancer deaths occurred. Hazard of mortality was not found associated with carbon black exposure. The results do not suggest that carbon black exposure is a human lung carcinogen. However, risk increased with duration of work in the lamp black producing department. This may indicate a history of exposure to gaseous polycyclic aromatic hydrocarbons.

17.4.2 Classification and Regression Trees

You may have noticed many methods in this text that predict a difficult-to-assess variable or a future outcome on the basis of relatively easily available or existing information. Mean ± 1.96SD limits for a measurement when computed on the basis of measurements in healthy subjects help to classify a new subject as normal or not normal for that measurement. Scoring systems can help in diagnostic classification and in assigning prognostic category. Standard regression can predict a quantitative variable on the basis of values of explanatory variables, and logistic regression can predict odds or the probability of a subject belonging to a particular category. Statisticians seem to have mastered this art although many predictions still remain questionable. Thus, it is not surprising that newer methods are devised. Classification and regression tree is one more method that can help in predicting the category of a subject. The dependent variable must be categorical for this method. The emphasis in this method is on simplification of a complex process without compromising the accuracy of prediction.

Consider predicting a subject with multiple injuries (in motor vehicle accident, earthquake, etc.) into a survivor with no disability, survivor with some disability, and death. The dimension of injuries can be observed as organs affected, severity of injury, number of fractures, etc. Individual's own characteristics such as age, gender, muscular strength, and bone density may be other determining factors for the outcome. Emergency help (time elapsed since injury), available medical facilities, and expertise of the medical care providers are also important. In place of saying that one factor will contribute 5%, another 10%, etc., would it not be nice to say that if head is injured, age is 60+, medical help is excellent but not available within 2 h after the injury, then the person is most likely to survive but will have residual disability? Classification and regression tree has precisely this function. It determines a set of logical *if-then* conditions instead of linear equation for predicting the category of a subject. Such statements are straightforward, easily understood, and intuitively appealing. The method is nonparametric and does not need Gaussian distribution of the underlying variables and does not need linearity either. The tree method is particularly suitable for data exploration where a priori information is scanty and where biological reasons are not immediately available for classification. The method can sometimes reveal simple relationships between just a few variables that can go unnoticed by other methods.

The tree algorithm, as the name suggests, devises split nodes that generate branches by using if-then rules. The adequacy of prediction is tested at each step. A fresh split node is not needed when the preceding node does not improve the accuracy of prediction. It is possible that a final tree so arrived still fails to correctly classify the adequate percentage of subjects. As in all other prediction methods, this failure would indicate that the predictors chosen for this purpose are not of the right kind. Also, if the tree becomes very large, it loses much of its operational utility.

The tree method is complex and is beyond the scope of this book. For details, see Denison et al. [7]. Briefly, the algorithm begins by identifying a predictor that splits the total subjects into two groups such that the subjects are similar within groups and dissimilar across groups, and the classification is correct for the largest percentage of subjects. The groups can be such that they have the strongest association with the response categories. If a predictor is quantitative, various cutoff points are tried and the one providing least within-groups sum of squares is chosen. For a qualitative predictor with K categories, all $(2^K - 1)$ possible splits are tried. Then, the process is repeated at second and subsequent steps with new predictors. The process stops when it does not add to the accuracy of classification.

It is customary in classification and regression trees to define a loss function for misclassification. Some misclassifications have more severe implications than others, and for them a larger loss is defined. The loss may be in terms of financial cost, inconvenience to the health agencies and the person concerned, or simply statistical in terms of increase in least squares or increase in proportional reduction in error (PRE). The last criterion is described later in this chapter in another context.

There is some similarity between classification trees and discriminant analysis. But in the latter case, all predictors are considered together to classify a subject, whereas in the trees, one or few predictors are considered in the first step, another one or more in second step, yet another set in third step, etc. More than three steps make the process too complex and generally not considered appropriate for classification trees.

17.4.3 Further Generalizations

As mentioned in Chapter 16, regression, analysis of variance (ANOVA), and analysis of covariance (ANCOVA) unify under general linear models for quantitative dependent. These along with logistic and some other regressions such as Poisson (not discussed in this book) also unify under, what is called, **generalized linear models**. This requires a link function for dependent that specifies that it is logistic or Gaussian or what. A further generalization is available under the name of **generalized estimating equations** (GEE). This dispenses with the requirement of independence of ys.

In GEE, you can have some ys belonging to the same person at different points of time, or measured at different parts of the body, or for persons belonging to the same cluster, so that they are correlated. Some ys, of course, will belong to different persons. The correlation structure among ys can be chosen as independent (i.e., no correlation, which returns to the classical structure), exchangeable (same correlation between values at time-1 and time-2 as between time-1 and time-3, etc.), autoregressive of specified lag (if lag = 2, correlation between values at time-1 and time-3 is the same as between time-2 and time-4, etc.), or unstructured (any correlation). GEE is popularly used for longitudinal data. This can have time-dependent regressors as well as time-independent covariates [8]. For example, if you want to examine how a particular daily exercise for 1 h affects BP level over a period of 3 months, the actual independent variable of interest is the cumulative exercise since the beginning. In this setup, age, sex, and social class are examples of time-independent covariates that can influence change in BP level, and BMI, smoking, and diet are examples of time-dependent covariates that can affect BP and can also change over the study period. Thus, the model for longitudinal quantitative data at time t could be

$$y_{it} = b_0 + b_1 t + \Sigma_j b_{2j} z_{ijt} + \Sigma_k b_{3k} x_{itk} + \Sigma_l b_{4l} r_{il} + e_{it};$$

$$i = 1, 2, \ldots, n; \quad t = 1, 2, \ldots, T; \quad j = 1, 2, \ldots, J; \quad k = 1, 2, \ldots, K; \quad l = 1, 2, \ldots, L;$$

where
  xs are the usual regressors
  zs are the time-dependent covariates
  rs are time-independent covariates

In our BP example, there is only one x of real interest, namely, the cumulative exercise. Time-dependent covariates are $z_1 = $ BMI, $z_2 = $ smoking, and $z_3 = $ diet.

Time-independent covariates are r_1 = age, r_2 = sex, and r_3 = social class. All these will be different for different subjects indexed by subscript i.

You can see how quickly a GEE model becomes complex. This does require quite some expertise for proper implementation and correct interpretation.

17.5 Strength of Relationship in Qualitative Variables

Whereas the strength of relationship between quantitative variables is easily obtained by computing correlation coefficient, it is not so simple for qualitative variables. The problem becomes more acute when one variable is quantitative and the other qualitative. The following are some of the more commonly used measures of the strength of relationship where at least one variable is qualitative. These are also known as distribution-free (nonparametric) measures as they do not require underlying values, even if metric, to follow any specific distribution such as Gaussian.

17.5.1 Both Variables Qualitative

The measures of the strength of relationship in qualitative variables can be discussed separately for variables with dichotomous and polytomous categories.

17.5.1.1 Dichotomous Categories

The most common situation in the case of qualitative variables is that of dichotomy. The bivariate binary observations in this case are generally summarized in the form of a 2 × 2 table of the type given in Table 17.4.

It is customary in the case of qualitative variables to use the term association in preference to correlation. The most widely used measure of the strength of association in such a setup is either RR or OR, depending on whether the design is prospective or retrospective/cross sectional. As mentioned in an earlier chapter, ln(RR) and ln(OR) are preferred because of the

TABLE 17.4

Cross-Classification of Subjects by Two Binary Variables

	Variable-1	
Variable-2	**Present**	**Absent**
Present	a	b
Absent	c	d

linearizing property of the logarithm in this case. This transformation also assigns a negative or positive sign to the relationship as appropriate, making the interpretation easy.

The conventional measures are the following dichotomy coefficients:

$$\text{Positive matching dichotomy coefficient: } S_1 = \frac{a}{a+b+c+d}.$$

This is the proportion of pairs with both values present.

$$\text{Jaccard's dichotomy coefficient: } S_2 = \frac{a}{a+b+c}.$$

This is the proportion of pairs with both values present given that at least one occurs.

$$\text{Simple matching dichotomy coefficient: } S_3 = \frac{a+d}{a+b+c+d}.$$

This is the proportion of pairs where the values of both variables agree.

$$\text{Anderberg's dichotomy coefficient: } S_4 = \frac{a}{a+2(b+c)}.$$

This is basically the same as S_2 but is standardized for all possible patterns of agreement and disagreement.

$$\text{Tanimoto's dichotomy coefficient: } S_5 = \frac{a+b}{a+2(b+c)+d}.$$

This is S_3 standardized for all patterns of agreement and disagreement.

It is not easy to choose one among these. You may have to carefully examine the data and the objective of the relationship that you want to measure. For example, S_2 and S_4 are appropriate when absence of an attribute in both variables (d in Table 17.4) does not convey any useful information.

Another popular measure of degree of association for dichotomous categories is

$$\text{Yule } Q: Q = \frac{ad-bc}{ad+bc}.$$

In terms of OR, this is $Q = (OR - 1)/(OR + 1)$. This lies between -1 and $+1$ and can be interpreted like a correlation coefficient for assessing the strength of relationship.

TABLE 17.5

Gender and Blindness in Patients Coming to a Cataract Clinic

Gender	Blind	Not Blind	Total
Male	101	419	520
Female	110	370	480
Person	211	789	1000

Example 17.5: OR as a Measure of Strength of Association between Gender and Blindness

Consider the data in Table 7.1 on age, gender, and visual acuity (VA) in the worse eye of 1000 subjects coming to a cataract clinic. Let the definition of blindness be VA < 1/60. When age is collapsed, Table 17.5 is obtained for gender and blindness.

What is the degree of association between gender and blindness? The answer lies in the OR. Since females have a higher rate of blindness in this case, it is better to compute OR in female relative to males. This is

$$OR = \frac{110 \times 419}{101 \times 370} = 1.23.$$

As far as these data are concerned, females are 1.23 times as likely to be blind as males, or the odds are nearly 5:4. You may wish to recompute OR for blindness (VA < 1/60) in persons of age 60 years and older relative to younger than 60 years (disregarding gender) on the basis of the data in Table 7.1 and confirm that the OR is 1.62. This translates into odds of nearly 8:5. It can thus be concluded that blindness in these subjects is apparently associated more with age than with gender.

SIDE NOTE: As an exercise you may like to calculate coefficients S_1 through S_5 for the data in Table 17.5 and interpret them.

17.5.1.2 Polytomous Categories: Nominal

If you continue to use the same VA categories (VA ≥ 6/60, 6/60 > VA ≥ 1/60, and VA < 1/60) as in Table 7.1, so that there are three categories and not two, then you need to compute OR for each pair of categories, that is, OR of category-1 versus category-2, of category-1 versus category-3, and of category-2 versus category-3. Such multiple ORs may be useful in some cases but can be confusing in many cases. Moreover, a measure of overall association will not be available.

An alternative is to compute the usual $\chi^2 = \Sigma(O-E)^2/E$ and use this as a measure of association. It is true that a higher degree of association will yield a higher value of chi-square. But the value of chi-square also depends heavily on the sample size n. It increases without bound as n increases. To counter this, the following measure is proposed:

$$\text{Phi coefficient: } \phi = \sqrt{\frac{\chi^2}{n}}, \tag{17.7a}$$

where the square root is taken to offset the effect of squares in chi-square. Note that the concept of negative association is not relevant in the case of polytomous *nominal* categories. The phi coefficient can exceed unity. Also, ϕ depends on the size of the table. A modification of this measure is

$$\text{Contingency coefficient: } C = \frac{\sqrt{\chi^2}}{\sqrt{\chi^2 + n}}. \tag{17.7b}$$

This cannot exceed unity but could also never be one, even when the association is perfect. The value of χ^2, and hence of ϕ and of C, can be severely affected by the cutoff points of categories when they are for a metric variable. Different cutoff points may give different values.

The most popular of such chi-square based measures is Cramer V:

$$\text{Cramer } V = \sqrt{\frac{\chi^2}{n * \min(R-1, C-1)}}. \tag{17.7c}$$

This provides a good measure for tables of any size and ranges between 0 and 1. For 2×2 tables, $V = \phi$. Cramer V is especially suitable for a square contingency table (i.e., when $R = C$).

These three measures of degree of association described earlier are, in fact, for nominal categories. Example 17.6 illustrates the calculations, but note that the categories in this example are metric. The calculations ignore the metric values or ordinal structure.

Example 17.6: Phi Coefficient and Contingency Coefficient for Association between Age and VA

Collapse gender in Table 7.1 and obtain Table 17.6. The value of χ^2 for this table is 37.14. If the cell frequencies are proportionately decreased to one-fifth, rounded off to the nearest integer, so as to have a total of 200, you get $\chi^2 = 7.39$. The large difference between this value of χ^2 for $n = 200$ and the previous value for $n = 1000$ illustrates that χ^2 is heavily dependent on n. A proportionate decrease (or increase) in cell frequencies does not affect the degree of association but affects the value of χ^2. For these data,

1. For $n = 1000$, as in the aforementioned table, $\chi^2 = 37.14$, $\phi = 0.19$, $C = 0.19$, and $V = 0.14$.
2. For $n = 200$ (proportionate cell frequencies), $\chi^2 = 7.39$, $\phi = 0.19$, $C = 0.19$, and $V = 0.14$.

TABLE 17.6

Age and VA in Patients Coming to a Cataract Clinic

Age-Group (Years)	VA			Total
	≥6/60	6/60–1/60	<1/60	
–49	19	69	22	110
50–59	39	142	29	210
60–69	46	325	89	460
70–79	21	98	51	170
80+	7	23	20	50
Total	132	657	211	1000

SIDE NOTE: Note that ϕ, C, and V for the two ns are the same while the value of χ^2 is very different. In this example, the values of ϕ and C seem equal for each of the two ns because of the rounding off, but this will not be generally the case.

17.5.1.3 Proportional Reduction in Error

A major objection to the measures ϕ, C, and V is that they lack underlying substantive meaning. A measure of strength of relationship with a useful interpretation is PRE. The primary purpose of PRE is to measure the utility of one characteristic in predicting the other. The higher the PRE, the more useful the predictor. This is directly dependent on the strength of the relationship. It ranges from 0 to 1 and can be interpreted from no association to perfect association. But there is no negative association. The basic principle underlying PRE is easily explained with the help of an example.

In Example 17.6, if a guess is to be made about the VA of a random person coming to the same cataract clinic, the best guess (with least error) is (6/60 > VA ≥ 1/60) because this is the most commonly (65.7%) occurring acuity in such subjects. This guess can be wrong in the other 100 – 65.7 = 34.3% of cases. This is the error of prediction. If we know that the age of the patient is between 60 and 69 years, then this guess is strengthened further because 325 out of 460 (70.7%) is a higher proportion in this acuity category than for any other age-group. The error is now reduced to 29.3%. The PRE by knowing the age in this case is (34.3 – 29.3)/34.3 = 0.15 or 15%. This measures the utility of age in predicting VA category. However, knowing age on the whole is not helpful in predicting VA category in the data of Table 17.6 because whatever the age, the most common VA category is always (1/60, 6/60). When all age-groups are put together, the PRE is zero. Example 17.7 explains PRE further.

Example 17.7: PRE for Predicting Neck Abnormality in Spermatozoa from Head Abnormality

A study was carried out on 80 subfertile men with varicocele on spermatozoal morphology with the objective of finding whether head abnormalities can be used to predict neck abnormalities in spermatozoa. The data obtained are shown in Table 17.7.

If head abnormality is present, the best prediction is that neck abnormality is doubtful because this is the most commonly observed (24 cases) category in the head abnormality group. This prediction will be wrong in 44 − 24 = 20 cases. Thus, Table 17.8 can be constructed.

Extent of error stated in the last column is obtained by subtracting the maximum in the row from the corresponding total. By knowing that the head abnormality is present or absent, the error is reduced from a total of 45 cases to a total of 20 + 10 = 30 cases. Thus,

$$PRE = \frac{(80 - 35) - [(44 - 24) + (36 - 26)]}{80 - 35} = 0.33.$$

Knowledge about the presence or absence of head abnormality reduces the error in predicting neck abnormality by 33%. This reduction is not really high in this example and shows that the association of head abnormality with neck abnormality in spermatozoa is not really strong.

TABLE 17.7

Head and Neck Abnormality in Spermatozoa in Subfertile Men

Head Abnormality	Neck Abnormality			Total
	Present	Doubtful	Absent	
Present	11	24	9	44
Absent	3	7	26	36
Total	14	31	35	80

TABLE 17.8

Calculation for PRE in Table 17.7

Head Abnormality	The Best Prediction for Neck Abnormality	Extent of Error
Present (44 cases)	Doubtful (24 cases)	44 − 24 = 20 cases
Absent (36 cases)	Absent (26 cases)	36 − 26 = 10 cases
If not known (total 80 cases)	Absent (35 cases)	80 − 35 = 45 cases

In terms of notations, the PRE in predicting column category when row category is known can be written as

$$\text{PRE} = \frac{\left(n - \overset{max}{C} O_{\bullet c}\right) - \Sigma_r(O_{r\bullet} - \overset{max}{C} O_{rc})}{n - \overset{max}{C} O_{\bullet c}}; \quad r = 1, 2, \ldots, R; \quad c = 1, 2, \ldots, C; \quad (17.8)$$

where
the order of the table is $R \times C$
O_{rc} is the frequency in the cell in the rth row and cth column
$O_{r\bullet}$ is the marginal total in rth row
$O_{\bullet c}$ is the marginal total in cth column
n is the total number of subjects

These notations are the same as used earlier for calculation of chi-square. The notation $\overset{max}{C}$ is for the maximum value among the columns.

For PRE in the formula given in Equation 17.8 to be interpretable, it is necessary that one variable is considered dependent on the other in the sense that one can be predicted by the other. In Example 17.7, neck abnormality is treated as dependent because it is sought to be predicted from head abnormality.

The formula given in Equation 17.8 assumes that the column category is to be predicted by the row category. The notations will change if row category is to be predicted by column category. The PRE considers all categories nominal. If categories are ordinal or metric, and if the order is important, then other measures of association should be used. These are discussed by Freeman [9]. Some of these are described next.

17.5.1.4 Polytomous Categories: Ordinal Association

Now, consider two ordinal characteristics such as severity of disease and obesity. One may have $R = 4$ categories and the other $C = 3$ categories. The interest is in measuring the strength of association between these two ordinal characteristics. The data can be arranged in a $R \times C$ table.

The association is high if higher category of one is more frequently seen with higher category of the other. Association between severity of disease and obesity is high if more severe cases are obese. This is called concordance. If less severe cases are mostly obese, this is discordance. Thus, in this case you need to consider all possible pair of pairs. If one pair is (x_1, y_1) and the other pair is (x_2, y_2), they are concordant if $x_1 < x_2$ and $y_1 < y_2$ or if $x_1 > x_2$ and $y_1 > y_2$, and disconcordant if $x_1 < x_2$ but $y_1 > y_2$, or $x_1 > x_2$ but $y_1 < y_2$. Also, pairs are tied for x if $x_1 = x_2$ irrespective of y and tied for y if $y_1 = y_2$ irrespective of x. In ordinal data, ties are quite common. For n

TABLE 17.9

Persons with Two Ordinal Characteristics

	Characteristic-1 (x)		
Characteristic-2 (y)	Low	Medium	High
Low	a	b	c
High	d	e	f

subjects, there are a total of $n(n-1)/2$ pairs since the pair $[(x_1, y_1), (x_2, y_2)]$ is considered same as $[(x_2, y_2), (x_1, y_1)]$. For a 3 × 2 table (Table 17.9), this can be explained as follows:

Total number of persons $= a + b + c + d + e + f = n$.

Total number of pairs of pair $= n(n-1)/2 = T$.

Concordant pairs $= a(e + f) + bf = P$.

Discordant pairs $= c(d + e) + bd = Q$.

Pairs tied on characteristic-1 (x) alone $= ad + be + cf = X_0$.

Pairs tied on characteristic-2 (y) alone $= a(b + c) + bc + d(e + f) + ef = Y_0$.

Pairs tied on both x and y $= a(a-1)/2 + b(b-1)/2 + c(c-1)/2 + d(d-1)/2 + e(e-1)/2 + f(f-1)/2 = (XY)_0$.

$(XY)_0$ does not contribute to the measure and ignored except for calculating total number of pairs.

Now, various measures of ordinal association can be defined as follows. You may find varying definitions in the literature.

$$\text{Kendall Tau-a: } \tau_a = \frac{P-Q}{n(n-1)/2}.$$

This is the surplus of concordant pairs over discordant pairs as proportion of the total pairs. If the agreement in pairs is perfect, $Q = 0$ and tau-a $= 1$ assuming no ties. If all are discordant pairs, $P = 0$ and tau-a $= -1$. Thus, this ranges from -1 to $+1$. If ties are present, use

$$\text{Kendall Tau-b: } \tau_b \frac{P-Q}{\sqrt{(P+Q+X_0)(P+Q+Y_0)}}.$$

The denominator is now partially adjusted for ties, and P and Q will also be automatically adjusted by definition. Tau-b works well for square tables where the number of categories for one characteristics is the same as for the other characteristics (i.e., $R = C$). Tau-b = +1 if the table is diagonal and Tau-b = –1 if all diagonal elements are zero. If the table is not square, the corresponding adjustment for the size is

$$\text{Kendall Tau-c: } \tau_c = \frac{2(P-Q)R}{n(n-1)(R-1)},$$

where R is the number of rows or columns, whichever is smaller. This is also called **Stuart Tau-c**.

Beside these variations of tau, there are two other popular measures of ordinal association.

$$\text{Goodman–Kruskal gamma: } \gamma = \frac{P-Q}{P+Q}.$$

This completely excludes ties from the numerator as well as from the denominator. This also ranges from –1 to +1. If the number of discordant pairs is the same as concordant pairs, $\gamma = 0$.

All these measures are symmetric in the sense that it does not matter which characteristic is in the rows and which in the columns. Both are treated in the same way. For directional hypothesis such that x predicts y, use

$$\text{Somer } d = \frac{P-Q}{P+Q+Y_0},$$

where Y_0 is the number of pairs tied for y. Note that only the pairs tied for x are excluded from the denominator.

Example 17.8: Association between Smoking and Drinking

Tai el al. [10] studied smoking and drinking as risk factors for esophageal cancer in Taiwanese women. They did not study association between smoking and drinking but suppose the Table 17.10 was obtained for cancer cases.

TABLE 17.10

Smoking and Drinking in Cancer Cases

Drinking	Nonsmokers	Smoking ≤3.5 Pack/Week	>3.5 Pack/Week	Total
Nondrinkers	35	4	1	40
Drinkers	6	2	3	11
Total	41	6	4	51

For this table, $n = 51$, and $T = 51 \times 50/2 = 1275$.

Concordant pairs $= P = 35(2+3) + 4 \times 3 = 187$.

Discordant pairs $= Q = 1(6+2) + 4 \times 6 = 32$.

Tied on x (smoking) $= X_0 = 35 \times 6 + 4 \times 2 + 1 \times 3 = 221$.

Tied on y (drinking) $= Y_0 = 35(4+1) + 4 \times 1 + 6(2+3) + 2 \times 3 = 215$.

Tied on both x and $y = (XY)_0 = \frac{1}{2}(35 \times 34 + 4 \times 3 + 1 \times 0 + 6 \times 5 + 2 \times 1 + 3 \times 2) = 620$.

These numbers give

$$\text{Tau-a} = \frac{187 - 32}{1275} = 0.12.$$

$$\text{Tau-b} = \frac{187 - 32}{\sqrt{(187 + 32 + 221)(187 + 32 + 215)}} = 0.35.$$

$$\text{Tau-c} = \frac{2 \times (187 - 32) \times 2}{51 \times 50 \times 1} = 0.24.$$

$$\text{Goodman–Kruskal } \gamma = \frac{187 - 32}{187 + 32} = 0.71.$$

$$\text{Somer } d = \frac{187 - 32}{187 + 32 + 215} = 0.36.$$

Note how widely different values are obtained by different measures for the same data. For this reason, many workers do not rely much on these measures.

17.5.2 One Qualitative and the Other Quantitative Variable

Risk of malaria dependent partly on the density of female anopheline mosquitoes in an area and diagnosis of primary biliary cirrhosis dependent on elevated serum alkaline phosphatase levels are examples in which the dependent variable is binary. You now know that logistic regression is used in this setup. The regressors in a logistic model can be either qualitative or quantitative or a mixture. One of the measures of strength of relationship in such cases is the OR although it has certain limitations as already mentioned.

In the reverse case, when dependent y is quantitative and the regressor x qualitative, the situation is the same as in ANOVA. The strength of relationship in this case as measured by R^2 is given as follows:

For quantitative y (dependent) and qualitative x,

$$R^2 = 1 - \frac{\text{Sum of squares due to error}}{\text{Total sum of squares}}. \tag{17.9}$$

This is the same as mentioned in the previous chapter. The quantity R^2 measures the proportion of total variation accounted for by the regressor. This takes values from 0 for no relationship to 1 for perfect relationship. Such extreme values rarely occur in empirical sciences. But R^2 is a very useful measure of the strength of relationship in the case of quantitative dependent whether the regressor is quantitative or qualitative.

The OR based on logistic regression and R^2 based on ANOVA are both applicable even when the number of regressor variables is more than one.

Example 17.9: R^2 as a Measure of Association between a Quantitative and Qualitative Variable

Example 15.5 describes rapid eye movement (REM) sleep time in rats that received different doses of an ethanol preparation. The ANOVA table in that example (Table 15.3) shows that

$$\text{Total sum of squares (SST)} = 7369.8$$

and

$$\text{Sum of squares due to error (SSE)} = 1487.4.$$

Thus,

$$R^2 = 1 - \frac{1487.4}{7369.8} = 0.798.$$

This means that 79.8% of the variation in REM sleep time among rats is due to differences in ethanol dosage. Thus, there is a fairly strong association between REM sleep time and ethanol dosage in this example.

17.5.3 Agreement in Qualitative Measurements (Matched Pairs)

Assessing optical disk characteristics by two or more observers, results of Lyme disease serological testing by two or more laboratories, and comparison of x-ray images with Doppler images are examples of the problem of qualitative agreement. The objective is to find the extent of agreement between two or more methods, observers, laboratories, etc. In some cases, for example in comparison of two laboratories, agreement has the same interpretation as **reproducibility**. In the case of comparison of observers, it is termed **interrater reliability**. In all these cases, only one group of subjects is assessed twice. Thus, this is a matched pair setup. Quantitative agreement was discussed in the previous chapter. Now consider qualitative agreement.

17.5.3.1 Meaning of Qualitative Agreement

For simplicity, restrict for the time being to the presence or absence of a characteristic assessed by two observers on the same group of subject. An example is presence or absence of a lesion in x-rays read by two observers. Suppose the observations are as given in Table 17.11.

TABLE 17.11

Presence or Absence of a Lesion Assessed by Two
Radiologists in X-Rays of 60 Suspected Cases

	Observer-1		
Observer-2	**Present**	**Absent**	**Total**
Present	29	7	36
Absent	13	11	24
Total	42	18	60

The two observers agree on a total of 40 cases in this example. This is the sum of the frequencies in the leading diagonal cells. In the other 20 cases, the observers do not agree. Apparently, the agreement = 40/60 = 66.7%. But part of this agreement is due to chance, which might happen if both are dumb observers and randomly allocate subjects to present and absent categories. This chance agreement is measured by the cell frequencies expected in the diagonal when the observer's ratings are independent of one another. These expected frequencies are obtained by multiplying the respective marginal totals and dividing by the grand total, as obtained for calculating the chi-square.

For the data in Table 17.11, the chance-expected frequencies are $36 \times 42/60 = 25.2$ and $24 \times 18/60 = 7.2$ in the two diagonal cells. The total of these two is 32.4. Agreement on so many cases is expected by chance alone. Thus, agreement in excess of chance is in only $40 - 32.4 = 7.6$ cases. The maximum possible excess is $60 - 32.4 = 27.6$. A popular measure of agreement is the ratio of the observed excess to the maximum possible excess, in this case $7.6/27.6 = 0.275$ or 27.5%. Thus, the two observers in this case do not really agree much on rating of x-rays for the presence or absence of lesion. Most of their agreement is due to chance.

17.5.3.2 Cohen Kappa

In terms of notations, the foregoing procedure to measure agreement between qualitative characteristics on the same scale is given by

$$\text{Cohen kappa: } \kappa = \frac{\Sigma O_{ii} - \Sigma(O_{i \bullet} O_{\bullet i}/n)}{n - \Sigma(O_{i \bullet} O_{\bullet i}/n)}, \tag{17.10}$$

where
   O_{ii} is the cell frequency in the ith row and ith column (diagonal element)
   $O_{i \bullet}$ is the marginal total in the ith row
   $O_{\bullet i}$ is the marginal total in the ith column
   n is the grand total

The first term in the numerator is the observed agreement and the second term is the chance agreement. Thus, the numerator is the agreement in excess of chance. The denominator is its maximum possible value. Kappa in the formula given in Equation 17.10 is for a general $I \times I$ table, that is, the rating is not necessarily restricted to present–absent type of dichotomy but could also be into three or four or a larger number of categories.

Kappa is used mostly to assess how close the agreement is to one rather than how far is it from zero. The following scale is suggested:

Kappa	Strength of Agreement
<0.3	Poor
0.3–0.5	Fair
0.5–0.7	Moderate
0.7–0.9	Good
>0.9	Excellent

Example 17.10: Cohen Kappa for Agreement between the Results of Two Laboratories

Detection of intrathecal immunoglobulin G (IgG) synthesis is important in patients with suspected multiple sclerosis. Isoelectric focusing is a method used for detection of intrathecal IgG synthesis. Let this be assessed as positive, doubtful, and negative by two laboratories on 129 patients. The results are shown in Table 17.12.

In this case, by Equation 17.10,

$$\kappa = \frac{(36 + 12 + 55) - (44 \times 44/129 + 25 \times 21/129 + 60 \times 64/129)}{129 - (44 \times 44/129 + 25 \times 21/129 + 60 \times 64/129)}$$

$$= \frac{54.155}{80.155} = 0.68.$$

SIDE NOTE: Generally speaking, a kappa value equal to 0.68 is adequate to conclude fair agreement but, in this example, the investigation is on reproducibility between laboratories. If both the laboratories are using standardized tools and methods, the agreement should be close to 1. Thus, the reproducibility of the method of isoelectric focusing between the laboratories for assessing intrathecal synthesis cannot be considered good in this case despite a not so disappointing value of kappa.

TABLE 17.12

Assessment of Intrathecal Synthesis by Two Laboratories

Laboratory-2	Laboratory-1			Total
	Positive	Doubtful	Negative	
Positive	36	5	3	44
Doubtful	7	12	6	25
Negative	1	4	55	60
Total	44	21	64	129

The following comments regarding Cohen kappa may be helpful:

1. There are some other measures of agreement for qualitative variables. These are discussed by Agresti [11]. He also describes other agreement assessment models such as log-linear, Rasch, and latent class models.

2. Cohen kappa is valid for nominal categories only. Ordinal or metric categories are considered nominal by this measure and the order is ignored. The other assumptions are that (a) the subjects are independent; (b) the observers, laboratories, or methods under comparison operate independently of one another; and (c) the rating categories are mutually exclusive and exhaustive. These conditions are easily fulfilled in most practical situations.

3. Rare though, you may sometimes find reference to **weighted kappa**. In this, the cells are assigned a weight according to the degree of disagreement they exhibit. Thus, cells in the diagonal, since they are in full agreement, get zero weight. Off-diagonal cells get varying weight depending upon either your perceived importance of the involved cells or on some objective criterion such as quadratic weight. All this makes kappa too complex and possibly not as useful.

4. The kappa value is +1 for complete agreement and 0 if the agreement is the same as expected by chance. However, the value does not become −1 for complete disagreement. Thus, *it is a measure of the extent of agreement but not of disagreement.*

5. For the formula of variance of kappa for large samples, see Fleiss et al. [12]. This variance can be used to construct a CI and to test a hypothesis on the value of kappa. Standard statistical softwares have provision to do these calculations.

6. Kappa values from different studies may not be comparable as the value also varies with the number of subjects in categories relative to the total. If one study has 30% subjects in category A and another study has 50%, the value of kappa will differ even if the extent of agreement is the same.

7. Kappa does not distinguish between +/− discordance and −/+ discordance. Both get the same weight. Thus, kappa can be 0.9 in 100 subjects when all discordances are +/− type and none of −/+ type.

References

1. Hosmer DW, Lemeshow S. *Applied Logistic Regression*, 2nd edn. New York: Wiley–Interscience, 2000.
2. Mehta CR, Patel NR. Exact logistic regression: Theory and examples. *Stat Med* 1995; 14:2143–2160.

3. Marquis GS, Habicht J, Lanata CF, Black RE, Rasmussen KM. Association of breastfeeding and stunting in Peruvian toddlers: An example of reverse causality. *Int J Epidemiol* 1997; 26:349–356.

4. Satchell M, Pati S. Insurance gaps among vulnerable children in the United States, 1999–2001. *Pediatrics* 2005; 116:1155–1161.

5. Kleinbaum DG, Klein M. *Survival Analysis: A Self-Learning Text (Statistics for Biology and Health)*, 2nd edn. New York: Springer, 2005.

6. Morfeld P, Buchte SF, Wallmann J, McCunney RJ, Piekarski C. Lung cancer mortality and carbon black exposure: Cox regression analysis of a cohort from a German carbon black production plant. *J Occup Environ Med* 2006; 48:1230–1241.

7. Denison DGT, Homes CC, Mallick BK, Smith AFM. *Bayesian Methods for Nonlinear Classification and Regression*. New York: John Wiley & Sons, Inc., 2002.

8. Twisk JWR. *Applied Longitudinal Data Analysis for Epidemiology: A Practical Guide*. Cambridge, U.K.: Cambridge University Press, 2003.

9. Freeman DH. *Applied Categorical Data Analysis*. New York: Marcel Dekker, 1987.

10. Tai SY, Wu IC, Wu DC, Su HJ, Huang JL, Tsai HJ et al. Cigarette smoking and alcohol drinking and esophageal cancer risk in Taiwanese women. *World J Gastroenterol* 2010; 16:1518–1521.

11. Agresti A. Modelling patterns of agreement and disagreement. *Stat Methods Med Res* 1992; 1:201–218.

12. Fleiss JL, Cohen J, Everitt BS. Large sample standard errors of kappa and weighted kappa. *Psychol Bull* 1969; 72:323–327.

18

Survival Analysis

Survival is complementary to mortality, but the statistical use of the term is for the duration of survival. The duration of survival of a patient with cancer after detection of malignancy and median survival time in cases of leukemia are common examples. Sometimes the interest is in the percent surviving for a specific duration such as 5 year survival rate of end-disease renal patients on dialysis.

In a generic sense, the method of survival analysis is used to analyze any data on duration. This can also be understood as time-to-event data. It could be duration of stay in the hospital, birth interval, duration of immunity, or any such duration.

Why cannot durations be analyzed as any other quantitative variable? One difficulty is that most durations do not follow a Gaussian pattern. Their distribution is highly skewed. This is not a big problem since this can be overcome by choosing a sample of sufficiently large size. The real problem arises when the duration for some subjects is not fully observed. The person moves out, dies due to an unrelated cause, refuses to cooperate after some follow-up, etc. The primary cause of such incomplete observation, however, is that the end point event does not occur during the follow-up period; the study is terminated after a fixed time. All these result in incomplete segments, called **censored observations**. The method of survival analysis is geared to meet this contingency, which can rarely be handled in any other manner. In addition, this chapter also describes some other methods to study the duration of survival.

It goes without saying that for any duration there must be a beginning (entry) point and an end (exit) point. In the case of diabetes, for example, the beginning point can be the time when the first symptom was noted, when the disease was first diagnosed, when blood glucose level showed a significant rise from the preexisting level, when a treatment was started, or any other point of time considered appropriate. The end point can be when blood glucose was first noted back to the normal level, when the treatment was discontinued, when the person is able to eat normally, or any other. Because of multiple choices, you can see that both the beginning point and the end point should be sharply defined for the duration to be observed without error. It is also important to decide whether the intervening period between the entry and exit has to be completely free of events and whether the occurrence of other events is to be disregarded.

This chapter: The most popular measure of duration of survival is expectation of life. This chapter begins with the methods to compute the life expectancy for a population. Section 18.1 describes a method for constructing life tables. This method forms the basis for discussing other life expectancy–related measures such as years of life lost because of a particular disease and healthy life expectancy. The methods of this section do not form the core of survival analysis but are described here to provide a foundation. Section 18.2 describes methods of survival analysis that include the life table method and the Kaplan–Meier (K-M) method. These are the core methods that tell you how to find a pattern of survival for a group on a particular regimen opposed to another group such as a control regimen and how to compare them. The methods of both these sections do not consider prognostic factors affecting the durations. The methods of assessing the role of various factors in affecting the duration are presented in Section 18.3.

For convenience of presentation, I am using the terms survival and mortality in this chapter but they really mean change from one state to the other. Survival time is any time-to-event duration.

18.1 Life Expectancy

The average number of years expected to be lived by individuals in a population is called expectation of life. This can be calculated as expectancy at birth or at any other age. The expectation of life at birth (ELB) can be crudely interpreted as the average age at death.

The ideal method for computing the expectation of life is by observing a large cohort (called a radix) of live births as long as any individual of the cohort is alive. This may take more than 100 years and thus is impractical unless records for over 100 years are complete. Such a **cohort life table** provides summaries of those born in a particular year and another cohort will be required for another year. This is also called a generation life table.

As a shortcut, a **current life table** is constructed. It assumes that the individuals at different ages are exposed to the *current* risks of mortality. Thus, the current age-specific death rates (ASDRs) are used on a preassumed cohort of, say, 100,000 persons. The average of life so obtained is the number of years a newborn is expected to live *at the current levels of mortality*. This may be characterized as cross sectional and provides a snapshot of the current mortality experience. This is more useful as it tells about the existing situation and can be computed immediately. There is no need to wait for 100 years. The calculations are done as follows. The method is also used in some other applications, as described shortly.

18.1.1 Life Table

The computation of ELB or at any other age becomes convenient by preparing what is called a life table. It utilizes the information regarding people alive at different ages, the ASDRs, and the years lived by segments of population of different ages. It is desirable to do the computation for each year of age, say, at age 46 years, 47 years, etc., but the mortality rates are generally available for age groups such as 45–49 years. When such groups are used in constructing a life table, it is called an **abridged life table**. When each single year of age is used, it is called a **complete life table**.

Table 18.1 is an abridged life table for urban females in a developing country for the block years 2007–2011. An example of a developing country provides an opportunity to discuss some features of a life table that do not appear in a life table constructed for developed countries. An explanation of the symbols in this table is as follows:

n_t is the length of the age interval beginning with age t, and

q_t is the probability of death in the age interval $(t, t + n_t)$ when the person is alive at age t.

TABLE 18.1

Abridged Life Table for Urban Females in a Developing Country for the Block Years 2007–2011

Age Interval (Years)	Probability of Death in the Interval	Expected Surviving Age t	Person-Years Lived in the Interval	Person-Years Lived Beyond t	Expectations of Life at Age t (Years)
$(t, t + n_t)$	q_t	l_t	L_t	T_t	e_t
0–1	0.04989	100,000	96,508	6,214,374	62.14
1–5	0.02453	95,011	375,149	6,117,866	64.39
5–10	0.00648	92,680	461,897	5,742,717	61.96
10–15	0.00489	92,079	459,270	5,280,820	57.35
15–20	0.00737	91,629	456,457	4,821,550	52.62
20–25	0.00874	90,954	452,782	4,365,093	47.99
25–30	0.01010	90,159	448,517	3,912,311	43.39
30–35	0.01089	89,248	443,810	3,463,794	38.81
35–40	0.01168	88,276	438,802	3,019,984	34.21
40–45	0.01480	87,245	432,997	2,581,182	29.58
45–50	0.02134	85,954	425,185	2,148,185	24.99
50–55	0.03522	84,120	413,192	1,723,000	20.48
55–60	0.05783	81,157	394,052	1,309,808	16.14
60–65	0.14623	76,464	354,367	915,756	11.98
65–70	0.15759	65,283	300,695	561,389	8.60
70+	1.00000	54,995	260,694	260,694	4.74

The latter is the deaths occurring during the interval as a proportion of the population at the *start* of the interval t. Generally, the population is available for the midpoint of the interval than at the start. Thus, q_t is estimated from the ASDR, which is based on the population at the midyear. This estimate is

$$q_t = \frac{n_t m_t}{1 + (1 - a_t) n_t m_t},\qquad(18.1a)$$

where

m_t is the ASDR for the age interval $(t, t + n_t)$
a_t is the average length of life lived per year in that interval by those who die in the interval

When $a_t = \frac{1}{2}$,

$$q_t = \frac{n_t m_t}{1 + \frac{1}{2} n_t m_t}.\qquad(18.1b)$$

The average $a_t = \frac{1}{2}$ in Equation 18.1b works well when the risk of death is uniform throughout the interval. This is generally true for age intervals (5–10), (10–15), ..., (65–70) years but not for the younger and higher age intervals. For example, it is seen that infant deaths are highly concentrated in the neonatal period. They are not uniformly distributed over (0–1) interval. In low-mortality areas (developed countries), those who survive the neonatal period tend to live almost as long as anybody else. In this case, for $t = 0$, the convention is to use $a_0 = 0.1$. For high-mortality areas (developing countries), $a_0 = 0.3$ is used. Such a higher value of a_0 for developing countries seems paradoxical, but infant deaths in developing countries tend to be relatively more evenly distributed as the deaths also continue to occur in the postneonatal period. This is rare in developed countries. In age interval (1–5) years, deaths are slightly more in (1–2) years than in (4–5) years. Thus for $t = 1$, $a_1 = 0.475$ can be used for all areas although 0.4 is advocated for developed countries. Additional adjustment is needed for the last age interval because its length is not known. Other notations are as follows:

l_t = Expected number of persons surviving the age t.

The radix is 100,000, which obviously survives age $t = 0$. The survivors at exact age $(t + n_t)$ are

l_{t+n} = Survivors at t $-$ dying in the interval $(t, t + n_t) = l_t - q_t l_t$.   (18.2)

For instance, in Table 18.1, expected survivors at age 35 years

$l_{35} = 89{,}248 - 0.01089 \times 89{,}248 = 88{,}276.$

L_t = Expected number of person-years lived in the interval $(t, t + n_t)$

$$= \left(\frac{n_t}{2}\right)(l_t + l_{t+n})\qquad(18.3)$$

TABLE 18.2

Estimation of L_0, L_1, and L_w for Developing and Developed Countries

Age Interval (Year)	Developing Country	Developed Country
0–1	$L_0 = 0.3l_0 + 0.7l_1$, for $a_0 = 0.3$	$0.1l_0 + 0.9l_1$, for $a_0 = 0.1$
1–4	$L_1 = 1.9l_1 + 2.1l_5$, for $a_1 = 0.475$	$1.6l_1 + 2.4l_5$, for $a_1 = 0.4$
Last	$L_w = l_w * \log_{10} l_w$	$L_w = l_w / M_w$

for all age intervals except the first two and the last. The last interval is denoted by w. For these three age intervals, L_t is given in Table 18.2. These are the adjustments made in appreciation of the fact that the deaths are not evenly distributed in these three intervals. Thus, these deaths cannot be averaged at the midpoint. Appropriate modifications can be made for a specific country according to its pattern of mortality. You may find different adjustments in the literature.

In Table 18.1,

$$L_w = 54{,}995 \times \log(54{,}995) = 260{,}694.$$

Now,

$$T_t = \text{Number of person-years lived beyond age } t$$

$$= L_t + L_{t+n_t} + \cdots + L_w$$

$$= L_t + T_{t+n_t} \tag{18.4}$$

and

$$T_w = L_w.$$

This means that the totaling is done from the bottom upward. Thus, the last two columns are calculated after the calculations for the previous columns are complete. Finally,

$$e_t = \text{Expectation of life at age } t$$

$$= \frac{\text{Number of person-years lived beyond age } t}{\text{Number of persons surviving age } t}$$

$$= \frac{T_t}{l_t}. \tag{18.5}$$

Table 18.1 is peculiar in the following ways: (1) For most of the intervals, as already stated, it is reasonable to assume that the deaths are uniformly distributed, and an average of half-length is lived by the persons dying in the interval. This, however, is not true for the intervals 0–1, 1–4, and 70+ years. Such a consideration leads to different values of L_0, L_1, and L_w as stated in Table 18.2 for developing and developed countries. (2) In this example, the expectation of life at 1 year (EL1) (e_1 = 64.39 years) is more than the ELB (e_0 = 62.14 years). Thus, a child of 1 year is expected to live for 64.39 more years, whereas at birth the expectation is only 62.14 years. This discrepancy is due to the high infant mortality rate (IMR). This is true for many developing countries. (3) This life table is for the block of a 5-year period, namely, 2007–2011. The annual estimates of ASDRs in many developing countries show considerable fluctuation because these are based on sample studies. The average of a 5-year period is fairly stable and is better in reflecting the trend at the midpoint of the block, which is the year 2009 in this case.

The following comments contain useful information regarding life tables:

1. All life tables assume that there are no sudden deaths owing to calamities such as famine and earthquake, or at most they are minimal that do not affect the general pattern of mortality.

2. Males and females in general have differential mortality patterns and, therefore, separate life tables are drawn for the two genders. A similar distinction may also be needed between other groupings such as rural–urban, low income–high income, and laborer–executive classes.

3. Age-specific mortality may change from year to year. The rates in the year 2012 are not necessarily the same as in the year 1995. Thus, a life table is prepared every year or at least once in 5 years.

4. ELB is severely affected by infant mortality. This is of particular concern to the developing nations, where IMR is high. For this reason, EL1 is sometimes preferred as an indicator of longevity.

The expectation of life is a very popular measure of health on the one hand, and of socioeconomic development on the other. It is considered a very comprehensive indicator because many aspects of development seem to reflect on the longevity of people. In the year 2009, it ranged from a low of 47 years in Malawi to a high of 83 years in Japan. It has gone up in China from nearly 40 years in 1950 to nearly 74 years in 2009. Recent evidence suggests that the maximum ELB attainable is 85 years [1]. Note that this is an *average* attainable by a population. Individuals are known to live for as long as 130 years.

18.1.2 Other Forms of Life Expectancy

The life table is a fairly general method and is used to compute many other expectancy-type indicators in a variety of situations. More common of these are listed in this section.

18.1.2.1 Potential Years of Life Lost

The death of a person at 55 years due to, say, a vehicular accident, who otherwise was expected to live for another 30 years, amounts to a loss of 30 potential years of life. Thus,

Potential years of life lost = Life expectancy at the age of premature death

$$- \text{actual age at death}$$

$$= e_t - t, \tag{18.6}$$

where
    t is the age at death
    e_t is expectation of life at age t

In place of using life expectancy for a country, the life expectancy in the most healthy population can be used to calculate potential years of life lost (PYLL). The concept has been in existence for a long time but was promoted by the World Bank in their 1993 report [2] and taken forward by the World Health Organization (WHO). They dropped the word "potential" and added features such as age weighting and discounting. Such years of life lost (YLL) is one of the two components of disability-adjusted life years (DALYs) that measures the burden of disease. Details were provided in Chapter 11.

YLL can be computed separately for each disease, for each group of people, for each gender, for each age group, etc. For the overall picture, YLL for all causes are added and computed per 1000 population for comparison between areas, groups, countries, etc. The WHO estimated that the YLL in the year 2004 per 1000 population per year of age was lowest in the age group 5–14 years, which is less than one-half those in the age group 15–29 years and less than one-fourth of those in the age group 0–4 years [3]. Does it sound strange to you?

18.1.2.2 Healthy Life Expectancy

It was realized a long time ago that increased longevity beyond a point does not generally translate into overall well-being of people. The advanced technology now available can delay death and prolong life in many cases. Many of these added years are, however, more painful or of increasingly restricted activity. Years are added to life but not life to years. Thus, the concept of healthy life expectancy was evolved. This is the average number of years that a person is expected to live in good health. Good health is subjective and defined variously by different workers. In the context of old age, this can be active life expectancy, that is, the period when a person can independently carry out the daily chores of life—walking, bathing, toileting, dressing, etc. One variation is **disability-free life expectancy** (DFLE). When considering all age groups in general, the period of disease and infirmity is excluded for calculating DFLE. This period can arise from acute diseases that occasionally occur for a short duration such as cough and cold, diarrhea, cholera, typhoid, and malaria, as well as from the chronic diseases such as ischemias, malignancies, tuberculosis, and ulcers that cause long-term sickness. Disabilities such as of vision, hearing, and movements are also counted. The procedure is basically the same as in Table 18.1 after such exclusion. However, estimating the duration of disease and infirmity in lifetime is a big challenge. A more popular version is the **health-adjusted life expectancy** (HALE) where the period of ill health is adjusted according to the severity of condition and equivalent years in good health lost are computed similar to the years of disability calculated for DALYs (Chapter 11). This is also popularly known as healthy life expectancy. Nearly 10%–15% of healthy equivalent of life is lost due to various conditions of ill health in different countries [4].

18.1.2.3 Application to Other Setups

As remarked earlier, the life table is a fairly general method and is used in a wide variety of situations. The method can be used for any **arrival–departure process**, of which the birth–death process is a particular case. The outcome of interest can be remission or recurrence or any such event. Note that you can be flexible with regard to the arrival time, which can vary from subject to subject. In Table 18.1, the population was fixed and subject to departures only. But the structure could be that subjects join the group at different chronological points of time, remain in the group for varying periods, and leave at different points of time. Tietze [5] proposed a life-table-like method for acceptors of temporary methods of birth control. For example, oral contraceptive users start taking the pill at different points of time, continue taking it for different periods, say, from 1 to 36 months, and then stop for a variety of reasons (accidental pregnancy, appearance of side effects, planned pregnancy, etc.). The life table method can be used to assess the expected duration of continuation of intake in this situation. Pharoah and Hollingworth [6] used the life table survival to workout the cost-effectiveness of statins in lowering serum cholesterol concentration in people at varying risk of fatal cardiovascular disease in the United Kingdom. The other important application of the life table method is in studying the pattern of survival. This is described next.

18.2 Analysis of Survival Data

Survival data arise when the aim is to study the time elapsed from some particular starting point to the occurrence of an event. Thus, a follow-up is an integral part of survival studies. The term "survival" presumes that the event of interest is death; in practice, this could be any failure, such as occurrence of metastasis, toxicity, or relapse, or any success, such as recovery, discharge from the hospital, or disappearance of a complaint. The term survival is used generically to measure time to any event under investigation. Other terms are waiting time, failure time, and transition time.

Although the duration of survival can be studied in terms of probability of outcome within a specified duration, such as chance of recurrence of sputum positivity in tuberculosis within, say, 6-month period after the treatment was discontinued in the middle of the course, this is a by-product of the survival analysis that models the entire pattern of survival from the beginning to the end point. When a single summary measure is required, the choice goes to median survival time instead of the mean. Median is favored because of highly skewed distribution of survival time in most situations. Several other features of survival are described in this section.

Besides modeling the survival pattern over a period of time, the other objectives of survival analysis are (1) to investigate factors that influence the duration of survival, (2) to compare two or more modalities for survival pattern, and (3) to estimate the future survival of individuals or groups with specified features.

Knowledge of survival pattern helps patients and physicians to decide which treatment or health strategy to prefer and when. For example, short-term survival may be better (in terms of percentage) with one regimen and long-term survival with another. The survival patterns are lessons to healthcare providers and seekers about what to expect in specific cases. The actual experience in individual cases would be different but not too much if the survival curves are valid and reliable.

18.2.1 Nature of Survival Data

Consider an example of the follow-up of breast cancer cases. If the follow-up is for a period of 6 years, there might be cases that are still alive at the end of the follow-up. Their duration of survival will not be known except that the survival is for *at least* 6 years. Also, in such a follow-up study, there is always a likelihood that some patients withdraw or are lost to follow-up. If a patient was alive at the 24-month visit and lost thereafter, the survival duration again is not known. These two situations give rise to censoring of the observations. These are also called **incomplete segments**. Because of such censoring, the classical method of life tables discussed in the previous section is not directly applicable. A modification is needed. The technique used to analyze such data is called **survival analysis**.

In the case of survival studies, not all the subjects of a cohort necessarily begin at the same point of time. The subjects can join and leave at any point. The joining time is considered time zero, and all durations are measured from this point. Other features of survival time can be described as follows.

18.2.1.1 Types of Censoring

The kind of censoring mentioned in the preceding example on breast cancer is called **right censoring**. In this case, the survival in censored cases is
known to be greater than the specified duration, but the exact duration is not
known because the event had not occurred when the study stopped. Right
censoring is the most common type of censoring for survival time so much
so that the word right is dropped and such values are referred to as just
censored. This kind of censoring after the durations are ordered from minimum to maximum leads to the situation as shown in Figure 18.1a. Hollow
diamond in Figure 18.1b represents right censoring. In cases D, F and H, this
occurred because the study has ended, whereas in case E, it is due to loss of
the subject for follow-up.

In **left censoring**, the beginning time is not known though the end time is
known. This also is an incomplete segment, but this kind of censoring is rare
in survival studies. This is shown for subject A in Figure 18.1a.

The third is **interval censoring**. This occurs when the time to event is
known to have occurred within a specified time interval but the exact time
is not known. Sometimes, continuous vigil on all subjects is impractical or
too expensive. If you are undertaking periodic visits to inspect a process,
the exact time of occurrence will not be known but only that it has occurred
sometime since the previous visit. If a group of diabetic persons are being
assessed quarterly for the development of retinopathy beginning with the
diagnosis of diabetes, the elapsed duration will be known only in quarterly
intervals. The exact day or week will not be known.

For the methods discussed in this chapter, the censoring scheme should
have nothing to do with the future survival or time to event. This means
censoring should not distort the information regarding duration in that
group. This condition is generally fulfilled in right-censored values

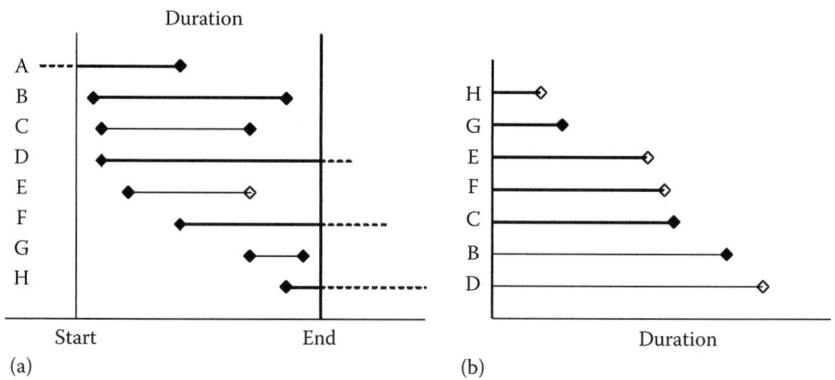

FIGURE 18.1
Censored and complete values (a) before ordering and (b) after ordering.

arising from the termination of the follow-up. But caution is required in those lost to follow-up for other reasons. In the case of prostate cancer, if a man is not traceable after a certain period, this should not be because he has moved to his family in the last moments of life or has dropped out of the study because he felt he has recovered. The implicit assumption in survival analysis is that the prospects of survival of censored cases are the same as that of uncensored cases. You should be convinced that this condition is met before using the methods of survival analysis discussed here.

18.2.1.2 Collection of Survival Time Data

Survival time data can be obtained in a variety of ways. The method of data collection can have important implications on the interpretation and sometimes on the method of analysis because different collection methods give rise to different kinds of censoring.

The most convenient method is to take a cross-sectional random sample of subjects who have the condition of interest and inquire when the condition started. This can be done when the starting point is known to subjects and they are capable of reporting it correctly. Thus, this method can be adopted when the beginning point is well defined, for example, by appearance of complaints. It can be defined by laboratory assessment when the concerned laboratory report is available with each subject. Any other record of interest may also help to identify the time of beginning of the condition. If there is no follow-up, the end point will remain unknown. One alternative is to consider the duration for all subjects as censored at the time of inquiry. This for *all* subjects is not admissible. In fact, incomplete segments for survival analysis should not be far too many, certainly not more than the complete segments. If there are no or very few subjects with complete segments, the pattern would not be known, and no worthy analysis can be performed.

The only way to assess the end point in case of a cross-sectional sample is to follow up at least half the subjects till the end point is reached. This should be a random subsample so that no bias creeps in. Because of the follow-up the strategy does not remain cross-sectional but hopefully this would still provide a representative picture of the target population.

The second method of collection of survival time data is to enquire from a sample that they ever (or in previous few years) had the condition of interest—when it started and when it finished. This entirely depends on proper recall and again can be adopted for those conditions only for which the beginning and the end points are known to the subjects.

The third method is to take a random sample of those who have just experienced the end point and find out when the condition started. For example, this can be adopted for death due to Alzheimer's disease. The time of start of the disease can be obtained by inspecting records. A great advantage of this method is that there are no incomplete segments. See if it is easier for you operationally to assess survival time in this manner.

The most satisfying but, perhaps, the most difficult method is to enroll people at the beginning of the condition and follow them up till the end point. If the end point is death, the total follow-up period could be quite long. Thus, the observation can be terminated after a prefixed reasonable time so that most subjects are able to reach the end point. Some, who do not, could be considered as censored values.

Before proceeding ahead with survival analysis, assess that there is no discernible pattern in the durations or in reaching to the end point when arranged by enrolment. For example, you can plot them as in Figure 18.1a. After reordering by date of entry, confirm that deaths and durations of survival are randomly distributed. Early recruiters should not have unusually high (or low) death rates or unusually long (or short) duration of survival.

18.2.1.3 Statistical Measures of Survival

Now that you have understood the censoring, you may realize the problems in computing mean survival time in this case. In Figure 18.1, the survival duration of subjects D, E, F, and H is not known. If these are ignored, the mean of the remaining three known durations can be unreliable and biased. Unreliable because the n available now is only three, whereas actually there are eight subjects in this study. Biased because the subjects excluded from this calculation due to incomplete observations could be very different. If this is duration of survival after a treatment, the persons alive at the end of follow-up are exactly those that may have benefited. If these censored observations are included with truncated values, the means would obviously be an underestimate. Survival duration is also known to have highly skewed distribution since some people tend to survive for long duration. And for such highly skewed values, you know mean is not a valid representative value.

If mean is not appropriate for survival, which summary measure do you use? The choice immediately falls on median, which is much less affected by such vagaries. **Median survival time** is indeed used as a summary measure to get feeling of duration of survival and to compare two or more groups, although you can see that this too is not completely free of bias because of incomplete segments.

For varying durations, you can think of person-years as the denominator and come up with an average such as 2.7 deaths per person-year. Although person-year can take care of the incomplete segments, this annoyingly considers first year after treatment on the same pedestal as, say, fifth year. Clearly, this is not true for survival. The risk of death increases as the time passes—one because of advancing age and two because the effect of treatment generally wanes. Thus, a measure such as death per person-year is also ruled out for survival studies.

Besides median survival time, the other appropriate measure is survival rate such as 5-year survival rate. This can be adjusted for censored values as discussed next and is a commonly used parameter in survival studies.

18.2.2 Survival Observed in Time Intervals: Life Table Method

Depending upon how time is measured, two methods are generally used in survival studies: the life table method and the K-M method. Both are nonparametric.

18.2.2.1 Life Table Method

The life table method of survival analysis is explicitly designed to handle the situation of interval censoring where the number of subjects at risk and those dying pertain to intervals of time. This will happen when, for example, you are visiting an experimental site each morning and counting the number of rats dead that were administered a particular dose of toxic substance. The exact time of death of rats will not be known but the duration of survival will be recorded as 0–1, 1–2, 2–3 days, and so on. Thus, time is divided into bands. Such observations in intervals may be cost-effective in some situations, particularly when n is large. Whereas uncensored values are easily handled, the difficulty arises for censored values in case of interval data. These pertain to those who vanish during the course of the study. You would not know whether they survived the interval or died by the time the interval ended.

In the life table method, lost-to-follow-up subjects are considered to have been lost at the midpoint of the interval. This is a crude adjustment but works well in situations in which the dropouts are uniform over the interval. Most practical situations conform to this requirement. Other incomplete segments are treated in a similar fashion. Thus, in this case, if there are a total of K intervals of observation,

$$p_k = \frac{n_k - d_k - c_k/2}{n_k - c_k/2}; \quad k = 1, 2, \ldots, K$$

$$= 1 - \frac{d_k}{n_k - c_k/2}, \tag{18.7}$$

where
  p_k is the proportion surviving the kth interval among those who survived $(k-1)$st interval (the estimated conditional probability)
  n_k is the number of subjects known to be alive at the beginning of the kth interval
  d_k is the number of deaths in the kth interval
  c_k is the number of subjects with incomplete segments in the kth interval whose fate is unknown

The numerator of the formula given in Equation 18.7 is the number of survivors and the denominator is the number of subjects at risk. Both contain half of those who have incomplete segments. Once this adjustment is done, the life table method is nearly the same as that already described for expectation of life.

The life table method is also called the **actuarial method** because actuaries are particularly interested in the probability of survival of a person the next one full year once the person has reached a particular age.

18.2.2.2 Survival Function

The cumulative probability of surviving the kth interval can be estimated using Equation 18.7. By-product rule for conditional probabilities, this is

$$s_k = p_k p_{k-1} \cdots p_2 p_1; \quad k = 1, 2, ..., K. \tag{18.8}$$

This probability of survival can be obtained for each of the K intervals. The formula given in Equation 18.8 is called the survival function. Its plot versus time k is called the survival curve. Since each $p_k \leq 1$, you can see that the survival probability s_k is nonincreasing and will eventually decline as time passes. Statistical software can easily do these calculations and plot the survival curve. Note that this method is nonparametric as it does not depend on any particular form of distribution of survival period.

Example 18.1: Survival following Mastectomy for Breast Cancer

Consider the following data on survival time of 15 patients following radical mastectomy for breast cancer. The study started in January 2008 and continued till December 2011. Thus, the maximum follow-up was 4 years. However, many patients joined the study after January 2008 as and when radical mastectomy was being performed.
Survival time (months):

6	8	20	20	20+	24+	25+	30+	35+	37	37	38+	40+	42	45+

where + means that the patients are lost to follow-up or not followed up after this period. They are the incomplete segments. Their exact survival time is not known but it is at least as many months as shown. The patients are deliberately ordered by survival time. This order is not important but makes the presentation simple. The sample size is small and there are many incomplete segments. This is not desirable for survival analysis, but the data are still adequate to illustrate the method.

In the case of life table method, the duration of survival is grouped. The calculations given in Table 18.3 are for 12 month grouping. Such a grouping is not needed in this case because the exact survival duration is known. But groups are used in this example to illustrate the life table method. The last column of this table gives the survival pattern at different time points.

TABLE 18.3

Illustration of Life Table Method for Calculation of Survivors

Interval Number k	Months after Surgery	Known Alive at the Start of the Interval n_k	Died during the Interval d_k	Patients with Incomplete Segments c_k	At Risk in the Interval $n_k - c_k/2$	Survivors $n_k - d_k - c_k/2$	Proportion Surviving the Interval $p_k{}^a$	Proportion Surviving since the Beginning $s_k{}^b$
1	0–11	15	2	0	15.0	13.0	$13/15 = 0.867$	0.87
2	12–23	13	2	1	12.5	10.5	$10.5/12.5 = 0.840$	0.73
3	24–35	10	0	4	8.0	8.0	$8.0/8.0 = 1.000$	0.73
4	36–47	6	3	3	4.5	1.5	$1.5/4.5 = 0.333$	0.24

[a] p_k as in the formula given in Equation 18.7.
[b] s_k as in the formula given in Equation 18.8 (e.g., $s_3 = 1.00 \times 0.840 \times 0.867 = 0.73$).

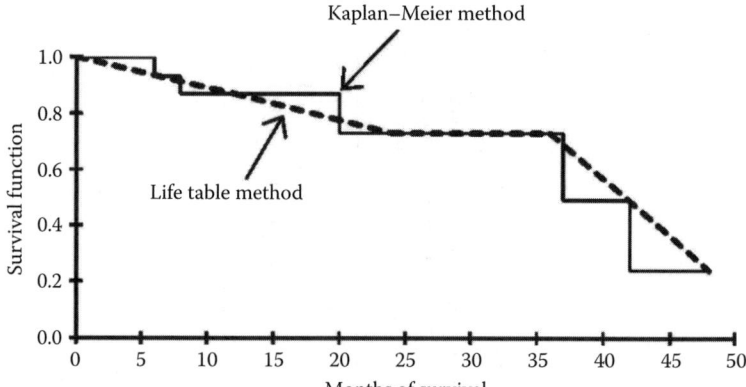

FIGURE 18.2
Survival function of breast cancer cases in Example 18.1: life table and K-M methods.

The survival curve is shown by dotted lines in Figure 18.2. Median survival time corresponds to survival function = 0.5. This is the same as the time at which 50% survive only if there are no censored observations. Thus, 50% survival and survival function = 0.5 can be different. More exactly, median is calculated as the smallest survival time at which survival probability is at least 0.5. From the figure, this can be worked out to nearly 43 months in this example.

SIDE NOTE: Since the duration of survival time is generally highly skewed, mean is not used. However, if needed, mean survival time is estimated by the area under the survival curve. Some softwares use only uncensored time points for calculation of mean but that tends to give a biased estimate.

For some conditions, such as peritonitis, deaths or failures are rapid and the survival curve shows quick decline. For others, such as kidney diseases, it may remain steady. Modeling survival rate on duration can sometimes help in understanding the underlying mechanism or in comparing the pattern in two or more conditions. This is discussed next under continuous time models. Failure distribution and hazard function are also discussed.

Software illustration: Appendix C contains an example on duration of breast feeding that illustrates how to do survival analysis by life table method using SPSS software.

18.2.3 Continuous Observation of Survival Time: Kaplan–Meier Method

Now, dispense with the restriction that survival is recorded in time intervals. Assume that continuous monitoring is done and the exact time of death is recorded. Analysis of such survival data is done with the help of the K-M method.

18.2.3.1 Kaplan–Meier Method

When exact duration of survival is recorded, there is no need of an adjust-ment of the type used in the life table method. Instead, the censored observa-tions are considered only till the time the subjects were last seen alive, and after that they are ignored. The probability of survival is computed for each observed duration in place of each interval. Survival rate at time t in this case is the proportion surviving longer than t. To estimate this, arrange the subjects according to the known duration of survival, including the censored duration. Corresponding to the formula given in Equation 18.7, now for T distinct time points,

$$p_t = \frac{n_t - d_t}{n_t}; \quad t = 1, 2, \dots, T; \tag{18.9}$$

where
  p_t is the proportion surviving the tth time point among those who sur-vived $(t-1)$st time point
  n_t is the number of subjects at risk of death at the tth time point, that is, those who are still being followed up (this excludes those with incom-plete segments, i.e., $n_{t+1} = n_t - c_t - d_t$); c_t is the number of subjects with incomplete segments
  d_t is the number who died at the tth time point

The estimated proportion surviving the tth time point is

$$s_t = p_t p_{t-1} \cdots p_2 p_1; \quad t = 1, 2, \dots, T. \tag{18.10}$$

This is the estimated survival function in this case. It is basically the same product as the formula given in Equation 18.8 but is now computed for each time point instead of the time interval. Only unique time points are considered. If a time point is applicable to two or more subjects, it is counted only once. This method requires the calculation of as many sur-vival rates by the product rule given in Equation 18.10 as there are events, unless several events occur at the same time. Hence, the K-M method is also called the **product-limit** method. The larger the T, the smoother the survival curve. Example 18.2 illustrates the method for the same subjects as in Example 18.1.

The K-M method for estimating survival probability at the last time point is usually very unreliable because of heavy censoring toward the end of the trial.

Example 18.2: Kaplan–Meier Survival of Breast Cancer Cases in Example 18.1

Consider the same duration of survival of breast cancer patients after radical mastectomy as in Example 18.1. Now there is no grouping of survival time. Calculations as per the K-M method are given in Table 18.4.

The patients with incomplete segments do not contribute to the calculation of s_t after they are lost. Thus, the rows such as those corresponding to the duration of 24–35 months in this case can be deleted. Note how p_t is calculated for survival period 20 months where there are two deaths and one incomplete segment. At this point, of the 13 patients, 11 survived and 2 died, and the patients with incomplete segments are not counted in this calculation. Thus, $p_3 = 11/13$. Similar are the other calculations. The survival curve is shown in Figure 18.2.

SIDE NOTE: Since time is continuously observed in this setup, many deaths at one time point are shown by a step ladder. The median duration corresponding to 50% survival (0.5 on y-axis) is nearly 37 months. This is very different from the 43 months arrived at by using the life table method.

Both the life table and K-M methods use all the censored durations but the life table method assumes half intervals for censored observations. Thus, of the two, the K-M method is considered a better method. In any case, as you can easily appreciate, observation of values in intervals rather than exact values implies loss of information.

The validity conditions of K-M are same as already noted, namely, (1) censoring is unrelated to prognosis or survival in comparison with the other subjects in the group, (2) chance of survival of subjects enrolled early in the study is the same as of those recruited late, and (3) duration is exactly recorded and not in intervals.

18.2.3.2 Using the Survival Curve for Some Estimations

When based on adequate sample, a survival curve can be used to estimate a variety of parameters. Figure 18.3 is a graphical representation of how duration with 80% survival, median survival time, and 5-year survival rate can be obtained. As always, the graphical method provides approximate values although they may be adequate for many practical applications.

18.2.3.3 Standard Error of Survival Rate

As you might have expected, if the expression for survival function is messy, the standard error (SE) cannot be simple. However, if deaths are not too many at any time point and if you started with sufficient number of subjects so that the number of survivors remains reasonably large at the end, the SE can be approximated by

$$\text{SE}(s_t) = \sqrt{\frac{s_t(1-s_t)}{n_t}}.$$

TABLE 18.4

Illustration of K-M Method for Calculation of Survivors

Patient Number	Months of Survival	Time Point t	No. of Patients at Risk (at the Beginning) n_t	Deaths at the ith Time Point d_t	Number of Survivors $n_t - d_t$	Proportion Survived ith Time Point p_t^a	Proportion Survived since the Beginning s_t^b	Number with Incomplete Segments c_t	Number of Known Survivors $n_t - c_t - d_t = n_{t+1}$
1	6+	1	15	1	14	14/15 = 0.933	0.93	0	14
2	8+	2	14	1	13	13/14 = 0.929	0.87	0	13
3,4,5	20+	3	13	2	11	11/13 = 0.846	0.73	1	10
6	24+	4	10	0	10	10/10 = 1	0.73	1	9
7	25+	5	9	0	9	9/9 = 1	0.73	1	8
8	30+	6	8	0	8	8/8 = 1	0.73	1	7
9	35+	7	7	0	7	7/7 = 1	0.73	1	6
10,11	37	8	6	2	4	4/6 = 0.667	0.49	0	4
12	38+	9	4	0	4	4/4 = 1	0.49	1	3
13	40+	10	3	0	3	3/3 = 1	0.49	1	2
14	42	11	2	1	1	1/2 = 0.50	0.24	0	1
15	45+	12	1	0	1	1/1 = 1	0.24	1	0

[a] p_t as in the formula given in Equation 18.9.

[b] s_t as in the formula in Equation 18.10 (e.g., $s_3 = (11/13) \times (13/14) \times (14/15) = 0.73$).

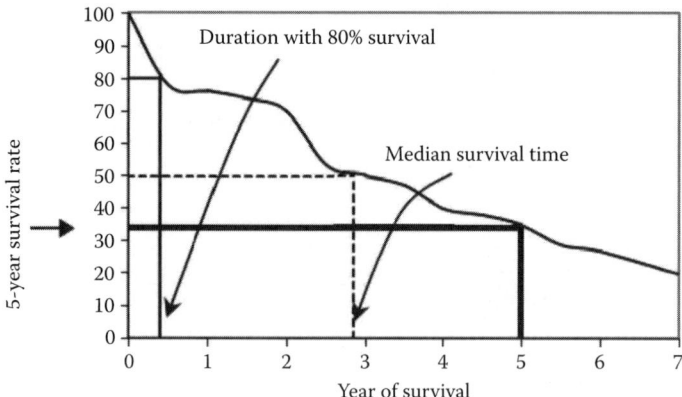

FIGURE 18.3
Using survival curve to estimate median survival time and duration with 80% survival.

This is the same as you have seen for proportions. The SE would be different for different points of time just as survival s_t and number of subjects n_t are. For life table survival also, the approximate SEs are the same—just that the subscript t is replaced by subscript k.

In view of small n_t in Table 18.4, you can see that this SE would not be applicable for these data. Even when n is large, note from SE that it increases (the precision declines) with time as the survival reduces and it is based on fewer and fewer subjects.

This SE can be used to build confidence interval around any survival rate.

18.2.3.4 Hazard Function

When the exact survival duration is considered, instead of an interval, for a large number of subjects, the survival curve can be modeled by a mathematical equation that will graphically translate into a smooth curve. In many situations, particularly for devising strategies to prevent mortality, failure function is considered more useful instead of survival function. Failure function is the complement of survival function.

If T is the notation for varying time of survival (different from T used earlier for the number of time points), failure function is $P(T < t)$. This is the probability that the time of survival is less than a specified time t. For example, after heart transplantation, $P(T \geq 2$ years) is the probability of survival for at least 2 years and $P(T < 2$ years) is the probability of death before 2 years. The latter is a cumulative probability and is nondecreasing. As time passes, the chance of failure increases and reaches 1 as t tends to become large.

Sometimes, the interest is not in the probability but in the rate of failure in a particular time interval. **Hazard** is the "limit" of this rate as this interval approaches zero. Statisticians use the concept of limit to obtain a mathematical form that works at a particular point of time when time is continuously observed.

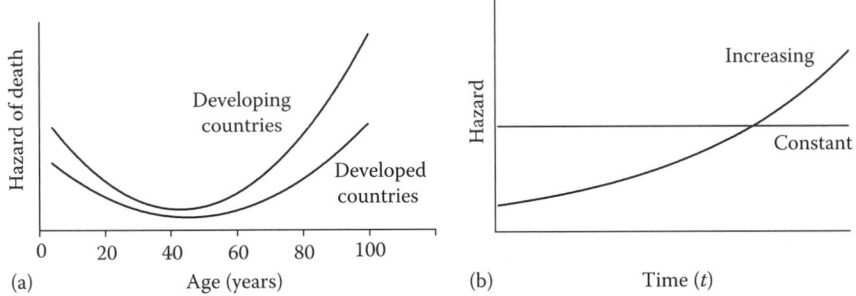

FIGURE 18.4
(a) Hazard of death at different age in general population. (b) Constant and increasing hazards.

The hazard at time t is the rapidity of failure next instant. This is obtained by dividing the conditional probability of failure next instant by the length of that instant. Thus, hazard is not the same as probability. It can exceed 1. Hazard function describes the failure rate per unit of time at different points of time.

You will appreciate hazard better in the context of general population where the hazard of death is high at the beginning of life (neonatal period), very low for age 5–59 years, and steeply increases thereafter. Thus, it has a bathtub shape (Figure 18.4a). The bend is more acute for developing countries than for developed countries.

As just mentioned, hazard of death is not uniform through the life in the general population. Though rare, if it is indeed uniform for some condition, such as for a disease that strikes young and the old at the same rate, the failure function $P(T < t)$ takes the form

$$\text{Exponential: } F(t) = 1 - e^{-\lambda t}. \tag{18.11}$$

In reference to hazard, it is easier to call failure function as the **cumulative hazard function**. The function given in Equation 18.11 is called an **exponential function**. A steeply increasing or gradually increasing form can be obtained by suitably choosing the parameter λ. In the case of exponential distribution, λ is the hazard at time t and it can be more than 1. Note that this is same at all time points in this case.

Since hazard is not uniform in most situations, the exponential form of cumulative hazard does not provide an adequate representation. Nonuniform hazard can be understood further by the example of an automobile whose failure rate in the beginning after coming out of the showroom is low and high in, say, the seventh year of running. In this case, the hazards rapidly increase as time passes. In the case of heart transplantation, the initial few days and weeks are critical. After this period, the hazard declines at least for some time. The cumulative hazard function that models varying hazards over time is

$$\text{Weibull: } F(t) = 1 - e^{-\lambda t^{\alpha}}. \tag{18.12}$$

When $\alpha = 1$, this reduces to exponential function. Thus, there is no need to consider exponential separately; it is just a special case of Weibull. Under this model, the hazard increases with time when $\alpha > 1$ and decreases when $\alpha < 1$. Depending upon the disease or the condition, appropriate values of the parameters λ and α can be chosen. For example, the Weibull model was used to conclude that the hazard of death in young patients of Ewing's sarcoma remains greatest but relatively constant over the first 2 years postdiagnosis and then declines to a lower value for the next 3 years before settling down to a low level at about 5 years [7].

The third commonly used cumulative hazard function is

$$\text{Log normal: } F(t) = 1 - \Phi\left[\frac{\ln t}{\sigma}\right], \tag{18.13}$$

where Φ is the cumulative probability of Gaussian distribution. This is appropriate when the duration of survival has a hugely skewed distribution to the right—long and very long durations are also present though mostly they are not very long. Life span of cells may have this pattern. Data from the Surveillance, Epidemiology, and End Results (SEER) revealed in a 27-year follow-up that the survival times of cancer patients who died of their disease can be considered to follow a log-normal distribution [8]. The parameters of the log-normal distribution were different for cancers of different organs. The survival duration for 97.75% cases of pancreatic cancer was less than 2.6 years and for cancer of the salivary gland less than 25.2 years. Note the striking difference. These are the durations one needs to consider while planning a survival study on these cancers if most cases are required to be followed-up till the end.

All these are parametric formulations. One rather simple nonparametric estimate of cumulative hazard function is

$$\hat{H}(t) = \sum_{T \leq t} \left(\frac{d_t}{n_t}\right).$$

Many workers feel that parametric models impose difficult-to-justify restrictions. Also, there are many options and prior information is rarely available to make a correct choice. You may have to try more than one model to get an adequate fit. Semiparametric model, particularly Cox discussed later, is more flexible and safer. This has become a standard choice for many applications.

If your interest in a parametric model persists, the estimates of the parameters can be generally obtained by least square regression. Luckily, the three commonly used parametric models just discussed can be converted to linear form by suitable transformation. For example, Weibull in Equation 18.12 reduces to

$$-\ln[\ln(1 - F(t))] = \ln\lambda + \alpha\ln t.$$

Thus, usual regression estimates can be obtained.

While analyzing survival data, if the hazard function is of interest, choose the one that meets its eligibility requirement. If you are examining hazard function used by someone else, view it in the light of the preceding discussion.

18.3 Issues in Survival Analysis

This section covers three unrelated topics. The first is the comparison of two survival curves obtained for the treatment group and the control group. Here you will find the log-rank test, which is so often used for this purpose. The second is the study of factors affecting survival pattern. The main method for studying this is the Cox proportional hazards model for survival. The third deals with the calculation of sample size for survival studies.

18.3.1 Comparison of Survival in Two Groups

The conventional method of comparison of survival curves is by comparing survival rates at specific times, for example, 2-year conception rate in subfertile women in one group treated with regimen-1 and another group treated with regimen-2. Regimen-2 could be no treatment or placebo (control). The other method is comparing the overall survival curves of the type shown in Figure 18.5 for a test and control group.

18.3.1.1 Comparing Survival Rates

Survival rates at any point of time are like proportions and the comparison of two groups can be easily done by chi-square when the usual validity conditions of this test are fulfilled. These survival rates are calculated

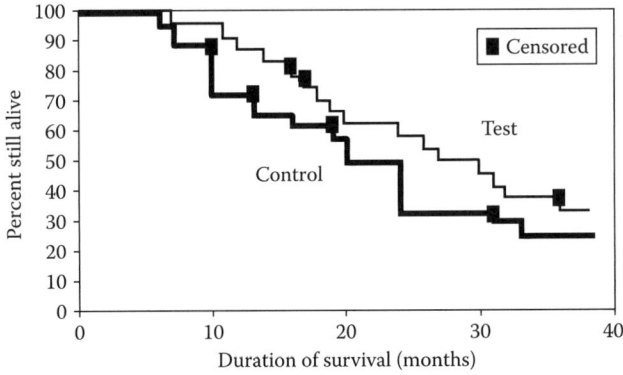

FIGURE 18.5
Survival curves with two regimens.

by the life table method when the observation of survival time is in the intervals and by K-M method when it is continuously observed. Survival–deaths in the two groups will give a 2 × 2 table, which will provide chi-square with 1 df as usual. The procedure remains essentially the same if there are more than two groups. A weakness of the procedure is that the time points chosen for calculating survival rate can be arbitrary and different time points can give different results. The procedure to compare the overall survival pattern is given next, but this procedure fails to detect if a big and significant difference exists at only one or two specific time points and not at other points.

Example 18.3: Five-Year Survival Rates in Gastric Cancer Patients with Different Types of Surgical Resection

Survival in gastric cancer patients depends on a large number of factors, but this example is restricted to the type of surgical operation. Zhang et al. [9] have presented 5-year survivals as shown in Table 18.5.

Censored observations are excluded. Thus, the last two columns of the table is a regular 3 × 2 contingency table and the statistical significance can be tested by the usual chi-square. In this case, $\chi^2 = 31.66$, which is highly significant ($P < 0.01$) for 2 df. Thus, there is sufficient evidence to conclude that gastric cancer cases with three types of operations have different 5-year survival rates.

SIDE NOTE: With such a large n as in this example, statistical significance is a foregone conclusion. The actual utility of this type of analysis is that the survival in proximal subtotal gastrectomy is nearly half of that in distal gastrectomy. Deaths are 79.9% in proximal and 62.8% in distal. This tends to quantify the hazard of death and says that the hazard ratio for death in proximal operation is nearly 1.3 relative to the distal operation (this is based on deaths and not on survivals). Not many studies of this type aim at this kind of a useful conclusion.

TABLE 18.5

Five-Year Survival Rate of Gastric Cancer Cases with Different Types of Operation

Type of Operation	Total Cases	Five-Year Survival		
		Rate (%)	Survived	Died
Distal subtotal gastrectomy	678	37.2	252	426
Total gastrectomy	466	32.4	151	315
Proximal subtotal gastrectomy	354	20.1	71	283
Total	1498	31.6	474	1024

18.3.1.2 Comparing Survival Experience: Log-Rank Test

The log-rank test is designed to compare the overall pattern of two survival curves without reference to any particular point of time. This is a large sample test and uses chi-square criterion to arrive at significance or nonsignificance of the difference. The categories for chi-square are defined by each of the *ordered* (ranked) failure times for the entire dataset. This is a nonparametric test and is not affected by the shape of the survival curves or distribution of survival times. It requires that censoring in both the groups is independent of survival process. Patients getting worse or better treatment should not dropout quicker than others. Also, log-rank method is valid when all time points are equally important. Sometimes, time points with higher number of subjects at risk are given more weight. In that case, another method such as **Breslow test** should be used. This is also known as Gehan–Breslow test and generalized Wilcoxon test. For details, see Hosmer and Lemeshow [10]. Since Breslow test gives more weight to the time points with higher number of subjects, initial time points tend to decide the statistical significance where the number of subjects at risk is higher. As time passes, subjects die or dropout and the numbers become smaller. As a middle path between log-rank and Breslow, some prefer **Tarone–Ware test** that gives weight proportional to $\sqrt{n_t}$, where n_t is the number at risk at time point t.

The null hypothesis is that the survival curves are identical in the two populations. This implies that the probability of survival (or of death) at *each* point of time is the same in one group as in the other. Under this null hypothesis, the expected number of deaths is calculated for each time using the combined experience in the two groups:

$$E_{1t} = \frac{n_{1t}}{n_{1t} + n_{2t}}(d_{1t} + d_{2t}) \quad \text{and} \quad E_{2t} = \frac{n_{2t}}{n_{1t} + n_{2t}}(d_{1t} + d_{2t}), \qquad (18.14)$$

where
n_{1t} is the number of subjects at risk at tth time point in the first group
n_{2t} is the number of subjects at risk at tth time point in the second group
d_{1t} and d_{2t} are the observed number of deaths at tth time point in the first and the second group, respectively

The method is illustrated in Example 18.4 and the calculations are shown in Table 18.6. When any survival time is censored, that individual is considered to be at risk of dying till the time of censoring but is ignored for subsequent time points. This is the same as done for obtaining the K-M curve. The sums ΣE_{1t} and ΣE_{2t} give the expected frequencies for use in chi-square.

TABLE 18.6

Calculation for Log-Rank Test for Example 18.4

Time (Months) T	Time Point[a] i	Number of Deaths			Number Censored		Number at Risk			Expected Frequency	
		Group-I d_{1i}	Group-II d_{2i}	Total $d_{1i}+d_{2i}$	Group-I c_{1i}	Group-II c_{2i}	Group-I n_{1i}	Group-II n_{2i}	Total $n_{1i}+n_{2i}$	Group-I E_{1i}	Group-II E_{2i}
1	1	0	1	1	0	0	137	79	216	$(137/216) \times 1 = 0.634$	$(79/216) \times 1 = 0.366$
2	2	0	1	1	0	0	137	78	215	$(137/215) \times 1 = 0.637$	$(78/215) \times 1 = 0.363$
3	3	1	0	1	0	0	137	77	214	$(137/214) \times 1 = 0.640$	$(77/214) \times 1 = 0.360$
4	4	0	1	1	0	0	136	77	213	$(136/213) \times 1 = 0.638$	$(77/213) \times 1 = 0.362$
5	5	0	3	3	0	0	136	76	212	$(136/212) \times 3 = 1.925$	$(76/212) \times 3 = 1.075$
8	6	0	1	1	0	0	136	73	209	$(136/209) \times 1 = 0.651$	$(73/209) \times 1 = 0.349$
9	7	0	1	1	0	0	136	72	208	$(136/208) \times 1 = 0.654$	$(72/208) \times 1 = 0.346$
10	8	0	1	1	0	0	136	71	207	$(136/207) \times 1 = 0.651$	$(71/207) \times 1 = 0.343$
12	9	2	0	2	2	0	136	70	206	$(136/206) \times 2 = 1.320$	$(70/206) \times 2 = 0.680$
13	10	0	1	1	4	2	132	70	202	$(132/202) \times 1 = 0.653$	$(70/202) \times 1 = 0.347$
14	11	0	1	1	0	2	128	67	195	$(128/195) \times 1 = 0.656$	$(67/195) \times 1 = 0.344$
16	12	0	0	0	3	0	128	64	192		
17	13	0	2	2	2	0	125	64	189	$(125/189) \times 2 = 1.323$	$(64/189) \times 2 = 0.677$
18	14	0	1	1	0	0	123	62	185	$(123/185) \times 1 = 0.665$	$(62/185) \times 1 = 0.335$
19	15	0	0	0	1	0	123	61	184		
20	16	0	1	1	2	2	122	61	183	$(122/183) \times 1 = 0.667$	$(61/183) \times 1 = 0.333$
21	17	0	0	0	2	0	120	58	178		
24	18	0	1	1	0	0	118	58	176	$(118/176) \times 1 = 0.670$	$(58/176) \times 1 = 0.330$
Total		3	16	19	16	6				12.390	6.610

Source: Adapted from Jeyaseelan, L. et al., *Natl. J. India*, 12, 172, 1999.

[a] At death or censoring in any of the two groups.

Example 18.4: Comparison of Survival with Different Antibody Levels in Patients with Renal Allograft

In a study of early posttransplant phase of renal allografts, peak antibody level was recorded to assess if it affects survival in the first 2 years [11]. Out of a total of 216 patients, the peak antibody level did not reach 15% in 137 patients (group-I). In the other 79 patients (group-II), the antibody level exceeded 15%.

Table 18.6 contains calculation for the two groups. Note how subjects with censored observations are excluded from the calculations. For the purpose of completeness, time = 16 months is shown but it does not contribute to the calculation as all the three subjects are censored at this time. A software will automatically do this for you. Under the null, the expected frequency (number of deaths) for group-I is 12.39 and for group-II is 6.61. The observed are 3 and 16, respectively. Subtract the expected frequencies from the observed frequencies for the two groups and get

$$(O_1 - E_1) = -9.39 \quad \text{and} \quad (O_2 - E_2) = +9.39$$

$$\chi^2 = \frac{(-9.39)^2}{12.39} + \frac{(+9.39)^2}{6.61}$$

$$= 20.46.$$

This has 1 df. The critical value for chi-square in Table B.3 is 3.84 at 5% level. Thus, $P < 0.05$. Using a software will give you $P < 0.001$. The difference in the survival pattern in the two groups is highly significant. The two survival curves are shown in Figure 18.6. You can easily see the difference.

The validity conditions of the log-rank test are the same as for the K-M curve. (1) Censored values have the same pattern of survival as the uncensored values. This means that they have same prognosis. (2) In case the

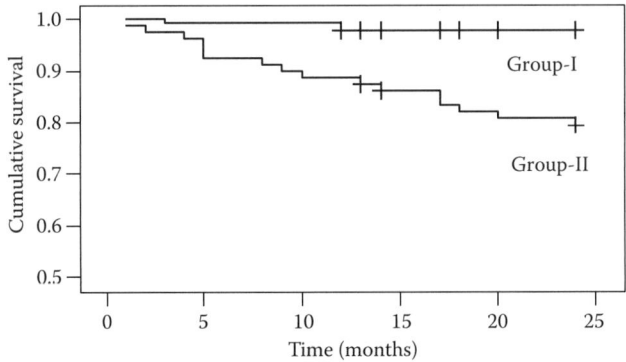

FIGURE 18.6
Survival curves for patients of renal allograft with peak antibody level <15% (group-I) and ≥15% (group-II) (+ sign for censored values).

subjects are enrolled sequentially over a period, those enrolled later have the same survival pattern as those enrolled early. This means that no innovation has occurred that may have a case mix. (3) The time of occurrence of events is exactly recorded and not in intervals. The test can be easily extended to more than two groups.

The log-rank test is most likely to succeed in detecting a difference between survival patterns in two groups when the chance of survival in one group is consistently higher or consistently lower at each time point than the other. If the survival curves cross one another, as can happen while comparing medical treatment with surgical operation, then the difference in the overall pattern may be masked. Also, this test only reveals whether the curves are different, without any implication regarding the magnitude of difference. One method for assessing the magnitude of difference is to compute ratio of the hazards in the two groups.

18.3.2 Factors Affecting Survival: Cox Model

You may have noticed that the discussion so far in this chapter has not dealt with the factors affecting the survival pattern. For example, the duration of survival after organ transplantation may depend on age of the patient, coexisting diseases, and the drug regimen followed. These factors are now called covariates. Because of highly skewed distribution and censored observations, the conventional regression approach is not applicable. Two approaches are available to study the effect of covariates on the survival duration.

18.3.2.1 Parametric Models

All of the parametric models discussed earlier, namely, the exponential, Weibull, and log-normal, can be extended to include covariates. For example, the exponential cumulative hazard model would look like

$$\text{Exponential model with covariates: } F(t) = 1 - e^{-(\beta_1 x_1 + \beta_2 x_2 + \cdots + \beta_K x_K)t}, \qquad (18.15)$$

where $x_1, x_2, \ldots, x_K$ are the covariates. Similarly, other models can also be parameterized to include the covariates. As mentioned earlier, these models work well when the cumulative hazard function indeed follows the modeled pattern. A priori this is difficult to assess, although the goodness of fit can be tested once the data are available.

A model that has found wide acceptability is the Cox model. This was introduced in Chapter 17. Now, examine its applicability to survival analysis with covariates where some observations are censored. Cox model is semiparametric since it is nonparametric for time and parametric for covariates. The following is an introductory discussion. For details, see Lee and Wang [12].

18.3.2.2 Cox Model for Survival

Recall that Cox regression mentioned earlier in Chapter 17 is for any binary outcome. When $\hat{\lambda}(t)$ is replaced by $h(t)$, this is as follows:

$$\text{Cox regression: } h(t) = h_0(t)e^{b_1 x_1 + b_2 x_2 + \cdots + b_K x_K}. \tag{18.16}$$

Consider now $h(t)$ in this equation as the hazard of death at time t and $h_0(t)$ is the baseline hazard at this time when all xs are zero. Note that $h_0(t)$ represents the effect of time alone without the covariates. Hazard has already been explained. The binary outcome is survive/dead. In other contexts, the binary outcome could be the end point reached or not reached. The model can be adapted to include censored data.

In the Cox regression (Equation 18.16), let the covariates be prognostic factors that affect survival. If you suspect that survival in males would be different than in females, sex would be a valid covariate. This type of modeling could identify the factors that have significant influence and their contribution can be quantified. Cox survival model does not require any specific form of survival over time but it does require that the hazard ratio remains unaltered over the period of follow-up (called proportional hazards) and cumulative hazard is multiplicative (additive on log scale). Note that Cox model is for hazard of death (or failure) and not for *duration* of survival.

18.3.2.3 Proportional Hazards

Proportional hazards imply that the hazard ratio remains same over the entire survival time. For example, if a person of age 60 years with hypertension has hazard of myocardial infarction (MI) 1.25 times the person without hypertension, this ratio should continue to be 1.25 at 70 years and at age 80 years if these ages are in your data. This means that the hazard of MI increases with age in the same proportion in hypertensives as in nonhypertensives. This may not hold, for example, for intricate surgery (compared with, say, medical treatment) because of high risk of death at early stages due to perioperative causes and relatively lower risk of mortality at later stages of recovery when the surgery is seen as successful. When proportionality holds, no correction is required for the varying length of follow-up of different subjects.

In terms of notations, consider two groups defined by indicator variables $x_1 = 0$ for control and $x_1 = 1$ for the treatment group. For simplicity, suppose this is the only variable under consideration. Equation 18.16 in this case gives $h(t) = h_0(t)$ for $x_1 = 0$ and $h(t) = h_0(t) e^b$ for $x_1 = 1$. Thus, the ratio of these two hazards is e^b for all t. It does not depend on time. The form of $h_0(t)$ does not matter for this ratio. This is what proportional hazards imply. This simple and useful interpretation has made hazard ratio so popular. It is because of this proportional property that the method is independent of the pattern of survival curve. Proportional hazards in the two groups would mean the plot of logarithm of hazards versus survival time would be parallel (Figure 18.7).

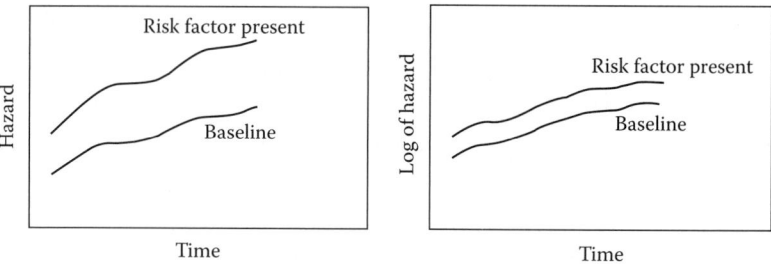

FIGURE 18.7
Hazards and proportional hazards in terms of logarithm.

Cox survival model also requires that the effect of covariates on hazard is multiplicative. This can also be stated as log-linear. To understand this, suppose now that another covariate x_2 is included. For two covariates, (x_1, x_2), Equation 18.16 gives

$$h(t) = h_0(t)e^{b_1x_1+b_2x_2}$$

$$= h_0(t)e^{b_1x_1} * e^{b_2x_2}.$$

Thus, the hazard is multiplicative of the effects of x_1 and x_2. If x_2 is also dichotomous, say, gender, with values $x_2 = 0$ for males and $x_2 = 1$ for females, this would imply that the effect of treatment in males is the same as in females. There is no interaction. If interaction exists, a third variable x_3 is introduced. This would be defined as $x_3 = x_1 * x_2$. This product has values $x_3 = 1$ when $x_1 = 1$ and $x_2 = 1$, otherwise zero. Such conditions are generally fulfilled when the covariate is time invariant.

As an illustration of how Cox model works, consider hazard of complications in case of peritonitis with different APACHE scores. If the coefficient b for APACHE is 0.0487, what does it imply? This means that if APACHE score increases by 1, the hazard ratio (relative to APACHE score = 0) of complication is exp(0.0487) = 1.050. If APACHE score increases by 8, the hazard ratio becomes exp(8 × 0.0487) = 1.476, that is, about 1½ times. In place of continuous APACHE, consider binary variable such as gender. If b for gender is 0.675 and the coding is 1 for men and 0 for women (so that women is the reference category), the hazard ratio of men to women is exp(0.675) = 1.96. Thus, men have nearly twice as much hazard of developing complication in peritonitis as in women. For these two risk factors, the model would be

$$h(t) = h_0(t)[\exp(0.0487) * \text{APACHE}] * [\exp(0.675) * \text{gender}] \quad \text{for all } t.$$

As in the case of logistic, any unusually large coefficient in Cox model or large SE should be regarded as wrong. It can arise either due to strong multicollinearity or due to extremely small number of subjects in a particular subgroup. Such instances highlight the importance of scrutinizing the data and exploring their suitability for a particular analysis. As in the regression models, the covariates in Cox model also should not have high multicollinearity.

Time invariance of covariates is quite a strict condition in the context of survival. Baseline values can always be incorporated as covariates because they remain same. Generally, life style and physiological variables change over time, while background characteristics such as gender, blood group, ethnicity, and urban/rural residence remain stable. If you are not sure, check that interaction between the covariate and time has statistically not significant coefficient. If significant, abandon the Cox proportion hazards model in favor of time-dependent covariates model (not discussed in this text).

For Cox survival model, all polytomous covariates are converted to a series of binary classes by using dummy variables. Most statistical softwares would do this once you specify that a particular covariate is polytomous.

When the aforementioned conditions are met, Cox model provides better estimates of the hazard at any point of time than the K-M method. Hazard ratio associated with a covariate is given by the exponent of its coefficient in the model. This will provide the estimate of the independent effect of that covariate after adjusting the effect of other covariates present in the model. The estimate would come from appropriate software.

Not many researchers test goodness of fit of Cox model. For test of this hypothesis, the deviance $-2\ln L$ is used as in the logistic regression. The test for the overall model is provided by the difference in deviance of the full model and the deviance for just $h_0(t)$, which measures the effect of time only. This will tell whether all covariates together are of any help or not in explaining the hazard. To assess the utility of any particular covariate, calculate $-2\ln L$ with and without that covariate and refer the difference to chi-square with 1 df. Statistics similar to R^2 in quantitative regression can also be used for this purpose. For testing significance of individual covariates, Wald statistic is preferred as it gives slightly better results. In this case, this is just $[b/SE(b)]^2$ and referred to chi-square at 1 df.

Example 18.5: Cox Model for Identification of Risk Factors of 15-Year Mortality after the Age 65 Years

Ahmad and Bath [13] obtained data from a nationally representative sample of 1042 community-dwelling people in the United Kingdom of age 65 years or more. Their survival time since 1985 was recorded with censoring in the year 2000. Data pertain to 460 independent variables on cognitive impairment, physical health, physical activity, psychological well-being, etc. Six Cox models were run that were asked to select 1, 2, 4, 8, 12, and 16 most important variables, respectively, to predict survival time. Besides age, the analysis found handgrip strength as an important marker of frailty in predicting early death. Pain in joints causing difficulty in carrying bags and self-rated activity compared to peers were important predictors of long-term mortality.

SIDE NOTE: With a sample of size 1042, the number 460 of independent variables is too large for any meaningful result.

Software illustration: For a software illustration of running a Cox model using SPSS, see Appendix C.

18.3.3 Sample Size for Survival Studies

For comparing survival pattern in one group with the other, the parameter of interest generally is the hazard ratio. Linearity is achieved by taking logarithm. Let the log of hazard ratio for detection be

$$\theta = \log\left(\frac{h_1}{h_2}\right), \tag{18.17}$$

where h_1 and h_2 are the hazards in the two groups. This is the ratio that you want to detect. Similar to the formula given in Chapter 15,

$$\text{Required number of deaths: } D = \frac{4(z_{\alpha/2} + z_\beta)^2}{\theta^2}. \tag{18.18}$$

Use this and death rate to obtain the total sample size. This will give

$$n = \frac{D}{P(\text{death})}.$$

Each group will have same sample size. Note that this assumes that the deaths will be less than the number of survivors. The formula given in Equation 18.18 is for two-sided tests. For one-sided tests, $z_{\alpha/2}$ will change to z_α.

If the most important covariate is quantitative, the formula changes to

$$D = \frac{4(z_{\alpha/2} + z_\beta)^2 \sigma_y^2}{\theta^2}, \tag{18.19}$$

where σ_y^2 is the variance of the covariate y. In case, there are several covariates such as gender and ethnicity, use

$$\text{Variance inflation factor: VIF} = \frac{1}{1-R^2}, \tag{18.20}$$

for variance, where R^2 is the proportion of variation in y explained by the covariates, same as defined in Chapter 16.

References

1. Olshanky SJ, Carnes BA, Cancel C. In search of Methusaleh: Estimating the upper limits to human longevity. *Science* 1990; 250:634–690.
2. World Bank. *World Development Report 1993: Investing in Health.* Oxford, U.K.: Oxford University Press, 1993, pp. 290–291.
3. WHO. *Global Health Observatory Data Repository, apps.who.int/ghodata* (last accessed on October 11, 2012).
4. WHO. *World Health Report 2004—Changing History.* Geneva, Switzerland: World Health Organization, 2004, pp. 126–131.
5. Tietze C. Intra-uterine contraception: Recommended procedures for data analysis. *Stud Fam Plann* 1967; No. 18 (Suppl):1–6.
6. Pharoah PDP, Hollingworth W. Cost effectiveness of lowering cholesterol concentration with statins in patients with and without pre-existing coronary heart disease: Life table method applied to health authority population. *Br Med J* 1996; 312:1443–1448.
7. Weston CL, Douglas C, Craft AW, Lewis IJ, Machin D. Establishing long-term survival and cure in young patients with Ewing's sarcoma. *Br J Cancer* 2004; 91:225–232.
8. Tai P, Yu E, Cserni G et al. Minimum follow-up time required for the estimation of statistical cure of patients: Verification using data from 42 cancer sites in the SEER database. *BMC Cancer* 2005; 5:48.
9. Zhang XF, Huang CM, Lu HS et al. Surgical treatment and prognosis of gastric cancer in 2,613 patients. *World J Gastroenterol* 2004; 10:3405–3408.
10. Hosmer DW, Lemeshow S, May S. *Applied Survival Analysis: Regression Modeling of Time to Event Data,* 2nd edn. New York: Wiley Interscience, 2008.
11. Jeyaseelan L, Walter SD, Shankar V, John GT. Survival analysis: An introduction. *Natl Med J India* 1999; 12:172–177.
12. Lee ET, Wang JW. *Statistical Methods for Survival Data Analysis,* 3rd edn. New York: Wiley Interscience, 2003.
13. Ahmad R, Bath PA. Identification of risk factors for 15-year mortality among community-dwelling older people using Cox regression and a genetic algorithm. *J Gerontol A Biol Sci Med Sci* 2005; 60:1052–1058.

19

Simultaneous Consideration
of Several Variables

In an example in an earlier chapter, "risk" of weaning a child from breast-feeding was considered to depend on weight-for-age, complementary food intake, and change in diarrheal incidence. In another example, birth weight of a child was considered dependent on the mother's weight and the father's weight. In both these examples, several variables were considered together but only one was considered a dependent variable. The others were predictors. But there are other situations where the dependent variable is a multifactorial entity in the sense that it is based on several measurements. Thyroid function is evaluated by simultaneous consideration of triiodothyronine (free T_3), thyroxine (free T_4), and thyroid-stimulating hormone (TSH). These might be dependent on age, diet, exercise, stress, etc. In another setup, outcome of a treatment is evaluated not just in terms of the extent of recovery but also in terms of time taken in recovery, nature of side effects, magnitude of discomfort experienced by the patient, convenience in administration of the regimen, etc. The predictors in this case could be treatment regimen, care provided, cooperation of the patient, severity of the condition at the time of admission, etc. An analysis of the extent of recovery alone as a dependent variable, ignoring the other aspects of outcome, or on each aspect separately, would not provide a holistic view. All aspects of outcome should be considered *together* for a total picture. Similarly, if severity of disease is an outcome of interest, this could be measured in terms of intensity and magnitude of complaints, extent of abnormality in laboratory and radiological evaluations, extent of disability, etc. All of them should be considered together. Thus, severity too is intrinsically a multivariate entity. It can be considered to depend on baseline physiological measurements of the patient, magnitude of infection, injury or toxicity, personality traits, etc.

Physical growth in infants is another example of a multivariate entity. It incorporates measurements such as weight, length, and head circumference. Dependence of these on maternal smoking during lactation could be a point of study (for one such study on Dutch infants, see Boshuizen et al. [1]). Several variables together considered dependent on another set of variables constitute one of the many situations that come under the domain of multivariate setup. There are several other types of situations that come in this domain. Some of them are discussed in this chapter.

This chapter: Mathematical aspects of a multivariate analysis are generally very complex. The attempt in this text as always is to explain the essentials in simple language. In the process, however, the intricacies tend to be oversimplified. The objective in this chapter is not to present the actual methods but only to apprise you of the situations where these methods could and should be used. The kinds of conclusions that can be reached by such methods are also indicated. It is hoped that you will then be able to identify a situation where multivariate methods can be helpful and be able to seek professional statistical assistance accordingly when needed. The contents of this chapter would not equip you with the skill to use these methods yourself unless you are well versed with statistical softwares. For details of multivariate methods see, for example, Hair et al. [2].

This chapter starts by discussing the scope of multivariate methods (Section 19.1) so that you can be confident when to expect the use of these methods. Section 19.2 considers situations where there is a set of independent variables and a set of dependent variables. It includes a brief description of multivariate multiple regression, path analysis, multivariate analysis of variance (MANOVA), and discriminant functions. These are the core multivariate methods. Section 19.3 is devoted to the search for a structure among subjects or among variables that can explain the observations. It includes the techniques of cluster analysis and factor analysis.

Analogous to the *F*-test in a univariate setup, the tests in a multivariate setup are based on a criterion such as Wilks lambda (Λ) and Pillai trace. The details are outside the scope of this chapter. Most standard statistical softwares give a *P*-value that you can directly use to draw conclusions.

A strategy sometimes used in a multivariate setup is to calculate the correlation between two *sets* of variables. This is generally done by obtaining a linear combination of variables in each set such that the usual product–moment correlation coefficient between these two combinations is maximum. This is called a **canonical correlation**. This also is not discussed in this text because of its extremely limited use in health and medicine. The interested reader is referred to Krzanowski [3].

There are several other multivariate methods not included in this chapter. Among these are correspondence analysis and conjoint analysis.

19.1 Scope of Multivariate Methods

Misuse of the term "multivariate" is quite common in medical literature. Quantitative multiple regression and logistic multiple regression are many times undesirably called multivariate. Thus, it is necessary to be clear what situation qualifies to be called multivariate and where to use multivariate methods. You should also be aware why separate methods are needed for multivariate setups.

19.1.1 Essentials of a Multivariate Setup

The distinction between a multivariate setup and a collection of univariate setups should be very clear. Consider the regression situation discussed in Chapter 16. The number of variables under simultaneous consideration in a regression situation is also more than one. In this sense, it is a multivariable setup. However, only one is dependent and the others are regressors. The regressors are considered "fixed" in that situation. Thus, they are deterministic and not stochastic. *All xs are considered known in regression. Only the response is subject to sampling fluctuation.* Although regression can be used in a cross-sectional study where both *y*s and *x*s were simultaneously observed (and thus both subject to fluctuation), it is interpreted as if *x*s were fixed. Since only one variable is considered stochastic, the regression in that chapter is essentially a univariate technique. For a genuine multivariate setup, it is essential that there are several stochastic variables.

The second requirement for a valid multivariate setup is that the stochastic variables are interrelated. Multivariate methods have special relevance when a joint conclusion is needed on several correlated variables, and this correlation should be fairly strong for multivariate methods to be really effective. Presence of statistically significant correlations between three or more variables can be examined by a test such as the one discussed by Srivastava and Carter [4]. Most statistical softwares would do this for you. If correlations exist, you can be convinced that multivariate methods are indeed needed.

In the case of growth of infants, weight, length, and head circumference are related to one another. If they are not, univariate analyses for weight, length, and head circumference can be separately done. But the conclusions so arrived at would be valid separately for weight, height, and head circumference but not jointly for growth. There are two reasons for this. First, each univariate conclusion, when based on a statistical test of hypothesis, is subject to the specified chance of Type-I error such as 0.05. The combined conclusion then would have a much higher chance of error. This situation is the same as that of multiple comparisons. Special methods are needed to ensure that the total probability of Type-I error does not become more than a predetermined threshold. The second and more important reason is that such variables are often correlated. Separate univariate analyses ignore these correlations. Special methods are required that give due consideration to the correlation structure. Multivariate methods take care of both these problems.

All the methods presented in this chapter assume that the variables under consideration are correlated. These methods generally use product–moment correlation that measures only the linear relationship. Though this includes the linear component, if any, of the nonlinear relationship also but excludes the genuine nonlinear part. Thus, if the variables are related in a nonlinear fashion, many methods of this chapter may not give adequate results.

However, in the case of a Gaussian form of distribution of several variables together, the relationship is always linear. Thus, in this case, there is no need to worry about the nonlinear form of relationships.

You may have noticed that efforts are often made to develop a composite index on the basis of several variables. Such an index is an attempt to express a multivariate phenomenon by a univariate quantity. Waist–hip ratio is an index based on waist and hip measurements. Apgar score is an index based on five measurements of the condition of a newborn. The hypothyroid diagnostic index tries to catch the essence of a large number of signs and symptoms. Many such indices were discussed in earlier chapters for assessing various aspects of health at the individual level and for assessing community health. Some of these indices have been validated to perform well. When an appropriate index is available, there is no need to consider those variables in a multivariate setup. Univariate analysis of the concerned index may be adequate to draw conclusions. In fact, it is often to avoid the complexity of a multivariate setup that such indexes are developed.

19.1.2 Statistical Limitation on the Number of Variables

A prerequisite for efficient application of multivariate methods is that the total number of subjects n is substantially more than the number of variables M. If the interest is in finding those signs–symptoms out of a list of 200 that can categorize thyroid subjects into hyperthyroid, euthyroid, and hypothyroid groups, then the number of subjects should be a minimum of 1000. Multivariate calculations for $M = 200$ variables on $n = 1000$ subjects require a computer with adequate capacity. Also, such a large n may not be easy to achieve in a single study of thyroid cases. If $n = 150$, then the number of variables that can possibly be studied simultaneously should not exceed 20. These different values of n are from my experience but no specific guidelines are available. It can be safely mentioned that n need not increase at the same rate as M. Experience suggests that the ratio of M to n for $M < 10$ should preferably be 1:10 for multivariate analysis to be efficient. This ratio could progressively decline for $M \geq 10$.

If the number of variables is really large, a statistical method called **principal components** can be used to reduce the dimension. These components are a linear combination of the variables that can often account for a large part of the information contained in the entire set of variables. M correlated variables reduce to a small number of K uncorrelated components. The loss of information in most cases would be small. But the principal components so derived are purely statistical and would not generally have any biological meaning. This can make interpretation very difficult. I do not normally advise use of principal components in a medical study.

The other commonly used method is to restrict multivariate analysis to the variables found statistically significant in univariate setups. This may be convincing to many but it fails to account for interdependence of variables that forms the core of multivariate methods.

19.2 Dependent and Independent Sets of Variables

Hypertension is defined in terms of both systolic and diastolic blood pressure (BP). It can be considered to depend on family history, dietary pattern, obesity, smoking, stress, etc. Thus, there are a dependent set and an independent set. Measuring quality of life of chronic disease patients consists of items such as sleep, appetite, sexual functions, social participation, and work performance. This can be considered dependent on age, gender, education, etc., or on a set of biochemical parameters. Health of a person can be measured in terms of hemoglobin (Hb) level, lung functions, oxidative stress, etc., and this can be considered to depend on age, calorie intake, type of diet, exercise, etc. These are many examples of situations where the variables can be grouped into a dependent set and an independent set. Mathematically, these sets are called **vectors**. The independent set may contain only one variable but the dependents must be more than one for a valid multivariate setup.

As usual, the methods for studying the relationship between these two sets of variables depend on whether the variables are quantitative or qualitative. If the dependent set is qualitative (independent set could be quantitative or qualitative) and the objective is to study the relationship, then we get a situation that is a multivariate extension of logistic regression. These methods are under development and are outside the scope of this book. If both sets contain quantitative variables, the relationship is studied by multivariate multiple regression. This is an extension of the usual quantitative regression and discussed first in this section as it may be fresh in your mind. The next situation discussed is dependent set quantitative and the independent set qualitative. This is studied by MANOVA, which is an extension of analysis of variance (ANOVA). The last situation discussed in this section is the classification of subjects into known groups on the basis of a set of measurements. This is done by discriminant functions.

19.2.1 Dependents and Independents Both Quantitative

In line with the desire of most statisticians, the first multivariate setup is when all variables under consideration are quantitative. Some of them are postulated to depend on the others. For example, you may like to study how lung functions (forced expiratory volume in one second [FEV_1], forced vital capacity [FVC], and peak expiratory flow rate [PEFR]) are affected by age, body mass index (BMI), and hours of heavy physical activity per week. All these variables are quantitative.

19.2.1.1 Multivariate Multiple Regression

Take the example of height, weight, head circumference, and chest circumference of a child dependent on parity, years of education of the mother, and per capita income of the family. Thus, there are four dependent and three independent variables in this case. All are quantitative. As an extension, now consider a general situation where there are J independent variables denoted by $x_1, x_2, \ldots, x_J$ and K dependent variables denoted by $y_1, y_2, \ldots, y_K$. In other words, the independent vector contains J components and the dependent vector has K components. Let the ys be correlated with one another so that they need to be considered simultaneously. The independent variables xs may or may not be related to one another. In fact, it is desirable that the xs are not highly correlated so that multicollinearity is not present. In any case, ys are related to xs, otherwise there is no point in studying their relationship. The dependent set should preferably be outcome or response of the independent set so that the relationship is plausible and interpretable.

As an extension of univariate quantitative regression, the relationship becomes **multivariate** when the number of dependent variables is more than one. As before, it becomes **multiple** when the independent set contains more than one variable.

Multivariate multiple regression gives the same regression equations as the univariate regressions. The regression coefficients do not change. Thus, there is no need for the multivariate method if the objective is to obtain regression equations only. However, if the objective is to check statistical significance of the regressions, then multivariate methods give the right answers for correlated dependents. This is explained with the help of an example.

Example 19.1: Multivariate Multiple Regression of Lung Functions on Age, Height, and Weight

Consider dependence of lung functions (FVC, FEV_1, PEFR, and total lung capacity [TLC]) on age, height (Ht), and weight (Wt) in apparently healthy males of age 20–49 years. The data for a random sample of 70 subjects are given in Table 19.1. Some information is not available for two subjects. These two have been deleted casewise. The univariate and multivariate results for the remaining 68 subjects are as follows:
Regressions (univariate and multivariate are the same):

$$FVC = 0.67 - 0.0016(age) + 0.0182(Ht) + 0.0011(Wt)$$

$$FEV_1 = 3.07 - 0.0220(age) - 0.0073(Ht) + 0.0332(Wt)$$

$$PEFR = 5.04 + 0.0022(age) - 0.0173(Ht) + 0.0661(Wt)$$

$$TLC = -5.21 + 0.0122(age) + 0.0596(Ht) - 0.0006(Wt)$$

Univariate tests of significance: Univariate P-values (each lung function investigated separately to depend on three regressors, that is, one dependent and three independent variables in the regression) are given in Table 19.2.

The scatter plot matrix, which plots each dependent versus each independent, is in Figure 19.1. No trend is immediately discernible although some P-values in Table 19.2 are less than 0.05. The plot also indicates that the independents chosen in this study do not have much predictive power. Discerning eye can pick up subject number 65 with age 52 years in Table 19.1, which should be between 20 and 49 years as stated for this example. This looks like data entry error about which I cautioned earlier.

TABLE 19.1

Lung Functions in a Sample of 70 Males of Age 20–49 Years

Subject Number	Age (Years)	Height (cm)	Weight (kg)	FVC (L)	FEV$_1$ (L)	PEFR (L/s)	TLC (L)
1	27	156	61	4.46	3.17	5.66	6.93
2	26	163	60	4.02	3.19	5.70	5.67
3	32	153	54	3.35	2.74	3.68	4.60
4	35	155	45	3.50	3.93	2.88	6.61
5	35	148	53	2.58	2.29	5.46	3.28
6	38	163	50	3.26	3.05	6.95	3.94
7	26	161	54	3.97	3.44	8.87	5.17
8	32	155	43	2.49	2.56	5.57	2.73
9	26	155	43	4.35	4.34	8.71	4.91
10	28	147	49	2.85	2.63	7.32	3.48
11	28	164	63	3.23	2.81	6.66	4.19
12	29	167	57	3.88	2.74	4.92	6.09
13	19	161	54	4.00	3.43	4.56	5.24
14	27	161	52	4.48	3.73	5.93	6.04
15	30	162	60	3.41	3.51	7.79	3.72
16	20	162	45	3.06	2.66	5.39	3.98
17	20	160	45	2.77	2.97	4.44	2.99
18	30	161	46	3.77	2.87	5.76	5.51
19	23	160	47	4.20	3.38	4.55	5.86
20	20	159	47	3.42	2.96	4.63	4.45
21	28	163	49	3.36	3.16	6.01	4.03
22	24	162	45	3.89	3.02	8.65	3.12
23	35	165	45	3.94	3.12	5.63	5.59
24	24	167	56	5.27	4.05	5.95	7.66
25	26	159	47	3.26	2.34	4.31	5.03
26	16	162	40	2.84	1.27	1.88	3.18
27	19	165	55	3.33	2.39	3.62	5.13
28	28	164	55	3.74	3.27	5.51	4.83
29	30	172	44	3.93	2.82	5.12	6.07
30	25	160	53	—	—	5.82	6.71
31	35	160	52	4.85	3.06	4.80	3.08
32	22	166	57	4.36	1.88	8.65	5.53
33	25	170	56	4.00	3.77	7.89	4.79
34	30	165	55	2.80	3.40	7.31	3.00
35	37	172	63	4.72	4.16	6.03	—
36	21	170	60	3.11	2.92	6.36	6.73
37	26	155	47	3.25	2.68	5.33	4.42
38	21	167	55	3.11	2.92	6.21	3.70

(*continued*)

TABLE 19.1 (continued)

Lung Functions in a Sample of 70 Males of Age 20–49 Years

Subject Number	Age (Years)	Height (cm)	Weight (kg)	FVC (L)	FEV$_1$ (L)	PEFR (L/s)	TLC (L)
39	23	163	54	3.80	3.12	6.33	5.19
40	25	174	59	2.02	3.76	4.64	4.84
41	38	155	44	3.16	1.48	2.31	3.10
42	34	168	53	5.22	4.52	5.12	4.61
43	30	163	56	3.23	2.70	5.30	4.33
44	40	161	60	2.17	1.76	5.18	2.99
45	23	159	48	3.79	3.60	5.44	6.01
46	25	162	61	3.46	4.66	5.02	3.09
47	30	163	60	3.78	3.29	6.50	4.90
48	32	161	52	3.03	2.69	5.64	4.32
49	30	163	59	3.60	3.02	6.03	4.81
50	34	166	50	3.34	1.68	5.10	6.72
51	27	162	42	4.57	2.04	6.29	2.89
52	23	161	62	3.83	3.35	5.97	4.93
53	24	156	49	3.50	3.30	6.22	4.18
54	40	158	54	3.50	3.01	8.01	4.57
55	24	157	52	3.84	3.15	4.74	5.26
56	25	146	45	3.48	3.12	5.96	4.39
57	22	169	77	4.26	4.68	6.11	4.81
58	27	170	60	4.54	1.28	5.91	5.42
59	40	156	55	3.22	2.77	8.63	4.22
60	30	170	58	4.38	3.51	4.64	6.12
61	41	169	55	4.18	2.14	3.15	8.32
62	26	161	64	3.45	4.43	5.17	4.33
63	50	157	53	3.60	2.89	5.72	5.02
64	25	160	60	3.80	3.30	8.20	5.45
65	52	162	76	3.62	2.99	6.78	4.92
66	25	175	68	2.77	3.99	8.25	3.41
67	33	174	82	3.85	3.31	8.37	5.05
68	25	163	47	3.20	2.61	3.82	4.41
69	34	162	50	3.85	3.31	8.32	6.12
70	28	159	55	4.84	3.51	6.78	3.81

First consider univariate simple regression results when Ht, Wt, and age are the individual regressors on individual lung functions. At $\alpha = 0.05$, Ht has a statistically significant effect on TLC ($P = 0.037$) but not on any other lung function (Table 19.2). On the other hand, Wt has a significant effect on FEV$_1$ ($P = 0.007$) and PEFR ($P = 0.014$) but not on FVC ($P = 0.924$) and TLC ($P = 0.975$). When Ht, Wt and age are considered together, they have significant effect on FEV$_1$ ($P = 0.023$) and marginally on PEFR ($P = 0.071$) but not on

TABLE 19.2

P-Values for Univariate Regression in Example 19.1

Lung Function	P-Values			All Three Together
	Age	Height	Weight	
FVC	0.899	0.250	0.924	0.580
FEV_1	0.090	0.661	0.007	0.023
PEFR	0.939	0.632	0.014	0.071
TLC	0.576	0.037	0.975	0.128

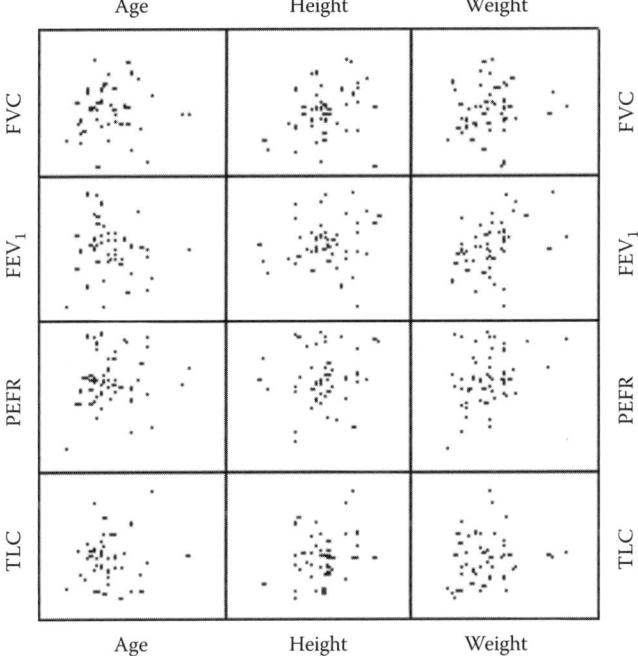

FIGURE 19.1

Scatter plot matrix of four dependent and three independent variables.

FVC ($P = 0.580$) and TLC ($P = 0.128$). In view of these conflicting results, what sort of conclusion can be drawn about the effect of Ht, Wt, and age on lung functions as a whole? The answer is given by the multivariate results.

Multivariate tests of significance: Multivariate *P*-values (Wilks Λ) (four lung functions investigated to depend on three regressors, that is, four dependent and three independent variables in the regression) are as follows:

	P-Values			
	Age	Ht	Wt	All Three Together
Lung functions (all four together)	0.636	0.159	0.012	0.056

Ht does not have a statistically significant ($P = 0.159$) influence on the "vector" of lung functions when assessed by these four measurements together, but Wt has a significant ($P = 0.012$) influence. Such a conclusion has less than 5% overall chance of being wrong. The multivariate method allows one to draw a conclusion on the basis of joint Type-I error for all four indicators of lung functions together. Multivariate multiple regression results show that their joint effect on lung functions is not statistically significant ($P = 0.056$) if the maximum tolerable error of Type-I is 0.05. If the level of significance is relaxed to 0.06, the conclusion in the multivariate setup is that the lung functions are significantly affected. Therefore, it can be concluded that age, Ht, and Wt together do influence the lung functions on the whole in the population from which this sample was drawn. If a threshold of 0.05 is used strictly, the conclusion is that of nonsignificance. The cautious conclusion is that the effect of age, Ht, and Wt on lung functions is *marginally significant*. In any case, the conclusion is different from the one obtained by univariate regressions.

The remarks on external validation that were made for univariate methods in earlier chapters apply to the multivariate methods as well. In the case of multivariate multiple regression, for example, the model may be an excellent fit to the data on which it is based but the evidence of its utility on new subjects comes when the model is tested on a new set of data.

19.2.1.2 Path Analysis

If you are studying role of heredity, diet, and obesity on size of prostate, the first method of choice is ordinary regression, provided they are all quantitative. Suppose you are able to develop a heredity score on the basis of proportion of male family members who have had prostate problem, and diet is measured for fat content. Obesity can be easily quantified by BMI. Ordinary regression may tell you, for example, that the net effect of diet on determining the size of prostate is 20% and the net effect of BMI is 15%. However, diet and BMI are related and the role of obesity through diet is obscure. Path analysis works on correlations of each predictor with the outcome and decomposes them in two components: (1) due to direct effect and (2) due to indirect effect through another predictor, such as of BMI through diet. Thus, a path is traced as illustrated in Figure 19.2. Although target variable may be one, all others are also considered stochastic in this setup.

Arrows and quantities in Figure 19.2 say that heredity is not affected by any of the predictors considered in this figure. But it is affecting both fat intake and BMI—almost twice (0.17) as much BMI as fat intake (0.09). BMI has little effect on dietary fat (0.04), but fat has substantial (0.33) effect on BMI. Fat intake affects size of prostate (0.45) more than BMI (0.28) does. The effect of heredity (0.25) and BMI (0.28) is nearly the same.

These values are **path coefficients** and interpreting them "effects" is little farfetched. But that is how the literature describes the results of path analysis.

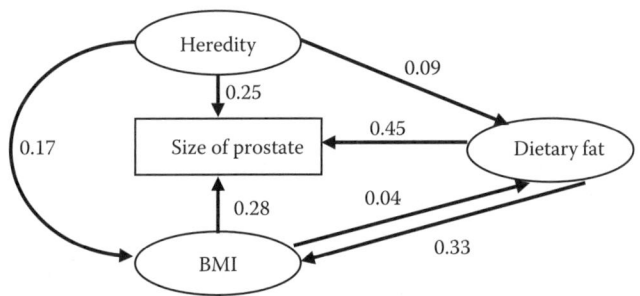

FIGURE 19.2
Example of path analysis of size of prostate affected by heredity, dietary fat, and BMI.

The path diagram is already looking complex with only three predictors, and it can soon become difficult to comprehend if there are more predictors.

You can see that it is for the researcher to postulate a right model for path analysis. In this example, the model is that the size of prostate is affected by dietary fat, BMI, and heredity, and nothing else. If postulated model is wrong, you cannot expect right results. Path analysis only provides algorithm for decomposing effect into direct and indirect within the postulated model. You can test how well the postulated model fits to the correlation structure. As in the case of regression, a good fit does not preclude other models with better fit.

Second, in Figure 19.2, ovals are used for "regressor" variables and rectangle for the dependent variable. Sometimes ovals are reserved for latent traits when those also are considered. The regressor variables are not considered fixed in this setup.

The actual procedure of path analysis is intricate and beyond the scope of this book. Also, there are many types of path models such as direct, independent, recursive, and nonrecursive. A good statistical software should be able to do this for you. This brief is just to apprise you of situations where you may like to try path analysis.

You may have noted that path analysis requires each variable to be quantitative. They must be actually observed so that measurements on them are available. What if some variables are qualitative and some others are unobserved underlying traits, also called latent traits of the type you will see in factor analysis? **Structural equation models** (SEM) consider all these and find out which variable to leave free and which of the others should be constrained or fixed. These models also are not discussed in this text. Details of path analysis and structural equation modeling are in Loehlin [5].

19.2.2 Quantitative Dependents and Qualitative Independents: Multivariate Analysis of Variance

You have seen ANOVA where a quantitative variable is dependent on qualitatives such as case and control group or different types of treatment.

Now consider a setup where the dependent is not one variable but a set of quantitative variables.

19.2.2.1 Regular MANOVA

As mentioned earlier, MANOVA is used when a set of correlated quantitative variables is considered dependent on a set of qualitative variables. For example, the dependent could be kidney functions (creatinine clearance, urea clearance, and diotrast or p-aminohippurate clearance) in persons of age 50–59 years. These persons may be grouped into those taking different diets (vegetarian, meat based, fish based, etc.) and grades of physical activity (sedentary, mild, moderate, and heavy). They may also be categorized into male and female. The dependent variables in this case are various kidney functions. These are quantitative. The independents are gender and the category of diet and physical activity. These are qualitative. Just as in ANOVA, the primary purpose of MANOVA in this case would be to test the null hypothesis of equality of means of kidney function parameters in people in different categories of independent factors. Interaction among these independent factors can also be investigated.

> **Example 19.2: MANOVA in Parkinson's Disease**
>
> To evaluate the effectiveness of an exercise intervention for people with early and midstage Parkinson's disease, 51 men and women with the disease were randomly allocated to two groups [6]. One group received individual instructions for exercise three times a week for 10 weeks and the other group remained in the usual care (control). In this study, 46 completed the trial. The outcome measures are functional axial rotation for spinal flexibility, functional reach, and the supine-to-standing time for measuring physical performance. Thus, there were three dependent variables. MANOVA performed for these three variables demonstrated a significant difference ($P < 0.05$) between the two groups. It was concluded that better improvements can be achieved by a 10-week exercise program for people in early and midstage Parkinson's disease.

Note that the conclusion in Example 19.2 is based on simultaneous consideration of the three outcome variables. When each of them is considered individually in a univariate setup, functional axial rotation and functional reach showed a significant difference but not the supine-to-sitting time. It can be argued that univariate analyses allow a more focused conclusion because they tell which particular outcome is affected and which is not. However, combining the univariate conclusions on correlated variables can sometimes be erroneous because the joint conclusion is subject to a relatively high Type-I error than is otherwise apparent. Also, in general, a combined conclusion cannot be drawn by univariate analyses when some outcomes show change and others do not. Multivariate analysis helps to draw a clear conclusion for all the variables together.

In the case of a univariate setup, Student t-test is a special case of the ANOVA F-test when the number of groups is two. The corresponding analogue of MANOVA for two groups is **Hotelling** T^2. In other words, MANOVA for two groups gives the same result as Hotelling T^2. Thus, this is not discussed separately.

Example 19.3: MANOVA of Lung Functions on Age and BMI Categories

Calculate BMI from height and weight given in Table 19.1 and divide it into three categories: low, BMI < 20; normal, $20 \le$ BMI < 25; and high, BMI $\ge$ 25. Also divide age into two groups: $\le$29 and $\ge$30 years. These two variables now become sort of qualitative from metric. Ignore the order in these categories. The dependent variables are FVC, FEV_1, PEFR, and TLC as before. Now the MANOVA results obtained from a statistical software are shown in Table 19.3.

Note how the conclusions change in some cases from the univariate to the multivariate setup. In Example 19.3, if the conclusion is drawn for lung functions as a whole, the interaction between BMI and age categories is not significant ($P > 0.10$) and the main effects also are not significantly different. Each of these conclusions can be drawn at an overall Type-I error not exceeding 5%.

TABLE 19.3

P-Values for MANOVA and ANOVA in Example 19.3

Multivariate setup (Wilks Λ)	
Effect of interaction of BMI and age categories on the lung functions	$P > 0.10$
Lung function differences between BMI categories	$P > 0.30$
Lung function differences between age categories	$P > 0.40$
Univariate results for FVC	
Interaction between BMI and age categories	$P > 0.10$
Differences between BMI categories	$P > 0.30$
Differences between age categories	$P > 0.30$
Univariate results for FEV_1	
Interaction between BMI and age categories	$P > 0.60$
Differences between BMI categories	$P > 0.20$
Differences between age categories	$P > 0.10$
Univariate results for PEFR	
Interaction between BMI and age categories	$P > 0.40$
Differences between BMI categories	$P < 0.05$
Differences between age categories	$P > 0.60$
Univariate results for TLC	
Interaction between BMI and age categories	$P < 0.01$
Differences between BMI categories	$P > 0.40$
Differences between age categories	$P > 0.80$

Univariate results show that interaction between BMI and age categories is significant for TLC, and differences in PEFR between BMI categories are significant. Examination of the data indicates that TLC is low when BMI and age are both high. But this univariate conclusion is different for FVC, FEV_1, and PEFR, and each has its own Type-I error. Note also that the P-values differ from one lung function to the other and that a joint P-value in multivariate setup can be very different. The correct joint conclusion on the lung functions as a whole is obtained by the multivariate test.

The following remarks about MANOVA may be helpful in clarifying certain issues:

1. The underlying assumption for a MANOVA test is a multivariate Gaussian pattern of the observations. This is not easy to verify. However, each variable separately can be checked for a Gaussian pattern using the methods described in an earlier chapter. When each is not far from Gaussian, there is a great likelihood that they are jointly multivariate Gaussian.

2. The other requirement for a valid MANOVA test is homogeneity of the dispersion matrices of ys in different groups. This matrix is the multivariate analogue of the variance. Methods such as Box M test are used for testing their homogeneity [7]. This, however, heavily depends on multivariate normality of data. You can depend on a statistical software for performing this test. When the same set of variables is measured in two or more groups, the dispersion matrices should be comparable, and there would be rarely any need to worry on this account.

3. The test criterion used for the MANOVA test is generally either Pillai trace or Wilks Λ as in the case of multivariate regression. Their distribution in most cases can be transformed, at least approximately, to the usual F as applicable to ANOVA. Statistical software would do this and provide the P-value.

4. Empty cells due to missing observations or otherwise are a more serious handicap in a multivariate setup than in a univariate setup. If information on just one variable is missing, the entire record is generally deleted. In Example 19.3, only FVC and FEV_1 could not be measured for one subject and TLC for another subject. Other information was available. But the analysis was done on 68 subjects after excluding these two subjects altogether.

5. These methods assume that all dependent variables have equal importance. If homocysteine and insulin levels both appear as dependents in the data for coronary artery disease (CAD) cases, both are given the same importance unless a weighting system is applied in the analysis. For example, it is possible to incorporate in the analysis that homocysteine level is six times as important as insulin level.

This will modify the method to some extent, but the major problem for a clinician would be to determine these weights objectively.

6. Example 19.3 stated that *P*-values as being more than or less than nice numbers. This is one way to report statistical results. The other, of course, is to state the exact *P*-values as shown in Table 19.2.

19.2.2.2 MANOVA for Repeated Measures

Another example where the dependent is a quantitative response is drug concentration in the blood at different points of time (repeated measures) after its administration to two or more groups of patients (e.g., experimental/control or control/drug-1/drug-2). The qualitative independent in this case is the group. The objective is to find whether or not the mean response at different points of time is different in various groups and not to explore the time trend of the response. The mean response in this case refers to the mean over the patients and not the mean over time. MANOVA would simultaneously compare several means, one at each time point, in one group with those in the other groups. Such simultaneous consideration obviates the need to examine univariate parameters such as time to reach the peak concentration (T_{max}) and the peak concentration (C_{max}) reached unless they are otherwise needed for evaluating pharmacological properties of the regimen. MANOVA in this case could be an alternative to the area under the (concentration) curve (AUC) used by many workers for this setup. As indicated later in Chapter 21, the AUC can lead to erroneous conclusions in some cases.

Repeated measures are naturally correlated and provide an apt situation for use of MANOVA whenever dependent is quantitative. Univariate repeated measures ANOVA was discussed earlier but MANOVA is considered better when the group sizes are equal (balanced design) because MANOVA is quite robust to assumptions such as of sphericity. Sphericity, as mentioned earlier, is contrasts (differences at various time points with the previous value) being independent (covariance = 0) and have same variance (homogeneity). This is tested by Mauchly test. These are rather strong assumptions for univariate repeated measures ANOVA but not as much for MANOVA. Univariate analysis is recommended for unbalanced design (unequal group sizes). In this case, as mentioned earlier, use *F*-test with dfs corrected by Huynh–Feldt epsilon.

If you are using MANOVA because of balanced design, care is needed in specifying the design regarding what factors are between subjects and what are within subjects. In repeated measures, time will always be within subjects but there might be other factors as well. Also, for MANOVA, prefer Pillai trace as the criterion instead of *F* for testing differences between groups, because Pillai trace is generally more robust to assumption violation. Tests between subjects are based on average over time—thus interpret them accordingly. The interaction between groups and time will indicate whether different groups have same time trend or not. Do not forget to test homogeneity of variance–covariance matrices between groups by Box M test.

Software illustration: Appendix C contains an illustration of use of SPSS software for running MANOVA for repeated measures.

19.2.3 Classification of Subjects into Known Groups: Discriminant Analysis

Move to a setup where the classification structure of n observations is known and this information is used to assign other observations whose classification is not known. Suppose you have information on clinical features and laboratory investigation for 120 thyroid cases. On the basis of extensive information and the response to therapy, they are divided into three groups, namely, the hyperthyroid, euthyroid, and hypothyroid patients. Assume that this division is almost infallible and there is practically no error. This is feasible in this case because the response to therapy is also known. The problem is to classify a new subject into one of these three groups on the basis of clinical features alone. This exercise would be useful when the facility for evaluation of thyroid functions is expensive. Response to therapy in any case would be available only afterward. What would be the best clinical criteria for classifying a new case with least likelihood of error? Note that this is also an exercise in management of uncertainties. The statistical method used to derive such classification criteria is called discriminant analysis. This can be viewed as a problem of finding combinations of independent variables that best separate the groups. The groups in this case define the dependent variable.

19.2.3.1 Discriminant Functions

The procedure to find the combinations of variables that best separate the groups is called discriminant analysis. The combinations so obtained are called discriminant functions. They may not have any biological meaning. These functions are considered optimal when they minimize the probability of misclassification. If there are K groups, the number of discriminant functions required is $(K - 1)$. The first is obtained in such a manner that the ratio of the between-group sum of squares to the within-group sum of squares is maximum. The second is obtained in such a manner that it is uncorrelated with the first and has the next largest ratio, and so on. If there are only two groups, only one discriminant function is needed. The nature of these functions is similar to the multiple regression equation, that is,

$$D_k = b_{0k} + b_{1k}x_1 + b_{2k}x_2 + \cdots + b_{Jk}x_J; \quad k = 1, 2, \ldots, (K-1); \quad (19.1)$$

but the x variables on the right side are stochastic in this setup. The function in Equation 19.1 is linear but other forms can also be tried. The x variables are standardized so that any particular x or few xs with large numerical values do not get unfair advantage. The method used to obtain the function as represented in Equation 19.1 is complex. Therefore, it is best to leave it to a software package. If J measurements $(x_1, x_2, ..., x_J)$ are available for each subject, it is not necessary to use all J of them. Simple discriminant functions with fewer variables are preferable, provided they have adequate discriminating power. Relevant variables can be selected by a stepwise procedure similar to the one explained for regression in Chapter 16. This procedure also helps to explore which variables are more useful for discriminating among groups. Sometimes just one variable may be enough to distinguish the groups.

19.2.3.2 Classification Rule

When values of $x_1, x_2, ..., x_J$ are substituted in the function in Equation 19.1, the value obtained is called the **discriminant score**. These scores for various sets of xs are used in Bayes rule to classify the cases. The probability required to use this rule depends on the distribution form of the variables. A multivariate Gaussian distribution is generally assumed. The rule also requires specification of the prior probability of a subject belonging to various groups. It is not necessarily equal. In some situations it is known from experience that subjects come more frequently from one group than from the others. This is generally estimated by the proportion of cases in different groups in the sample, provided there is no intervention that could distort sampling. If in 120 consecutive thyroid cases coming to a clinic, 30 are found to be hyperthyroid, 70 euthyroid, and 20 hypothyroid, then the estimates of prior probabilities are $30/120 = 0.25$, $70/120 = 0.58$, and $20/120 = 0.17$, respectively. If the subjects are deliberately chosen to be in a certain proportion in your sample such as being equal, then the prior probabilities are specified in accordance with the actual group prevalence in the target population. For example, you may wish to have 50 cases each of hyperthyroidism, euthyroidism, and hypothyroidism for your discriminant analysis, but the prior probabilities would continue to depend on their respective proportion in all the incoming cases that are planned to be the subjects for the future classification. Many published reports seem to ignore this aspect and assume equal probabilities. The results then can be fallacious. Also do not try to temper prior probabilities to get a better classification.

In case of two groups, the threshold for classification is

$$d = \frac{D_{\mathrm{I}} + D_{\mathrm{II}}}{2} + \frac{\ln(\text{prior probability of group-II}/\text{prior probability of group-I})}{D_{\mathrm{I}} - D_{\mathrm{II}}},$$

(19.2)

where
  D_{I} is D in Equation 19.1 evaluated at the mean of the xs in group-I
  D_{II} is D evaluated at the mean of the xs in group-II

The labeling is such that $D_I > D_{II}$. For a particular subject, if $D > d$ then that subject is assigned to group-I, otherwise to group-II. This is called the classification rule. If the prior probabilities are equal, the second term on the right-hand side of Equation 19.2 becomes zero.

19.2.3.3 Classification Accuracy

The exercise of classification is first used on the existing cases. For each case, a predicted class is obtained on the basis of the discriminant score for this case. Its actual class is already known. A cross-classification of cases by actual and predicted classification thus obtained is called a **classification table**. This is used to find the percentage correctly classified, called the **discriminating power**. For discriminant analysis to be successful, this power should be high, say, exceeding 80%. If a set of discriminant functions cannot satisfactorily classify the cases on which it is based, it certainly cannot be expected to perform well on new cases. When the percentage correctly classified is high, it is still desirable to try the discriminant functions on another set of cases for which the correct classification is known. It is only after such external validation that the hope for its satisfactory performance on new cases is high.

> **Example 19.4: Discriminant Function for Survival in Cervical Cancer Cases**
>
> Consider 5-year survival in cases of cervical cancer investigated as depending on age at detection (AGE) and PARITY. Suppose data are available for $n = 53$ cases, of whom 33 survived for 5 years or more but 20 died before this. Can AGE and PARITY be effectively used to predict whether the duration of survival is going to be at least 5 years? Suppose the data summary is as given in Table 19.4.
>
> The difference between the groups in mean age at detection of cancer is not significant ($P > 0.50$) but the difference in mean parity is ($P < 0.05$). The ratio of survivors to nonsurvivors in this group is $33/20 = 0.623/0.377$. Assuming that the same ratio will continue in the future, these can be used as prior probabilities for running the discriminant analysis. Then, the following discriminant function is obtained by a software:
>
> $$D = -0.327 - 0.0251(\text{AGE}) + 0.774(\text{PARITY}).$$
>
> **TABLE 19.4**
>
> Age and Parity in Cases of Cervical Cancer by 5-Year Survival
>
> | 5-Year Survival | Number of Subjects | Age Mean (SD) | Parity Mean (SD) |
> |---|---|---|---|
> | No | 20 | 43.2 (6.55) | 2.6 (1.43) |
> | Yes | 33 | 42.6 (6.84) | 1.4 (1.27) |

For classification of subjects, you need the threshold d. This is obtained from Equation 19.2 as follows.

Substitution of group means in the discriminant function D gives for nonsurvivors $D_I = -0.327 - 0.0251 \times 43.2 + 0.774 \times 2.6 = 0.6011$ and for survivors $D_{II} = -0.327 - 0.0251 \times 42.6 + 0.774 \times 1.4 = -0.3127$.

Thus,

$$d = \frac{0.6011 - 0.3127}{2} + \frac{\ln(0.623/0.377)}{0.6011 + 0.3127} = 0.1442 + \frac{0.5023}{0.9138}$$
$$= 0.69.$$

If $D > 0.69$ for a particular subject, then assign the subject to group-I (surviving less than 5 years), if not to group-II. The discriminant function in this case can be represented by a line as shown in Figure 19.3. This shows that the chance of survival for at least 5 years is less if age at detection and parity are high. Some points in Figure 19.3 overlap.

SIDE NOTE: Variations and uncertainties play their role, and many subjects are misclassified. The actual numbers are shown in Table 19.5. Only 40 out of 53 (75.5%) could be correctly classified. As many as 8 out of 20 (40%) nonsurvivors are classified as survivors by this discriminant function. When the search is restricted to linear functions, no better discrimination criterion can be achieved. For better discrimination, either look for nonlinear functions (not included in this text) or include more variables in addition to AGE and PARITY. Perhaps, a more plausible

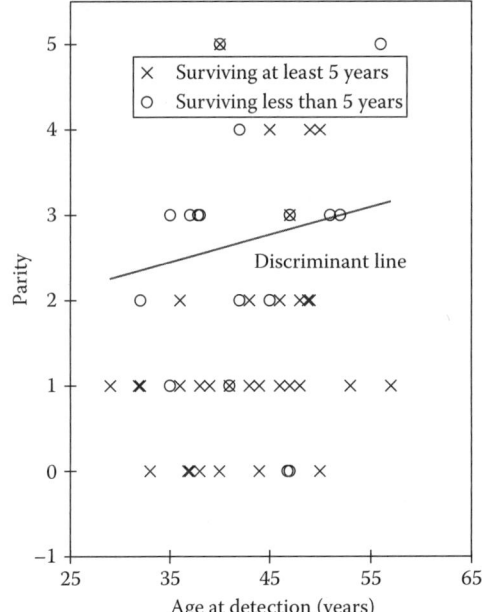

FIGURE 19.3
Scatter and discriminant line for surviving less than 5 years and at least 5 years.

TABLE 19.5

Classification Table Based on Discriminant Function in Example 19.4

| | Predicted Group | | |
Observed Group	Survivors[a]	Nonsurvivors	Total
Survivors[a]	28	5	33
Nonsurvivors	8	12	20
Total	36	17	53

[a] For 5 years or more.

conclusion is that AGE and PARITY by themselves are not sufficient to predict 5-year survival in cervical cancer cases. You can take the view that correct classification in three-fourths of the cases on the basis of such mundane variables as AGE and PARITY is good enough for practical applications. If this view is accepted, the discriminant function must be externally validated before it is used on new cases.

Example 19.5: Discriminant Function to Determine Sex from Patellar Dimensions

Introna et al. [8] studied the right patella of 40 male and 40 female Italian skeletons with respect to seven measurements: maximum height, maximum width, thickness, height and width of external facies articularis, and height and width of the internal facies articularis. Through a stepwise procedure, they found that only two measurements, maximum width and thickness, could discriminate gender correctly in 83.3% of cases. It was concluded that gender can be predicted by patellar dimensions when no other suitable remains of a human skeleton are available for gender determination. The prior probabilities in this example are 0.5 each because the two genders have same proportions in the population.

Note the following regarding discriminant functions:

1. Discriminant functions divide the universe of subjects into sectors corresponding to different groups. Each sector defines the group to which a subject is most likely to belong.

2. You may have noticed earlier that logistic regression is used when the target variable is binary. Logistic for polytomous-dependent variables exists but that also generally considers two categories at a time. Discriminant analysis can be used for all categories together when the target variable is polytomous. The interpretation of a discriminant function, even when there are two groups, is not exactly the same as that of a logistic regression but it serves the purpose

of predicting the category provided the distribution is multivariate Gaussian. The associated probability can also be calculated although no details are provided in this text. For the calculation of these probabilities, see Everitt and Dunn [9]. Most statistical softwares calculate these probabilities.

3. Discriminant analysis is sometimes used to find whether or not a particular variable or a set of variables has sufficient discriminating power between groups. Kilcoyne et al. [10] studied the renin–angiotensin system in 146 Black patients with essential hypertension in the United States. They divided these patients into low, normal, and high renin groups. It was observed that none of the variables they studied, including incidence of cerebrovascular and cardiovascular events or age, had adequate discriminating power.

4. As stated earlier, the foregoing discriminant analysis assumes that the variables are jointly multivariate Gaussian. This necessarily implies that the distribution of each variable is Gaussian. Mild deviations do not do much harm. In Example 19.4, the parity distribution is not Gaussian yet the result may still be valid for large n. If one or more variables are binary, then the foregoing procedure is questionable unless n is really large. In that case, use a logistic discriminant function. This is briefly discussed by Everitt [11].

5. The other important requirement for discriminant analysis is equality of the dispersion matrices in various groups. This is checked by the Box M test. If the dispersion matrices are really very different, the linear discriminant function is not adequate in separating the groups. A more complex, quadratic discriminant function [11], may be helpful in this case.

19.3 Identification of Structure in the Observations

There were two groups of variables in the setup considered in the previous section: the dependent and the independent. Now consider a setup where all the variables have the same status and there is no such distinction. Suppose you obtain information on a large number of health variables of a set of individuals and want to divide them into two or more groups such that the individuals within each group have similar health. The number of groups is not predetermined, and there is no dependent variable in this case. Dividing cases into diagnostic categories on the basis of various clinical features and laboratory investigations is a classical example of this type of exercise. It was observed in the past, for example, that acute onset of fever, hemorrhagic manifestations, thrombocytopenia, and rising hematocrit occur together in some cases, and this entity was called dengue hemorrhagic fever under certain conditions. Similarly, fever, splenomegaly, and hepatomegaly accompanied by anemia and weight loss were given a name, kala-azar, particularly in sandfly infested areas. All diagnostic categories have evolved from the observation of such common features. Subjects with the same features are put together into one "cluster." In these examples, clustering of cases is with respect to the clinical features or etiology. The structure is not known a priori, and there is no target variable.

The second type of structure discussed in this section concerns the constructs, if any, underlying a set of observations. These constructs are common to many observations and generally are unobservable or unmeasurable entities. These are popularly called factors underlying the observations. Such factors can be identified in some cases by suitable statistical methods. The term "factors" in this context has entirely different meaning from what has been used earlier.

19.3.1 Identification of Clusters of Subjects: Cluster Analysis

The problem addressed now is the division of subjects or units into an unspecified number of affinity groups. The statistical method that discovers such natural groups is called cluster analysis. This was briefly described in Chapter 8 in the context of graphics. The grouping is done in such a manner that units similar to one another with respect to a set of variables belong to one group and the dissimilar ones belong to another group. Thus, natural groupings in the data are detected. This is, for instance, unwittingly done in assigning grades to students in a course where instructor looks for rational cutoff points. Sometimes only two grades A and B are considered enough; sometimes they go up to E. In another setup, the method can also create taxonomic groups for future use. An example already cited is of cases falling into various diagnostic groups on the basis of their clinical features. A name is subsequently assigned to these groups depending on their etiology or features. Although overlapping clusters can be conceived so that one or more units belong simultaneously to two or more clusters, this chapter is restricted to exclusive clusters. Cluster analysis is a nonparametric procedure and does not require values to follow a Gaussian or any other pattern.

19.3.1.1 Measures of Similarity

Division of subjects into a few but unknown number of affinity or natural groups requires that proximity between subjects is objectively assessed. Those with high proximity go into one group and those with low proximity are assigned to some other group. Thus, the subjects resembling one another are put together in one group. As many groups are formed as needed for internal homogeneity and external isolation of the groups (Figure 19.4). The points plotted in Figure 19.4b are the same as in Figure 19.4a but now the affinity groups are shown.

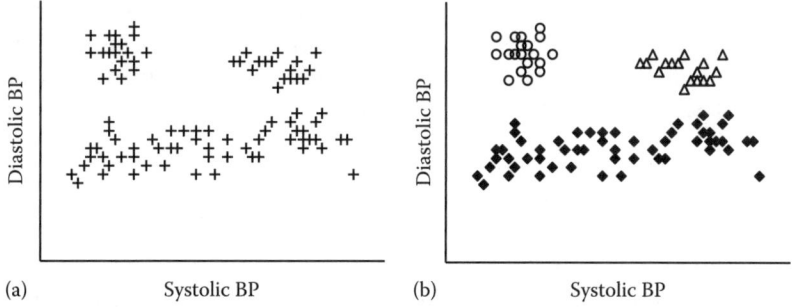

FIGURE 19.4
Scatter plot of diastolic and systolic levels of BP: (a) affinity not shown and (b) one form of affinity groups.

TABLE 19.6

Matches and Mismatches in Variables in Two Subjects Measured on K Binary Variables

	*i*th Subject		
*j*th Subject	Yes	No	Total
Yes	k_{11}	k_{12}	$k_{11} + k_{12}$
No	k_{21}	k_{22}	$k_{21} + k_{22}$
Total	$k_{11} + k_{21}$	$k_{12} + k_{22}$	K

Similarity between two subjects can be measured in a large number of ways. The methods are different for qualitative than for quantitative variables. In the case of binary qualitative variables, the affinity can be shown in the form of a 2 × 2 table as in Table 19.6. There is a total of K variables out of which matches between the *i*th and the *j*th subject occur in $k_{11} + k_{22}$ variables. A popular measure of similarity in this case is the

$$\text{Simple matching coefficient: } s_{ij} = \frac{k_{11} + k_{22}}{K}; \quad i, j = 1, 2, \ldots, n. \quad (19.3)$$

Note that s_{ij} is the ratio of the number of variables on which the *i*th and the *j*th subjects match to the total number of variables. This can be calculated for each pair of subjects. Similar measures are available for polytomous variables [12]. This reference also discusses strategies for mixed sets (quantitative and qualitative) of variables.

In the case of quantitative variables, the usual Pearsonian correlation can be used as a measure of similarity. If $y_{i1}, y_{i2}, \ldots, y_{iK}$ are K quantitative measurements on the *i*th subject and $y_{j1}, y_{j2}, \ldots, y_{jK}$ on the *j*th subject, the correlation between the two can be directly calculated. Note that this is being used here for assessing similarity between subjects whereas the earlier use was to measure strength of correlation between variables. A more acceptable method in this case is to compute a **measure of dissimilarity** instead of similarity. This, between the *i*th and the *j*th subjects, can be measured by

$$\text{Euclidean distance: } d_{ij} = \sqrt{\Sigma_k(y_{ik} - y_{jk})^2}; \quad i, j = 1, 2, \ldots, n. \qquad (19.4)$$

This is calculated after standardization (Z-score) of the variables so that the scales do not affect the value. Otherwise the variables with larger numerical values will mostly determine distance. This distance can also be calculated for a setup with one variable ($K = 1$). In this case, this reduces to a simple difference between the values.

On the basis of such measurement of dissimilarity, the subjects are classified into various groups using one of the several possible algorithms. A very popular one is described next.

19.3.1.2 Hierarchical Agglomerative Algorithm

With hierarchical algorithm, two units (or subjects) that are most similar (or least distant) are grouped together in the first step to form one group of two units. This group is now considered as one entity. Now the distance of this entity from other units is compared with the other distances between various pairs of units. Again, the closest are joined together. This hierarchical agglomerative process goes on in stages, reducing the number of entities by one each time. The process is continued until all units are clustered together as one big entity. Described later is a method to decide when to stop the agglomerative process so that natural clusters are obtained. This process is graphically depicted by a dendrogram of the type shown in Figure 8.14. Note for this method that subsequent clusters completely contain previously formed clusters.

It may not be immediately clear how to compute the distance between two entities containing, say, n_1 and n_2 units, respectively. Several methods are available. First is to consider all units in an entity centered on their average. A second is to compute the distance of the units that are farthest in the two entities. A third is to base it on the nearest units. There are several others. Depending on how this distance is computed, names such as centroid, complete linkage, single linkage, etc., are given to these methods. Different methods can give different results. Jain et al. [13] have studied relative merits and demerits of some of these methods. No specific guideline can be given, but a method called average linkage has been found to perform better in many situations. This method uses the average of the distances between units belonging to different entities as the measure of distance between two entities.

19.3.1.3 Deciding on the Number of Natural Clusters

The most difficult decision in the hierarchical clustering process is regarding the number of clusters naturally present in the data. The decision is made with the help of a criterion such as pseudo-r or the cubic clustering criterion [14]. These values should be high compared with the adjacent stages of the clustering process. Another criterion could be the distance between the two units or entities that are being merged in different stages. If this shows a sudden jump, it is indicative of a very dissimilar unit joining the new entity. Thus, the stage where the entities are optimal in terms of internal homogeneity and external isolation can be identified. The entities at this stage are the required natural clusters.

The following comments regarding cluster analysis may be helpful:

1. The algorithm just described is a hierarchical agglomerative algorithm. You can use a hierarchical divisive algorithm in which the beginning is from one big entity containing all the units, and divisions are made in subsequent stages. However, this is rarely favored because agglomeration is considered a natural clustering process.

2. The other algorithm is nonhierarchical. This can be used when the number of clusters is predetermined. I do not recommend this algorithm because it does not adequately meet the objective of discovering an unspecified number of natural clusters.

3. Cluster analysis methods have the annoying feature of "discovering" clusters when, in fact, none exists. A careful examination of the computer output for cluster analysis, particularly with regard to the criteria for deciding the number of clusters, should tell you whether or not natural clusters really exist.

4. Since there is no target variable, the clusters so discovered may or may not have any medical relevance.

5. Different clustering methods can give different clusters. One strategy to overcome this problem is to obtain clusters by several different methods and then look for consensus among them. Such consensus clusters are likely to be stable. The consensus may be difficult to identify in a multivariate setup. Indrayan and Kumar [15] have given a procedure to identify consensus clusters in the case of multivariate data.

6. Cluster analysis has developed into a full subject by itself. Details of the procedures just described and of several other cluster procedures are available in Everitt et al. [16].

7. The clustering just mentioned is different from the clustering of subjects with disease in population units such as households. For this, papers by Mantel [17] and Fraser [18] may be helpful.

Example 19.6: Clustering of Countries for Expectation of Life at Birth

Cluster analysis can also be done on just one variable. Figure 19.5 shows the dendrogram obtained for Pan American countries when clustered on the basis of 2004 values of expectation of life at birth (ELB). The method followed is average linkage. The distance on the horizontal axis is rescaled in proportion to the actual distances between entities. The algorithm detected four clusters as shown from 1 to 4 in the first clustering in the figure. These are as follows

Cluster	Expected Life at Birth (Years)
1	80–75
2	73–68
3	64
4	52

The fourth cluster has only one country, namely, Haiti, which has very low ELB. Note the gap in ELB between clusters that makes these clusters distinct.

Example 19.7: Clusters of *Pseudomonas aeruginosa* by Whole-Cell Protein Analysis by SDS–PAGE

See Figure 8.14 for a dendrogram that shows hierarchical clustering of 20 sodium dodecyl-sulfate–polyacrylamide gel electrophoresis (SDS–PAGE) groups on the basis of the whole-cell protein profile in strains of *Pseudomonas aeruginosa* [19]. The authors did not study the number

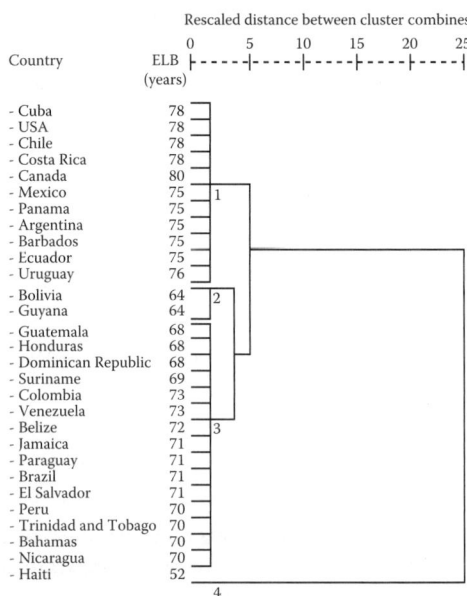

FIGURE 19.5
Dendrogram of Pan American countries for ELB (2004): average linkage method.

of natural clusters in this case but the affinity can be seen in the figure. For example, groups 5 through 8 have been found alike and they are close to group 9 (note the small vertical distance between group 9 and groups 5 through 8; the vertical distance from group 4 is larger). Similarly, groups 18 and 19 are alike. There are many such affinity groups in this example.

19.3.2 Identification of Unobservable Underlying Factors: Factor Analysis

How do you assess health of a child? You will see his height, weight, skinfold, eyes, nails, etc. Health is a directly unobservable factor and it is measured through a host of variables. Conversely, you may have a host of variables and you want to know what factors are underlying those variables. There are situations where the observed variables can be considered to be made up of a small number of unobservable, sometimes abstract, factors or constructs. Health in its comprehensive form (physical, social, mental, and spiritual) can be measured for adults by a host of variables such as the kind and severity of complaints if any, obesity, lung functions, smoking, sexual behavior, marital relations, job satisfaction, unmet aspirations, income, and education. It is not apparent how much of, say, job satisfaction is ascribable to physical health, how much to each of the other components (social, mental, and spiritual) of health, and how much is the remainder that cannot be assigned to any of these components. For a variable, such as obesity, the physical component may be dominant, and for a variable, such as education, the social component may be dominant. The technique called exploratory factor analysis can be sometimes used to identify such underlying factors.

The theory of factor analysis presumes that the variables share some common factors that give rise to correlation among them. Actual interest in this analysis is in identifying relatively few underlying factors, and the variables actually measured are now better understood as **surface attributes** because these are considered manifestations of the factors. The underlying factors are the **internal attributes**. The primary objective of factor analysis is to determine the number and nature of the underlying internal attributes, and the pattern of their influence on the surface attributes. In a successful factor analysis, a few factors would adequately represent relationships among a relatively large number of variables. These factors can then be used subsequently for other inferential purposes. An example described later should clarify the meaning and purpose of such factors.

19.3.2.1 Factor Analysis

How do you measure physical health of a community? You may like to assess it by expectation of life at 1 year, absence of various diseases, disability-free life years, etc. All these are kind of surrogates. You may like to combine all these and call it index of physical health. Note that whereas observed variables are surface attributes, the real interest is in an intrinsic attribute—health in this example, and health cannot be directly measured. This is a **factor** in the terminology of factor analysis. Factors are unobservable underlying attributes and can be abstract.

Now add other attributes such as beds and doctors per 1000 population, per capita health expenditure by the government and out of pocket, specialists available, smoking, divorces, crimes, and suicides, so that the spectrum enlarges to include other aspects of health. If a mix of these measurements is available, and you want to discover the unknown intrinsic attributes, you may find one set of measurements comprising those mostly concerned with health resources and second set with mental health, in addition to the set concerned with physical health.

Factor analysis is easy to understand through principal components. Consider two variables—height and weight—whose essence is captured by BMI for many applications. This combined variable provides most of the information contained in two different but correlated variables. BMI is not a linear combination of height and weight but there are situations where much of the information contained in two or more variables can be captured by one linear combination of them. This linear combination is called **principal components** (PCs). If substantial information remains uncaptured, you can try to find second PC. This would be independent of the first and may capture much of this remainder. If the remainder is substantial even after the second, a third PC can be obtained, and so on. When really successful, few PCs would be able to capture most of the information in a large number of variables. Now you do not need to work with the original large set and can work with few PCs without much loss. This is called data reduction—more appropriately dimensionality reduction.

In principal component method of factor analysis, first linear component is discovered that is able to account for maximum variation among observed variables. Second linear combination is obtained from the remaining variation, and so on. First principal component may be able to take care of 62% of total variation, second 23% and third only 8%. A strategy called rotation of axes is used that can modify PCs such that first PC has high weight for only few variables, second for another set of few variables, etc. In our example, although the dataset comprises all the measurements, but the first principal component may have high weight for variables pertaining to physical health (low weight for others), second principal component with high weight for variables pertaining to health resources, and third set with high weight for variables pertaining to mental health. Some variation would still remain unaccounted irrespective of rotation.

The statistical purpose of factor analysis is kind of reverse. It is to obtain each observed variable as a combination of a few unobservable factors, that is,

Observed value of a variable = linear combination of factors + error.

If the observed variables are $x_1, x_2, ..., x_K$, the factor analysis seeks the following:

$$\left.\begin{aligned}
x_1 &= a_{11}F_1 + a_{12}F_2 + \cdots + a_{1M}F_M + U_1, \\
x_2 &= a_{21}F_1 + a_{22}F_2 + \cdots + a_{2M}F_M + U_2, \\
&\vdots \quad\quad \vdots \quad\quad \vdots \quad\quad\quad\quad \vdots \quad\quad \vdots \\
&\vdots \quad\quad \vdots \quad\quad \vdots \quad\quad\quad\quad \vdots \quad\quad \vdots \\
x_K &= a_{K1}F_1 + a_{K2}F_2 + \cdots + a_{KM}F_M + U_K,
\end{aligned}\right\}$$

(19.5)

where

$F_1, F_2, ..., F_M$ are the M unobservable factors common to x_ks

U_ks ($k = 1, 2, ..., K$) are called unique factors

A schematic representation is in Figure 19.6 where the number of x variables is six but the number of factors is two. Unique factors represent the part that remains unexplained by the factors.

The coefficients a_{km} ($k = 1, 2, ..., K; m = 1, 2, ..., M; M \ll K$) are estimated by the factor analysis procedure. These coefficients are called **loadings** and measure the importance of the factor F_m in the variable x_k on a scale 0–1. When a loading is very small, say, less than 0.20, the corresponding factor is dropped from Equation 19.5. Thus, different variables may contain different sets of factors. Some factors would overlap and would be present in two or more variables. In Figure 19.6, factor F_1 is common to $x_2, x_3, x_5,$ and x_6, and F_2 is common to $x_1, x_3,$ and x_4. Each x has its own specific U. Factors are also called constructs.

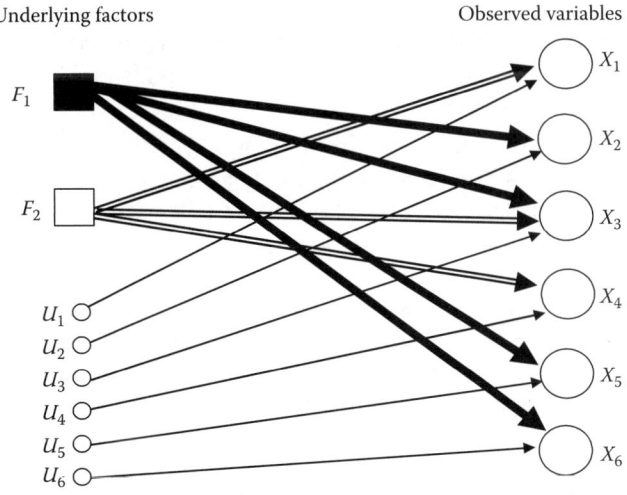

FIGURE 19.6
Schematic representation of factors.

Equation 19.5 differs from the usual multiple regression equations because the F_ms are not single independent variables. Instead, they are labels for combination of variables that characterize these constructs. They are obtained in such a manner that they are uncorrelated with one another. Statistical method of principal components is generally used for this purpose. Details of the method are complex and are beyond the scope of this book. You may like to see Kline [20] for details. Statistical softwares would easily provide an output, but they may ask you to specify different aspects of the methodology. You should not undertake this exercise yourself unless you understand the intricacies. Do not hesitate to obtain the help of a biostatistician when needed.

19.3.2.2 Steps for Factor Analysis

Basic steps for doing factor analysis are as follows:

Step 1. Obtain the correlation matrix of all x-variables you have observed.

Step 2. Check that correlations are sufficiently high among subsets of variables. You can use Bartlett test to test the null hypothesis that the correlation matrix is an identity. Identity matrix means that the diagonal elements are one (correlation of x_k with x_k is always one) and the off-diagonal elements are zero.

Step 3. Check that the variables indeed share some common factors that are probably giving rise to these correlations. This can be checked as follows for linear correlations:

    a. Multiple correlation of each x_k with other xs should be generally high.

    b. The partial correlations, which are correlations between two variables after effect of others is adjusted, should be generally low. This is tested by Kaiser–Meier–Olkin (KMO) measure.

Step 4. Once you are convinced from steps 1 through 3 that you have an appropriate setup to try factor analysis, enter correlation matrix into factor analysis software. The output will give you factor loadings and a host of other information.

Depending on the correlation structure among variables, it is often possible to identify meaningful underlying constructs. For proper interpretation, it is necessary to group variables that have large loading for the same factor. A suitable name can accordingly be assigned to these factors. Such an exercise can sometimes lead to a better understanding of the structure of the interaction among variables. However, the procedure based on correlations as just described assumes *linear* relationship. Other forms of relationships may require another set of procedures that are not discussed in this chapter. The following example may clarify the process and meaning of factor analysis.

Example 19.8: Underlying Diet Patterns in Colon Cancer

Colon cancer is considered to be associated with several nutrients and foods. Detailed dietary intake data can be examined to find dietary patterns related to colon cancer. Slattery et al. [21] surveyed a large number of subjects for intake of over 800 individual food items. These items were categorized into 35 food groups such as processed meat, red meat, fish, eggs, butter, margarine, liquor, coffee, fresh fruits, dry fruits, fruit juice, green salads, grains, and desserts. These served as input variables for factor analysis. The underlying factors in this study are the dietary patterns. Six dietary patterns for males and seven for females were identified in control subjects after performing factor analysis. Depending on the patterns receiving higher loadings in specific food groups, they were given a name. For example, the pattern that received high loadings in processed meat, red meat, eggs, refined grain, added sugar, etc., was called a "Western" diet. The pattern that received high loadings in fruit juice, potatoes, salads, vegetables, and whole grain was called a "prudent" diet. The pattern receiving high loadings in fresh liquor and wine was called a "drinker diet."

SIDE NOTE: Evaluation of these dietary patterns with regard to the risk of colon cancer showed that the Western-type diet increased the risk of colon cancer in both men and women. The prudent diet pattern was associated with decreased risk.

19.3.2.3 Features of a Successful Factor Analysis

The details are omitted but there are some steps in factor analysis that are not considered fully scientific. Despite this, the technique is popular with social scientists. It is now making inroads into medical sciences as well. But the technique sometimes fails to identify meaningful factors. Its success is assessed on the basis of the following considerations:

1. One of the steps in factor analysis is breaking down the total variation among variables into variations accountable by different factors. The analysis is considered successful when few factors are able to account for a large part of the total variation, say, more than 70%. In Example 19.8, the total variation explained by the six factors together in males was only 36.9% and by seven factors in females only 34.3% [21]. This is low. Such a low percentage indicates that there were factors other than the identified dietary patterns that were responsible for intake of specific type of food. Also, in this example, the percentage of variation explained by individual factors ranged from only 4 to 10.

This is also low. In many situations, you may find that the first factor alone explains something like 20% of the total variation and the others 15% or 10% each. This did not happen in this example.

2. A criterion for successful factor analysis is to be able to find some factors with very high loadings (close to ±1) and the others very low loadings (close to 0). Also the factors should be largely distinct or nonoverlapping. To achieve this, a technique called rotation of axes is sometimes adopted. One popular rotation is called **varimax** as it aims to maximize the variance of that factor. Others are quartimax and equamax. This sometimes helps to extract factors with selectively high loadings and thus to obtain interpretable factors.

3. In a successful factor analysis, the number of identified factors is very small relative to the number of variables. In Example 19.8, the number of factors is six or seven and the number of variables is 35. Thus, the number of factors is indeed low.

A basic requirement to achieve successful factor analysis is that the underlying common factors are really present. In Example 19.8, food intake was likely to be governed by the dietary patterns. It was thus possible to identify the patterns on the basis of the kind of food consumed. The presence of underlying factors would mean at least a fair degree of correlation between most variables. If correlations are not present, it is futile to try factor analysis. Bartlett test [7] and the one discussed by Srivastava and Carter [4] are available to test the significance of the correlations. These correlations are stated in the form of a matrix in the case of a multivariate setup.

The success of factor analysis, as for all multivariate methods, also depends on the proper choice of variables to include in the analysis. In Example 19.8, the authors divided more than 800 food items and nutrients into 35 food groups. Some of these could be arbitrary. The results can be different if the grouping of food items is different or the number of groups is 50 instead of 35.

Perhaps, the most important decision in factor analysis is the choice of the number of factors to be extracted. The statistical procedure theoretically is such that as many factors can be extracted as the number of variables. But only a few factors would be important. This importance is generally assessed by the **eigenvalues** for the factors. Eigenvalues depend on the correlation structure among the variables. Factors with eigenvalues greater than 1.0 can be considered important because they explain more variance than is explained by a single variable. When common factors are indeed present, only a few factors are likely to have this property. If the number of factors so identified is more than you think it should, the threshold of the eigenvalue can be raised from 1.0 to 1.5 or any other suitable number.

In the example on dietary patterns, the authors used a threshold of 1.25. It is on the basis of this cutoff that they came up with six factors for males and seven for females.

The factors identified after this analysis are not necessarily exclusive. In the example on dietary patterns, the factors might correlate well with lifestyle (physical activity, smoking, use of drugs such as aspirin, etc.) and socioeconomic status. These are the confounders in this case. It cannot be immediately concluded that dietary patterns per se were responsible for differential risks of colon cancer. The lifestyle and socioeconomic status may also have contributed irrespective of the diet pattern.

Another term talked about in the context of factor analysis is **communality**. Note that the basic premise of factor analysis is that various variables have one or more common factors. In accordance with its social meaning, the proportion of variance of the variable x_k that is shared with other xs is called its communality. The remainder is unique to x_k. Generally, the value of R^2 obtained when x_k is regressed on other xs is regarded as communality of x_k. There are other methods also. High communality of several xs indicates that the factor analysis is likely to be successful.

The method described earlier is mostly for exploratory purposes. This method does not assume any preconceived structure of the factors. If you want to assess whether a set of variables conforms to a *known* structure of the factors, then confirmatory factor analysis is needed. One of the methods for this is path analysis. This is discussed earlier in this chapter. For details, see Cattel [22].

19.3.2.4 Factor Scores

As a reverse process, it is possible to obtain factors as a linear combination of the variables, that is,

$$F_m = b_{m1}x_1 + b_{m2}x_2 + \cdots + b_{mK}x_K; \quad m = 1, 2, \ldots, M. \qquad (19.6)$$

The coefficients b_{mk} ($m = 1, 2, \ldots, M$; $k = 1, 2, \ldots, K$) are called factor score coefficients. When the values of ($x_1, x_2, \ldots, x_K$) for a subject are substituted in Equation 19.6, the quantity obtained is called the factor score for that subject. This measures the importance of the factor for that individual. If the factor score for the third factor (F_3) is high for the sixth subject and relatively low for the tenth subject, it can be concluded that F_3 is influencing the sixth subject more than the tenth subject. Such factor scores for each factor can be obtained for each subject. These scores can be used for a variety of purposes. Chandra Sekhar et al. [23] used these scores to develop an index of need for health resources in different states of India, and Riege et al. [24]

used them to study the age dependence of brain glucose metabolism and memory functions.

After the discussion of a wide variety of multivariate methods, it might be helpful to have a glance at all these methods. This is given in **Table S.7** at the beginning of the book.

References

1. Boshuizen HC, Verkerk PH, Reerink JD et al. Maternal smoking during lactation: Relation to growth during the first year of life in a Dutch birth cohort. *Am J Epidemiol* 1998; 147:117–126.
2. Hair JF, Black B, Babin B, Anderson RE, Tatham RL. *Multivariate Data Analysis*, 6th edn. London, U.K.: Prentice Hall, 2005.
3. Krzanowski WJ. *Principles of Multivariate Analysis: A User's Perspective*. Oxford, U.K.: Oxford University Press, 1988, pp. 432–455.
4. Srivastava MS, Carter EM. *An Introduction to Applied Multivariate Statistics*. New York: North Holland, 1998, pp. 332–333.
5. Loehlin JC. *Latent Variable Models: An Introduction to Factor, Path and Structural Equation Analysis*, 4th edn. Philadelphia, PA: Psychology Press, 2003.
6. Schenkman M, Cutson TM, Kuchibhatla M et al. Exercise to improve spinal flexibility and function for people with Parkinson's disease: A randomized controlled trial. *J Am Geriatr Soc* 1998; 46:1207–1216.
7. Morrison DF. *Multivariate Statistical Methods*, 4th edn. New York: Duxbury Press, 2004.
8. Introna F Jr, Di Vella G, Campobasso CP. Sex determination by discriminant analysis of patella measurements. *Forensic Sci Int* 1998; 95:39–45.
9. Everitt BS, Dunn S. *Applied Multivariate Data Analysis*. London, U.K.: Hodder Arnold, 2001.
10. Kilcoyne MM, Thomson GE, Branche G et al. Characteristics of hypertension in the black population. *Circulation* 1974; 50:1006–1013.
11. Everitt BS. *Statistical Methods in Medical Investigations*, 2nd edn. Sevenoaks, London, U.K.: Edward Arnold, 1994.
12. Romesberg MR. *Cluster Analysis for Researchers*. Lulu.com, 2004.
13. Jain NC, Indrayan A, Goel LR. Monte Carlo comparison of six hierarchical clustering methods on random data. *Pattern Recognit* 1986; 19:95–99.
14. Milligan GW, Cooper MC. An examination of procedures for determining the number of clusters in a data set. *Psychometrika* 1985; 50:159–179.
15. Indrayan A, Kumar R. Statistical choropleth cartography in epidemiology. *Int J Epidemiol* 1996; 25:181–189.
16. Everitt B, Landau S, Lease M. *Cluster Analysis*. London, U.K.: Hodder Arnold, 2001.
17. Mantel N. Re: Clustering of disease in population units: An exact test and its asymptotic version. *Am J Epidemiol* 1983; 118:628–629.
18. Fraser DW. Clustering of disease in population units: An exact test and its asymptotic version. *Am J Epidemiol* 1983; 118:732–739.

19. Khan FG, Rattan A, Khan IA, Kalia A. A preliminary study of fingerprinting of *Pseudomonas aeruginosa* by whole cell protein analysis by SDS-PAGE. *Indian J Med Res* 1996; 104:342–348.
20. Kline P. *An Easy Guide to Factor Analysis.* New York: Routledge, 1994.
21. Slattery ML, Boucher KM, Caan BJ, Potter JD, Ma KN. Eating patterns and risk of colon cancer. *Am J Epidemiol* 1998; 148:4–16.
22. Cattel RB. *The Scientific Use of Factor Analysis in Behavioral and Life Sciences.* New York: Plenum, 1978.
23. Chandra Sekhar C, Indrayan A, Gupta SM. Development of an index of need for health resources for Indian states using factor analysis. *Int J Epidemiol* 1991; 20:246–250.
24. Riege WH, Metter EJ, Kuhl DE, Phelps ME. Brain glucose metabolism and memory functions: Age decrease in factor scores. *J Gerontol* 1985; 40:459–467.

20

Quality Considerations

Quality in a statistical sense means meeting a specified standard. It is always desirable to aim at perfection, but that is quite often elusive. For example, it would be wonderful to have a hospital that discharges fully cured patients all the time without any death or disability, but that is impossible to achieve. Thus, a lower standard of, say, a 90% cure rate in casualty cases and 98% in routine admissions may have to be fixed. Limitations of knowledge and cost considerations often determine this feasibility level. This applies to statistical methods also. Once a standard—high or not so high—is fixed, everything possible should be done to achieve it. This is best done by controlling errors.

Errors are distinct from variation. The latter is endogenous whereas the former are exogenous. Variability remains; it is only their impact that can be minimized. But the errors themselves can be largely avoided by being sufficiently careful, although it may not be easy to eliminate them fully. A standardized mercury manometer would still give variable readings in different individuals, but occurrence of air bubbles in the mercury column produces an error. Clinicians may genuinely differ on the dose of a drug to be given to a particular patient, but giving the drug two times when the prescription says three times a day is an error. In statistical methods too, misinterpretation of an incidental correlation, occurring mostly due to uncontrolled confounding factors, as an indication of a cause–effect relationship is an avoidable error. On the other hand, not being able to reject a false null hypothesis due to, for example, a small sample is not an error in that sense.

This chapter: The illustrations just given indicate that errors, in our context, are primarily of three kinds: (1) errors concerned directly with the provision of medical care, (2) errors related to the measurements and to the medical tools used to obtain data, and (3) errors concerning statistical results. Statistical methods that can help to control the first two types of errors are readily available. These are discussed in Sections 20.1 and 20.2, respectively. Control of the third type requires assessment of robustness. Some methods of this assessment are introduced in Section 20.3. The third type of error is further discussed in Chapter 21.

20.1 Statistical Quality Control in Medical Care

Medical care is a big industry in many parts of the world. Providing good care is a prerequisite for this industry to maintain and increase clientele and perhaps profits. In an activity that deals with the life and health of people, the primary concern is with the humanitarian aspect of doing the best to prolong life and reduce suffering. In this context, quality of medical care assumes importance much more than in other industries. Perhaps, millions of episodes of illness and deaths around the world every year can be attributed to medical errors. According to one estimate, nearly one in three patients encounter problems due to medical errors during their hospital stay.

A patient sometimes has to undergo an array of steps while seeking treatment. The patient describes complaints and responds to questions about the history of the problem, environmental exposures, extent of disabilities, etc. The attending clinician gears questions to meet the requirement of the situation in terms of the condition of the patient, the type of complaints, and the intelligence level of the patient. The clinician also examines the patient as required and obtains data on body temperature, heart rate, blood pressure (BP), and weight. An interim therapy is sometimes started. Then, laboratory and radiological investigations are ordered, if considered desirable. The laboratory and radiological units carry out these investigations and report findings. The clinician reassesses and sometimes the cycle restarts. In a hospital setup, outpatients receive their supply of medicines from a pharmacy according to the prescription and ingest the drug. Inpatients are administered prescribed dosages by the nursing staff. Some patients undergo surgery. A series of steps are taken in the preoperative ward, operation theater, and postsurgical care unit. A hospital does all this for a large number of patients day after day. It is unrealistic to expect that all steps will be correctly done in all cases all the time. Errors do occur. The question is whether these errors are far too many or too large, or are within a tolerance limit. This tolerance has a statistical nature because it is determined by the experience. Before this is discussed, a brief overview of steps is given that could enhance the quality of medical care.

Edward Deming is among those who were outstanding in promoting quality control. He suggested many steps to keep a check on quality. Some of these, in the context of medical care, are as follows [1]:

1. Prepare operational definitions of the services to be provided. Specify the standards of service and formulate clear guidelines about the identification of patients to be served and to be excluded by referral or otherwise.

2. Refuse to accept a higher level of mistakes in diagnosis and treatment for any reason including inadequate instruments; inappropriate technology; lack of expertise; or substandard drugs, chemicals, etc.

3. Instead of depending on ad hoc inspection, depend on statistical evidence of quality of incoming material such as pharmaceuticals, testing kits, and blood. The meaning of statistical evidence is given later in this chapter.

4. Implement a system of self-detection of errors and do not wait for complaints.

5. Continually update the doctors and technicians on rapidly growing developments and encourage them to acquire new skills.

6. Have as the objective not of finding cases of fault but of diagnosing the fault so that remedial steps can be taken.

7. Drive out fear so that employees feel free to make suggestions. Assign failure to the lacuna in the system rather than to the individuals.

8. Instead of setting numerical targets, set work standards in terms of quality. Thus, do not count the patients treated but assess the percentage of patients who were cured and satisfied.

9. Top management must be committed to quality and should earnestly implement the preceding suggestions.

A large number of issues related to quality of medical care can be discussed. The focus in this text is on statistics-based issues rather than medicine-based issues. Among the statistics-based issues, the following discussion is limited to some specific aspects that *illustrate* the help statistical methods can provide in improving the quality of medical care. The same kind of methodology can be used to control other kinds of errors.

20.1.1 Statistical Control of Medical Care Errors

If a patient with a disease, the treatment of which is known, is not cured, it would be generally correct to assume that this failure is a consequence of error on the part of the medical care facility. This can be due to carelessness or inadequate expertise of the attending personnel or to faulty tools and equipment. The present discussion is restricted to deviation such as from the prescription or the procedures and calls such errors as medical care errors. In a hospital, these could be in terms of administration of doses at times other than specified, in the wrong amount, omission of a dose or giving an extra dose, administration of an unprescribed drug, etc. Such errors can also occur with other procedures such as obtaining the history; establishing the diagnosis, measurements, and assessments; attending to the complaints of the patients; acting on warning signals; consulting a specialist; carrying out a laboratory investigation; and identifying side effects.

20.1.1.1 Adverse Patient Outcomes

Medical care errors are difficult to identify. These come to notice mostly when the outcome of a therapy is a sudden deterioration of the condition of the patient rather than gradual improvement. Listed next are adverse patient outcomes [2] that indicate that an error has occurred somewhere in patient management, including a clinician's error in judgment about the patient's condition:

1. Admission due to adverse results of outpatient management
2. Admission for complication of a problem on previous admission
3. Perforation, laceration, or injury to an organ incurred during invasive procedure
4. Adverse reaction to a drug or transfusion
5. Unplanned return to operation theater
6. Infection developing subsequent to admission, including infection occurring after operation
7. Transfer from general care to special care unit
8. New complications occurring after the start of therapy

This list may convince you that errors in medical care are not uncommon. It is preferable to keep separate track of each type of adverse outcome so that corrective steps can be taken accordingly.

20.1.1.2 Monitoring Fatality

The adverse outcomes just listed assume that the patient is alive. Some events are bound to be fatal in a hospital despite the best care. But some fatalities may be avoided. **Case-fatality** per 1000 patients admitted can be an indicator of the quality of care in a hospital. Because some hospitals are specialized to treat serious and terminal cases, interhospital comparison on the basis of gross fatality may not be valid. The same is true for different units of a hospital. However, trend within a hospital can certainly be monitored over time using this indicator. Mortality within 48 h in many cases is determined by the condition of the patient at the time of admission instead of the quality of medical care. When this is excluded, the rate is called the **net fatality rate**. This rate is relatively more valid for comparison between units or between similar hospitals. Case-fatality rate, which is calculated separately for each disease or condition, may also provide a good indication of the quality of care in different departments of a hospital and may reflect the ability to manage critical patients with different diseases. The other useful indicators are general anesthesia death rate, which is deaths related to general anesthesia per 1000 patients administered anesthesia, postoperative death rate, hospital maternal death rate (maternal deaths per 100,000 antenatals admitted), hospital early neonatal death rate, etc.

It is often difficult to identify a single error that has caused a particular adverse outcome. But if the management is sufficiently careful and the staff cooperative, analysis of each case of adverse outcome may successfully pinpoint the major cause. Thus, errors of different types can indeed be monitored.

20.1.1.3 Limits of Tolerance

Because the number and type of errors would vary from hospital to hospital and in a hospital from time to time, there is a need to handle these errors through statistical methods. It is the responsibility of the hospital managers to develop a system to keep track of errors of different types and of the opportunities where these errors can occur. The incidence of errors is generally computed in terms of rate per 10,000 opportunities for that type of error. Depending on the load of patients, availability of staff, and desire of management to provide quality care, a tolerable average proportion of medical care errors is fixed, say, at a level of 1 per 10,000 opportunities (i.e., $\pi = 0.0001$). An error-free environment is not possible even in a top-grade hospital. Again, because of various factors, fluctuations in errors would occur from day to day. These fluctuations, as you now know, are measured in terms of the standard error (SE). For number of errors, $SE = \sqrt{n\pi(1-\pi)}$ where n is the number of opportunities. These can be counted per day, per week, or per month, depending on convenience. For sufficiently large n,

$$\text{Upper limit of tolerance of error} = n\pi + 1.645 * SE. \qquad (20.1)$$

This is a one-sided (one-tailed) limit based on the Gaussian approximation with $\alpha = 0.05$. The number n would necessarily be large in a hospital of reasonable size. Although $n\pi$ could be small, large n may be able to take care of the Gaussian approximation. If so, there is only about 5% chance that the errors would exceed this threshold when the true rate is π. If errors exceed, action to rectify is warranted. In the context of control of errors, the lower limit of errors is of no consequence. Thus, the limit in Equation 20.1 is one sided.

Example 20.1: Tolerance for Medication Errors

In a 500-bed hospital with 80% occupancy, the number of medication opportunities per month at an average of four per patient-per day is $n = 500 \times 0.80 \times 4 \times 30 = 48,000$. If the long-range average medication errors are to be kept below 1 per 10,000 opportunities, then $\pi = 0.0001$. For $n = 48,000$ and $\pi = 0.0001$, the upper limit of tolerance is approximately

$$48,000 \times 0.0001 + 1.645\sqrt{48,000 \times 0.0001 \times 0.9999} = 8.4.$$

If in any particular week the errors are more than this limit, that is, nine or more, the cause should be investigated and remedial steps taken. When this is done, the error rate in the long run would not exceed the targeted average of 1 per 10,000 opportunities.

SIDE NOTE: In this case $np = 4.8$, which is less than 8. Thus the Gaussian approximation is suspect despite such a large n. Adjusting 8.4 to 9 should take care of this approximation.

Example 20.1 concerns medication errors. These can be more during night time for admitted patients than in day time and can be more with resident doctors than specialists. Thus, a break up may be helpful. Such error can be easily detected and can be kept at a very low level in a good hospital. However, errors in clinical judgment would not be so uncommon. For these errors, perhaps even $\pi = 0.01$ would be considered very good. Thus, the tolerable error rate would differ from one type of error to another.

It may not be easy to count the number of opportunities of error in a hospital. Special efforts may be needed to get an adequate count. The count is easier when the opportunities are divided into segments such as diagnostic opportunities, medication opportunities, and minor surgery opportunities. The breakdown also helps in identifying corrective action when needed.

20.1.2 Quality of Lots

Quality of the incoming material such as drugs, chemicals, and blood products in a hospital was mentioned earlier. When purchased from a reputed company, these may have already been subjected to a rigorous quality check. Yet, it is sometimes desirable to check their quality at the time of receiving them. This can be done by using a lot quality assurance scheme (LQAS). This scheme is also applied to the assessment of health services and to disease surveillance.

20.1.2.1 Lot Quality Method

The total material to be inspected for quality is first divided into lots of nearly equal size and a small sample of units from each lot is checked for quality. If the number of defective units exceeds a predetermined tolerance, the whole lot is rejected, otherwise it is accepted.

Suppose a hospital received 80 boxes, each box containing 10 laboratory kits. These can be divided into five lots of 16 boxes each. Each lot will now have 160 kits. Take a random sample of, say, $n = 15$ kits from each lot, and check these for a predetermined quality standard. If the number of kits not meeting the standard exceeds the tolerance level, reject the whole lot. The tolerance level for large n is determined by expression similar to Equation 20.1 depending on the acceptable rate of error. For small n, an exact threshold [3] based on the binomial distribution is needed.

20.1.2.2 LQAS in Health Assessment

The lot quality method has been used in disease surveillance, nutrition programs, assessing women's health services, and, most of all, immunization coverage assessment [4].

Example 20.2: Tolerance for the Level of Nonimmunization

Suppose coverage of at least 90% by polio vaccine is required to build herd immunity so that the disease is not able to transmit itself. Thus, not more than 10% of children should remain uncovered ($\pi = 0.10$). The target population is children below the age of 3 years in a district. The area of the district is divided into, say, 20 zones that will serve as a "lots" in LQAS. Within each zone, a random sample of $n = 80$ children is examined for noncoverage by polio vaccine. The tolerance threshold in this case is

$$80 \times 0.10 + 1.645\sqrt{80 \times 0.10 \times 0.90} = 12.4.$$

If in any sample of size 80, the number of nonimmunized children is 13 or more, reject that lot. That lot—zone in this case—can be considered to have almost surely not reached the level of 90% coverage required for herd immunity because the probability of this happening is less than 0.05 when the number of nonimmunized in a sample of 80 children from that zone is 13 or more.

The following comments regarding LQAS may be helpful:

1. The division of the total material into lots should preferably be such that each lot is homogeneous while the lots themselves are different from one another. In Example 20.2, the zones may be such that one consists mostly of a slum or underprivileged population and another of a population of high socioeconomic status. LQAS can then better identify the zones with low coverage.

2. The defectives in the case of commodities transform to immunization noncoverage in Example 20.2. The problem can be stated in terms of immunization coverage in place of noncoverage. The tolerance threshold will then change to ($n\pi - 1.645 * SE$), where π is now the proportion coverage of 0.90 required for herd immunity in this example.

3. Sometimes, lots with a high proportion of defectives can also be wrongly accepted under this scheme (for that matter, in any statistical decision). This is the same as Type-II error under the testing of hypothesis procedure. Methods are available to devise an LQAS such that the probability Type-II error is under control. For details, see Robertson et al. [4].

20.1.3 Quality Control in a Medical Laboratory

Laboratories differ in their methods, chemicals, skills of the staff, etc., and thus results for aliquots of same specimen may differ from laboratory to laboratory. Part of this variability can be eliminated by standardization across laboratories. Differences also occur within the laboratory from time to time. If such differences are substantial, it shakes the clinician's confidence in the values reported. Quality control helps to keep a check and maintain a high level of performance. This can be done with the help of a control chart.

20.1.3.1 Control Chart

For quality control in medical laboratories, a specimen with known composition is analyzed at least once everyday. Such a specimen is called a control. This is preserved under standard conditions for repeated analysis. The daily readings for this control specimen are plotted on a graph called the control chart. This is prepared by adopting the following steps:

Step 1. Carefully analyze the control specimen at least 20 times under standard conditions. This will delineate the random error and should be small in this set up. They can occur due to the factors that vary in the operation of method such as in timing, temperature, humidity, staining, and actual measurement. However, this may be restricted to the factors operating in short span of period. It fails to consider long-term variation such as in different seasons, even between night and day. For better control, the random errors should be studied separately "within day" and "between days." Also, the random error must be estimated for the same solution (aqueous, serum, etc.) as proposed for final use.

Step 2. Calculate the mean and standard deviation (SD) of these 20 readings. Because the same specimen is being repeatedly analyzed, the distribution pattern would be Gaussian and SD would be small.

Step 3. The tolerance range is

$$(\text{mean} - 2\text{SD}, \text{mean} + 2\text{SD}). \tag{20.2}$$

The justification for the tolerance range is the same as for the normal range. A useful property of this range is that, on average, only 1 in 20 samples is expected to fall outside the range, if no systematic error is committed.

Step 4. Draw the lines corresponding to mean, (mean − 2SD), and (mean + 2SD) on a chart as in Figure 20.1. Your control chart is ready.

Step 5. Plot the readings for the control specimen on this chart every morning before using the laboratory for other incoming specimens. If a reading on any day is outside the tolerance range, consider this as an indication that the laboratory requires scrutiny. Take corrective measures as revealed by this scrutiny before starting the analysis of other specimens.

Example 20.3: Control Chart for Measuring Serum Glucose Level

Consider the following data on serum glucose level on repeated analysis of a control specimen with a known level of 80 mg/dL:

Repetition number	1	2	3	...	...	20
Serum glucose level (mg/dL)	78	80	84	...	...	82

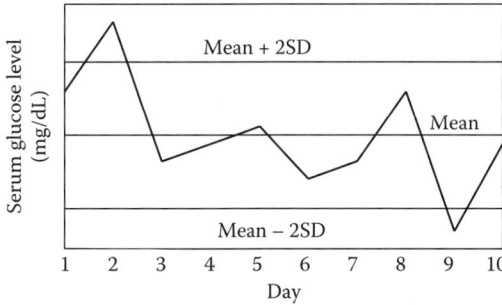

FIGURE 20.1

Quality control chart.

TABLE 20.1

Serum Glucose Readings in Control Specimen

	Day									
	1	2	3	4	5	6	7	8	9	10
Serum glucose level (mg/dL)	83	87	79	80	81	78	79	83	75	80
Difference from 80	+3	+7	−1	0	+1	−2	−1	+3	−5	0
Cumulative sum of difference from 80	+3	+10	+9	+9	+10	+8	+7	+10	+5	+5

The data for intermediary days are not shown. Suppose mean = 80.5 mg/dL and SD = 2.1 mg/dL. Thus mean ± 2SD limits are (76.3–84.7). The lines corresponding to mean, (mean − 2SD), and (mean + 2SD) are shown in Figure 20.1. This completes the first four steps just outlined. Step 5 is taking readings of the control specimen on each day. Suppose these readings for 10 days are as shown in Table 20.1. These values are plotted on the control chart (Figure 20.1). Any reading above or below the respective tolerance limits is considered a potential error and is investigated further for reasons that produced that kind of unusual value. In most such cases, an assignable reason can be identified and corrective steps can be taken before actual specimens from patients are analyzed. In this example, the value on day 2 is unacceptably high and the value on day 9 is unacceptably low. When corrective steps are regularly taken to minimize the occurrence of such outliers, the laboratory's performance is considerably improved.

SIDE NOTE: The value 2.1 of SD is rather high in this example considering that it is for variation in control specimen in standard conditions. In a good laboratory, this would be quite low.

20.1.3.2 Cusum Chart

In addition to outliers, an increasing or decreasing trend in the plot may indicate a need to check the functioning of the laboratory. Some subtleties

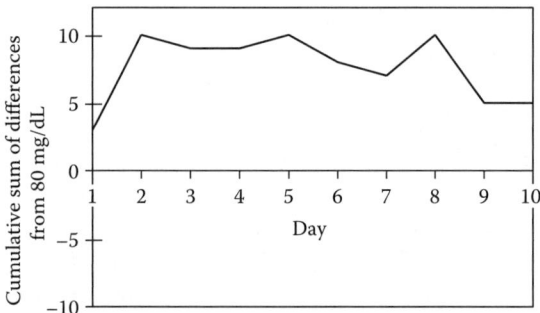

FIGURE 20.2
Cusum chart for the data in Example 20.3.

of the daily change in the values are revealed much better by what is called a cusum (cumulative sum) chart. For this, the cumulative differences over days from the average (or standard value) are plotted as in Figure 20.2 for the data in Table 20.1.

Even though no trend in this plot is discernible, all values are on the positive side of zero. This shows that there is some tendency to overestimate the value. It is, thus, worthwhile to examine the laboratory procedures and functioning for possibility of a positive bias. This drift is more easily detected by a cusum chart than by the conventional control chart.

A quality control chart takes care of one aspect of quality, namely the performance relative to a standard. The other aspect is the extent of repeated investigations ordered by clinicians either with the same laboratory or with another laboratory. Reordering of a laboratory investigation indicates that the attending clinician lacks confidence in the results. The discrepancy can occur when the results do not adequately match the clinical condition of the patient. This does not necessarily arise because of lack of quality of the laboratory. Nevertheless, the magnitude of reordering can be considered to measure the extent of confidence in a laboratory.

20.1.3.3 Other Errors in a Medical Laboratory

The discussion so far was on errors in analysis of the specimen. Many other errors can occur in a medical laboratory. These include incomplete preparation of the patient before sampling, collection of specimen in inappropriate container, mislabeling or mishandling, delay in transport, improper storage, mistakes in data entry, and above all misinterpretation of the physician's order for the test. All these can add up to a formidable chance of error.

Try to imagine the cost of these errors to the patient and medical care system. A wrong reporting can cause immense discomfort and can even cost a life. Ethics is compulsive in this activity as in all others involving health and well-being of people. If you are interacting with a laboratory where a large number of specimens are processed everyday, see if the laboratory is keeping a count of such errors and has a mechanism to address the quality issues.

Because of widespread computerization, much of the data required for quality control are available online these days. Programs are prepared to issue warning signals when the system seems to derail. This, as well as issues related to multivariate control of quality, is discussed by MacGregor [5], although in the context of manufacturing industry. The application to medical setups is immediate.

20.1.3.4 Six-Sigma Methodology

As stated earlier, health care is an industry and quality issues cannot be sidelined. Six-Sigma methodology has saved billions of dollars for different industries by improving efficiency and profitability. The concept applies to the whole process and requires cross-cutting approach involving various departments such as administrators, managers, physicians, and technicians.

You know that sigma (σ) is the notation for SD. You also know that three-sigma on either side of mean in case of Gaussian distribution covers 99.7% and leaves out only 0.3% probability. Use it for errors and realize that these limits leave out only 0.3% chances of error as uncovered. Extend it to six-sigma on either side of mean. Table B.1 does not go that far but it is known that 4.5σ limits tolerate only 3.4 errors per million. Notice how intolerant this is compared to the usual 95% confidence intervals (CIs) defined by ± 1.96SE that leaves out 5% chances of error. When the possibility of long term shift of 1.5σ is allowed, this becomes 6σ.

If you are in an industry, you can easily realize how difficult it is to achieve error rate of less than 3.4 per million. A motivated team is required to identify areas of improvement and devise strategies to tackle them. Customer satisfaction surveys would be integral part of this exercise. Statistically, this may require studies on cause and effect, and development of tree diagrams and flow charts, besides study of error distributions with its mean and variance, CIs, and tests of significance.

Whereas Six-Sigma concept has given a definite boost to statistical thinking, a caution is required when nonexperts use the concept. They may have unacceptably low numbers for monitoring and this could jeopardize Gaussian application. Statistical errors such as ill-placed control limits and wrongly chosen factors can confound the results and may push some manager to consider it as a nuisance. When properly used with well-meaning intentions, Six-Sigma has considerable potential to improve various aspects of health care.

20.1.3.5 Nonstatistical Issues

Statistical methods just discussed help to identify the errors. However, quality is achieved by taking effective steps: first to prevent recurrence and second by anticipating and preventing new occurrence. Spear and Schmidhofer [6] discuss how high degree of specification together with rapid response to individual problems can improve health care. The other key is to make this a cross-departmental so that the quality control exercise is owned by all and results implemented as required for Six-Sigma efforts.

20.2 Quality of Measurements

You may be familiar with the garbage-in garbage-out (GIGO) syndrome. Even the most immaculate statistical treatment of data cannot lead to correct conclusions if the data are basically wrong. Before starting an analysis, everything possible should be done to ensure that the information is correct. This mostly depends on the validity and reliability of the tools and instruments (including manpower) used for the collection of information and on the quality of data actually obtained through these instruments.

You have earlier seen an account of the large number of steps that a patient or a subject has to pass through after entering a health-care web. For example, in a clinic setup, health is assessed by eliciting the history, conducting a physical examination, carrying out laboratory or radiological investigations, and, sometimes, longitudinally monitoring the progress of the subject. Expertise and attention of the person doing this assessment are critical. The health assessment requires use of a questionnaire or schedule for systematically eliciting and recording information; medical instruments such as stethoscope, BP instrument, and weighing machine; laboratory chemicals, reagents, and methods including machines such as analyzers; x-ray, ultrasound and other imaging devices, and all that go along with them; diagnostic criteria and definitions including scoring systems, if any; and prognostic indicators. All these generically are instruments in statistical sense because it is through them that the data are obtained. Various procedures for patient management such as surgery and treatment regimens also come under this general term. Under certain conditions, these procedures can also be called tests instead of instruments. As emphasized earlier, no instrument is perfect in practice because the performance of the instrument also depends, to a degree, on the human input. The quality of medical decisions depends very substantially on their dependability of these instruments in actual use. This is assessed in terms of the validity and reliability of the instruments.

20.2.1 Validity of Instruments

Validity refers to the ability of an instrument to measure correctly what it is supposed to measure. What signs and symptoms or hematological parameters (such as hematocrit) can correctly identify cases of dengue fever in early stages because serological evidence takes so much time? Is BP > 140/90 mmHg the right definition for hypertension or is BP > 160/95 more correct? Both the definitions are used. (See Refs [7,8], incidentally these are nearly consecutive articles in the same journal.) This means that those with BP = 150/92 are considered hypertensive by one definition but not by the other. Is waist–hip ratio (WHR) an appropriate measure of obesity or is body mass index (BMI) more appropriate? If it is BMI, should the threshold be 25 or 30 or somewhere in between, such as 27? How can one distinguish, without error, a hypothyroid condition from euthyroid goiter? How can one correctly ascertain cause of death in an inaccessible rural area where a medical doctor is not available? How can age be assessed when a dependable record is not available? Can the diet content of an individual be adequately assessed by taking a 3-day history? How can physical activity be validly measured?

Instruments of different hues are used, and their validity is never perfect. Validity is measured against a gold standard. Such a standard, in most cases, is elusive. As a consequence, validity suffers. Once an acceptable gold standard is available, indices such as sensitivity, specificity, and predictive values are calculated to measure validity. These measures have been discussed in Chapter 9.

In the context of a questionnaire or a schedule, validity is assessed in terms of, for example, its ability to distinguish between a sick and a healthy subject or a very sick and a not-so-sick subject. This means that the scores received or the responses obtained from one group of subjects should be sufficiently different from those of another group when the two groups are known to be different with respect to the characteristic under assessment. Consider a questionnaire containing 20 items measuring quality of life. If the response on one particular item, say, on physical independence, from the healthy subjects is nearly the same as from the mentally retarded subjects, then this item is not valid for measuring quality of life in those types of subjects. This item can be deleted. A shorter questionnaire is always better than a longer questionnaire.

20.2.1.1 Types of Validity

Validity is an intricate concept that changes according to the setup. Important types of validity for the purpose of medical applications are face, criterion, concurrent, content, and construct validity.

Face validity is an apparent correspondence between what is intended and what is actually obtained. Grossly speaking, ovarian cancer cannot occur in males and prostate cancer cannot occur in females. Cleft lip, deafness, and dental caries cannot be causes of death. If any data show such inconsistency, these are not face valid. Face validity is also violated when a patient is shown as discharged from a hospital and cause of death within the hospital is also recorded. Current age cannot be less than the age at first childbirth. Thus, face validity is achieved when the observations look just about right. If a man tells his age as 20 years but looks like a 40 year old, the response is not face valid. If a family reports consuming food amounting to 2800 cal per unit per day but the children are grossly undernourished, then again the response or the assessment is not face valid.

An instrument or a method is called **criterion valid** if it gives nearly the same information as a criterion with established validity. Carter et al. [9] reported criterion validity of Duke Activity Status Index with respect to standard pathologic work capacity indices in patients of chronic obstructive pulmonary disease. Whenever a new index or a score or any other method is developed, it is customary that its criterion validity is established by comparing its performance with a standard. Review of literature suggests that correlation is the generally used statistical method for this purpose, whereas the right assessment of criterion validity is obtained by evaluating predictivity. If negative and positive predictivities are high, the tool can be considered criterion valid.

It is not uncommon in medical setup that no validated standard is available. For example, no valid measure is available to assess physical balance in people with vestibular dysfunction. The Berg Balance Scale is often used but is not considered a fully valid measure. A relatively new tool is Dynamic Gait Index. The two can be investigated for agreement. Again, this is different from correlation. If the agreement is good, the two methods can be called **concurrently valid**. This means that both are equally good (or, equally bad!). This, however, does not establish superiority of one over the other. The superiority can be inferred only when two methods under evaluation are compared with a known gold standard.

The fourth type is **content validity**. This is based on the domain of the content of the measurement or the device. Would you consider kidney function tests such as urea clearance and diodrast clearance content valid for assessing overall health of kidneys? Such tests restrict to specific aspects and do not provide a complete picture. Content validity is often established through qualitative expert reviews. Wynd and Schaefer [10] explain how content validity of osteoporosis risk assessment tool was established through a panel of experts.

The last is the **construct validity**. This seeks agreement of a device with its theoretical concept. BMI is construct valid for overall obesity but not for central obesity, whereas WHR is construct valid for central obesity and not for overall obesity. Construct validity is sometimes assessed by the statistical method of factor analysis that reveals "constructs." If these constructs correspond well with the ones that are otherwise theoretically expected, the tool is considered construct valid.

20.2.2 Reliability of Instruments

You know that a clinical thermometer is a reliable tool for measuring body temperature. It behaves the same way under a variety of conditions. Reliability of an instrument is its ability to be consistent across varying measurements in the sense of providing same results when used by different observers or in varying conditions. The performance of a reliable tool is not affected much by the external environment in which the tool is used. Unlike validity, the concept of reliability is not related to any gold standard. Interrater reliability and interlaboratory reliability were discussed in an earlier chapter. The latter is also called reproducibility. Both can be measured by the extent of agreement. This is applicable to one item at a time.

An instrument, such as a questionnaire, can have multiple items. How does one assess the reliability of such an instrument? Reliability of a multiple-item instrument, which is now called a "test" for convenience, has two components: (1) internal consistency and (2) stability across repeated use. The first concerns the consistency of responses across various items in the same test. The second is self-explanatory and is also called repeatability.

20.2.2.1 Internal Consistency

Consider a test in the form of a questionnaire containing several questions or items on, say, a scale of 0–5. An example is the assessment of ability to perform activities of daily living (ADL) by a geriatric population. Items such as ability to dress, to bathe, to walk around, and to eat can be scored from zero for complete inability (full dependence) to five for complete independence requiring no assistance. It is expected that the ability score on one item would correspond to score on other items. If there were a difference, it would generally persist across subjects. In fact, all items are different facets of the same entity—the ADL in this case. That the underlying construct is uniform across all items of a test is an important prerequisite for measuring internal consistency. If that is so, the responses will be consistent with one another, provided the items are properly framed and the questions appropriately asked.

One way to measure internal consistency is to split the test into ostensible halves where feasible. This is generally possible for a questionnaire by randomly dividing questions into two parts. The other method is to put odd-numbered questions in one-half and even-numbered questions in the other half. The product–moment correlation coefficient between total scores in the two parts across several subjects is a measure of internal consistency. This is called **split-half consistency**. Note that there is no one-to-one correspondence between one item in one half of the test and an item in the other half of the test. Thus, intraclass correlation cannot be used in this case. The usual product–moment correlation between total scores is used to assess split-half consistency.

Split-half implies that the number of items available for calculating the total score is one-half of the items in the whole test. Such reduction in number of items adversely affects reliability assessment. An adjustment, called the

Spearman–Brown coefficient, can be used to estimate the reliability if the test were to consist of all the items. For details, see Carmines and Zeller [11].

20.2.2.2 Cronbach Alpha

The second method for assessing internal consistency is

$$\text{Cronbach alpha} = \frac{M\bar{r}}{1+(M-1)\bar{r}}, \quad (20.3)$$

where
  M is the number of items in the test
  $\bar{r}$ is the average of $M(M-1)/2$ correlations between each pair of M items

Again, product–moment correlation is used. Thus, this alpha is applicable only when the response to various items is on a metric scale such as the graded response on ADL in the previous example. A basic assumption is that the variances of the scores on individual items are equal. If this is not the case, the measure in Equation 20.3 needs modification as indicated by Carmines and Zeller [11].

Equation 20.3 shows that the value of Cronbach alpha depends on the number of items in the test. When $\bar{r}$ is 0.3, alpha = 0.68 for $M = 5$ and alpha = 0.87 for $M = 15$. Thus, the Cronbach reliability of a test can be increased by simply increasing the number of items, provided that the average correlation does not deteriorate. The reason for such behavior is not too hard to find. Cronbach alpha assumes that the items included in the test are a random sample from a universe of many possible items. The higher the number of items, the better is the representation.

The following comments give some more information regarding Cronbach alpha:

1. Cronbach alpha is based on all items of the test and thus is a better indicator of reliability than the split-half coefficient. The latter is based on only half the items. Also, split-half is based on the sum total of the scores on the items in the half-test, whereas Cronbach alpha considers each item separately and uses the average correlation.

2. Cronbach alpha measures how well the items are focused on a single idea or an entity or a construct. All items are supposed to be focused on the same entity such as disability in the ADL example. But it does not tell you how well the entity is covered. A high value of this alpha is possible even when the breadth of the construct is covered only partially.

3. Cronbach alpha measures the internal consistency component of reliability of an instrument as a whole and not of individual items. However, correlations of individual items with one another can be examined to find whether or not any particular item is adequately related to the others.

4. Cronbach alpha can also be viewed as the correlation between the test and all other possible tests containing the same number of items on the same entity or construct. In the ADL example, the questions can be on being able to sit on a chair instead of walking around or can be on being able to climb stairs. If the test is reliable, the alpha will not materially change when an item is replaced by another equivalent item.

5. All measures of reliability, including Cronbach alpha, range from 0 to 1. Thus, these measures are easy to interpret. A value less than 0.4 indicates poor reliability, between 0.4 and 0.6 toss-up, between 0.6 and 0.8 acceptable, and a value more than 0.8 excellent. However, an exceedingly high value such as >0.90 suggests some repetition and redundancy.

6. If the response to items is binary (yes/no, true/false, or present/absent type), then Cronbach alpha reduces to what is called the **Kuder–Richardson (K-R) coefficient**. For details, see Carmines and Zeller [11].

7. All measures of internal consistency are affected by length of test, homogeneity of test items, heterogeneity of the subjects in the sample to whom the test was administered, and objectivity of the test items.

In addition, the conditions of administration of the test such as time allowed for completing the test and observer–subject interaction affect reliability assessment.

Example 20.4: Internal Consistency of an AIDS Questionnaire

A questionnaire consisting of four items on knowledge of acquired immunodeficiency syndrome (AIDS) was administered to 80 students of grade VII. These are as follows:

Item 1: Knowledge about mode of spread of human immunodeficiency virus (HIV) infection

Item 2: Knowledge about its prevention

Item 3: Knowledge about social handling of AIDS patients

Item 4: Knowledge about symptoms of AIDS

Each item is graded on a four-point scale from zero for no or completely wrong knowledge to three for perfect knowledge. Suppose the correlations are as shown in Table 20.2. These give $\bar{r} = 0.563$. Thus,

$$\text{Cronbach alpha} = \frac{4 \times 0.563}{1 + 3 \times 0.563} = 0.84.$$

The internal consistency, thus, is "excellent," despite the fact that the fourth item has poor correlation with the other items. Low correlation of this item with others is an indication that knowledge of symptoms is probably not part of the same entity that is being measured by the other three items.

TABLE 20.2

Correlation between Items of Knowledge
about AIDS

Item	Item			
	1	2	3	4
1		0.87	0.62	0.45
2			0.75	0.33
3				0.36

20.2.2.3 Test–Retest Reliability

The second component of reliability is repeatability. This is the stability of the response with repeated use of the instrument. The measurement of repeatability essentially involves administering the test twice to the same subjects. This is therefore known as test–retest reliability. It can be calculated for an instrument that does not provide any learning to the respondent, and the second-time responses are not affected by the first-time responses. The scores or the responses obtained on the two occasions are checked for agreement. These can be checked itemwise, but generally the total score is compared. You now know that the extent of agreement is measured by intraclass correlation if the responses are quantitative and by Cohen κ (kappa) if they are qualitative.

Suppose you want to find out whether laboratory-A is more reliable than laboratory-B in measuring plasma glucose levels. One strategy for this could be to split each blood sample into six parts. Send three parts to laboratory-A and three to laboratory-B. Do this for, say, 25 subjects. Keep the laboratories blind regarding sending aliquots of the same sample. Calculate the intraclass correlation for each laboratory. If the laboratory is reliable, the measurement on the different parts of the same sample will give nearly the same values, yielding a high value of the intraclass correlation coefficient. The laboratory with higher intraclass correlation is more reliable. Note that the reliability may be higher even when there is a constant bias in the measurement. It is possible that a laboratory has an intraclass correlation coefficient of 0.90 but consistently reports values lower or higher than the actual value. Thus, reliability does not ensure validity.

Example 20.5: Repeatability of Vision in Keratoconus

Keratoconus is a disease in which the eye lens bulges and becomes conical in shape. It is believed that these patients generally report variable vision on different occasions. Gordon et al. [12] measured visual acuity (VA) in 134 keratoconus patients on two visits separated by a median of 90 days. The examiner was same in the two visits for some subjects but different for other subjects. High- and low-contrast VA was measured with the patient's habitual visual correction and with the best correction monocularly. Thus, four measurements were available for each patient for each visit.

The intraclass correlation between VA on the two visits ranged from 0.76 to 0.85. The VA score was somewhat higher at the repeat visit than at the baseline visit when the examiners were different between visits. It was concluded that the VA in this sample was very repeatable, and repeatability was slightly poorer when different examiners tested VA at the baseline and repeat visits.

Example 20.6: Reliability of SF-36 in Chronic Schizophrenia

Measurement of the changes in physical and role functioning in outpatients of mental health clinics has been a challenge. Russo et al. [13] tested a 36-item health survey form (SF-36) on 36 outpatients with chronic schizophrenia (incidentally, in this example $n = 36$ and $M = 36$ are equal). The form was completed in writing as well as orally by each patient. The test–retest reliability as well as internal consistency showed that SF-36 is an appropriate outcome measure for schizophrenic outpatients.

20.3 Quality of Statistical Models: Robustness

Models by definition are relatively simple statements of complex processes. They are man-made manifestations of the nature. By their very nature, models are imperfect representation of real world and at best only approximations of the actual situation. Statistical models are no different. Given the complex interrelated nature of health and disease, the models also tend to become complex. They go much beyond the simple ±2SE thresholds. Efforts are made to keep parsimony intact so that the utility is not compromised. Because of this limitation, most statistical models used in medicine and health are fragile. Often they work only in very specific situations, and slight variation in the conditions can produce weird results. Antonym to fragility is **robustness**. This means consistency, sustainability, and stability of results in less than ideal conditions. It refers to insensitivity of the overall conclusion to various limitations of data, limited knowledge, and varying analytical approaches. By showing endurance of results to withstand the pressure of minor variations in the underlying conditions, robustness establishes their wider applicability.

The results that risk of coronary disease is higher in persons with higher BP and risk of lung cancer is higher in persons smoking heavily are robust. The presence or absence of other factors does not affect these results much. Also, this is seen in all population segments. On the other hand, the conclusion that urinary sodium excretion is dependent on salt intake is not robust since it is so easily affected by metabolism, creatinine excretion, and BMI. The result that risk of lung cancer increases by 1% by smoking 100 additional cigarettes-years is also not robust. This chance can easily be 2% or 0.5% depending on the person's nutrition, exposure to kitchen smoke, etc.

Robustness is relatively a new concept and is not in regular practice yet in medicine. Only good researchers use this strategy to rule out challenges from critics. Robustness is difficult to ascertain. The evaluation could be done for the set of underlying conditions as a whole but is generally done for its components such as robustness to choice of variables, to instrumentation, and to the methodology. Individual conditions are altered by turn to examine if the broad conclusion still remains the same. If you are using a published result for your practice, examine if the result is sufficiently robust to the altered conditions under which you intend to apply it. This also is a way to minimize the impact of uncertainties.

You know that most biological phenomena are very complex and a large number of factors need to be included if these are to be expressed in terms of a mathematical expression. For example, a prerequisite for a regression model to be useful is that all relevant regressors are included. This may be hindered by lack of knowledge about the relevant regressors. For example, if it were not adequately known which factors influence the duration of survival after detection of malignancy, any model on survival would be tentative.

Other limitations of statistical models are as follows:

1. Models may not help—even mislead—if the underlying processes are not fully specified.
2. A critical step is to decide whether or not a model is credible, plausible, and consistent with existing knowledge and actual observations. This step was being ignored in earlier literature but is now almost ritually followed.
3. Models are accepted in the absence of better alternative despite aleatory and epistemic uncertainties they contain.
4. Models are based on past experience but are used for predicting future. Past trends may change in future due to interventions such as improved strategies. Thus, guard against overconfidence.
5. Beware of butterfly effect. As in theory of chaos, a tiny error in the beginning can end up in tornado. Forecast can go haywire if there is even minor error in the input values.
6. Many models simulate the present well, yet disagree in the magnitude of changes they predict for future.

The caveats do not imply that model building is a worthless exercise—only that adequacy of model for future projections is difficult to assess as the underlying uncertainties can be very unpredictable. Also modeling is a continuous process—it evolves as the new evidences emerge. Less effective components are replaced or modified, whereas the successful components are retained.

Nevertheless, statistical models do approximate biological processes fairly well when prepared with sufficient care and when the missing links are not too many. But their adequacy must be fully assessed before they are actually used. For this, three types of approaches are available: (1) external validation, (2) sensitivity and uncertainty analysis, and (3) resampling. You would soon notice that all these approaches answer questions regarding the population settings, outcome variables, and antecedents to which the model can be generalized. Opposed to this, as you might have noticed, the concern in most statistical methods discussed in this book is on internal validity. Internal validity primarily answers questions regarding the effect being really present rather than arising from extraneous considerations.

20.3.1 External Validation

A model may turn out weak when datasets for developing it is the same as to evaluate it. This is a kind of circular reasoning. External validation overcomes this drawback. Although statistical significance implies that the results are replicable, trial of results on another group of subjects is desirable. Agreement on repeated application means that the model is more likely to be adequate. To examine this kind of robustness, the first step is to test results on another group from the same target population. If found valid, the next step is to test the results on a group of subjects from a different but similar population. Sometimes, the settings and measurement scales are varied to assess external validity. New sample of subjects can be taken in a variety of ways.

20.3.1.1 Split-Sample Method

The split-sample method can be applied where a large number of eligible subjects are available for the study. This method requires that the information is collected on all subjects but only random half is used for statistical analysis. If the result fails statistical significance in this half-sample, just draw lessons for another study and do not draw any substantive conclusion regarding the question under research. If significance is achieved, use the second half group of subjects and examine if the results validate. The first half-sample is called **training dataset** and the second half **test dataset.** If the test dataset supports the model derived from training dataset, combine the two and obtain more reliable (due to increased *n*) estimates of the parameters of the model.

Suppose you want to develop a model to predict the gestation age on the basis of anthropometric measurements at birth in situations where the date of last menstrual period is not known as can happen with less-educated women. If a large sample such as 1800 is available, random 900 of them can be used to develop the model, and the other 900 to test it. If testing gives correct gestation such as within ±1 week in at least 80% cases, the model can be considered robust for application.

Split sample method may look like more for internal validation than external validation as the same sample is used. Some consider split sample a weaker procedure as it is based on just one sample.

20.3.1.2 Another Sample Method

Another sample method requires a new sample at another point of time, possibly in other setting. If this sample also gives the same result, utility increases tremendously. In the gestational age example in the preceding paragraph, the validation sample can be taken from some other hospital located in the same area or in some other area with same socioeconomic milieu. Many researchers leave such validation onto the future researchers. McGill pain score has been validated in a variety of settings by different researchers, yet its utility in largely illiterate population remains questionable. Try to use properly validated results in your professional practice and research.

> **Example 20.7: Validation of Automatic Scoring Procedure for Discriminating between Breast Cancer Cases and Controls**
>
> Micronucleus test (MNT) in human lymphocytes is frequently used to assess chromosomal damage. Varga et al. [14] developed MNT on the basis of computerized image analysis. In a test sample of 73 persons (27 breast cancer cases and 46 controls) in Germany, the automated score gave OR = 16. This is a substantial improvement over OR = 4 for the conventional score based on visual counting. The improvement was confirmed in a validation sample of 41 persons (20 cases and 21 controls) that gave OR = 11.
>
> **SIDE NOTE:** In this study, the variation in OR in test and validation sample is high. This tends to reduce the confidence in the result.

20.3.2 Sensitivity Analysis and Uncertainty Analysis

Models, particularly statistical models, are not unique. Different sets of parameters may reasonably reproduce the same sample values. Thus, agreement between model and the observed data does not imply that the model assumptions accurately describe the underlying process. Only that it is one of the plausible explanations and empirically adequate. Thus, the model needs to be checked under varying conditions. The sensitivity analysis refers to the study of effect of changes in the basic premise such as individual and societal preferences or assumptions made at the time of model development. These assumptions are made to plug the knowledge gaps.

Against this, uncertainty analysis is the process to measure the impact on the result of changing values of one or more key inputs about which there is uncertainty. Thus, this mostly arises due to use of sample estimates. The inputs are varied over a reasonable range that can practically occur. In essence, sensitivity analysis is primarily for epistemic uncertainties, whereas uncertainty analysis is mostly for aleatory uncertainties.

20.3.2.1 Sensitivity Analysis

In the calculation of disability adjusted life years (DALYs), the assumption that a death in young age is much more important than death in childhood or old age is a value choice. In addition to such assumptions, variables used for developing model also are basic to the model. These also depend on the choice of the investigator. A simple example is that you model systolic BP in healthy subjects to depend on age, sex, and BMI. Another choice could be age, sex, and socioeconomic status. These choices are deterministic rather than stochastic. More accurate measurement or scientifically sound methodology does not help alleviate this uncertainty. Sensitivity analysis is varying these basic conditions and verifying that the broad conclusion still remains the same. Thus, this also is an exercise in external validation.

Risk of coronary disease can be modeled to depend upon the presence or absence of diabetes, hypertension, and dyslipidemia. This model might be able to correctly predict 10-year risk in 62% cases. However, addition of smoking and obesity can increase this to 70%. This addition of 8% is substantial. Thus, predictivity of coronary disease is *sensitive* to the choice of risk factors. The first model is based on three risk factors and the second on five factors. Had the contribution of smoking and obesity been only 2% or 3%, the conclusion would be that prediction of coronary disease is insensitive to smoking and obesity when diabetes, hypertension, and dyslipidemia are known. Then, robustness would have established.

The sensitivity analysis investigates how the result is affected when the basic premise is altered. Unlike uncertainty analysis, this has nothing to do with the variation in the input values. Input factors themselves are changed. Sensitivity analysis deals with uncertainty in model structure, assumption, and specification. Thus, it pertains to model uncertainty.

Example 20.8: Sensitivity Analysis of Shock for Predicting Prolonged Mechanical Ventilation in Intensive Care Unit

Estenssoro et al. [15] compared Argentinean patients on mechanical ventilation in intensive care unit (ICU) for more than 21 days ($n = 79$) with those with less ventilation ($n = 110$) for severity score, worst PaO_2/FIO_2 fraction, presence of shock on ICU admission day, length of stay in ICU, and length of stay in the hospital. Logistic regression identified shock on ICU admission day as the only significant predictor with an OR = 3.10. This analysis excluded patients who died early. It was not known whether this result would or would not hold for patients dying early. To bridge this epistemic gap, the authors conducted a sensitivity analysis by including 130 patients who died early. Shock remained a powerful predictor. The conclusion is that the only prognostic factor for prolonged mechanical ventilation is shock on ICU admission day irrespective of early or late death (or survival). Perhaps, shock itself is a by-product of severity of illness and hypoxemia.

The purpose of sensitivity analysis is to examine whether the key result continues to point to the same direction when the underlying structure is altered within plausible range. The process involves identifying key outcome in the first place, and then the basic inputs that can affect this outcome. For example, in a clinical trial setup, the outcome could be mortality or the length of hospital stay. Both should generally lead to nearly the same conclusion. Thus, the outcome measure can also be changed to see if the results are still same. Patients' inclusion and exclusion criteria can be relaxed, the method of assessment can be altered, and even the data analysis methods can be changed to see if this affects the final result. Intention-to-treat (ITT) analysis can be tried in addition to the regular per protocol analysis that excludes the missing or distorted data. If the results do not materially change, the confidence in the results strengthens.

20.3.2.2 Uncertainty Analysis

The study of the effect of varying the value of parameters included in the analysis is called uncertainty analysis. Thus, this pertains to parameter uncertainty. Examples of such parameters used in calculation of DALYs are incidence, prevalence, and duration of disease and mortality rates for various health conditions. The value of DALYs will naturally change if any of these are changed. Correct estimation of these parameters is vital to the validity of the DALYs obtained. The uncertainty surrounding these parameters can be reduced through more accurate measurement and by adopting a scientifically sound methodology of estimation. This is not possible for the preferences mentioned for sensitivity analysis.

In medicine, uncertainty analysis is generally done for a model that relates an outcome with its antecedents. For example, urinary excretion of creatinine can be predicted by age, dietary constituents, and lean body mass. If age is reported approximately in multiples of 5, such as 45 years instead of exact 43, will the model would still be able to predict nearly the correct value of urinary creatinine? If this example is not convincing, consider the effect of changes in diet from day to day. If the outcome prediction remains more or less same, the model is considered robust.

Uncertainty analysis incorporates three types of aleatory variations in the inputs: (1) Random measurement errors such as approximate age in the creatinine example. (2) Natural variation that can occur in the input parameters such as dietary constituents can change from day to day. If the model is based on the average diet over a month, the model may or may not be able to reflect the effect of daily changes. If it is based on diet of the previous day, what happens if the diet changes the next day? (3) Variation in the multipliers used in the prediction. If ln(creatinine in mmol/day) is predicted as $(0.012 \times$ height in cm $- 0.68)$ in children, what happens if the multiplier of height is 0.013 or 0.011. The multiplier 0.012 is an estimate, which is subject to sampling fluctuation. This multiplier can change depending upon the subjects happen to be in the sample.

Example 20.9: Uncertainty Analysis of Estimates of Health-Adjusted Life Expectancy

Calculation of health-adjusted life expectancy (HALE) for any population requires three inputs: (1) life expectancy at each age, (2) estimates of prevalence of various nonhealthy conditions at each age, and (3) a method of valuing nonhealthy period in comparison to full health. All three are estimates and have built-in sampling and other variation. When the life expectancy of females at birth in Brazil is estimated as 72 years in 2012, it could actually be 71 or 73 years. Life expectancy is relatively robust but such variation is more prominent in prevalence rates of various diseases. Many of these prevalences are not known, and they are estimated by indirect method. A statistical distribution can be imagined around all such estimates. Salomon et al. [16] have described a method to calculate uncertainty interval around HALE for each member country of the World Health Organization. They used computer simulations for generating statistical distributions around input values and thus propagate uncertainty intervals.

An uncertainty interval is very different from the CI. In Example 20.9, CI would be based exclusively on SE of HALE without considering the variation in input values. Uncertainty interval would be much larger as it incorporates the effect of variation in each of the inputs. This variation depends on repeatability of instruments, number of measurements, and other sources of variation that can contribute to disagreement between the predicted and the actual result.

An uncertainty analysis is recommended when it is necessary to disclose the potential bias associated with models that use single value of the parameters, particularly when your calculations indicate the need for further investigation before taking any action. Also, do uncertainty analysis if the proposed model has serious consequences.

The preceding discussion is in the context of health outcomes but the major application of uncertainty analysis has been in the context of costing. Healthcare cost can substantially vary depending on the quantity and quality of various inputs, and uncertainty analysis helps to delineate the range of costs for varying inputs.

20.3.3 Resampling

When the opportunity for studying another sample does not exist, the results can still be validated by using, what is called, resampling. Under this method, subsamples from your existing sample are drawn and results checked if they replicate on such subsamples. Thus, this method assesses reproducibility or reliability of the results. Two popular methods of resampling are bootstrapping and jackknifing. The following is a brief and introductory account. For details, see Good [17].

20.3.3.1 Bootstrapping

You know that the distance ±1.96SE in 95% CI comes from the Gaussian distribution. This can be used for CI for mean from non-Gaussian distribution also when n is large because then the central limit theorem comes to the rescue. This does not happen for estimating nonadditive parameters of interest such as median (e.g., ED_{50}) and quartile; neither for mean of highly non-Gaussian distribution. For these situations, ±1.96SE cannot be used. The method of bootstrap was originally devised to find the CI in such situations. This is done by generating pseudodistribution using thousands of samples with replacement drawn by computer from the available sample, and thus create a proxy universe. This universe is used for finding the CI. The method is nonparametric.

The actual procedure of bootstrap is something like this. If you have a sample of size $n = 18$, select one at random from this sample. Replace this value so that you again have 18 values. Select another from these 18. This is your second value in the generated pseudosample. Since the sampling is with replacement, the same value can be selected again. Do this 18 times and you have a pseudosample of size 18. In a rare case, same value can occur 18 times in one pseudosample. You can repeat the process and build thousands of pseudosamples from one sample. Each of these samples can give mean, median, SD, or whatever statistic is required. You will have thousands of these values and can generate a pseudodistribution of sample mean, sample median, etc. You can use this distribution, for example, to find a range such that 2.5% medians are less than the lower limit of this range and 2.5% are more than the upper limit of this range. This is your 95% bootstrap CI for median. This can be done for any parameter of interest.

Note how the available data are used to tell more about the data. No extra help is needed—thus the name bootstrap.

The concern in this section is with robustness of results. For this, bootstrapping involves taking several subsamples from the same group that was actually studied and examine if the results replicate. This is slightly different from what is just described. For checking robustness, the method should be applied only when the available sample is large. If your result is based on a group of 200 subjects, take several random subsamples of, say, 150 subjects from this sample of 200. Of course, these are not exclusive samples but drawn after replacement. Each subsample will have some different subjects but most will be those that appear in the previous subsample. This overlap does not matter. Although such resampling can be done many times, generally three or four subsamples of this type would be enough to assess robustness.

20.3.3.2 Jackknife Resampling

The only difference between bootstrap and jackknife is that in jackknife one or more values are serially dropped at a time and results recalculated. Thus, in jackknife resampling, the subsamples are not necessarily random. This method can be used even when the sample is relatively small such as 10 or 15. For dropping two at a time, the procedure would be as follows. If you have a sample of $n = 12$ subjects, drop numbers 1 and 2 and recalculate result on the basis of numbers 3–12. Then, drop numbers 2 and 3 and recalculate result based on numbers 1 and 4–12, and so on. In jackknifing, generally, all possible subsamples are studied.

The method was originally devised to detect outliers that have major effect on the results. You can see that jackknifing is very effective in detecting outliers. It can be used for assessing robustness also. For robustness, there is no need to study all possible samples—possibly 5 or 6 or 10 subsamples are enough. If the results replicate in repeated subsamples, there is evidence that they are robust. Example 20.10 given next illustrates another use of jackknife resampling that derives various validity measures of a new method of classification of breast cancer cases.

Example 20.10: Jackknife Testing of Computer-Aided Classification of BI-RADS Category-3 Breast Lesions

Breast Imaging Reporting and Data System (BI-RADS) is a conventional method of mammographic interpretation. Buchbinder et al. [18] used a computer-aided classification (CAC) that automatically extracted lesion-characterizing quantitative features from digitized mammograms. This classification was evolved on the basis of 646 pathologically proved cases (323 malignant). Jackknife method (100 or more subsamples by computer) was used to calculate that the sensitivity of CAC is 94%, specificity is 78%, positive predictivity is 81%, and area under the receiver operating characteristic (ROC) curve is 0.90.

SIDE NOTE: The authors also used CAC on 42 proved malignant lesions that were classified by conventional method as probably benign and found that CAC correctly upgraded the category in 38 (90%) of them. The implication is that CAC is better than the conventional method for categorizing breast lesions.

Resampling methods can be used for a variety of purposes and obviate the need to depend on Gaussian distribution. Many enthusiasts believe that there would be a paradigm shift and statistical methods would be primarily based on resampling instead of the conventional methods. These methods show promise today as much as they did 25 years ago but have not been able to make much headway so far.

20.4 Quality of Data

Valid and reliable instruments can give erroneous results when not used with sufficient care. In addition, some errors creep in inadvertently due to ignorance. Sometimes, data are deliberately manipulated to support a particular viewpoint. Also, remember what Tukey said. Availability of set of data and aching desire for an answer do not ensure that a reasonable answer would be extracted.

20.4.1 Errors in Measurement

Errors in measurement can arise due to several factors as listed in the following.

20.4.1.1 Lack of Standardization in Definitions

If it is not decided beforehand that an eye will be called practically blind when VA < 3/60 or VA < 1/60, then different observers may use different definitions. When such inconsistent data are merged, an otherwise clear signal from the data may fail to emerge, leading to a wrong conclusion. An example was cited earlier of the variable definition of hypertension used by different workers. In addition, special attention is required for borderline values. Chambless et al. [7] use BP $\geq$ 140/90 for hypertension, whereas Wei et al. [19] use BP > 140/90. One considers BP = 140/90 hypertensive; the other considers this normotensive. The other such example is age. This can be recorded in completed years (age last birthday) or as on the nearest birthday. These two are not necessarily the same. A difficulty might arise in classifying an individual as anemic when the Hb level is low but the hematocrit is normal. Guidelines on such definitions should be very clear.

20.4.1.2 Lack of Care in Obtaining or Recording Information

There can be a lack of care in obtaining or recording information when, for example, sufficient attention is not paid to the appearance of Korotkoff sounds while measuring BP by sphygmomanometer or to the waves appearing on a monitor for a patient in critical condition. This can also happen when responses from patients are accepted without probing and some of them may not be consistent with the response obtained on other items. If reported gravidity in a woman does not equal the sum of parity, abortions, and stillbirths, then obviously some information is wrong. A person may say that he does not know anything about AIDS in the early part of an interview but states sexual intercourse as the mode of transmission in the latter part of the interview. The observer or the interviewer has to exercise sufficient care so that such inconsistencies do not arise.

Then is the question of correct transfer of datasheet to the spreadsheet. I have often detected errors in data entry when intimately checked after a suspicion arose. This I do only for studies where I am intimately involved, else the analysis is done on the submitted data. It is mostly for investigators themselves to carefully check the data. Here is a sample of what I could detect.

In one instance, codes for males and females were reversed for some subjects. Such errors are difficult to detect unless you find a pregnant male!

In another case, the pretest values were unwittingly swapped with posttest values for some cases at the time of data entry. In yet another instance, the values for case number 27 were exactly same as for case number 83. This may have occurred due to repeat recording of the same subject by two different workers or by wrong renumbering of forms.

To give you a few more practical examples, at the time of entering change from pre- to postvalues, minus sign was inadvertently omitted. In another instance, calculation of scores was based on wrong columns F, G, H, M, N in spreadsheet where the columns actually needed were F, G, H, N, O, and the scores turned out that did not arouse suspicion. Only when a third person examined the data for some other purpose, this error was detected. All these are actual instances and can happen in best of setups. Thus, proper scrutiny of data is a must for valid results. Some such errors may have crept in this book also and I will fake it as nothing serious has happened.

Now some tips on how to check the data. For large datasets, double entry by two independent workers and matching may help in detecting errors. Second method is the range check. If hemoglobin level is typed as 3.2 mg/dL, birth weight as 6700 g, or age at menopause 67 years, you know these are improbable values and need to be double checked. Whether qualitative or quantitative, frequency tabulation can help in detecting such outliers. The difficulty is in detecting age 32 years typed as 23 years—when both are equally plausible. If the stakes are high, you may like to double check all the entries with the forms (assuming that information in forms is correct). If the entries are direct online, one source of errors is eliminated but the chance of detecting any error in entry steeply declines. Do whatever you can since you cannot disown such errors later on when detected.

20.4.1.3 Inability of the Observer to Get Confidence of the Respondent

This inability can be due to language or intellectual barriers if the subject and observer come from widely different backgrounds. They may then not understand each other and generate wrong data. In addition, in some cases such as in sexually transmitted diseases (STDs), part of the information may be intentionally distorted because of the stigma or the inhibition attached to such diseases. An injury in a physical fight may be ascribed to something else to avoid legal wrangles. Some women hesitate to divulge their correct age. Some may refuse physical examination, forcing one to depend on less valid information. Correct information can be obtained only when the observer enjoys full confidence of the respondent.

20.4.1.4 Bias of the Observer

Some agencies deliberately underreport starvation deaths in their area as a face-saving device and overreport deaths due to calamities such as flood, cyclone, and earthquake to attract funds and sympathy. Improvement in condition of a patient for reasons other than therapy can be wrongly ascribed to the therapy. Tighe et al. [20] studied observer bias in the interpretation of dobutamine stress echocardiography. They concluded that the potential for observer bias exists because of the influence of ancillary testing data such as angina pectoris and ST-segment changes. Lewis [21] found that psychiatric assessments of anxiety and depression requiring clinical judgment on the part of the interviewer are likely to suffer from observer bias. These examples illustrate some situations in which observer bias can occur.

20.4.1.5 Variable Competence of the Observers

Quite often, an investigation is a collaborative effort involving several observers. Not all observers have the same competence or the same skill. Assuming that each observer works to his fullest capability, faithfully following the definitions and protocol, variation can still occur in measurement and in assessment of diagnosis and prognosis because one observer may have different acumen in collating the spectrum of available evidence than the others.

Many inadvertent errors can be avoided by imparting adequate training to the observers in the standard methodology proposed to be followed for collection of data and by adhering to the protocol as outlined in the instruction sheet. Many investigations do not even prepare an instruction sheet, let alone address adherence. Intentional errors are, however, nearly impossible to handle and can remain unknown until they expose themselves. One approach is to be vigilant regarding the possibility of such errors and deal sternly with them when they come to notice. Scientific journals can play a responsible role in this respect. If these errors are noticed before reaching the publication stage, steps can be sometimes taken to correct the data. If correction is not possible, the biased data may have to be excluded altogether from analysis and conclusion.

It is sometimes believed that bad data are better than none at all. This can be true if sufficient care is exercised in ensuring that the effect on the conclusion of bias in bad data has been minimized, if not eliminated. This is rarely possible if the sources of bias are too many. Also, care can be exercised only when the sources of bias are known or can be reasonably conjectured. Even the most meticulous statistical treatment of inherently bad data cannot lead to correct conclusions.

20.4.2 Missing Values

Two kinds of missing values occur in medical setup. First, some of the subjects included in the sample are not available for some reason right from the beginning and for them no information is available at all. Hopefully they are random and not introducing any bias. For these misses, the study protocol should provide for an unbiased mechanism to replace and select other subjects from the same target population and use them as replacement under certain conditions. Second, the subjects drop out after initially consenting. For them, some initial or baseline information is available. They should not be replaced but need to be handled separately.

It is quite common that part of the information is not available for some subjects—even under the best-designed and monitored studies. This can happen for reasons beyond the control of the investigator. (1) The subjects may cease to cooperate after giving consent earlier and may not show up or may leave a hospital ward against medical advice. Some subjects may unexpectedly die in the midst of the investigation. Some may change their medical care provider. Some may develop side effects and drop out. Some subjects may refuse to divulge part of the information that they want to keep secret. Some may have urgent work at home/office or may be ill so that the appointment is rescheduled. (2) In some cases, a particular investigation in the laboratory is not done because the kit or the reagent is not available at that time, or the sample of blood, urine, etc., is not adequate or is ruined in storage or transportation. The equipment may not be functioning properly. If the information is to be obtained from records, these may be incomplete. Some laboratory results may be unbelievably wrong and would be discarded. The person responsible for collection of data may not be sufficiently careful. He may goof up and not collect some of the information as per the protocol. Some reports may be misplaced. When the collection of data involves handwriting, it may not be legible in some cases.

The opportunities of missing values due to errors enumerated in reason (2) have substantially declined after the advent of automated and online systems but continue to occur, perhaps with increased frequency, for personal reasons listed in (1). If the number of subjects investigated is 250, complete and correct data may be available only for 220. How to account for the missing 30?

20.4.2.1 Approaches for Missing Values

It has been seen time and again that the missing values are not *random* subsample. These tend to follow a different pattern. Missing subjects may be those who have experienced considerable improvement in their condition or those who see no hope. If they have moved out, this may be related to their condition. If some have died, this again reflects on their condition. Thus, the effect of missing values on conclusion is not just due to reduced sample size but, more importantly, because the available sample is no longer representative.

Missing values can be especially dangerous where several variables are under consideration. Multiple regression and multivariate methods are particularly vulnerable. Suppose 15 variables are studied for 60 individuals. If one variable value is missing for two subjects and another variable value for another three subjects, these five subjects will be deleted from the regression calculations. In practice, it is possible that one or two values are missing for 40 out of 60 in the sample. Thus, n reduces to only 20. As stated earlier, this is called casewise deletion or **listwise deletion** and results in a big loss. Most statistical packages use this as default. Methods are available, particularly for regression, which use **pairwise deletion**. This means that all available values for each *pair of variables* are included for calculation. This results in much less loss but one analysis may be based on $(n - 3)$ subjects and the other on $(n - 7)$ subjects. Thus, the reliability will differ. And this may cause problems.

Few missing values may not affect the findings and the available data can be analyzed without any adjustment. If the missing values are far too many, such as more than 50%, asses the utility of available data because such deficient data can lead to distorted results. There is a view though that imputation or adjustment is better than throwing away the data. Three alternative approaches are available:

1. Simplest but undesirable approach is to ignore the missing values and pretend that nothing is lost. Only the available data are analyzed. This approach may work well in a random nonresponse situation. But, as you are now aware, nonresponse is seldom random. In addition, the sample size will in any case reduce, which will compromise the reliability of the results. Thus, this approach is not advisable unless the missing values are really sparse, say, less than 5%.

2. Second approach is to compare the baseline information of the non-respondents with the respondents and adjust results accordingly if the baselines are found different. This approach is illustrated in Example 20.11 and works well if baseline characteristics are making a difference in the outcome, and nothing else is much important.

3. Third approach is to make up the lost data by imputation.

You should carefully examine the data regarding pattern of missing values—whether some information is missing for some specific types of subjects or are they dispersed, does it look random or fall into a pattern. If same variable values are missing for a large proportion of subjects, consider deleting this variable from the analysis.

Example 20.11: Adjustment in Incidence of Cancer Based on Age Structure of Respondent and Nonrespondent Women

In a study to evaluate the risk of breast or endometrial cancer in women treated with menopausal estrogen, a cohort of 5000 women was randomly chosen from those who received the prescription from pharmacies across a country. (The study would have an equivalent control without estrogen therapy but leave that outside the purview of this example.) At the end of 10-year follow-up period, information on 1237 women was not available for a variety of reasons. The broad distribution of the 3763 women whose data were available is shown in Table 20.3. There is

TABLE 20.3

Incidence of Cancer in Menopausal Women Treated with Estrogen

	\<40	40–44	45–49	50+	Total
			Age at Menopause (Years)		
Respondents					
Number of women	240	1888	1534	101	3763
Percent of women	6.38	50.17	40.77	2.68	100.00
Cancer cases	6	29	19	2	56
Incidence rate (%)	2.50	1.54	1.24	1.98	1.49
Nonrespondents					
Number of women	638	412	162	25	1237
Percent of women	51.58	33.31	13.10	2.02	100.00

apparently some gradient of incidence rate with age at menopause in this study. The overall rate of 1.49% is unadjusted for the nonresponse.

Basic data are available for the nonrespondents as well. Their distribution by age at menopause is also shown in the table. Note that women with early menopause predominated among the nonrespondents. Since incidence seems related with this age, an adjustment is required. Assuming that the age-wise incidence among nonrespondents is same as among the respondents, the cases among the nonrespondents would be 0.025 × 638 = 16 in the first age group, similarly 6 in the second, 2 in the third, and 0 in the fourth group. This makes a total of 24 in this group and a total of 80 in the total of 5000 women. The incidence rate, adjusted for nonrespondents by age at menopause, is now 80/5000 = 1.60%.

SIDE NOTE: The adjustment may look small in absolute sense but can make material difference when the total number of women with the risk of breast or endometrial cancer is projected for the whole country or when the odds ratio or its significance is computed.

20.4.2.2 Handling Nonresponse

The question of handling nonresponse arises only when the nonresponse is present. See if you can reduce the opportunity of nonresponse to begin with so that the problem does not arise. A proper groundwork with the sample subjects may help. Explain the benefits of the study to the subjects so that they cooperate. The investigations should be such that these cause least inconvenience or distress to the respondents. These should have no adverse financial implications. In fact, it may be necessary in some cases to provide some sort of compensation to the subjects so that they extend full cooperation.

The problem of nonresponse can be particularly severe in mailed questionnaire, primarily due to disinterest of the respondents. In the case of personal contacts, this can occur for reasons already enumerated that include non-availability of the selected subjects (due to unexpected death, migration, or visit to other places), refusal to cooperate even after providing informed consent including patients leaving against medical advice due to unsatisfactory service or change of mind, unexpected development of a side effect, or inclusion of houses that are not occupied; or the selected subject may fail to meet the criteria for inclusion in the study. It is only in the last two cases that the subjects can be replaced by fresh subjects, that too in accordance with the procedures laid down in the protocol before starting the sampling. In most other cases, two or three attempts may have to be made to contact the subject and get his cooperation. Those who remain nonrespondents even after such efforts could be different from the respondents with regard to some of the characteristics under study. They may either be disproportionately normal or disproportionately abnormal. For example, those moved out may be the ones who are very healthy to take up another job or can be those who visualized death and want to spend time with the family.

In a study on the effect of an exercise regimen on cardiac patients, those who do not show up for a recheck may be the ones who did not follow the regimen. This could be because they felt healthy otherwise or are too sick to perform the exercise or perhaps are very busy or careless. Some may have even died. If some have moved out, those may be among the relatively healthier patients. Thus, the nonrespondents would not be a random subsample, and the results derived from only the respondent can be distorted. In this case, an adjustment may be required to make results valid for the entire group.

One way to find whether the nonrespondents are different from the respondents or not is to callback a random subsample of nonrespondents. Intensive efforts are made to track down this subsample and get their response on the regular format. This exercise is easily said than done because some of the nonrespondents in the subsample also may not be traceable or refuse to cooperate. In case the intensive efforts are successful, the pattern of responses obtained from this subsample can be compared with that of the respondents. This comparison may indicate in what respect, if at all, the two groups are different and what kind of adjustment is required.

Whether adjustment is done or not, the extent of nonresponse must always be stated in the report so that the reader can decide how much confidence he should place in the results. The method of adjustment should also be stated.

Although the methods described in Example 20.11 tend to correct the imbalance due to bias among nonrespondents, they do not make up for loss in efficiency in statistical inference arising from smaller n. To compensate, imputation can be used.

20.4.2.3 Imputations

The deletion methods restrict calculations to complete cases. Imputation methods replace the missing data with most plausible values. When successfully done, imputation tends to make the data complete for analysis as though nothing is lost.

As stated earlier, one view is that imputation can be used even when many values are missing. Imputation allows full use of the available data. Data are expensive to collect and discarding the available data on any subject just because part of the information is missing can be unethical.

Rarely though, you can deduce the missing value from the available data such as missing age from age at starting work in a mine and the duration of work (when there is no interruption). Note in this case that these three are linearly related and possibly only two should be used. If age and duration of work are highly correlated, one more can be dropped.

Most conservative method is to replace all missing observations by the most pessimistic values. If the result is still statistically significant, a safe conclusion can be drawn. The second is to use imputation technique [22]. This technique tries to fill a missing value in such a manner that it fits into the trend otherwise seen in the data. Imputation can be done by group means, closest match, or last-value-carried-forward methods as well as by some other methods.

Replacing missing values by the mean of the available values introduces an artifact. Surely, all missing values would not be at the center as assumed by this method. You can easily see that this method would reduce the variance, which can contribute to false statistical significance. Among many algorithms for closest match, the regression algorithm is popular. In this method, regressions are obtained of the variables with missing values on the variables without missing values. These regressions are used to predict the missing values for each subject separately. The last-value-carried-forward approach can be used for serial observations. The previous value is assumed to remain same for the next occasion. Among other methods are (1) replacement of missing value by generating a random value from the distribution of the variable when known and (2) adding a random residual to the respective mean values. Both these methods work only when the missing values are random. Yet another method is to determine classes of similar subjects and replace missing values with a random value from the same class.

Imputing a missing value by its estimate fails to account for uncertainty about the missing value. Analysis based on such single imputations, as it is called, treats this imputed value as though this is the actual observed value. To account for uncertainty about the actual missing value, sometimes one missing value is replaced by several plausible values. It is like a random sample of missing values. Such multiply imputed datasets are then analyzed as usual. The process results in relatively more valid inference. For details, see Rubin [22].

Adjustments done by imputation or otherwise certainly reduce the validity of the data, but that is the best one can do when values are missed. Remember that imputation is an exercise in salvaging the damage done by missing values. These help but do not fully rectify. The only better course is to do everything possible to obtain complete data.

20.4.2.4 Intention-to-Treat Analysis

ITT is an ingenious strategy to partially circumvent the limitation imposed by specific kind of distortion in the data. This strategy is particularly advocated for randomized controlled trials. Consider a typical trial for a new drug for vertigo in which randomized patients are given a tablet and told to take one if and when an attack occurs. Some subjects will have vertigo and take the tablet, and some would not have the attack and would not take the tablet. Should the analysis be based on those who had the attack or on all those who were randomized? There might be other situations where patients randomized to receive the intended treatment shift loyalty in between and go for another treatment. Sometimes the assigned regimen is not given in full due to side effects, or one or more subject receive incorrect regimen in error. Deviation from protocol can occur for a variety of other reasons also. These are the examples of some situations where ITT analysis can be helpful. Since this can occur in practice as well, ITT simulates pragmatic trials. For a complete ITT analysis, the outcome must be known for all the patients including those who shifted to other treatments. There is no consensus on how to handle nonresponse in ITT analysis.

ITT strategy requires that all participants be analyzed according to their original group assignment regardless of what occurred subsequently. This reduces bias due to nonrandom loss or shift of participants. Efficacy of a regimen may be overestimated if ITT is not done. ITT analysis is conservative—this provides more confidence.

An appropriate situation for ITT is given in Table 20.4 where medical and surgical treatments are the two treatments under comparison for stable

TABLE 20.4

ITT and Other Analyses to Compare Medical and Surgical Treatments for Stable Angina Pectoris

	Treatment—Allocated/Actual			
	Medical/ Medical	Medical/ Surgical	Surgical/ Surgical	Surgical/ Medical
	A	B	C	D
Number of patients	323	50	368	26
Number of deaths	27	2	15	6
Mortality	8.4%	4.0%	4.1%	23.1%
ITT analysis (A + B vs. C + D)	29/373 = 7.8%		21/394 = 5.3%	
As per protocol (A vs. C)	27/323 = 8.4%	—	15/368 = 4.1%	—
As per actual treatment (A + D vs. B + C)	33/349 = 9.5% (A + D)	17/418 = 4.1% (B + C)		

Source: Adapted from European Coronary Surgery Study Group, *Lancet*, i, 889, 1979.

angina pectoris [23]. The outcome measure is 2-year mortality. Some patients switched from medical to surgical and some from surgical to medical under compelling circumstances.

A total of 373 patients were allocated to receive medical treatment (groups A and B) as first-line treatment but 50 of these ended up getting surgery (second-line treatment) (group B). Similarly, out of 394 patients allocated for surgery as first line treatment (groups C and D), 26 got medical treatment (second-line treatment) (group D). The ITT analysis would be (A + B) versus (C + D). Two-year mortality in these groups was 7.8% and 5.3%, respectively. Opposed to this, if the protocol is to be strictly followed, the comparison would be of group A with group C, and the deviant groups B and D would be ignored. If the patients actually receiving medical (A + D) and surgical (B + C) treatments are compared, the 2-year mortality was 9.5% and 4.1%, respectively.

Note how different approaches can provide different results. ITT analysis provides a pragmatic estimate of the effect or of difference in actual situations and not the potential difference in ideal situations. Thus, this analysis mirrors clinical decisions and avoids bias in most situations. However, there are exceptions such as safety of a regimen.

ITT fails to take care of fallacies such as participant tossing the drug or placebo in a toilet, or when some take four doses one day after missing them on previous three occasions. The latter might come out from the interview of the subjects, the former may never surface.

Whereas efficacy of a regimen can be evaluated for clinical realities where some patients end up in unintended group, common sense dictates that safety analysis should be on "as treated" basis. It would be unfair to ascribe a serious side effect to placebo where the patient supposed to receive placebo actually received the test drug. For not so serious side effects such as headache and nausea, which can occur due to nonpharmacological reasons and can occur in a placebo group, ITT analysis can still be adopted. In case of efficacy, major concern is with avoidance of Type-I error, while for safety avoidance of Type-II matters more—as in equivalence studies.

20.4.3 Lack of Standardization in Values

If apples should be compared with apples, how to compare the level of abnormality in health revealed by systolic level with the level revealed by cholesterol level? The admissible range of normality of BP is narrow and of cholesterol level wide. Also, the cholesterol level is numerically higher. Similarly, average duration of smoking in young adults is not comparable with the average in the older population. Older population has had more years to smoke than the young. Answer to such discrepancies is standardization. Some methods of standardization have already been discussed as listed next, and some new methods are described later in this section.

20.4.3.1 Standardization Methods Already Described

Z-scores: As described in Chapter 9, Z-score for variable x is obtained for each subject by subtracting its mean and by diving by its SD. For sample values, $Z = (x - \bar{x})/s$. If Z-score for systolic level for a person is 0.7 and for cholesterol level 1.3, you can compare and say that cholesterol is almost twice as different from mean as the systolic level in that person. Thus, such standardization provides mechanism to compare values of different variables that can be on widely different scale.

Normalization: This term is generally used when the values are converted to (0–1) scale. This can be easily done by the transformation (value – minimum)/(maximum – minimum). Thus, a normalized value 0.5 indicates that the value is in the middle of the actual range of values. You have seen such normalization in Chapter 11 for calculating, for example, life expectancy index for use in human development index.

Standardized death rate: This standardization is described in Chapter 11 as a strategy to make two populations with different age structure comparable for death rate. Since old age people are expected to die anyway, a higher crude death rate (CDR) in a population comprising large percentage of old people is not as bad as a high CDR in a population with predominantly child population. Standardized death rate removes this discrepancy and brings the two populations on common age structure. See Chapter 11 for direct and indirect methods of standardization.

Some other methods of standardizations are described next.

20.4.3.2 Standardization for Calculating Adjusted Rates

Standardization as discussed in Chapter 11 in the context of death rates is a fairly general method and can be used in a variety of setups. For example, it can be used to standardize means, although then this is called the adjusted mean. One such application is illustrated in the following.

Suppose in a survey of 300 adults, the mean serum folate level (nmol/L) in current, former, and never smokers of different age groups is as given in Table 20.5.

The mean in column D is calculated with due consideration to the different numbers of subjects in different age groups as is needed for grouped data. For example, mean serum folate level for current smokers is

$$\frac{6.5 \times 80 + 7.0 \times 40 + 8.5 \times 10}{80 + 40 + 10} = 6.8 \text{ nmol/L.}$$

Of the 70 former smokers, 40 are in the age group of 60+ years. This number is only 10 out of 130 in the current smokers. This differential can affect the mean serum folate level because the level depends on age. This mean in column D is **unadjusted** in the sense that the difference in age distribution

TABLE 20.5

Mean Serum Folate Level by Age and Smoking Status

	Age (Years)			Unadjusted Mean	Age-Adjusted
	20–39	40–59	60+	for All Adults	Mean
Smoking Status	A	B	C	D	E
Current smokers	6.5	7.0	8.5	6.8	7.1
	(80)	(40)	(10)	(130)	
Former smokers	7.5	8.5	9.0	8.6	8.1
	(10)	(20)	(40)	(70)	
Never smokers	7.0	8.0	9.0	7.6	7.7
	(50)	(40)	(10)	(100)	
Total	6.8	7.7	8.9	7.5	
	(140)	(100)	(60)	(300)	

Note: The number of subjects is in parentheses.

is overlooked. The difference in these unadjusted means across smoking status categories is not necessarily real but could be partially due to the difference in their age structure. The effect of age disparity can be removed by calculating the **age-adjusted mean**. For this, a standard age distribution is required. The age distribution of total subjects (last row) can serve as the standard. When this standard is used on the unadjusted means, the following is obtained:

Age-adjusted mean serum folate level for current smokers is

$$\frac{6.5 \times 140 + 7.0 \times 100 + 8.5 \times 60}{140 + 100 + 60} = 7.1 \, \text{nmol/L}.$$

This is obtained by applying the age-specific rates in the group on to the standard age structure. This procedure is similar to the procedure used earlier to calculate the directly standardized death rate. Similar age-adjusted mean can be calculated for the former smokers and the never smokers. The adjusted means so obtained are shown in column E of Table 20.5.

Note that the large difference between the unadjusted means in current and former smokers decreased considerably after the age adjustment. Much of this difference was due to the differential in age structure of the subjects in these two categories. This adjustment brought the groups to a common base with respect to age and made them comparable.

20.4.3.3 Standardized Mortality Ratio

The standardized mortality ratio (SMR) is the ratio of the number of observed deaths in a study group to the number that would be expected if the study group had the same specific rates as in a standard group. This procedure is the same as indirect method of standardization. The ratio is sometimes multiplied by 100 for expressing it in terms of percentage:

$$\text{SMR} = \frac{\text{Observed number of deaths}}{\text{Expected number of deaths}} * 100, \qquad (20.4)$$

where the denominator is based on the specific rates in the chosen standard population. These are not necessarily age specific but could be gender specific, exposure specific, or specific for any other categorization. An SMR greater than 100 is interpreted as indicating that the study group has excess mortality relative to the standard. The study and the standard group can be disease and control groups in a case–control study or general populations of two types or any other groups of interest. In this method, more stable rates of the larger population are applied to the smaller study group to obtain the expected number of deaths. SMR gives a measure of the likely excess or reduction in mortality in the study group. Note that two SMRs can be compared only when they are based on same standard population.

Example 20.12: SMR for Textile Workers

Stress of work, exposure to fiber dust, and other factory environmental factors are known to cause excess mortality in the staff of a textile mill. Their age distribution and calculation of SMR are shown in Table 20.6.

Expected deaths are obtained by applying the national age-specific death rates (ASDRs) to the number of textile workers in different age groups. Since the national rate for the age group 25–34 years is 3.0 per 1000 population, the expected deaths in 400 workers of this age group are 3.0 × 400/1000 = 1.2. Similarly for other age groups. If the total number of observed deaths is 10 and the expected deaths based on the national rates is 6.8, then from the formula given in Equation 20.4, SMR = (10/6.8)100 = 147.

TABLE 20.6

Calculation of SMR in Textile Workers

Age Group (Years)	ASDR in the National Population per 1000	Textile Workers	
		Number	Expected Deaths
25–34	3.0	400	1.2
35–44	5.0	300	1.5
45–54	8.0	200	1.6
55–64	25.0	100	2.5
Total		1000	6.8

This shows that the mortality level of textile workers (in this mill) was 147% of the national average. This is 47% higher than that experienced by the national population.

SIDE NOTE: Excess mortality in textile workers was never in doubt, but SMR delineates the exact magnitude of this excess.

For a similar but more extensive application of SMR, see Johansen and Olsen [24]. They investigated mortality from amyotrophic lateral sclerosis (ALS) and other chronic disorders among male employees of electric supply companies in Denmark. The 21,236 men included in the study accrued nearly 303,000 person-years of follow-up. They observed 14 deaths from ALS when only 6.9 deaths were expected on the basis of the national ASDRs. This yielded an SMR of nearly 200. This means that these employees are dying from ALS nearly twice as much as the general population. These excess deaths can be attributed to the risks present in the company environment.

References

1. Deming WE. *Quality, Productivity and Competitive Position*. London, U.K.: Prentice Hall, 1978, pp. 240–245.
2. Demos MP, Demos NP. Statistical quality control's role in health care management. *Qual Prog* 1989; 22:85–89.
3. Montgomery DC. *Introduction to Statistical Quality Control*. New York: John Wiley & Sons, 1985, pp. 149, 351–373.
4. Robertson SE, Anker M, Roisin AJ, Macklai N, Engsirom K, LaForce FM. The lot quality technique: A global review of applications in the assessment of health services and disease surveillance. *World Health Stat Q* 1997; 50:199–209.
5. MacGregor JF. Using on-line process data to improve quality: Challenges for statisticians. *Int Stat Rev* 1997; 65:309–323.
6. Spear SJ, Schmidhofer M. Ambiguity and workarounds as contributors to medical error. *Ann Intern Med* 2005; 142:627–630.
7. Chambless LE, Shahar E, Sharrett AR et al. Association of transient ischemic attack/stroke symptoms assessed by standardized questionnaire and algorithm with cerebrovascular risk factors and carotid artery wall thickness. The ARIC study, 1987–1989. *Am J Epidemiol* 1996; 144:857–866.
8. Dwyer JH, Li L, Dwyer KM, Curtin LR, Feinleib M. Dietary calcium, alcohol and incidence of treated hypertension in the NHANES I epidemiologic followup study. *Am J Epidemiol* 1996; 144:828–838.
9. Carter R, Holiday DB, Grothues C, Nwasuruba C, Stocks J, Tiep B. Criterion validity of the Duke Activity Status Index for assessing functional capacity in patients with chronic obstructive pulmonary disease. *J Cardiopulm Rehabil* 2002; 22:298–308.

10. Wynd CA, Schaefer MA. The osteoporosis risk assessment tool: Establishing content validity through a panel of experts. *Appl Nurs Res* 2002; 15:184–188.
11. Carmines EG, Zeller RA. *Reliability and Validity Assessment.* London, U.K.: Sage Publications, 1979.
12. Gordon MO, Schechtman KB, Davis LJ, McMahon TT, Schornack J, Zadnik K. Visual acuity repeatability in keratoconus: Impact on sample size: Collaborative Longitudinal Evaluation of Keratoconus (CLEK) Study Group. *Optom Vis Sci* 1998; 75:249–257.
13. Russo J, Trujillo CA, Wingerson D et al. The MOS 36-item Short Form health survey: Reliability, validity and preliminary findings in schizophrenic outpatients. *Med Care* 1998; 36:752–756.
14. Varga D, Johannes T, Jainta S et al. An automated scoring procedure for the micronucleus test by image analysis. *Mutagenesis* 2004; 193:391–397.
15. Estenssoro E, Gonzalez F, Laffaire E et al. Shock on admission day is the best predictor of prolonged mechanical ventilation in ICU. *Chest* 2005; 127:598–603.
16. Salomon JA, Mathers CD, Murray CJL, Ferguson B. Methods for life expectancy and healthy life expectancy uncertainty analysis. Global Programme on Evidence for Health Policy Working Paper No. 10. Geneva, Switzerland: World Health Organization, 2001.
17. Good P. *A Practitioner's Guide to Resampling for Data Analysis, Data Mining, and Modeling.* London, U.K.: Chapman and Hall, 2011.
18. Buchbinder SS, Leichter IS, Lederman RB et al. Computer-aided classification of BI-RADS category 3 breast lesions. *Radiology* 2004; 230:820–823.
19. Wei M, Mitchell BD, Haffner SM, Stern MP. Effects of cigarette smoking, diabetes, high cholesterol, and hypertension on all-cause mortality and cardiovascular disease mortality in Mexican Americans: The San Antonio Heart Study. *Am J Epidemiol* 1996; 144:1058–1065.
20. Tighe JF Jr, Steiman DM, Vernalis MN, Taylor AJ. Observer bias in the interpretation of dobutamine stress echo cardiography. *Clin Cardiol* 1997; 20:449–454.
21. Lewis G. Observer bias in the assessment of anxiety and depression. *Soc Psychiatry Psychiatr Epidemiol* 1991; 26:265–272.
22. Rubin DB. *Multiple Imputation for Nonresponse in Surveys.* New York: Wiley Interscience, 2004.
23. European Coronary Surgery Study Group. Coronary–artery bypass surgery in stable angina pectoris: Survival at two years. *Lancet* 1979; i:889–893.
24. Johansen C, Olsen JH. Mortality from amyotrophic lateral sclerosis, other chronic disorders, and electric shocks among utility workers. *Am J Epidemiol* 1998; 148:362–368.

21

Statistical Fallacies

Fallacies are anomalies that considerably reduce the credibility of a report. Statistical fallacies are common in medical literature. This chapter enumerates many such fallacies, hoping to create awareness of situations when these can occur. Consider the following stimulating examples.

Statistics show that more people die in hospital than at home. Also, there is a strong association between dying and being in bed. Can intelligence quotient (IQ) be related to the first names of people? The absurdity of such associations is apparent. No one would advocate avoiding hospitals or beds to prolong life. These might seem like extreme examples, but the same sorts of errors of logic sometimes pass unrecognized in the medical literature [1].

If a child cuts his finger with a sharp knife, blame the child or the knife? If a person takes an excessive dose of sleeping pills and dies, the pills cannot be blamed. Abuse or misuse of statistics has precisely the same consequence. An *abuse* occurs when the data or the results are presented in a distorted form with the intention to mislead. Sometimes part of the information is deliberately suppressed to support a particular hypothesis. A *misuse* occurs when the data or the results of analysis are unintentionally misinterpreted because of lack of comprehension. The fault in either case cannot be ascribed to statistics. It lies with the user. Correct statistics tell nothing but the truth. The difficulty, however, is that the adverse effects of wrong biostatistical methods are slow to surface. And this makes these methods even more vulnerable to wrong use. The other difficulty is that the learning curve of medical professionals for statistical methods, particularly for intricate methods, is shallow and daunting—thus errors become inevitable. Few researchers, if at all, are able to engage a qualified statistician beginning to end. They tend to depend on their colleagues, their own experience, and software "help" files. These helps are incomplete, to say the least. At the same time, many biostatisticians also are not above the board. Besides biostatistical competence, their appreciation of medical implications is a severe constraint.

This chapter: While presenting various statistical methods in previous chapters, my regular advice was to exercise sufficient caution regarding the validity of assumptions such as those of Gaussian pattern, independence of observations, and equality of variances before using those methods. The purpose is not to repeat all that here. But there are several other aspects that have not been sufficiently covered earlier. These are fallacies arising from intentional and unintentional biases and errors in the data, in their presentation, and in the interpretation of statistical results. Some of these fallacies can occur despite the use of quality tools in collection of data and despite the availability of accurate data. Problems arising from inadequate design and samples are discussed in Section 21.1. Section 21.2 describes fallacies due to inadequate analysis and errors in presentation of findings are discussed in Section 21.3. Section 21.4 deals with misinterpretation of data. Some of these problems are also briefly mentioned in earlier chapters but they are reiterated here and discussed in more detail in view of their importance. Thus, you may notice some repetition. The list of fallacies in this chapter is not exhaustive but may be sufficient to illustrate the kind of problems that need attention.

21.1 Problems with the Sample

The investigation should provide sufficient and valid data to take a decision. Sometimes this does not happen for various reasons.

21.1.1 Biased Sample

A sample is considered unbiased when it truly represents the target population. A frequent source of error in statistical conclusions is a biased sample. This can happen even when the selection is random or even when random allocation is made in experimental studies. A large sample tends to magnify these errors rather than control them.

21.1.1.1 Survivors

Consider the relationship of lung functions with age after the age of 50 years. It is well known that lung functions decline because of biological degeneration but the gradient differs from population to population. How do we find the exact effect of age in a particular population? Those with lower lung functions are likely to be in poor health and thus with less longevity. Any estimation based on survivors is bound to be biased. It may not be easy to find the correct gradient in this case. Perhaps, a more valid picture can be obtained if lung function parameters of those dying at varying ages are available in records and are used for adjustment. Availability of such records is a severely restrictive condition for most populations. In the absence of such adjustment, the lung functions among the survivors are not a true reflection of the age-based gradient. The actual gradient could be a sharper decline. Similarly, a study based only on hospital cases will exclude those who die before admission. Serious cases, those residing in remote areas, and those who are poor to afford hospitalization tend to be excluded. Similar bias occurs in a variety of other situations where the study is on prevalent cases rather than on incident cases.

The same sort of caution is required in some serial procedures in medicine. In some cancers, surgery is done when radiotherapy fails and the patient remains operable at the conclusion of radiotherapy. The patients not responding to radiotherapy are not necessarily uniform in their grade of malignancy or age, nor in their resolve to fight the disease. Thus, additional caution is required at the time of drawing conclusions in such cases.

21.1.1.2 Volunteers

Early phases of clinical trials are often done on volunteers. They self-select themselves as the subjects. Volunteers tend to be very different from the general class of subjects. Many of them are either hopeless terminal cases or people with exceptional courage. Both affect the response. The results, thus, are not applicable to the general class of patients. Notwithstanding this limitation, volunteer studies have a definite place in medicine as they do provide important clues on the toxicity of the regimen under test, the dose level that can be tolerated, and the potential for further testing of the modality.

Even for nonvolunteers, medical ethics requires that the subject's consent be taken for participation in a study. Real consent, after fully explaining the underlying uncertainties, is difficult to obtain without an inducement. Again, the subjects self-select themselves by providing consent, and bias is likely. Some of this is eliminated by random allocation. This is feasible for trials but not, for example, for cross-sectional surveys. Even for trials, the generalizability suffers due to self-selection of subjects. Conclusions based on such studies can be fallacious. Not many professionals appreciate this limitation imposed by consent-based selection, and they wonder later why their results cannot be reproduced in practical situations.

The same kind of problem, although of less severity, arises with the surveys based on mailed questionnaires, subjects listed in a telephone directory, and Internet users. None of these sources are adequate for an unbiased sample of the general population. Because of the high nonresponse in many such surveys, the results are rarely valid even for the restricted class they represent.

The solution in all these cases is proper adjustment of the results. This is easily said than done. The adjustment can be done only when information is available on at least some subjects who are truly representative of the target population. In the absence of this, the alternative is to use the results only for a restricted type of population that these subjects represent. The third is to interpret the results as indicative of what *might* be occurring in the total population.

21.1.1.3 Clinic Subjects

Clinic and hospital subjects too form a biased sample, as they tend to have a more severe form of disease and include mostly those who can afford these services. Mild cases tend to ignore their condition or self-treat themselves. An interesting example is migraine, which was once believed to be more common in the intelligent professional class [2]. An epidemiological study on a random sample could not substantiate such a relationship when the role of confounding factors was eliminated. Those in the more intelligent professional class perhaps seek medical assistance in the early phases of the disease and with greater frequency than their nonprofessional counterparts. Despite such limitation, clinic-based studies do give important information on the presenting symptoms, their correspondence with laboratory and radiological findings, response to various therapeutic procedures, prognostic features, etc. But the results are seldom applicable to the type of cases that do not show up in clinics. It is sometimes believed that consecutive cases coming to a clinic would be free from bias. Such cases could be true representatives of clinic subjects, but the general bias in all clinic-based subjects still remains.

A similar bias occurs when the sample is restricted to people in employment. Such persons are healthier than the general population. The results based on employed people cannot be generalized to the other segments of the population.

21.1.1.4 Publication Bias

Meta-analysis that tends to provide more valid and reliable results by pooling results from different publications is an accepted statistical technique. However, two points deserve attention in this regard. First, the literature is generally unduly loaded with positive results. Negative or indifferent results are either not sent for publication or not published by the journals with the same frequency. Any conclusion based on commonality in publications thus can magnify this bias. Second, individual studies with small sample sizes may give statistically not significant results but when combined can give significant results because the sample size is so much increased by pooling. This pooling by itself is not a danger, provided different methodologies used in different publications are properly accounted for.

21.1.1.5 Inadequate Specification of Sampling Method

Certain terms in statistics have specific meaning. "Systematic sampling" is a particular method that includes every kth subject in the sample. But the term has also been used for another sampling [3] that does not match with such accepted use. The term "cluster sampling" is often confused with multistage sampling [4] where whole clusters are not included in the sample. Such use, which is at variance with the accepted meaning, can create confusion regarding the actual method adopted for sampling.

21.1.1.6 Abrupt Series

If you are examining or otherwise dealing with a data series, check for abrupt changes. A sudden break or unexpected changes may be hiding important information. Silence between numbers may also be mysterious. When a series lacks random variation or starts to show a pattern that happen to match the target levels, reality may be disguised. Always suspect credibility when a series shows such abnormal patterns.

21.1.2 Inadequate Size of Sample

Size of the sample is an important issue so much so that this sometimes defines the size and adequacy of a study. Despite all the attention it gets from researchers, problems may exist.

21.1.2.1 Size of Sample Not Adequate

Chapter 12 explained why statisticians prefer large samples. It also cautioned that a large sample could become counterproductive in some cases. The number of subjects in a study should be adequate to generate sufficient confidence in the results. Considering this aspect, a larger sample is not harmful but a small sample can be a waste of resources. A small sample may fail to detect the difference when really present in the target population. There is a growing concern among the medical community regarding the failure of many randomized controlled trials (RCTs) in detecting a medically relevant difference because these studies do not have sufficient power. The main driving force behind such studies is the size of the trial and the interindividual variability. During the 1970s, Freiman et al. [5] observed after studying 71 negative RCTs on new therapeutic procedures that the sample size was too small in most of them for power 90% to detect a 25% improvement in outcome. In nearly three-fourths of these trials, the sample was inadequate to detect a 50% improvement. In other words, even if there is a 50% improvement in efficacy of the new treatment compared with the old, the trial results were still statistically not significant. It would have most likely become significant had there been a bigger trial. Quarter-century later, Dimick et al. [6] reported the same for surgical trials. This once again underscores the need to be careful about the size of trial—the number of subjects must be adequate to inspire confidence that a medically relevant difference would not go undetected.

In addition, a small sample has high likelihood of being not fully representative and thus of producing biased results. Statistical methods have an inbuilt provision to take care of the larger sampling fluctuations in small samples, but they are not equipped to take care of the lack-of-representativeness bias that is more likely to creep in small samples.

Sample size becomes critical for rare outcomes. With thorough aseptic conditions now maintained in operation theaters, chance of infection is negligible in routine surgeries such as for hernia. However, antibiotics are still given in some settings, albeit for a short duration. If a trial on the effect of antibiotics is done on 100 cases in test group and another 100 controls, the trial is doomed from the start since no group may develop infection, or just one or two infections may occur in one group. This cannot lead to any reliable conclusion.

A lot of debate goes on the sample size required for multivariable situations. This is different from multivariate setup where the random variable has several components. In a logistic regression with K independents, conservatives generally advocate that at least $10 * K$ subjects should be observed for the rarest outcome in the analysis. If $K = 4$, and mortality is the outcome of interest, there must be at least 40 deaths in your sample according to this rule. If mortality is 8%, the total sample must be 500. My advice is not so strict. In most situations, if no category has frequency less than 5, the results should be fairly reliable.

In a rare situation, the problem could be the reverse—an exceedingly large sample. As stated earlier, Schnitzer et al. [7] studied the prescribing pattern for rofecoxib and celecoxib in 47,935 patients with osteoarthritis (OA) and 10,639 patients with rheumatoid arthritis. The objective was to find the most frequently prescribed dose. They realized that statistical test of significance for such a large sample has no relevance since clinically unimportant small difference would turn out to be statistically significant. Despite such a big sample, their conclusion had limitations because of lack of clinical information, inability to ascertain actual use, and potential for selection bias.

21.1.2.2 Problems with Calculation of Sample Size

Sample size and power have become critical issues in planning a study. But the formulas for many situations are just not available or are based on several assumptions. Gaussianity is taken for granted in sample size calculation although you now know that many measurements in sick people particularly do not follow this pattern. Recourse to central limit theorem for large sample helps in many cases but not all. Transformations such as log, square, and inverse help in some other cases but the interpretation distorts. No perfect remedy is available yet. This is an area where biostatisticians may like to research.

Sample size calculation needs value of standard deviation (SD) or some such parameters, which had to be estimated from a similar previous study or a pilot study. In both situations, the estimate is tentative and thus the sample size calculation also remains tentative. My earlier advice to upwardly revise the calculated size was because of this uncertainty. Exact calculation is elusive. For this reason, you may still not get the aimed power.

Next problem with sample size calculation is in specifying the minimum medically important difference that you want your study to have power to detect. Many researchers are fiercely independent and they tend to differ from one another. Thus, improvement regarded as medically important by one may be dismissed as trivial by another researcher. Some researchers tend to approach a statistician for advice to determine this too though the difference to be detected should be based mainly on clinical judgment.

The last problem is specification of required power. This is not so serious. Most tend to agree that power of 0.80 is reasonably good and 0.90 is excellent. This consensus though is not as much as is for 0.05 for level of significance.

For all these reasons, a study based on perfectly calculated sample size may still fail to deliver the result with expected validity. Just beware.

21.1.3 Incomparable Groups

If sufficient care is not exercised with regard to various features of the groups under comparison, fallacies can occur easily. For comparability, the control group in a clinical trial should undergo the same medical maneuvers, such as transfer from one unit to another and duration and frequency of examination, which are used for the treatment group. This is what makes blinding very important. If a group of women receiving oral contraceptives are being observed for thromboembolism, the same set of observations should be done on a group of women using mechanical devices for contraception to obtain a valid conclusion on excess incidence of thromboembolism in the oral contraceptives group. In practice, the group of women using mechanical devices may not be observed for thromboembolism with similar attention. Such fallacies due to incomparable groups are described in the following paragraphs.

21.1.3.1 Differential in Group Composition

Matching or randomization is advised as a strategy to minimize bias in results based on comparison of groups. Comparability should not be merely for age and gender but should be for all those prognostic factors that might possibly affect the outcome. These include severity of disease, coexisting diseases, care of the subjects, etc. This equivalence is difficult to achieve in practice. Also, all prognostic factors may not be fully known. Randomization is a good strategy to average out the factors and to obtain two equivalent groups, but it works well in the long run. It may fail in a particular study. It is necessary that similarity of groups is checked as a post hoc procedure even when the groups are formed by genuine randomization. This similarity has to be cautiously interpreted because a large difference could be statistically not significant if the sample size is small.

See Example 6.2 in Chapter 6 on effect of education on anxiety coping skills in women with abnormal mammogram [8]. Controls unmatched for age created a lacuna in the study and the conclusions were questionable. The difficulty that incomparable groups can sometimes create is also illustrated by the following two examples.

TABLE 21.1

Response Rate in Mild and Severe Hyperthyroid Cases

Group	Number of Subjects	Number Responded	Response Rate (%)
I. Treatment group	40	32	80.0
Mild	30	26	86.7
Severe	10	6	60.0
II. Control group	40	25	62.5
Mild	8	7	87.5
Severe	32	18	56.2

Example 21.1: Fallacious Unequal Response Rates due to Incomparable Groups

Consider a clinical trial on a new therapeutic regimen that is expected to improve the response rate in patients with hyperthyroidism within a stipulated period. A total of 40 cases with varying grades of disease were put on trial and an equal number receiving the existing therapy were used as controls. The allocation of subjects into treatment and control groups was randomized. The data in Table 21.1 were obtained when the subjects were subsequently divided according to the severity of their condition.

The response rate in the treatment group was 80.0% and in the control group 62.5%. This difference is statistically significant ($P < 0.05$). When the subjects were divided according to the severity of the disease, it turned out that the therapy has a differential effect in mild than in severe cases. Randomization did not match the control group with the treatment group for severity of the disease in the subjects. When mild and severe cases from the control group were compared with those of the treatment group, respectively, the new treatment was found to be of no significant value.

Example 21.2: Fallacious Equal Death Rates due to Incomparable Groups

Consider the data in Table 21.2 where case fatality in patients in two hospitals is the same on aggregate but glaring differences emerge when it is broken down by age and severity of the patients.

If one looks at the top row for the total alone in this table, it would seem that the two hospitals have nearly the same case-fatality rates. The breakdown by age in the next two rows shows that both the hospitals have nearly the same case-fatality rates in each age group also. But further breakdown by severity of illness in the subsequent rows shows that hospital A has less fatality in every group. Why then the rates in the aggregate are the same? The reason is that hospital A is catering much more to serious cases and to older cases. These have higher fatality although less than in hospital B for such patients. The breakdown clearly establishes better performance of hospital A. This was otherwise concealed by the aggregate rate in top three rows of the table.

TABLE 21.2

Case Fatality in Two Hospitals by Age of the Patients
and Severity of Disease

Severity and Age	Hospital A			Hospital B		
	n	Deaths	Rate[a] (%)	*n*	Deaths	Rate[a] (%)
Total	800	158	19.8	600	120	20.0
−60 years	280	38	13.6	340	48	14.1
60+ years	520	120	23.1	260	72	27.7
Mild	200	28	14.0	350	62	17.7
−60 years	60	5	8.3	190	21	11.1
60+ years	140	23	16.4	160	41	25.6
Moderate	300	51	17.0	200	41	20.5
−60 years	120	13	10.8	120	19	15.8
60+ years	180	38	21.1	80	22	27.5
Serious	300	79	26.3	50	17	34.0
−60 years	100	20	21.0	30	8	26.7
60+ years	200	59	29.5	20	9	45.0

[a] Case-fatality rate.

The reverse of what is seen in Example 21.2 can also happen. There might be a large difference in aggregate mortality rates in two hospitals falsely showing that the quality of care in one hospital is better than in the other. This might vanish when adjustment for severity of cases is made. Silber and Rosenbaum [9] have discussed this aspect in detail.

21.1.3.2 Differential Definitions

In a study on trend in contribution of coronary artery disease (CAD) to total mortality over the past 50 years, an important factor is the change in definition and detection of CAD from 1950 to the present. The disease is understood much better now than 60 years ago. The diagnostic procedures have greatly improved and the awareness is high. Thus, the detection rate has increased. This will surely affect the trend. For a simple measurement such as blood pressure (BP), the changing definition of hypertension has already been cited. From a conservative threshold of 160/95 mmHg, it now seems to have settled at 140/90 mmHg. But different workers use different definitions even today. Thus, comparability suffers.

Definitions do matter. When counting cases of dengue in an outbreak, should a person who is weakly positive for the presence of anti-dengue IgG antibody be included? If a person contracts the infection a second time, is he to be counted again? If a person dies of renal failure arising out of dengue hemorrhagic fever, should he be counted among dengue deaths? The serology test in early phases of disease, say, initial 5–6 days after the infection, is often negative. If a person dies during this period, how can he be counted among dengue deaths?

21.1.3.3 Differential Compliance

In a clinical trial setup, subjects in the treatment group may drop out more because of discomfort or poor taste of the drug, even when the placebo looks like the drug and the trial is randomized. On the other hand, the subjects in the treatment group may stay if they see improvement in their condition whereas the placebo group can become noncompliant. The compliance rate in this case is related to the efficacy of the regimen and the comparison can be jeopardized.

A related situation occurs when different groups do not have the same opportunity of being screened, similar to the one mentioned earlier on thromboembolism in oral contraceptive users. Diabetes and gallstones may appear to be associated because persons with diabetes are regularly checked for gallstones but persons without diabetes are rarely checked.

A similar situation arises when the two groups under comparison have differential responses. Gorey and Trevisan [10], while discussing the secular trend in Black/White hypertension prevalence ratio in the United States, observed that the respondents tend to be younger and of higher socioeconomic status than the nonrespondents. Because socioeconomic status is associated with other risk factors of hypertension, the nonresponse may affect the prevalence rate ratio.

21.1.3.4 Variable Periods of Exposure

Because of oversight or otherwise, the differential period of exposure is sometimes ignored and an erroneous conclusion reached. This is illustrated in Example 21.3, which is similar to the one given by Hill [2].

> **Example 21.3: Effect of Variable Periods**
> **of Exposure on Vaccine Efficacy**
>
> A district with an eligible child population of 8000, hitherto uncovered by any typhoid vaccine, was chosen for a vaccine trial starting on the first of January of a year. The campaign was done in three phases of 4 months each. Only 500 children could be vaccinated between January and April, 1000 between May and August, and 1200 between September and December. Thus, a total of 2700 received the vaccine. The other 5300 remained unvaccinated. During the year, 6 contracted typhoid among the vaccinated group and 40 among the unvaccinated group. At the end of the year, the attack rates were $6 \times 1000/2700 = 2.2$ per 1000 for the vaccinated group and $40 \times 1000/5300 = 7.5$ per 1000 for the unvaccinated group.
>
> The rate in the unvaccinated group is more than three times the rate in the vaccinated group. This is the gross position at the end of the year. The actual position is shown in Table 21.3. Assume for the sake of simplicity that all were vaccinated at the middle of their respective phase.

TABLE 21.3

Vaccinated and Unvaccinated Children Exposed to Infection in Different Months

Months[a]	Number of Months	Number of Vaccinated Children Exposed to Infection	Number of Unvaccinated Children Exposed to Infection
January–February	2	0	8000
March–June	4	500	7500
July–October	4	1500	6500
November–December	2	2700	5300

[a] Children vaccinated between January and April are assumed to have been vaccinated on the first of March and similarly for other phases.

Person-months of exposure

1. In vaccinated children after vaccination = $(0 \times 2) + (500 \times 4) + (1,500 \times 4) + (2,700 \times 2) = 13,400$.
2. In unvaccinated children = $(8,000 \times 2) + (7,500 \times 4) + (6,500 \times 4) + (5,300 \times 2) = 82,600$.

Attack rate per 1000 person-months of exposure

1. In vaccinated group = $6 \times 1,000/13,400 = 0.45$.
2. In unvaccinated group = $40 \times 1,000/82,600 = 0.48$.

The attack rate among the vaccinated children is not much different from that in those not vaccinated. Thus, there is no evidence that the vaccine is successful in preventing typhoid that year. This conclusion is entirely different from the one arrived earlier by looking at the gross rates.

A different example on variable period of exposure can be cited. Consider the association of lung cancer with smoking in American women compared with that in French women. American women started smoking six or seven decades ago and the French only about three or four decades ago. The effect would obviously be more pronounced in American women and may not be so evident in French women. When a comparison between smokers and non-smokers is made, the duration of exposure tends to be ignored. This might affect the results in some cases.

21.1.3.5 Improper Denominator

Fractions based on the sample subjects can be sometimes misleading. For example, the finding that 40% of HIV-positive cases are heterosexual promiscuous (HTPR) males, whereas only 10% are sex workers (SWs) (in bold in Table 21.4) does not mean that the seropositivity rate in HTPR males is four times that in SWs. There is a risk of coming to this wrong

TABLE 21.4

HIV-Positive Cases in Different Risk Groups in an HIV-Infected Area

HIV Risk Group	HIV-Positive Cases	Population in the Risk Group	Seropositivity Rate (%)
Sex workers	63 (9.94%)	1,270	4.96
HTPR males	254 (40.06%)	17,033	1.49
Intravenous drug users	37 (5.84%)	528	7.01
Blood donors	56 (8.83%)	8,605	0.65
Blood recipients	182 (28.71%)	12,376	1.47
Others	42 (6.62%)	16,468	0.26
Total	634 (100.00%)		

conclusion, if the last two columns in Table 21.4 are suppressed. There is a high number of HTPR males among HIV positives mostly because the HTPR population is high. As the last two columns reveal, the prevalence measured by seropositivity is actually very low in HTPR males and high in SWs in these data.

If the distribution of peptic ulcer in blood groups O, A, B, and AB is 45%, 35%, 15%, and 5%, respectively, it would be naive to conclude that peptic ulcer occurs more commonly in subjects with blood group O. For a correct conclusion, this distribution should be compared with the proper denominator, which in this case is the blood group distribution in the concerned population.

Another kind of problem with the denominator occurs when organs or episodes are counted instead of individuals. Modern medicine indeed seems to have fragmented human body into a conglomerate of different parts, organs, tissues, etc. An ophthalmologist tends to count the number of eyes with a specific problem such as blindness and not as individuals. Although this might be adequate in some instances, such as in restoration of sight after intraocular lens (IOL) implantation in cataractous eyes, it may be inadequate to assess the magnitude of the problem of blindness. A similar problem arises when a subject is repeatedly counted in case of recurrent attacks of a disease. This can happen, for example, in angina pectoris and in asthma. Repeated infections can possibly be considered independent, but recurrent attacks of angina pectoris and asthma tend to have the same origin. Extra caution may be required in stating and interpreting results based on such repeated count of individuals.

21.1.4 Mixing of Distinct Groups

Example 21.2 shows how the superiority of one hospital was concealed when the data on different age groups and different severity of cases were combined. Mixing of distinct groups can give fallacious results in a variety of other ways. Two of them are illustrated here.

21.1.4.1 Effect on Regression

Mixing of two distinct groups can give a false sense of regression as illustrated in Example 21.4.

> **Example 21.4: Trend When Two Groups Are Mixed**
>
> A survey was carried out on boys of grade IX belonging to three randomly selected schools from each of the two strata—schools in slums catering to a low socioeconomic class and schools in posh localities catering to an upper class. The objective was to find health correlates. Among the measurements made were hemoglobin (Hb) level and body mass index (BMI). When these are plotted, a scatter of the type shown in Figure 21.1 is obtained. This contains two distinct clusters of points: the lower left belonging to the low SE class and the upper right belonging to the upper SE class.
>
> The points within each cluster are randomly scattered and indicate no relationship between Hb level and BMI in either class. Thus, there is no evidence that a high BMI in either class is accompanied by a high level of Hb. When the data for the two SE classes are mixed, a distinct relationship appears. But this is false because neither group has that kind of relationship unless one wants to draw a composite conclusion for the two groups combined.

An annoying feature of regression (and correlation) is that a single outlier can influence it. The position is shown in Figure 21.2. If the outlier is excluded, there is no relationship but if it is included it can produce a fairly high degree of relationship.

You may wish to generate such data and see yourself whether false regression and correlation appear in these situations. The lesson learnt is that a

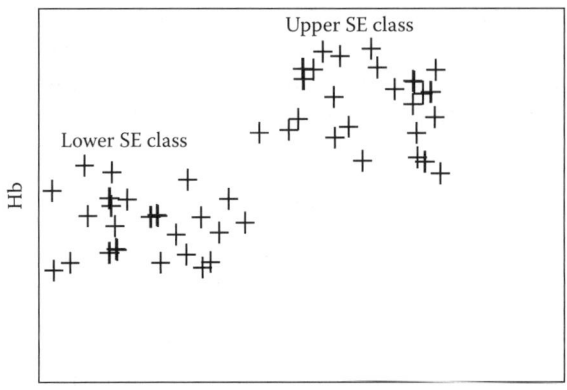

FIGURE 21.1
False relationship when two groups, each with no relationship, are merged.

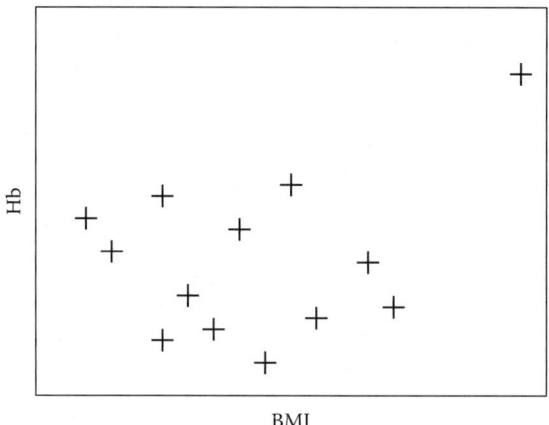

FIGURE 21.2
Even one outlier can yield a significant relationship.

scatter diagram should be examined for data exploration because otherwise such fallacies may never be noticed.

21.1.4.2 Effect on Shape of the Distribution

The pattern of distribution of the serum cholesterol level in two distinct groups as well as jointly in their mixture is shown in Figure 21.3. The group of CAD patients has mean 280 mg/dL and SD 50 mg/dL. There are 100 such subjects. The non-CAD group contains 500 subjects and its mean is 220 mg/dL (SD = 50 mg/dL). The pattern is Gaussian in both the groups. When these two groups are combined, the resultant again is a Gaussian curve with mean 230 mg/dL and SD 50 mg/dL as before. This is shown by the thick curve in Figure 21.3. The composite curve is nice and masks the fact that this actually is made up of two distinct groups. When only the composite curve is available, it is extremely difficult to make out that two very different groups constitute the population. It is only through extraneous information that any clue to the presence of such distinct groups would be available.

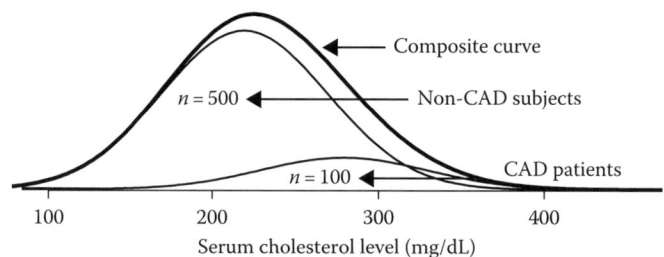

FIGURE 21.3
Mixing of two distinct groups resulting in a nice composite curve.

21.1.4.3 Lack of Intragroup Homogeneity

Consider cases of cervical cancer that are followed up for survival from the time of detection. The time of detection may look an appropriate starting point but the stage of disease among those detected in screening by Pap smear may be very different from those who were diagnosed after appearance of signs and symptoms. This difference in lead time can cause systematic error in the results when groups with such differentials are combined.

Consider a regimen found effective in 78% of children with chronic diarrhea. If at the time of data analysis you discover that the efficacy is 97% in well-nourished children but only 59% in malnourished children, and in your sample if they happen to be equally represented, the finding regarding overall efficacy is suspected. Most cases of chronic diarrhea occur in malnourished children. Thus, the composite sample was biased to begin with. Also note that most well-nourished children get well even without treatment. The need is to find a regimen that helps malnourished children. This regimen does not meet this requirement despite apparent efficacy of 78%.

21.2 Inadequate Analysis

Among the most common sources of fallacies in data-based conclusions is the use of inappropriate method of analysis. Cautions against most such misuses have already been advised in different chapters, but they are reiterated here with an altered focus because of their importance in drawing valid conclusions.

21.2.1 Ignoring Reality

Most statistical methods take a simplified view of the complex biological process. Because computers are available for intricate calculations, the statistical need to simplify is really not as great now as it used to be in the precomputer era. But simplification is still required for easy comprehension. This should not be done to an extent that could distort the essential features of a biological process. Consider this problem in the context of the following common situations.

21.2.1.1 Looking for Linearity

There is no doubt that hardly any relationship in medicine is linear. Yet, a linear relationship is the most commonly studied form of relationship in health and medicine. This simplification seems to work fairly well in many situations but can destroy an otherwise very clear relationship in others. There is an example in Chapter 16 of the rise and fall of lung function with increase in age. This is aptly represented by a parabolic curve but the relationship vanishes if only linearity is considered. Another example is the relationship between glomerular filtration rate and creatinine level (Example 16.4). The linear relationship is medically unsatisfactory despite a high $R^2 = 0.81$. These examples show that there is a definite need to curb the tendency to linearize a clear nonlinear relationship.

Note that the concern in the preceding paragraph is with the narrow sense of linearity that defines a line (or a hyperplane in multiple dimensions). Statistical linearity also includes a parabola because the coefficient is still linear, but square or such other terms of the regressors are rarely included in medical investigations. And this can cause fallacies. Such terms should be included where needed.

As stated in Chapter 16 on regression, investigators tend to look for one equation—either linear, curvilinear or whatever—for the entire range of a measurement. Left ventricular ejection fraction (LVEF) ranges from 10% to 90%—less than 50% is considered a coronary risk and more than 60% does not provide any benefit. If coronary risk is plotted against LVEF, it will be flat after 60%. One model over the entire range 10%–90% would not capture this phenomenon. For measurements that show one pattern for $x \leq a$ and another $x > a$, a **spline function** is needed as stated in the regression chapter. Without this the results could be fallacious. Realize, however, that this could be used only if you know, at least suspect, that the patterns could be different over different ranges of x. If this is not known, the fallacy can continue for long time.

21.2.1.2 Overlooking Assumptions

A Gaussian form of distribution of various quantitative measurements is so ingrained in the minds of some workers that they take it for granted. Many examples are provided in this text of medical measurements that do not follow a Gaussian pattern. For large n, the central limit theorem can be invoked for inference on means but nonparametric methods should be used when n is small and the distribution is far from Gaussian. At the same time, note also that most parametric methods, such as t and F, are quite robust to mild deviation from the Gaussian pattern. Their use in such cases does a limited harm so long as the distribution has a single mode. Special attention is required if the distribution looks bimodal.

Transformations such as logarithm and square root sometimes help to "Gaussianize" a positively skewed distribution. But these also make interpretation difficult and unrealistic. If the logarithm of the mean of lipoprotein(a) level does not differ significantly in males and females, what sort of conclusion can be drawn for the lipoprotein(a) level itself? Despite this limitation, such transformations are in vogue and seem to lead to correct conclusions in many cases, particularly if n is not too small. Special statistical methods for many non-Gaussian distributions, such as exponential and gamma, can be developed, but experience suggests that the gains are not commensurate with the efforts. However, the need for such statistical methods cannot be denied.

The assumptions of independence of observations and of homoscedasticity are more important than that of a Gaussian pattern. You now know that independence is threatened when the measurements are serial or longitudinal. Homoscedasticity or uniformity of variance is lost when, for example, SD varies with mean. And this is not so uncommon. Systolic levels are higher in postmenopausal females and so are their SDs. Persons with higher BMI also tend to exhibit greater variability in BMI than those with lower BMI. Thus, care is needed in using methods that require uniformity of variance.

In the case of chi-square for proportions, the basic assumption is that expected frequency in most cells is five or more. This restriction is sometimes overlooked. Fisher exact test for a 2×2 table has become an integral part of most statistical packages of repute but the multinomial test required for larger tables with many expected frequencies less than five has not found a similar place. It is for you to locate a statistical package that is right for your problem.

21.2.1.3 Selection of Inappropriate Variables

More than 100 signs–symptoms of hypothyroidism can be identified. These include puffy face, cold intolerance, thin hair, and brittle fingernails. This probably is true for many medical conditions. Two kinds of fallacy can arise in such cases. One is due to a priori (before modeling) selection of variables. If you cannot study all because of limitation of sample size or otherwise, you must have sufficient reasons for including some and excluding others. In many situations, such reasons do not adequately exist and subjective preference plays a role. This obviously will affect the results. If all known variables are entered into the model, the sample size must be enormously large. Sometimes one variable at a time is considered and the statistically significant ones are included in a multivariable setup. To reiterate what was stated earlier, this ignores the interdependence—thus the choice is statistically difficult. You should keep this limitation in mind while making an inference. Many times this is forgotten.

Final wrinkle is epistemic gap about which also a remark was made earlier in the context of regression models. Quite often the factors affecting the outcome are not fully known. Obviously only known or at best suspected factors can be studied. Large unexplained variation is one indicator of inadequacy of model but unknown factors can be important even when unexplained part is small. This can happen when an unknown factor is closely linked to one or more of the known factors.

You know that all models are necessarily simplified version of the reality. They many times present a false aura of precision due to mathematical formulation. Limitations just stated increase vagaries. For this reason, even statistically adequate models may fail to live up to the expectations in practice.

21.2.1.4 Area under the Concentration Curve

There is a practice in animal experiments and clinical trials to note down the intensity of response at different time points. This could be in terms of concentration of drug after administration, intensity of pain after analgesia, physical stress during an exercise, or any other response. One popular method generally used to compare the *pattern* of response is the area under the concentration (AUC) curve. For example, Gaudreault et al. [11] evaluated truncated AUC curve as a measure of relative extent of bioavailability. This area is different from area under the receiver operating characteristic (ROC) curve. The difficulty is that the AUC curve has limited physical meaning. When used in isolation, it may fail to provide an adequate assessment of the difference in trend, even on average, because it is neither specific nor sensitive to the changes in the patterns. Curves with markedly different trends can give the same area (Figure 21.4). AUC curve is not a valid measure of bioequivalence. Also, as the time passes, some patients dropout or get cured—thus average response at different points of time is based on different n. This is forgotten while studying AUC curve.

AUC curve can give valid results when the response in one group is better than that in the other at almost every time point. If they crisscross then the conclusion can be very wrong. In this case, the investigator may have to be more specific on what exactly he is looking for. Merely stating a trend or pattern does not help. Depending on what is most relevant for the objective of the study, particularly in pharmacokinetic studies, it could be the time taken to reach the peak (T_{max}), time taken to return to the initial level, the peak level attained (C_{max}), the response level after a specific time gap, etc. In many situations, the AUC curve needs to be considered in conjunction with other parameters such as T_{max} and C_{max} for a valid conclusion. In a study on treatment

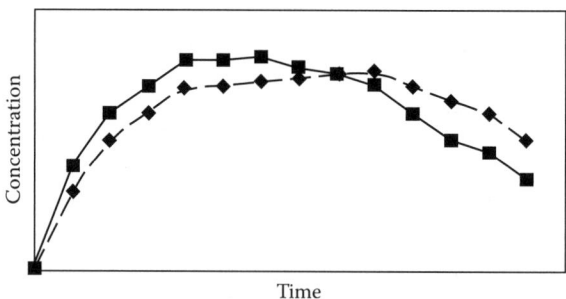

FIGURE 21.4
Markedly different curves with the same AUC.

efficacy, the interest may be in the value at the end relative to the initial value. If the study is on effectiveness of an analgesic, the interest may be in T_{max}, C_{max}, and possibly time to reach to some specified critical level. No matter what parameter is used, it should always be decided beforehand on the basis of clinical utility and not after inspecting the data.

21.2.1.5 Further Problems with Statistical Analysis

At the cost of repetition but in the interest of completeness, the following problems in analysis may be reminded:

1. Categorizing a continuous variable causes loss of information and can result in bias, loss of power, and increase in Type-I error. Different categorizations can lead to different conclusions. In some cases a clever investigator can choose categories after seeing the data that connives to reach to a preconceived result.

2. Mean and SD are inappropriate for variables with highly skewed distributions. For such variables, median and inter-quartile range are more appropriate. You know that inter-quartile range comprises middle half of the subjects. The difference (median $- Q_1$) when compared with ($Q_3 -$ median) gives an idea of the extent of skewness. Inter-quartile range also rules out the "absurd" looking statement such as mean lipoprotein(a) level is 8 ± 20 since this cannot be negative. The distribution of lipoprotein(a) is highly positively skewed and the SD is very high. Most values are around 5 mg/dL.

3. Real anomaly arises when, for example, too many zeros occur in a dataset (such as runs by batters in baseball games). In medicine, this happens when, for example, you are studying smoking among adolescents. If you record duration as zero for nonsmokers and actual duration on continuous scale for smokers, neither mean nor median will work well because of large number of zeros. If you categorize this as zero and more than zero, or duration in three or four categories, the analysis can be appropriately done by using proportions.

4. Many researchers take Gaussianity for granted, and use Gaussian-based methods even for small samples. Remember that small samples would not be able to provide evidence against Gaussianity because of lack of power. The tests such as Kolmogrov–Smirnov, Anderson–Darling, and Shapiro–Wilk work well with large samples and not small samples. For small samples you should have some extraneous evidence (from experience or literature) that the distribution is Gaussian. If the plot suggests non-Gaussian pattern, it might be safe to use nonparametric methods for small samples.

On the other hand, some researchers are nonparametric enthusiasts and use these methods even when the sample size is large. This could result in loss of power. So far, fortunately, the number of such enthusiasts is not high.

5. In paired comparisons, change such as from pre to post is a natural outcome of interest. Since baseline (pre) values can affect the amount of change, many times percent change is calculated. This works well provided (1) negative and positive changes are not canceling on averaging, and (2) percent change is independent of baseline. These limitations tend to be overlooked.

21.2.1.6 Anomalous Person-Years

It is customary to calculate person-years of smoking and use this for various inference purposes. The calculation presumes that smoking 5 cigarettes a day for 20 years has the same implication as smoking 50 cigarettes a day for 2 years. For many diseases, this obviously is not true. Person-years of exposure is a valid epidemiological tool only when each year of exposure has the same risk. Smoking first 2 years may carry more risk than additional risk of 2 years of smoking after 10 years of smoking when the damage is already done. Calculation of mortality rate per thousand person-years after an episode of fracture in patients on dialysis [12] is valid only when the risk of mortality in the first year after the fracture is the same as in, say, the tenth year after fracture. This obviously is not true. If nothing else, aging will make an impact in this case. Thus, use the person-years tool with abundant caution. Most of the literature using person-years seems to be missing this point.

21.2.1.7 Problems with Intention-to-Treat Analysis and Equivalence

Use of intention-to-treat (ITT) strategy in superiority trials is preferred, but in equivalence or noninferiority trials, ITT is suspect. When the patients switch from one group to the other, ITT may show equivalence where none really exist. The other fallacy occurs when a failed superiority trials is touted as evidence of equivalence or noninferiority. A review of 88 superiority studies over a 5-year period found that the claim was not properly supported in 67% studies [13]. Failed superiority is not evidence of equivalence or noninferiority. Fresh trial is needed to confirm this.

On the other hand, per-protocol (PP) analysis is based on those patients who stick to the allocated regimen. If there are many who frittered away, reduced sample size will affect the power to detect clinically important difference. At the same time, often smaller variability is associated with PP analysis that can compensate loss of power due to small n. Perhaps the best strategy is to do both types of analysis. If they do not conflict, you are done. If not, there might be lessons on where the problem lies.

You now know that equivalence of a regimen can be concluded when its efficacy is different by not more than a specified margin. If an existing regimen has 79% efficacy and if 3% difference is your tolerance, a regimen with 76% efficacy can be considered equivalent. This, however, means that the standard can slip over time. Now a new regimen could be evaluated for 76% efficacy in place of 79%. You may like to guard against such a fallacy. Also make sure that equivalence or noninferiority is not due to mischievous execution of the study such as a lot of dropouts and noncompliance of regimen by either group, particularly the group with standard regimen.

21.2.2 Choice of Analysis

Large number of fallacies can be cited that arise due to inappropriate choice of analysis but this section contains two everyday examples that illustrate how varying conclusion can be reached by one kind of analysis opposed to another kind. Aura of *P*-values is so much that a novice can even be happy using an inapplicable statistical method that happens to yield the lowest *P*-value.

21.2.2.1 Mean or Proportion?

The blame lies mostly with statisticians who fancy quantity than quality. Although quantitation in many cases does help in achieving exactitude in thinking and in drawing conclusions, it can sometimes suppress important findings. Example 21.5a illustrates one such situation. This is similar to Example 15.13 in Chapter 15.

Example 21.5a: Conclusion Based on Mean Can Be Different from the Conclusion Based on Proportion

Given in the following is the rise or fall in Hb levels in a random sample of eight adolescent girls after iron supplementation for 2 weeks:

Rise in Hb level (g/dL)	0.4	0.7	−0.9	0.3	0.1	0.5	0.9	0.2

$$\text{Mean rise} = 0.275 \text{ g/dL}$$

$$\text{SD of rise} = 0.542 \text{ g/dL}$$

$$t_7 = \frac{0.275}{0.542/\sqrt{8}} = 1.44$$

Use of the *t*-test is valid in this case (small *n*) when the rise follows a Gaussian pattern at least approximately. The null hypothesis is $H_0: \mu = 0$, where μ is the mean rise in the target population. The critical value of *t* for 7 df at 5% level is 1.895 for a one-tailed test. Note in this case that the possibility of iron supplementation reducing the mean Hb level is ruled out, so the test is one sided. Since the calculated value of *t* is less than the critical value, infer that the rise is not statistically significant and conclude that iron supplementation for 2 weeks in adolescent girls is not sufficient to produce a rise on average.

The fact also is that Hb levels in seven out of eight girls showed a rise. If the supplement is not effective, then the chance of rise is the same as that of a fall. Assume that Hb level will not remain exactly at the same level. Thus, H_0: $\pi = 1/2$ and H_1: $\pi > 1/2$, where π is the probability of a rise. Because of small n, you need to use the exact binomial distribution to obtain the P-value. Under H_0, this is given by

$$P(x \geq 7) = {}^8C_7 \left(\frac{1}{2}\right)^7 \left(\frac{1}{2}\right)^1 + {}^8C_8 \left(\frac{1}{2}\right)^8 \left(\frac{1}{2}\right)^0$$

$$= 0.035$$

where x is the number of girls showing a rise in Hb level. Thus, the chance of seven or more girls showing a rise out of eight when $\pi = 1/2$ is less than 0.05. Reject H_0 and conclude that the supplement is effective. This conclusion is different from the one reached on the basis of the mean.

Both analyses in Example 21.5a are right. If one wants to draw a conclusion about the rise in *mean*, then the answer is that it is not statistically significant. For the *proportion* of girls showing rise, the conclusion is that a statistically significant number show a rise. The first is based on the quantity of rise and the second disregards the exact quantity. Even a rise of 0.1 g/dL is a rise and a fall of 0.9 is a fall, irrespective of the quantity. Depending on the purpose of the investigation, the finding can be presented either way. It is for you to be on guard and check that the conclusion drawn indeed emerges from the data and that it is appropriately answering the intended question.

21.2.2.2 Forgetting Baseline Values

Although the rise in Example 21.5a is calculated over the baseline values, a rise of 0.3 over 8.7 g/dL is considered same in this example as this rise over 13.7 g/dL. In this sense, the example does not consider the actual Hb level in the subjects before the start of the supplementation. Now examine what can happen if the initial levels are also considered.

Example 21.5b: Consideration of Baseline Values Can Alter the Conclusion

Given in the following are the Hb levels (g/dL) of eight girls in Example 21.5a before the supplementation:

10.1	8.5	13.8	9.7	12.4	9.0	8.6	12.1

On critical examination of the data, you will find that the girls with higher level of Hb showed a smaller rise. In fact, a girl with an Hb level of 13.8 g/dL showed a fall. The relationship can be examined by running a simple linear regression of rise on the initial level. The regression equation is given by

$$\text{Rise} = 2.85 - 0.244 \text{ (initial Hb)}$$

This gives $R^2 = 0.81$ and $r = 0.90$. The regression shows that the rise, on average, is higher when the initial values are lower and becomes negative in this case when the initial Hb level exceeds 11.7 g/dL. Since fall can be ruled out in the case of iron supplementation, it might be concluded that iron supplementation in such girls is possibly not useful when Hb level is already 11.7 g/dL or higher.

The lesson from Example 21.5b is that baseline values can be important. In the case of comparison of two groups, if baseline values differ, the analysis may have to be geared to adjust for this difference. In many cases, change from baseline, rather than the final outcome, would be more adequate in coming to a valid conclusion.

Example 21.5b also illustrates the possible adverse effect of small n on the conclusion. The finding that Hb level substantially falls in some cases after iron supplementation is medically untenable and possibly indicates that other uncontrolled factors were playing spoil sport. This finding would most likely reverse if the number of girls in the trial were large. In this example, only one girl exhibited a fall, and this is able to vitiate the regression because of small n. Also, the objective of the study was not to find a threshold beyond which iron supplementation is not useful. To achieve this objective, a larger trial is required anyway.

21.2.3 Misuse of Statistical Packages

Computers have revolutionized the use of statistical methods for empirical inferences. Methods requiring complex calculations are now done in seconds. This is a definite boon when appropriately used but is a bane in the hands of nonexperts. Understanding of the statistical techniques has not kept pace with the spread of their use. This is particularly true for medical and health professionals. The danger arises from misuse of sophisticated statistical packages for intricate analysis without fully appreciating the underlying principles. The following are some of the common misuses of statistical packages.

21.2.3.1 Overanalysis

Popularly termed as torturing the data until they confess, data are sometimes overanalyzed, particularly in the form of post hoc analysis. A study may be designed to investigate the relationship between two specific measurements but correlations between pairs of a large number of other variables, which happen to be available, are calculated and examined. This is easy these days because of the availability of computers. If each correlation is tested for statistical significance at $\alpha = 0.05$, the total error rate increases enormously. Also, $\alpha = 0.05$ implies that 1 in 20 correlations can be concluded to be significant when actually it is not. If measurements on 16 variables are available, the total number of pairwise correlations is $16 \times 15/2 = 120$. At the error rate of 5%, 6 of these 120 can turn out to be falsely significant. Hofacker [14] has illustrated this problem with the help of randomly generated data.

Any result from post hoc analysis should be considered indicative and not conclusive. Further study should be planned to confirm such results. This also applies to post hoc analysis of various subgroups that were not part of the original plan. There is always a tendency to try to find the age–sex or severity groups that benefited more from the treatment than the others. Numerous such analyses are sometimes done using drop down menu in the hope of finding some statistical significance somewhere. Again, such detailed analysis is fine for searching a ground to plan a further study but not for drawing a definitive conclusion. However, this is not to deny the existence of Americas because it was not in Columbus' plan. This is not empirical. In most empirical cases though, it may be sufficient to acknowledge that the results are based on post hoc analysis and possibly need to be confirmed.

21.2.3.2 Data Dredging

Because of availability of software packages, it is now easy to reanalyze data after deleting some inconvenient observations. Valid reasons, such as presence of outliers, for this exercise are sometimes present, but this can be misused by excluding some data that do not fit the hypothesis of the investigator. It is extremely difficult to get any evidence of this happening in a finished report. Integrity of the workers is not and cannot be suspected unless evidence to the contrary is available. Thus, data dredging can go unnoticed.

21.2.3.3 Quantitative Analysis of Codes

Most computer programs, for the time being, do not have the capability to distinguish numeric codes from quantitative data unless you define them categorical. If disease severity is coded as 0, 1, 2, 3, and 4 for none, mild, moderate, serious, and critical conditions, respectively, statistical calculations may treat them as the usual quantitative measurements. As cautioned in Chapter 9, this runs the risk of considering three mild cases as equal to one serious case, and so on. This means codes can be mistreated as scores. This can happen even with nominal categories such as signs and symptoms when they are coded as 1, 2, 3, etc. Extra caution is needed in analyzing such data so that codes do not become quantities unless you want them to be so.

21.2.3.4 Soft Data versus Hard Data

The data to be analyzed statistically are mostly entered in terms of numerics in a computer. There may be some features of the health spectrum that are soft in the sense that they can only be understood and are difficult to put on paper. This applies particularly to psychological variables such as depression and frustration. Even if put on paper, they may defy coding, particularly if it is to be done before collection of data. A precoded pro forma is considered desirable these days because it makes computer entry so easy. But it should be ensured in this process that the medical sensibility of the information is not lost.

21.3 Errors in Presentation of Findings

Out of ignorance or deliberate, the presentation of data in medical reports sometimes lack propriety. This can happen in a variety of ways.

21.3.1 Misuse of Percentages and Means

Some fallacies occurring due to misuse of proportions and means are described next.

21.3.1.1 Misuse of Percentages

Percentages can mislead if calculations are (1) based on small n or (2) based on an inappropriate total. If two patients out of five respond to a therapy, is it correct to say that the response rate is 40%? In another group of five patients, if three respond, the rate jumps to 60%. A difference of 20% looks like a substantial gain, but, in fact, the difference is just one patient. This can always occur due to sampling fluctuation. Thus, the percentage based on small n can mislead. It is preferable to have $n \geq 100$ for valid percentages, but the following is my subjective guideline:

1. State only the number of subjects without percentages if $n < 30$.
2. For $n \geq 30$, percentages can be given but n should always be stated.

The second misuse of percentages occurs when they are calculated on the basis of an inappropriate group. Example 21.6 illustrates this kind of misuse.

Example 21.6: Correct Base for Calculating Percentages

Consider a group of 134 cases of heart bypass surgery who are followed for postsurgical complications. The data obtained are presented in Table 21.5. This is similar to Table 7.5 in Chapter 7.

Information was not available for seven patients. Since 83 did not experience any significant complication, it would be wrong to calculate the complication rate on the basis of the 44 patients who experienced complications. It unnecessarily magnifies the problem. The correct base for the complication rate is 127. The fact is not that 20.5% had excessive bleeding but that 7.09% of the patients had this problem. It would also be wrong to include seven patients in this calculation for whom the data are not available. For a nonresponse rate, however, the correct base is 134. This can be separately stated as shown in parentheses in the table.

Example 21.6 also illustrates the calculation of percentages in the case of multiple responses. A patient can have two or more complications. Thus, the percentages are not additive in this case.

Another example of use of a wrong base is as follows. Suppose 300 subjects are asked about their preference for more convenient, less expensive, but less efficacious medical treatment versus less convenient, more expensive, but more efficacious surgical removal of a benign tumor. Only 60 (20%) reported preference for medical intervention and 84 (28%) preferred surgery. The remaining 156 (52%) were noncommittal. If this finding is reported as 28% preferring surgery, it has the risk of being interpreted as saying that the remaining 72% preferred medication. Obviously this is wrong. The nonrespondents or the neutrals should always be stated so that no bias occurs in the interpretation. For an example from everyday life,

TABLE 21.5

Complications in Cases of Heart Bypass

Complication	Number of Cases	Wrong Percentage (Out of 44)	Correct Percentage (Out of 127)	
Excessive bleeding	9	20.5	7.09	
Chest wound infection	15	34.1	11.81	
Other infections	8	18.2	6.30	
Breathing problems	10	22.7	7.87	
Blood clot in the legs	13	29.5	10.24	
Others	11	25.0	8.66	
Any complication	44	100.00	34.65	
No significant complication	83		65.35	
Total (data available)	127		100.00	(94.78)
Data not available	7			(5.22)
Grand total	134			(100.00)

consider the interpretation of the forecast of 30% chance of rain. In fact, then, the chance of no rain is 70%, which is more than two times. Here there is no neutral category.

21.3.1.2 Misuse of Means

A popular saying by detractors of statistics is, "Head in an oven, feet in a freezer, and the person is comfortable, on average!" There is no doubt that an inference on mean alone can sometimes be very misleading. It must always be accompanied by the SD so that an indication is available about the dispersion of the values on which the mean is based. This text has emphasized this from time to time. Sometimes the standard error (SE) is stated in place of SD. This might mislead unless its implications in the context are fully explained. Also, n must always be stated when reporting a mean. These two, n and SD, should be considered together when drawing any conclusion based on mean. Statistical procedures such as confidence intervals (CIs) and test of significance have a built-in provision to take care of both of them. A mean based on large n naturally commands more confidence than the one based on small n. Similarly, a smaller SD makes the mean more meaningful.

The general practice is to state mean and SD with a "±" sign in between such as given by Park et al. [15]. Opinion is now generating against the use of the ± sign because it tends to give a false impression. If mean serum bilirubin is reported as 1.1 ± 0.4 mg/dL, it gives the impression that the variation is from 0.7 to 1.5 mg/dL. In fact, the variation is much more, even more than ±2SD limits 0.3–1.9 mg/dL. Thus, it is better to state clearly that mean is 1.1 mg/dL and SD is 0.4 mg/dL without using a ± sign.

You should also evaluate whether the mean is an appropriate indicator for a particular data set. Averages are not always what they seem. If in a group of 10 persons, 9 do not fall sick and 1 is sick for 40 days, how correct is it to say that the average duration of sickness in this group is 4 days per person? If extreme values or outliers are present, mean is not a proper measure. Either use the median or recalculate mean after excluding the outliers. If exclusion is done, this must be clearly stated. Else consider if proportions are more adequate than mean.

21.3.1.3 Unnecessary Decimals

There is a tendency to overuse decimals in the reporting of results, creating a false sense of accuracy. Perhaps, higher number of decimals is considered to give a scientific look. A large number of decimal places should be used for intermediary calculations (a computer will automatically do that), but the final result should have only an appropriate number of decimal places. A rule for percentages is to have only as many decimal places as are needed to retrieve the original number. As an extra precaution, one more decimal place can be used. For the percentage of subjects, this can be obtained by the following rules:

In percentages,

Maximum one decimal place if $n \leq 99$

Maximum two decimal places if $100 \leq n \leq 999$

Maximum three decimal places if $1000 \leq n \leq 9999$

There is a concept of **significant digits**. This is applicable particularly to the reporting of calculated values with an extremely large denominator. The leading zeros after decimal are ignored while counting the significant digits. The value 0.00720 and the value 0.458 both have three significant digits and have the same accuracy. The values 3.06 and 0.069 have two significant digits each. All reporting should have the same number of significant digits. This concept is particularly useful when the reported values are meant to be used for further calculations. I have not followed this rule in this text.

In expressing quantities other than percentages, such as mean and SD, the following rule is generally adequate:

Report one decimal place more than in the original measurements from which mean and SD are calculated.

If the last digit to be rounded off is 5, make the previous digit even, that is, 1.15 is rounded off as 1.2 and 3.45 as 3.4. Or, you can decide to stick to the odd digit in place of even digit. The idea is that 5 should go up half the time and down half the time because it is exactly midway.

It is sometimes desirable to use the same number of decimal places uniformly throughout a report even if n varies from one section of the report to another. The number of decimal places in this case would depend on the highest n. If there are 260 subjects belonging to 80 families, the percentages for subjects as well as for families can go up to two decimal places each.

The number of decimal places in a coefficient, such as in a regression equation, would depend on the numerical magnitude of the quantity with which it is multiplied. The coefficient 1.07 for birth weight in kilograms has the same accuracy as 0.00107 for birth weight in grams.

Sometimes original measurements are also made with excess accuracy. It would be a waste of resources to measure survival time in cancer patients in days and report it as 3.7534 years, or insulin level in a subject to two decimal places. A minute difference in such readings does not affect the validity of the measurement. The same sort of logic can be applied to percentages and means also. It does not matter whether the mean diastolic BP level of a group is 86.73 or 86.54 mmHg. Both could be rounded off to 87 for medical interpretation. Thus, the decimal places in this case do not serve any useful purpose. Instead they complicate the presentation and interpretation.

Many rates can be expressed as percentage but for others a different multiplier is used for convenience. For example, the death rate due to cervical cancer is stated as 8 per 100,000 women. This is the same as 0.00008 per woman or 0.008%. Siegrist [16], however, found that rates expressed as a frequency (8 per 100,000) are perceived differently than rates expressed as a probability (0.00008).

21.3.2 Problems in Reporting

Among many problems that can occur with the reporting, two requiring special attention are incomplete reporting and overreporting. The other is misuse of graphs.

21.3.2.1 Incomplete Reporting

All reports should state not only the truth but the whole truth. There is a growing concern in the medical fraternity that part of the information in reports is sometimes intentionally suppressed and sometimes unknowingly missed. If so, the reader gets a biased picture. This is easily illustrated by the bias of many medical journals for reporting positive findings and ignoring the negative reports. While knowledge about the side effects of a therapeutic regimen is definitely important, knowledge about their absence cannot altogether be ignored. Both should be reported in a balanced manner. Similarly, properly designed studies that do not reject a particular null hypothesis deserve a respectable place in the literature. That, at present, is sadly lacking, although awareness for a balanced approach is increasing. If you are scanning the literature to update your knowledge, examine if you are getting balanced information.

It is not an uncommon practice in the medical literature not to give full details of the methodology, of the sources of subjects, or of the data. For example, reports on trials sometimes say that randomization was done but exactly how it was achieved will not be stated. Sometimes the source of the subject is not clearly stated. Among many, one such example is the study of infants and young children reported by Kirjavainen et al. [17] on monitoring of respiration during sleep. The source of the subjects is obscure. If the subjects are from a mix of heterogeneous sources, then the results in some cases can be suspected. Often the target population to which the results are intended to be extrapolated remains obscure. There are examples of articles in reputed journals that report that data were statistically analyzed without stating the particular statistical methods that were used for different inferences. Some remain absolutely silent about statistical methods. See, for example, an article by Belgorosky et al. [18]. Another problem is with regard to the data. Although complete data are desirable, it may not be feasible to reproduce them in many cases. However, they can be placed on the web. In any case, the reports should contain enough details for a suspicious reader to verify the results.

A serious problem is reporting only those parts of a study that support a particular hypothesis. The other parts are suppressed. An example is a series of studies on the carcinogenic effect of asbestos. According to an analysis of different studies provided by Lilienfeld [19], deliberate attempts were made by the industry to suppress information on the carcinogenicity of asbestos that affected millions of workers.

If 12 frequent users of mobile phones in an area are found to develop cancer, how big is this warning sign? For media, this is a big catch. If there are 90,000 other users who did not develop cancer, can you say that mobile phones protect from cancer? Such reports deserve to be fully investigated to discover the truth before they are sensationalized.

While discussing critique of a medical report in Chapter 1, the advice was to ensure that the results do indeed emanate from the data. Inference must always be accompanied by evidence. Some reports do not live up to expectation in this respect.

21.3.2.2 Overreporting

The statement of results in a report should generally be confined to the aspects for which the study was originally designed. If there are any unanticipated findings, these should be reported with caution. These can be labeled as "interesting" or "worthy of further investigation" but not presented as conclusions. A new study, specifically designed to investigate these findings, should be conducted.

If a drug is found marginally effective in improving vision in an isolated small-size trial, it is likely to be blown up by the hasty media because such a drug is the need of the public. Thus, unproved treatments can get undue promotion.

If one subgroup showed large effect of a regimen, it can be highlighted by more detailed analysis and by speaking extensively about it. Sometimes even the primary focus may change from the one stated in the protocol. Any violation of protocol leaves strong ground for suspicion. Also, beware of the "other" and "unknown" category of the responses in a finished report. They may be hiding important information on cases adverse to the hypothesis of the investigator.

21.3.2.3 Selective Reporting

Researchers around the world tend to be selective in reporting results of their work. Whereas some of this can be justified to save journal space and complexity, at other times such filtering can introduce bias. If you started with 1000 subjects, all these must be fully accounted for in your report. Statements such as CONSORT and STROBE require this in full measure. But filtering can be intentionally or unsuspectingly done at other levels too. These are (1) selecting the "favorable" end point for reporting; (2) reporting result for three treatments while the study had four treatments; (3) follow-up was 2 years but the results truncated to, say, 6 months because they are as per your liking till this point of time; (4) you studied 15 predictors but restrict results to 6 convenient predictors; (5) trying different cut points for your continuous variables and choosing those for reporting that provide favorable results; (6) using categories such as none, mild, moderate, and serious (e.g., for hypertension) where actual levels (of BP in this case) should be used; and (7) reporting two favorable results out of dozens of attempted analyses.

What is the way out? Do not compromise integrity of your work. Fully disclose what variables you started with, what analyses were tried, and what part of the results you are reporting and why. Why other results are not included should also be expressly stated. This does not necessarily remove the bias but makes the reader aware of the limitations of the results.

21.3.2.4 Self-Reporting versus Objective Measurement

Self-perception of health may be very different from an assessment based on measurements. A person with an amputated leg may consider himself absolutely healthy. Besides such discrepancies, it has been observed, for example, that people tend to report lower weight than their actual weight but when it comes to reporting height, they actually report higher values. This could make a substantial difference when BMI is calculated. The percentage of subjects with BMI ≥ 25 may thus become much lower than obtained by actual measurements. Thus, only those characteristics should be self-reported that are so required for the fulfillment of the objectives of the study. All others should be objectively measured.

21.3.2.5 Misuse of Graphs

While describing graphs and diagrams in Chapter 8, it was mentioned that some fallacies could occur due to an inadequate choice of scale. A steep slope can be represented as mild and vice versa. Similarly, a wide scatter may be shown as compact. Variation is shown as $\pm 1SD$ whereas actually it is much more. Also, means in different groups or means over time can be shown without corresponding SDs. They can be shown to indicate a trend that really does not exist or is not statistically significant.

One of the main sources of fallacies in graphs is their insensitivity to the size of n. A mean or a percentage based on $n = 2$ is represented in the same way as the one based on $n = 100$. The perception, and possibly cognition, received from a graph is not affected even when n is explicitly stated. One such example is box-and-whiskers plots drawn by Koblin et al. [20] for the time elapsed between cancer and AIDS diagnoses among homosexual men with cancer diagnosed before or concurrently with AIDS in San Francisco during 1978–1990. This is reproduced in Figure 21.5. Five lines (minimum, Q_1, median, Q_3, and maximum) are shown on the basis of only four cases of

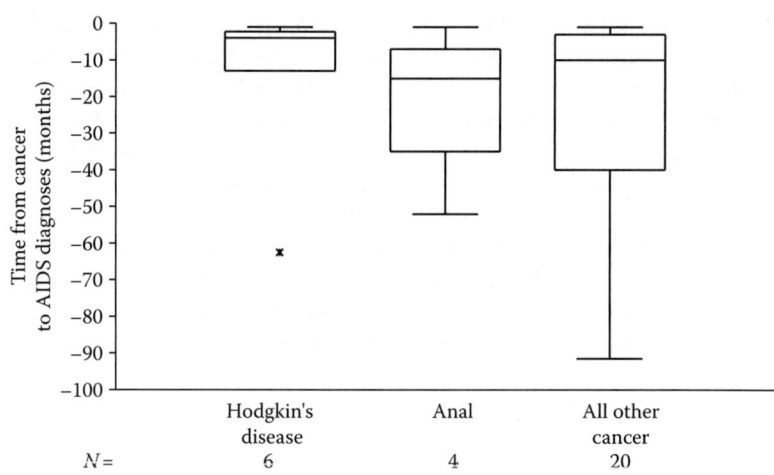

FIGURE 21.5
Time between cancer and AIDS diagnoses among homosexual men with cancer diagnosed before or concurrently with AIDS, San Francisco, CA, 1978–1990. Asterisk indicates outlier. (Reproduced from Koblin, B.A. et al., *Am. J. Epidemiol.*, 144, 916, 1996. With permission of the Johns Hopkins University School of Hygiene and Public Health.)

anal cancer. It was concluded on the basis of the figure that "Hodgkin's disease cases occurred relatively close to AIDS diagnoses." This may have merit but is stated without checking statistical significance and in disregard of the small number of available cases.

21.4 Misinterpretation

Misinterpretation of statistical results is mostly due to failure to comprehend them in their totality and inability to juxtapose them with the realities of the situation. This can happen either because many medical professionals have limited knowledge of statistical concepts [21] or because of inadequate understanding of medical issues by the statisticians associated with medical projects. Some of these are discussed in this section.

21.4.1 Misuse of *P*-Values

Statistical *P*-values seem to be gaining acceptance as a gold standard for data-based conclusions. However, biological plausibility should not be abandoned in favor of *P*-values. Inferences based on *P*-values can also produce a biased or incorrect result.

21.4.1.1 Magic Threshold 0.05

A threshold of 0.05 in Type-I error is customary in health and medicine. Except for convention, there is no specific sanctity of this threshold. There is certainly no cause for obsession with this cutoff point. A result with $P = 0.051$ is statistically almost as significant as one with $P = 0.049$, yet the conclusion reached would be very different if $P = 0.05$ is used as the threshold. Borderline values always need additional precaution.

A value close to the threshold such as $P = 0.06$ can be interpreted both ways. If the investigator is interested in showing the presence of difference, he might argue that this P approaches significance. If the investigator is not interested, this can be easily brushed aside as not indicating significance at $\alpha = 0.05$. It is for the reader to be on guard to check that the interpretation of such borderline P-values is based on objective consideration and not driven by bias. I generally interpret $P = 0.06$ or 0.07 as encouraging though not quit good enough to conclude a difference. They can be called marginally significant. If feasible, wait for some more data to come and retest the null in this situation.

The second problem with threshold 0.05 is that it is sometimes used without flexibility in the context of its usage. In some instances, as in the case of a potentially hazardous regimen, a more stringent control of Type-I error may be needed. Then, $\alpha = 0.02, 0.01$, or 0.001 may be more appropriate. It is not necessary to use $\alpha = 0.01$ if a value less than 0.05 is required. The value $\alpha = 0.02$ can also be used if you can justify this. In some other instances, as in concluding the presence of differences in social characteristics of the subjects, a relaxed threshold $\alpha = 0.10$ may be appropriate. For most physiological and pathological conditions, however, the conventional $\alpha = 0.05$ works fine, and that is why it has stayed as a standard for so long. This text has used this α in most of the examples.

As stated in Chapter 12, the practice now generally followed is to state exact P-values so that the reader can draw his own conclusion. A value of P around 0.10 can possibly be considered weak evidence against the null hypothesis and a small P, say, less than 0.01, as strong evidence.

21.4.1.2 One-Tail or Two-Tail P-Values

Example 21.5a on rise in Hb level is a situation where a one-tailed test is appropriate because biological knowledge and experience confirm that iron supplementation cannot reduce Hb level. Use of the two-tailed test in this case makes it unnecessarily restrictive and makes rejection of H_0 more difficult. However, in most medical situations, assertion of one-sided alternative is difficult and a two-sided test is needed. Most statistical packages provide two-tailed P-values as a default and many workers would not worry too much about this aspect. Scientifically, a conservative test does not do much harm, although some real differences may fail to be detected when a two-tailed test is used instead of a one-tailed test. My advice is to use a one-tailed test only where a clear indication is available but not otherwise.

21.4.1.3 Multiple Comparisons

This point has already been explained in the paragraph on overanalysis of Section 21.2. It would be unwise to repeat the argument but needs mention here also for completeness. Several tests of statistical significance, each at $\alpha = 0.05$, would make the total error much higher than acceptable. Specific statistical methods, such as the Tukey procedure, that control α to 0.05 or any other specified level should be used in such cases. A wiser approach is to limit the number of P-values you calculate to the original question. Do not give too many P-values that you cannot handle.

The concept of multiple comparisons and adjustment of P-values is complex. Many P-values may have been considered to arrive at a conclusion. The procedure just mentioned is valid for the stated situations but in practice you may have tried many statistical tests before reaching to the final ones you decide to report. Hardly anybody will make adjustment for such "behind-the-scene" statistical tests although they also affect the final P-value.

There might be two or more publications on the same set of data with different focus or with different outcome variables. Each publication may be complete within itself for multiple comparisons but would be oblivious to the comparisons made for the other papers. The argument can be extended to Type-I errors committed by other workers in similar studies and possibly in lifetime. For some researchers, ignoring accumulation of Type-I errors is one reason that the statistically significant results fail to reproduce. Just be on guard for such fallacies.

21.4.1.4 Dramatic P-Values

Attempts are sometimes made to dramatize the P-values. James et al. [22] stated $P < 0.00000000001$ for difference in seroconversion rate against Epstein–Barr virus between patients with lupus and controls. It is pulling out a number from a computer output without being careful about its implications. Such accuracy is redundant. It really does not matter whether $P < 0.001$ or $P < 0.000001$ as far as its practical implication is concerned. Many statistical software packages rightly stop at three decimal places and give $P = 0.000$ when it is exceedingly small. It only means that $P < 0.0005$, and this is enough to draw a conclusion. Further decimal places are rarely required. But do not interpret $P = 0.000$ as $P = 0$.

21.4.1.5 P-Values for Nonrandom Sample

This text has repeatedly emphasized that all statistical methods of inference such as CI and test of significance require random sample of subjects. Patients coming to a clinic during a particular period can be considered a *random* sample from the population of patients that are currently coming to that clinic. But this limited definition of population is sometimes forgotten and generalized conclusions are drawn on the basis of *P*-values. This is quite frequent in medical literature and mostly accepted without question. An example is a study [20] of a New York City cohort of homosexual men who participated in two studies of hepatitis B virus infection in the late 1970s and a San Francisco city clinic cohort of homosexual men recruited from the municipal sexually transmitted disease (STD) clinics from 1978 through 1990. Neither cohort is a random sample. Yet *P*-values for statistical significance of difference between the two cohorts have been calculated. These *P*-values are valid only when the cohorts are considered a random sample from a hyperpopulation of homosexuals in the two cities. The hyperpopulation exists in concept but not in reality and may include the past and, more importantly, homosexuals arising in immediate future. The implication of the result, which in this case is the increased incidence of cancer among homosexual men, is a warning to the future homosexual men more than to the present homosexual men.

Two more points need stress in this context. First, a sample can rarely be fully representative of the target population in a true sense. Thus, the risk of a wrong conclusion at $\alpha = 0.05$ is, in fact, slightly more in many cases than it is in case of truly random samples. This probability can also be affected by, for example, non-Gaussian distribution of the measurement under consideration. Second, *P*-values have no meaning if no extrapolation of findings is stipulated. Conclusions based on a sample *for the sample subjects* can be drawn without worrying about *P*-values.

21.4.1.6 Assessment of "Normal" Condition Involving Several Parameters

Consider individual persons instead of groups. As explained earlier, the reference range for most quantitative medical parameters is obtained as mean ± 2SD of healthy subjects. These statistical limits carry a risk of excluding 5% healthy subjects who have levels in the two extremes. When such limits are applied on several parameters, it becomes very unlikely that all parameters in any person are within such statistical range, even if one is fully healthy. This anomaly is sometimes forgotten while evaluating health parameters of a person or while devising inclusion and exclusion criteria for subjects in a research study.

This fallacy is similar to a multivariate conclusion based on several univariate analyses, and the one inherent in multiple comparisons.

21.4.1.7 Absence of Evidence Is Not Evidence of Absence

I have tried to make a distinction all through this book that a large *P*-value is not interpreted as evidence of absence. When $P > 0.05$ for a difference between two groups, the only valid conclusion is that the data are not sufficient to provide evidence of difference being present. This does not imply that there is no difference. Yet, when the sample size is reasonably large to have enough power for a specified unimportant difference, the conclusion can be that the difference is less than the specified value—thus practically there is no difference. This kind of argument is generally forwarded to assert that the two groups are not different at baseline. Remember though that the sample size must be large to arrive at this conclusion.

21.4.2 Correlation versus Cause–Effect Relationship

Some professionals feel that an RCT is the only mode to provide scientific evidence of a cause–effect relationship. A moment's reflection will convince you that this is not so. Most accept that cigarette smoking *causes* lung cancer but no trial has ever been conducted. The controlled experiments do provide direct evidence but evidence from observational studies can be equally compelling when confounders are really under control and the results replicate in a variety of settings.

Keep in mind though that an association or a correlation in health measurements can arise because of a large number of intervening factors. It is rarely a cause–effect type of relationship unless established by a carefully designed study that rules out the possibility of a significant role of any confounding factor. McLaughlin et al. [23] report from epidemiological studies that women in the United States with cosmetic breast implants have two- to threefold risk of suicides compared to the general population. This is statistically significant. However, this by itself does not support a cause–effect relationship. They speculate that psychopathological condition of women may be a confounding factor since this condition can contribute to the need for a breast implant as well as suicide. They also remark that any study aiming to investigate this relationship for cause–effect should also rule out year of birth, family history of psychiatric admission, etc., as factors. More important threats to cause–effect inference are not known confounders but the unknown confounders.

A strong correlation between heights of siblings in a family exists not because one is the cause of the other but because both are affected by parental height. Similarly, correlation between visual acuity and vital capacity in subjects of age 50 years and above is not of the cause–effect type but arises because both are products of the same degeneration process. No one expects vision to improve if vital capacity is improved by some therapy. Maternal mortality rate declined in Mexico between 1960 and 2010, and the proportional mortality from coronary diseases increased. These too have a strong negative correlation but these are not cause-and-effect type of relationships. Counterfactuals provide useful armory to refute causality.

An unusual confounding factor may provide useful information in some cases. Persons with depressed moods may have an elevated risk of lung cancer because of a third intervening factor, namely, smoking [24]. Depressiveness seems to modify the effect of smoking on lung cancer either by a biological mechanism or by affecting smoking behavior. This example illustrates how a seemingly nonsense correlation can sometimes lead to a plausible hypothesis. Another example is the parental age gap influencing the sex ratio of the firstborn children [25]. At the same time, our existing knowledge tells us that no drug can change the blood group. There is no need of empirical evidence to confirm this.

21.4.2.1 Criteria for Cause–Effect

For an association or a correlation to indicate a cause and effect, the following points should be carefully examined:

1. Association should be present not only in the prevalence but also in incidence. This should continue to be statistically significant when the effect of other confounding factors is eliminated by means of a method such as multiple regression. If feasible, confirm this by an experiment.

2. Association or correlation must be consistent across various groups, periods, and geographical areas. Wherever or whenever the postulated cause is present, the effect must also be present. If an association is found between lipoprotein(a) and incidence of CAD in Canada, Singapore, England, and South Africa, which has persisted since this protein was first discovered, then this is likely to be a cause of CAD, although this could be one of the many causes.

3. There must be a dose–response kind of relationship in the sense that if the cause is present in higher amounts or in greater intensity then the chance of effect should also be high. The more one smokes, the higher is the risk of lung cancer.

4. Relationship should also be specific, that is, if the postulated cause is absent then the effect should also be absent. However, make a distinction between a necessary cause and a sufficient cause.

5. Relationship must be biologically plausible, which means explanation linking the two in a causal way is available. For example, cigarette smoke has cotinine that easily affects lung cells and causes aberrant cell behavior. Thus, a biological explanation is available.

6. It is generally stated for a cause–effect relationship that the degree of association or the correlation should be high. This may be so when the factor under investigation is a dominant cause. The correlation between parental levels and children's level of cholesterol is not high but this does not exclude genetic influence as a cause of raised cholesterol level. When a correlation, albeit low, is consistently present, a causal relationship can still be inferred but that would be one of the many causes. It should, however, be statistically significant so that sampling fluctuation is adequately ruled out as a likely explanation. In addition, also note that the correlation coefficient can be small just because small range of values has been observed.

The causal role of elevated serum cholesterol in atherosclerotic CAD, for example, fulfills all these criteria. A consistent association has been found in prevalence as well as in incidence in a large number of studies carried out in populations with different backgrounds. A gradient in risk of the disease has been seen with rising cholesterol level. In controlled clinical trials, cholesterol-lowering drugs and diet therapy have been shown to result in a reduction of coronary events. Angiographic regression of coronary lesions has also been demonstrated. Biological plausibility has been established from animal and human studies that have demonstrated cholesterol deposition in atheromatous plaques.

21.4.2.2 Other Considerations

Distinction may be made between a necessary cause and a sufficient cause. Sexual intercourse is necessary for initiating pregnancy in natural course but it is not sufficient. In fact, the correlation between number of intercourses and number of pregnancies, even without barriers, is negligible.

In this era of diseases with multifactorial etiology, one particular factor may not be sufficient to cause the disease. Presence or absence of one or more of the others may also be needed. Thus, a factor could be a contributing cause in the sense that it is a predisposing, enabling, precipitating, or reinforcing factor. For a detailed discussion of causal relationships in medicine, see Elwood [26].

The aforementioned discussion assumes that the role of chance and of bias has been minimized if not eliminated. Chance due to sampling fluctuation is adequately ruled out by demonstrating statistical significance. Bias, as discussed in Chapter 2, can occur due to a host of factors such as the selection process, observational methods, differential recall of past events, and suppression of information. Measures such as standardization of methods, training, randomization, and matching have been already discussed to minimize the bias. Cause–effect hypothesis is strengthened when all alternative explanations are also studied and shown to be not tenable. Larger sample size does help in increasing confidence but its role in causal analysis beyond statistical significance is marginal.

21.4.3 Sundry Issues

The list of statistical fallacies mentioned earlier is not complete. Several others are mentioned in various chapters. Some of these need to be reiterated.

21.4.3.1 Diagnostic Test Is Only an Additional Adjunct

Moons et al. [27] make a forceful plea and rightly so that hardly ever a diagnostic test is used in isolation. In almost every situation, the patient's history and physical examination findings are already available before a test is used.

In fact, a test is advised only on the basis of the clinical picture. Thus, the utility of the test should be examined only as an additional adjunct instead of solely on its performance. They illustrate this by calculating the area under the ROC curve for 140 patients suspected of pulmonary embolism who had an inconclusive ventilation-perfusion lung scan. The tests they evaluated are partial pressure of oxygen in arterial blood (PaO_2), x-ray film of thorax, and leg ultrasound. The ROC area for history and physical examination was 0.75 that rose to 0.77 when only PaO_2 test was added but to 0.81 when only thorax x-ray was added and was also the same when ultrasound was added. There was no clear-cut advantage in choosing between the x-ray and ultrasound from this viewpoint; however, the x-ray is preferred because of convenience. Against this, single-test evaluations showed that ultrasound was the most informative. The authors rightly concluded that single-test evaluations could be very misleading. Also, note in this example that the additional utility of the tests once clinical picture is available is very limited. The gain is only 0.02–0.06 in the AUC. Such low gain over the clinical picture is often forgotten when the utility of tests are evaluated.

Medical tests are often ordered to get an objective evidence of what has been subjectively evaluated using the clinical picture. A clinician may be already confident but corroborative evidence may be more convincing to the patient. Also, tests are often considered essential to wriggle out legal hassles. Statistically though, tests in some situations do not sufficiently add to the already arrived clinical picture–based probability. In these cases, posttest probability of a disease is not much different from the pretest probability.

21.4.3.2 Medical Significance versus Statistical Significance

As emphasized in an earlier chapter, there is a need to scrupulously maintain the distinction between statistical significance and medical significance. A decline of 1 mmHg in systolic BP can be statistically significant if n is large but may not have any medical significance in terms of condition of the patient or in terms of management of the condition. This text tried to make it clear in many examples that statistical methods check statistical significance only, and medical significance needs to be examined separately. For example, in a test for bioequivalence of two pharmaceutical preparations, if the difference is less than a specified amount, you can conclude that the groups are essentially equivalent even if the difference is statistically significant.

21.4.3.3 Interpretation of Standard Error of p

As explained in Chapter 12, standard error $SE(p)$ has two very different meanings depending upon this is measured in an absolute sense or in a relative sense. If $p = 0.03$, then $SE(p) = 0.022$ is exceedingly high because the 95% CI for π is (0, 0.07) for large n. If $p = 0.40$, even a double $SE(p) = 0.044$ is low because 95% CI (0.31, 0.49) is still narrow relative to the value of p.

21.4.3.4 Univariate Analysis but Multivariate Conclusions

For simplicity in calculations and for easy understanding, statistical analysis is often done for one variable at a time. This is adequate as long as the conclusion too is univariate. But a multivariate conclusion on the basis of several univariate analyses can be wrong. Consider, for example, the relationship of weight, length, and head circumference of infants to antenatal smoking and feeding practices. If weight is affected but length and head circumferences are not affected, no definite conclusion can be drawn for growth as a whole. Growth is a multidimensional variable that includes all three measurements if not more. If development is also to assessed, and one milestone is properly achieved and the others are not, the conclusion again can be drawn for that milestone by univariate analysis but not for the development in general. The interest may be in the composite answer. If so, multivariate analysis becomes essential.

There is another dimension to the problem. It is quite often possible that three or four variables, when considered separately, exhibit statistical significance but become not significant when considered together in a multivariate setup. The opposite can also happen, although this is rare. Such instances occur because multivariate analysis gives due consideration to the interrelations of the variables, which a set of univariate analyses would not do.

A classical example of a bivariate entity is hypertension. It requires simultaneous consideration of both systolic and diastolic BP. One of them in isolation is rarely enough to draw a valid conclusion. Similarly, efficacy of a treatment regimen is assessed not just in terms of cure rate but also in terms of speed of relief, side effects, cost, convenience of administration, etc. The cure itself may be multivariate in some situations consisting of symptomatic response, functional change, laboratory results, etc., and they may not be of equal importance. A practicing clinician may be interested in a composite answer about whether or not the regimen *on the whole* is beneficial to his patients. This conclusion will be necessarily multivariate and requires simultaneous consideration of the factors involved. Realization of the importance of multivariate analysis is common but its use in practice is rare, probably because of the intricacies involved in interpretation. The situation may change in the course of time.

21.4.3.5 Limitation of Relative Risk

No risk can exceed one. The 10-year risk of death in leukemia may be 0.99 and in anemia only 0.02. Then the relative risk (RR) of death in leukemia is nearly 50. Opposed to this, if the comparison is with breast cancer cases where the 10-year risk of death is 0.60, the RR is only $0.99/0.60 = 1.65$. This is as high as it can get in this situation. If risk at baseline is 0.5 or more, RR cannot exceed 2.0. Note the physical limitation imposed by high risk in the reference group. The value of the RR is greatly influenced by the risk in the control group. This aspect is many times forgotten while interpreting the value of RR.

21.4.3.6 *Misinterpretation of Improvements*

A gain of 2 years in survival duration after an intervention has different meaning for a disease affecting in old age such as Alzheimer's than for a disease affecting young people such as motor vehicle injuries. A gain of 2 mg/dL in Hb level is easy when baseline is 6.2 mg/dL than when baseline is 14.7 mg/dL. An improvement of 1.5/1000 in case fatality in case of Stage IV cancer has different implication than a similar improvement in cases of peritonitis. Such gains may look statistically same but have different social implications. You should be careful in interpreting such improvements. Put value to the outcome and interpret accordingly.

Another kind of misinterpretation is cited by Vickers [28] in the context of result of a trial on prostectomies. This trial found that 53 (15.3%) out of 347 patients of prostate cancer died in the prostectomy group, of which 16 from prostate cancer. In the control group of 348 patients, 62 (17.8%) died of which 31 from prostate cancer. The conclusion was that prostectomy reduced deaths from prostate cancer but not overall deaths. How that is possible when the cases are all of prostate cancer and this is the primary cause of death? What other factors were operating in the prostectomy group to cause nearly same number of deaths? This example illustrates that statistical nonsignificance in overall deaths (15.3% vs. 17.8%, $P > 0.10$) should not be considered as no difference. It is possible that overall mortality was different but not picked up in this study. The second possibility is that prostectomy reduced prostate cancer deaths but propped up other factors to cause death. This is not unlikely. The third, as emerged later on, is statistical. In this case, the difference $17.8\% - 15.3\% = 2.5\%$ in overall mortality is lower than the difference $31/348 - 16/347 = 4.3\%$ in prostate cancer–specific mortality. Statistical significance is more difficult to reach for same difference in two large percentages compared with two small percentages. That is, it is easier to detect a difference between, for example, a rate of 8% and 4% cancer-specific mortality than a similar difference between 28% and 24% in overall mortality. Difference of 4% in the former is one-half of the higher percentage whereas this is one-seventh in the latter. How many of us really look at the data so critically?

21.4.4 Final Comments

No decision is more important than one concerning the life and health of people. The medical fraternity has a tall order. They are expected to prolong life and reduce suffering that occurs as a consequence of complex and often poorly understood interaction of a large number of factors. Some of these factors are explicit but many remain obscure, and some behave in a very unpredictable manner. Uncertainties in health and medicine are indeed profound. A main reason behind the success achieved so far on the health front is the ability of some professionals to learn quickly from their own experience or from the experience of the others and articulately collate the experience gained in laboratories, clinics, and the field. Successful empiricism is proper discernment of trends from fluctuation, segregation of focus from turbulence, and order from chaos. Knowingly for some but unknowingly for many, statistical methods play a vital role in all empirical inferences. Know the limitations also. Nothing in medicine can be predicted with certainty, no matter how sophisticated is the method. However, uncertainties can be controlled and estimated with reasonable confidence. That helps in some but not in all cases.

Some researchers (see, e.g., Vickers [28]) have passionately and convincingly pleaded that inappropriate use of statistical methods and consequent wrong decisions can risk life and health of *many* people. A surgeon is trained for years, yet a mistake is dreaded as it can cost life of the patient. Statistical methods seem to belong to everyone. Sufficiently trained or not, almost anybody can comment on statistical data and can carry out statistical analysis. Vickers illustrates how a mistake in statistical analysis and interpretation may have resulted in 750 premature deaths due to prostate cancer in Scandinavia. This indeed is serious and you should take all possible precautions that right method and right software are used and results are correctly interpreted.

It should be clear that statistics could be a dangerous tool when used carelessly. Realization of cultivating statistics-based thinking in medical professionals is recent. I have tried to demonstrate in this book that biostatistics is about rational thinking and it does not require fancy mathematics. It is necessary to use an appropriate method for the problem in hand, and for this, mathematics is not all that necessary.

As with all experts, statisticians too sometimes differ on the method to be adopted in a particular setup, but there is a wide agreement on the basic methods. The choice of a basic method is thus not much of a problem. Statistical packages do not yet have the inbuilt expertise to decide the correct method although they sometimes generate a warning message when the data are not adequate. The user of the package decides the method. If you are not sufficiently confident, do not hesitate to consult an expert biostatistician. This consultation is much more effective when done at the planning stage than at the data analysis stage.

Statistics seems to have become too important for some of us. Some statisticians may allow themselves to be manipulated when pressed to do so. Senn [29] narrates how this can happen in some cases. Biostatisticians should never indulge in losing soul for pecuniary gains—they must let the data speak. All data and analysis may be put on the Internet for anybody to check.

Medical journals too have a responsibility to ensure that the results of dubious quality are not published. Statistical refereeing is a norm for some journals but some are lax on this issue. There is also a need to realize that biostatisticians too need some basic training in medicine or health before being admitted to this fraternity. Some statisticians win the label of a biostatistician because they happen to work in a medical or health institution. Some universities require sufficient grounding in health or medicine before offering degrees in biostatistics, but most universities around the world are not so particular. Many statistical fallacies occur due to the inability of biostatisticians to comprehend the medical aspects of the problem. They are thus not able to provide sufficiently valid consultation in some cases.

My last advice is not to rely solely on statistical evidence. Statistical tools are surely good as an aid but rarely as a master. Do not allow it to dominate other ways of thinking. Statistical methods do not still incorporate skilled clinical judgment, which remain hallmark of clinical practice. In addition, like a diagnostic test, a statistical test can be falsely positive or falsely negative. In the case of diagnosis, decisions are sometimes made against the test results when other evidence is overwhelmingly against. This can be done against statistical tests too when sufficient evidence is otherwise available. You may like to distinguish between statistical evidence and scientific conclusion. Previous knowledge, medical evidence, and biological plausibility must remain driving considerations in reaching to a conclusion. The context cannot be divorced and numbers by themselves seldom provide infallible evidence. The approach should be holistic rather than isolated and graded instead of binary. Rely on your intuition more than science. If scientific results fail intuitional judgment, look for gaps. They would most likely lie with science than with intuition.

References

1. Ludwig EG, Collette JC. Some misuses of health statistics. *JAMA* 1971; 216:493–499.
2. Hill AB. *A Short Textbook of Medical Statistics*. London, U.K.: The English Language Book Society, 1977, p. 261.
3. Fernandez WG, Mehta SD, Coles T, Feldman JA, Mitchell P, Olshanker J. Self-reported safety belt use among emergency department in Boston, Massachusetts. *BMC Pub Health* 2006; 6:111.
4. Vitolo MR, Canal Q, Campagnolo PD, Gama CM. Factors associated with risk of low folate intake among adolescents. *J Pediatr (Rio J)* 2006; 82:121–126.
5. Freiman JA, Chalmers TC, Smith H Jr, Kuebler RR. The importance of beta, the type II error, and sample size in the design and interpretation of the randomized control trial: A survey of 71 "negative" trials. *N Engl J Med* 1978; 299:690–694.
6. Dimick JB, Diener-West M, Lipsett PA. Negative results of randomized clinical trials published in the surgical literature: Equivalency or error? *Arch Surg* 2001; 136:796–800.
7. Schnitzer TJ, Kong SX, Mitchell JH et al. An observational, retrospective cohort study of dosing patterns for rofecoxib and celecoxib in the treatment of arthritis. *Clin Ther* 2003; 25:3162–3172.
8. Barton MB, Morley DS, Moore S et al. Decreasing women's anxieties after abnormal mammograms: A controlled trial. *J Natl Cancer Inst* 2004; 96:529–538.
9. Silber JH, Rosenbaum PR. A spurious correlation between hospital mortality and complication rates: The importance of severity adjustment. *Med Care* 1997; 35 (10 Suppl):OS77–OS92.

10. Gorey KM, Trevisan M. Secular trends in the United States black/white hypertension prevalence ratio: Potential impact of diminishing response rates. *Am J Epidemiol* 1998; 147:95–99.

11. Gaudreault J, Potvin D, Lavigne J, Lalonde RL. Truncated area under the curve as a measure of relative extent of bioavailability: Evaluation using experimental data and Monte Carlo simulations. *Pharm Res* 1998; 15:1621–1629.

12. Danese MD, Kim J, Doan QV, Dylan M, Griffiths R, Chertow GM. PTH and the risks for hip, vertebral and pelvic fractures among patients on dialysis. *Am J Kidney Dis* 2006; 47:149–156.

13. Greene WL, Concato J, Feinstein AR. Claims of equivalence in medical research: Are they supported by the evidence? *Ann Intern Med* 2000; 132:715–722.

14. Hofacker CF. Abuse of statistical packages: The case of the general linear model. *Am J Physiol* 1983; 245:R299–R302.

15. Park SW, Kim JY, Lee SW, Park J, Yun YO, Lee WK. Estimation of smoking prevalence among adolescents in a community by design-based analysis. *J Prev Med Pub Health* 2006; 39:317–324.

16. Siegrist M. Communicating low risk magnitudes: Incidence rates expressed as frequency versus rates expressed as probability. *Risk Analysis* 1997; 17:507–510.

17. Kirjavainen T, Cooper D, Polo O, Sullivan CE. The static-charge-sensitive bed in the monitoring of respiration during sleep in infants and young children. *Acta Paediatr* 1996; 85:1146–1152.

18. Belgorosky A, Chahin S, Rivarola MA. Elevation of serum luteinizing hormone levels during hydrocortisone treatment in infant girls with 21-hydroxylase deficiency. *Acta Paediatr* 1996; 85:1172–1175.

19. Lilienfeld DE. The silence: The asbestos industry and early occupational cancer research—A case study. *Am J Pub Health* 1991; 81:791–800.

20. Koblin BA, Hessol NA, Zauber AG et al. Increased incidence of cancer among homosexual men, New York City and San Francisco, 1978–1990. *Am J Epidemiol* 1996; 144:916–923.

21. Wulff HR, Andersen B, Brandenhoff P, Guttler F. What do doctors know about statistics? *Stat Med* 1987; 6:3–10.

22. James JA, Kaufman KM, Farris AD, Taylor-Albert E, Lehman TJA, Harley JB. An increased prevalence of Epstein-Barr virus infection in young patients suggests a possible etiology for systemic lupus erythematosus. *J Clin Invest* 1997; 100:3019–3026.

23. McLaughlin JK, Lipworth L, Tarone RE. Suicide among women with cosmetic breast implants: A review of the epidemiologic evidence. *J Long Term Eff Med Implants* 2003; 13:445–450.

24. Knekt P, Raitasalo R, Heliovaara M et al. Elevated lung cancer risk among persons with depressed mood. *Am J Epidemiol* 1996; 144:1096–1103.

25. Hakko H, Rasanen P, Jarvelin M, Tiihonen J. Parental age-gap and child sex ratio—Fact or fiction? *Int J Epidemiol* 1998; 27:929–930.

26. Elwood JM. *Causal Relationship in Medicine: A Practical System for Critical Appraisal.* Oxford, U.K.: Oxford University Press, 1992.

27. Moons KG, van Es GA, Michel BC, Buller HR, Habbema JD, Grobbee DE. Redundancy of single diagnostic test evaluation. *Epidemiology* 1999; 10:276–281.

28. Vickers A. Interpreting data from randomized trials; the Scandinavian prostatectomy study illustrates two common errors. *Nat Clin Pract Urol* 2005; 2:404–405.

29. Senn S. Sharp tongues and bitter pills. *Significance* 2006; 3:123–125.

Appendix A: Statistical Softwares

In view of the medical-professional-friendly nature, this text avoided the mathematical intricacies of many statistical methods and left the computations to a statistical software. It is thus important that some information is provided on the currently available statistical softwares with regard to their features. The following presentation is divided into information on general purpose software and the specific purpose software. The description is brief. For further details, see the respective manuals.

The software scenario is changing very quickly. Some of those described here may discontinue and new ones may appear. The web is the best source to find out what is coming up in the market and whether it meets your need. Many useful statistical resources are available free on the web. Most provide free trial versions for a limited period. Before you invest, it is a good idea to try it for some time and judge if it meets your requirement.

Availability of wide range of statistical softwares has indeed made life easy both for statisticians and medical professionals as the complex calculation can be done by a click. However, this also has encouraged irresponsible use, even misuse of statistical methods. I have been sounding caution from time to time in this book regarding inappropriate use of statistical methods in general and the role of easy availability of software in this in particular. Remember that the conclusions arrived in your study are expected to be used on the larger groups of patients and the population. These conclusions pertain to the health and survival—thus, an undetected error can risk life of many. In addition, beware of the following limitations of statistical softwares:

1. Statistical softwares still have not been provided intelligence to decide which methods to use where, nor do they enhance the appreciation of the investigator of the statistical subtleties of his work, except for few researchers who carefully examine various statistical options and try to understand them.
2. Computer programs are dumb—they will do what you ask them to do. Important is that you ask right thing for the data in hand considering the design followed for collecting that data and the objectives you stipulated.
3. When you start using a software, consult the manual regularly and understand the intricacies. Do not leave this consultation to a situation where everything else fails.

4. Some universities encourage self-use of statistical softwares by students and faculty of medical professions, whereas others advise to take help of expert biostatisticians to analyze the data. There are dangers in the first approach and dependence in the second. Middle path is that the researcher understands and does the basics himself and leaves complexities to the statistical expert.

5. One of the reasons for medical professionals directly reaching to the statistical software is that good biostatisticians are scarce (as for almost any other profession). Thus, softwares tend to be misused as a substitute for statistical consultation.

6. Outputs of statistical softwares are not uniform—they do not make life simple. Different softwares can give different results for the same set of data depending upon the algorithm they use. They can use different methods also—for example, one using Gaussian confidence interval (CI) for odds ratio (OR) and the other using nonparametric method. Thus, a dependable software should be used. For any serious, high-stake conclusion, analysis should be done in at least two different statistical packages. If they give identical results, you are safe.

7. Realize that no software is free of bugs. There is a chance, extremely small though, that you get weird output for perfectly valid data. For this reason, confidence in many free softwares is low. Even the market-established softwares have to be checked if they are really performing as well as expected of them.

A.1 General Purpose Statistical Softwares

Several general purpose statistical softwares are available. Many features of these softwares are common to all, but some features are typical that may be of specific interest to the user.

SAS: Originally called the statistical analysis system, SAS is just about the most comprehensive statistical system available at present and most dependable. The library of this system contains a very large variety of statistical methods such as analysis of variance (ANOVA), regression, categorical data analysis, survival analysis, multivariate analysis, and nonparametric methods. Many users believe that it is the ultimate answer to statistical computations. The system is modularly designed to give flexibility to license only the needed functionalities. As the needs grow, additional components can be added in an integrated manner.

Among the SAS components are Base SAS, SAS/STAT, and SAS/GRAPH. Other modules that may be of interest are SAS/GIS for geographical information system and SAS/QC for quality control measures. SAS/INSIGHT includes a dynamic tool for visualizing data that can uncover trends, spot outliers, describe data distributions, and fit exploratory models. SAS/Model Manager helps in tracking model performance while it is being developed.

SAS also has an extensive SQL procedure, allowing SQL programmers to use the system with little additional knowledge. SAS is supported by a series of technical reports that gives detailed description of the newer methods. This is a very useful feature of this system. SAS Learning Edition is an inexpensive version with impressive features and this may be able to fulfill your initial requirement. Included procedures are MIXED, GENMOD, ARIMA, GLM, REG, LOGISTIC, and PHREG.

SPSS: Initially called the statistical package for social sciences, it was once revised to stand for statistical product and service solutions, but now SPSS stands alone. Among the modules available for data analysis are base, tables, regression models, advanced models, conjoint, categories, trends, and exact tests. Additionally, other modules include missing value analysis that helps to uncover missing data patterns and complex samples that is for more accurate analysis of data from surveys using two or more sample procedures in the same survey. Thus, SPSS now is among the most comprehensive statistical packages, although the statistical methods it uses are not among the most advanced.

SYSTAT: The core strength of SYSTAT is its wide variety of analytical graphs. These are not so readily available in other softwares. Its output organizer combines formatted statistical outputs with publication quality graphs. The output can be saved to HTML for publication on the web. SYSTAT has provisions for bootstrap, classification and regression trees, path analysis, and ridge regression, in addition to other common statistical methods.

S-Plus: This also is a comprehensive package for data exploration, analysis, and modeling. Among its special features are nearly 4000 functions for data analysis. Besides routine analysis, S-Plus also has capability to run time-series models, Cox regression, bootstrap and Jackknife estimation, and fuzzy clustering. It has a powerful programming language that can be used to create new programs for data analysis or to extend the existing ones. Graphic facility is good. Check for its free student version if you are enrolled in a university.

Minitab: Minitab statistical software has provision for data, graphics, and macros. It has strong graphics for displaying data, including multiple graphs and three-dimensional (3D) graphs. You can write macros for immediate execution. Among statistical tools, Minitab has a provision for generating quality control charts besides the usual tests, regression, ANOVA, and multivariate methods. It has programs for response optimization and Taguchi designs.

NCSS: Number cruncher statistical system (NCSS) is a cost-competitive, user-friendly, general statistical software. It incorporates some of those advanced statistical methods that are difficult to implement with some other softwares. These include repeated measures ANOVA, ridge and principal component regressions, area under the curve, construction of design of experiments, survival and reliability with a large number of features, life tables, accelerated life testing, repeatability and reproducibility, spectral analysis, and linear programming. Many of these methods are not discussed in this text.

NCSS output is easily transferred to popular word processors and presentation software such as PowerPoint. NCSS Probability Calculator is a freeware for Windows that calculates probabilities and quantiles of 14 statistical distributions such as Gaussian, gamma, exponential, and Weibull.

Stata: This is an integrated package and not a collection of modules. Among the methods available are general linear models, logistic and Poisson regression, random and fixed effects models, multilevel mixed models, nonparametric methods, different kinds of survival analyses, time-series analysis, multivariate methods, and simulations. Publication quality graphs can be generated. Thus, this also is quite a comprehensive package.

GraphPad: This has three products. GraphPad Prism is user friendly to nonstatisticians as it guides through each analysis. However, its statistical capabilities are limited to day-to-day methods. GraphPad InStat is an even less cumbersome alternative. It helps to pick an appropriate test by asking questions about your data. Results also are produced in easy-to-understand format. GraphPad StatMate calculates sample size or power. This also is text based for the user rather than formula based.

MedCalc: Specially designed for medical professionals, MedCalc has most of the tools you would need to analyze routine statistical data. It has provisions for various graphs and diagrams, correlation and regression including logistic, chi-square for association and trend, survival analysis, Cox regression, agreement assessment, ROC curve, quality control charts, and calculation of sample sizes for simple situations. This software does not have tools for multivariate analysis.

R: This is more of a data analysis language than a statistical system. R is in public domain and its users are increasing by the day. Large collection of packages is available with R that includes classical as well as much of the modern methods, including generalised linear models (GLM) and generalised estimating equations (GEE). You may have to do some work to find the right package for your problem. As of now, R is not available as click radiobutton tool—commands are written instead. Thus, expertise is required. However, it is similar to S-plus and many S codes will run unchanged.

Other free downloads: Several statistical softwares are available as free downloads. Some of these are (1) VISta that performs univariate and multivariate visualization and data analysis; (2) OpenStat, which is a general package with SPSS user interface and contains a wide variety of procedures for data analysis; (3) Dap package that contains commonly needed data management, analysis, and graphics; and (4) MYSTAT, which is a limited version of SYSTAT available free to the students and teachers. There are several others.

Excel: As part of MS Office, Excel has provision to calculate mean, standard deviation (SD), regression, correlation, etc., but users warn that the results may not be accurate. I do not advise Excel for statistical applications.

StatCrunch: This is a useful online statistical software that allows users to perform complex analysis anywhere with Internet access. Reports with interactive graphics can be generated. However, the software is geared for students than professionals.

STATGRAPHICS: This also performs and explains basic and advanced methods with interactive graphics and is available online also. The online version works with fast Internet connections. StatGraphics is easy-to-use and easy-to-learn package but has not found extensive use in practical applications.

A.2 Special Purpose Statistical Software

These are basically of two types: the graphics and the advisory. Some of these are briefly described in the following:

SigmaPlot: This is popular for its capacity to draw exact graphs. Its 3D features are particularly attractive. Multiple axes can be inserted. Graph attributes can be customized. Any mathematical function can be plotted. More than 80 2D and 3D graph types are available. The graphs can be shared on the web. SigmaPlot also incorporates some data analysis tools, and it can be used within Excel.

TableCurve 3D: A very special feature of this software is that it can find the ideal equation to describe 3D empirical data. Once the fit is complete, TableCurve presents with a statistically ranked list of the best-fit equations. This can help in choosing an equation that best meets the requirement. Besides linear, it has provisions to try polynomials, logarithmic and exponential curves, nonlinear curves, and user-defined curves. You can graphically review the surface fit results.

SigmaStat: This software wants the user to run through a series of simple questions to be able to recommend the best statistical test to run. Thus, it has a structure similar to an expert system. The software can study the features of a dataset and can recommend a statistical procedure accordingly. It also handles missing and unbalanced data. Among the statistical methods available are repeated measures analysis, survival analysis, regression and correlation, and power or sample size calculations.

EquivTest/PK: EquivTest provides a range of statistical tests and CIs for the analysis of equivalence studies. One-sided and two-sided tests can be done for difference or ratio of means and difference or ratio of proportions, including OR. Both parametric and nonparametric tests are available. The added feature is pharmacokinetic calculations such as C_{max} and T_{max}.

nQuery Advisor: This software is particularly helpful in planning a research study. It determines the sample size to meet the research objective with specified precision or power. Sample sizes for survival studies, sampling for finite population, cluster sampling, repeated measures, crossover design, and equivalence studies can be obtained. Both parametric and nonparametric procedures are available.

StatXact: Most of the methods available in general purpose software are for large n. StatXact provides exact methods to analyze small sample data. Most of these methods are nonparametric and generally distribution free. This software could be very useful to those who deal with small datasets. This software is widely perceived to provide valid results under varying degrees of uncertainty regarding underlying assumptions. More than 130 procedures are now available in this software.

LogXact: Specially designed for logistic regression, LogXact provides tools to fit a logistic model under a variety of conditions. For example, it can fit logistic when some values of discrete covariates are missing. CIs can be obtained that use a newly developed profile likelihood method. It has provision for analyzing small datasets. LogXact can reside in the SAS environment and thus is available to SAS users without opening another application.

EpiInfo: Developed primarily for epidemiologists, it offers basic general statistics and mapping capabilities with EpiMap.

PASS: This is an acronym for Power Analysis and Sample Size. This is among the most popular packages for sample size calculations. Sample size for more than 150 statistical procedures can be calculated with ease. It produces charts, graphs, and text that you can copy and paste to your document.

PiFace: An interesting Java applet with graphical interface that provides instant sample size for your selected procedure in terms of sliders. Worth giving a try.

Appendix B: Some Statistical Tables

The following tables may be useful for medical professionals when relevant statistical software is not available or for doing exercises for better understanding of the statistical procedures.

TABLE B.1

Probability from the Standard Gaussian Distribution

z	0.00	0.01	0.02	0.03	0.04	0.05	0.06	0.07	0.08	0.09
0.0	0.5000	0.4960	0.4920	0.4880	0.4840	0.4801	0.4761	0.4721	0.4681	0.4641
0.1	0.4602	0.4562	0.4522	0.4483	0.4443	0.4404	0.4364	0.4325	0.4286	0.4247
0.2	0.4207	0.4168	0.4129	0.4090	0.4052	0.4013	0.3974	0.3936	0.3897	0.3859
0.3	0.3821	0.3783	0.3745	0.3707	0.3669	0.3632	0.3594	0.3557	0.3520	0.3483
0.4	0.3446	0.3409	0.3372	0.3336	0.3300	0.3264	0.3228	0.3192	0.3156	0.3121
0.5	0.3085	0.3050	0.3015	0.2981	0.2946	0.2912	0.2877	0.2843	0.2810	0.2776
0.6	0.2743	0.2709	0.2676	0.2643	0.2611	0.2578	0.2546	0.2514	0.2483	0.2451
0.7	0.2420	0.2389	0.2358	0.2327	0.2296	0.2266	0.2236	0.2206	0.2177	0.2148
0.8	0.2119	0.2090	0.2061	0.2033	0.2005	0.1977	0.1949	0.1922	0.1894	0.1867
0.9	0.1841	0.1814	0.1788	0.1762	0.1736	0.1711	0.1685	0.1660	0.1635	0.1611
1.0	0.1587	0.1562	0.1539	0.1515	0.1492	0.1469	0.1446	0.1423	0.1401	0.1379
1.1	0.1357	0.1335	0.1314	0.1292	0.1271	0.1251	0.1230	0.1210	0.1190	0.1170
1.2	0.1151	0.1131	0.1112	0.1093	0.1075	0.1056	0.1038	0.1020	0.1003	0.0985
1.3	0.0968	0.0951	0.0934	0.0918	0.0901	0.0885	0.0869	0.0853	0.0838	0.0823
1.4	0.0808	0.0793	0.0778	0.0764	0.0749	0.0735	0.0721	0.0708	0.0694	0.0681
1.5	0.0668	0.0655	0.0643	0.0630	0.0618	0.0606	0.0594	0.0582	0.0571	0.0559
1.6	0.0548	0.0537	0.0526	0.0516	0.0505	0.0495	0.0485	0.0475	0.0465	0.0455
1.7	0.0446	0.0436	0.0427	0.0418	0.0409	0.0401	0.0392	0.0384	0.0375	0.0367
1.8	0.0359	0.0351	0.0344	0.0336	0.0329	0.0322	0.0314	0.0307	0.0301	0.0294
1.9	0.0287	0.0281	0.0274	0.0268	0.0262	0.0256	0.0250	0.0244	0.0239	0.0233
2.0	0.0228	0.0222	0.0217	0.0212	0.0207	0.0202	0.0197	0.0192	0.0188	0.0183
2.1	0.0179	0.0174	0.0170	0.0166	0.0162	0.0158	0.0154	0.0150	0.0146	0.0143
2.2	0.0139	0.0136	0.0132	0.0129	0.0125	0.0122	0.0119	0.0116	0.0113	0.0110
2.3	0.0107	0.0104	0.0102	0.0099	0.0096	0.0094	0.0091	0.0089	0.0087	0.0084
2.4	0.0082	0.0080	0.0078	0.0075	0.0073	0.0071	0.0069	0.0068	0.0066	0.0064
2.5	0.0062	0.0060	0.0059	0.0057	0.0055	0.0054	0.0052	0.0051	0.0049	0.0048
2.6	0.0047	0.0045	0.0044	0.0043	0.0041	0.0040	0.0039	0.0038	0.0037	0.0036
2.7	0.0035	0.0034	0.0033	0.0032	0.0031	0.0030	0.0029	0.0028	0.0027	0.0026
2.8	0.0026	0.0025	0.0024	0.0023	0.0023	0.0022	0.0021	0.0021	0.0020	0.0019
2.9	0.0019	0.0018	0.0018	0.0017	0.0016	0.0016	0.0015	0.0015	0.0014	0.0014
3.0	0.0013	0.0013	0.0013	0.0012	0.0012	0.0011	0.0011	0.0011	0.0010	0.0010

Notes: (1) For $z \geq 3.10$, the probability $P(Z \geq z)$ is less than 1 in 1000 that can be taken as almost 0 for most practical applications. (2) The tabulated value is $P(Z \geq z)$. Because of symmetry, $P(Z \leq z) = P(Z \geq z)$. (3) Also, $P(Z \leq z) = 1 - P(Z > z) = 1 - P(Z \geq z)$.

TABLE B.2

Critical Values of Student *t*

	Probability				
df	0.10	0.05	0.025	0.01	0.005
1	3.078	6.314	12.706	31.821	63.656
2	1.886	2.920	4.303	6.965	9.925
3	1.638	2.353	3.182	4.541	5.841
4	1.533	2.132	2.776	3.747	4.604
5	1.476	2.015	2.571	3.365	4.032
6	1.440	1.943	2.447	3.143	3.707
7	1.415	1.895	2.365	2.998	3.499
8	1.397	1.860	2.306	2.896	3.355
9	1.383	1.833	2.262	2.821	3.250
10	1.372	1.812	2.228	2.764	3.169
11	1.363	1.796	2.201	2.718	3.106
12	1.356	1.782	2.179	2.681	3.055
13	1.350	1.771	2.160	2.650	3.012
14	1.345	1.761	2.145	2.624	2.977
15	1.341	1.753	2.131	2.602	2.947
16	1.337	1.746	2.120	2.583	2.921
17	1.333	1.740	2.110	2.567	2.898
18	1.330	1.734	2.101	2.552	2.878
19	1.328	1.729	2.093	2.539	2.861
20	1.325	1.725	2.086	2.528	2.845
21	1.323	1.721	2.080	2.518	2.831
22	1.321	1.717	2.074	2.508	2.819
23	1.319	1.714	2.069	2.500	2.807
24	1.318	1.711	2.064	2.492	2.797
25	1.316	1.708	2.060	2.485	2.787
26	1.315	1.706	2.056	2.479	2.779
27	1.314	1.703	2.052	2.473	2.771
28	1.313	1.701	2.048	2.467	2.763
29	1.311	1.699	2.045	2.462	2.756
30	1.310	1.697	2.042	2.457	2.750
40	1.303	1.684	2.021	2.423	2.704
50	1.299	1.676	2.009	2.403	2.678
60	1.296	1.671	2.000	2.390	2.660
120	1.289	1.658	1.980	2.358	2.617
∞	1.282	1.645	1.960	2.327	2.576

Note: The value tabulated is *c* such that $P(t \geq c)$ is equal to the probability shown at the top. This is one-tailed probability. Two-tailed probability is the double of the one-tailed probability.

TABLE B.3

Critical Values of Chi-Square

df	$\alpha = 0.10$	$\alpha = 0.05$	$\alpha = 0.01$
1	2.706	3.841	6.635
2	4.605	5.991	9.210
3	6.251	7.815	11.345
4	7.779	9.488	13.277
5	9.236	11.070	15.086
6	10.645	12.592	16.812
7	12.017	14.067	18.475
8	13.362	15.507	20.090
9	14.684	16.919	21.666
10	15.987	18.307	23.209
11	17.275	19.675	24.725
12	18.549	21.026	26.217
13	19.812	22.362	27.688
14	21.064	23.685	29.141
15	22.307	24.996	30.578
16	23.542	26.296	32.000
17	24.769	27.587	33.409
18	25.989	28.869	34.805
19	27.204	30.144	36.191
20	28.412	31.410	37.566
21	29.615	32.671	38.932
22	30.813	33.924	40.289
23	32.007	35.172	41.638
24	33.196	36.415	42.980
25	34.382	37.652	44.314
30	40.256	43.773	50.892
35	46.059	49.802	57.342
40	51.805	55.758	63.691
45	57.505	61.656	69.957
50	63.167	67.505	76.154
60	74.397	79.082	88.379
70	85.527	90.531	100.425
80	96.578	101.879	112.329
90	107.565	113.145	124.116
100	118.498	124.342	135.807

Note: The value tabulated is c such that $P(\chi^2 \geq c) = \alpha$.

TABLE B.4

Critical Values of F for $\alpha = 0.05$

Denom-inator df (v_2)	Numerator df (v_1)									
	1	**2**	**3**	**4**	**5**	**6**	**7**	**8**	**9**	**10**
1	161.45	199.50	215.71	224.58	230.16	233.99	236.77	238.88	240.54	241.88
2	18.51	19.00	19.16	19.25	19.30	19.33	19.35	19.37	19.38	19.40
3	10.13	9.55	9.28	9.12	9.01	8.94	8.89	8.85	8.81	8.79
4	7.71	6.94	6.59	6.39	6.26	6.16	6.09	6.04	6.00	5.96
5	6.61	5.79	5.41	5.19	5.05	4.95	4.88	4.82	4.77	4.74
6	5.99	5.14	4.76	4.53	4.39	4.28	4.21	4.15	4.10	4.06
7	5.59	4.74	4.35	4.12	3.97	3.87	3.79	3.73	3.68	3.64
8	5.32	4.46	4.07	3.84	3.69	3.58	3.50	3.44	3.39	3.35
9	5.12	4.26	3.86	3.63	3.48	3.37	3.29	3.23	3.18	3.14
10	4.96	4.10	3.71	3.48	3.33	3.22	3.14	3.07	3.02	2.98
11	4.84	3.98	3.59	3.36	3.20	3.09	3.01	2.95	2.90	2.85
12	4.75	3.89	3.49	3.26	3.11	3.00	2.91	2.85	2.80	2.75
13	4.67	3.81	3.41	3.18	3.03	2.92	2.83	2.77	2.71	2.67
14	4.60	3.74	3.34	3.11	2.96	2.85	2.76	2.70	2.65	2.60
15	4.54	3.68	3.29	3.06	2.90	2.79	2.71	2.64	2.59	2.54
16	4.49	3.63	3.24	3.01	2.85	2.74	2.66	2.59	2.54	2.49
17	4.45	3.59	3.20	2.96	2.81	2.70	2.61	2.55	2.49	2.45
18	4.41	3.55	3.16	2.93	2.77	2.66	2.58	2.51	2.46	2.41
19	4.38	3.52	3.13	2.90	2.74	2.63	2.54	2.48	2.42	2.38
20	4.35	3.49	3.10	2.87	2.71	2.60	2.51	2.45	2.39	2.35
21	4.32	3.47	3.07	2.84	2.68	2.57	2.49	2.42	2.37	2.32
22	4.30	3.44	3.05	2.82	2.66	2.55	2.46	2.40	2.34	2.30
23	4.28	3.42	3.03	2.80	2.64	2.53	2.44	2.37	2.32	2.27
24	4.26	3.40	3.01	2.78	2.62	2.51	2.42	2.36	2.30	2.25
25	4.24	3.39	2.99	2.76	2.60	2.49	2.40	2.34	2.28	2.24
26	4.23	3.37	2.98	2.74	2.59	2.47	2.39	2.32	2.27	2.22
27	4.21	3.35	2.96	2.73	2.57	2.46	2.37	2.31	2.25	2.20
28	4.20	3.34	2.95	2.71	2.56	2.45	2.36	2.29	2.24	2.19
29	4.18	3.33	2.93	2.70	2.55	2.43	2.35	2.28	2.22	2.18
30	4.17	3.32	2.92	2.69	2.53	2.42	2.33	2.27	2.21	2.16
40	4.08	3.23	2.84	2.61	2.45	2.34	2.25	2.18	2.12	2.08
60	4.00	3.15	2.76	2.53	2.37	2.25	2.17	2.10	2.04	1.99
120	3.92	3.07	2.68	2.45	2.29	2.18	2.09	2.02	1.96	1.91
∞	3.84	3.00	2.60	2.37	2.21	2.10	2.01	1.94	1.88	1.83

Note: The value tabulated is c such that $P(F \geq c) = 0.05$.

TABLE B.5

Critical Values of the Studentized Range Criterion (Q) for 2–10 Groups for $\alpha = 0.05$

Error df	Number of Groups								
	2	3	4	5	6	7	8	9	10
1	18.0	27.0	32.8	37.1	40.4	43.1	45.4	47.4	49.1
2	6.1	8.3	9.8	10.9	11.7	12.4	13.0	13.5	14.0
3	4.50	5.91	6.82	7.50	8.04	8.48	8.85	9.18	9.46
4	3.93	5.04	5.76	6.29	6.71	7.05	7.35	7.60	7.83
5	3.64	4.60	5.22	5.67	6.03	6.33	6.58	6.80	6.99
6	3.46	4.34	4.90	5.31	5.63	5.89	6.12	6.32	6.49
7	3.34	4.16	4.68	5.06	5.36	5.61	5.82	6.00	6.16
8	3.26	4.04	4.53	4.89	5.17	5.40	5.60	5.77	5.92
9	3.20	3.95	4.42	4.76	5.02	5.24	5.43	5.60	5.74
10	3.15	3.88	4.33	4.65	4.91	5.12	5.30	5.46	5.60
11	3.11	3.82	4.26	4.57	4.82	5.03	5.20	5.35	5.49
12	3.08	3.77	4.20	4.51	4.75	4.95	5.12	5.27	5.40
13	3.06	3.73	4.15	4.45	4.69	4.88	5.05	5.19	5.32
14	3.03	3.70	4.11	4.41	4.64	4.83	4.99	5.13	5.25
15	3.01	3.67	4.08	4.37	4.60	4.78	4.94	5.08	5.20
16	3.00	3.65	4.05	4.33	4.56	4.74	4.90	5.03	5.15
17	2.98	3.63	4.02	4.30	4.52	4.71	4.86	4.99	5.11
18	2.97	3.61	4.00	4.28	4.49	4.67	4.82	4.96	5.07
19	2.96	3.59	3.98	4.25	4.47	4.65	4.79	4.92	5.04
20	2.95	3.58	3.96	4.23	4.45	4.62	4.77	4.90	5.01
24	2.92	3.53	3.90	4.17	4.37	4.54	4.68	4.81	4.92
30	2.89	3.49	3.84	4.10	4.30	4.46	4.60	4.72	4.83
40	2.86	3.44	3.79	4.04	4.23	4.39	4.52	4.63	4.74
60	2.83	3.40	3.74	3.98	4.16	4.31	4.44	4.55	4.65
120	2.80	3.36	3.69	3.92	4.10	4.24	4.36	4.48	4.56
∞	2.77	3.31	3.63	3.86	4.03	4.17	4.29	4.39	4.47

Source: From the *Biometrika Tables for Statisticians*, Vol. 1, 2nd edn., 1958, Table 29. With permission of the Biometrika Trustees.

Appendix C: Software Illustrations

Given next are illustrations of how some of the methods described earlier can be implemented with the help of a software. The first is on receiver operating characteristic (ROC) curve, which uses MedCalc software. All others use SPSS ver 17.0. This can be run by clicks on the menu, but commands are shown in the outputs in illustrations so that you can see what is going on. These illustrations are for relatively intricate methods such as repeated measures analysis of variance (ANOVA), various types of regressions, analysis of covariance (ANCOVA), logistic regression, survival analysis, and Cox model. The last on Cox model also illustrates unpaired t-tests and how it is affected by inequality of variances as judged by Levene test. Regression examples contain commands for obtaining correlations.

The figures in these illustrations are not in the sequence in which they are obtained in the software outputs. All other relevant outputs are reproduced as such.

C.1 ROC Curves

Studies have established that visceral adiposity is a strong determinant of growth hormone (GH) secretion. GH deficiency is associated with increased body fat and decreased lean body mass. However, visceral adiposity is difficult to evaluate as it requires CT scan. A study was carried out to find if high body fat itself can be a good indicator of GH deficiency opposed to visceral adiposity. Largest waist circumference was used as a surrogate for body fat. The data were obtained from 60 subjects: 28 with GH deficiency and 32 without this deficiency.

The following results are from MedCalc software for largest waist circumference.

Variable	Largest_waist_circumference__Cms_
	Largest waist circumference (Cms)
Classification variable	Deficiency
Select	1

Sample size		60
Positive group :	GHDeficiency = 1	28
Negative group :	GHDeficiency = 0	32

Disease prevalence (%)	unknown

Area under the ROC curve (AUC)	0.865
Standard Error [a]	0.0450
95% Confidence Interval [b]	0.752 to 0.939
z statistic	8.113
Significance level P (Area=0.5)	0.0001

[a] DeLong et al., 1988.

[b] Binomial exact.

Criterion values and coordinates of the ROC curve

Criterion	Sensitivity	95% CI	Specificity	95% CI
>=84	100.00	87.7–100.0	0.00	0.0–10.9
>98	100.00	87.7–100.0	53.13	34.7–70.9
>99 *	89.29	71.8–97.7	65.62	46.8–81.4
>100	75.00	55.1–89.3	71.87	53.3–86.3
>101	64.29	44.1–81.4	84.37	67.2–94.7
>102	50.00	30.6–69.4	96.87	83.8–99.9
>104	21.43	8.3–41.0	96.87	83.8–99.9
>105	14.29	4.0–32.7	100.00	89.1–100.0
>107	0.00	0.0–12.3	100.00	89.1–100.0

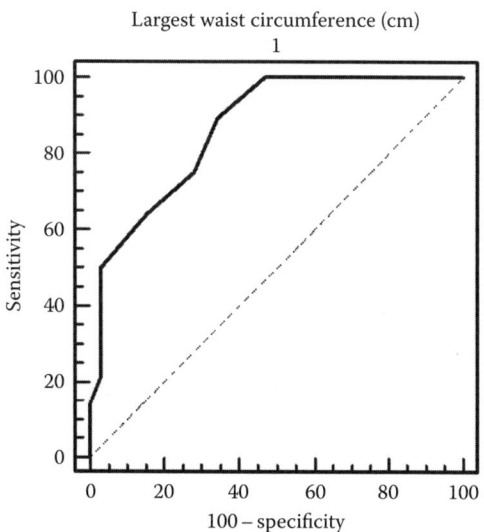

FIGURE C.1

ROC curve for largest waist circumference.

The best cutoff for largest waist circumference is 99 cm (marked by asterisk sign in the output). The sensitivity and specificity at this cutoff are 89% and 66%, respectively. The area under the curve (AUC) for largest waist circumference is 0.865. The ROC curve is in Figure C.1.

The following output shows for visceral adiposity that the best cutoff is 9960 mm² with sensitivity = 89% and specificity = 91%. Thus, visceral adiposity is more specific (gives negative result when GH deficiency not present). The area under the ROC curve (Figure C.2) for visceral adiposity is 0.904. Apparently, it is higher for visceral adiposity as expected.

Variable	Visceral_adiposity__mm2_
	Visceral adiposity (mm2)
Classification variable	GH Deficiency
Select	1

Sample size		60
Positive group :	GHDeficiency = 1	28
Negative group :	GHDeficiency = 0	32

Disease prevalence (%)	unknown

Area under the ROC curve (AUC)	0.904
Standard Error [a]	0.0455
95% Confidence Interval [b]	0.800 to 0.965
z statistic	8.876
Significance level P (Area=0.5)	0.0001

[a] DeLong et al., 1988.

[b] Binomial exact.

Criterion values and coordinates of the ROC curve [Hide]

Criterion	Sensitivity	95% CI	Specificity	95% CI
>=8259	100.00	87.7–100.0	0.00	0.0–10.9
>8886	100.00	87.7–100.0	9.38	2.0–25.0
>9017	96.43	81.7–99.9	9.38	2.0–25.0
>9485	96.43	81.7–99.9	25.00	11.5–43.4
>9528	92.86	76.5–99.1	25.00	11.5–43.4
>9785	92.86	76.5–99.1	53.13	34.7–70.9
>9794	89.29	71.8–97.7	53.13	34.7–70.9
>9960 *	89.29	71.8–97.7	90.62	75.0–98.0
>9994	78.57	59.0–91.7	90.62	75.0–98.0
>10004	78.57	59.0–91.7	93.75	79.2–99.2
>10089	71.43	51.3–86.8	93.75	79.2–99.2
>10148	71.43	51.3–86.8	96.87	83.8–99.9
>10289	53.57	33.9–72.5	96.87	83.8–99.9
>10295	53.57	33.9–72.5	100.00	89.1–100.0
>11714	0.00	0.0–12.3	100.00	89.1–100.0

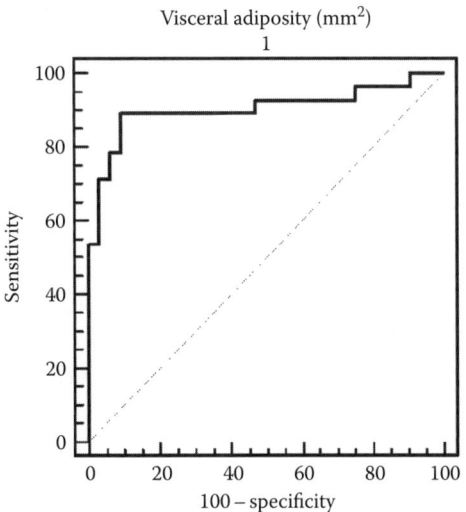

FIGURE C.2
ROC curve for visceral adiposity.

The following output gives test of significance of difference between AUC for largest waist circumference and visceral adiposity (Figure C.3). It is a pairwise comparison as the subjects are the same. The *P*-value is 0.387 and the difference is not statistically significant. Thus, these data do not provide evidence of one being better than the other for assessing GH deficiency. Note that despite failing to reject the null of equality, we cannot say that largest waist circumference is as good as visceral adiposity—only that these data could not say that these two indices have differential performance.

Variable 1	Largest_waist_circumference__Cms_
	Largest waist circumference (Cms)
Variable 2	Visceral_adiposity__mm2_
	Visceral adiposity (mm2)
Classification variable	GH Deficiency
Select	1

Sample size		60
Positive group :	GHDeficiency = 1	28
Negative group :	GHDeficiency = 0	32

	AUC	SE [a]	95% CI [b]
Largest_waist_circumference__Cms_	0.865	0.0450	0.752 to 0.939
Visceral_adiposity__mm2_	0.904	0.0455	0.800 to 0.965

[a] DeLong et al., 1988.

[b] Binomial exact.

Pairwise comparison of ROC curves

Largest_waist_circumference__Cms_ ~ Visceral_adiposity__mm2_	
Difference between areas	0.0391
Standard Error [a]	0.0451
95% Confidence Interval	−0.0494 to 0.127
z statistic	0.866
Significance level	P = 0.387

[a] DeLong et al., 1988.

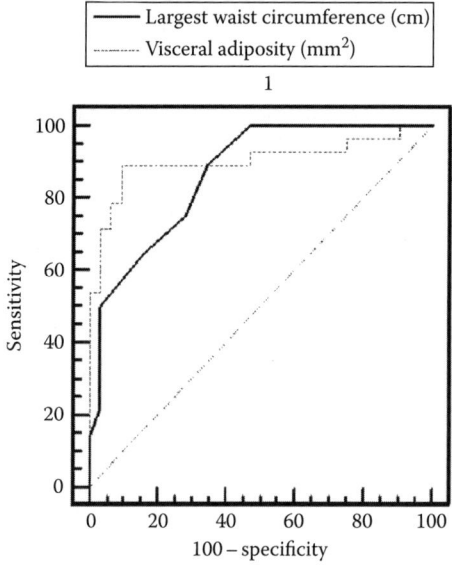

FIGURE C.3
Comparison of ROCs obtained for the two methods.

The following illustrations are based on SPSS ver 17.0.

```
Basic SPSS command to open a SPSS file:
GET FILE = <Drive name:\folder name\file name>.
```

C.2 Repeated Measures ANOVA

A study was conducted to find the effect of yoga on lowering the total serum cholesterol level in newly diagnosed patients of hypercholesterolemia. Thirty patients were selected having total serum cholesterol between 200 and 250 mg/dL and not taking any drug to lower the cholesterol. Each subject was given yoga for 3 months and cholesterol level measured at five time points, namely, baseline, after 15 days, 1 month, 2 months, and 3 months. The objective of the study is to find overall effect of yoga on lowering the cholesterol level and whether the mean cholesterol level is significantly different from baseline to 15 days, 15 days to 1 month, and so on.

```
Command to run one factor repeated measures ANOVA:
GLM baseline days_15 month_1 month_2 month_3
/WSFACTOR = cholesterol 5 repeated
/METHOD = SSTYPE(3)
/EMMEANS = TABLES(cholesterol) COMPARE ADJ(BONFERRONI)
/PRINT = DESCRIPTIVE
/CRITERIA = ALPHA(.05)
/WSDESIGN = cholesterol.
```

Observations at different time points are within subjects (WS) and have to be so specified in the aforementioned command. The command gives output for both univariate and multivariate analysis of the repeated measures analysis. It runs Mauchly test for sphericity by default and provides Huynh–Feldt correction for the degrees of freedom (dfs). The output gives certain other corrections also. The aforementioned command also asks for means at different time points and their Bonferroni comparison.

The following confirms five repeated measures WS and gives the mean and SD of total cholesterol level at different time points.

Within-Subjects Factors

Measure: MEASURE_1

cholesterol	Dependent Variable
1	baseline
2	days_15
3	month_1
4	month_2
5	month_3

Descriptive Statistics

	Mean	Std. Deviation	N
baseline	226.67	14.337	30
15 days	222.13	13.733	30
1month	218.27	13.318	30
2month	213.83	13.613	30
3month	206.57	13.234	30

The following table gives results of multivariate tests and shows there is significant ($P < 0.001$) difference in mean cholesterol levels at different time points.

Multivariate Tests (b)

Effect		Value	F	Hypothesis df	Error df	Sig.
cholestero 1	Pillai's Trace	.795	25.220 (a)	4.000	26.000	.000
	Wilks' Lambda	.205	25.220 (a)	4.000	26.000	.000
	Hotelling's Trace	3.880	25.220 (a)	4.000	26.000	.000
	Roy's Largest Root	3.880	25.220 (a)	4.000	26.000	.000

a Exact statistic.

b Design: Intercept
   Without Subjects Design: Cholesterol.

Mauchly test is significant ($P = 0.000 < 0.001$) as given in the following table. Sphericity assumption is violated. Correction to the df is required.

Mauchly's Test of Sphericity[b]

Measure: MEASURE_1

Within Subjects Effect	Mauchly's W	Approx. Chi-Square	df	Sig.	Epsilon[a] Greenhous e-Geisser	Huynh-Feldt	Lower-bound
cholesterol	.154	51.310	9	.000	.559	.608	.250

a May be used to adjust the degrees of freedom for the averaged tests of significance. Corrected tests are displayed in the Tests of Within-Subjects Effects table.

b Design: Intercept
   Within Subjects Design: cholesterol.

Huynh–Feldt correction to the dfs is 0.608. With this correction, the numerator df of F-test is $4 \times 0.608 = 2.432$ and denominator df is $116 \times 0.608 = 70.528$. These automatically come in the following table although slightly different due to decimal approximation. This test finds significant ($P < 0.001$) difference in mean cholesterol level at different time points.

Tests of Within-Subjects Effects

Measure: MEASURE_1

Source		Type III Sum of Squares	df	Mean Square	F	Sig.
cholesterol	Sphericity Assumed	7171.960	4	1792.990	62.011	.000
	Greenhouse-Geisser	7171.960	2.237	3206.204	62.011	.000
	Huynh-Feldt	7171.960	2.433	2948.128	62.011	.000
	Lower-bound	7171.960	1.000	7171.960	62.011	.000
Error(cholesterol)	Sphericity Assumed	3354.040	116	28.914		
	Greenhouse-Geisser	3354.040	64.870	51.704		
	Huynh-Feldt	3354.040	70.549	47.542		
	Lower-bound	3354.040	29.000	115.657		

Various types of comparisons can be made between time points. The following compares mean level at any time point with its value at preceding time point as per the objective of the study. All these are statistically highly significant ($P < 0.001$).

Tests of Within-Subjects Contrasts

Measure: MEASURE_1

Source	cholesterol	Type III Sum of Squares	df	Mean Square	F	Sig.
cholesterol	Level 1 vs. Level 2	616.533	1	616.533	21.872	.000
	Level 2 vs. Level 3	448.533	1	448.533	16.774	.000
	Level 3 vs. Level 4	589.633	1	589.633	27.878	.000
	Level 4 vs. Level 5	1584.133	1	1584.133	38.805	.000
Error(cholesterol)	Level 1 vs. Level 2	817.467	29	28.189		
	Level 2 vs. Level 3	775.467	29	26.740		
	Level 3 vs. Level 4	613.367	29	21.151		
	Level 4 vs. Level 5	1183.867	29	40.823		

The following table shows the mean, standard error (SE), and 95% confidence interval (CI) of mean cholesterol level at each time point:

Estimates

Measure: MEASURE_1

cholesterol	Mean	Std. Error	95% Confidence Interval	
			Lower Bound	Upper Bound
1	226.667	2.618	221.313	232.020
2	222.133	2.507	217.005	227.261
3	218.267	2.432	213.294	223.240
4	213.833	2.485	208.750	218.917
5	206.567	2.416	201.625	211.508

The following table shows the comparison of mean cholesterol level at each time point with all other time points using Bonferroni adjustment:

Pairwise Comparisons

Measure: MEASURE_1

(I) cholesterol	(J) cholesterol	Mean Difference (I–J)	Std. Error	Sig.[a]	95% Confidence Interval for Difference[a]	
					Lower Bound	Upper Bound
1	2	4.533*	.969	.001	1.588	7.478
	3	8.400*	1.353	.000	4.289	12.511
	4	12.833*	1.690	.000	7.700	17.967
	5	20.100*	1.911	.000	14.295	25.905
2	1	−4.533*	.969	.001	−7.478	−1.588
	3	3.867*	.944	.003	.998	6.735
	4	8.300*	1.429	.000	3.959	12.641
	5	15.567*	1.690	.000	10.434	20.699
3	1	−8.400*	1.353	.000	−12.511	−4.289
	2	−3.867*	.944	.003	−6.735	−.998
	4	4.433*	.840	.000	1.882	6.984
	5	11.700*	1.465	.000	7.250	16.150
4	1	−12.833*	1.690	.000	−17.967	−7.700
	2	−8.300*	1.429	.000	−12.641	−3.959
	3	−4.433*	.840	.000	−6.984	−1.882
	5	7.267*	1.167	.000	3.723	10.811
5	1	−20.100*	1.911	.000	−25.905	−14.295
	2	−15.567*	1.690	.000	−20.699	−10.434
	3	−11.700*	1.465	.000	−16.150	−7.250
	4	−7.267*	1.167	.000	−10.811	−3.723

Based on estimated marginal means

[a] Adjustment for multiple comparisons: Bonferroni.

* The mean difference is significant at the .05 level.

All differences are statistically significant. Yoga has been able to make a significant change in cholesterol level at each time point of observation as far as these data are concerned.

C.3 One-Way ANOVA and Tukey Test

To investigate that high-density lipoprotein (HDL) levels in three groups, normal, coronary artery disease patients, and diabetes patients, are different or not, 50 subjects in each group were randomly selected from their respective population and their HDL level recorded.

Tests to be used: one-way ANOVA F-test, followed by Tukey test at 5% level of significance for pairwise comparisons.

One-way ANOVA has two main assumptions, namely, homogeneity of variances across the groups and independence of observations. Independence in this case is assured as the observations belong to separate subjects. Homogeneity of variances can be tested using Levene test. For pairwise comparisons, Tukey test is used. Gaussianity is assumed as the number of subjects in each group is 50, which is reasonably large.

```
Command to run the one-way ANOVA, Levene test, and Tukey test:
ONEWAY HDL BY Group
/STATISTICS DESCRIPTIVES HOMOGENEITY
/MISSING ANALYSIS
/POSTHOC = TUKEY ALPHA(.05).
```

This command displays the mean, SD, SE, and 95% CI for each group, Levene test for homogeneity of variance, and post hoc Tukey test at 5% level. Controls are the normal subjects.

Descriptives

HDL

	N	Mean	Std. Deviation	Std. Error	95% Confidence Interval for Mean		Minimum	Maximum
					Lower Bound	Upper Bound		
Control	50	43.9800	8.09507	1.14482	41.6794	46.2806	32.00	64.00
CAD	50	33.9600	6.24650	.88339	32.1848	35.7352	22.00	50.00
Diabetic	50	41.5400	7.12343	1.00741	39.5155	43.5645	23.00	56.00
Total	150	39.8267	8.33022	.68016	38.4827	41.1707	22.00	64.00

Test of Homogeneity of Variances

HDL

Levene Statistic	df1	df2	Sig.
1.644	2	147	.197

The Levene test is not significant (P = 0.197), which indicates that condition of homogeneity of variances is not violated. The following is the ANOVA table:

ANOVA Table

Dependent Variable: HDL

Source of variation	Sum of Squares	df	Mean Square	F	Sig.
Between Groups	2730.173	2	1365.087	26.371	.000
Within Groups	7609.320	147	51.764		
Total	10339.493	149			

The value of F shows that between-group mean square is more than 26 times the value of within-group mean square. This is highly significant ($P < 0.001$ which is written as .000 in the preceding table) and indicates that group means are very different. The following Tukey test shows where the difference lies:

Tukey HSD

Group (I)	Group (J)	Mean Difference (I–J)	Sig.
Normal	CAD Patients	10.02000(*)	.000
	Diabetic Patients	2.44000	.210
CAD Patients	Diabetic Patients	−7.58000(*)	.000

* The mean difference is significant at the 0.05 level by Tukey test.

Mean HDL of CAD patients is significantly different from the mean of normal subjects as well as from the mean of diabetic patients. Mean HDL of diabetic patients is not significantly different from normal subjects. The statistical conclusion is that HDL level is affected in CAD patients but not in diabetes patients.

C.4 Stepwise Multiple Linear Regression

In many developing countries, especially in underprivileged sections, last menstrual period in pregnant women is not known and thus gestational age (GA) of pregnancy also is not known. The aim of this study is to explore the possibility of predicting GA from the anthropometrical measurements of the newborn. These anthropometry are birth weight, crown heal length (CHL), head circumference (HC), mid-arm circumference (MAC), foot length (FL), hand length (HL), and calf circumference (CaC). The data were obtained on 800 newborns.

Multiple linear regression method using stepwise selection is applied. To build a good model, assumptions required to be tested are linearity, outliers, and multicollinearity. Gaussianity of error term will be tested after fitting the model.

The linearity can be explored by drawing the scatter plot.

```
Command for the scatter plot:
GRAPH
/SCATTERPLOT(MATRIX)=GA BirthWeight CHL HC MAC FL HL CaC
/MISSING=LISTWISE.
```

Scatter plot in matrix form is shown in Figure C.4.

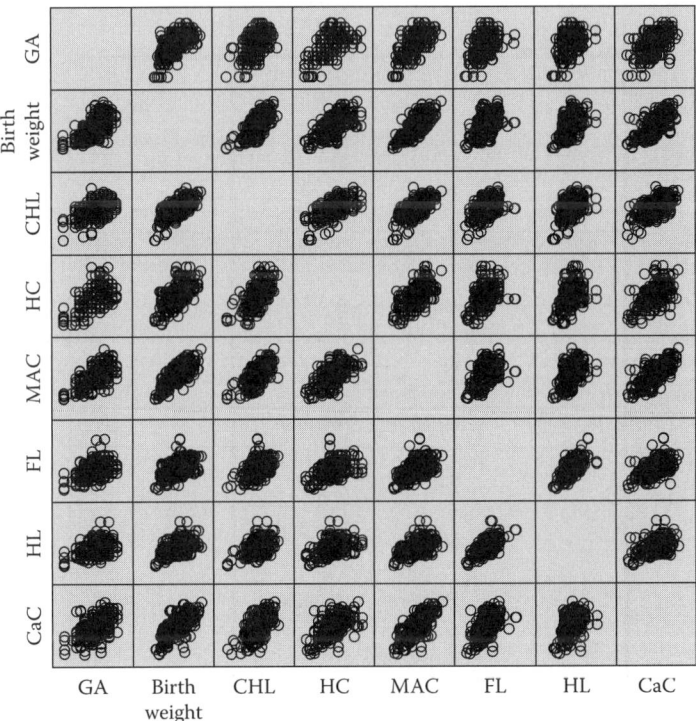

FIGURE C.4
Scatter matrix of all the variables under consideration.

The scatter plot matrix gives the visual assessment of linearity. Plots in the first row indicate that most variables are linearly related with the GA. Birth Weight may have quadratic relationship as indicated by slightly curved plot but keep it linear for the present example. Apparently there are no outliers. This plot also suggests strong linear relationship (multicollinearity) among some independents. This is examined in more details slightly later in this output. To numerically check linearity, compute Pearson correlations.

```
Command to run the Pearson correlations:
CORRELATIONS
/VARIABLES=GA with BirthWeight CHL MAC FL HL CaC HC
/PRINT=TWOTAIL NOSIG
/MISSING=PAIRWISE.
```

This command prints the Pearson correlation with two-tailed *P*-values and excludes the pair value if either GA or other independent variable has missing value.

		GA
Birth Weight	Pearson Correlation	.679**
	Sig. (2-tailed)	.000
	N	800
CHL	Pearson Correlation	.565**
	Sig. (2-tailed)	.000
	N	800
MAC	Pearson Correlation	.643**
	Sig. (2-tailed)	.000
	N	800
FL	Pearson Correlation	.432**
	Sig. (2-tailed)	.000
	N	800

** $P < 0.01$

HL	Pearson Correlation	.407**
	Sig. (2-tailed)	.000
	N	800
CaC	Pearson Correlation	.552**
	Sig. (2-tailed)	.000
	N	800
HC	Pearson Correlation	.517**
	Sig. (2-tailed)	.000
	N	800

The GA is significantly linearly correlated with all the seven independent variables. GA is highly correlated with Birth Weight and MAC. The strength of correlation of baby's anthropometry with GA varies from 0.407 to 0.679.

To numerically assess the collinearity among the independent variables, again Pearson correlation can be used.

```
Command to run the Pearson correlation:
CORRELATIONS
/VARIABLES=BirthWeight CHL HC MAC FL HL CaC
/PRINT=TWOTAIL NOSIG
/MISSING=PAIRWISE.
```

This command displays the matrix with correlation coefficients, P-values (two tailed), and number of pairs considered to calculate the correlation.

Correlations

		Birth Weight	CHL	HC	MAC	FL	HL	CaC
Birth Weight	Pearson Correlation	1	.702**	.625**	.757**	.507**	.501**	.715**
	Sig. (2-tailed)		.000	.000	.000	.000	.000	.000
	N	800	800	800	800	800	800	800
CHL	Pearson Correlation	.702**	1	.607**	.695**	.432**	.400**	.599**
	Sig. (2-tailed)	.000		.000	.000	.000	.000	.000
	N	800	800	800	800	800	800	800

HC	Pearson Correlation	.625**	.607**	1	.607**	.439**	.333**	.528**
	Sig. (2-tailed)	.000	.000		.000	.000	.000	.000
	N	800	800	800	800	800	800	800
MAC	Pearson Correlation	.757**	.695**	.607**	1	.500**	.458**	.720**
	Sig. (2-tailed)	.000	.000	.000		.000	.000	.000
	N	800	800	800	800	800	800	800
FL	Pearson Correlation	.507**	.432**	.439**	.500**	1	.588**	.485**
	Sig. (2-tailed)	.000	.000	.000	.000		.000	.000
	N	800	800	800	800	800	800	800
HL	Pearson Correlation	.501**	.400**	.333**	.458**	.588**	1	.434**
	Sig. (2-tailed)	.000	.000	.000	.000	.000		.000
	N	800	800	800	800	800	800	800
CaC	Pearson Correlation	.715**	.599**	.528**	.720**	.485**	.434**	1
	Sig. (2-tailed)	.000	.000	.000	.000	.000	.000	
	N	800	800	800	800	800	800	800

** Correlation is significant at the 0.01 level (2-tailed).

Some of the correlations among the independent variables are more than 0.7 and indicate that some of variables have high collinearity. These should be automatically excluded by stepwise procedure. The aforementioned is only to illustrate the computer procedure. We are now ready to run multiple stepwise regression.

```
Command to run stepwise regression:
REGRESSION
/MISSING LISTWISE
/STATISTICS COEFF OUTS CI(95) R ANOVA
/CRITERIA=PIN(.05) POUT(.10)
/NOORIGIN
/DEPENDENT GA
/METHOD=STEPWISE BirthWeight CHL HC MAC FL HL CaC
/RESIDUALS HIST(ZRESID).
```

This command runs stepwise multiple linear regression after excluding the cases for whom variable values are missing. The STATISTICS option provides the 95% CI on regression coefficients; CRITERIA option specifies that a variable be entered into the model if $P < 0.05$ and to remove it from the model is $P > 0.10$. Same criteria are used for all the variables in this example although you can specify different criteria for different variables. The RESIDUALS command is to draw a histogram and superimpose Gaussian curve to check the Gaussianity of error terms.

The following table displays the sequence of entering and removing of variables according to the criteria specified in the command. In this case, no variable was removed.

Variables Entered/Removed[a]

Model	Variables Entered	Variables Removed	Method
1	Birth Weight	.	Stepwise (Criteria: Probability-of-F-to-enter <= .050, Probability-of-F-to-remove >= .100).
2	MAC	.	Stepwise (Criteria: Probability-of-F-to-enter <= .050, Probability-of-F-to-remove >= .100).
3	HC	.	Stepwise (Criteria: Probability-of-F-to-enter <= .050, Probability-of-F-to-remove >= .100).
4	FL	.	Stepwise (Criteria: Probability-of-F-to-enter <= .050, Probability-of-F-to-remove >= .100).

[a] Dependent Variable: GA.

Following table shows the model summary. Birth Weight is the first to enter and explained about 46.1% of variation in GA. Next is MAC that explained additional 3.9% variation. Then came HC and FL that explained 0.4% and 0.3% additional variance, respectively. All these are statistically significant. No other variable contributed significantly. Total variance of GA explained by four anthropometric variables is 50.7%. This is not high and the regression is not likely to provide accurate prediction. This is the best one can obtain by these data. Perhaps, a better prediction can be obtained by considering nonlinear regression or by including other predictors. This was not done in the present exercise.

Model Summary[a]

Model	R	R Square	Adjusted R Square	Std. Error of the Estimate
1	.679[b]	.461	.460	1.602
2	.707[c]	.500	.498	1.544
3	.710[d]	.504	.503	1.538
4	.712[e]	.507	.505	1.534

[a] Predictors: (Constant), Birth Weight.

[b] Predictors: (Constant), Birth Weight, MAC.

[c] Predictors: (Constant), Birth Weight, MAC, HC.

[d] Predictors: (Constant), Birth Weight, MAC, HC, FL.

[e] Dependent Variable: GA.

Following table depicts the ANOVA that provides results of tests of the overall significance of the model at each stage of the stepwise procedure.

ANOVA[a]

Model		Sum of Squares	df	Mean Square	F	Sig.
1	Regression	1749.125	1	1749.125	681.313	.000[b]
	Residual	2048.695	798	2.567		
	Total	3797.820	799			
2	Regression	1897.676	2	948.838	397.983	.000[c]
	Residual	1900.144	797	2.384		
	Total	3797.820	799			
3	Regression	1915.975	3	638.658	270.145	.000[d]
	Residual	1881.845	796	2.364		
	Total	3797.820	799			
4	Regression	1926.782	4	481.695	204.671	.000[e]
	Residual	1871.038	795	2.354		
	Total	3797.820	799			

[a] Dependent Variable: GA.

[b] Predictors: (Constant), Birth Weight.

[c] Predictors: (Constant), Birth Weight, MAC.

[d] Predictors: (Constant), Birth Weight, MAC, HC.

[e] Predictors: (Constant), Birth Weight, MAC, HC, FL.

The following table shows steps in building the model with unstandardized or standardized coefficients of variables included in the model. Out of seven initial variables, only four are entered into the model, others are not significant. All coefficients are positive that indicates these variables are positively associated with GA. The standardized coefficients are unitless and are useful in ranking of variable according to their importance. For example, in this case, Birth Weight is more than six times as important as FL (see Beta for Model 4). Step 4 gives the final model equation.

The regression equation from the last block of the following table is

$$GA = 24.734 + 1.508(\text{birth weight}) + 0.414(\text{MAC}) + 0.080(\text{HC}) + 0.177(\text{FL})$$

Coefficients[a]

Model		Unstandardized Coefficients		Standardized Coefficients	t	Sig.	95.0% Confidence Interval for B	
		B	Std. Error	Beta			Lower Bound	Upper Bound
1	(Constant)	30.226	.246		122.681	.000	29.743	30.710
	Birth Weight	2.571	.098	.679	26.102	.000	2.378	2.764
2	(Constant)	27.595	.409		67.431	.000	26.792	28.398
	Birth Weight	1.702	.145	.449	11.711	.000	1.417	1.987
	MAC	.483	.061	.303	7.894	.000	.363	.603
3	(Constant)	25.390	.891		28.483	.000	23.640	27.139
	Birth Weight	1.566	.153	.414	10.257	.000	1.267	1.866
	MAC	.437	.063	.274	6.923	.000	.313	.561
	HC	.090	.032	.092	2.782	.006	.026	.153
4	(Constant)	24.734	.941		26.295	.000	22.887	26.580
	Birth Weight	1.508	.155	.398	9.744	.000	1.204	1.812
	MAC	.414	.064	.260	6.480	.000	.288	.539
	HC	.080	.033	.082	2.473	.014	.017	.144
	FL	.177	.083	.064	2.143	.032	.015	.340

[a] Dependent Variable: GA.

Residual statistics in the following table show that the standardized residuals range from −3.520 to +3.143 and the SD is nearly 1. This is acceptable.

Residuals Statistics[a]

	Minimum	Maximum	Mean	Std. Deviation	N
Predicted Value	31.18	41.29	36.49	1.553	800
Residual	−5.400	4.821	.000	1.530	800
Std. Predicted Value	−3.419	3.095	.000	1.000	800
Std. Residual	−3.520	3.143	.000	.997	800

[a] Dependent Variable: GA.

The histogram of standardized residuals, superimposed with Gaussian curve (Figure C.5), shows an approximate Gaussian distribution of error term. Thus, this assumption is fulfilled and the model can be believed.

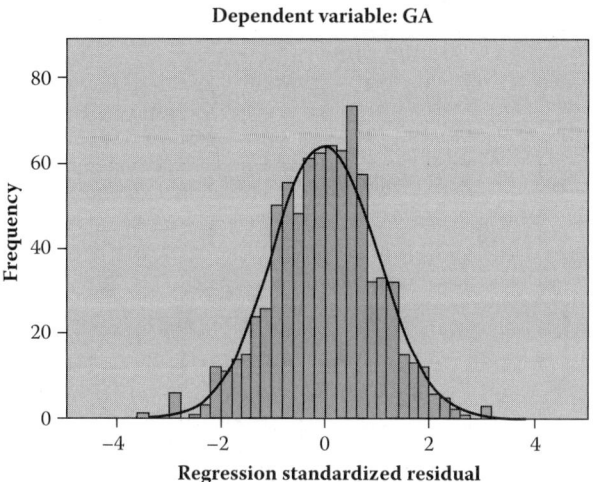

FIGURE C.5
Histogram of standardized residuals with superimposed Gaussian curve.

C.5 Curvilinear Regression

The aim of the study is to find an adequate relationship between birth weight and GA on the basis of observations on 800 births in a hospital in a developing country. These data are the same as in Section C.4. The following command fits several types of regressions, namely, linear, logarithmic, inverse, quadratic, power, growth, and exponential so that a choice can be made.

```
Command for fitting regressions of various types:
CURVEFIT /VARIABLES=BirthWeight WITH GA
  /CONSTANT
  /MODEL=LINEAR LOGARITHMIC INVERSE QUADRATIC POWER GROWTH
  EXPONENTIAL
  /PLOT FIT.
```

The following table provides the model summary and parameter estimates of various types of relationships. The one with maximum R^2 is statistically most adequate. There is a marginal difference among linear, logarithmic, inverse, and quadratic regressions. In such a situation, you can choose any that is biologically more plausible. Power, growth, and exponential models have higher R^2.

Only quadratic regression has two parameters for which b2 appears in the following table. All other models have only one parameter each (other than constant).

Model Summary and Parameter Estimates

Dependent Variable: BirthWeight

Equation	Model Summary					Parameter Estimates		
	R Square	F	df1	df2	Sig.	Constant	b1	b2
Linear	.461	681.313	1	798	.000	−4.102	.179	
Logarithmic	.462	686.184	1	798	.000	−20.283	6.319	
Inverse	.461	682.550	1	798	.000	8.497	−220.330	
Quadratic	.463	343.713	2	797	.000	−8.840	.448	−.004
Power	.514	843.637	1	798	.000	3.73E-005	3.075	
Growth	.506	816.918	1	798	.000	−2.304	.087	
Exponential	.506	816.918	1	798	.000	.100	.087	

The independent variable is GA.

Figure C.6 depicts the scatter plot and various types of regression between the birth weight and GA.

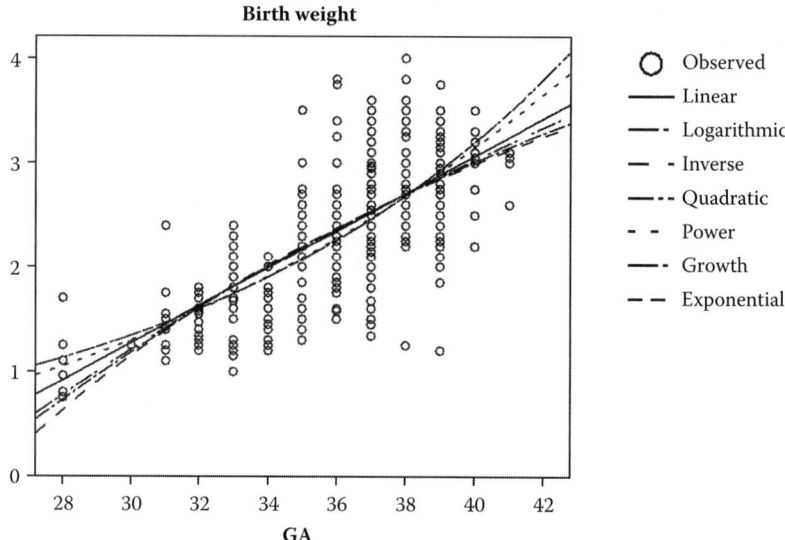

FIGURE C.6
Scatter plot and various regressions.

This figure does not provide much of a clue in this case. However, the preceding table shows that power relationship has maximum R^2 although not particularly high to inspire confidence. But that is the best among these choices that one can obtain with these data. The general form of power equation is

$$\text{Birth weight} = \beta_0 * \text{GA}^{\beta_1}$$

This has two parameters, namely, β_0 and β_1. This is curvilinear as logarithm converts this to a linear form:

$$\text{Log}_{10}(\text{Birth weight}) = \text{Log}_{10}(\beta_0) + \beta_1 \text{Log}_{10}(\text{GA})$$

The parameter estimates shown in the preceding table for power regression gives the fitted regression equation:

$$\text{Birth weight (kg)} = 0.0000373 \times \text{GA (weeks)}^{3.075}$$

To get more information on this model such as CI, now fit this model directly to the data.

```
Command to fit linearized form of the power equation:
COMPUTE Log10GA = LG10(GA).
COMPUTE Log10Birtweight = LG10(BirthWeight).
REGRESSION
/MISSING LISTWISE
/STATISTICS COEFF OUTS CI R ANOVA
/CRITERIA=PIN(.05) POUT(.10)
```

```
/NOORIGIN
/DEPENDENT Log10Birtweight
/METHOD=ENTER Log10GA
/RESIDUALS NORM(ZRESID).
```

Model summary now is as follows, which is the same as obtained earlier:

Model Summary(b)

Model	R	R Square	Adjusted R Square	Std. Error of the Estimate
1	.717(a)	.514	.513	.08048

a Predictors: (Constant), Log10GA.
b Dependent Variable: Log10Birtweight.

ANOVA(b)

Model		Sum of Squares	df	Mean Square	F	Sig.
1	Regression	5.464	1	5.464	843.637	.000(a)
	Residual	5.169	798	.006		
	Total	10.633	799			

a Predictors: (Constant), Log10GA.
b Dependent Variable: Log10Birtweight.

The following table displays the unstandardized and standardized coefficients and the 95% CI. The coefficient is statistically highly significant ($P = 0.000 < 0.001$).

Coefficients[a]

Model		Unstandardized Coefficients B	Std. Error	Standardized Coefficients Beta	t	Sig.	95% Confidence Interval for B Lower Bound	Upper Bound
1	(Constant)	−4.428	.165		−26.788	.000	−4.753	−4.104
	Log10GA	3.075	.106	.717	29.045	.000	2.867	3.283

[a] Dependent Variable: Log10Birtweight.

The regression equation is

$$\mathrm{Log}_{10}(\text{Birth weight}) = -4.428 + 3.075 \times \mathrm{Log}_{10}(\text{GA})$$

The antilog of −4.428 is 0.0000373 and the equation is the same as obtained before.

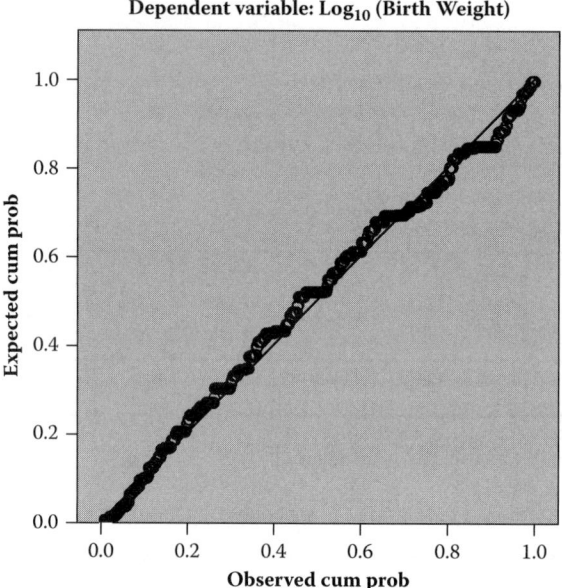

Dependent variable: Log$_{10}$ (Birth Weight)

FIGURE C.7
Gaussian P-P plot of standardized residuals.

Figure C.7 displays the Normal P-P plot to study the Gaussianity of the residuals. There is some deviation around 0.9 otherwise all the points lie on the expected line. This confirms that the deviation from Gaussianity is not large and establishes that the tests of significance and CIs are not invalid.

```
Command to calculate the observed and predicted mean birth
weight at each gestational age and to draw the graph:
AGGREGATE
  /OUTFILE=*
  MODE=ADDVARIABLES
  /BREAK=GA
  /BirthWeight_mean_1 = MEAN(BirthWeight).
COMPUTE predicted_BW = 0.0000373*GA ** 3.075.
EXECUTE.
GRAPH
  /SCATTERPLOT(OVERLAY)=GA GA WITH BirthWeight_mean_1
  predicted_BW (PAIR)
  /MISSING=LISTWISE.
```

To avoid complexity of plot of 800 points, Figure C.8 shows only the mean birth weight for each gestation age. The fitted regression curve shows faster than linear rise in birth weight as the gestation age increased. Examine biological admissibility of this possibility before accepting the regression curve.

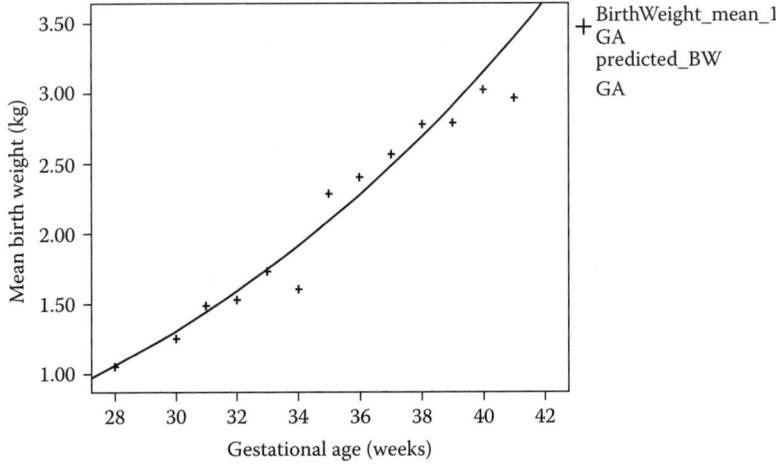

FIGURE C.8
Power regression of birth weight on GA.

C.6 Analysis of Covariance (ANCOVA)

Uric acid is measured in three groups of subjects, namely, normal (group-1), diabetic (group-2), and cardiovascular disease (group-3). Sample size is 30 in each group. The purpose is to find if three groups have same mean level of uric acid. However, age pattern was different in the three groups. It is suspected that age would influence the uric acid level. ANCOVA is the method that would eliminate the linear effect of age when age is considered as a covariate.

First confirm that uric acid and age are indeed correlated. If not correlated, there is no need to run ANCOVA.

```
Command for Pearson correlation:
CORRELATIONS VARIABLES = uricacid age
/PRINT=TWOTAIL NOSIG
/MISSING=PAIRWISE.
```

This command produces the Pearson correlation with two-tailed *P*-value after excluding the pair value if either uric acid or age is missing.

Correlations

		uric acid	age
uric acid	Pearson Correlation	1	.476(**)
	Sig. (2-tailed)		.000
	N	90	90
age	Pearson Correlation	.476(**)	1
	Sig. (2-tailed)	.000	
	N	90	90

** Correlation is significant at the 0.01 level (2-tailed).

Correlation coefficient between level of uric acid and age is 0.476 ($P < 0.001$). This is statistically significant (cannot be considered zero). Thus, it is in order to run ANCOVA and adjust for effect of age. However, ANCOVA is valid only if (1) error variance of uric acid across groups is nearly equal (homoscedasticity) and (2) the gradient of relationship between uric acid and age is the same across the groups. To check these two assumptions, run the following ANCOVA command:

```
UNIANOVA uricacid BY group WITH age
/METHOD = SSTYPE(3)
/INTERCEPT = INCLUDE
/PRINT = HOMOGENEITY
/CRITERIA = ALPHA(.05)
/DESIGN = group age age*group.
```

The DESIGN command includes interaction age * group also to test equality of gradients.

Levene Test of Equality of Error Variances(a)

Dependent Variable: uric acid

F	df1	df2	Sig.
1.523	2	87	.224

a Design: Intercept+group+age.

The preceding table shows the result of Levene test for equality of error variance using F-test. Since P-value > 0.05, evidence is not enough to reject the null hypothesis. We are safe in assuming homoscedasticity. The following table shows that the interaction term between the age and group is not significant ($P = 0.571$). Thus, there is no evidence of violation of the equal gradient assumption. Assumptions of ANCOVA are fulfilling and we can run ANCOVA.

Test of Between-Subjects Effects

Dependent Variable: uric acid

Source	Type III Sum of Squares	df	Mean Square	F	Sig.
Corrected Model	61.468(a)	5	12.294	8.723	.000
Intercept	89.012	1	89.012	63.162	.000
group	.996	2	.498	.353	.703
age	10.426	1	10.426	7.398	.008
group * age	1.588	2	.794	.563	.571
Error	118.377	84	1.409		
Total	2413.276	90			
Corrected Total	179.845	89			

```
a R Squared = .342 (Adjusted R Squared = .303).
```

```
Command to run analysis of covariance:
UNIANOVA uricacid BY group WITH age
/METHOD = SSTYPE(3)
/INTERCEPT = INCLUDE
/EMMEANS = TABLES(group) WITH(age=MEAN) COMPARE
ADJ(BONFERRONI)
/PRINT = DESCRIPTIVE OPOWER PARAMETER HOMOGENEITY
/CRITERIA = ALPHA(.05)
/DESIGN = group age.
```

The DESIGN command now excludes interaction as it is not significant. The following output gives the group means and SDs:

Descriptive Statistics

Dependent Variable: uric acid

group	Mean	Std. Deviation	N
1	4.003	.9234	30
2	5.181	1.3736	30
3	5.760	1.3431	30
Total	4.982	1.4215	90

Some of these differences across groups could be due to differences in age structure of the subjects in the three groups. The following tests the null hypothesis that the error variance of the dependent variable is equal across groups, now for ANCOVA model.

Levene Test of Equality of Error Variances[a]

Dependent Variable: uric acid

F	df1	df2	Sig.
1.480	2	87	.233

a Design: Intercept+group+age.

Since P-value of Levene test is >0.05, we do not have enough evidence to reject the null hypothesis. The following is the output for ANCOVA model.

Test of Between–Subject Effects

Dependent Variable: uric acid

Source	Type III Sum of Squares	df	Mean Square	F	Sig.	Noncent. Parameter	Observed Power(a)
Corrected Model	59.880(b)	3	19.960	14.309	.000	42.927	1.000
Intercept	100.193	1	100.193	71.826	.000	71.826	1.000
group	19.097	2	9.549	6.845	.002	13.690	.913
age	11.796	1	11.796	8.456	.005	8.456	.820
Error	119.965	86	1.395				
Total	2413.276	90					
Corrected Total	179.845	89					

a Computed using alpha = .05.

b R Squared = .333 (Adjusted R Squared = .310).

The preceding table shows sources of variation in the uric acid. Effect of both age and group are significant, indicating that mean uric acid is different in different groups even after adjusting for age differentials, and age effect by itself is also significant. However, $R^2 = 0.333$ is not high and the model is not sufficiently good for prediction.

Parameter Estimates

Dependent Variable: uric acid

Parameter	B	Std. Error	t	Sig.	95% Confidence Interval		Noncent. Parameter	Observed Power[a]
					Lower Bound	Upper Bound		
Intercept	4.309	.544	7.925	.000	3.228	5.389	7.925	1.000
[group=1]	−1.268	.348	−3.643	.000	−1.960	−.576	3.643	.950
[group=2]	−.415	.310	−1.337	.185	−1.031	.202	1.337	.262
[group=3]	0[b]	.	.	.	.	.	.	.
age	.029	.010	2.908	.005	.009	.049	2.908	.820

[a] Computed using alpha = .05.

[b] This parameter is set to zero because it is redundant.

The preceding table shows regression parameter estimate of uric acid on age. This is $B = 0.029$. B-coefficients for groups are used to estimate mean uric acid in the three groups after removing the effect of age. This is calculated at mean age as shown in the following. Reference group in this case is group = 3 (cardiovascular diseases).

Estimates

Dependent Variable: uric acid

group	Mean	Std. Error	95% Confidence Interval	
			Lower Bound	Upper Bound
1	4.274[a]	.235	3.807	4.741
2	5.128[a]	.216	4.698	5.558
3	5.543[a]	.228	5.089	5.996

[a] Covariates appearing in the model are evaluated at the following values: age = 42.37.

At average age = 42.37 years, mean uric acid is 5.543 − 5.128 = 0.415 mg/dL higher in cardiovascular disease group than diabetic group and 5.543 − 4.274 = 1.269 mg/dL higher than control group (group = 1). Mean of which group is significantly different from the others is found by the following pairwise comparisons using Bonferroni adjustment.

Pairwise Comparisons

Dependent Variable: uric acid

(I) group	(J) group	Mean Difference (I–J)	Std. Error	Sig. [a]	95% Confidence Interval for Difference [a]	
					Lower Bound	Upper Bound
1	2	−.854*	.325	.030	−1.647	−.061
	3	−1.268*	.348	.001	−2.118	−.418
2	1	.854*	.325	.030	.061	1.647
	3	−.415	.310	.554	−1.172	.343
3	1	1.268*	.348	.001	.418	2.118
	2	.415	.310	.554	−.343	1.172

Based on estimated marginal means

[a] Adjustment for multiple comparisons: Bonferroni.

* The mean difference is significant at the .05 level.

The results show that there is a significant difference in diabetic and cardio-vascular groups compared with normals. However, there no significant difference between the diabetic and cardiovascular subjects after linear effect of age is eliminated.

C.7 Logistic Regression

The aim of this study is to explore the predictors of malnutrition in children from a deprived section of society. One hundred malnourished and 100 normal children of age between 1 and 5 years—matched for age within ±6 months and sex—were considered for the study. The potential predictors were decided either from the previous knowledge or significant predictors from univariate analysis. The dependent variable is group (malnutrition = 1, normal = 0) and the predictors are maternal education (illiterate = 1, literate = 0); daily income of parents (low = 1, not low = 0); immunization (no = 1, yes = 0); colostrum (not given = 1, given = 0); breast feeding till 6 months (no = 1, yes = 0); and mode of feeding (bottle = 1, others = 0). Consider only these six predictors for this exercise.

Backward elimination (BSTEP(LR)) method was applied with criteria $P < 0.05$ to enter and $P > 0.10$ to remove a predictor from the model. Hosmer–Lemeshow goodness of fit was applied to find appropriateness of model. Pseudo-R^2 was also calculated to assess the variability explained by the predictors (pseudo-R^2 is analog to the R^2 used in multiple linear regression).

CONTRASTs define the indicator variables. Note that all the predictors are binary in this example.

```
Command to run backward elimination logistic regression:
LOGISTIC REGRESSION VARIABLES Group
  /METHOD = BSTEP(LR) maternal_education Breast_feed_6months
  mode_feeding
    immunization_status daily_income colostrumgiven_code
  /CONTRAST (maternal_education)=Indicator(1)
  /CONTRAST (Breast_feed_6months)=Indicator(1)
  /CONTRAST (mode_feeding)=Indicator(1)
  /CONTRAST (immunization_status)=Indicator(1)
  /CONTRAST (daily_income )=Indicator(1)
  /CONTRAST (colostrumgiven)=Indicator(1)
  /PRINT = GOODFIT SUMMARY CI(.95)
  /CRITERIA = PIN(.05) POUT(.10) ITERATE(20) CUT(.5)  .
```

This command runs the backward elimination stepwise(likelihood ratio) procedure. CI(.95) displays the 95% CI of odds ratio (OR). PIN and POUT specify the probability to enter and remove a predictor from the model. Logistic will predict the probability of being malnourished or normal on the basis of the given value of predictors. CUT(.5) is the cutoff point of probability chosen to classify children as malnourished or normal on the basis of the fitted model. The following confirms the coding for the dependent variable:

Dependent Variable
Encoding

Original Value	Internal Value
0	0
1	1

The next table displays variables, their categories, frequency, and coding scheme. For example, daily income of parents is divided into two categories: Low category has 120 children and not low has 80 children. Category perceived as risk for malnourishment is coded as 1 and the other as 0.

Categorical Variables Codings

		Frequency	Parameter coding (1)
daily_income	Not low	80	.000
	Low	120	1.000
Breast_feed_6months	Yes	126	.000
	No	74	1.000
colostrumgiven_code	Given	140	.000
	Not given	60	1.000
mode_feeding	Other types	119	.000
	Bottle feeding	81	1.000
immunization_status	Immunized	150	.000
	Not immunized	50	1.000
maternal_education	Literate	110	.000
	Illiterate	90	1.000

The following two tables give result when no predictor is in the model.

Blocl 0: Beginning Block

Classification Table[a,b]

		Predicted		
		Group		Percentage Correct
Observed		0	1	
Step 0 Group 0		0	100	.0
	1	0	100	100.0
Overall Percentage				50.0

[a] Constant is included in the model.

[b] The cut value is .500.

Variables in the Equation

		B	S.E.	Wald	df	Sig.	Exp(B)
Step 0	Constant	.000	.141	.000	1	1.000	1.000

The following table displays the initial univariate significance of each predictor. Score in this table is similar to chi-square. Mode_feeding is marginally significant and all others are highly significant when considered one at a time.

Variables not in the Equation

			Score	df	Sig.
Step 0	Variables	maternal_education(1)	13.657	1	.000
		Breast_feed_6months(1)	21.965	1	.000
		mode_feeding(1)	3.507	1	.061
		immunization_status(1)	24.000	1	.000
		daily_income(1)	10.083	1	.001
		colostrumgiven_code(1)	18.667	1	.000
	Overall Statistics		65.333	6	.000

The following table displays the steps of backward elimination method and corresponding significance. In this method, since the number of predictors serially reduces, chi-square value at each subsequent step either decreases or remains same.

Omnibus Tests of Model Coefficients

		Chi-square	df	Sig.
Step 1	Step	77.950	6	.000
	Block	77.950	6	.000
	Model	77.950	6	.000

Step	Step	−.666	1	.414
2(a)	Block	77.284	5	.000
	Model	77.284	5	.000
Step	Step	−2.110	1	.146
3(a)	Block	75.174	4	.000
	Model	75.174	4	.000

a A negative Chi-squares value indicates that the Chi-squares
value has decreased from the previous step.

The following table is showing the model summary. Last two columns are two versions of pseudo-R^2. The value of pseudo-R^2 is decreasing because of backward elimination method. This output does not say which variables have been eliminated.

Model Summary

Step	−2 Log likelihood	Cox & Snell R Square	Nagelkerke R Square
1	199.309(a)	.323	.430
2	199.975(a)	.321	.427
3	202.085(a)	.313	.418

a Estimation terminated at iteration number 5 because
parameter estimates changed by less than .001.

The next table shows the goodness of fit of the model at each step. At the final step, the P-value is 0.091. This is not significant. So we can not reject the null hypothesis that there is significance difference in observed and expected values. In other words, the model can be considered adequate.

Hosmer and Lemeshow Test

Step	Chi-square	df	Sig.
1	6.001	8	.647
2	8.319	7	.305
3	12.306	7	.091

The following table shows the observed and expected frequencies at each step of elimination in normal and malnutrition categories. These are divided into 10 strata as required for Hosmer–Lemeshow test. These start with nearly 20 subjects in each stratum. Some frequencies are less than 5 and that raises questions about validity of this test for these data.

Contingency Table for Hosmer-Lemeshow Test

		Group = Normal		Group = Malnutrition		Total
		Observed	Expected	Observed	Expected	Observed
Step 1	1	19	17.244	0	1.756	19
	2	21	20.765	3	3.235	24
	3	15	15.451	5	4.549	20
	4	11	12.959	9	7.041	20
	5	11	10.897	9	9.103	20
	6	10	9.019	12	12.981	22
	7	6	6.558	14	13.442	20
	8	2	4.095	17	14.905	19
	9	4	2.521	17	18.479	21
	10	1	.492	14	14.508	15
Step 2	1	31	27.784	0	3.216	31
	2	13	13.403	3	2.597	16
	3	14	16.790	9	6.210	23
	4	11	14.091	12	8.909	23
	5	11	9.469	9	10.531	20
	6	7	6.838	11	11.162	18
	7	6	5.864	14	14.136	20
	8	4	3.874	16	16.126	20
	9	3	1.887	26	27.113	29
Step 3	1	44	41.108	3	5.892	47
	2	14	16.914	9	6.086	23
	3	5	7.895	8	5.105	13
	4	14	11.178	7	9.822	21
	5	11	8.686	9	11.314	20
	6	3	6.239	17	13.761	20
	7	5	5.141	16	15.859	21
	8	3	2.427	19	19.573	22
	9	1	.412	12	12.588	13

The following table displays the classification of subjects as observed and predicted by the model. Logistic regression is used to calculate probability of each subject based on the model and subjects with probability more than 0.5 were classified as malnourished. The final logistic model was able to correctly predict 75% of the subjects. This is not high but that is the best achievable by these data when linear combination of these six predictors is considered.

Classification Table (a)

Observed			Predicted		
			Group		Percentage Correct
			Normal	Malnutrition	
Step 1	Group	Normal	72	28	72.0
		Malnutrition	25	75	75.0
	Overall Percentage				73.5
Step 2	Group	Normal	72	28	72.0
		Malnutrition	27	73	73.0
	Overall Percentage				72.5
Step 3	Group	Normal	77	23	77.0
		Malnutrition	27	73	73.0
	Overall Percentage				75.0

a The cut value is .500.

Following table displays the coefficient (B = log-OR), their SE, Wald test, P-value, OR = exp(B), and its 95% CI.

Variables in the Equation

		B	S.E.	Wald	df	Sig.	Exp(B)	95.0% C.I. for EXP(B)	
								Lower	Upper
Step 1[a]	maternal_education(1)	.848	.362	5.477	1	.019	2.335	1.148	4.749
	Breast_feed_6months(1)	1.779	.381	21.802	1	.000	5.925	2.808	12.504
	mode_feeding(1)	.530	.358	2.195	1	.138	1.698	.843	3.423
	immunization_status(1)	1.833	.461	15.800	1	.000	6.256	2.533	15.449
	daily_income(1)	.312	.382	.669	1	.413	1.366	.647	2.886
	colostrumgiven_code(1)	1.529	.405	14.252	1	.000	4.616	2.086	10.212
	Constant	−2.284	.411	30.850	1	.000	.102		
Step 2[a]	maternal_education(1)	.923	.351	6.921	1	.009	2.517	1.265	5.006
	Breast_feed_6months(1)	1.825	.378	23.324	1	.000	6.202	2.957	13.007
	mode_feeding(1)	.515	.356	2.090	1	.148	1.674	.833	3.364
	immunization_status(1)	1.926	.449	18.381	1	.000	6.859	2.844	16.541
	colostrumgiven_code(1)	1.527	.405	14.233	1	.000	4.603	2.083	10.176
	Constant	−2.156	.375	33.068	1	.000	.116		
Step 3[a]	maternal_education(1)	.920	.349	6.956	1	.008	2.510	1.267	4.975
	Breast_feed_6months(1)	1.813	.375	23.420	1	.000	6.130	2.941	12.775
	immunization_status(1)	1.946	.450	18.691	1	.000	7.001	2.897	16.917
	colostrumgiven_code(1)	1.506	.399	14.262	1	.000	4.511	2.064	9.858
	Constant	−1.943	.335	33.711	1	.000	.143		

a Variable(s) entered on step 1: maternal_education, Breast_feed_6months, mode_feeding, immunization_status, daily_income, colostrumgiven_code.

The predictors remained significant after elimination are maternal education, breast feeding till 6 months, immunization, and colostrum given. The other two were eliminated as not significant (see table given later). When these four variables are in the model, income and feeding by bottle or otherwise do not make any significant contribution to the prediction of malnourishment. Since our coding is 1 for the risk present and 0 for absent for all the predictors, positive values of the coefficients indicate that these are positively associated with the risk of malnourishment. OR of maternal education represents that illiterate women were 2.510 times likely to have malnourished child compared to literate women as far as this model is concerned. Similarly, those mothers who did not breast feed till 6 months has 6.130 times likely to have malnourished child compared to mother who had breast feed more than 6 months. Highest OR is for immunization. These ORs are adjusted for other terms in the model and measure independent contribution of these predictors.

The coefficients stated in the aforementioned table can be used to construct the logistic model. This is

$$\ln[p/(1-p)] = -1.943 + 1.506(\text{colostrum}) + 1.946(\text{immunization})$$
$$+ 1.813(\text{breast feeding}) + 0.920(\text{maternal education})$$

For example, when all risk factors are present (all predictors = 1),

$$\ln[p/(1-p)] = -1.943 + 1.506 + 1.946 + 1.813 + 0.920 = 4.242$$

OR for this child to be malnourished is 4.242 relative to the one with no risk factor. This equation gives $p/(1-p) = e^{4.242} = 69.547$ and $p = 69.542/(1+69.547) = 0.986$. The estimated probability of a child with these risk factors being malnourished is 0.986.

The following table displays the status of variable excluded from the model.

Variables not in the Equation

			Score	df	Sig.
Step 2(a)	Variables	daily_income(1)	.671	1	.413
	Overall Statistics		.671	1	.413
Step 3(b)	Variables	mode_feeding(1)	2.109	1	.146
		daily_income(1)	.562	1	.453
	Overall Statistics		2.765	2	.251

a Variable(s) removed on step 2: daily_income.

b Variable(s) removed on step 3: mode_feeding.

C.8 Survival Analysis (Life Table Method)

Duration of exclusive breast feeding was recorded for 64 children by monthly visit to their homes. Some were not available at the time of the visit and the data for them was incomplete. Termination 0 means that the duration is incomplete and 1 means complete. Let us use life table method to obtain survival curve for duration of exclusive breast feeding and the median duration of exclusive breast feeding in these children. Life table method is applied because time is aggregated in the equal intervals of 1 month each.

```
Command to run the life table method:
SURVIVAL TABLE=Month_ending
/INTERVAL=THRU 9 BY 1
/STATUS=Termination(1)
/PRINT=TABLE
/PLOTS (SURVIVAL)=Month_ending.
```

This command calculates the median duration of exclusive breast feeding and prints the survival curve.

Survival Analysis

Survival Variable: Duration of breast feed (months)

Life Table[a]

Interval Start Time	Number Entering Interval	Number Withdrawing during Interval	Number Exposed to Risk	Number of Terminal Events	Proportion Terminating
0	64	0	64.000	0	.00
1	64	2	63.000	1	.02
2	61	3	59.500	2	.03
3	56	3	54.500	11	.20
4	42	1	41.500	14	.34
5	27	2	26.000	11	.42
6	14	0	14.000	7	.50
7	7	2	6.000	3	.50
8	2	0	2.000	1	.50
9	1	0	1.000	1	1.00

[a] The median duration of exclusive breastfeed is 5.01 months.

Life Table[a]

Interval Start Time	Proportion Surviving	Cumulative Proportion Surviving at End of Interval	Std. Error of Cumulative Proportion Surviving at End of Interval	Probability Density	Std. Error of Probability Density
0	1.00	1.00	.00	.000	.000
1	.98	.98	.02	.016	.016
2	.97	.95	.03	.033	.023
3	.80	.76	.06	.192	.052
4	.66	.50	.07	.256	.059
5	.58	.29	.06	.213	.056
6	.50	.15	.05	.145	.050
7	.50	.07	.04	.073	.039
8	.50	.04	.03	.036	.032
9	.00	.00	.00	.000	.000

[a] The median duration of exclusive breastfeed is 5.01 months.

The median duration of exclusive breast feeding is 5.01 months. The survival curve in Figure C.9 has been obtained by Excel using the survival probabilities given in the table mentioned earlier. The survival curve obtained by SPSS is a step function as shown in Figure C.10. Survival plot depicts cumulative probability of exclusive breast feeding according to the month. The median 5.01 months corresponds to the cumulated probability 0.50.

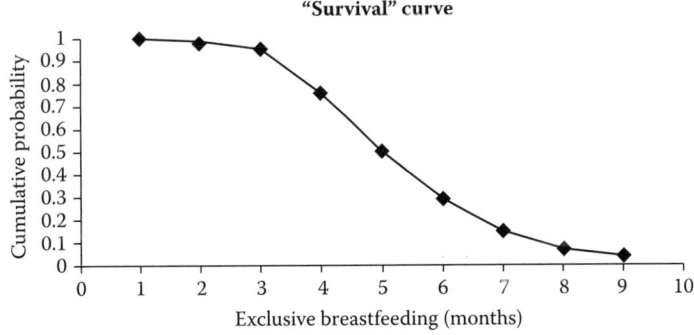

FIGURE C.9
Life table survival curve by Excel for duration of exclusive breast feeding.

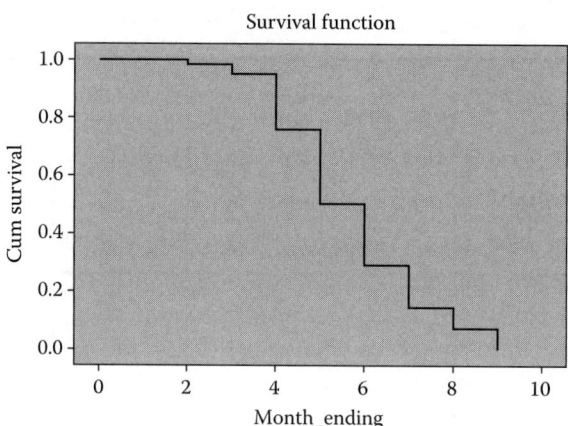

FIGURE C.10
Survival curve obtained by SPSS.

C.9 Cox Proportional Hazards Model

In a study of peritonitis disease, survival duration (in years) was compared between two regimens (control and experimental). The survival has been taken to be affected by severity of disease at admission (assessed by APACHE score), sex, and age in the two groups. These are the covariates in this example. Since measured at baseline, these are time independent. The uncensored values denote death within the follow-up period, whereas censored denotes both lost-to-follow-up and alive subjects at the end of the study. The follow-up was for 6¼ years.

Basic assumption of survival analysis is that censoring should have nothing to do with the survival pattern—both censored and uncensored persons should have similar characteristics at baseline. To test this, first unpaired *t*-tests should be applied to compare the mean APACHE score in censored and uncensored subjects.

```
Command to run unpaired t-test to compare the APACHE score in
censored and uncensored groups:
T-TEST GROUPS=Censored(0 1)
/VARIABLES=APACHEatadmission
/CRITERIA=CI(.95).
```

This command runs the unpaired *t*-test to compare the mean APACHE score in censored (coded as 0) and uncensored (coded as 1); CRITERIA= CI(.95) calculates the 95% CI for mean difference. The following table displays the mean, SD, SE of mean of APACHE score in censored and uncensored subjects and the next table provides results of *t*-test.

Group Statistics

	Censored	N	Mean	Std. Deviation	Std. Error Mean
APACHE at admission	uncensored	63	23.84	9.185	1.157
	censored	17	25.65	7.689	1.865

Independent Samples Test

		Levene's Test for Equality of Variances		t-test for Equality of Means						
									95% Confidence Interval of the Difference	
		F	Sig.	t	df	Sig. (2-tailed)	Mean Difference	Std. Error Difference	Lower	Upper
APACHE at admission	Equal variances assumed	.981	.325	−.743	78	.460	−1.806	2.432	−6.648	3.036
	Equal variances not assumed			−.823	29.563	.417	−1.806	2.195	−6.291	2.679

Levene test for equality of variances is applied to check the equality of variances in APACHE score in censored and uncensored groups before deciding which *t*-test result is applicable. Levene test, not significant ($P = 0.325$), represents no violation of homogeneity of variance assumption. The *P*-value of *t*-test for equal of variances found no significant ($P = 0.460$) difference in mean APACHE score in censored and uncensored values.

Chi-square test is applied to compare proportion of males and females in censored and uncensored subjects. TIMER(5) sets a limit of 5 minutes for recursive calculations of exact test. These calculations can take substantial time.

Command to run the chi-square test:

```
CROSSTABS
/TABLES=Censored BY Sex
/FORMAT=AVALUE TABLES
/STATISTICS=CHISQ
/CELLS=COUNT
/COUNT ROUND CELL
/METHOD=EXACT TIMER(5).
```

This command displays the following contingency tables between sex and censoring and runs the chi-square and other similar tests. Cell frequencies are not small and exact test is not required. Chi-square results can be believed. No chi-square is significant.

Censored * Sex Crosstabulation

Count

		Sex		
		0	1	Total
Censored	uncensored	28	35	63
	censored	9	8	17
Total		37	43	80

Chi-Square Tests

	Value	df	Asymp. Sig. (2-sided)	Exact Sig. (2-sided)	Exact Sig. (1-sided)	Point Probability
Pearson Chi-Square	.389[a]	1	.533	.591	.362	
Continuity Correction[b]	.122	1	.727			
Likelihood Ratio	.388	1	.533	.591	.362	
Fisher's Exact Test				.591	.362	
Linear-by-Linear Association	.384[c]	1	.536	.591	.362	.178
N of Valid Cases	80					

[a] 0 cells (.0%) have expected count less than 5. The minimum expected count is 7.86.

[b] Computed only for a 2 x 2 table.

[c] The standardized statistic is −.620.

Similar comparison is required for age at onset also.

Command to run unpaired t-test to compare the age at onset:

```
T-TEST GROUPS=Censored(0 1)
/MISSING=ANALYSIS
/VARIABLES=AgeatonsetYears
/CRITERIA=CI(.95).
```

Group Statistics

	Censored	N	Mean	Std. Deviation	Std. Error Mean
Age at onset (Years)	uncensored	63	50.00	10.526	1.326
	censored	17	49.53	13.816	3.351

Independent Samples Test

		Levene's Test for Equality of Variances		t-test for Equality of Means							
										95% Confidence Interval of the Difference	
		F	Sig.	t	df	Sig. (2-tailed)	Mean Difference	Std. Error Difference		Lower	Upper
Age at onset (Years)	Equal variances assumed	2.063	.155	.153	78	.879	.471	3.083		-5.667	6.608
	Equal variances not assumed			.131	21.270	.897	.471	3.604		-7.018	7.959

Average age at onset in censored observations is 49.53 years and in uncensored is 50.00 years. Levene test is not significant—thus equal variances can be assumed. The difference in mean age at onset is not statistically significant ($P = 0.879$).

All the three covariates are not different in subjects with censored and uncensored survival. Thus, this basic assumption is fulfilled.

As an additional inquiry, you can test whether APACHE score at admission differed between males and females. For this the command is as follows:

```
T-TEST GROUPS=Sex(0 1)
/MISSING=ANALYSIS
/VARIABLES=APACHEatadmission
/CRITERIA=CI(.95).
```

Group Statistics

	Sex	N	Mean	Std. Deviation	Std. Error Mean
APACHE at admission	Female	37	23.54	9.197	1.512
	Male	43	24.81	8.650	1.319

Independent Samples Test

		Levene's Test for Equality of Variances		t-test for Equality of Means						
									95% Confidence Interval of the Difference	
		F	Sig.	t	df	Sig. (2-tailed)	Mean Difference	Std. Error Difference	Lower	Upper
APACHE at admission	Equal variances assumed	.062	.803	−.638	78	.526	−1.273	1.997	−5.250	2.703
	Equal variances not assumed			−.635	74.609	.528	−1.273	2.007	−5.271	2.724

Although APACHE score is higher in males, it is not statistically significant. Just to explore the data further, find correlation between APACHE score and age at onset of the disease.

```
Command to run the Pearson correlation between the APACHE
score and age at onset:
CORRELATIONS
/VARIABLES=APACHEatadmission AgeatonsetYears
/PRINT=TWOTAIL NOSIG
/MISSING=PAIRWISE.
```

This command displays the correlation coefficient with two-tailed P-value and the number of subjects after excluding pair values if either APACHE score or age is missing. Nothing is missing in this example. The correlation is 0.132, which is not high and not significant ($P = 0.243$).

		APACHE at admission	Age at onset (Years)
APACHE at admission	Pearson Correlation	1	.132
	Sig. (2-tailed)		.243
	N	80	80
Age at onset (Years)	Pearson Correlation	.132	1
	Sig. (2-tailed)	.243	
	N	80	80

Now, it is a good idea to explore whether the two groups are similar with respect to the baseline characteristics. The following is the test for age at onset.

```
Command to run unpaired t-test for group differences in age at
onset:
T-TEST GROUPS=Group(0 1)
/MISSING=ANALYSIS
/VARIABLES=AgeatonsetYears
/CRITERIA=CI(.95).
```

Following table gives means, SDs, and SEs:

Group Statistics

	Group	N	Mean	Std. Deviation	Std. Error Mean
Age at onset (Years)	0	40	48.90	10.068	1.592
	1	40	50.90	12.293	1.944

Following tables display the results of *t*-test. In this case, Levene test for equality of variances is significant and tells that homogeneity of variance assumption is violated by these data. The *P*-value ($P = 0.429$) of *t*-test for unequal variances found no significant difference in mean age at onset in the two groups.

Independent Samples Test

		Levene's Test for Equality of Variances		t-test for Equality of Means						
									95% Confidence Interval of the Difference	
		F	Sig.	t	df	Sig. (2-tailed)	Mean Difference	Std. Error Difference	Lower	Upper
Age at onset (Years)	Equal variances assumed	4.066	.047	−.796	78	.428	−2.000	2.512	−7.002	3.002
	Equal variances not assumed			−.796	75.086	.429	−2.000	2.512	−7.005	3.005

```
Command to run unpaired t-test to compare the APACHE score in
the two groups:
T-TEST GROUPS=Group(0 1)
/MISSING=ANALYSIS
```

```
/VARIABLES=APACHEatadmission
/CRITERIA=CI(.95).
```

Group Statistics

	Group	N	Mean	Std. Deviation	Std. Error Mean
APACHE at admission	0	40	24.48	9.021	1.426
	1	40	23.98	8.830	1.396

Independent Samples Test

		Levene's Test for Equality of Variances		t-test for Equality of Means							
										95% Confidence Interval of the Difference	
		F	Sig.	t	df	Sig. (2-tailed)	Mean Difference	Std. Error Difference	Lower	Upper
APACHE at admission	Equal variances assumed	.030	.863	.251	78	.803	.500	1.996	−3.474	4.474
	Equal variances not assumed			.251	77.964	.803	.500	1.996	−3.474	4.474

APACHE score at admission in the two groups also is not significantly different ($P = 0.803$).

A chi-square test for equality of sex distribution should also be done but that is not shown here. These tests establish that the two groups were not different to begin with. Assuming that all other factors not in the model are also same in the two groups, any difference in survival can be attributed to the difference in regimens.

After this exploration, ground is set for **Cox proportional hazard model**. First is to confirm that the hazards are proportional or not.

```
Command to run whether the hazard is proportional between the
two groups:
COXREG DurationofsurvivalYears
/STATUS=Censored(0)
/STRATA=Group
/PLOT HAZARDS LML.
```

This command graphically tests the proportionality of hazard assumption between the two groups. The option STATUS = Censored(0) represents event has happened (death). Many would like it to be coded as 1 but that does not matter. The variable Group was declared as STRATA instead of covariate so that this comparison can be done. PLOT command displays log-minus-log (LML) hazard for each group to visually check the proportionality assumption.

The following table gives the data summary. Seventeen out of 80 observations are censored and these are nearly equally divided in the two groups.

Stratum Status[a]

Stratum	Event	Censored	Censored Percent
0	31	9	22.5%
1	32	8	20.0%
Total	63	17	21.3%

[a] The strata variable is : Group.

Figure C.11 displays the cumulative hazard for both the groups. Cumulative hazard at 5 years is obviously much more than at 1 year. The difference between the two hazard curves remains proportional over the time. This proportionality is easily seen in the Figure C.12 on LML hazard. These plots are almost parallel for the groups in this figure, including proportionality.

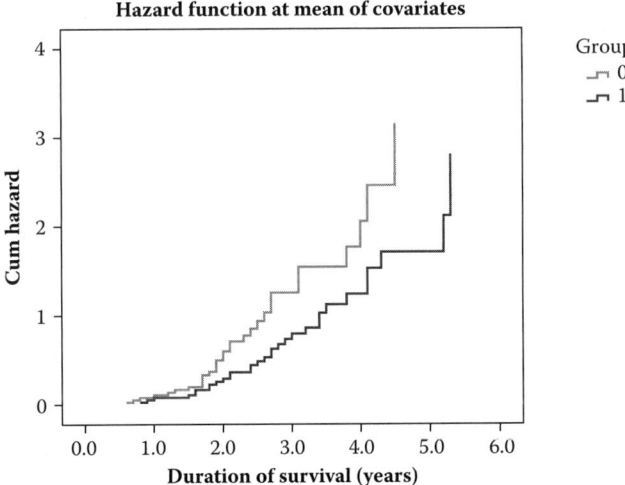

FIGURE C.11
Cumulative hazard in the two groups.

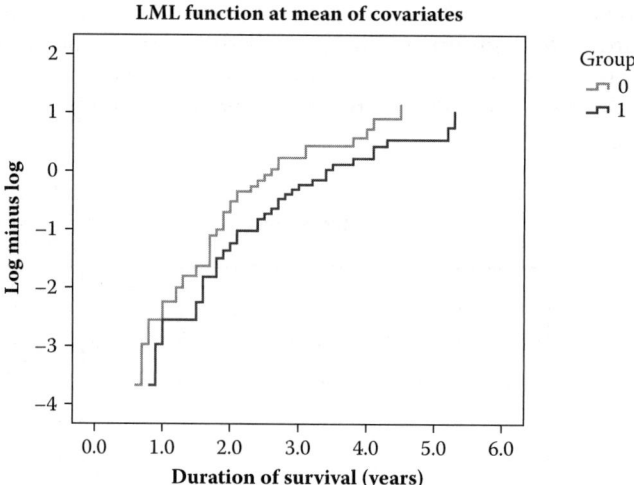

FIGURE C.12
LML plot of cumulative hazard in the two groups.

Thus, proportionality is not violated and Cox proportional hazard model is suitable for these data.

```
Command to run the Cox-regression model:
COXREG DurationofsurvivalYears
/STATUS=Censored(0)
/PATTERN BY Group
/CONTRAST (Sex)=Indicator(1)
/CONTRAST (Group)=Indicator
/METHOD=ENTER AgeatonsetYears APACHEatadmission Sex Group
/PLOT SURVIVAL
/PRINT=BASELINE.
```

This command runs Cox regression for survival considering survival duration as the dependent variable. STATUS defines censoring with zero for death. PATTERN By Group option plots the survival curve for both the regimens separately. CONTRAST options are for binary variables Sex and Group. For Sex, first category (female) was selected as reference category, and for Group, 0 is selected as reference category. Age at onset and APACHE score are entered as continuous variable in the model. METHOD option is for the method of selecting the variable—here, ENTER method is used that forced all the four independents in to the model. PLOT option displays the survival curve. The PRINT = BASELINE option displays the table for baseline hazard $h_0(t)$ at different time points and $h_1(t)$ value at same time points at the mean values of the covariates.

Cox Regression

The following table displays the frequencies and coding scheme of the categorical variables. Control regimen was taken as reference and for sex female was taken as reference.

Categorical Variable Codings [a,b]

		Frequency	(1)[c]
Group[d]	Test	40	1
	Control	40	0
Sex[d]	Female	37	0
	Male	43	1

[a] Category variable: Group.

[b] Category variable: Sex.

[c] The (0,1) variable has been recoded, so its coefficients will not be the same as for indicator (0,1) coding.

[d] Indicator Parameter Coding.

The following is value of $-2\ln L$ at the beginning without any covariate.

Block 0: Beginning Block

Omnibus Tests of

Model Coefficients

–2 Log Likelihood
439.542

The following table shows that when all the four covariates are entered, $-2\ln L$ reduces by 15.038 and this is statistically significant at 4 df.

Block 1: Method = Enter

Omnibus Tests of Model Coefficients[a,b]

-2 Log Likelihood	Overall (score)			Change From Previous Step			Change From Previous Block		
	Chi-square	df	Sig.	Chi-square	df	Sig.	Chi-square	df	Sig.
424.137	15.038	4	.005	15.404	4	.004	15.404	4	.004

[a] Beginning Block Number 0, initial Log Likelihood function: –2 Log likelihood: 439.542.

[b] Beginning Block Number 1. Method = Enter.

The following table displays the coefficients, their SE, relative risk $\exp(B)$, and its 95% CI.

Variables in the Equation

	B	SE	Wald	df	Sig.	Exp(B)	95.0% CI for Exp(B)	
							Lower	Upper
AgeatonsetYears	.000035	.011	.000	1	.997	1.000	.979	1.021
APACHEatadmission	.051	.016	9.929	1	.002	1.052	1.019	1.086
Sex	.207	.262	.622	1	.430	1.230	.735	2.057
Group	−.672	.273	6.036	1	.014	.511	.873	.299

The APACHE score and Group are significant predictor for mortality. $\exp(B)$ is the relative risk and represents the change in risk with each unit change in the value of the covariate. For example each unit increase in the APACHE score is expected to increase the risk of mortality by 5.2% relative to control because $\exp(B) = 1.052$. Similarly, test regimen decreases the risk of mortality by nearly one-half compared with control since $\exp(B) = 0.511$. The Cox model can be written as

$$h_1(t) = h_0(t) \times e^{[0.000035 \times \text{age} + 0.051 \times \text{APACHE} + 0.207 \times \text{sex} - 0.672 \times \text{group}]}$$

The Cox equation for control (Group = 0) is

$$h_1(t) = h_0(t) \times e^{[0.000035 \times \text{Age} + 0.051 \times \text{APACHE} + 0.207 \times \text{sex} - 0.672 \times 0]}$$

and the equation for test regimen (Group = 1) is

$$h_1(t) = h_0(t) \times e^{[0.000035 \times \text{age} + 0.051 \times \text{APACHE} + 0.207 \times \text{sex} - 0.672 \times 1]}$$

The following table displays the baseline cumulative hazard rate at different time (t), survival probability, its SE, and cumulative hazard rate $h_1(t)$ at mean values of the covariates.

Survival Table

Time	Baseline Cum Hazard	At mean of covariates		
Years		Survival	SE	Cum Hazard
.6	.002	.989	.010	.011
.7	.004	.979	.015	.022
.8	.008	.956	.021	.044
.9	.011	.945	.024	.056
1.0	.015	.922	.028	.081
1.2	.018	.911	.030	.094
1.3	.020	.899	.032	.107
1.5	.025	.874	.036	.134
1.6	.030	.850	.039	.162
1.7	.041	.802	.044	.221
1.8	.051	.763	.047	.271
1.9	.064	.709	.051	.343
2.0	.076	.667	.053	.406
2.1	.093	.607	.055	.499
2.3	.098	.591	.056	.525
2.4	.116	.539	.058	.618
2.5	.128	.504	.059	.684

2.6	.142	.469	.059	.757
2.7	.172	.398	.059	.920
2.8	.181	.381	.058	.966
2.9	.190	.362	.058	1.016
3.0	.200	.343	.058	1.069
3.1	.222	.306	.057	1.184
3.2	.234	.287	.056	1.250
3.4	.265	.242	.055	1.417
3.5	.282	.221	.054	1.509
3.8	.325	.176	.050	1.736
4.0	.349	.155	.048	1.866
4.1	.436	.097	.038	2.331
4.3	.471	.081	.035	2.518
4.5	.514	.064	.031	2.747
5.2	.717	.022	.024	3.831
5.3	1.065	.003	.006	5.692
6.1	.	.000	.	.

For example, at survival duration 4 years, the baseline cumulative hazard rate $h_0(t) = 0.349$ and cumulative hazard rate at mean values of covariates is

$$h_1(4 \text{ years}) = 0.349 \times e^{[0.000035 \times 49.90 + 0.051 \times 24.225 + 0.207 \times 0.538 + 0.672 \times 0.5]}$$
$$= 0.349 \times \exp(1.684) = 0.349 \times 5.387 = 1.88$$

The value of h_1 at 4 years in the table is 1.866. This is close to the 1.88. The minor difference is due to rounding off in calculations.

You can calculate the cumulative hazard for each regimen separately and for various combinations of values of the covariates by substituting the values in the Cox model.

Figure C.13 gives the survival curve for the two groups.

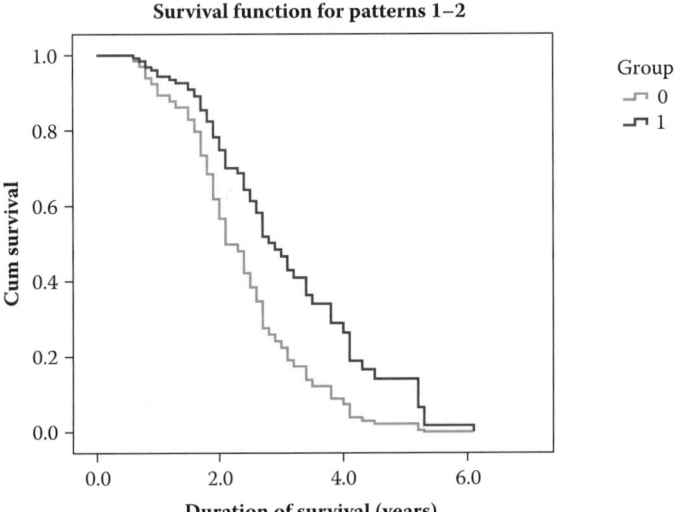

FIGURE C.13
Survival curve for the two groups as per the Cox model.

Now we need to test whether the difference between the two survival curves is statistically significant or not. This is done by log-rank test.

```
Command to run the log rank test:
KM DurationofsurvivalYears BY Group
/STATUS=Censored(0)
/PRINT MEAN
/PLOT SURVIVAL
/TEST LOGRANK.
```

This command runs the Kaplan–Meier method to obtain the two survival curves and test the significance of difference between the two groups using log-rank method. PRINT option displays the mean and median of each group. PLOT option displays the survival curve for both the groups.

Case Processing Summary

Group	Total N	N of Events	Censored N	Percent
0	40	31	9	22.5%
1	40	32	8	20.0%
Overall	80	63	17	21.3%

The next table displays the mean and median duration of survival in two regimen groups, and their SEs and CIs. The median survival duration for test regimen (group = 1) is 2.9 years which is higher than 2.1 years for control group. Note that mean is much higher than median.

Means and Medians for Survival Time

Group	Mean[a] Estimate	Std. Error	95% Confidence Interval Lower Bound	95% Confidence Interval Upper Bound	Median Estimate	Std. Error	95% Confidence Interval Lower Bound	95% Confidence Interval Upper Bound
0	2.469	.195	2.086	2.852	2.100	.266	1.580	2.620
1	3.156	.242	2.681	3.631	2.900	.291	2.330	3.470
Overall	2.839	.166	2.514	3.165	2.600	.142	2.321	2.879

[a] Estimation is limited to the largest survival time if it is censored.

The following table displays the results of log-rank test. The difference between the survival patterns in the two groups is statistically significant ($P = 0.032$).

Overall Comparisons

	Chi-Square	df	Sig.
Log Rank (Mantel-Cox)	4.584	1	.032

Test of equality of survival distributions for the different levels of Group.

Figure C.14 gives the two survival curves again, this time with + sign for censored values. These are slightly different from the curves obtained earlier by Cox method.

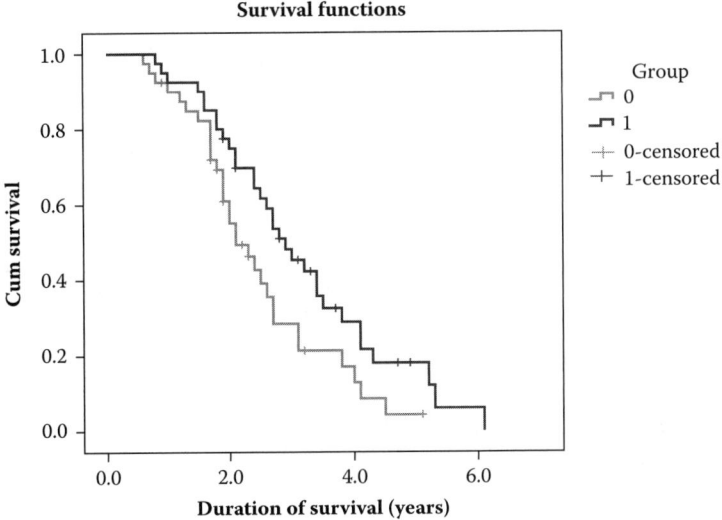

FIGURE C.14
Survival curves in the two groups by Kaplan–Meier method.

The survival function of test group descends more slowly than the survival function of control group.

Index

NOTE: For your convenience, this index is much more comprehensive than is available in most other books. In view of encyclopedic nature of contents of this book, many terms are configured in multiple ways in which a reader may like to search. The terms are arranged alphabetically after ignoring spaces, hyphens, propositions, etc. I hope the reader will find the index useful and exploit it fully. Terms followed by *see* and *see also* is only where you can find more information on similar or alternative terms. Numbers in **bold** face indicate the page where the term is explained or defined.